Photodetectors

Woodhead Publishing Series in Electronic and Optical Materials

Photodetectors

Materials, Devices and Applications

Second Edition

Edited by

Bahram Nabet

Electrical and Computer Engineering Department, Drexel University, Philadelphia, PA, United States

Woodhead Publishing is an imprint of Elsevier
50 Hampshire Street, 5th Floor, Cambridge, MA 02139, United States
The Boulevard, Langford Lane, Kidlington, OX5 1GB, United Kingdom

Copyright © 2023 Elsevier Ltd. All rights reserved.

No part of this publication may be reproduced or transmitted in any form or by any means, electronic or mechanical, including photocopying, recording, or any information storage and retrieval system, without permission in writing from the publisher. Details on how to seek permission, further information about the Publisher's permissions policies and our arrangements with organizations such as the Copyright Clearance Center and the Copyright Licensing Agency, can be found at our website: www.elsevier.com/permissions.

This book and the individual contributions contained in it are protected under copyright by the Publisher (other than as may be noted herein).

Notices
Knowledge and best practice in this field are constantly changing. As new research and experience broaden our understanding, changes in research methods, professional practices, or medical treatment may become necessary.

Practitioners and researchers must always rely on their own experience and knowledge in evaluating and using any information, methods, compounds, or experiments described herein. In using such information or methods they should be mindful of their own safety and the safety of others, including parties for whom they have a professional responsibility.

To the fullest extent of the law, neither the Publisher nor the authors, contributors, or editors, assume any liability for any injury and/or damage to persons or property as a matter of products liability, negligence or otherwise, or from any use or operation of any methods, products, instructions, or ideas contained in the material herein.

ISBN: 978-0-08-102795-0 (print)
ISBN: 978-0-08-102876-6 (online)

For information on all Woodhead Publishing publications
visit our website at https://www.elsevier.com/books-and-journals

Publisher: Matthew Deans
Acquisitions Editor: Stephen Jones
Editorial Project Manager: Michael Nicholls
Production Project Manager: Fizza Fathima
Cover Designer: Miles Hitchen

Typeset by MPS Limited, Chennai, India

Cover image, by Kiana Montazeri, is a MXene-Semiconductor-MXene photodetector on a GaAs/AlGaAs heterojunction substrate with embedded two-dimensional electron gas

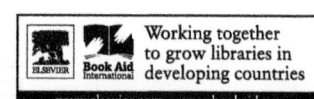

Contents

List of contributors		xiii
Preface		xvii

1 Metrology of thin-film photodetectors   1
Canek Fuentes-Hernandez

1.1	Introduction	1
1.2	Basic radiometry	3
	1.2.1 Point sources	5
	1.2.2 Extended sources	7
	1.2.3 Spectral characteristics of light sources	9
1.3	Ideal measurement of physical quantities	10
1.4	The responsivity	12
	1.4.1 Frequency-dependent responsivity	13
1.5	Electrical signals	14
	1.5.1 Photogenerated current	14
	1.5.2 Photogenerated voltage	19
1.6	Noise equivalent power	20
1.7	Measurement and noise	21
	1.7.1 Measuring electronics	24
	1.7.2 Effective noise bandwidth	25
	1.7.3 Sources of noise	27
	1.7.4 Thermal noise	27
	1.7.5 Shot noise	28
	1.7.6 Generation-recombination noise	29
	1.7.7 1/f or flicker noise	29
1.8	Specific detectivity	30
1.9	Photodiodes	30
Acknowledgments		34
References		35

2 Silicon based single-photon avalanche diode technology for low-light and high-speed applications   37
Daniel Durini, Uwe Paschen, Werner Brockherde and Bedrich J. Hosticka

2.1	Introduction	37
2.2	Single-photon counting in silicon	42
	2.2.1 From Avalanche photodiodes to SPAD: Geiger-mode operation and different readout modes	44
	2.2.2 Fabrication issues in CMOS-based SPAD arrays	49

2.3		Definition of SPAD figures of merit and general aspects of SPAD characterization	57
	2.3.1	Photon detection efficiency	58
	2.3.2	Dark count rate	58
	2.3.3	Afterpulsing probability	58
	2.3.4	SPAD dead time	59
	2.3.5	Timing jitter	59
	2.3.6	Crosstalk	59
	2.3.7	Fill factor	60
	2.3.8	Homogeneity of SPAD parameters in an array	60
	2.3.9	General aspects of SPAD characterization	60
2.4		SPAD arrays in CMOS technology: architecture overview	61
2.5		Active illumination based ToF 3D imaging and ranging with SPAD imaging arrays	62
2.6		Conclusions and outlook	67
		References	68

3 Organic photodetectors — 73
Vincenzo Pecunia, Dario Natali and Mario Caironi

3.1		Introduction	73
3.2		Organic semiconductors	74
	3.2.1	Charge photogeneration	75
	3.2.2	Deposition techniques	77
3.3		Device structure and operation mechanisms	83
	3.3.1	Photodiodes operation mechanism	84
	3.3.2	Photoconductors	86
	3.3.3	Reduction of dark currents	86
3.4		Photoactive materials and detectors for different spectral regions	91
	3.4.1	Visible and NIR photodetectors	91
	3.4.2	Wavelength selective OPDs	94
	3.4.3	UV detectors	103
3.5		All-printed organic photodetectors fabricated by means of scalable solution-based processes	108
	3.5.1	Inkjet-printed organic photodetectors	108
	3.5.2	Organic photodetectors fabricated by spray-coating	111
3.6		Applications of organic photodetectors	113
	3.6.1	Integrated transceivers	113
	3.6.2	Imaging applications	115
	3.6.3	Other applications	121
3.7		Conclusions	123
3.8		Note to the second edition	124
		References	124

| 4 | **Nanowires for photodetection** | **139** |

Badriyah Alhalaili, Elif Peksu, Lisa N. Mcphillips,
Matthew M. Ombaba, M. Saif Islam and Hakan Karaagac

	4.1	Introduction	139
	4.2	Nanowires photodetector fabrication themes	141
		4.2.1 Direct nanowire integration	141
		4.2.2 Transfer-printing/pick-and-place techniques	144
		4.2.3 Transfer printing of horizontally oriented semiconductor nanowires	145
	4.3	Recent device demonstrations	147
		4.3.1 Demonstrations of direct-growth photodetectors	147
		4.3.2 Waveguide coupled photodetectors	148
		4.3.3 Plasmonic photodetectors	150
		4.3.4 Photoactive oxide devices	150
	4.4	Device design challenges	170
		4.4.1 Impediments on contact formation	170
		4.4.2 Hybrid contacts for print transferred nanowires	172
		4.4.3 Control of nanowires doping	172
		4.4.4 Nnanowire photo-trapping enhancement	174
		4.4.5 Additional challenges	177
	4.5	Towards development of integrated multispectral nanowires photodetectors	177
	4.6	Conclusions and perspectives	179
	Acknowledgments		180
	References		180

| 5 | **High-speed InAs quantum dot photodetectors for data/telecom** | **199** |

Adriano Cola, Gabriella Leo, Annalisa Convertino, Anna Persano,
Fabio Quaranta, Marc Currie and Bahram Nabet

	5.1	Introduction	199
	5.2	InAs quantum dots: epitaxial growth advances	201
	5.3	Electrical and photoelectrical properties of quantum dots	206
		5.3.1 Electrical properties of quantum dots in vertical device structures	207
		5.3.2 In-plane electrical properties of quantum dots	212
		5.3.3 Stark effect in quantum dots	213
	5.4	Photodetectors for optical communication	213
	5.5	InAs quantum dot-based photodetectors: trends and performance	215
	5.6	InAs quantum dots photodetectors: the way ahead	221
	References		222

| 6 | **Advances in chip-integrated silicon-germanium photodetectors** | **233** |

Daniel Benedikovič

	6.1	Introduction	233
	6.2	Photodetection material systems: standard semiconductors and beyond	236

		6.2.1	Silicon photodetectors	236
		6.2.2	III/V compound photodetectors	238
		6.2.3	Germanium photodetectors	238
		6.2.4	Germanium-tin photodetectors	239
		6.2.5	New materials beyond semiconductors	239
	6.3	Processing methods and integration opportunities		240
		6.3.1	Growth methods: overview and recent trends	241
		6.3.2	Detector-waveguide integration	241
		6.3.3	On-chip schemes for waveguide coupling	242
		6.3.4	Si-complementary metal-oxide-semiconductor and Si-foundry integration	244
	6.4	Contemporary advances and state-of-the-art silicon-germanium photodetectors		245
		6.4.1	Metal-semiconductor-metal diodes	245
		6.4.2	PIN diodes	246
		6.4.3	Avalanche photodiodes	249
	6.5	Conclusion		256
	Funding			256
	References			257

7	**Ultraviolet detectors for harsh environments**	**267**
	Ruth A. Miller, Hongyun So, Thomas A. Heuser,	
	Ananth Saran Yalamarthy, Peter F. Satterthwaite and Debbie G. Senesky	
	7.1 Introduction	267
	7.2 Photodetector parameters	270
	7.3 III–nitride-based ultraviolet photodetectors	272
	7.4 SIC-based ultraviolet photodetectors	279
	7.5 Other types of ultraviolet photodetectors	283
	7.6 Conclusions	286
	References	286

8	**Low-temperature grown gallium arsenide (LT-GaAs) high-speed detectors**		**293**
	Marc Currie		
	8.1	Introduction	293
	8.2	Attributes of low-temperature-grown photodetectors	294
		8.2.1 Growth temperature	294
		8.2.2 Antisite defects and carrier traps	295
		8.2.3 Annealing	297
		8.2.4 Mobility and resistivity	298
		8.2.5 Carrier lifetime	300
		8.2.6 Optical absorption at mid-gap states	303
	8.3	Material systems	305
	8.4	Principle of operation for LT-GaAs photodetectors	306
	8.5	Photodetector technologies	306

	8.5.1	PIN photodiodes	306
	8.5.2	Photoconductive switches	308
	8.5.3	MSM photodetectors	309
	8.5.4	Waveguide photodetectors	311
8.6	Photodetector performance		313
8.7	Applications		317
	8.7.1	Sub-bandgap absorption photodetectors	317
	8.7.2	Nonlinear optics	317
	8.7.3	Pulsed magnetic spin experiments	318
	8.7.4	THz emitters and receivers	318
8.8	Conclusions and future trends		321
References			321

9 Faster than electron speed: photodetectors with confined 2D charge plasma overcome transit-time limit — 327
Bahram Nabet, Fabio Quaranta, Adriano Cola, Pouya Dianat and Marc Currie

9.1	Introduction		327
9.2	Device structure		330
9.3	Current-voltage relationship		333
9.4	Time response		337
	9.4.1	Components of total temporal response	339
9.5	Analysis and modeling		345
9.6	Conclusions		348
References			348

10 Plasmonic photodetectors — 353
Arash Ahmadivand, Mustafa Karabiyik and Nezih Pala

10.1	Introduction to surface plasmon resonances		353
	10.1.1	Physics of plasmon resonances	353
	10.1.2	Plasmonic devices	358
10.2	Photodetectors		359
	10.2.1	Semiconductor photodetectors	360
	10.2.2	Grating-coupled plasmonic photodetectors	362
	10.2.3	Plasmonic detectors with metallic nanoparticles	366
	10.2.4	Waveguide detectors	372
	10.2.5	Graphene plasmonics for photodetection	374
	10.2.6	Plasmonic metamaterials	377
References			382

11 CMOS-integrated waveguide photodetectors for communications applications — 391
Shiyang Zhu and Guo-Qiang Lo

11.1	Introduction	391
11.2	Waveguide-integrated Ge-on-Si photodetectors	393

		11.2.1	Selective hetero-epitaxy of Ge-on-Si and CMOS integration	393

   11.2.1 Selective hetero-epitaxy of Ge-on-Si and CMOS
          integration   393
   11.2.2 Waveguide-integrated PIN Ge-on-Si photodetectors   395
   11.2.3 Surface plasmon enhanced Ge-on-Si photodetectors   398
   11.2.4 Waveguide-integrated avalanche photodetectors   400
 11.3 Waveguide-integrated silicide Schottky-barrier photodetectors   403
   11.3.1 NiSi$_2$ film and absorption   403
   11.3.2 NiSi$_2$/p-Si and NiSi$_2$/n-Si Schottky-barrier
          photodetectors   404
   11.3.3 Metal-semiconductor-metal Schottky-barrier
          photodetectors   406
   11.3.4 Schottky-barrier collector phototransistors   408
   11.3.5 Schottky-barrier detector with embedded silicide
          nanoparticles   411
 11.4 Conclusions   415
 References   415

12 Photodetectors for silicon photonic integrated circuits 419
Molly Piels and John E. Bowers

 12.1 Introduction   419
 12.2 Technology   420
   12.2.1 Germanium   420
   12.2.2 Hybrid III/V-silicon   421
   12.2.3 Other technologies   421
 12.3 Optical properties of Si-based WGPDs   422
 12.4 Demonstrated WGPDs on silicon   423
   12.4.1 Vertically coupled PIN photodiodes in Si/Ge and
          InGaAs   424
   12.4.2 Butt-coupled PIN photodiodes in Ge   427
   12.4.3 Metal-semiconductor-metal photodetectors   427
   12.4.4 Separate absorption charge and multiplication
          avalanche photodiodes   428
   12.4.5 Si/Ge uni-traveling carrier photodiodes   428
   12.4.6 Hybrid III/V-silicon uni-traveling carrier
          photodiodes   431
 12.5 Conclusions and future outlook   433
 References   433

13 Efficient surface nano-textured CMOS-compatible photodiodes for Optical Interconnects 437
Soroush Ghandiparsi, Ahmed S. Mayet, Cesar Bartolo-Perez and M. Saif Islam

 13.1 Introduction   437
   13.1.1 Global IP traffic trend and forecast   437

		13.1.2 Optical communication in datacenters	437
		13.1.3 Chapter outline	439
	13.2	Theory and design	440
		13.2.1 Motivation	440
		13.2.2 Light trapping theory and background	440
	13.3	Vertical PIN silicon-based photodiode for short-reach communication	444
		13.3.1 Device design	444
		13.3.2 Simulation and optimization	445
		13.3.3 Fabrication	446
		13.3.4 Experimental results	448
		13.3.5 Equivalent photodiode model (system verification)	453
		13.3.6 Discussion and future roadmap	459
	References		462
14	**Photodetectors for microwave photonics**		**467**
	Tadao Nagatsuma		
	14.1	Signal generation	467
		14.1.1 Schemes	467
		14.1.2 Optical signal generation	468
		14.1.3 Fundamental characteristics	469
	14.2	Signal detection	471
		14.2.1 Schemes	471
		14.2.2 Photonic local oscillator for mixers	472
		14.2.3 Photonic mixers	472
	14.3	Applications	472
		14.3.1 Wireless communications	472
		14.3.2 Spectroscopy	475
		14.3.3 Electric-field measurement	477
		14.3.4 Imaging	478
	References		481
Index			**485**

List of contributors

Arash Ahmadivand Department of Electrical and Computer Engineering, Florida International University, Miami, FL, United States

Badriyah Alhalaili Nanotechnology and Advanced Materials Program, Kuwait Institute for Scientific Research, Kuwait City, Kuwait

Cesar Bartolo-Perez Department of Electrical and Computer Engineering University of California, Davis, CA, United States

Daniel Benedikovič Department of Multimedia and Information-Communication Technologies, University of Žilina, Žilina, Slovakia

John E. Bowers Department of Electrical and Computer Engineering, University of California, Santa Barbara, CA, United States

Werner Brockherde Fraunhofer-Institute for Microelectronic Circuits and Systems IMS, Duisburg, Germany

Mario Caironi Center for Nano Science and Technology @PoliMi, Istituto Italiano di Tecnologia, Milano, Italy

Adriano Cola IMM-CNR, Institute for Microelectronics and Microsystems, Unit of Lecce, National Research Council, Lecce, Italy

Annalisa Convertino Institute for Microelectronics and Microsystems (IMM)-CNR, Rome Unit, Rome, Italy

Marc Currie Optical Sciences Division, Naval Research Laboratory, Washington, DC, United States

Pouya Dianat Electrical and Computer Engineering Department, Drexel University, Philadelphia, PA, United States

Daniel Durini National Institute of Astrophysics, Optics and Electronics INAOE, Puebla, Mexico

Canek Fuentes-Hernandez Department of Electrical and Computer Engineering, Northeastern University, Boston, MA, United States

Soroush Ghandiparsi Department of Electrical and Computer Engineering University of California, Davis, CA, United States

Thomas A. Heuser Department of Materials Science and Engineering, Stanford University, Stanford, CA, United States

Bedrich J. Hosticka Fraunhofer-Institute for Microelectronic Circuits and Systems IMS, Duisburg, Germany

M. Saif Islam Department of Electrical and Computer Engineering University of California, Davis, CA, United States; Integrated Nanodevices and Nanosystems Research, Electrical and Computer Engineering, University of California, Davis, CA, United States

Hakan Karaagac Physics Department, Istanbul Technical University, Istanbul, Turkey

Mustafa Karabiyik Department of Electrical and Computer Engineering, Florida International University, Miami, FL, United States

Gabriella Leo Institute for the Study of Nanostructured Materials (ISMN)-CNR, Rome, Italy

Guo-Qiang Lo Institute of Microelectronics, The Agency for Science, Technology and Research (A*STAR), Singapore

Ahmed S. Mayet Department of Electrical and Computer Engineering University of California, Davis, CA, United States

Lisa N. Mcphillips Integrated Nanodevices and Nanosystems Research, Electrical and Computer Engineering, University of California, Davis, CA, United States

Ruth A. Miller Department of Aeronautics and Astronautics, Stanford University, Stanford, CA, United States; NASA Ames Research Center, Mountain View, CA, United States

Bahram Nabet Electrical and Computer Engineering Department, Drexel University, Philadelphia, PA, United States

Tadao Nagatsuma Graduate School of Engineering Science, Osaka University, Osaka, Japan

Dario Natali Dipartimento di Elettronica, Informazione e Bioingegneria, Politecnico di Milano, Milano, Italy; Center for Nano Science and Technology @PoliMi, Istituto Italiano di Tecnologia, Milano, Italy

Matthew M. Ombaba Integrated Nanodevices and Nanosystems Research, Electrical and Computer Engineering, University of California, Davis, CA, United States

Nezih Pala Department of Electrical and Computer Engineering, Florida International University, Miami, FL, United States

Uwe Paschen Westfälische Hochschule, Gelsenkirchen, Germany

Vincenzo Pecunia School of Sustainable Energy Engineering, Simon Fraser University, Surrey, BC, Canada

Elif Peksu Integrated Nanodevices and Nanosystems Research, Electrical and Computer Engineering, University of California, Davis, CA, United States; Physics Department, Istanbul Technical University, Istanbul, Turkey

Anna Persano IMM-CNR, Institute for Microelectronics and Microsystems, Unit of Lecce, National Research Council, Lecce, Italy

Molly Piels Department of Electrical and Computer Engineering, University of California, Santa Barbara, CA, United States

Fabio Quaranta IMM-CNR, Institute for Microelectronics and Microsystems, Unit of Lecce, National Research Council, Lecce, Italy

Peter F. Satterthwaite Department of Electrical Engineering and Computer Science, Massachusetts Institute of Technology, Cambridge, MA, United States

Debbie G. Senesky Department of Aeronautics and Astronautics, Stanford University, Stanford, CA, United States; Department of Electrical Engineering, Stanford University, Stanford, CA, United States

Hongyun So Department of Mechanical Engineering, Hanyang University, Seoul, South Korea

Ananth Saran Yalamarthy Department of Mechanical Engineering, Stanford University, Stanford, CA, United States

Shiyang Zhu Institute of Microelectronics, The Agency for Science, Technology and Research (A*STAR), Singapore

Preface

Every bit of information that circulates the Internet across the globe is a pulse of light, which at some point will need to be converted to an electric signal in order to be processed by the electronic circuitry in our data centers, computers, and cell phones. Photodetectors (PDs) perform this conversion with ultrahigh speed and efficiency and are ubiquitously present in many other devices, ranging from the mundane TV remote controls to ultrahigh resolution instrumentation used in Laser Interferometer Gravitational Wave Observatory that extends our reach to the edge of the universe and measures gravitational waves. The second edition of "Photodetectors" fully updates the popular first edition with current material covering the state of the art in modern PDs.

The second edition starts with basic metrology of PDs and common figures of merit to compare various devices. It follows with chapters that discuss single-photon detection with avalanche photodiodes, organic PDs that can be inkjet printed, and silicon-germanium PDs popular in the burgeoning field of silicon photonics. Internationally recognized experts contribute chapters on one-dimensional, nanowire, PDs, as well as high-speed zero-dimensional, quantum dot, versions that increase the spectral span as well as the speed and sensitivity of PDs and can be produced on various substrates. Solar-blind PDs that operate in harsh environments such as deep space, or rocket engines, are reviewed and new device designs in GaN technology. Novel plasmonic PDs, as well as devices which employ microplasma of confined charge in order to overcome the speed limitation of transfer of electronic charge, are covered in other chapters. Using different, novel technologies, CMOS compatible devices are described in two chapters, and ultrahigh-speed PDs that use low-temperature-grown GaAs to detect fast THz signals are reviewed in another chapter. PDs used in the application areas of silicon photonics and microwave photonics are reviewed in the final chapters of this book.

All chapters are of a review nature with extensive list of references, providing a perspective of the field before concentrating on particular advancements. As such, the book should appeal to a wide audience that ranges from those with general interest in the topic to practitioners, graduate students, and experts who are interested in the state of the art in photodetection.

The editor is grateful to all the contributors who shared their considerable expertise despite much demand on their time and hopes that they find our collective work to be informative, thorough, in-depth, current, and relevant. We greatly appreciate the professional and capable support of Kayla Dos Santos, Mariana L. Kuhl, and Fizza Fathima of Elsevier.

Bahram Nabet

Metrology of thin-film photodetectors

Canek Fuentes-Hernandez
Department of Electrical and Computer Engineering, Northeastern University, Boston, MA, United States

1.1 Introduction

Many physical processes can be used to detect light. Generally, this means to transduce an optical signal into an electric one. This chapter focuses on semiconductor photodetectors, wherein the absorption of light yields free-charge carriers that change the semiconductor's conductivity (i.e., photoconductive effect) and/or its electric potential (i.e., photovoltaic effect). These changes, enable transduction of the optical power carried by light impinging onto a photodetector's surface, into an electrical signal—current or voltage—which can be detected by an external circuit. The characteristics of this external circuit are important and will influence measurement accuracy and overall photodetector performance, particularly at the limit of detection, where its response is limited by the electronic noise.

Semiconductor photodetectors have become the preferred technology to detect light, or electromagnetic radiation, across many spectral regions of techological interest. This book, presents advances in semiconductor photodetector technologies from the ultraviolet (Chapter 7), to the microwave (Chapter 14) regions of the electromagnetic spectrum. In scientific and commercial applications, semiconductor photodetectors offer key advantages over other photodetector technologies, among them: (1) seamless integration with integrated circuits used to amplify, process, store and communicate the information generated by the photodetector; (2) fast response time; (3) the possibility of achieving high internal gain, and (4) ease of integration into imaging arrays, with optical communication technologies (Chapters 11 and 13) and silicon photonics (Chapter 12).

Semiconductor photodetectors are optoelectronic devices commonly made of inorganic crystalline semiconductors such as Si (Chapter 2), InAs (Chapter 5), 6H-SiC and GaN (Chapter 7), GaAs (Chapter 8), Ge, SiGe (Chapter 6), among many others. Also, these materials can enable nanowire photodetectors (Chapter 4), photodetectors with confined 2D charge plasmas (Chapter 9) and plasmonic photodetectors (Chapter 10). Remarkable progress in research institutions and the semiconductor industry has yielded semiconductor photodetectors that are both fast and sensitive, enabling single photon detection as well as arrays that integrate millions of pixels into a few millimeters-squared. Standard fabrication of semiconductor photodetectors typically involves high processing temperatures using epitaxial growth techniques and photolithography. High processing temperatures are needed, because the strong interatomic bonds of the raw materials require heating to sufficiently high-temperature values to enable long-range order to be achieved through the controlled solidification of liquid or vapor phases of the raw materials, into crystalline solids with a controlled density of defects. Epitaxial growth requires using highly ordered rigid substrates to seed crystalline growth; limiting the range of substrates onto which high-quality crystalline semiconductors can be grown. The applications of these inorganic semiconductor photodetectors are generally well defined, and the metrics used to characterize their performance, are well aligned to assess their usefulness and to compare their performance in that context.

As photodetectors continue to make inroads towards integration with everyday objects of the world, including those of the biological world, mechanical compliance and a need for reducing photodetector cost while maintaining high performance, has fueled a quest for semiconductor technologies that could enable next-generation flexible or stretchable high-performance photodetectors fabricated onto large areas at a reduced cost when compared to their commercially-available counterparts. The performance of thin-film photodetector technologies based on organic semiconductors (Chapter 3), 2D materials and other emerging semiconductor platforms has significantly improved in recent years. In some cases, with figures-of-merit (FOM) that are arguably comparable to those found in well-established photodetector technologies. However, these FOM may not be adequate to evaluate photodetector performance in the context where they may be relevant. The performance of a photodetector is typically evaluated through metrics such as responsivity ($\Re$), noise equivalent power (*NEP*), detectivity (*D*), specific detectivity (D^*), dark current and others. $\Re$ is used to quantify the efficiency of the transduction of an optical signal into an electrical signal. *NEP* is used to quantify the optical power needed to produce an electrical signal equal to the photodetector's electronic noise. *D* is defined as the inverse of the *NEP*, and D^* as $D \times (A_{\text{det}} B)^{1/2}$ to enable a comparison of photodetectors of different area (A_{det}) and evaluated at different measurement bandwidths (*B*). These metrics relate to measurable physical quantities, but the conditions under which they need to be measured, depend upon the type of photodetector, electronic circuits used to assess their performance, as well as on their intended application.

This last point is emphasized in one of R. Clark Jones seminal papers (Jones, 1949), where he states: "….when it is necessary to choose a detector for a given task, a decision must be made. The writer proposes to assist in this decision by defining a factor of merit which depends upon, and only upon, the relative signal-to-noise ratios obtainable with the various detectors." Photodetector metrics are not physical properties of a system but figures-of-merit that assist scientists and engineers in choosing the most appropriate photodetector for a specific application. For instance, the specific detectivity is only relevant if a photodetector will be used to detect faint optical signals and close to the limit where its performance is limited by its electronic noise. However, its value is irrelevant to compare the performance of two photodetectors operating under conditions wherein the electronic noise is negligibly small compared to the background noise (i.e., the background light detected that is different from the signal), as it happens in applications such as free-space optical communications, remote sensing, and others. In reviewing fundamental concepts used to define these metrics, it is expected that the reader will have the tools necessary to conduct accurate measurements of the relevant physical quantities, and to critically evaluate if the metrics of performance used, align well, and are relevant within the context of a targeted application.

In this chapter, we will use the term emerging semiconductors to encompass a wide range of semiconductor materials with strong optical absorption that enables the realization of thin-film photodetector architectures using photoactive layers that are no thicker than a couple of micrometers. Moreover, is used to refer to semiconductors that can be processed from solution or through low-temperature vacuum evaporation processes that are compatible with large-area, flexible, or stretchable substrates. Emerging semiconductors include synthetic organic compounds, perovskites, nanocrystals, nanowires, and two-dimensional (2-D) semiconductors among others (Jansen-van Vuuren et al., 2016; de Arquer et al., 2017; Natali and Caironi, 2016; Ombaba et al., 2016; Song, 2016). In contrast to inorganic crystalline semiconductors, these novel semiconductors exhibit a wide range of atomic and/or molecular order, from amorphous to poly- and para-crystalline order (Noriega et al., 2013). Morphological differences and differences in elemental composition and strength of interatomic or intermolecular interactions, produce differences in the physical processes leading up to the photogeneration, transport, and recombination of charge carriers with respect to those observed in inorganic crystalline semiconductors. These differences may cause parasitic effects generally not seen in the optoelectronic performance of photodetectors made with inorganic crystalline semiconductors and may suggest that different physical theories are needed to describe their performance. While this is true to some extent, it is also true that advances in device physics and engineering have resulted in thin-film photodetectors with a reduced number of parasitic effects, which reminds us, that the limit to their performance remains bound by general thermodynamic principles (de Arquer et al., 2017; Fuentes-Hernandez et al., 2020a).

In this chapter, basic concepts of radiometry and measurement will be reviewed. Photodetector performance parameters defined and techniques used to measure the average values and fluctuations of the physical quantities needed for calculation of relevant FOMs are discussed. These discussions are not meant to constitute a comprehensive review, but to provide sufficient background for readers to understand the origin of a photodetector's performance FOMs and to appreciate the assumptions made in their definition and evaluation.

1.2 Basic radiometry

Photodetectors are designed to transduce optical signals, i.e., light, emitted by incoherent natural or artificial light sources such as sunlight, light-emitting diodes (LEDs), and many others. The propagation through space of the radiant energy emitted by these sources is typically described within the conceptual framework of radiometry. Radiometry describes light propagation within the limits of geometrical optics. It assumes ray-like propagation of incoherent, and generally, unpolarized light on a macroscopic scale. It neglects any effects arising from the interference or diffraction of light through apertures and other optical elements, as well as during its propagation through space.

The approximations used in radiometry work surprisingly well for most applications, even if for instance, all real light sources have some degree of temporal coherence (Born and Wolf, 1980). Temporal coherence is indeed behind many difficulties that arise in connecting the quantities used in radiometry to describe radiant transport, and particularly the radiance, with a corresponding description based on electromagnetic fields and their statistical properties (Wolf, 1978). The temporal coherence of a light source quantifies a characteristic time wherein the phase difference between two electromagnetic waves remains constant. Within its coherence time (τ_c), a light source can produce visible interference fringes. The coherence time is inversely proportional to the frequency bandwidth $(\tau^{-1} \approx \Delta\nu_c)$. Since $\nu = c/\lambda_0$ it follows that $l_c = c\tau_c = \overline{\lambda}_0^2/\Delta\lambda$ where $\overline{\lambda}_0$ is the mean wavelength and l_c is defined as the coherence length. As an example, filtered sunlight with wavelength values between 400 and 700 nm, and a mean wavelength of 550 nm has an l_c of ~1 μm. Sunlight filtered with a 10 nm spectral width around 550 nm has an l_c of ~30 μm. Although these l_c values are first-order approximations, they clearly illustrate the fact that even for "incoherent" light sources, interference effects cannot be neglected over short distances, such as when light propagates through a semiconductor layer of thickness $d \leq 1$ μm. In such thin layers, the absorption of light can not be accurately described by Beer's law: $1 - \exp(-\alpha d)$, where α is the absorption coefficient and d the sample's thickness. Instead, multiple-beam interference from internally reflected beams will determine the total absorptance within the semiconductor layer (Born and Wolf, 1980). Interference effects become more pronounced when the spectral characteristics of a photodetector are measured using nearly monochromatic light sources, such as lasers, gas-discharge lamps, or spectrally filtered white-light sources, consequently, it is always important to consider how they may impact experimental results.

The central quantity used in radiometry to describe the propagation of radiant energy, hereon referred to as optical energy (Q) through space, is a directional quantity called radiance $L(\theta, \varphi)$. The radiance, represents the optical power, $\phi = dQ/dt$, per unit area A and per unit projected solid angle Ω, emitted from a surface element in a given direction. It has units of Wm²/sr in the international system of units (SI). The radiance emitted by the source element dA_s shown in Figure 1.1A in the direction defined by the polar angle θ, and the azimuthal angle φ, over a solid angle $d\omega$ is given by:

$$L(\theta, \varphi) = \frac{d^2\phi(\theta, \varphi)}{dA_s \cos\theta d\omega} = \frac{d^2\phi(\theta, \varphi)}{dA_s d\Omega} \tag{1.1}$$

Now, let's consider the radiative transfer between two surfaces, for example, between a light source and photodetector surface, as shown in Figure 1.1B.

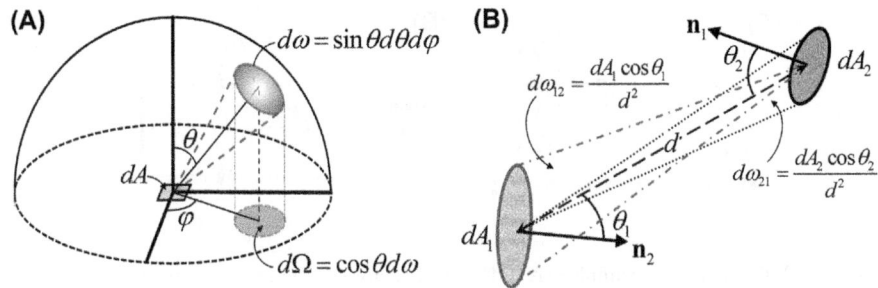

Figure 1.1 (A) Representation of solid angle and projected solid angle in spherical coordinates. (B) Relevant areas and solid angles considered when solving a radiative transfer problem between them.

In the absence of transmission losses due to atmospheric absorption or reflection at optical elements, and assuming the incoherent superposition of the power emitted by all surface elements from all directions, that is, a perfectly incoherent source (Wolf, 1978), Eq. (1.1) allows writing the optical power in its integral form as:

$$\begin{aligned}\phi(\theta,\varphi)_{1\to 2} &= \int_{A_1}\int_{\omega_2} L(\theta,\varphi)dA_1\cos\theta_1 d\omega_{21} \\ &= \int_{A_1}\int_{A_2} \frac{L(\theta,\varphi)\cos\theta_1\cos\theta_2}{d^2}dA_1 dA_2 \\ &\approx \frac{L(\theta,\varphi)A_1\cos\theta_1 A_2\cos\theta_2}{d^2}\ ;\ \text{if}\ d^2 >> A_1\ \text{or}\ A_2\end{aligned} \quad (1.2)$$

Note that if $\phi(\theta,\varphi)_{2\to 1}$ is calculated, indices can be interchanged in Eq. (1.2), yielding the same final result. Since $\phi(\theta,\varphi)_{1\to 2} = \phi(\theta,\varphi)_{2\to 1}$, it can be concluded that energy is conserved during radiometric transfer. This symmetry also implies that $A_1\Omega_{2,1} = A_2\Omega_{1,2}$. This is, the throughput: $A_i\Omega_{j,i}$, is conserved in a radiometric system. If surfaces are not located in the same media, the conservation of the throughput is written in its more general form as: $n_1^2 A_1\Omega_{2,1} = n_2^2 A_2\Omega_{1,2}$, where n_i are the refractive index of the two different media.

1.2.1 Point sources

Point sources are mathematical idealizations. However, commonly used light sources in photodetector characterization such as LEDs or arc gas discharge lamps, when placed at a sufficiently large distance can be treated as point sources.

As shown in Figure 1.2A, now let's consider a photodetector of area: A_{det}, with its normal vector **n** forming an angle θ_{det} with respect to the direction of a point source. This direction, representing the shortest path between source and detector, defines the optical axis of this problem. Hence, the photodetector is at a distance d from the point source. Let's assume that the point source has a radiance: $L(\theta,\varphi)$.

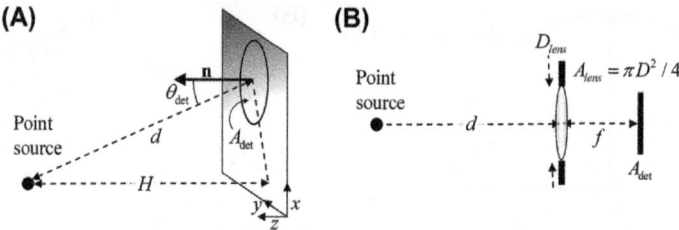

Figure 1.2 Point source examples (A) Off-axis detector (B) On axis-detector with a lens.

Since a point source does not have an area, it is useful to rewrite Eq. (1.2) in terms of the intensity of the source, defined as $I = \int L\, dA$ in SI units of W/sr. This is,

$$\begin{aligned}
\phi_{det} &= \int_{\omega_{det}} I\, d\omega_{det} \\
&= \int_{A_{det}} \frac{I \cos\theta_{det}}{d^2} dA_{det} \\
&\approx \frac{I A_{det} \cos\theta_{det}}{d^2} \quad \text{if} \quad d^2 \gg A_{det}
\end{aligned} \quad (1.3)$$

Although intensity is often used as synonymous with power or irradiance, in radiometry this use should be avoided. Irradiance is defined as: $E = \int L\, d\Omega$ in SI units of W/m. From Eq. (1.3), the irradiance at the detector is: $E_{det} = \phi_{det}/A_{det} \approx I\cos\theta_{det}/d^2$. If we take H as the minimum distance between the photodetector's plane in its normal direction, and the point source, then $d = H/\cos\theta$ and $E_{det} = I\cos^3\theta_{det}/H^2$, which shows that the irradiance falls off as: $\cos^3\theta$, causing an inhomogeneous power distribution over the photodetector plane. If $d^2 \approx A_{det}$, the irradiance fall off will be significant over the photodetector area and the integral in the third row of Eq. (1.3) would need to be solved to accurately calculate the total optical power collected. This situation should be avoided during the characterization of a semiconductor photodetector because fundamentally, the electrical signal generated and its linearity as a function of optical power, will depend on the peak irradiance of the power collected at its surface. This is because the rates of certain physical processes, such as the photogeneration and charge recombination rates, are dependent on local density values of photons and free charge carriers, respectively. Hence, two different illumination conditions having different peak irradiance values could yield different electrical signal values, even if the total power collected by the photodetector is equal under both conditions. This situation is common when a lens is used to collect more optical power by focusing it into a photodetector's area. In such case, the peak irradiance could change significantly if the distance between the lens and photodetector varies.

Let's then analyze the situation when a lens is used in front of a photodetector as shown in Figure 1.2B. Assume that all power collected by the lens is focused onto the photodetector. For the radiometric calculation, the lens becomes the relevant surface onto which power is transferred from the point source. Strictly speaking, the entrance pupil of the lens, that is, the limiting aperture of the optical system as seen from the source plane, is the relevant area that should be considered in these radiometric calculations. However, when the focusing optics comprises many lenses and apertures, for example a camera lens, the entrance pupil may not be a physical aperture but a virtual image in space. For simplicity and in relation to Figure 1.2B, the entrance pupil will be considered to be the area A_{lens} of the lens. The lens will be assumed to have a transmittance $\tilde{T}_{lens}$ that is smaller but close to unity. To avoid confusion with other physical quantities used in this chapter, a tilde is used to denote transmittance, reflectance, and absorptance, this is, $1 = \tilde{T} + \tilde{R} + \tilde{A}$. We will also assume that all surfaces are properly aligned so that all the surface normal vectors are parallel to the optical axis. In such a case, all cosine terms in Eqs. (1.2) and (1.3) equal one. Assuming that the distance $d \geq f$, where f is the focal distance of the lens, the total optical power collected by the lens is: $\phi_{det/lens} \approx \tilde{T}_{lens} A_{lens} I / d^2$. Hence, when placed at a distance d, the lens will increase the optical power collected by the photodetector, under free space propagation, by a factor: $\tilde{T}_{lens} A_{lens} / A_{det}$. Let's note that since the lens is an imaging system, the values of f and d will determine the position of the source image z_{image} behind the lens according to the lens equation: $d^{-1} + z_{image}^{-1} = f^{-1}$. If $d = f$ then $z_{image} \to \infty$ and the beam is collimated in the image space; a situation that is ideal for photodetector characterization since it will produce a uniform power distribution across the photodetector area, bar a $\cos^3\theta$ fall-off depending on the lens radius. However, when $d > f$ and if the photodetector is placed at the image conjugate plane, z_{image}, then the irradiance of the point source's image predicted by geometrical optics will be infinite. In the real world, point sources do not exist and diffraction and aberrations will cause the point source's image to have a finite area, $A_{beam@det}$, typically much smaller than the photodector area. Since $d > f$, $A_{beam@det} \leq A_{det} \leq A_{lens}$, and the irradiance at the photodetector: $E_{det/lens} = \phi_{det/lens} / A_{beam@det}$ will increase with respect to the free-space propagation case we first analyzed by a factor of $\tilde{T}_{lens} A_{lens} / (A_{beam@det})$.

In general, to minimize irradiance variations, it is desirable that $A_{beam@det} \approx A_{det}$, suggesting that point sources should be placed in front of the lens, at a distance that is close to its focal length. In practice, when a point source is placed at the front focal point of a lens, a stop aperture must also be used to minimize the $\cos^3\theta$ irradiance fall off across the lens area, to minimize the detrimental effects on the image quality, that is, the power distribution across the photodetector, produced by so-called third-order optical aberrations such as spherical aberration, distortion, field curvature, and others and to match the beam area with the photodetector area.

1.2.2 Extended sources

Extended sources are those that completely fill the field-of-view (FOV) of a photodetector. Extended sources such as integrating spheres are rarely used in photodetector characterization. Extended sources generally require solving the integral in Eq. (1.2), a situation that will not be discussed here but can be found in reference (Palmer and Grant, 2009). However, many sources of light having relatively large areas, display properties that can be approximated as being Lambertian (e.g., the sky).

A Lambertian source is that for which the radiance is independent of the direction, that is, $L(\theta, \varphi) \equiv L$. For a Lambertian source, Eq. (1.2) is reduced to:

$$\phi_{1\to 2} = LA_1\Omega_{2,1} \tag{1.4}$$

where $\Omega_{2,1}$ denotes the projected solid angle of the second surface as seen from the first surface having an area A_1. Hence, Eq. (1.4) reduces the radiative transfer problem to solving the geometrical problem of calculating the projected solid angle subtended by the source as seen from the photodetector.

As a first example, consider an on-axis Lambertian disk source as shown in Figure 1.3A. In such a case, the projected solid angle is the right circular cone formed by the disk and is given by $\Omega = \pi\sin^2\Theta_{1/2}$. This solid angle yields a collected optical power equal to $L_s A_{det} \pi\sin^2\Theta_{1/2}$, and an irradiance of $\pi L_s \sin^2\Theta_{1/2}$. If the disk is located at a large distance from the source we have that $\Omega = \pi\sin^2\Theta_{1/2} \approx \pi r_{disk}^2/d^2$, and consequently $\phi_{det} \simeq L_s A_{det} A_{disk}/d^2$.

Now, let's consider the situation depicted in Figure 1.3B, wherein an extended Lambertian source is placed in front of a photodetector of the area A_{det}, having an aperture area A_{ap} placed at a distance f in front of it. In such case, since the extended source overfills the aperture, the total collected power is determined by the projected solid angle of the aperture as: $\phi_{det} = L_s A_{det} \pi\sin^2\Theta_{1/2} \simeq L_s A_{det} A_{ap}/f^2$; equivalent to having a Lambertian disk of the area A_{ap} at a distance f from the photodetector. If now a lens with $\tilde{T}_{lens} < 1$, entrance pupil $A_{EP} = A_{ap}$, and focal distance f, is placed at the aperture plane, the collected optical power becomes: $\tilde{T}_{lens} L_s A_{det} \pi\sin^2\Theta_{1/2}$. Which is smaller than the power collected without the lens due to transmittance losses. Hence, for Lambertian sources, it is generally better not to use a lens.

Apertures can play different and important roles in an optical system. The aperture stop and the field stop in a radiometer are the most important in radiometric calculations. The aperture stop limits the amount of light that reaches the photodetector. The field stop limits the angular FOV of the photodetector. In radiometers that use a lens, the lens defines a right circular cone with a projected solid angle $\Omega = \pi\sin^2\Theta_{1/2}$ where $\Theta_{1/2}$ is the cone's half angle. The numerical aperture is then defined as: $NA = \sin\Theta_{1/2}$ and can be related to the f-number of the lens as:

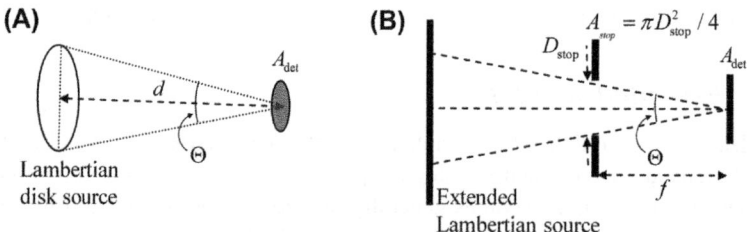

Figure 1.3 Extended sources (A) On-axis Lambertian disk and photodetector (B) Extended Lambertian source and photodetector.

$f/\# = f/D = (2NA)^{-1}$, where D are the entrance pupil diameter and f the focal length of the lens. For the same circular cone, the solid angle is given by $\omega = 2\pi(1 - \cos\Theta_{1/2})$. The FOV is defined by the half angle of the maximum projected solid angle subtended by the photodetector as seen from the entrance pupil, that is, the lens aperture in a single lens radiometer.

1.2.3 Spectral characteristics of light sources

From a practical perspective, radiometry is extended to account for the fact that real light sources emit light at different wavelengths (λ) by replacing the radiance function with a spectral radiance function: $L(\theta, \varphi) \to L(\theta, \varphi, \lambda)$. Hence, the optical power function is also replaced with a spectral optical power function: $\phi(\theta, \varphi) \to \phi(\theta, \varphi, \lambda)$. From a fundamental perspective, and beyond issues relating to temporal and spatial coherence that arise when nearly monochromatic light propagates through space, it is generally difficult to fully reconcile the radiometric view that energy flows through particular points in space and at particular directions, with standard electromagnetic theory or with the quantum theory of radiation (Foley and Wolf, 1991; Fante, 1981). Consequently, deriving a radiance function from first principles presents significant challenges and won't be discussed here. Despite these difficulties, radiometry offers an excellent approximation to solve radiative transfer problems encountered during photodetector characterization.

Light sources used for photodetector characterization are electrically powered and display a spectral radiance that can be considered as being constant over the period in which experiments are typically carried out. Calibrated spectrometers, perhaps offer the easiest and fastest way to measure the spectral radiance, spectral irradiance, spectral intensity, or spectral optical power of a light beam. Spectrometers are complex optical systems that use dispersive elements such as diffraction gratings or prisms to spatially separate the spectral components of a light beam. The spatial distribution of optical power at various wavelengths, created by dispersive elements, is typically measured using a one-dimensional photodetector array. The size of the individual pixels in the photodetector array and their location determines the spectral bandwidth $\Delta\lambda$ captured by each element and offers a rapid way to produce a snapshot of the spectral characteristics of the light entering the instrument. However, knowledge of the spectral radiance of a light source is not sufficient to determine the electrical output of a semiconductor photodetector because its responsivity is wavelength-dependent.

To determine the spectral response of a photodetector, the most common approach is to use a spectrally broad light source, such as a gas discharge arch lamp, and to spectrally filter it using an instrument known as a monochromator. As in a spectrometer, a monochromator uses dispersive elements such as diffraction gratings or prims to separate in space the different spectral components of a light beam. In contrast to a spectrometer, a monochromator mechanically redirects the spatially distributed spectral components of light into a slit at its output. The geometry of the monochromator and the properties of the dispersive elements along with the width of the slit determine the spread of wavelengths $\Delta\lambda$ passing through the slit. Narrower slits yield smaller $\Delta\lambda$ but decrease the amount of optical power available for photodetector characterization. To avoid averaging effects $\Delta\lambda$ must be selected to be smaller than the full-width-at-half-maximum (FWHM) of the narrowest spectral peak in the light source's spectrum, which can be challenging with gas discharge arch lamps that have narrow spectral emission lines. If some averaging is acceptable, because the responsivity of a semiconductor photodetector such as a silicon photodiode is roughly constant for tens of nm and if the spectral shape of the light source is symmetrical, for example, a Gaussian or Lorentzian profile with maximum power at λ_0, determination of the optical power only requires knowledge of the responsivity at λ_0 since the overall effect of multiplying the spectral power distribution by a linear function in λ around λ_0 will increase one side of the distribution in the same proportion than the other side decreases. This is useful because with a calibrated photodetector, the only knowledge needed to determine the total power of a light source having a broad but symmetric emission spectrum, such as a LED, is its peak emission wavelength.

1.3 Ideal measurement of physical quantities

The characterization of thin-film semiconductor photodetectors requires measurements of time-averaged physical quantities, such as optical power, current, or voltage, over integration times wherein all effects related to their quantized nature can be generally neglected. Thin film photodetectors have response time values that are generally too slow, for example, >10 ns, and responsivity values that are too small, for example, <0.5 A/W at visible wavelengths, to generate electrical signals with a magnitude that is larger than that of the electronic noise and consequently, capable of detecting the arrival of individual photons. However, the statistical properties of fluctuations arising from the quantized nature of photons and electrons, allow their collective effect to be measurable even when long integration times are used. A more detailed discussion will be given in Section 1.7.

Here, let's consider a physical quantity $x(t)$, where t represents the instantaneous time. In general, we would only be interested in measuring running time average values of the physical quantity $x(t)$, which are given by: $\langle x(t) \rangle = \frac{1}{\tilde{\tau}} \int_t^{t+\tilde{\tau}} x(t') dt'$, where the time interval $\tilde{\tau}$, is negligibly short when compared to the integration time of standard measurement equipment used to sample $\langle x(t) \rangle$. To determine $\langle x(t) \rangle$, measurements at discrete times t_m are taken using an integration time $\tau = t_{m+1} - t_m >> \tilde{\tau}$, and corresponding to certain measurement bandwidth $B \leq \tau^{-1}$. The sampling process produces a set of time-averaged values calculated as

$$x_B(t_m) = B \int_{t_j}^{t_j + B^{-1}} \langle x(t') \rangle dt'. \tag{1.5}$$

According to the fundamental sampling theorem (a.k.a. Nyquist-Shannon theorem), to adequately reconstruct the temporal profile of $\langle x(t) \rangle$, B must be at least equal, but preferably larger than $2B_x$, where B_x represents the highest frequency component on the $\langle x(t) \rangle$ frequency spectra. This means that the Fourier transform: $\mathcal{F}\{\langle x(t) \rangle\} \equiv \int \langle x(t) \rangle e^{-i2\pi ft} dt \equiv X(f)$, is assumed to be bound, that is $X(f) = 0$ for $f > B_x$. If the average value of $\langle x(t) \rangle$ is constant over time, the physical system is said to be in a steady-state and consequently, $X(f) = \delta(f)$, where $\delta(f)$ is the delta function, and B could be made arbitrarily small without aliasing (distortion) effects. In the real world, even in a steady state, physical quantities fluctuate around an average or mean value. These fluctuations, known as noise, prevent $X(f)$ being zero for any reasonable value of B. Hence, the act of measuring the average value of a physical quantity will always result in cropping of its noise spectra.

In photodetector characterization, we are interested in estimating the value of $\langle x(t) \rangle$, in or near steady-state conditions, through the average value of the discrete set of measurements described in Eq. (1.5), this is:

$$\overline{x_B(t_m)} = N^{-1} \sum_{m=1}^{N} x_B(t_m). \tag{1.6}$$

The overbar is used to denote a discrete time average. In steady-state conditions, $\overline{x_B(t_m)} = \overline{x_B(t_k)}$ for any j or k, so the steady-state average value can simply be written as $\bar{x}$. To simplify the notation, hereon let's redefine $x_B \equiv x_B(t_m)$ and $\overline{x_B} \equiv \overline{x_B(t_m)}$; keep the subscript B as a reminder that these are discrete time-dependent values.

In addition to the value of $\bar{x}$, we will be interested in measuring the magnitude of the temporal fluctuations around this value. To do this, x_B values are assumed to be randomly drawn from a probability distribution determined by the stochastic nature of the physical processes that give rise to them. According to the Central Limit Theorem, the sum of m random samples drawn from a probability distribution will tend to be Gaussian distributed around the expectation value of the sum as $m \to \infty$. The width of the Gaussian distribution is quantified by the standard deviation, σ, corresponding to the root square of the sample's variance:

$$\sigma_{x_B}^2 = \frac{1}{N} \sum_{m=1}^{N} (x_B - \overline{x_B})^2 \equiv \overline{(x_B - \overline{x_B})^2}. \tag{1.7}$$

The root mean squared value, $x_{rms,B}$, is then defined as:

$$x_{rms,B} \equiv \sigma_{x_B} = \sqrt{\overline{(x_B - \overline{x_B})^2}}, \tag{1.8}$$

and represents an average value of the "noise." Therefore, the signal-to-noise ratio (SNR) is given by:

$$SNR = \frac{\overline{x_B}}{x_{rms,B}}. \tag{1.9}$$

In steady-state, even large values of $\tau = B^{-1}$, a few seconds long (e.g., $B < 1$ Hz), will have little impact on the value $\overline{x_B}$, but could strongly impact $x_{rms,B}$ depending on its spectral properties. As will be discussed in Section 1.8, defining the measurement bandwidth could be complex since it depends on the spectral properties of the electrical circuits used to measure a signal as well as those intrinsic to the sources of noise. Here, the term "spectral" is used to refer to frequency values that are well below 10^{12} Hz and not those associated with the frequency at which the electromagnetic field of light oscillates.

1.4 The responsivity

The responsivity $\mathfrak{R}_{y_B}(\phi_{\text{det},B})$ of a semiconductor photodetector is a figure-of-merit defined in terms of its spectral responsivity function $R_{y_B}(\phi_{\text{det},B}, \lambda)$ by:

$$\mathfrak{R}_{y_B}(\phi_{\text{det},B}) = \frac{y_B}{\phi_{\text{det},B}} = \frac{\int R_{y_B}(\phi_{\text{det},B}, \lambda) \phi_{\text{det},B}(\lambda) \, d\lambda}{\int \phi_{\text{det},B}(\lambda) \, d\lambda}, \tag{1.10}$$

where y_B denotes either a current or a voltage. Note that the spectral responsivity $R_{y_B}(\phi_{\text{det},B}, \lambda)$, does not represent a derivative quantity expressed per-unit wavelength, but the magnitude of the responsivity, this is, the ratio between the electrical output of the photodetector and the input optical power of a monochromatic light source, at a particular wavelength; in SI units of amps (volts) per W. The responsivity can also be defined as a function of other optical inputs. For example as an irradiance responsivity $\mathfrak{R}_{y_B}(E_{\text{det},B})$ in amps (volts) per W/m^2 or as a radiance responsivity $\mathfrak{R}_{y_B}(L_{\text{det},B})$ in amps (volts) per W/m^2/sr, with their corresponding spectral responsivity functions $R_{y_B}(E_{\text{det},B}, \lambda)$ and $R_{y_B}(L_{\text{det},B}, \lambda)$. The choice, depends on the photodetector's physical properties, as well as on the application of interest.

If we consider that the electrical signal produced by the photodetector is given by:

$$Signal_{electrical} = \int R_{y_B}(x_{\text{det},B}, \lambda) x_{\text{det},B}(\lambda) \, d\lambda, \tag{1.11}$$

in response to an optical signal defined by:

$$Signal_{optical} = \int x_{\text{det},B}(\lambda) d\lambda, \tag{1.12}$$

where $x_{\text{det},B}$ can be optical power, irradiance, radiance, photon flux, etc. at the photodetector, an alternative definition of the responsivity can be given as:

$$\mathfrak{R} \equiv \frac{Signal_{electrical}}{Signal_{optical}}. \tag{1.13}$$

The emphasis provided by this definition is on the role of responsivity as a figure-of-merit of the transduction efficiency from a signal of one kind, into a signal of a different kind. It forces consideration of what is meant by an optical signal, that is, what information is of interest in a particular context. Note that the spectral responsivity is a photodetector property but it lacks relevance as a photodetector's figure-of-merit. On the other hand, as a figure-of-merit, the responsivity is relevant only in relation to the signal(s) of interest, or in the context of a given application.

As an example, consider an ideal photodetector (PD1) with a spectral responsivity that as a function of wavelength (λ), expressed in nm, is given by: $8\lambda \times 10^{-4}$ A/W from 495 nm to 505 nm, and zero elsewhere. Consider a second photodetector (PD2) with a spectral responsivity function of $4\lambda \times 10^{-4}$ A/W from 400 nm to 600 nm and zero elsewhere. Let's now assume that both photodetectors are used to capture a signal corresponding to a 1 mW laser emitting at 500 nm, the electrical signal produced by PD1 will be 400 µA, corresponding to a responsivity of 0.4 A/W. These values are $2\times$ bigger than those for PD2, producing an electrical signal, ca. 200 µA corresponding to a responsivity of 0.2 A/W. Now, consider that the signal is a 1 mW (total power) LED with peak emission at 500 nm and a Gaussian emission profile with a 50 nm FWHM bandwidth (peak power 12.78 µW). In this case, PD2 will produce an electrical signal of 200 µA and a corresponding responsivity of 0.2 A/W; much bigger than the corresponsing values produced by PD1, yielding an electrical signal of 75 µA and a responsivity of 0.0745 A/W. Finally, consider a situation wherein the 1 mW laser signal at 500 nm is embedded on the 1 mW LED (peak emssion at 500 nm) background. In this situation, PD1 will yield an signal-to-noise ratio, SNR = 400/75 = 5.3, while for PD2 SNR = 1. Based on their spectral responsivity values, can any of the photodetectors be claimed to be superior? Clearly not. Hence, a photodetector's performance, its responsivity (*i.e.*, as defined by Eqs (1.10) or more explicitly (1.13)) depends on what is considered a signal and what is not. Hence, it is inherently related to an application. This distinction is often overlooked in the literature.

1.4.1 Frequency-dependent responsivity

As defined by Eq. (1.10), the responsivity is a function of the time-dependent optical power at the photodetector. If the optical power varies in time, or is modulated at a specific frequency f, the amplitude of the photodetector signal will change depending on its characteristic response time τ_0. The response time of a photodetector may be dominated by intrinsic (i.e., τ_{det} if related to charging carrier dynamic in the photodetector) or extrinsic (i.e., $\tau_{circuit}$ if related to an amplifying circuit) properties. For reasons later explained, the responsivity often displays a frequency response that corresponds to that of a first-order or single pole low pass filter, given by:

$$\Re_{y_B}\left(\phi_{det,B}, f\right) = \frac{\Re_{y_B}\left(\phi_{det,B}, f = 0\right)}{1 + 2\pi f \tau_0} \tag{1.14}$$

When the photocurrent is measured as the voltage drop across a load resistor R_{load}, that is, $V = i_{ph,B} R_{load}$, and if $\tau_{circuit} = R_{load} C > \tau_{det}$, then $\tau_0 = \tau_{circuit}$. Correspondingly, if $\tau_{circuit} < \tau_{det}$ then $\tau_0 = \tau_{det}$. When $\tau_{circuit} \approx \tau_{det}$, Eq. (1.14) is no longer valid, and needs to be replaced by the functional dependence of a second-order low pass filter with equal time constants.

Understanding the frequency dependence of $\mathfrak{R}_{y_B}(\phi_{det,B}, f)$, which always depends on the measuring circuit even if it is not limited by it, is of critical importance to determine the optimum conditions for the evaluation of a photodetector's performance. If the photodetector's responsivity reaches a maximum value $f_m \neq 0$, proper selection of the B value becomes critical to maximize the photodetector output and properly evaluate its performance metrics (Jones, 1953).

1.5 Electrical signals

1.5.1 Photogenerated current

Photodetectors are commonly engineered to have a constant responsivity. A constant responsivity as a function of optical power enables photodetectors to be used for metrology of optical signals in applications such as imaging. However, for applications such as motion sensing, bar code recognition, and others, linearity may not be necessary.

Semiconductor photodetectors generally display a constant responsivity over a wide range of optical power values because the photogeneration of free charge carriers is proportional to the number of photons absorbed by the material. A photon of light is absorbed if the photon's energy, $\varepsilon_{ph} = hc/\lambda$, is equal or greater than the energy difference between an occupied and an optically allowed empty electronic state. In a semiconductor, the minimum energy at which this electronic transition occurs is known as the optical bandgap, ε_{opt}. The value of ε_{opt} is generally different from the value of the so-called transport bandgap, $\varepsilon_{transport}$, defined as the minimum energy difference needed to produce a free charge carrier, a free electron, or hole. If IE is the ionization energy of the semiconductor, that is, the minimum energy required to remove an electron, and EA is its electron affinity, that is, the energy gained by adding a negative charge carrier, then $IE - EA = \varepsilon_{transport}$. The difference between these two energies is known as the exciton binding energy, $\varepsilon_B = \varepsilon_{transport} - \varepsilon_{opt}$, which characterizes the strength of the electrostatic interaction between the hole left in the ground state and the optically excited electron in a higher energy electronic state. The value ε_B is of fundamental importance for understanding charge generation in molecular solids with low dielectric constant values. In inorganic crystalline semiconductors, the value ε_B is comparable to the thermal energy and consequently, $\varepsilon_{opt} \approx \varepsilon_{transport}$ at room temperature. In other semiconductors, this equivalency should not be taken for granted. Beyond direct electronic excitations described, light can be absorbed through other physical phenomena such as photon-phonon interactions, which in indirect gap semiconductors leads to phonon-assisted absorption.

Regardless of the mechanism of photon absorption, the photogeneration of free charge carriers increases the density of holes (p) and electrons (n). Under steady-state conditions the total density of carriers can be expressed as:

$$n = n_{dark} + \Delta n_{ph,n} = n_{dark} + G_{ph}\tau_n$$
$$p = p_{dark} + \Delta n_{ph,p} = p_{dark} + G_{ph}\tau_p, \qquad (1.15)$$

where G_{ph} is the photogeneration rate of electrons (holes) per unit time and unit volume, and $\tau_{n(p)}$ is the average lifetime of electrons (holes), defined as:

$$\tau_n(t) = \frac{\Delta n}{\partial \Delta n(t)/\partial t} = \frac{\Delta n}{R}$$
$$\tau_p(t) = \frac{\Delta p}{\partial \Delta p(t)/\partial t} = \frac{\Delta p}{R},$$
(1.16)

where R is the rate of recombination of excess charge carriers, which may arise from one or more physical processes, and is generally proportional to the product of the n and p carrier densities, this is: $R \sim np$. In steady-state and under continuous monochromatic illumination, the generation rate in Eq. (1.15) can be written as:

$$G_{ph} = \eta_{ext} \frac{E_{0,ph}}{d}$$
(1.17)

where η_{ext} is the external quantum efficiency, d is the sample's thickness and $E_{0,ph} = \phi_{det}/(\varepsilon_{ph} A_{beam})$ the incident photon irradiance. Note that η_{ext} can be written as $\eta_{ext} = \eta_A \eta_{int}$. Where η_A corresponds to the photoactive layer absorptance: $\eta_A = \tilde{A}_{PhL}$, and the internal quantum efficiency where η_{int} can be expressed as the product of the photogeneration efficiency η_{phgen}, and the charge collection efficiency $\eta_{collection}$. This allows the external quantum efficiency to be defined as:

$$\eta_{ext} = \tilde{A}_{PhL} \eta_{phgen} \eta_{collection}$$
(1.18)

This equation contains a few modifications with respect to the ones often found in the literature in the context of crystalline inorganic semiconductor photodetectors:

1. The more general term $\tilde{A}_{PhL}$ replaces the commonly used term $(1 - \tilde{R}_0)(\alpha d)$, representing the absorptance of the photoactive layer based on the incoherent transmission of light. As discussed, the propagation of light in thin film photodetectors cannot be considered purely incoherent. Interference effects arising from light being partially coherent at those scales and the fact that the photoactive layer is often in a cavity structure can substantially change the amount of energy absorbed by a thin-film photodetector. Although close solutions can be found to calculate $\tilde{A}_{PhL}$ (Kishino et al., 1991), in general, it is a nontrivial calculation that requires knowledge of the refractive index values of the individual layers in a device and of its geometry. However, if the photodetector transmittance can be assumed to be zero, that is, $\tilde{T}_{PD} = 0$, a measurement of a device's reflectance can be used to approximate $\tilde{A}_{PhL}(\lambda) \approx \tilde{A}_{PD}(\lambda) = 1 - \tilde{R}_{PD}(\lambda)$ since the total absorptance of the device can be calculated by adding up the absorptance of the individual layers: $\tilde{A}_{PD} = \tilde{A}_{PhL} + \tilde{A}_{transparent\_electrode} + \ldots$, and in most photodetectors, absorption losses in all other layers will be negligible compared to those in the photoactive layer.
2. In contrast to most inorganic crystalline semiconductors, in many emerging semiconductors η_{phgen} is often a function of the temperature T and the applied voltage V. This is because in excitonic materials, ε_B can be substantially larger than $k_B T$, the thermal energy at room temperature. Consequently, excitons will require extra energy to be dissociated

into free-charge carriers. This energy can be supplied by an externally-applied electric field, by increasing the temperature or, for organic bulk heterojunctions, by engineering the frontier orbitals of electron-donating and accepting molecules to promote the formation of charge-transfer states having reduced exciton binding energy values.
3. In inorganic crystalline semiconductors, high charge carrier mobility and low trap density values reduce recombination losses. In contrast, emerging semiconductors often have a large degree of positional and energetic disorder which may yield low charge mobility and high trap density values that lead to substantial recombination losses in the bulk and/or at the charge collecting electrodes. Hence, the collection efficiency $\eta_{\text{collection}}(V, T, \phi_{\text{det},B})$ in emerging semiconductor devices is generally a complex function of: i) the temperature, since charge mobility and recombination are thermally activated phenomena; ii) of applied voltage, because the charge mobility is generally electric field dependent when carriers move by incoherent hopping transport; and iii) of optical power, because trap filling and recombination losses may depend on the density of photogenerated charge carriers.

Since the conductivity is $\sigma = n\mu_n + p\mu_p$, where $\mu_{n(p)}$ is the electron or hole mobility, a change of charge density leads to a differential change of the semiconductor's conductivity given by

$$\begin{aligned}\sigma_{ph} &= q\left(\Delta n \mu_n + \Delta p \mu_p + n\Delta \mu_n + p\Delta \mu_p\right) \\ &= q\left(G_{ph}\tau_n \mu_n + G_{ph}\tau_p \mu_p + n\Delta \mu_n + p\Delta \mu_p\right),\end{aligned} \quad (1.19)$$

where the terms $\Delta\mu_{n(p)}$ arise because, in many disordered solids, the mobility can be charge density dependent. The total conductivity is: $\sigma = \sigma_{ph} + \sigma_{dark}$. Therefore, the total current density ($j = \sigma F$, where F is the electric field) can be separated as:

$$j(t) = j_{ph}(t) + j_{dark}(t). \quad (1.20)$$

From Eq. 1.17 and Eq. 1.19 is clear that j_{ph} is proportional to the total number of absorbed photons in a given volume and, consequently, linearly proportional to the irradiance, and consequently to the optical power; for as long as charge recombination processes in the material do not change. However, Eq. (1.16) also shows that the average carrier lifetime scales linearly with the excess density of free charge carriers and is inversely proportional to the recombination rate. Recombination processes in a semiconductor do not remain constant over the many orders of magnitude in which the optical power can vary in any given photodetector application. Hence, the internal quantum efficiency, and consequently the responsivity, may change at different irradiance values. This is shown in Figure 1.4A using a logarithmic plot of the photocurrent versus irradiance (optical power) measured five times in a low-noise SiPD. A clear dispersion of the data is observed at low optical power values. These changes are better seen when the responsivity is plotted versus irradiance or optical power, as in Figure 1.4B, also showing the statistical nature of these changes, since repeated measurements on the same device yield a wide spread of values. Deviations from linearity, have been observed in inorganic and organic photodiodes (Fuentes-Hernandez et al., 2020a) and are expected to be of concern

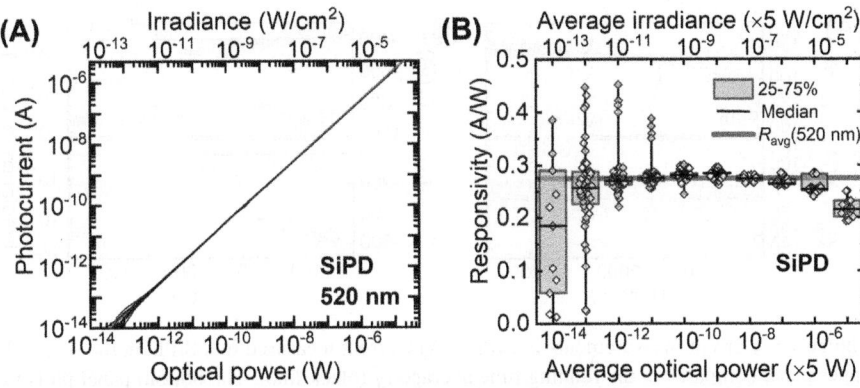

Figure 1.4 Responsivity and linearity (A) Six measurements (black) of the steady-state photocurrent measured in SiPD as a function of optical power and irradiance. The Gray line is the extrapolated constant responsivity. (B) Box plots showing responsivity values measured as a function of the average irradiance and optical power. Average power/irradiance taken per decade.

for other thin-film semiconductor photodetector technologies (Fang et al., 2019), so it is necessary that statistical analysis and careful measurements over illumination conditions varying many orders of magnitude, especially close to darkness, is carried out. Data on low noise SiPDs is used here to show that these effects are of general concern.

Using Eq. (1.10), the photocurrent ($i_{ph} = j_{ph} A_{det}$) in a semiconductor photodetector can be defined as:

$$i_{ph,B}(\phi_{det,B}, V, T) = \int R_{i_B}(\phi_{det,B}, V, T, \lambda) \phi_{det,B}(\lambda) d\lambda \quad (1.21)$$

The photocurrent can be measured directly as the average difference between the total current under illumination $\overline{i_{ill,B}}(\phi_{det,B}, V, T)$ and current in the dark $\overline{i_{dark,B}}(V, T)$. Hence,

$$\overline{i_{ph,B}}(\phi_{det,B}, V, T) \equiv \left| \overline{i_{ill,B}}(\phi_{det,B}, V, T) - \overline{i_{dark,B}}(V, T) \right| \quad (1.22)$$

At high optical power values $\overline{i_{ill,B}}(\phi_{det,B}, V, T) \gg \overline{i_{dark,B}}(V, T)$, the dark current can be neglected. However, when $\overline{i_{ill,B}}(\phi_{det,B}, V, T) \approx \overline{i_{dark,B}}(V, T)$, careful consideration of the dark current and its dynamics is needed.

In the dark, semiconductor photodetectors can display very slow current dynamics because of deep trapping states. Figure 1.5A shows the temporal evolution of the dark current in a SiPD. An accurate determination of $\overline{i_{ph,B}}(\phi_{det,B}, V, T)$ and of its statistical behavior, requires subtracting slow varying current components from the measured current i_B. Because dynamical behavior of i_B is slow and generally well approximated as an exponential decay, an exponential fit of the data yields residuals

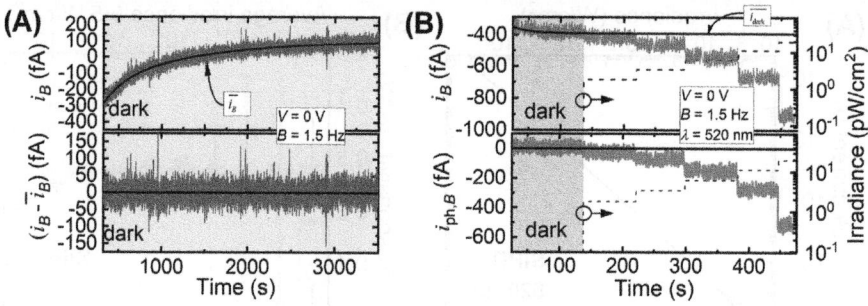

Figure 1.5 Current measurements in SiPDs (A) Current measured in a SiPD in the dark. The upper panel compares i_B the running time average $\overline{i_B}$ (black line). The bottom panel plots the error between them. (B) Upper panel plots the measured current in an experiment wherein the optical power of a 520 nm LED was varied from 0 to 20 pW/cm^2. Shaded areas in both graphs correspond to data measured in the dark.

that are described as random fluctuations around a zero mean, as shown in the bottom panel of Figure 1.5B. As we will discuss in Section 1.7, only then, do the stochastic properties of these residuals, or errors, allow measured fluctuations to become suitable for statistical analysis. Using the same method, $\overline{i_{ph,B}}(\phi_{det,B}, V, T)$ can be determined by measuring $i_B(V, T)$ as the illumination over the sample is changed from complete darkness, as shown in Figure 1.5B. Note that $\overline{i_{dark}}$ values are extrapolated to time values where the sample, experimentally, is being illuminated. Because illumination can potentially change the magnitude $\overline{i_{dark}}$, this extrapolation contributes to the error in estimating the noise equivalent power value. However, it also allows a direct measurement of the electronic noise in the dark, that is, $i_{rms,B}$, and during different conditions of illumination, enabling direct measurements of the *SNR* as a function of irradiance or optical power. Because a low-noise electrometer is needed to resolve very small current values (Fuentes-Hernandez et al., 2020a,b), this approach has a general limitation in that the measurement bandwidth is small compared with other approaches such as using lock-in measurements (Fang et al., 2019). Unfortunately, if the electronic noise of the photodetector is below the lock-in amplifier electronic noise, this approach cannot be used to estimate the *SNR*.

From past discussions, the spectral responsivity is related to the spectral external quantum efficiency as:

$$\eta_{ext}(\phi_{det,B}, V, T, \lambda) = \frac{hc}{q\lambda} R_{i_B}(\phi_{det,B}, V, T, \lambda) \quad (1.23)$$

where h is Planck's constant, q is the elementary charge and c the speed of light in a vacuum. Figure 1.6A shows a comparison of the spectral responsivity measured in silicon photodiodes (SiPDs) and a thin-film photodetector, here organic photodiodes (OPDs) are described in (Fuentes-Hernandez et al., 2020a). Continuous lines are plots of Eq. (1.23) where it has been assumed that the external quantum

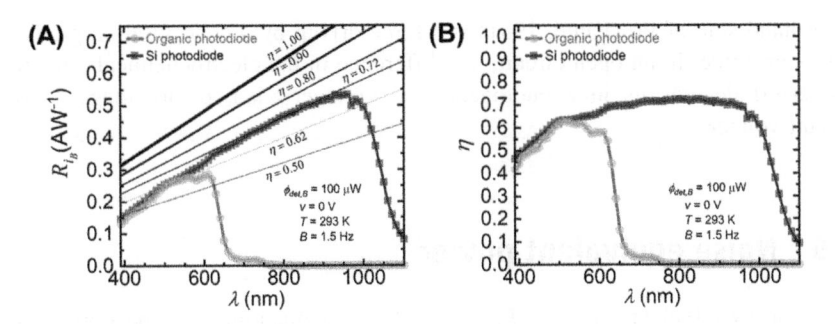

Figure 1.6 Spectral responsivity and external quantum efficiency (A) Comparison of the spectral responsivity of Si photodiode and organic photodiode. Lines represent the ideal responsivity dependence for different constant external quantum efficiency values (lines). (B) Comparison of spectral external quantum efficiency of Si photodiode and organic photodiode. Insets display experimental conditions.

efficiency is constant. In Figure 1.6B, a comparison of the external quantum efficiency of both types of photodiodes is shown.

1.5.2 Photogenerated voltage

The photovoltage produced by a photodetector can be defined by the same expression used to define the photogenerated current. This is:

$$V_{ph,B}(T) = \int R_{V_B}(\phi_{det,B}, T, \lambda) \phi_{det,B}(\lambda) d\lambda \tag{1.24}$$

where $V_{ph,B}(T) = |V_{ill,B}(T) - V_{dark,B}(T)|$. For most devices of interest, $V_{dark,B}(T) = 0$. A change of voltage in the photodetector reflects an increased charge given by $\Delta Q = V_{ph,B} C_{det}$, where the photodetector capacitance density is given by: $C_{det} = \varepsilon \varepsilon_0 A_{det}/d$ and d is the photoactive layer thickness.

Photogenerated charge carriers in a semiconductor rapidly thermalize and give rise to the emergence of quasi-Fermi levels across the material. This is an electrochemical potential, with regions of low and high charge carrier concentrations that allow free charge carriers to drift, and diffuse from regions of high carrier concentration to regions of lower carrier concentration. According to classical mechanics, the force field associated with the electrochemical potential is:

$$\begin{aligned} \nabla \varepsilon_{fn}(\mathbf{r}, T) &= \nabla E_c(\mathbf{r}) + kT \frac{\nabla n_c(\mathbf{r}, T)}{n_c(\mathbf{r}, T)} \\ \nabla \varepsilon_{fp}(\mathbf{r}, T) &= \nabla E_v(\mathbf{r}) - kT \frac{\nabla p_v(\mathbf{r}, T)}{p_v(\mathbf{r}, T)}. \end{aligned} \tag{1.25}$$

The first term on the right-hand side of Eq. (1.25) can be identified as an electric force: $q\mathbf{F} = \nabla E_{c,v}(\mathbf{r})$, where $\mathbf{F}$ is the local electrical field. The second term on the right-hand side of Eq (Fante, 1981) is a force driven by a concentration gradient, a diffusion force. In an open circuit, the difference in the electrochemical potential at the electrodes results in a photogenerated voltage, a.k.a. photovoltage or open-circuit voltage.

1.6 Noise equivalent power

The noise equivalent power (*NEP*) quantifies the magnitude of the optical signal needed to produce an electrical signal of the same magnitude as the photodetector's electrical noise. This is an electrical *SNR* = 1. The *NEP* is defined as:

$$\begin{aligned} NEP_B &= \frac{Signal_{optical}}{SNR_{electrical}|_{SNR=1}} \\ &= Signal_{optical} \frac{Noise_{electrical}}{Signal_{electrical}} \\ &= \int x_{\det,B}(\lambda)\, d\lambda \frac{y_{rms,B}}{\int R_{y_B}(x_{\det,B}, \lambda) x_{\det,B}(\lambda)\, d\lambda} \end{aligned} \qquad (1.26)$$

where y_B is either a time-varying current or voltage, and $y_{rms,B}$ is the electronic noise defined through Eq. (1.8). Note that the $NEP_B \to 0$ when $SNR_{electical}$ values are much larger than one. However, the NEP_B is only meant to be evaluated at $SNR_{electical} = 1$. Since the *SNR* is unit less, the NEP_B has the same units than $x_{\det,B}$, typically an optical power in SI units of W. Definitions in Eq. (1.26), also convey two simple messages that can be overlooked: (1) all physical quantities are measured with the same measurement bandwidth; (2) the magnitude of the optical signal that produces a $SNR_B = 1$, enters this equation not only as a multiplicative factor but also as a variable that determines the amplitude of the electrical signal. Also note that Eq. (1.26) is commonly written in the literature as:

$$NEP_B = \frac{y_{rms,B}}{\mathfrak{R}_{y_B}(x_{\det,B})} \qquad (1.27)$$

However, the functional dependences highlighted in Eq. (1.26), could and are often lost through this definition because $\mathfrak{R}_{y_B}(x_{\det,B})$ is treated as a constant parameter with respect to $x_{\det,B}$. The NEP_B can also be defined as a spectral quantity $NEP_B(\lambda)$, by removing the integration over wavelength from Eq. (1.26).

The NEP_B is the key metric that characterizes the physical limit to the performance of a photodetector since it represents the minimum amount of optical input that is detectable. However, it has an idiosyncratic characteristic in that better photodetectors achieve smaller NEP_B values. Therefore, the detectivity was defined as: $D = NEP_B^{-1}$, to have the reassuring quality of being larger for better-performing photodetectors.

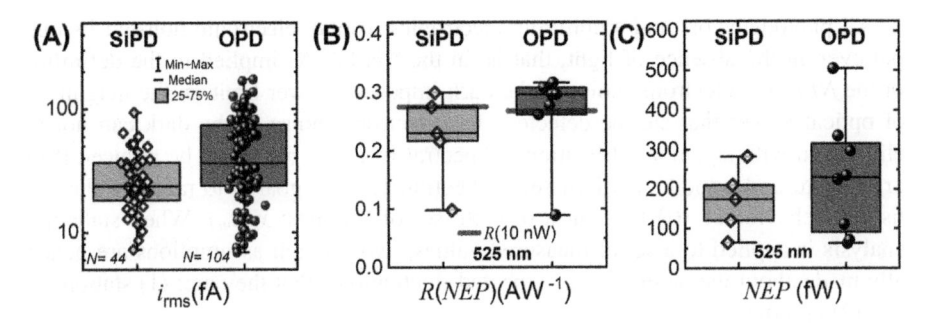

Figure 1.7 Noise and Noise equivalent power in SiPDs and OPDs (A) Electronic noise box charts. (B) Responsivity box chart. Gray lines are responsivity values at 10 nW. (C) Noise equivalent power box chart.

To estimate the *NEP* it is commonly assumed that the responsivity remains constant as the optical power is reduced. Consequently, the photocurrent is extrapolated to reach the estimated or measured value of the electronic noise in a logarithmic plot such as the one shown in Figure 1.4A. Even when the noise is measured, as shown in Figure 1.5, Figure 1.7A clearly shows that there is a widespread distribution of values so extrapolations should at minimum account for the statistical spread of these measurements. Furthermore, when the *NEP* is measured from direct measurements of the *SNR* at low optical powers as shown in Figure 1.5, lower *NEP* values may not be found in devices or during experimental conditions yielding smaller electronic noise values. This is because, as shown in Figure 1.7B, the responsivity values also change at the *NEP*, acquiring values below or above the mean value measured at higher optical powers; even when the same device is measured multiple times. Hence, measured values of *NEP* shown in Figure 1.7C, reflect complex dynamic physical processes in the devices and need to be better understood. What is clear from these experiments, is that extrapolated *NEP* values will generally be an unreliable estimator of the *NEP* even in commercial photodiode technologies such as low-noise SiPDs (Fuentes-Hernandez et al., 2020a).

1.7 Measurement and noise

In a seminal paper, Callen and Welton stated (Callen and Welton, 1951): "The parameters which characterize a thermodynamic system in equilibrium do not generally have precise values, but undergo spontaneous fluctuations." Noise is fundamental to nature, and can be broadly defined as the random fluctuations of a physical quantity $x(t)$ around a mean value. These fluctuations prevent the value of $x(t)$ to be known at any time. Deterministic physical theories are developed only in relation to temporal means of physical quantities, that is, $\langle x(t) \rangle$. Under certain circumstances, the noise amplitude can be described by the magnitude of the temporal mean-variance $\langle x^2(t) \rangle$, that is, the square mean value of $x(t)$.

All photodetectors and amplifying electronics display electronic noise at its output even in the absence of light, that is, in the "dark." As implied in the definition of the *NEP*, the electronic noise in the dark imposes a lower limit on the magnitude of optical power that can be detected. The electronic noise in the dark can not be eliminated with signal amplification or spectral filtering, but it can be reduced if its origins are understood. As discussed in Section 1.3, the goal is to measure the statistical behavior of $x(t)$ through a discrete set of samples: $x_B(t_m)$. When statistical analysis is applied to a set of measured values, two implicit assumptions are generally made about the nature of the expected fluctuations, that they are: (1) stationary; and (2) ergodic.

The stationary assumption implies that its statistical properties are time-independent. This is the main reason behind the need to subtract the slow varying average $\overline{i_B}$ from i_B in Figure 1.5. For simplicity, let's assume that $\overline{x_B}(t_m) = 0$. By stationary, it is also implied that the time-averaged autocorrelation function, defined as $c_{xx}(t_m, t_m + t_n) \equiv \overline{x_B(t_m)x_B(t_m + t_\kappa)}$ has a value that does not depend on t_m but only on the time difference t_κ if averaged over a long enough period. Hence, $c_{xx}(t_m, t_m + t_\kappa) \equiv c_{xx}(t_\kappa)$. Since t_κ is an integer multiple of the integration time, that is, $t_\kappa = \kappa\tau$ with κ a positive or negative integer, and since only the relative time difference is of importance, we can write the time-dependent autocorrelation function as $c_{xx}(\kappa)$.

Ergodicity implies that over a very long period, the physical quantity $x_B(t_m)$ will take all the possible values of an, in principle unknown, statistical distribution. Consequently, its time average should approach the ensemble average as the integration time goes to infinity. This is a critical assumption, because the statistical properties of the ensemble, that is, the discrete set of measurements, are assumed to accurately represent those of the random fluctuations of $x_B(t_m)$, and implicitly, $x(t)$ over time.

The Wiener–Khintchine theorem demonstrates that if $x_B(t_m)$ is stationary and ergodic, then its power density spectrum is given by:

$$S_x(f) = \sum_{\kappa=-\infty}^{\infty} c_{xx}(\kappa) \, e^{-i2\pi\kappa f}, \quad |f| < \frac{1}{2}. \tag{1.28}$$

Corresponding to the discrete-time Fourier transform of the set $c_{xx}(\kappa)$, and having an inverse Fourier transform given by:

$$c_{xx}(\kappa) = \int_{-1/2}^{1/2} S_x(f) \, e^{i2\pi\kappa f} df. \tag{1.29}$$

If $\tau \to \tilde{\tau}$, $\overline{x_B(t)} \to \langle x(t) \rangle$ the autocorrelation function becomes a continuous time average:

$$c_{xx}(\tau) \equiv \lim_{T \to \infty} \frac{1}{2T} \int_{-T}^{T} x(t')x(t' + \tau) \, dt' = \langle x(t')x(t' + \tau) \rangle. \tag{1.30}$$

An important property that can be derived from either form of the autocorrelation function is that $|c_{xx}(\tau)|$ is always smaller than or equal to $c_{xx}(0) = \overline{x_B^2}(t) = \langle x^2(t) \rangle$. This is to say that the value of the autocorrelation is bound and reaches its maximum at $\tau = 0$. Using Eq. (1.30), $S_x(f)$ and $c_{xx}(\tau)$ become Fourier transform pairs:

$$\begin{aligned} S_x(f) &= \int_{-\infty}^{\infty} c_{xx}(\tau)\, e^{-i2\pi\tau f}\, d\tau \equiv \mathcal{F}\{c_{xx}(\tau)\} \\ c_{xx}(\tau) &= \int_{-\infty}^{\infty} S_x(f)\, e^{i2\pi\tau f}\, df \equiv \mathcal{F}^{-1}\{S_x(f)\}. \end{aligned} \qquad (1.31)$$

Furthermore, using Eqs. (1.30) and (1.31), and since $\mathcal{F}\{\langle x(t+\tau)x(t)\rangle\} = \mathcal{F}\{\langle x^2(t)\rangle\} = \langle X^2(f)\rangle$, we have that

$$S_x(f) = \lim_{T \to \infty} \frac{1}{2T} \int_{-T}^{T} X^2(f)\, dt' = \langle X^2(f) \rangle. \qquad (1.32)$$

Therefore, the power spectral density $\overline{x_B}(t)$ can be determined as the time average of the square of its Fourier transform. From this definition, it can be shown that $S_x(f)$ is a strictly positive, even function of f. Also, that at $f = 0$, $S_x(0) = \int_{-\infty}^{\infty} c_{xx}(\tau)d\tau = \langle X^2(0)\rangle$, while at $\tau = 0$ $\langle x^2(t) \rangle = c_{xx}(0) = \int_{-\infty}^{\infty} S_x(f) df = \int_{-\infty}^{\infty} \langle X^2(f) \rangle df$. This last equation corresponds to Parseval's theorem, which establishes an equivalence between the total area of the fluctuations in the time domain and the total area of its power density spectra in the frequency domain, which is a reassuring quality for energy conservation. Because $x(t)$ typically represents either a current or a voltage, $x^2(t)$ is proportional to a dissipated power or energy (e.g., the electrical power dissipated through a resistor R'_Ω is $P = i^2 R'_\Omega = V^2/R'_\Omega$). To avoid infinite energy, $x(t)$ must be a function for which the time average $\langle x^2(t) \rangle < \infty$.

Now consider a signal of the form $x(t) = \sum_n y_n(t)$, where $y_n(t)$ are different stochastic contributions to the signal. Because the Fourier transform is a linear operator we have that

$$\begin{aligned} \langle x^2(t) \rangle &= \left\langle \mathcal{F}\left\{ \left(\sum_n y_n(t)\right)^2 \right\} \right\rangle \\ &= \left\langle \mathcal{F}\left\{ \sum_n y_n^2(t) + \sum_{n \neq m} y_n(t) y_m(t) \right\} \right\rangle \\ &= \sum_n \langle Y_n^2(t) \rangle + \sum_{n \neq m} \langle Y_n(t) Y_m(t) \rangle \end{aligned} \qquad (1.33)$$

If $y_n(t)$ are statistically independent then $\langle Y_n(t) Y_m(t) \rangle = 0$ for all n and m, leading to the very important result,

$$\langle x^2(t) \rangle = \sum_n S_{y_n}(t). \qquad (1.34)$$

This relation will be used to calculate the total noise in a photodetector as the sum of statistically independent contributions.

1.7.1 Measuring electronics

Electrical circuits are used to measure physical quantities that are transduced into electrical signals. They amplify and in some cases, transduce one type of electrical signal into another; for example, current to voltage or vice versa. An amplification circuit, or amplifier, is characterized by its gain, defined as:

$$gain = 20\log(y_{out}/y_{in}) \tag{1.35}$$

where y_{in} represents the magnitude of the current or voltage signal before the amplifier, and y_{out} after the amplifier. Although the *gain* is a unit less quantity, when expressed as in Eq. (1.35) it is typically referred to as given in decibels or dB, as is most commonly referred to in electronics.

The effect of a measuring circuit on a signal, on the time and frequency domain, can be analyzed using linear filter theory, through the use of an impulse response function $g(t)$ and its corresponding transfer function $G(f) = \mathcal{F}\{g(t)\}$, heron referred to as power gain.

The impulse response actuates on an input signal $y_{in}(t)$ through its correlation to produce an output signal equal to

$$y_{out}(t) = y_{out}(t) * g(t) \tag{1.36}$$

Since the Fourier transform transforms the autocorrelation function into a product, it becomes easier to appreciate the effects of the circuit on a signal by considering that the input signal spectra $Y_{in}(f)$ will produce an output spectra given by:

$$Y_{out}(f) = Y_{in}(f)G(f) \tag{1.37}$$

In a circuit, the power gain corresponds to its total impedance $Z(f)$. Since the impedance for a resistor $Z_R = R_\Omega$ and for a capacitor is $(j2\pi fC)^{-1}$ for the simple RC circuit shown in Figure 1.8A, as in a voltage divider the total impedance is

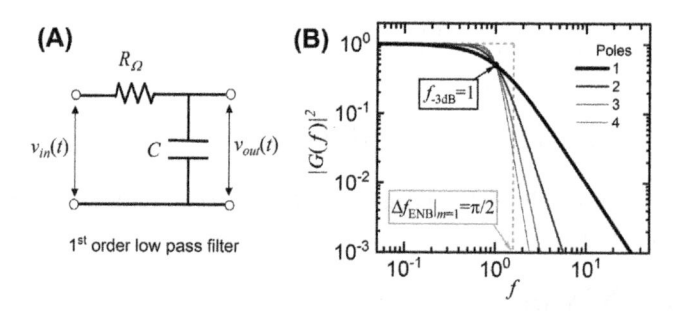

Figure 1.8 *RC* low pass filter. (A) *RC* circuit of first order band pass filter. (B) Power gain frequency dependence for 1 to 4 pole low pass filters with $f - _{3dB} = 1$ (continuous lines). Dashed lines represent a brick-wall filter with an effective noise bandwidth of $\Delta f_{ENB} = \pi/2$, corresponding to a first order (1 pole) low pass filter.

$Z_C(Z_C+Z_R)^{-1}$, and its power gain is given by

$$G(f) = \frac{V_{out}(f)}{V_{in}(f)} = \frac{1}{1+j2\pi f R_\Omega C} \tag{1.38}$$

Since $V_{in}(f) = I_{in}(f)R_\Omega$ the current to voltage gain function can be expressed in terms of the input current $I_{in}(f)$ spectra as:

$$G_{IV}(f) = G(f)R_\Omega = \frac{V_{out}(f)}{I_{in}(f)} \tag{1.39}$$

Using Eq. (1.32) the power spectral density of the output voltage noise produced by stochastic fluctuations of a current at the entrance of the circuit is given by

$$\begin{aligned} S_{V_{out}}(f) &= \langle V_{out}^2(f) \rangle \\ &= \langle I_{in}^2(f) |G_{IV}(f)|^2 \rangle \\ &= S_{I_{in}}(f) |G_{IV}(f)|^2 \\ &= \frac{S_{I_{in}}(f)}{1+(2\pi f R_\Omega C)^2} \end{aligned} \tag{1.40}$$

1.7.2 Effective noise bandwidth

To quantify the spectral filtering effect of a measuring circuit characterized by a power gain function $G(f)$, the effective noise bandwidth is defined as:

$$\Delta f_{ENB} = \int \left| \frac{G(f)}{G(f_{max})} \right|^2 df \tag{1.41}$$

which arises from equating the area underneath the normalized power gain function with that of a "brick-wall" filter of width Δf_{ENB}, as shown in Figure 1.8B. Eq. (1.41) assumes that the input noise is white, that is, frequency independent.

In electronics, a filter's bandwidth is typically specified in terms of the so-called -3dB bandwidth (f_{-3dB}). The -3dB bandwidth corresponds to the frequency at which the gain has decreased to 0.707 from its maximum value $G(f_{max})$, $20\log_{10}(0.707) \simeq -3dB$ for a voltage or current gain amplifier. For a power gain, that is, $G_{power} = 10\log(P_{out}/P_{in})$, $10\log_{10}(0.5) \simeq -3dB$ so the bandwidth corresponds to the frequency range to which the power drops $0.5 \times G(f_m)$. For a first-order low pass filter, it can be shown that $\Delta f_{ENB} = \pi f_{-3dB}/2$. For a second-order low pass filter with equal time constants, it is $\Delta f_{ENB} = 1.22 \times f_{-3dB}$. Hence, for an increased number of poles, the response of the filter becomes increasingly more rectangular and $\Delta f_{ENB} \to f_{-3dB}$, as shown in Figure 1.8B. For a passband filter comprising a single pole high pass filter with -3dB bandwidth $f_{-3dB,High}$ and a single pole low pass filter with -3dB bandwidth $f_{-3dB,Low}$, $\Delta f_{ENB} = \pi(f_{-3dB,High} - f_{-3dB,Low})/2$.

In the presence of frequency-dependent noise sources, for example, flicker noise, the effective noise bandwidth is given by

$$\Delta f_{ENB} = \frac{1}{S_{I_{in}}(f_{max})|G(f_{max})|^2} \int S_{I_{in}}(f)|G(f)|^2 df \tag{1.42}$$

The effective noise bandwidth is very important because it is the bandwidth that needs to be used to calculate the specific detectivity D^*. This is a direct consequence of (1.40) since the SNR in the NEP definition is measured through measurements of $\overline{v^2_{out,ENB}}(t)$ or $\overline{i^2_{out,ENB}}(t)$. Hence, from Parseval's theorem,

$$\begin{aligned}\overline{v^2_{out,ENB}}(t) &= \int_{-\infty}^{\infty} S_{V_{out}}(f) df \\ &= \int_{-\infty}^{\infty} S_{I_{in}}(f)|G(f)|^2 df \\ &= S_{I_{in}}(f_{max})|G(f_{max})|^2 \Delta f_{ENB}\end{aligned} \tag{1.43}$$

and consequently, if the ultimate interest of these measurements is to estimate the photodetector's current noise

$$\overline{i^2_{in,ENB}}(t) = \frac{\overline{v^2_{out,ENB}}(t)}{|G(f_{max})|^2} = S_{I_{in}}(f_{max})\Delta f_{ENB} \tag{1.44}$$

As it should be clear from these definitions, what we have previously referred to as the "measurement bandwidth" corresponds to the effective noise bandwidth, hence $B \equiv \Delta f_{ENB}$. To emphasize this point hereon Δf_{ENB} is used.

The effective noise bandwidth is not commonly discussed in the context of thin-film photodetector characterization. This is in great part because technological advances and widespread commercialization of automated measurement tools have hidden all the details of the amplifying electronics and signal processing needed to measure physical quantities, particularly, when their magnitude is very small. Indeed, Eqs. (1.41) and (1.42) present a number of difficulties:

1. The power gain function of commercial measurement equipment may not be known.
2. The power noise spectra of a new photodetector are generally unknown.
3. For any nonwhite noise source, the integral must be solved to estimate the ENB.

A potential approach to estimate the effective noise bandwidth is to use the same measurement equipment used for photodetector characterization, for example, an electrometer, to measure the current or voltage variance of an electrical element, such as a resistor, having a known spectral power distribution, and to estimate the effective noise bandwidth from those measurements, as done in (Fuentes-Hernandez et al.,

2020a). The use of lock-in measurement techniques (Fang et al., 2019) offers an alternative to more accurately control and significantly reduce the effective noise bandwidth; which is typically inversely proportional to the averaging time (Libbrecht et al., 2003). In either case, the goal of photodetector characterization is to measure the intrinsic sources of electronic noise, so it is important to minimize noise sources associated with the experimental setup by proper electrical grounding, shielding, and reducing parasitic capacitances, etc. If the f_{3dB} bandwidth of the amplifier is known, another approach may be to use f_{3dB} as a low bound to the value of Δf_{ENB}, as it is shown in Figure 1.8B.

1.7.3 Sources of noise

Noise is inherent to nature. Using quantum mechanics, Callen and Welton (1951) demonstrated that any physical system capable of absorbing energy when subjected to a time-periodic perturbation will produce random fluctuations with a power spectral density $S_y(f)$ given by:

$$S_y(f) = 4k\mathfrak{J}(f)R(f), \tag{1.45}$$

where $R(f)$ is a dissipation function associated with a force y driving the perturbation over time and

$$\mathfrak{J}(f) = \frac{hf}{k_B} \frac{1}{e^{hf/k_B T} - 1}, \tag{1.46}$$

is a generalized temperature function. For frequency values such that $k_B T > hf$, $\mathfrak{J} = T$. For example, at 300K, $\mathfrak{J} = T$ up to $f = 10^{10}$ Hz and falls to 0K when $f > 10^{14}$ Hz, which means that for most electronic applications it can be considered a constant, but still, formally fulfills the condition: $\int_{-\infty}^{\infty} S_x(f) df < \infty$

1.7.4 Thermal noise

Eq. (1.45) is referred to as the general Nyquist formula since it reduces to the well-known result derived by Nyquist and Johnson for the case wherein $R(f) = R_\Omega$ is resistance and $y = v$ voltage. Hence the spectral density for the so-called thermal or Nyquist-Johnson noise can be written as:

$$S_{thermal} = 4k_B T R_\Omega \tag{1.47}$$

as implied by (1.44), the maximum variance of the voltage is given by

$$\overline{v_{thermal}^2} = 4k_B T R_\Omega \Delta f_{ENB} \tag{1.48}$$

If now these fluctuations are measured using the $R_\Omega C$ circuit in Figure 1.8 and described by Eq. (1.40)

$$\overline{v^2_{thermal}} = \int \frac{4k_B T R_\Omega}{1 + (2\pi f R_\Omega C)^2} df$$
$$= \frac{k_B T}{C} \qquad (1.49)$$

Which is expected since $f_{-3dB} = (2\pi R_\Omega C)^{-1}$ and consequently $\Delta f_{ENB} = \pi f_{-3dB}/2 = (4R_\Omega C)^{-1}$.

Similarly, the rms current produced by thermal noise in a resistor is given by

$$\overline{i^2_{thermal}} = \frac{4k_B T}{R_\Omega} \Delta f_{ENB} \qquad (1.50)$$

and if measured through a $R_\Omega C$ circuit it results in

$$\overline{i^2_{thermal}} = \frac{k_B T}{R_\Omega^2 C} \qquad (1.51)$$

1.7.5 Shot noise

Random uncorrelated discrete events in time follow a Poisson distribution given by:

$$P(N) = \frac{(\xi T)^N}{N!} e^{-\xi T} \qquad (1.52)$$

The mean value of this distribution is given by $\overline{N} = \xi T$ and most importantly: $\sigma_N^2 = \overline{N}$, that is, the mean is equal to its variance. Now, let's assume that each event produces a charge in the photodetector, using (1.5) yields $\overline{i_T} = q\overline{N} T^{-1}$. By definition of variance: $\overline{i_T^2} = q^2 \sigma_N^2 T^{-2} = q(q\overline{N} T^{-1})T^{-1} = q\overline{I} T^{-1}$. Because of the fundamental sampling theorem, $T^{-1} = 2\Delta f_{ENB}$, that is, shot noise will be sampled at twice the effective noise bandwidth frequency, therefore

$$\overline{i^2_{shot}} = 2q\overline{I_{ENB}} \Delta f_{ENB} \qquad (1.53)$$

corresponding to a power spectral density

$$S_{shot}(f) = 2q\overline{I_{ENB}} \qquad (1.54)$$

For very large numbers a Poisson distribution is expected to converge to Gaussian distribution with a mean equal to its variance: $\sigma_N^2 = \overline{N}$. This is the key property that gives rise to Eqs. (1.53) and (1.54), but not all Gaussian distributed random variables lead to these equations. Indeed, they will lead to expressions that could be derived from Eq. (1.45). This is a key observation, since not all currents measured in a circuit will inherently be shot noise.

The key assumption behind Poisson statistics, is indeed the uncorrelated nature of their random events. However, it is important to note that this assumption holds true not in the context of a physical quantity, but in the context of the physical processes that lead to the detection of a physical quantity. As an example, it must be noted that photon emission is not inherently uncorrelated. For instance, photons emitted by a black body obey Bose-Einstein statistics, and lead to an emission that has a power spectral density given by the expression (Jones, 1953)

$$S_{photon-noise}(f) = 8k_B T \sigma_{SB} T^4 \tag{1.55}$$

where σ_{SB} is the Stefan-Boltzmann constant. However, if the process of detection is a random process such as the random passage of charge carriers across a potential barrier (Ziel, 1970), then shot noise could still exist independent of the correlated nature of the photons arriving at the detector.

1.7.6 Generation-recombination noise

The random thermal generation and recombination, trapping, and detrapping of charge carriers, as well as variations in carrier lifetime produced by charge density changes, produce fluctuations in semiconductor photodetectors. The power spectral density distribution depends on the specific detector but for a photoconductor, it can be given by (Palmer and Grant, 2009)

$$S_{G-R}(f) = \frac{4R_\Omega^2 i^2 \tau_l}{\overline{N}(1 + 4\pi^2 f^2 \tau_l^2)} \tag{1.56}$$

where R_Ω is the photoconductor's resistance, τ_l is the carrier lifetime, and $\overline{N}$ is the average number of carriers in a volume. In this case, the power spectral density is shaped as a Lorentzian curve.

1.7.7 1/f or flicker noise

An important class of noise phenomena show power spectral density distributions that depend on the inverse of the frequency and are present in all photodetectors. Multiple physical phenomena have been associated with the emergence of so-called 1/f noise, flicker noise, or colored noise. For example, the superposition of Lorentzian spectra arising from generation and recombination noise sources but having different characteristics and times yields

$$S_{G-R}(f) \propto \int_0^\infty \frac{d\tau_l}{(1 + 4\pi^2 f^2 \tau_l^2)} \propto \frac{1}{f} \tag{1.57}$$

Other phenomena that can lead to $1/f$ noise in semiconductor devices (Ziel, 1988), include cases wherein the power spectral density is proportional to the dc current flowing in a device. In general, $1/f$ the noise will have a functional dependence given by

$$S_{1/f}(f) = K \frac{I^\delta}{f^{-\alpha}} \tag{1.58}$$

where α is a constant typically close to 1, δ is generally between 1 and 2 and K is a frequency and current independent constant.

1.8 Specific detectivity

The electronic noise in most photodetectors depends on the product $\sqrt{A_{\text{det}} \Delta f_{ENB}}$. Hence, to compare the performance of different photodetectors it is useful to define a new metric, the specific detectivity as:

$$D^* = D\sqrt{A_{\text{det}} \Delta f_{ENB}} = \frac{\sqrt{A_{\text{det}} \Delta f_{ENB}}}{NEP_B} \tag{1.59}$$

D^* is in SI units of cm•Hz$^{1/2}$W^{-1}, which are also referred to as Jones, in honor of R. Clark Jones who originally proposed this metric (Jones, 1947). The spectral $D^*(\lambda)$ is defined by replacing NEP_B with $NEP_B(\lambda)$ in Eq. (1.59). As recently discussed (Fuentes-Hernandez et al., 2020a; Fang et al., 2019) and discussed here, deficiencies in the determination of NEP values have caused important inaccuracies in the determination of D^* values. Critical evaluation of how D^* values are determined is needed before values measured for different technologies can be compared.

1.9 Photodiodes

Photodiodes are an important class of photodetector architecture in which emerging semiconductors are incorporated. Accurate characterization of current versus voltage characteristics in the dark and under illumination is needed to evaluate the photodetector metrics of the performance of a photodiode. As discussed, the goal of these measurements is to extract statistically relevant information to estimate the average and the variance of current or voltage values. To relate the statistical properties of measured data with the photodetector metrics of performance and theoretical descriptions of a photodiode, a central assumption is that the system is in a steady-state. Consequently, close attention needs to be placed on insuring that measured values correspond to steady-state. As shown in Figure 1.5a this can be challenging because the dark current dynamics can be extremely slow, particularly in low-noise photodiodes in the dark.

Accurate measurement of the current versus voltage characteristics of a photodiode allow extraction of important device parameters: (1) the electronic noise in the dark and as a function of voltage; (2) the responsivity of the photodiode under calibrated illumination; (3) extraction of the shunt, series resistance, reverse saturation currents and ideality factor values by fitting the data to an equivalent circuit model; (4) direct measurement of the noise equivalent power and estimation of the specific detectivity.

The current-voltage characteristics can be modeled using the equivalent circuit developed by Prince to describe silicon solar cells and later adopted for other photovoltaic technologies (Potscavage et al., 2008; Prince, 1955). Prince's equivalent circuit is shown in Figure 1.9A, accounts for parasitic effects attributed to the shunt resistance (R_p) and series resistance (R_s). In this model, hereon referred to as the P-model, the steady-state current density (j_{PM}) is given by:

$$j_{PM}(v,T) = \frac{R_p}{R_p + R_s}\left[j_0\left\{\exp\left(\frac{v - jR_sA_{PD}}{n_{id}k_BT/q}\right) - 1\right\} - \left(j_{ph} - \frac{v}{R_pA_{PD}}\right)\right]; \quad (1.60)$$

where k_B is the Boltzmann constant, n_{id} is the ideality factor, and j_{ph} is the photogenerated current density as described in Eq. (1.21). In the limit $R_p \to \infty$ and $R_s \to 0$, the Prince's model reduces to the well-known Shockley diode (S-model)

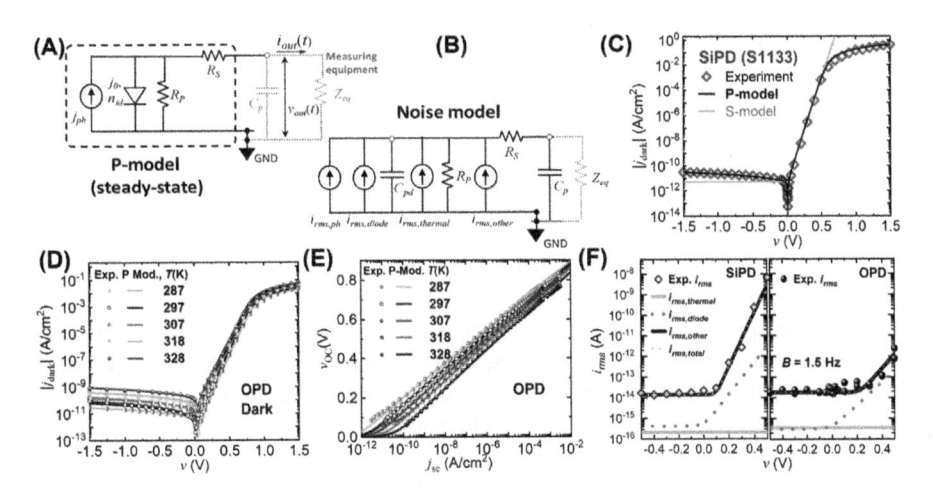

Figure 1.9 Photodiode modeling and noise (A) Equivalent circuit corresponding to P-model, also showing the impedance of the measuring equipment and a parasitic capacitance. (B) Equivalent circuit used for noise calculations. (C) Comparison between measured current densities versus voltage values and fits to the S-model and P-model in SiPD. (D) Comparison of measured versus voltage values and fits to P-model at different temperatures in OPD. (E) Comparison of measured and modeled open circuit voltage values and P-model predictions at different temperatures. (F) Measured electronic noise current in SiPDs and OPDs and comparison with calculated contributions to the electronic noise.

model (Shockley, 1949):

$$j_{SM}(v,T) = j_0\left[\exp\left(\frac{qv}{n_{id}k_BT}\right) - 1\right] - j_{ph} \tag{1.61}$$

In both models the reverse saturation current j_0, arises from charge carrier generation and recombination in the dark. The S-model reflects the principle of detailed balance (Shockley and Queisser, 1961), which states that in thermodynamic equilibrium, that is, in the dark and at $V = 0$ V, the average value of j_0 associated with the thermal generation and thermal recombination of charge carriers are equal but of opposite sign. Therefore, j_0 is an equilibrium current density that is proportional to the recombination or generation rate associated with the dominant charge recombination process in the dark, such as band-to-band, surface or trap-assisted recombination. The reverse saturation current can be described as a thermally activated process across the transport bandgap of the material, this is

$$j_0 = j_{00}\exp\left(\frac{q\varepsilon_{transport}}{n_{id}k_BT}\right) \tag{1.62}$$

where j_{00} is a multiplicative factor, and the ideality factor $n_{id} \approx 2\delta^{-1}$ is inversely proportional to the recombination order (δ) in the dark (Kirchartz et al., 2011; Göhler et al., 2018). The recombination order is commonly interpreted as the number of charge carriers available for recombination. Hence, for band-to-band recombination $\delta = 2$ yields $n_{id} = 1$, whereas for Shockley-Read-Hall recombination $\delta = 1$ yields $n_{id} = 2$. For trap-assisted recombination through an exponential trap distribution (Urbach tail) model, the ideality factor becomes temperature dependent and given by: $n_{id}(T)^{-1} = n_o^{-1} + (k_BT/2\varepsilon_U)$; where ε_U is the characteristic Urbach energy (van Berkel et al., 1993) and n_o is an ideality factor associated with the thermal activation of carrier lifetime (Foertig et al., 2012).

If now an electrical bias is applied, electrically injected carriers thermalize to produce distinct electron (n) and hole (p) thermal distributions having quasi-Fermi levels F_n and F_p, respectively; a condition that is defined as quasiequilibrium. The emergence of the quasi-Fermi levels gives rise to an electric potential $qv = |F_n\text{-}F_p|$ that represents a measure of how far from equilibrium is the semiconductor. In quasiequilibrium, the principle of detailed balance is once more valid, leading to a recombination/generation current that, using the Boltzmann approximation (i.e., $|F_n|$ and $|F_p| > 3k_BT$), can be simplified to the form:$j_{quasi-eq} = j_0\exp(qv/n_{id}k_BT)$. The total current density is then equal to: $j = j_{quasi-eq} - j_0$, which is equal to Eq. (1.61) in the absence of a photogenerated current density. The photogenerated current density enters the model as a field-independent quantity, a condition that due to the large binding energy displayed by some semiconductors may not always be held, but can easily be incorporated into the S or P models. It should be stressed that the S-model is rooted in fundamental thermodynamic principles which should apply to any photodiode in the absence of parasitic effects, other than those accounted for by the P-model through the shunt and series resistance. In the dark, these effects generally relate to changes of the ideality factor value with respect to the applied voltage, which reflect changes in the dominant recombination mechanism as injection of charge carriers into the devices increases under forward bias (Kirchartz et al., 2013).

The use of these equivalent circuit models to analyze the photodetector properties of SiPDs and OPDs was described in (Fuentes-Hernandez et al., 2020a). A summary of the results is presented as an example and to highlight key features of the analysis. Figure 1.9C shows the current voltage characteristics of a low noise SiPD and corresponding fits to the P-model using $j_0 = 5 \, \text{pA/cm}^2$, $n_{id} = 1.05$, $R_p = 786 \, \text{G}\Omega$ and $R_s = 30 \, \Omega$, and to the S-model for comparison. As it is clear from these fits, the P-model yields an excellent fit to measured steady-state data. The effects produced by the series resistance, in reducing the current in forward bias, and by the shunt resistance, in increasing the current in reverse bias, are clearly seen when compared with the values predicted by the S-model. In the real-world, the shunt resistance of a diode will not only prevent its dark current reverse bias from reaching its limiting J_0 value, but introduces an uncorrelated source of noise to the diode's output current, thermal noise. Figure 1.9D and E show that the P-model can also be used to accurately fit the current density versus voltage in the dark and the open-circuit versus short circuit current characteristics under illumination in the OPDs described in (Fuentes-Hernandez et al., 2020a) by using: $j_0 = 1.2 \, \text{pA/cm}^2$, $n_{id} = 1.515$, $R_p = 217 \, \text{G}\Omega$ and $R_s = 165 \, \Omega$ at 297K. For these OPDs, the temperature dependence of J_0 follows Eq. (1.62) with parameters $j_{00} = 650 \, \text{A/cm}^2$ and $\varepsilon_{transport} = 1.31$ eV which were found consistent with values for the materials used in these OPDs (Fuentes-Hernandez et al., 2020a). The ideality factor remains relatively constant as a function of temperature and in the range between Eqs. (1.52) and (1.51), while a best fit to the temperature dependence of the shunt resistance yields: $R_p(T) = (209 \, \text{G}\Omega) \times \exp\left[-(T-298)/12.6\right]$.

From steady-state dark current measurements the electronic noise can be calculated as an rms current value, i_{rms}; see Eq. (1.8). Figure 1.9F shows a comparison of the i_{rms} values measured in SiPDs and OPDs as a function of applied voltage. To rationalize i_{rms} values, parameters derived from fits to the equivalent circuit model can be used to calculate the power spectral densities of thermal and shot noise contributions. Figure 1.9b shows the equivalent circuit used for noise analysis, wherein the diode noise contribution is taken as uncorrelated from those of the parasitic resistances. Consequently, their variances or power spectral values add as in Eq. (1.34).

Even at zero volts, the contribution to the photodiode current flowing through the shunt V/R_p will fluctuate with a thermal noise power spectral density determined by: $S_{thermal} = 4k_B T/(R_p + R_s) \simeq 4k_B T/R_p$. In addition to this current, the diode current, determined by $j_{SM}(v,T)$ in Eq. (1.61) and not the total diode current in Eq. (1.60), is expected to fluctuate with two shot-noise power spectral density contributions arising from its equilibrium current $i_{eq} = j_0 A_{PD}$, and from its nonequilibrium current, that is, $i_{non-eq} = j_{non-eq} A_{PD} = [j_{SM}(v,T) + j_0] A_{PD}$ (Van Der Ziel, 1970). Consequently, the total shot-noise contribution of the diode is given by:

$$S_{diode} = 2q\left[i_{non-eq} + i_{eq}\right] = 2q j_{SM}(v,T) + 4q j_0 \qquad (1.63)$$

This expression is thermodynamically consistent (Van Der Ziel, 1970; Stratonovich, 1992; Wyatt and Coram, 1997) in that it avoids a nonphysical consequence of considering that the shot noise of produced by a photodiode is proportional to the total current, that is, $S_{diode} = 2qJ_{PM}A_{PD}$ which unphysically vanishes at $V = 0$ V, instead of converging to the value $4qj_0$ as implied by Eq. (1.63). A formal derivation of a diode's power spectral density in the context of nonlinear nonequilibrium thermodynamics (Stratonovich, 1992) yields:

$$S_{diode} = 2q\coth\left(\frac{qV}{2n_{id}k_BT}\right)j_{SM}(v,T) \tag{1.64}$$

which can be reduced to Eq. (1.63). It is important to note that this is also equivalent to previous expressions derived in ref (Van Der Ziel, 1970) to account for recombination noise in p-n junction diodes, showing the clear connection that exists between the ideality factor and the physics of recombination in the photodiode. The noise model also consider the shot noise produced by the absorption of light, with power spectral density: $S_{diode} = 2qi_{ph}$ which vanishes in the dark.

Using the parameters derived from fitting data to the P-model, in the dark we can estimate the magnitude of other noise contributions by

$$i_{rms}^2 = \left(S_{thermal} + S_{shot-noise} + \sum S_{others}\right)\Delta f_{ENB} \tag{1.65}$$

The results of these calculations are shown in Figure 1.9F, which demonstrate that neither the thermal nor the shot noise contributions dominate measured i_{rms} values in SiPDs or OPDs at low bandwidth values of 1.5 Hz.

Although not discussed here, to estimate the transient response of the photodiode, its internal capacitance C_{pd}, parasitic capacitances in the setup, at the contact with the instrument, C_p, as well as the instrument's impedance Z_{eq}, along with the photodiode's intrinsic response time will have to be combined to determine the measured response time. For this reason, they have been included in the equivalent circuits shown in Figure 1.9.

Acknowledgments

The author acknowledges financial support from the Department of Energy National Nuclear Security Administration (NNSA), award no. DE-NA0003921 through the Consortium for Enabling Technologies and Innovation (ETI). He also thanks Prof. Bernard Kippelen and students in his group whom through years of fruitful discussions have contributed and motivated efforts to better understand photodiode characterization. He finally thanks his family, Kate, Galia and Lucas, for their patience and loving support.

References

Born, M., Wolf, E., 1980. Principles of Optics: Electromagnetic Theory of Propagation, Interference and Diffraction of Light. Elsevier Science & Technology, Saint Louis.

Callen, H.B., Welton, T.A., 1951. Irreversibility and generalized noise. Phys. Rev. 83 (1), 34–40.

de Arquer, F.P.G., et al., 2017. Solution-processed semiconductors for next-generation photodetectors. Nat. Rev. Mat. 2, 16100.

Fante, R.L., 1981. Relationship between radiative-transport theory and Maxwell's equations in dielectric media. J. Opt. Soc. Am. 71 (4), 460–468.

Fang, Y., et al., 2019. Accurate characterization of next-generation thin-film photodetectors. Nat. Photonics 13 (1), 1–4.

Foertig, A., et al., 2012. Shockley equation parameters of P3HT:PCBM solar cells determined by transient techniques. Phys. Rev. B 86 (11), 115302.

Foley, J.T., Wolf, E., 1991. Radiance functions of partially coherent fields. J. Mod. Opt. 38 (10), 2053–2068.

Fuentes-Hernandez, C., et al. In low-noise large-area organic photodiodes. In: Organic and Hybrid Sensors and Bioelectronics XIII, International Society for Optics and Photonics. 2020b; p 114750N.

Fuentes-Hernandez, C., et al., 2020a. Large-area low-noise flexible organic photodiodes for detecting faint visible light. Science 370 (6517), 698.

Göhler, C., et al., 2018. Nongeminate recombination in organic solar cells. Adv. Electron. Mater. 4 (10), 1700505.

Jansen-van Vuuren, R.D., et al., 2016. Organic photodiodes: the future of full color detection and image sensing. Adv. Mater. 28 (24), 4766–4802.

Jones, R.C., 1949. Factors of merit for radiation detectors*. J. Opt. Soc. Am. 39 (5), 344–356.

Jones, R.C., 1947. The ultimate sensitivity of radiation detectors. J. Opt. Soc. Am. 37 (11), 879–890.

Jones, R.C., 1953. Performance of detectors for visible and infrared radiation. In: Marton, L. (Ed.), Advances in Electronics and Electron Physics, Vol. 5. Academic Press, pp. 1–96.

Kirchartz, T., et al., 2011. Recombination via tail states in polythiophene:fullerene solar cells. Phys. Rev. B 83 (11), 115209.

Kirchartz, T., et al., 2013. On the differences between dark and light ideality factor in polymer:fullerene solar cells. J. Phys. Chem. Lett. 4 (14), 2371–2376.

Kishino, K., et al., 1991. Resonant cavity-enhanced (RCE) photodetectors. IEEE J. Quantum Electron. 27 (8), 2025–2034.

Libbrecht, K.G., et al., 2003. A basic lock-in amplifier experiment for the undergraduate laboratory. Am. J. Phys. 71 (11), 1208–1213.

Natali, D., Caironi, M., 2016. 7—Organic photodetectors. In: Nabet, B. (Ed.), Photodetectors. Woodhead Publishing:, pp. 195–254.

Noriega, R., et al., 2013. A general relationship between disorder, aggregation and charge transport in conjugated polymers. Nat. Mater. 12 (11), 1038–1044.

Ombaba, M.M., et al., 2016. 4—Nanowire enabled photodetection. In: Nabet, B. (Ed.), Photodetectors. Woodhead Publishing:, pp. 87–120.

Palmer, J.M., Grant, B.G., 2009. Art of Radiometry. SPIE, Bellingham.

Potscavage, W.J., et al., 2008. Origin of the open-circuit voltage in multilayer heterojunction organic solar cells. Appl. Phys. Lett. 93 (19), 3.

Prince, M.B., 1955. Silicon solar energy converters. J. Appl. Phys. 26 (5), 534–540.
Shockley, W., 1949. The theory of p-n junctions in semiconductors and p-n junction transistors. Bell Syst. Tech. J. 28 (3), 435–489.
Shockley, W., Queisser, H.J., 1961. Detailed balance limit of efficiency of p-n junction solar cells. J. Appl. Phys. 32 (3), 510–519.
Song, Y.W. 2016. 3—Carbon nanotube and graphene photonic devices. In: Nabet, B. (Ed.), Photodetectors. Woodhead Publishing, pp 47–85 (**This chapter was first published as Chapter 3 'Carbon nanotube and graphene photonic devices: nonlinearity enhancement and novel preparation approaches' by Y.-W. Song in Carbon Nanotubes and Graphene for Photonic Applications, (ed.) S. Yamashita, Y. Saito and J.H. Choi, Woodhead Publishing Limited, 2013, ISBN: 978-0-85709-417-9).
Stratonovich, R.L., 1992. Nonlinear Nonequilibrium Thermodynamics I: Linear and Nonlinear Fluctuation-Dissipation Theorems. Springer-Verlag, Berlin, New York.
van Berkel, C., et al., 1993. Quality factor in a-Si:H nip and pin diodes. J. Appl. Phys. 73 (10), 5264–5268.
Van Der Ziel, A., 1970. Noise in solid-state devices and lasers. Proc. IEEE 58 (8), 1178–1206.
Wolf, E., 1978. Coherence and radiometry. J. Opt. Soc. Am. 68 (1), 6–17.
Wyatt, J.L.; Coram, G.J., Thermodynamic validity of noise models for nonlinear resistive devices. In: Proceedings of IEEE International Symposium on Circuits and Systems. Circuits and Systems in the Information Age ISCAS, 1997, 2, 889–892.
Ziel, A.v d, 1970. Noise in solid-state devices and lasers. Proc. IEEE 58 (8), 1178–1206.
Ziel, A.v d, 1988. Unified presentation of 1/f noise in electron devices: fundamental 1/f noise sources. Proc. IEEE 76 (3), 233–258.

Silicon based single-photon avalanche diode technology for low-light and high-speed applications

Daniel Durini[1], Uwe Paschen[2], Werner Brockherde[3] and Bedrich J. Hosticka[3]
[1]National Institute of Astrophysics, Optics and Electronics INAOE, Puebla, Mexico, [2]Westfälische Hochschule, Gelsenkirchen, Germany, [3]Fraunhofer-Institute for Microelectronic Circuits and Systems IMS, Duisburg, Germany

2.1 Introduction

Once photodetectors ceased to be used exclusively for creating images aimed to conserve the human memory or communicating ideas and were transformed into measuring devices, they almost immediately found their application in many different areas of science and industry. For an increasing number of applications, mainly related to material inspection and research, automation through optical sensing, particle physics, biology, or medicine, one of the main goals that photodetector technology has been striving for over the last decades is the ability to detect single photons over a variety of photon wavelengths, measuring their exact time of arrival to the detectors' active area and reconstructing their accurate spatial path. In other words, the "holy grail" of photodetection became offering the ability to measure as many characteristics as possible of a single photon in absence of other quanta of radiation. This proved to be a quite difficult task.

Over several decades, the internal photo effect, described by Albert Einstein in 1905 (Einstein, 1905), on which many of the modern photodetector technologies are based, has been successfully exploited by photocathode-based devices, for example, photomultiplier tubes (PMT) (Hamamatsu, 2007) or image intensifiers (I2) (Nützel, 2011). Photocathodes are materials, normally alkaline or multi alkaline metals, in which the impinging photons (normally in the near ultraviolet (UV), visible (VIS), and near infra-red (NIR) ranges, that is, with wavelengths (λ) between 250 and 1400 nm) generate electrons by means of the photoelectric effect. Once generated, these electrons are transported via electrical fields toward a certain sense-node (SN) of the device, where they can be accumulated over time and eventually read out.

The first problem encountered when using photocathode-based devices to detect very low photon fluxes is the very low amount of electrons that are photogenerated in the photocathode, so small indeed that the flux of these photogenerated electrons induced by the applied electrical field (the photocurrent) is almost impossible to be measured with presently available electronics. So, the first idea is to try to increase the number of photogenerated electrons by different means, that is, try to implement some kind of internal gain into the photodetector (Brennan, 1999).

In the case of vacuum tubes introduced in the first half of the 20th century, and soon attached to photocathodes to produce the so-called phototubes (see Figure 2.1A), the photogenerated electrons are transported through the vacuum tube to the biasing anode and eventually read out. The phototubes could not be used at extremely low irradiances, so dynodes (electron multiplication sites acting as

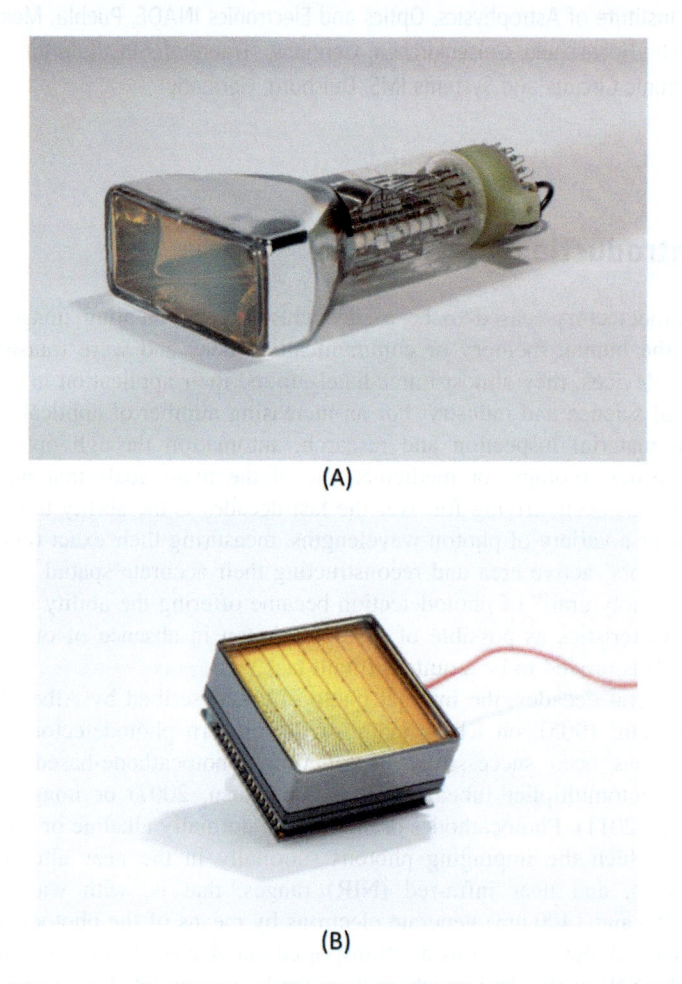

(A)

(B)

Figure 2.1 (A) An example of a photomultiplier tube (PMT); (B) an example of a flat-panel PMT.
Source: (a) Photograph taken by H. Timmermanns, Forschungszentrum Jülich—ZEA-2;
(b) Photograph taken by H. Timmermanns, Forschungszentrum Jülich—ZEA-2.

separate anodes) were introduced within the vacuum tube, biased at consequently increasing high voltages. The resulting device was the PMT. In a PMT the electrons photogenerated at the photocathode travel in a vacuum in the direction of the anode undergoing avalanche multiplication processes at every dynode, being consequently multiplied in a process called secondary emission with multiplication factors of around 10 in modern PMTs (Hamamatsu, 2007). Having on average between 5 and 7 dynodes per PMT, these devices normally deliver a total of 10^5-10^7 electrons per single impinging photon (if the photoelectric effect has taken place within the photocathode). To achieve this, these devices operate at hundreds to even thousands of volts and must be sometimes cooled down below 0°C to achieve acceptable dark-signal performance, that is, to reduce the number of electrons generated in complete darkness through the thermal energy of the material itself. All of this makes the electromechanical complexity of these devices quite high. Additionally, the PMTs cannot operate in presence of magnetic fields, as these would deflect the electrons during their motion across the vacuum tube causing them to miss the dynodes and the end-anode. Nevertheless, the time resolution of the PMTs in the range of nanoseconds and the extremely high sensitivity in low-level radiation environments make them, despite quite poor spatial resolution (of several mm^2), even today the instrument of choice in many applications.

To solve the problem of very poor spatial resolution, several solutions have been recently proposed, such as the flat-panel PMTs (Hamamatsu, 2007), an example of which can be observed in Figure 2.1B, where compact packaged PMTs have a reduced dead area between them, and a square detector pitch of between 2.8 and 5.8 mm (Pani et al., 2007, Hamamatsu, 2007).

The PMT are normally used also in experiments dealing with particle physics and in medicine or biology to detect electron clouds being generated by different particles reacting with some kind of scintillator material. Higher spatial resolutions (of approximately 1 mm^2 or lower) than those enabled by using flat-panel PMTs can be achieved if, for example, Anger camera concepts (Anger, 1958) are applied for signal processing. Hal O. Anger was a US electrical engineer and biophysicist who worked a long time at the Donner Laboratory of the University of California, Berkeley. In 1957 he fabricated a scintillator-based camera to image gamma radiation emitting radioisotopes to view and analyze images of the human body or the distribution of medically injected, inhaled, or ingested radionuclides emitting gamma rays. In this approach (Anger, 1958), the Gamma rays are emitted from the subject, some of which travel through the aperture (pinhole) in a lead shield and continue traveling in straight lines until they impinge on the scintillating crystal, where they decay into secondary particles that in a secondary reaction finally produce visible photons. The light that is produced in the scintillator is emitted isotopically to illuminate a set of photodetectors (e.g., PMTs) placed in a hexagonal configuration. The cloud of photons generated by the Gamma-ray will be distributed among the photodetectors; those placed exactly beneath the scintillation event receiving most light. The actual position of the impinging gamma-photon can be determined by processing the output voltage signals from all the illuminated photodetectors. In simple terms, the location can be found by weighting the position of each photodetector by the strength of its signal and then calculating a mean position from the weighted positions. A common method used for such processing tasks is based on the so-called Center-of-Gravity determination; although this method does not work properly if the detectors yield strong inhomogeneities in which case alternative methods must be considered (Kemmerling et al., 2001).

Another solution was proposed in form of the so-called multi-channel plates (MCP) (Nützel, 2011), schematically depicted in Figure 2.2 (Durini and Arutinov, 2014). These devices are based on a typically 2 mm thick highly resistive lead (Pb) glass carrier material with a regular array of tiny cylinders (microchannels) of about 15 μm in diameter leading from one face to the opposite one, which is densely distributed over the whole surface and attached to a photocathode (see Figure 2.2A). The walls of these micro-channels are typically covered by a thin layer of lead (Pb) to provide electrical conductivity to the cell covered by an even thinner layer of silicon oxide which is used to cause the secondary emission of electrons. The applied bias voltage will cause, due to the MCP internal resistance, a certain strip current of some microamperes to start flowing across the microchannel that will at the same time evenly distribute the voltage along it. When the first photon-generated electron hits the input of a microchannel, it will create a few secondary electrons that will get accelerated in the direction of the microchannel output. After these secondary electrons have gained sufficient kinetic energy, they will collide again with the microchannel wall creating even more secondary electrons, as can be observed in Figure 2.2B. The probability of the created electrons hitting the microchannel walls that creates the secondary electron emission is increased due to a small inclination (of some 10 degrees) of these walls with respect to the photocathode. This inclination reduces simultaneously the ion or light feedback within the microchannels. At the end of the cascaded process, several hundreds of thousands of electrons emerge from the bottom side of the MCP. The biasing voltages required for the proper functioning of an MCP are typically in hundreds of volts, and the gain factors achieved are similar to those achieved by the PMT.

Image intensifiers (I2) are based on another principle. The photogenerated electrons emitted by the photocathode are here once again converted into multiple photons by a phosphor screen, which provides the signal gain if biased at accelerating voltages of 12−20 kV (Nützel, 2011). Many developments are based on an MCP device located between the photocathode and the phosphor screen to increase even further the overall gain factor by taking advantage of electron multiplication on the one side, and secondary photon amplification on the other. The three active components require typical voltages of 200−1000 V for the photocathode, 600−1000 V for the MCP, and 4−6 kV for the anode. These devices have been used so far principally for night-vision applications.

In the described technologies, single-photon counting and spatial resolutions of only several micrometers nevertheless remain a big challenge. Their obvious electro-mechanical complexity and their rather poor spatial resolution, accompanied by the operating restriction in presence of magnetic fields, opened the door to a new approach enabled through the significant developments in microelectronics carried out throughout the second half of the last century: the integrated silicon-based imaging sensors.

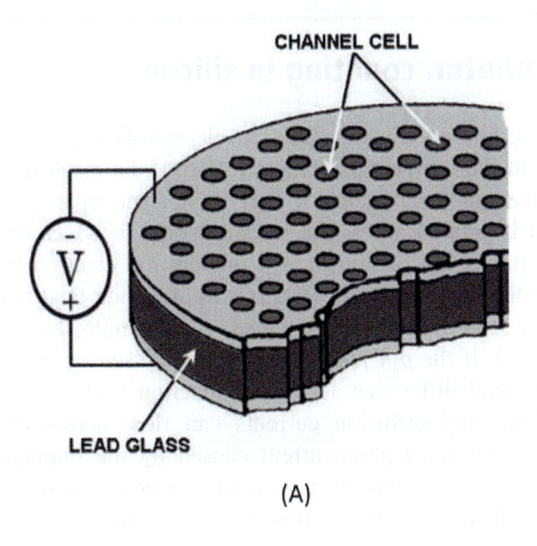

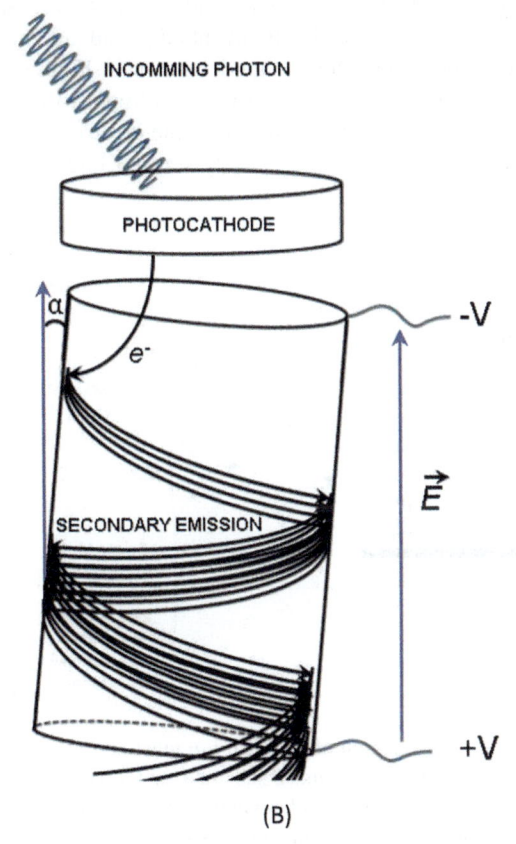

Figure 2.2 (A) A cutaway view of an MCP (Durini and Arutinov, 2014), and (B) a cross-section of the MPC cell causing the electron multiplication process due to secondary emission of electrons (Durini and Arutinov, 2014).

2.2 Single-photon counting in silicon

Once an impinging photon has generated an electron-hole pair (EHP) in a silicon crystal by means of the intrinsic photoelectric effect, the electron and its accompanying hole will eventually recombine if they do not get separated. Nevertheless, if an electrical field is applied across the silicon crystal in which this EHP has been generated by the impinging photon, it will prevent the electron from recombining immediately with the hole, directing electrons to an anode (and holes to a cathode) electrode. This electric field can be induced, for example, by a *p-n* junction (as shown in Figure 2.3). If the *p-n* junction is reverse biased by an external bias voltage U_{bias}, the potential difference across the junction will increase, so that under thermal equilibrium only diffusion currents can flow across the junction (Sze, 2002). Under illumination, a photocurrent caused by the impinging photons will start flowing across the *p-n* junction, now used as a conventional photodiode (PD), in addition to the diffusion currents flowing in thermal equilibrium (Sze, 2002). The electrons will be separated from the holes within the space-charge region (SCR) created at the junction and will eventually end up in the region of most positive electrostatic potential (the n^+ anode shown in Figure 2.3). They will then constitute the signal charge that is directly proportional to the number of photons impinging on the *p-n* junction. If the change in the electrical properties of the PD is to be measured, the junction must be initially set to a known electrostatic potential

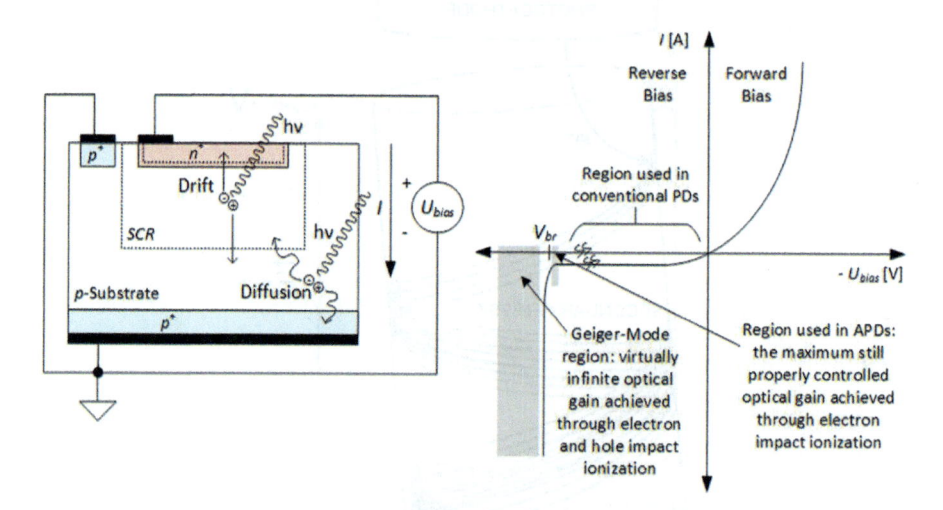

Figure 2.3 Schematic representation of a typical *p-n* junction (left) fabricated in a CMOS planar technology having a *p*-type grounded substrate, and a typical *I-V* characteristic (right) of the same *p-n* junction, used here as a photodiode, with its different regions of operation: forward bias, reverse bias, with the region without internal gain is used in standard applications, the region slightly below the breakdown voltage (V_{br}) is used in APDs, with a minimum controlled (but highly unstable) internal gain caused by impact ionization, and the Geiger-mode region, that is, the region above the breakdown voltage V_{br}.

level, that is, loaded to a certain defined initial (reset) voltage in such a way that the change of that initial voltage can be measured as generated signal and correlated to the properties of the radiation impinging on the photodetector. Once the PD is no longer electrically coupled to any fixed potentials, the change in the total amount of charge collected in it will cause the floating PD potential to change, and this change is interpreted as the PD output signal (Janesick, 2001).

The first problem here is the magnitude of this change in the electrical properties of the semiconductor translated into the electrical output signal proportional to the amount of photogenerated EHP and the sensitivity of the available devices that are required to measure it. Nevertheless, even if we manage to measure the generated signal, the second problem we encounter is the statistical variation of the accumulated charge due to additional sources of electrons (noise) present in the device and the ability of the readout circuit to discriminate between the usable and the spurious signals. To solve both problems, photogenerated electrons are normally accumulated at a certain sense-node (SN)—even the same PD—during a certain charge collection (or photocurrent integration, T_{int}) time. As soon as enough charge carriers have been collected to measure the change in the electrical properties of the SN, the readout can be carried out. The number of photons that impinges upon the photodetector during a given T_{int} can be calculated considering the achieved quantum efficiency (QE) of the photodetector. The quantum efficiency is a figure of merit expressed in percent (%) that describes the probability one photon (the energy of which can be calculated by multiplying its frequency ν by the Planck constant h) has to create an EHP, separate the electrons from the holes to avoid recombination losses, and successfully transport and finally collect the photogenerated electrons at the SN for readout.

So it can be concluded that if the photogenerated charge has to be accumulated during a certain time, and it has to generate an output signal that is higher than the noise-signal (consisting of the statistical variation of the amount of photons impinging the PD emitted by a source of constant optical power and in the same time interval that obeys the Poisson probability distribution function and is called photon shot noise, the variation of thermally generated EHP at a constant temperature in the same time window that also follows the Poisson probability distribution function and is called dark shot noise with additional EHP generated out of traps added to it, as well as all the additional EHP generated during the operation of all the additional electronic devices used for charge readout called read noise) generated at the same time through readout and reset operations, in order to perform single-photon counting there are only two main options available: to minimize the signal noise, or to increase the actual signal generated by a constant impinging radiation. As reducing the read noise has its physical limits (Janesick, 2001), internal amplification of the photogenerated signal appears to be a very good possibility. In both cases, the time-resolution of the photodetector remains rather poor compared to PMTs, reaching in the best case some hundreds of microseconds, but always at a cost of its spectral responsivity; which means that single-photon detection and high time resolution remain a very complicated combination of goal parameters.

2.2.1 From Avalanche photodiodes to SPAD: Geiger-mode operation and different readout modes

If a reverse-biased *p-n* junction is used as a photodiode, and the bias voltage increases, as it approaches its breakdown voltage (U_{BD}) it starts causing impact ionization of the generated electrons, that is, carrier multiplication (Sze, 2002) will take place and an equivalent increase in the current flowing across the junction can be measured, as shown in Figure 2.3. Basically, if biased well below the breakdown voltage (V_{br}), the *p-n* junction will operate with only diffusion current flowing through the junction. But, as soon as the first photo- or thermally generated charge carrier is introduced into the SCR generated at bias voltages approaching V_{br}, it will create a self-sustained avalanche of carriers (Brennan, 1999) caused by the effect of impact ionization. The expected number of these carriers will vary linearly with the amount of impinging photons (Fishburn, 2012) and the magnitude of the electrical field generated in the photoactive (SCR) region, as shown in the right side part of the Figure 2.3, where an *I-V* diagram of a *p-n* junction under different biasing conditions is depicted. Normally, the amplification factor induced in this way is around 10 (10 electrons generated by impact ionization at relatively low electrical fields for each electron entering the SCR of the PD). The first photodetector structures with internal signal amplification that appeared on the market were based on this property and are called APD.

The current response of an APD depends on the number of secondary EHP created from the original photogenerated charge carriers. The bias voltage of the APDs is held directly below the *p-n* junction breakdown voltage V_{br}, maintaining a sustained and well-controlled amplification of charge carriers. The biggest problem with APDs has been the high dependence of their gain factor on temperature changes and applied bias as well as other operating conditions. Namely, an increase in temperature causes a larger amount of EHP to be thermally generated increasing the carrier concentration in the SCR or photoactive region, which as a direct effect causes an increase in the local potential and, consequently, an increase in the magnitude of the electrical field induced. So, to reach the same critical electrical field necessary to initiate the avalanche process, we would need a decreasing V_{br} proportional to the amount of increment in temperature. This dependence between the magnitude of the electrical field and temperature holds in the opposite direction as well. If the gain factor is known and is held stable throughout the operation of the APD, the signal can be read out in an analog manner and the initial amount of impinging photons can be calculated. Moreover, if the impact ionization condition is sustained too long, the diode will start to heat up until it eventually gets permanently damaged.

Another idea appeared in the early 1960s, further encouraged by the theoretical prediction formulated in (McIntyre, 1961), and then also by the experimental proof reported in (Haitz, 1964), based on the fact that uniform *p-n* junctions may show microplasma (avalanche or impact ionization) performances when the bias exceeds the breakdown voltage, which is basically a spurious effect in the APDs (McIntyre, 1966). If biased far above the V_{br}, the electric field induced beneath the anode of the PD is so strong that both electrons and holes start causing significant impact ionization processes (Charbon and Fishburn, 2011), as shown in Figure 2.4. The photodiodes based on this principle of operation are also known as Geiger-mode APDs (GAPDs), named in this way due to the similarity in their principle of operation with the Geiger counter devices: they detect the presence of an avalanche multiplication process (at first, independently of the origin of the charge carrier that started it) and relate it to a visible photon impinging the photodetector; the amount of impinging photons can thus be counted similarly to what happens in a Geiger counter. The idea is to sense the appearance of the avalanche current flowing across the p-n junction biased above V_{br} and then lower this bias voltage (stop the impact ionization) until the excess carriers get drained away from the photodiode. The quenching operation avoids permanent damage to the avalanche device and significantly improves its time resolution.

The quenching operation can be carried out in a passive manner, where a so-called "quenching" resistor limits the bias voltage and is responsible for draining the avalanche multiplication region of excess charge carriers, or in an active manner by connecting the p-n junction to an active quenching circuit (Cova et al., 1981; Ghioni et al., 1996). So, the GAPDs will create a current pulse following the avalanche process normally started by single electrons and/or holes appearing within the avalanche or multiplication region of the PD (see Figure 2.4), the length of which will be controlled either by the quenching resistor or by an active quenching circuit (Schaart et al., 2016). Thus, if the impinging photon has succeeded in creating an EHP within the multiplication region (the region with the maximum electrical field), and the

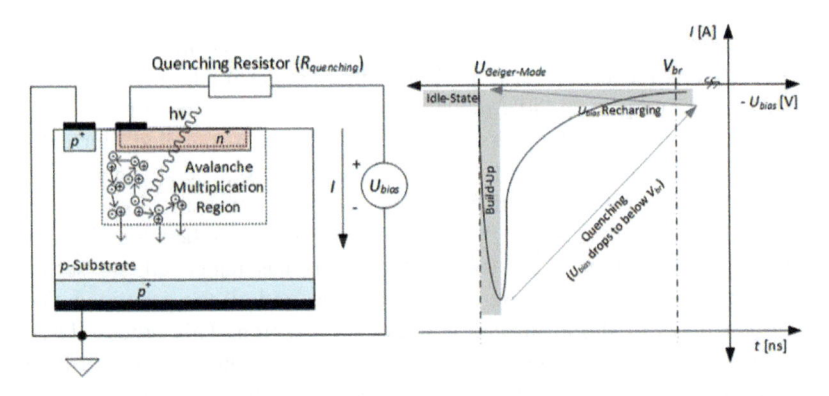

Figure 2.4 Schematic representation of a typical GAPD device fabricated in CMOS planar technology (left) showing a reverse biased p-n junction, the avalanche (multiplication) region, and a quenching resistor ($R_{quenching}$). The time-dependent current schematic diagram during an avalanche event, in which the GAPD remains in an idle condition until an electron starts the avalanche process in the avalanche region generating a short current peak, which then gets immediately quenched, can be observed on the right diagram. Superimposed, the three main U_{bias}-dependent operation modes can be observed on the same diagram (right). Here, U_{bias} drops below the breakdown voltage V_{br} due to GAPD quenching, and recharges back to the original $U_{Geiger\text{-}Mode}$, once all the generated electrons have been drained away from the GAPD.

avalanche process has created the first current peak that can be sensed at the PD anode, then it can be stated that the device is capable of detecting single impinging photons. The latter depends, of course, on the quantum efficiency of the device which suggests that to detect a single photon with a wavelength of 550 nm, it would be required to have at least two impinging photons for a quantum efficiency between the 50% and 80%. Additionally, it must be taken into account that thermally generated electrons can as well start an avalanche process during the duration of which the GAPD will not be able to detect any additional impinging photons. Thus, the single-photon counting (SPC) ability results are quite relative, and we normally speak of near SPC. If specially tailored for this task, this kind of device is also called a single-photon avalanche diode (SPAD), as proposed in (Cova et al., 1981).

By properly optimizing the quenching resistor and through proper calibration of the device, a gain factor can be considered, and the device can be read out very similarly to a common PMT, considering its analog output voltage (the amount of generated charge). This is the normal readout technique applied in most of the so-called silicon photomultipliers (SiPM) that will be separately addressed later on. On the contrary, if only the amount of avalanche events taking place is taken into consideration, the readout would consist of digital information reporting the amount of these "events," which makes the notion of quantified internal amplification of the internal flow of electrons meaningless. This second approach opened the way for use of small area diodes for single photon detection (Cova et al., 1981). Here, the SPAD basically becomes a "trigger" detector that provides a standard pulse every time one photon (or thermally generated dark electron) is detected (Cova et al., 1996).

If the output current of a GAPD is monitored, the time-dependent current change will be observed as the one schematically depicted in Figure 2.4 (*right*). At first, there is no current to be measured as no electron has started the impact ionization and U_{bias} remains far above V_{br} (i.e., at a certain Geiger-mode bias voltage U_{GM}): the SPAD is operating in an "idle mode." As soon as the avalanche process starts, a current peak of several mA appears at the output, during a so-called "current build-up" phase (see Figure 2.4, *right*). The same current starts being quenched (by a quenching resistor or through an active quenching circuit) and lowers its value causing at the same time the U_{bias} to drop below V_{br}, as schematically indicated with gray arrows in Figure 2.4 (*right*). Once this has happened, no second (or impact) ionization of charge carriers can take place, and all the generated electrons can be drained out of the avalanche multiplication region. This process lasts typically several hundreds of nanoseconds in case of passive quenching or several tens of nanoseconds if active quenching is used, in which the device remains "blind." After the charge carriers have been drained out from the GAPD, U_{bias} starts rising until it reaches its original Geiger-mode value again, and the GAPD is once again prepared to detect a new photon or thermally generated electron.

In the beginning, the so-called reach-through single GAPDs or SPADs were fabricated containing a thick portion of intrinsic silicon, similar to a *p-i-n* (or PIN) photodiode (Fishburn, 2012), in which the SCR extended over hundreds of μm. This was done to maximize the photon detection efficiency (PDE) of these devices, especially in the red and near-infra-red parts of the spectrum. But, just as it happened with PIN photodiodes, there was a desire to fabricate arrays of such devices to cover broader photoactive areas and overcome the quenching-related issues of these devices that were limiting their photodetection efficiency as well as their repetition rates.

One solution appeared in the late 1980s (Gasanov et al., 1989) and experienced a revival in the 2000s, and is based on the already mentioned silicon photomultipliers (SiPM). The idea is simple but very effective. If several GAPD or SPAD devices can be fabricated in a way that their anodes followed by a quenching resistor get connected to a common output anode (see Figure 2.5), then the quenching time required by one photosensitive SPAD cell does not affect the overall

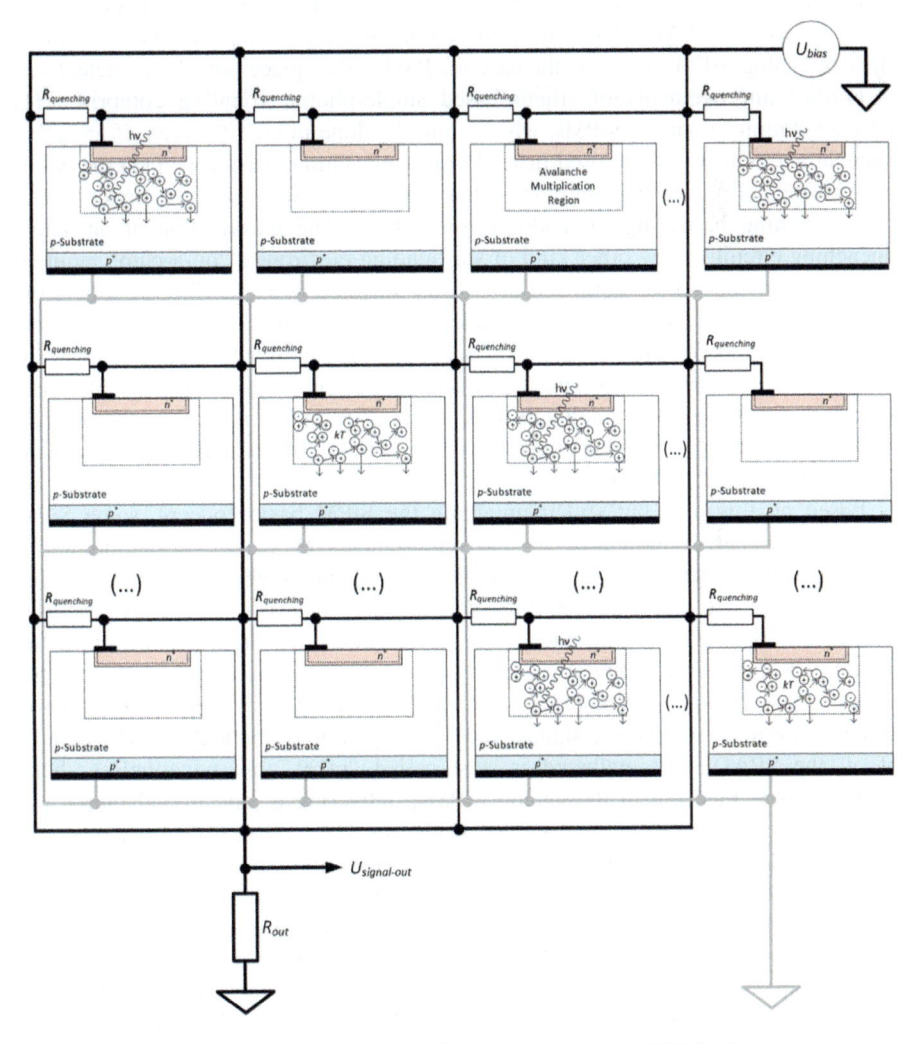

Figure 2.5 Schematic representation of a silicon photomultiplier (SiPM), showing an array of individual photosensitive GAPD or SPAD cells, each with an individual quenching resistor, and short-circuited anodes. The generated charge from all the photosensitive cells is accumulated and analog read out at the output resistor R_{out}.

performance of the entire "pixel" where thousands of such devices are placed and interconnected. The charge generated by all the photosensitive cells is accumulated on the common anode output. The output node is connected to a common readout resistor, as shown in Figure 2.5. If the average amount of charge per avalanche event is known (obtained through characterization), the amplification (gain) factor of the SiPM can be established and used to statistically determine the number of photons impinging the SiPM. This means that all the photons impinging on the photodetector that cannot be detected by a single SPAD device due to the quenching ("dead") time during which the detector remains "blind" (i.e., the so-called Geiger-mode limitation), can now be detected by a SiPM: an array of independent passively quenched SPAD cells with short-circuited analog outputs. The readout of such an analog SiPM, just as is the case for PMT, takes place outside the detector.

Further improvements of silicon-based single-photon counting comprised (1) time-resolution of analog SiPMs, which could be done more effectively if an active quenching circuit would be used instead of the quenching resistor, and (2) the possibility of signal-processing and analog-to-digital conversion (or even better, direct pulse counting based digital readout) "on-a-chip." The introduction of an active quenching circuit or any other kind of surrounding electronics "on-a-chip" required the SPAD devices to be fabricated in a planar technology (e.g., CMOS), where the surrounding circuit and the photodetectors would share the same silicon substrate (Zappa et al., 2003). This would even enable the possibility to fabricate arrays of SPAD cells, each containing the active quenching circuit, digital counters, and even signal processing circuits that could be read out as frames if necessary and even be employed in imaging-based applications (Zappa et al., 2003; Charbon and Fishburn, 2011; Schaart et al., 2016).

Based on this idea but still maintaining the SiPM basic concept, some SiPM structures appeared, motivated by the requirements of gamma-ray scintillation-based detector systems used in Positron-Emission Tomography (a nuclear medicine imaging technique that produces three-dimensional images of functional processes in a human or animal body), which use active quenching circuits and digitize the SPAD cell outputs for a digital readout (Philips, 2013; Schaart et al., 2016). In the case of these digital SiPMs, the detectors themselves deliver readout information and even perform some basic signal processing operations "on a chip." The outputs of all the SPAD cells, gathered in a so-called "pixel," are nevertheless short-circuited as is the case in the analog SiPMs, delivering a joint digital output per pixel. These so-called Digital Photon-Counting Devices (DPC), developed by Philips (Philips, 2013), yield much better time resolutions than analog SiPMs and are, as it otherwise could not be possible, fabricated in CMOS technology.

Other digital SiPM approaches are being currently developed on a research level that uses actively quenched CMOS technology-based SPADs, maintain the basic SiPM approach, but allow for an independent readout of each SPAD cell. In the digital SiPM, developed in the 0.35 μm CMOS-based SPAD-array technology, designed by the University of Heidelberg and fabricated by Fraunhofer IMS (Fischer et al., 2014), digital counting of avalanche events taking place in each SPAD cell delivers 1-bit digital information about it. Due to the fact that this information gets individually read out, an independent digital readout of each SPAD forming part of the SiPM is enabled. This additionally allows for flexible space resolution definitions, as the outputs of different SPAD cells can be combined in many different ways, for example, by binning different amounts of SPAD cells together.

SPAD or SiPM technology unfortunately has also other drawbacks. Basically, the avalanche processes in a SPAD can be also triggered by thermally generated electrons, which translates into spurious counts expressed through the dark count rate (DCR) of the device. Additionally, if two photons impinge the photodetector on the same spot, there is no possibility of detecting the second one. Finally, there is a good chance of the avalanche process getting restarted by the remaining excess charge carriers (after-flow) which translates into additional spurious counts. Finally, if thousands of individual SPAD cells are to be interconnected to build a single pixel structure, the minimum size of such a pixel would have to be around 1 mm^2; a fact that limits the spatial resolution of these photodetector devices and eliminates the possibility of using them for imaging-based tasks. One of the main advantages, however, of the SiPM resides in the fact that the average gain factor achieved nowadays by a typical analog SiPM is about 10^6, which is relatively the same as the gain achieved by most of the PMT, but at significantly lower biasing voltages: typically, between 20 and 60 V (see Table 2.1). Also, the time-jitter performance is quite impressive, especially in digital SiPMs. These devices are additionally immune to the influence of magnetic fields. All of this can be observed in Table 2.1, where the main figures of merit of nowadays commercially available analog and digital SiPMs have been summarized based on an example manufacturing series. Many of these figures of merit will be discussed more in detail in the following section.

In summary, if an application requires high spatial resolution, the short response time (time-jitter), and additional signal processing and analog-to-digital conversion (ADC) on a chip, whether PMT, MCP, I2 nor analog SiPM technologies would be able to become the technology of choice. If additionally, a GAPD array is to be used in imaging-based applications, digital SiPMs could also not be used, firstly due to the size of their "pixels" and secondly due to the readout technique they use, which is not used for frame-wise readout. Due to all these issues, during the last 10 years, several research groups have been working on the design and fabrication of SPAD arrays, in which each photosensitive cell could be read out independently from all the other cells but the entire array could be read out frame-wise, reaching ps time resolutions through active quenching circuits, and in-pixel signal processing and ADC, as it will be described in the following sections. As it will be shown, this opened the possibility of using this technology also for time-of-flight (TOF) based three-dimensional (3D) imaging, fluorescence lifetime imaging, and many other applications.

2.2.2 Fabrication issues in CMOS-based SPAD arrays

Although it worked wonderfully for single SPAD devices and analog SiPMs to be fabricated using dedicated fabrication processes, the idea emerged almost immediately to profit from the maturity of the CMOS technology to expand the

Table 2.1 Typical figures of merit of commercially available SiPM technologies (with analogue and digital outputs).

SiPM parameter	Hamamatsu MPPC S13360 series for precision measurements (Hamamatsu, 2019)	On-semiconductor J-Series MicroFJ – 30035 (Sens, 2019)	(Philips, 2013)	(Fischer et al., 2014)
Effective module photosensitive area, mm^2	3 × 3 (available also 1.3 × 1.3 and 6 × 6)	3 × 3 (available also 4 × 4 and 6 × 6)	32.6 × 32.6 (x 0.75, tile fill-factor)	4.96 × 4.96 (+ 1.5 mm, for readout circuitry)
No. of detector units per module	/	/	8 × 8	/
Detector unit (pixel) dimension (H x V), mm^2	/	/	4 × 4	/
Photosensitive cell size, μm^2	25, 50, 75 (square)	20, 35 (square)	59.4 × 32 (also available: 59.4 × 64)	56.44 × 56.44
No. of photosensitive cells per module	14400 (size: 25 μm) 3600 (size: 50 μm) 1600 (size: 75 μm)	14410 (size: 20 μm) 5676 (size: 35 μm)	6396 pixel (3200 / pixel)	88 × 88 (=7,744)
Geometrical fill-factor, %	47 (photosensitive cell size: 25 μm) 74 (photosensitive cell size: 50 μm) 82 (photosensitive cell size: 75 μm)	62 (photosensitive cell size: 20 μm) 75 (photosensitive cell size: 35 μm)	54 (pixel)	38 (photosensitive cell)
Breakdown voltage (V_{br}), V	53 ± 5	24.2–24.7	24 ± 0.5	27.5 ± 2
Typical operating voltage (V_{bias}), V	S13360–3050PE: V_{br} + 3	V_{br} + 2.5	27 ± 0.5	V_{br} + 3
Temperature coefficient of V_{bias} ($\Delta T \cdot V_{bias}$), mV/°C	54	21.5	tbm	~ 38

Readout mode	Analog	Analog	Digital (one output per pixel, no possibility of discrimination between photosensitive cells)	Digital (each photosensitive SPAD can be read out individually)
Output signal gain @ typ. V_{bias}	S13360–3050PE: 1.7×10^6	2.9×10^6 (anode-cathode @ 2.5V OV)	/	/
Spectral response range ($\Delta\lambda$), nm	320–900	200–900	380–700	300–1000
Peak sensitivity wavelength (λ_p), nm	450	420	420	430
Max. photon detection efficiency (PDE) @ λ_p, recommended typ. V_{bias}, and no after-pulsing or cross-talk, %	S13360–3050PE: 40	50	30	35
Dark Count Rate (DCR) per mm^2 @ T = 25°C, typ. V_{bias}, kcps	S13360–3050PE (3×3 mm^2): 55.5–166.7 @ terminal capacitance of 320 pF	(0.23 µA @ 2.5V OV – 3 µA @ 6 V OV) → 55–718	< 312.5 (< 5×10^3 per pixel divided by pixel size of 4×4 mm^2)	75–100
Time resolution (FWHM) for a single-photon, ps	/	1500	44	< 200
Photoactive cell recovery (RC) time constant, ns	/	45	/	90 (after-pulsing probability < 1%)
After-pulsing probability @ typ. V_{bias}, %	/	0.75	/	< 1

performance of the photodetectors. Using CMOS technology would also enable the fabrication of readout and quenching circuits on the same substrate and enable on-chip signal processing, analog-to-digital conversion, etc. This is how the so-called planar SPAD array solutions appeared. The space charge region (SCR), where the impinging photons can be converted into EHPs and immediately start avalanche multiplications, was in this kind of development of the order of hundreds of nanometers or some micrometers at the most. This limited the photodetection efficiency (PDE) of the devices in the red and near infra-red parts of the spectrum, but also significantly improved the SPAD time responses (i.e., lowered their time-jitter) due to the short distances between the avalanche (SCR) region and the anode (readout node) of the SPAD devices, a distance the generated electrons have to cross to get read out. Also, the silicon resistivity diminished accordingly, causing the RC constant to drop, having the same result: faster devices. In the past years, SPAD structures have been successfully implemented in CMOS processes belonging to very different technology generations. The integration of a SPAD module for single photon counting and timing into a 0.8 μm CMOS technology using only the standard wells was described by Zappa et al. (2004). Much work was done on SPAD integration in 0.35 μm CMOS processes (Goll et al., 2018, Jahromi and Kostamovaara, 2018) due to their widespread availability, cost efficiency, and the capability to handle the required high voltages. SPADs were also integrated into CMOS processes with much smaller feature sizes, for example, 90 nm (Webster et al., 2012, Karami et al., 2011) and even 65 nm (Karami et al., 2011; Nolet et al., 2018) and 40 nm (Dutton et al., 2018). In these advanced technologies, however, the breakdown voltages tend to be smaller due to increased doping levels (even though there can be large variations for different technologies belonging to the same node). Breakdown voltages were 14.9 V and 10.4 V for the two 90 nm technologies (Webster et al., 2012, Karami et al., 2011), and 9.5 V and 9.9 V for the 65 nm technologies (Karami et al., 2011; Nolet et al., 2018). The low breakdown voltages lead to a strong increase in the dark count rate due to tunneling processes.

So the idea was wonderful, but the standard CMOS processes were not designed for the fabrication of optimal SPAD devices. Thus, very soon it became quite clear that to be able to fabricate operational SPAD devices with good performances, additional implantation steps had to be incorporated into the already existing CMOS process flows. This was not possible in the case of many foundries, as the predicted numbers of fabricated chips were not high enough to be commercially attractive for them. This situation changed during the last couple of years. Nowadays many foundries in addition to dedicated CIS (CMOS Image Sensor) processes also offer SPAD devices in their portfolio of technologies and devices.

The second immediate problem, having SPAD imagers in mind, was the one of dealing with a single SPAD-pixel fill-factor: if an array of SPAD pixels is to be fabricated that can be read out frame-wise, this means that each SPAD pixel has to be read out independently from the rest of the pixels, that is, every pixel must contain at least one active quenching circuit, one digital counter required to count the avalanche events, and any additional circuitry used for "in-pixel" signal processing. If a 0.8 μm, 0.35 μm, or even a 0.25 μm CMOS process is used, this yields small pixel fill factors. The solutions pursued so far have been using CMOS processes capable of always smaller geometrical sizes (Karami et al., 2011; Webster et al., 2012), using microlenses on top of the pixels (Waddie et al., 2013, Mata Pavia et al., 2014), developing back-side illuminated SPAD (BackSPAD) arrays (Moser et al., 2007; Choong and Holland, 2012; Durini et al., 2012), where the pixel circuitry is placed on an additional chip connected by through-silicon via—vertical interconnect access—(TSV) or similar technologies to the photosensitive chip, an approach followed down to 45 nm 3D stacked back-side illuminated (BSI) CMOS SPAD arrays as reported in (Lee et al., 2017).

As mentioned above, CMOS integrated SPADs, in contrast to stand-alone SPADs fabricated in dedicated technologies, offer the possibility to integrate analog and digital circuitry such as counters, signal processing, active quenching, or time-to-digital converters with the SPADs in one device. This enables compact integrated SPAD arrays for example for photon counting or time of flight applications. It also allows special techniques like, for example, using coincidence information of the signals from different SPADs to separate signals from the background (Beer et al., 2017) in time of flight sensors. However, combining the requirements for CMOS processes with those for the SPADs leads to several restrictions and trade-offs. Basically there exist two fundamental possibilities to integrate SPADs into a CMOS, although many variations in detail are employed to improve specific properties in the different actual realizations:

1. The CMOS substrates are typical of p-type. As indicated in Figure 2.6A, one can therefore implement SPADs by employing an n^+ region at the surface and using the diode formed by the n^+-p − substrate junction.
2. As a second implementation possibility, one can employ the CMOS n-well to isolate the SPAD structures from each other (and the rest of the circuitry) and thus prevent electrical crosstalk between neighboring SPAD cells. In this case, a p^+ doping at the surface, typically the Source/Drain doping of a PMOS transistor, is used as the anode and the n-well is used as the cathode (Figure 2.6B). Due to additional design rules for the n-well isolation, this second realization involves larger distances between the SPAD cells and thus reduces the fill factor of the pixels, that is, the optical sensitive percentage of the total pixel area. However, this can be remedied to some extent by the use of an array of microlenses that concentrate the incoming light on the optically sensitive area of each pixel.

If the breakdown voltage of SPADs constructed according to Figure 2.6 is too high to handle without problems on the chip, an additional implant can be used to reduce the breakdown voltage as shown in Figure 2.7.

As discussed above, the SPAD function relies on biasing the p-n junction with voltages beyond the breakdown voltage of the planar junction (Geiger mode). To enable this, one must prevent a premature breakdown at the periphery of the device. This is achieved by an additional implant that acts as a guard ring at the edge of the diode as indicated in Figures 2.6 and 2.7. In contrast to the implant that determines the breakdown voltage of the SPAD, this implant is less critical so that usually already existing implants of the base (standard) CMOS process can be used without the restriction of the SPAD performance.

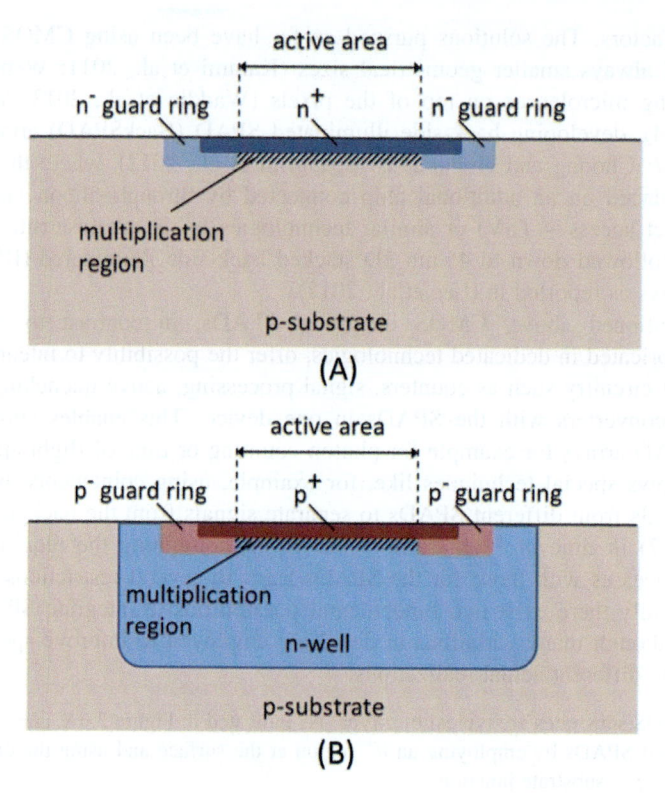

Figure 2.6 (A) A SPAD structure directly integrated into the *p*-substrate; and (B) an SPAD structure integrated inside a CMOS *n*-well.

Modern CMOS processes already have several implants available (e.g., source and drain implantations, *n*-well or *p*-well structures used for the fabrication of CMOS MOSFETs, etc.) that can in principle be used for the creation of SPAD devices. In many commercial foundry processes, the designer is restricted to using only these implants without the possibility to introduce additional ones. In that case, however, the breakdown voltage of the SPAD cannot be freely chosen but is restricted to the combination of the existing doping profiles, and the risk is high that the breakdown event happens in the horizontal direction, parallel to the silicon surface, between the supposed SPAD anode and a substrate contact placed on a side of the SPAD instead of it happening in the desired vertical direction, normal to the silicon substrate, in the foreseen multiplication region located beneath the anode electrode of the SPAD (see Figures 2.6 and 2.7).

Ideally, the foundry that processes the wafers offers the flexibility to add device-specific implants (or, as mentioned above, already offers dedicated SPAD implants and even optimized SPAD devices) which enables more freedom in the SPAD development. In this case, typically one additional implant is employed to optimize the multiplication region (see Figure 2.7) and to achieve a specific breakdown

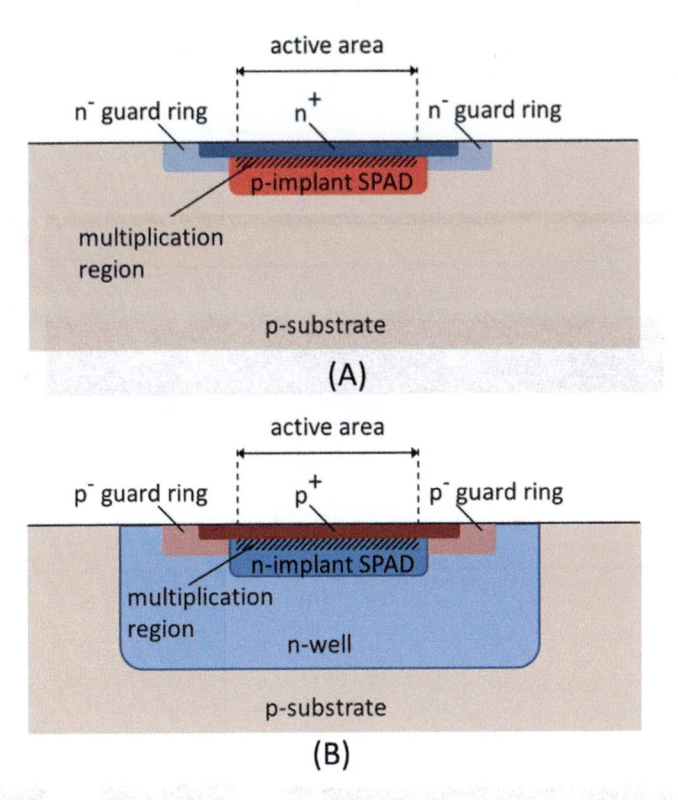

Figure 2.7 SPAD structures with an additional implant introduced to reduce the breakdown voltage V_{br}: (A) according to the SPAD structure proposed in Figure 2.6A and B according to the SPAD structure proposed in Figure 2.6B.

voltage. It must be noted, that the temperature steps of a given CMOS process are fixed, and no additional annealing or diffusion steps can be introduced without detrimental effect on the CMOS devices. Thus, the diffusion of the additional SPAD implant can only be influenced by defining the position in the process flow where it is introduced. Together with implantation dose and energy, this gives three degrees of freedom for the SPAD device optimization. Usually, initial parameters for implantation dose and energy as well as the position where the additional implant is introduced in the process flow are determined by 2D process and device simulations (see Figure 2.8). This enables the fabrication of optimized devices without too many "trial and error" iterations in the actual fabrication. However, in most cases, the calibration of the simulation setup does not describe the reality of the effects that are relevant for the SPADs perfectly, so these simulation results only give a good initial clue as to where to start with respect to implant dose and energy. Typically, processing splits will be required to reach the desired optimal performance. An example of SPAD integration in a 0.35 μm CMOS process with the optimized performance achieved by adding a special SPAD implant is given in (Villa et al., 2014).

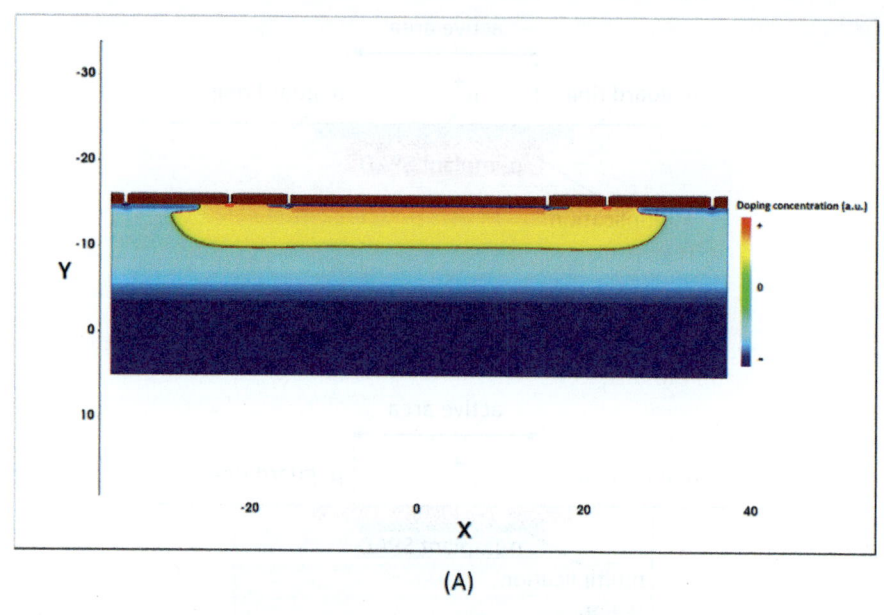

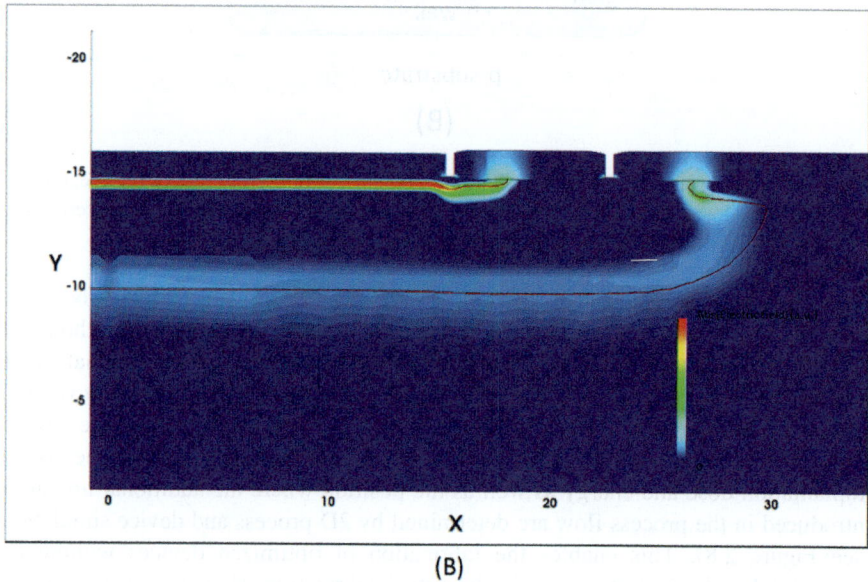

Figure 2.8 Two-dimensional (2D) simulation of the SPAD structure fabricated in a 0.35 μm CMOS process as reported in (Villa et al., 2014): (A) doping concentration; (B) electrical field distribution under bias.

Choosing the CMOS technology node for the implementation of SPAD structures typically involves a trade-off between various influences. While there is a logical trend to integrate SPADs into deep submicron CMOS processes to implement advanced functionality into each pixel (quenching, counter, and time-to-digital converters) there are several drawbacks regarding the SPAD performance of such technologies compared to the "older" technologies with design rules of about 0.35 μm and above. The breakdown voltage of the SPADs should be significantly beyond 10 V, otherwise band-to-band tunneling occurs and increases the dark count rate significantly (Migdall et al., 2013, p. 89; Webster et al., 2012). Modern deep submicron CMOS processes, however, have increasingly higher doping levels to enable the reduced design rules. This is tantamount to small breakdown voltages and high electrical fields, thus presenting potential band-to-band tunneling issues. CMOS processes with increasingly smaller feature sizes also tend to have higher defect concentrations because the annealing temperatures are getting smaller in these technologies. In addition, CMOS technologies at the 250 nm node and below use trench isolation instead of LOCOS (Local Oxidation of Silicon) isolation which also leads to increased defect concentrations and increased dark count rate if the trench is close to the multiplication region (Henderson et al., 2009). The high defect density can also lead to an increased after-pulsing probability by trapping carriers and releasing them after some time.

Deep submicron CMOS processes frequently employ substrates with a rather thin epitaxial layer (where the devices are integrated) on top of the highly doped substrate. This is a measure to eliminate latch-up effects. This, however, means that charge carriers that are generated below the epitaxial layer in the highly doped substrate recombine quickly without reaching the multiplication region of the SPAD device and thus the photon detection efficiency in the long wavelength range of the spectrum is reduced.

As mentioned above, an important class of applications for SPADs is 3D ranging with LIDAR systems, for example in the context of advanced driver-assistance systems or autonomous driving. For this kind of application, the sensitivity of SPAD devices in the NIR region of the spectrum is highly important. A high PDE in that region, typically starting at wavelengths around 800 to 850 nm, enables a large detection range with a given laser power, which is limited by eye safety restrictions. The product of PDE and the fill factor (i.e., the fraction of the pixel area that is sensitive to light) is the relevant quantity here. For example, Takai et al. (2016) and Gulinatti et al. (2016) describe measures to develop SPAD devices with a high fill factor and high PDE for wavelengths of impinging radiation of 800 nm and beyond in the near infra-red (NIR) part of the spectra. Careful optimization of the layout and dedicated doping profiles employed for the construction of the SPAD is the key elements for the optimization of such devices. Extended depletion regions are required to provide the necessary collection of carriers generated deep in the silicon, high electric fields ensure a small timing jitter by pulling the carriers quickly into the multiplication zone.

2.3 Definition of SPAD figures of merit and general aspects of SPAD characterization

In this section, we present a brief overview of the main performance parameters of SPADs, (mostly) applicable both for stand-alone and CMOS integrated SPAD devices (see also Table 2.1). For a comparison of typical values of several of these parameters for SPAD arrays fabricated in a large variety of CMOS technologies see for example (Bronzi et al., 2013; Bronzi et al., 2016). Here, also some additional figures of merit have been derived from these parameters for photon counting and photon timing applications, respectively, to enable a simple comparison of SPADs fabricated in different technologies. We also comment on characterization techniques and issues arising from the transition from using SPADs mostly in scientific measurements in a well-controlled laboratory environment to new applications, most notably in the automotive field.

2.3.1 Photon detection efficiency

This parameter specifies the fraction of incoming photons that result in a detected pulse at the SPAD. The photons must be absorbed creating an EHP. If the absorption takes place outside the multiplication region of the SPAD the charge carriers must travel into this region and start an avalanche breakdown with a certain probability which is detected by the electronics (see Figure 2.3). The PDE value depends on the wavelength of the impinging light because the penetration depth is wavelength dependent. Measurement is performed by determining the number of counts per number of incoming photons as a function of wavelength. In LiDAR applications, for example, the PDE in the infrared part of the spectrum is important (Takai et al., 2016), other applications like spectroscopy require a sufficient PDE over a broad range of the spectrum from IR to UV.

2.3.2 Dark count rate

The avalanche of the SPAD is triggered by charge carriers reaching the multiplication region, no matter if they are generated by the absorption of photons or by other mechanisms. Even in complete darkness, several mechanisms can provide charge carriers that lead to dark counts: thermally excited charge carriers, carrier generation at defects (contamination, lattice dislocations), and band-to-band tunneling in regions of very high electric fields. Processing in a very contamination-free environment, high annealing temperature, not too high doping concentration, and low defect wafer material are means to keep this parameter as low as possible. The dark count rate is measured under zero light conditions. It increases with temperature due to thermally excited carriers and depends on the quality of the manufacturing process (number of defects). A relatively large dead time is used during the measurement of this parameter to eliminate counts due to afterpulsing.

2.3.3 Afterpulsing probability

Some of the charge carriers created during the avalanche process can be trapped at a defect, kept there for some time, and then be released again when they can trigger a new avalanche. The number of recorded afterpulses depends on the defect (trap) density and the waiting time (dead time) between quenching the avalanche and recharging the SPAD. Measurements of this parameter are based on the statistics of the time between consecutive pulses under illumination. Devitations from the ideal Poisson distribution can be attributed to afterpulsing events (Piemonte et al., 2012; Xu et al., 2017).

2.3.4 SPAD dead time

After ignition, the avalanche must be quenched, an intentional waiting time might be introduced to reduce the after-pulsing probability and then the *p-n* junction must be recharged. During this time interval (dead time) the SPAD cannot detect other photons. The dead time limits the maximum photon rate that can be detected and is thus a compromise between suppression of unwanted afterpulsing events and loss of "real" counts of photons that arrive during the dead time of the SPAD.

2.3.5 Timing jitter

This parameter reflects the time spread of the interval between the arrival of the photon and the detected signal output. In addition to the timing jitter that originates from the detection electronics, also the specific position in the multiplication region where the avalanche starts to develop has an impact on the exact dynamics and timing behavior of the avalanche process (Villa et al., 2014). The diffusion time of charge carriers created outside the multiplication region also adds to the timing uncertainty. Measurement of this parameter is performed by employing very short laser (ps) pulses and measuring the statistics of the timing of the SPAD signal relative to the excitation pulse.

2.3.6 Crosstalk

The avalanche ignited in one SPAD can lead to stray charge carriers that reach the multiplication region of a neighboring SPAD and ignite an avalanche there (electrical crosstalk). Additionally, high-energy charge carriers during avalanche can generate secondary photons that lead to signals in neighboring cells (optical crosstalk). Rech et al. (2008) describe a method to determine this parameter experimentally based on correlation measurements of the signals of pairs of SPADs. Since these measurements are rather time-consuming, an alternative (much faster) method was also employed, in which one of the two SPADs was used as an "emitter" by driving a constant electrical current through it and measuring the count rate of the second SPAD ("detector") of the pair. Crosstalk in principle can be reduced by providing deep trench isolation between the SPADs of an array. However, deep trench isolation is not available in most foundry processes. In addition, due to mechanical stress, trenches can lead to detrimental effects like higher defect density and thus higher DCR and afterpulsing probabilities.

2.3.7 Fill factor

Each pixel in a SPAD array contains some area that is not optically sensitive due to minimum distance rules required for the process, guard rings, or electronics included in the pixel. The fill factor is the ratio of the optically active area and the total pixel area. The fill factor can be maximized by layout optimization but is ultimately limited by the required electrical isolation between neighboring pixels and crosstalk requirements.

2.3.8 Homogeneity of SPAD parameters in an array

In arrays, all pixels should behave as similarly as possible. However, in practice, several important SPAD parameters tend to have statistical variations across the array, most notably the dark count rate, photon detection efficiency, and breakdown voltage (Perenzoni et al., 2016). Ideally, all SPAD cells in an array should have very similar parameters. Concerning the dark count rate, there is always a fraction of SPADs that have significantly higher dark count rates than the others due to local defects. The dark count rate of SPAD pixels in a single array can span several orders of magnitude and show characteristic distributions depending on the underlying physical mechanism as discussed in detail (Perenzoni et al., 2016).

2.3.9 General aspects of SPAD characterization

In the past, SPADs were frequently used for scientific measurements, often in a controlled lab environment. Today, many applications are envisioned where SPADs are used in less controlled and more challenging environments. The probably most demanding application field (with the prospect of very large numbers of systems), both concerning the harshness of the environment and required reliability is in the automotive sector. LiDAR (Light Detection And Ranging) will probably be an important part of advanced driver-assistance systems and autonomous driving, where a 3D mapping of the area in front of the car is mandatory. Currently, much effort is made worldwide to develop small, cheap, and highly reliable LiDAR systems based on SPADs. The use conditions for this kind of application are very harsh and demanding, especially concerning temperature. Thus, device characterization has to be carried out over the whole automotive temperature range, reaching far below 0°C on the one side and up to 100°C and more (Takai et al., 2016) on the other. This wide temperature range strongly influences vital SPAD parameters such as the dark count rate, the breakdown voltage and also the failure rate, and the potential long-term parameter degradation of the SPAD devices. Generally, applications in the automotive field require very mature technologies, extreme reliability, close to zero failures, and of course also extreme cost efficiency.

2.4 SPAD arrays in CMOS technology: architecture overview

The co-integration of SPADs and CMOS circuitry enables a bunch of new architectures and capabilities. First, we can have a look at the SiPM technologies, figures of merit of which are listed in Table 2.1, which are traditionally implemented in special silicon processes optimized for the fabrication of these devices. The integration of SPADs in standard CMOS processes allows for the co-integration of the trans-impedance-amplifier (TIA) on the same die (Schwinger et al., 2017). This has some advantages concerning the noise and speed behavior of SPAD-based pixels since the capacitance of the interface between the photodetector and its readout circuitry can be lowered drastically.

A second approach to improve the SiPM performance is in form of the already mentioned "digital SiPM" or dSiPM, figures of merit of which are also presented in Table 2.1 compared to typical analog SiPM (aSiPM) technologies, which incorporates electronic circuitry in each cell of the SiPM array. This circuitry may be used for example, to shut down hot elements that dominate the DCR and hence, lower the overall DCR of the SiPM. Another feature of a dSiPM can be random access of each element and thus the possibility to detect the spatial distribution of the impinging photons.

This is already a solution close to the CMOS SPAD arrays or CSPAD arrays which, as already explained, stand for spatial resolving or imaging devices with one-bit digital output per SPAD cell. Besides the dimensional classification — linear or 2-D arrays — we distinguish between the CSPAD array for "photon timing" and for "photon counting." A CSPAD array for photon counting typically exhibits a time-to-digital converter (TDC) circuit unit in every pixel. These arrays are used for example, for time-of-flight applications like flash LiDAR or time resolving spectroscopy or other time-resolving measurements such as those used in fluorescence lifetime imaging spectroscopy (FLIM), to mention just one example. Photon counting CSPAD arrays are used for low-light and very high-speed applications. These arrays combine the spatial resolution of up to picoseconds with single photon resolution. These detectors need to have the lowest DCR, high QE, and high fill factors. Some multipurpose CSPAD arrays allow both regimes: photon counting and timing.

Common to all these CSPAD arrays is the need for complex pixel electronics. Apart from the active quenching and reset (AQR) circuits, these pixels incorporate respectively digital counters and registers for photon counting, or time-to-digital converters and registers for photon timing. The complex pixel electronics can be placed outside the photo-sensitive area for linear detector arrays, but for 2D arrays, the resulting fill factor will be very low. As already explained above, this drawback can be mitigated by applying micro-lenses on top of the array, using more advanced CMOS submicron processes of even 3D integration (or stacking).

Recently, 3D integration is becoming increasingly attractive. This technology is based on wafer bonding, thinning, and connecting the wafers vertically using fine-pitch "through silicon vias" (TSVs) (Hamamatsu, 2019). The lower wafer bears the readout electronics (ROIC) and the upper one the SPADs. These backside-illuminated hybrid devices allow for independent selection and optimization of the two tiers. The ROIC can be implemented in an advanced CMOS process with a low feature size, whereas the process of the SPAD wafer can be optimized for DCR and spectral QE. As a result, the pixel size can be drastically smaller than a pixel in a single CMOS tier, allowing either smaller chip sizes or higher pixel counts.

To illustrate the performance these sensors have recently reached, (Zhang et al., 2018) a 252×144 SPAD pixel sensor called *Ocelot* is reported, fabricated in the 180 nm CMOS technology featuring 1728 12-bit TDCs with a 48.8 ps resolution. In this development, every 126 pixels are connected to six TDCs, which enables effective sharing of resources and a fill-factor of 28% with a pixel pitch of 28.5 μm (Zhang et al., 2018). On the other hand, (Lee et al., 2017) reports on the back-side illuminated 3D-stacked SPAD structure achieving a dark count rate (DCR) of 55.4 cps/μm^2, a maximum photon detection efficiency (PDE) of 31.8% for 600 nm of impinging (red) light, and timing jitter of 107.7 ps at 2.5 V of excess bias (over the breakdown voltage to operate the SPAD in the Geiger mode). The development of these first SPAD structures in a 45 nm BSI 3D stacked technology opens a path toward high-performance SPAD-based sensors that should be solving many currently present issues in this technology.

2.5 Active illumination based ToF 3D imaging and ranging with SPAD imaging arrays

Three-dimensional (3D) imaging systems based on the contactless time-of-flight (ToF) principle have been entering a multitude of markets in the last decade. With the cost-efficient manufacturing possibilities of CMOS-based ToF sensor technology it is feasible to realize low-cost scanners 3D cameras, which address numerous mass market applications, such as gesture control for consumer electronics (e.g., smart TVs or gaming stations) or assisted driving systems for the automotive domain (e.g., based on LIDAR). Moreover, various high-end applications utilizing ToF sensors have been emerging in the fields of industrial imaging, machine vision and robotics, and security and surveillance, just to name a few.

The main principle of a ToF system consists of an active light source (laser diode or LED, typically in the near-infrared regime), a ToF sensor, and the diffusive optics at the emitter and the receiver optics, as shown in Figure 2.9. Here, being synchronized with the ToF sensor, the pulse emitted from the active light source illuminates the scene under investigation and travels at the light velocity c through the distance z twice before it impinges on the ToF sensor. This leads to a

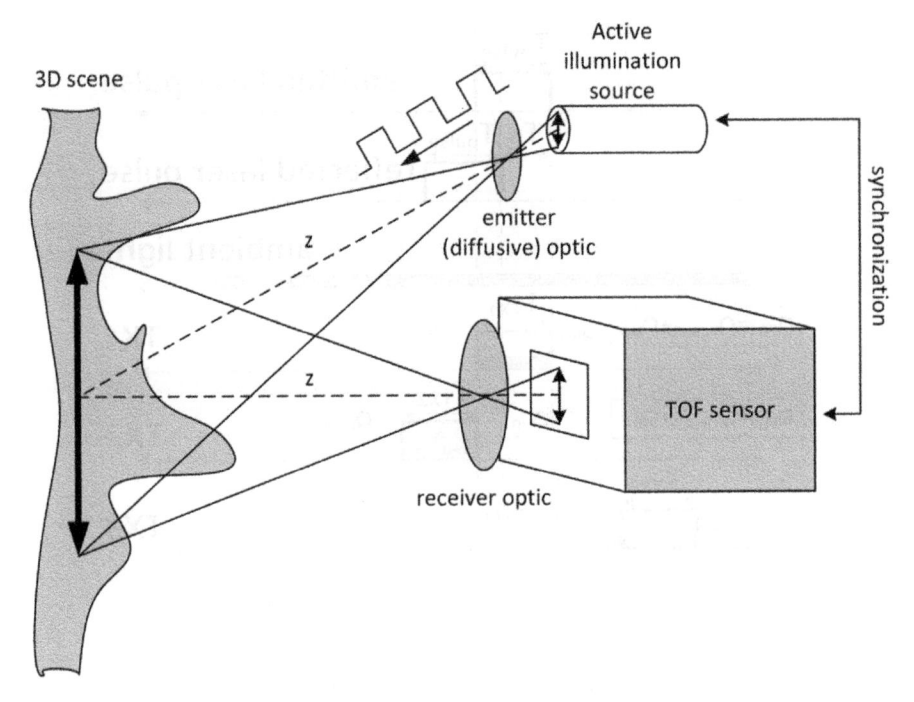

Figure 2.9 Schematic representation of the Time-of-flight (ToF) principle used for 3D imaging and ranging.

simple relation between the distance z and the round trip time of the light T_{tof} is given by Eq. (2.1).

$$z = \frac{c}{2} T_{tof} \tag{2.1}$$

In general, two distance measurement techniques can be used in modern ToF systems. The first approach is to use a highly accurate stop-watch mechanism to acquire the time-of-flight T_{tof} that elapses between the emission of the light pulse and the detection of the first photons belonging to the incoming pulse at the ToF sensor that returned reflected by a distant object. This approach is typically called direct-ToF (dToF) and requires ToF sensors with photodetector structures providing a very high temporal resolution. Because of their superior time resolution in the picosecond range, SPADs are perfect candidates for the dToF approach.

Another way of performing ToF measurements is based on time-gated light intensity acquisition and the indirect reconstruction of the ToF information. This approach is called indirect-ToF (iToF) and is typically used in pulsed modulated (PM) or continuous-wave modulated (CW) active illumination configurations. The timing schemes and the corresponding equations for distance reconstruction are summarized in Figure 2.10.

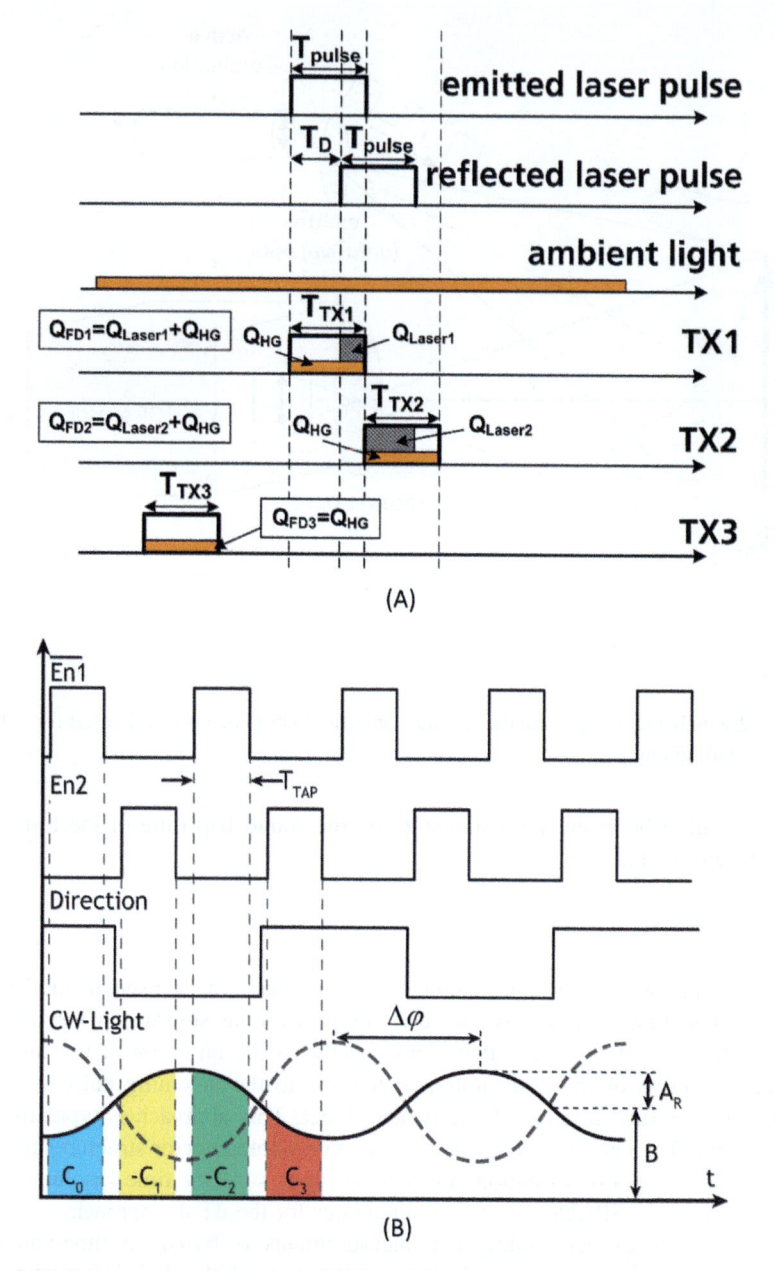

Figure 2.10 Schematic representation of the modulation of the active illumination in (A) pulse-modulated indirect time-of-flight (PM iToF) 3D imaging and ranging systems, and (B) continuous-wave modulated indirect time-of-flight (CW iToF) 3D imaging and ranging systems.

With its picosecond time resolution, SPAD arrays are also well suitable for the iToF approaches, used to acquire the temporal position of the light pulse in the PM case and the phase shift of the sinusoidal wave in the CW case caused by the travel time. In both cases, SPAD-based ToF imaging systems can achieve distance measurements with millimeter accuracy. Another important feature of SPAD arrays is their high single photon sensitivity, which makes it possible to carry out ToF measurements under very low light conditions, for example in long-range measurements or ranging of scenes with very low reflectance of the objects present.

The monolithic integration of SPADs in CMOS technologies combines the features of an array of photodetectors with near single-photon counting possibilities, and highly integrated readout and signal processing capabilities. Typically, a SPAD-based CMOS ToF sensor consists of the SPAD array including reset and quenching circuits, time-to-digital-converters (TDC), digital counters, and circuits for XY-addressing of the SPAD matrix.

In recent years, various SPAD-based sensors fabricated in CMOS technologies, used in applications employing both the dToF and the iToF technique, have been developed. These developments include systems that rely on scanning sensor technology as well as on scannerless approaches. The sensors described in this section represent the state-of-the-art performance currently achievable with SPAD-based technology in scannerless ToF applications.

A 64×32 pixel SPAD array for iToF-based depth-imaging applications has been designed, developed, and characterized in high-voltage 0.35 μm CMOS technology (Zappa and Tosi, 2013). It features circular SPADs with a 30 μm diameter and a low median DCR of 100 cps at room temperature. Each pixel also contains an analog avalanche sensing and quenching circuit in addition to the digital processing circuitry. A fill factor (FF) of 3.14% was achieved at 150 μm pixel pitch, due to the large in-pixel area reserved for electronics. The processing circuit contains 3 synchronous 9-bit counters based on Fibonacci LFSRs (linear feedback shift registers), 2 of which are bidirectional, and each connected to an in-pixel memory register. Storing the contents of the counters in the memory registers, and reading the array in global shutter mode, allows for continuous image acquisition, effectively preventing motion artifacts in the processed images. The sensor implements CW-based iToF ranging with multiple photon counters. By exploiting their bidirectional mode, only 2 in-pixel counters are required to calculate the distance to the measured object. Using this technique, a precision of 85 cm was achieved over a measuring range of 20 m at 100 frames per second (fps) (Bronzi et al., 2014). This precision can be further improved by lowering the frame rate.

As another example, the Fraunhofer Institute of Microelectronic Circuits and Systems in Duisburg has recently developed a 1×80 pixel sensor for flash dToF in a high-voltage 0.35 μm CMOS process (Beer et al., 2017). The sensor chip employs rounded square SPADs with a 20 μm size and pixel pitch of 100 μm (see Figure 2.11). This yields a FF of 19% with the circuitry placed at the periphery. Each pixel contains 4 SPADs which allows the implementation of coincidence detection to achieve high rejection of background-generated photons. The sensor

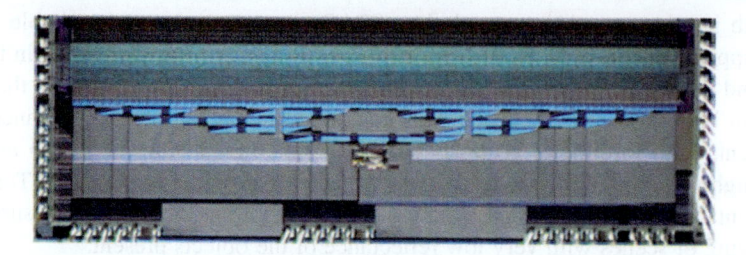

Figure 2.11 Chip photomicrograph of the 1 × 80-pixel ToF sensor fabricated in 0.35 μm HV CMOS technology at the Fraunhofer IMS in Germany, reported in (Beer et al., 2017).

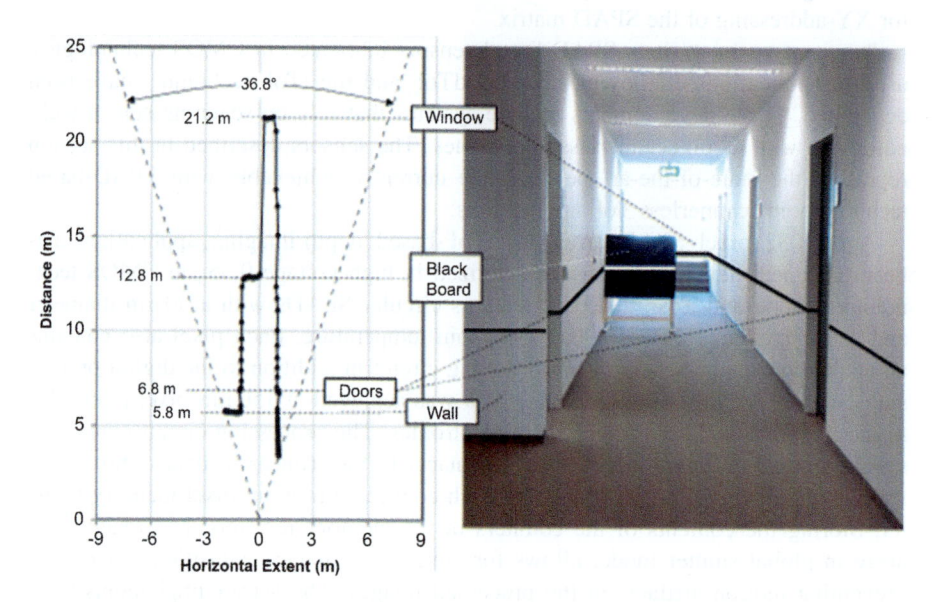

Figure 2.12 Measurement performed using the ToF sensor shown in Figure 2.11, reported in (Beer et al., 2017), under indoor conditions. On the right-hand side, the target scene is shown. The solid line marks the sensor field-of-view (FOV). On the left side the measured distance of each pixel is shown in polar coordinates.

operates with a 75 W pulsed laser diode providing 400 pulses per frame, each 25 ns wide, at 905 nm wavelength and 10 kHz repetition rate. Note this yields an average power of 18.75 mW, since the duty cycle is only 0.025%, and 25 fps. A precision of ca. 5 cm at the range of 12 m has been achieved in presence of background illumination of 90 klux. Figure 2.12 shows an example of an indoor 3D measurement. The performance of the sensors mentioned in this section is compared in Table 2.2.

Table 2.2 System performance of the ToF sensors described in this section (Bronzi et al., 2014; Niclass et al., 2013; Beer et al., 2017).

	Scannerless iToF (Bronzi et al., 2014)	Scanning dToF (Niclass et al., 2013)	Scannerless iToF (Beer et al., 2017)
Tech. node	350 nm	180 nm	350 nm
Pixel resolution	64 × 32 pixel	340 × 96 pixel	1 × 80 pixel
SPAD size (form)	30 μm diameter (circular)	< 25 μm (rounded square)	20 μm (rounded square)
Pixel pitch	150 μm	25 μm	100 μm
Pixel fill factor	3.14%	70%	19%
DCR	100 cps (median)	2650 cps (mean)	100 cps (median)
ToF technique	iToF (CW)	dToF	dToF (PM)
Illumination	800 mW @ 850 nm	40 mW @ 870 nm	25 ns pulse 75 W @ 905 nm
FoV (H × V)	40 × 20 degrees	170 × 4.5 degrees	36.8 × 0.46 degrees
Range	20 m	100 m	12 m
Precision	85 cm	10 cm	5 cm
Background light	0.45 klux	80 klux	90 klux
Target reflectivity	70%	100%	80%
Frame rate	100 fps	10 fps	25 fps

2.6 Conclusions and outlook

The concept, operational principle, and performance results of different SPAD-based technologies achieved so far have been presented and discussed in this chapter, comparing them to other single-photon counting technologies, such as the PMT, MCP, or I2. It was shown that each of these approaches has several advantages but also some drawbacks. Thus, the photodetector technology to be employed and the counting technique used depend to a very high degree on the particular application.

In the case of SPADs or SPAD arrays (analog and digital SiPMs and SPAD-based imagers), the main advantages of implementation in CMOS technology, not encountered in other technologies considered here, are the near single-photon counting capability, very high time resolution (of some picosecond in contrast to several tens of nanoseconds yielded by the other technologies), a high spatial resolution of the SPAD arrays with pixel sizes down to 20–50 μm (a huge advantage when compared to detector sizes of several millimeters offered by MCPs or PMTs), and the possibility of on-chip realization of active quenching and recharging, analog-to-digital conversion (if required), and signal and data processing. The main drawback of the SPAD technology in its current state is its still somewhat high DCRs (because these developments are normally based on standard CMOS technologies without much room for technological enhancement), and low pixel FFs, which may be a problem in high-resolution 2D SPAD-based imagers. There are several approaches under development to deal with these drawbacks, such as the use of microlenses or even back-side illumination and 3D integration technologies. Especially the latter approach is quite promising for future high-performance high-speed applications.

References

Anger, H.O., 1958. Scintillation camera. Rev. Sci. Instrum. 29 (1), 27−33.
Beer, M., Schrey, O.M., Hosticka, B., Kokozinski, R. 2017. Coincidence in SPAD-based time-of-flight sensors. PRIME 2017. In: Proceedings of the Thirteenth Conference on Ph.D. Research in Microelectronics and Electronics. Conference Proceedings: 12th−15th June 2017, Giardini Naxos—Taormina, Italy, pp. 381−384.
Brennan, K.F., 1999. The Physics of Semiconductors with Applications to Optoelectronic Devices. Cambridge University Press, UK.
Bronzi, D., Villa, F., Bellisai, S., Tisa, S., Ripamonti, G., Tosi, A., 2013. Figures of merit for CMOS SPADs and arrays. Proc. SPIE 8773, 877304-1-877304-7.
Bronzi, D., Villa, F., Tisa, S., Tosi, A., Zappa, F., Durini, D., et al., 2014. 100 000 Frames/s 64x32 single-photon detector array for 2-D imaging and 3-D ranging. IEEE J. Sel. Top. Quantum Electron. 20 (6), 1−10.
Bronzi, D., Villa, F., Tisa, S., Tosi, A., Zappa, F., 2016. SPAD figures of merit for photon-counting, photon-timing, and imaging applications: a review. IEEE Sens. J. 16 (1), 3−12. Available from: https://doi.org/10.1109/JSEN.2015.2483565.
Charbon, E., Fishburn, M.W., 2011. Monolithic single-photon avalanche diodes: SPADs. In: Seitz, P., Theuwissen, A.J.P. (Eds.), Single-Photon Imaging. Springer Series in Optical Sciences 160, Springer Verlag Berlin-Heidelberg, pp. 123−157. (2011).
Choong, W.-S., Holland, S.E., 2012. Back-side readout silicon photomultiplier. IEEE Trans. Electron. Devices 59 (8), 2187−2191. Available from: https://doi.org/10.1109/TED.2012.2200684.
Cova, S., Longoni, A., Andreoni, A., 1981. Towards a picosecond resolution with single-photon avalanche diodes. Rev. Sci. Instr. 52 (3), 408−412.
Cova, S., Ghioni, M., Lacaita, A., Samori, C., Zappa, F., 1996. Avalanche photodiodes and quenching circuits for single-photon detection. Appl. Opt. 35, 1956−1963.
Durini, D., Arutinov, D., 2014. Fundamental principles of photosensing. In: Durini, D. (Ed.), High Performance Silicon Imaging: Fundamentals and Applications of CMOS and CCD Sensors. Woodhead Publishing Series in Electronic and Optical Materials, *Woodhead Publishing Ltd. an imprint of Elsevier*, UK, pp. 3−24. (2014).
Durini, D., Weyers, S., Stühlmeyer, M., Goehlich, A., Brockherde, W., Paschen, U., et al., 2012. BackSPAD—back-side illuminated single-photon avalanche diodes: concept and preliminary performances. IEEE NSS/MIC/RTSD. Anaheim, California, 27 October−3 November 2012.
Dutton, N.A.W., Al Abbas, T., Gyongy, I., Della Rocca, F.M., henderson, R.K., 2018. High dynamic range imaging at the quantum limit with single photon avalanche diode-based image sensors,. Sensors 18 (4), 1166. Available from: https://doi.org/10.3390/s18041166.
Einstein, A., 1905. Über einen für die Erzeugung und Verwandlung des Lichtes betreffenden heuristischen Gesichtspunkt. Annal. Phys. 17 (6), 132−148.
Fischer, P., Armbruster, T., Blanco, R., Ritzert, M., Sacco, I., Weyers, S., 2014. A dense SPAD array with full frame readout and fast cluster position reconstruction. In: Proceeding of IEEE Nuclear Science Symposium, Seattle, WA.

Fishburn, M.W. 2012. Fundamentals of CMOS single-photon avalanche diodes. (Ph.D. thesis), Techn. Univ. Delft, NL.

Gasanov, A., Golovin, V., Sadygov, Z., Yusipov, N., 1989. Avalanche Detector. Russian Patent #1702831, application from 09/11/1989.

Ghioni, M., Cova, S., Zappa, F., Samori, C., 1996. Compact active quenching circuit for fast photon counting with avalanche photodiodes. Rev. Sci. Instrum. 67, 3440–3448.

Goll, B., Hofbauer, M., Steindl, B., Zimmermann, H., 2018. A fully integrated SPAD-based CMOS data-receiver with a sensitivity of −64 dBm at 20 Mb/s. IEEE Solid-State Circuits Lett. 1 (1), 2–5. Available from: https://doi.org/10.1109/LSSC.2018.2794766.

Gulinatti, A., Ceccarellia, F., Recha, I., Ghionia, M., 2016. Silicon technologies for arrays of single photon avalanche diodes. Proc. SPIE Int. Soc. Opt. Eng. 9858. Available from: https://doi.org/10.1117/12.2223884. 2016 April 17.

Haitz, R.H., 1964. Model of the electrical behavior of a microplasma. J. Appl. Phys. 35 (5), 1370–1376.

Hamamatsu. 2007. Photomultiplier Tubes. Basics and Applications, third ed. <http://sales.hamamatsu.com/assets/pdf/catsandguides/PMT_handbook_v3aE.pdf>.

Hamamatsu. 2019. <https://www.hamamatsu.com/resources/pdf/ssd/s13360_series_kapd1052e.pdf>, September 06.

Henderson, R.K., Richardson, J., Grant, L. 2009. Reduction of band-to-band tunneling in deep-submicron CMOS single photon avalanche photodiodes. In: Proceeding International Image Sensor Workshop, Bergen (Norway).

Jahromi, S., Kostamovaara, J., 2018. Timing and probability of crosstalk in a dense CMOS SPAD array in pulsed TOF applications. Opt. Express 26 (No.6), 20622–20632. Available from: https://doi.org/10.1364/OE.26.020622.

Janesick, J.R., 2001. Scientific Charge-Coupled Devices. SPIE Press, USA.

Karami, M., Yoon, H.-J., Charbon, E. 2011. Single-photon avalanche diodes in sub-100nm standard CMOS technologies. In: Proceeding of the International Image Sensor Workshop (IISW), Onuma, Hokkaido.

Kemmerling, G., et al., 2001. A new two-dimensional scintillation detector system for small angle neutron scattering experiments. IEEE Trans. Nucl. Sci. 48 (4), 1114–1117.

Lee, M.J., Ximenes, A.R., Padmanabhan, P., Wang, T.J., Huang, K.C., Yamashita, Y., et al. 2017. A back-illuminated 3D-stacked single-photon avalanche diode in 45 nm CMOS technology. In: Proceeding IEEE International Electron Devices Meeting (IEDM), Dec. 2–6, 2017, San Francisco, USA, pp. 405–408.

Mata Pavia, J., Wolf, M., Charbon, E., 2014. "Measurement and modeling of microlenses fabricated on single-photon avalanche diode arrays for fill factor recovery. Opt. Express 22 (4), 4202–4213. Available from: https://doi.org/10.1364/OE.22.004202.

McIntyre, R.J., 1961. Theory of microplasma instability in silicon. J. Appl. Phys. 32 (6), 983–995.

McIntyre, R.J., 1966. Multiplication noise in uniform avalanche diodes". IEEE Trans. Electron. Devices 13 (1), 164–168.

Migdall, A., Polyakov, S., Fan, J., Bienfang, J., 2013. Single-Photon Generation and Detection: Physics and Applications. Academic Press.

Moser, H.-G. et al. 2007. Development of back illuminated SiPM at the MPI semiconductor laboratory. In: Proceeding International Workshop on New Photon-Detectors PD07, June 27–29 2007, Kobe University, Kobe, Japan.

Niclass, C., Soga, M., Matsubara, H., Kato, S., Kagami, M., 2013. A 100-m range 10-frame/s 340 96-pixel time-of-flight depth sensor in 0.18-CMOS. IEEE J. Solid-State Circuits 48 (2), 559–572.

Nolet, F., Parent, S., Roy, N., Mercier, M.-O., Charlebois, S.A., Fontaine, R., et al., 2018. Quenching circuit and SPAD integrated in CMOS 65 nm with 7.8 ps FWHM single photon timing resolution. Instruments 2 (4), 18. Available from: https://doi.org/10.3390/instruments2040019.

Nützel, G., 2011. Single-photon imaging using electron multiplication in vacuum. In: Seitz, P., Theuwissen, A.J.P. (Eds.), Single-Photon Imaging. Springer Series in Optical Science 160, Springer-Verlag Berlin-Heidelberg, pp. 73–102. (2011).

Pani, R., et al., 2007. Factors affecting hamamatsu H8500 flat panel PMT calibration for gamma ray imaging. IEEE Trans. Nucl. Sc. 54 (3), 438–443.

Perenzoni, M., Pancheri, L., Stoppa, D., 2016. Compact SPAD-based pixel architectures for time-resolved image sensors. Sensors 16 (5), 745. Available from: https://doi.org/10.3390/s16050745.

Philips. 2013. Digital silicon photomultiplier. <http://www.digitalphotoncounting.com/wp-content/uploads/dSiPM-Leaflet_A4_2013-11_A4.pdf>.

Piemonte, C., Ferri, A., Gola, A., Picciotto, A., Pro, T., Serra, N., et al. 2012. Development of an automatic procedure for the characterization of silicon photomultipliers. In: Proceedings of the IEEE Nuclear Science Symposium and Medical Imaging Conference Record (N S S/MIC), pp. 428–432. <https://doi.org/10.1109/NSSMIC.2012.6551141>.

Rech, I., Ingargiola, A., Spinelli, R., Labanca, I., Marangoni, S., Ghioni, M., et al., 2008. Optical crosstalk in single photon avalanche diode arrays: a new complete model. Opt. Express 16 (12), 8381–8394. Available from: https://doi.org/10.1364/OE.16.008381.

Schaart, D., Charbon, E., Frach, T., Schulz, V., 2016. Advances in digital SiPMs and their application in biomedical imaging. Nucl. Instr. Meth. Phys. Res. A 809, 31–52. Available from: https://doi.org/10.1016/j.nima.2015.10.078.

Schwinger, A., Brockherde, W., Hosticka, B.J., Vogt, H. 2017. CMOS SiPM with integrated amplifier. In: Proceedings of the SPIE Conference on Optical Components and Materials XIV, San Francisco, January 2017.

Sens, L., 2019. <https://www.onsemi.com/pub/Collateral/MICROJ-SERIES-D.PDF> (accessed 6.09.19.).

Sze, S.M., 2002. Semiconductor devices, Physics and Technology, second ed. *John Wiley & Sons, Inc*, USA.

Takai, I., Matsubara, H., Soga, M., Ohta, M., Ogawa, M., Yamashita, T., 2016. Single-photon avalanche diode with enhanced NIR-sensitivity for automotive LIDAR systems. Sensors 16 (4), 459. Available from: https://doi.org/10.3390/s16040459.

Villa, F., Bronzi, D., Zou, Y., Scarcella, C., Boso, G., Tisa, S., et al., 2014. CMOS SPADs with up to 500 μm diameter and 55% detection efficiency at 420 nm. J. Mod. Opt. 61, 102–115.

Waddie, A.J., McCarthy, A., Buller, G.S., Tisa, S., Taghizadeh, M.R. 2013. Diffractive and refractive microlens integration with single photon detector smart pixels. In: Proceeding The European Conference on Lasers and Electro-Optics, Munich Germany, May 12–16, 2013.

Webster, E.A.G., Richardson, J.A., Grant, L.A., Renshaw, D., Henderson, R.K., 2012. A single-photon avalanche diode in 90-nm CMOS imaging technology with 44% photon detection efficiency at 690 nm. IEEE Electron. Device Lett. 33, 694–696.

Xu, H., Pancheri, L., Dalla Betta, G.-F., Stoppa, D., 2017. Design and characterization of a p + /n-well SPAD array in 150nm CMOS process. Opt. Express 25 (No.11), 12765–12778. Available from: https://doi.org/10.1364/OE.25.012765.

Zappa, F., Tosi, A., 2013. MiSPIA: microelectronic single-photon 3D imaging arrays for low-light high-speed safety and security applications. Proc. SPIE 8727, 87270L–87270L–11.

Zappa, F., Lotito, A., Giudice, A.C., Cova, S., Ghioni, M., 2003. Monolithic active-quenching and active-reset circuit for single-photon avalanche detectors. IEEE J. Solid-State Circ. 38, 1298–1301.

Zappa, F., Tisa, S., Gulinatti, A., Gallivanoni, A., Cova, S. 2004. Monolithic CMOS detector module for photon counting and picosecond timing. In: Proceedings of the Thirty-Fourth European Solid-State Device Research Conference, ESSDERC 2004, 341–344.

Zhang, C., Lindner, S., Antolovic, I.M., Pavia, J.M., Wolf, M., Charbon, E., 2018. A 30-frames/s, 252×144 SPAD Flash LiDAR with 1728 dual-clock 48.8-ps TDCs, and pixel-wise integrated histogramming. IEEE J. Solid-State Circuits. Available from: https://doi.org/10.1109/JSSC.2018.2883720. Dec. 2018.

Organic photodetectors

Vincenzo Pecunia[1], Dario Natali[2,3] and Mario Caironi[3]
[1]School of Sustainable Energy Engineering, Simon Fraser University, Surrey, BC, Canada, [2]Dipartimento di Elettronica, Informazione e Bioingegneria, Politecnico di Milano, Milano, Italy, [3]Center for Nano Science and Technology @PoliMi, Istituto Italiano di Tecnologia, Milano, Italy

3.1 Introduction

Organic semiconductors (Natali, 2020) have become an appealing class of materials, with several applications being developed at the R&D and industrial level, supported by a continued research effort. Not only their optoelectronic properties can be tuned by means of chemical tailoring, but also they can be deposited by means of cost-effective printing methods (Schmidt et al., 2011; Dong et al., 2012). Photodetectors based on organic semiconductors (OPD) can thus be realized with a spectral sensitivity that—spanning from UV to the near-infrared (NIR)—can be made panchromatic or selectively tuned to specific wavelengths. Direct or indirect sensing schemes can expand detectors' sensitivity to higher energy radiation, such as X-rays and γ-rays. The low-temperature processability over large areas and on virtually every substrate, and the possibility of integrating different organic devices [organic light-emitting diodes (OLEDs) (Thejo Kalyani and Dhoble, 2012), organic field-effect transistors (Sirringhaus, 2009), organic photovoltaics (Lin et al., 2012; Nelson, 2011), organic memories (Cho et al., 2011), and organic sensors (Angione et al., 2011; Briand et al., 2011)], makes it possible to target innovative optoelectronic systems such as large-area imagers and scanners, short-range optoelectronic transceivers, integrated sensors for Lab-on-a-chip to cite a few. Such integrated, lightweight, and even conformable systems have the potential to pervade sectors such as healthcare, home, and industrial automation, homeland security, automotive, industrial diagnostic, and gaming. While demonstrators and first products are already being proposed, some of the fundamental properties of photoactive organic materials and photodetection mechanisms are curiously not fully understood yet, thus the lack of generalized quantitative models. It is therefore expected that the ongoing research efforts will help to further expand the field with a full deployment of the potentiality of OPDs.

This chapter aims at introducing these aspects and it is organized as follows. Organic semiconductors are introduced in Section 3.2 focusing on charge photogeneration and deposition techniques. Section 3.3 deals with OPDs architecture and operation mechanisms. Section 3.4 reports exemplary realizations to provide an overview of the achieved specifications, and it is organized based on the targeted region of the spectrum. Section 3.5 is dedicated to photodetectors fabricated by means of scalable, printing processes. Finally, Section 3.6 contains an overview of realizations going beyond the single device and targeting more complex optoelectronic systems.

3.2 Organic semiconductors

Organic semiconductors (Wudl, 2014) are materials based on sp^2 hybridized carbon atoms. Hybridized orbitals give rise to three strongly localized, covalent σ−bonds which constitute the molecular backbone, whereas the remaining half-filled p orbitals give rise to π orbitals that delocalize over the entire molecule (π-conjugation). p-orbitals are the frontier orbitals and are responsible for the optoelectronic properties of a molecule. Oversimplifying, the highest occupied molecular orbital (HOMO) and the lowest unoccupied molecular orbital (LUMO) play somehow the role of the valence and conduction band edges respectively of inorganic semiconductors. Organic semiconductors can be classified as single small molecules, oligomers (given by the repetition of few monomer units), and polymers (given by the repetition of many monomer units). In the solid state, they give rise to *molecular solids*, which are characterized by intermolecular bonds, usually of the Van der Waals type. The latter has dissociation energy of about 10 kcal/mol, which is low if compared to the dissociation energy of the covalent Si−Si bond (78 kcal/mol). This implies that the tendency towards the formation of ordered structures is mild because the gain in enthalpy is low. From the processing point of view, this is advantageous because the involved thermal budget is relatively low (typically well below 200°C); in addition, many semiconductors can be functionalized to be soluble in suitable solvents: this enables the adoption of solution-based deposition protocols and opens the way to high throughput, low-cost techniques borrowed from graphical arts (Section 3.2.2). Depending on the type and parameters of the deposition technique and on the nature of the molecule, aggregation and packing in the solid-state can result in amorphous materials, inhomogeneous materials with nano- and microcrystalline regions interspersed in an amorphous matrix, up to a single crystal (obtained by means of dedicated vapor-phase techniques or, in exceptional cases also directly from solution).

The lack of long-range order, together with a strong electron-phonon coupling, causes charge carrier localization on a single molecule or very few neighboring molecules at best, in contrast to Bloch waves typical of crystalline, inorganic semiconductors. On the one hand, this implies a relatively ineffective transport of charge carriers, which can be modeled as a thermally activated tunneling from molecule to molecule termed hopping. The resulting mobility is low, often in the $10^{-4}-10^{-1}$ cm^2/Vs range. Only in the latest generation of high-performing polymers (Olivier et al., 2014) remarkable carrier mobilities exceeding 10 cm^2/Vs have been recorded (but only at a high carrier density of about 10^{19}/cm^3, as those reached in the accumulation channel of transistors): there is an ongoing debate regarding the transport mechanism in these cases (Venkateshvaran et al., 2014). On the other hand, carrier localization makes this class of material far more tolerant to impurities than its inorganic counterpart, thus drastically simplifying their processing (Kawase et al., 2003).

To understand device behavior, it is important to note that, unlike inorganic counterparts, *doping* of organic semiconductors (Lussem et al., 2013; Jacobs and Moulé, 2017; Salzmann et al., 2016) is not a very well mastered aspect yet. Dopants are conjugated molecules with suitable electronic levels: to perform p-type (n-type) doping, the dopant must have a high electron affinity (EA) [or low ionization potential (IP)]. The main issues are environmental stability of the dopant molecule; effectiveness of dopant ionization, due to the relatively low dielectric constant of organic semiconductors ($\varepsilon_r \cong 3-4$); and thermodynamic stability of dopants, related to their interstitial nature. There are some notable exceptions belonging to the category of molecularly doped polymers, such as poly(3,4-ethylene dioxythiophene) polystyrene sulfonate PEDOT: PSS, where the semiconducting polymer PEDOT is p-doped by sulfonyl groups carried by PSS. PEDOT:PSS is quite stable and its conductivity can be as high as 100–1000 S/cm.

Doping unreliability, together with a generally negligible thermal generation (the energy gap of many materials is around 2 eV), implies that excess carriers are always a large perturbation with respect to thermodynamic equilibrium. An additional consequence is that to engineer *metal-organic semiconductor contacts* to be non-injecting or ohmic, one has to rely on the alignment of metal Fermi level and molecular HOMO/LUMO levels. Charge injection has a strong dependence on the height of the Schottky barrier φ_B, where φ_B (assuming that the metal Fermi level lies within the semiconductor energy gap) for holes (or electrons) is the difference between the metal Fermi level (semiconductor LUMO) and the semiconductor HOMO (metal Fermi level). The lower (or higher) φ_B, the closer is the contact to be injecting (or non-injecting). Indeed, charge injection in low mobility, undoped, disordered semiconductors is a delicate issue, and going beyond the role of φ_B is outside the scope of this chapter: the interested reader is referred to Natali and Caironi (2012) and Braun et al. (2009).

To adjust φ_B various techniques have been devised, most of them relying on the insertion of suitable interlayers bearing intrinsic dipoles which can enlarge or reduce the metal work function (Natali and Caironi, 2012).

3.2.1 Charge photogeneration

In organic semiconductors, a neutral and spinless excited state is formed upon (single) photon absorption (Kippelen and Bredas, 2009). This excitation is a singlet, molecular, Frenkel-type *exciton*, and can be roughly viewed as a Coulombically bound electron-hole (e/h) pair with a relatively large binding energy of a few hundreds of meV, due to low dielectric constant, disorder, electron-electron correlation and electron-phonon coupling peculiar to π-conjugated compounds. Since a non-negligible potential barrier separates the excitonic state from the charge-pair state, the spontaneous dissociation of the exciton into free e/h pairs at room temperature is far less efficient than in inorganic semiconductors [pristine single-crystals representing an exception thanks to their high degree of order (Najafov et al., 2008)].

To enhance photogeneration yield, it is common to exploit the phenomenon of *photoinduced charge separation* occurring at donor (D)−acceptor (A) interfaces (Figure 3.1). Donor (or acceptor) molecules are characterized by a low IP (high EA). Upon photoexcitation of D (or of A), an exciton reaching a D/A interface decays into a charge transfer (CT) state, with the electron residing on the LUMO of A and the hole residing on the HOMO of D. For this process to be effective, an offset between the donor LUMO and acceptor LUMO of 0.3−0.4 eV is required (Hendriks et al., 2014). Photoinduced charge separation is an ultrafast process taking place on a sub-100 fs time scale, thus being very competitive with respect to other exciton deactivation paths. In the framework of photogeneration, the advantage of CT states is that e/h pairs are less coulombically bound and hence more prone to dissociation. High photogeneration yields require that the generation of excitons occurs within an exciton diffusion length from a D/A interface, stressing the importance of a high degree of intermixing between D and A phases. On the other hand, it must be taken into account that after being generated holes (electrons) have to be collected, which requires the continuity of D (A) phases, thus suggesting that a certain degree of demixing between D and A is mandatory to avoid dead-end transport paths and to ensure effective charge collection. From a device point of view, various solutions have been adopted, from bi- (or multi) D/A layers (Yu and Heeger, 1995; Someya et al., 2005) to bulk heterojunctions (BHJs) (Ruderer and Muller-Buschbaum, 2011; Brady et al., 2011). In the latter case, the D/A blend nano-morphology has a critical impact on device performances.

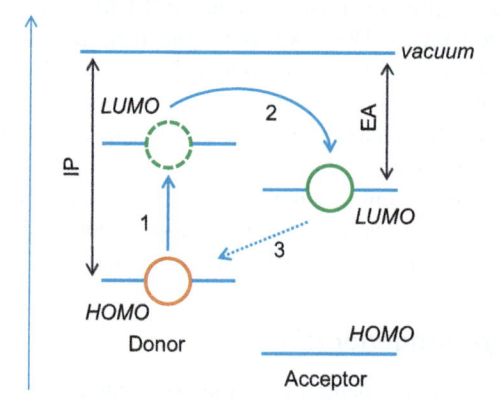

Figure 3.1 Simplified picture of a D/A interface. Process 1 refers to exciton generation upon photon absorption in the donor; process 2 refers to ultrafast charge separation at the interface; process 3 to the usually scarcely effective recombination between the electron on the acceptor and the hole on the donor. Also shown are the IP of the donor and the EA of the acceptor.
Source: Reproduced with permission from Baeg et al. (2013). Copyright 2013, Wiley.

Organic photodetectors

Once a CT state is formed, the generation of free e/h pairs has to compete against other CT deactivation paths such as electron (hole) back transfer to D (A) molecule thus regenerating an exciton on D (A) moiety, radiative or nonradiative recombination to the ground state (also termed geminate recombination) (Deibel et al., 2010). Going beyond this simplified description of D/A interfaces is outside the scope of this chapter. Nevertheless, we just point out that the CT state formed at a D/A interface still has a non-negligible binding energy of about a few tenths of eV and that the mechanism ruling its dissociation in free charges is still under debate (Deibel et al., 2010; Arkhipov et al., 2003; Verlaak and Heremans, 2007; Clarke and Durrant, 2010; Grancini et al., 2013; Morteani et al., 2004; Gao and Inganas, 2014, Hedley et al., 2017).

3.2.2 Deposition techniques

In this section, we give a short introduction to the deposition techniques most commonly adopted to fabricate OPDs. It is not intended to be comprehensive of all the many different techniques reported in the literature, but it has the aim to provide the reader with some basic knowledge of the approaches specific to OPDs. We first briefly revise vacuum thermal evaporation, which dominates part of the literature on OPD and has industrial applications, and spin-coating, the most widely adopted solution-based process in research. We then move to solution-based deposition processes enabling scalability, and we focus on inkjet printing, aerosol jet printing, and spray coating since notable examples of OPDs fabricated by means only of these techniques (Section 3.5) have been proposed.

3.2.2.1 Vacuum thermal evaporation

Vacuum thermal evaporation (Klauk, 2006) has largely dominated the first examples of organic photodetectors, and it is a very well-known and adopted technology in organic optoelectronics. The deposition system consists of a vacuum chamber, which in typical laboratory equipment is designed to achieve high vacuum levels ($\sim 10^{-6}$ mbar). The chamber is equipped with a series of thermal sources, to allow either multiple or codeposition processes. Each source can allocate a suitable crucible, where the powder of the molecules to be deposited is placed. The sources are typically temperature controlled so that the deposition can occur at an optimal temperature for the evaporation/sublimation of the target material, as opposed to the typical power control adopted in the deposition of metallic layers. The chamber is equipped also with different shutters which allow protecting the samples during ramping up or cooling down of the sources, avoiding uncontrolled deposition. Other shutters are present to isolate each source.

Vacuum thermal evaporation can be adopted for the deposition of those organic semiconductors which can withstand the evaporation/sublimation process without being degraded. Such constraint limits the range of materials to small molecules and oligomers. The great advantage of vacuum deposition, with respect for example to solution-based processes, is the possibility of achieving a nanometer control of each active layer thickness, the capability of easily stacking multilayers, and high large-area uniformity. One issue that has to be kept under control is the material usage, which may not be efficient due to the lack of directionality of the evaporation cone in standard tools. Control of uniformity requires large chambers since the source has to be positioned far away from the target. Such a high level of control over large areas is one of the main reasons that allowed the industrial development of commercial organic semiconductors-based devices, that is, OLED. Roll-to-roll processes have also been developed by the company Heliatek GmbH (http://www.heliatek.com) for the vacuum deposition of organic solar cells, sharing a very similar architecture with OPDs. Therefore, the same advantages can favor the development of highly uniform light detectors over large areas for imaging and diagnosis. This technique allows the deposition and integration of pristine semiconducting layers as well as photoactive blends (by means of coevaporation), with the additional feature of adding either vertical or lateral composition gradients.

Not only the photoactive blend can be vacuum processed, but the full photodetecting device can be produced by evaporating all the functional layers. Besides evaporated metallic electrodes, typically gold, aluminum, and silver, also interlayers playing the role of charge extraction layers and/or exciton blocking layers (BLs), can be evaporated, including small molecules, oxides, and salts (e.g., fluorides, carbonates). Patterning of photodetectors can be easily achieved by the use of shadow masks with feature sizes that can easily reach approximately 100 μm with chemically etched masks. Finer resolution can be obtained by means of laser patterned shadow masks, though requiring very thin masks to avoid shadowing effects, thus losing intrinsic rigidity.

3.2.2.2 Spin-coating

Spin-coating (Klauk, 2006) is probably the most widely adopted solution-based deposition technique for organic materials. It is derived from micro-electronics processes, where typically photoresists are deposited by spin-coating on wafers. The technique is very well-known and studied, and it is based on a rotating chuck where the substrate to be coated is held in place thanks to a vacuum. The solution is simply dispensed on the substrate, which is then accelerated to the desired final speed rotation. The acceleration and the regime speed rotation are two critical parameters, as well as the spin duration. In the case of organic semiconductors, a drying step usually follows shortly after to remove the residual solvent from the films. This technique allows the deposition of uniform layers with very precisely controlled thicknesses, ranging from a few tens of nanometers to microns. The final film thickness can be controlled through a series of parameters, as indicated by the two following equations which can be derived under the assumption that the evaporation of the solvent is neglected in the initial spin-off stage (Hall et al., 1998):

$$h_\text{f} = (1 - x_1^0)h_\text{w}$$

$$h_{\rm w} = \left[\left(\frac{3\eta_0}{2\rho\omega^2}\right)k\left(x_1^0 - x_{1\infty}\right)\right]^{1/3}$$

where h_f is the final film thickness, ρ is the liquid density, ω is the spin speed, η_0 is the initial solution viscosity, k is the mass transfer coefficient, x_1^0 is the initial solvent mass fraction in the coating solution, and $x_{1\infty}$ is the solvent mass fraction that would be in equilibrium with the solvent mass fraction in the gas phase.

Being a coating technique, it does not allow lateral patterning. Patterning of the photoactive materials is not strictly required in case bottom electrodes or contact pads of a backplane are prefabricated (Section 3.6). When patterning of the functional layers is required, this has therefore to be obtained by postetching or other kinds of processing. The maximum covered areas are those typical of wafer processing. For this reason, while being an excellent laboratory research and development tool, it presents limits in scalability and the realization of large-area detector arrays. Waste of materials is very high since most of the functional formulations dispensed on the substrate are lost at the initial spin-off.

3.2.2.3 Ink-jet printing

Inkjet printing derives from the observation of fluid jet instabilities, first observed in the nineteenth century by Savart (1833) and later described mathematically by Lord Rayleigh (1878). The first practical inkjet printing devices appeared in the mid of the twentieth century (Elmqvist, 1951). The first adopted apparatus was the so-called "continuous" inkjet printing (Figure 3.2A) (Gili et al., 2009), where a continuous stream of droplets is produced by applying an electrostatic field to the nozzle where a fluid, that is, the ink, is pumped to. The droplets, which are electrostatically charged, are subsequently deflected by an electric field according to the patterning needs, and either reach the substrate moving underneath or are recirculated through a gutter. This scheme is very simple and allows very fast printing of droplets with a typical diameter of approximately 100 μm thanks to a typical jetting frequency of 80–100 kHz, with maximum frequencies up to 1 MHz. This scheme is very suitable for low precisions industrial applications, such as fast labeling of packages and goods, but it is not flexible enough for printed electronics. For industrial applications requiring a higher resolution and for research activities on functional electronic materials, comprising organics, the drop-on-demand (DOD) approach is the most widely adopted (Figure 3.2B) (Hansell, 1950). The apparatus is simpler and the inks are more efficiently used since a droplet is produced on demand only when needed, either by a thermal ("bubble jet" first adopted in Canon printers) or piezoelectric actuator (adopted in Epson printers) coupled with the nozzle. Piezoelectric transduction is most commonly preferred for functional inks since it can be applied to a vast set of formulations, imposing fewer constraints on the fluids than thermal actuation in terms of solvent vaporability. In piezoelectric systems the nozzle is compressed or expanded by means of a voltage pulse, thus

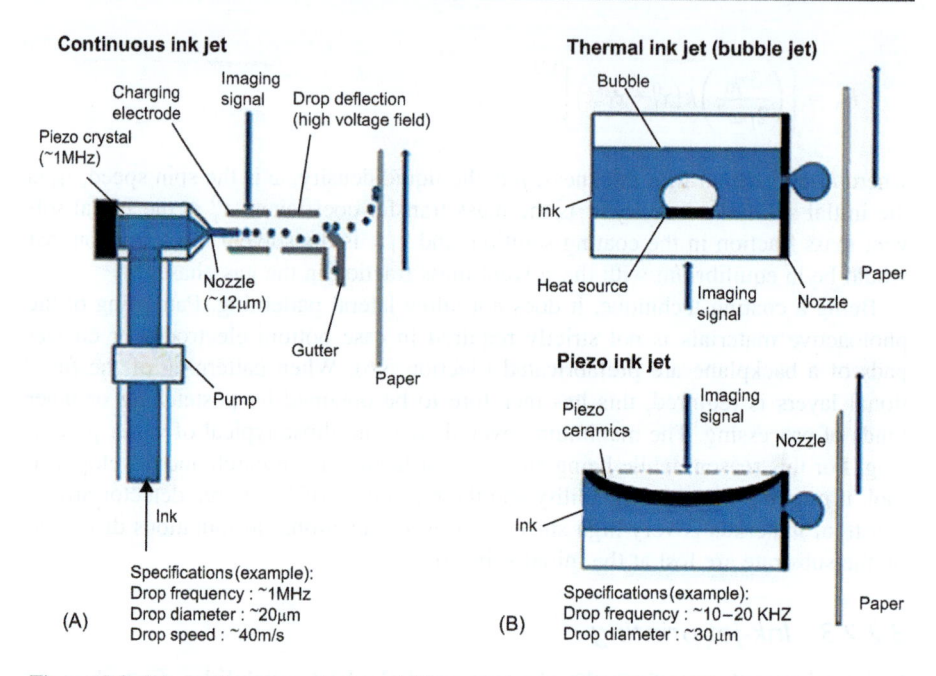

Figure 3.2 Schematic representation of (A) a continuous and (B) DOD inkjet printing apparatus: thermal inkjet on top and piezo inkjet on the bottom.
Source: Reproduced with permission from Weng et al. (2010). Copyright 2013 Royal Society of Chemistry.

producing an acoustic wave that, if properly set, can produce the ejection of a droplet. State-of-the-art inkjet printing tools for organic printed electronics, specifically adopted to inkjet print OPDs, are typically based on piezoelectric DOD systems. Several commercial tools by many different companies have become available, covering all the way from desktop, research printers equipped with single or a limited set of nozzles, to prototyping and industrial printers covering large areas and equipped with thousands of nozzles. Inkjet is a noncontact technique and the jetting distance is usually shorter than 1 mm. Printing heads have been developed to withstand typical solvents adopted for organic materials, insulators, and metallic ink formulations (Caironi et al., 2012). Inkjet is therefore very flexible, and it has the advantage of depositing a controlled volume of ink at specific places on the substrate. This is ideal for patterning functional inks to realize arrays of optoelectronic devices, in a direct-writing approach, without the need for any mask and by reducing materials waste to a minimum. It has therefore been employed in various fabrication processes of organic optoelectronics devices (Jung et al., 2014; Gili et al., 2012; Singh et al., 2010; Mandal et al., 2015; Noh et al., 2007; Li et al., 2013, 2014; Minemawari et al., 2011; Sirringhaus et al., 2000; Torrisi et al., 2012). The technique is suitable for low viscosity formulations, typically within the 1 to 20 mPas range. Other ink parameters to be controlled are the surface tension and the volatility of the solvents composing the formulation: surface tension should be in the $20-70 \times 10^{-3}$/Nm range, with pure water-based dispersions being at the limit of the jettable range, and dominant solvents should typically have a boiling point higher than 140°C to avoid fast evaporation at the nozzle orifice and clogging. Clogging is indeed one of the main problems to overcome to achieve stable jetting over time and it is one of the usual difficulties encountered when developing new formulations at the research level. Single droplets with a volume as small as 1 pL can be obtained with commercial tools, corresponding to a droplet diameter of roughly 10 μm. Smaller sizes cannot be achieved for rheological issues related to the inks. The printed patterns reflect such limitation, achieving a lateral features size not smaller than 20 μm in most cases. The film thickness achievable is typically below 1 μm if no suitable containing wells or trenches are preformed onto the substrate. Inkjet can be integrated into roll-to-roll lines, something which has already been done for a long in the graphic art industry, however with a limited throughput with respect to roll-to-roll workhorses such as gravure or flexography, likely achieving a linear web speed in the tens m/min range.

3.2.2.4 Aerosol-jet printing

Aerosol-jet printing is a very interesting alternative to inkjet. It shares the same advantages, being a noncontact, digital, direct-writing technique while presenting some additional features, in terms of the lateral resolution, which can be as low as 10 μm, and expanded flexibility in terms of jettable inks, being compatible with higher fluid viscosity, up to thousands of mPas (Mette et al., 2006).

The ink to be deposited is fed to an atomizer, where an aerosol is produced (Figure 3.3). Different atomizers are adopted to handle materials with different viscosities. Low viscosity materials (<10 mPas) are atomized by means of an

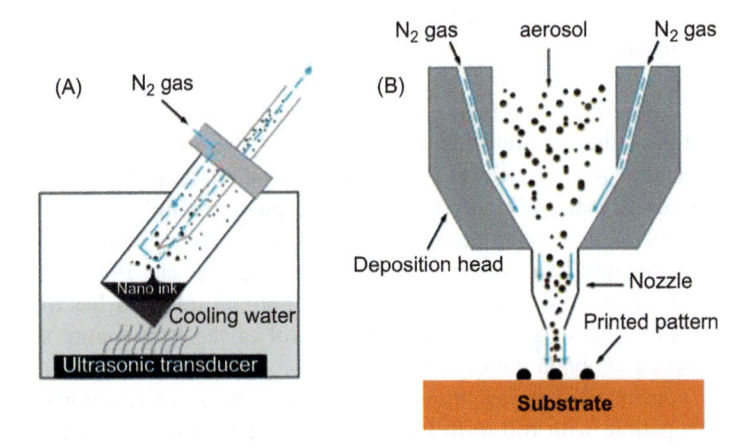

Figure 3.3 Sketches of an aerosol jet apparatus: scheme (A) of the ultrasonic transducer and (B) of the nozzle, with the aerosol being directed by a focusing gas.
Source: Adapted from permission from Zhao et al. (2012). Copyright 2012 Elsevier.

ultrasonic transducer producing high-frequency pressure waves. Higher viscosities (>10 mPas) can be handled with a pneumatic atomizer based on the action of a high-velocity gas stream to generate the mist. Viscosities higher than 1000 mPas may require suitable heating and stirring. The aerosol is then fed to a nozzle where it is aerodynamically focused thanks to a sheath gas and directed towards a substrate. The ink does not get into contact with the inner nozzle surface, thus solving at the root the nozzle clogging issue and enabling smaller jet diameters. A mechanical shutter is adopted to produce patterns, actually limiting the process throughput because of a response time in the range of milliseconds. The aerosol jet is more collimated than the inkjet and it can be used to pattern functional inks from longer distances, up to a few mm, simplifying printing on 3D-shaped surfaces. Also in this case single and multinozzle tools are available, with industrial equipment commercially available (Optomec. Aerosol jet 300 datasheet. *Datasheet available at* http://optomec.com). Aerosol jet printing has been successfully adopted to pattern a range of carbon-based materials and produce organic circuits (Hong et al., 2014).

3.2.2.5 Spray-coating

Spraying (Krebs, 2009) is an interesting approach for very high throughput deposition of functional inks. It can be thought of as a simple version of aerosol jetting, implying strongly reduced costs, where the atomized ink is much less focused. Spray can be generated with different methods: (1) pneumatic systems, with pressurized gas, (2) ultrasonic systems (Tenent et al., 2009), and (3) electrospraying systems. Air and ultrasonic assisted approaches are the preferred ones for the fabrication of organic devices, and consist of a gun with a mounted nozzle to where the ink mist, created by an air/gas stream is fed to. In ultrasonic tools, which are less liable to nozzle clogging, a piezoelectric transducer produces the atomizing vibrations, with a much finer capability of tuning the droplets' dimensions and distribution.

The quality of coated patterns critically depends on the size and distribution of the particles in the spray, and their coalescence once on the substrate. Viscosity and surface tension of the ink are critical parameters for the quality of the atomization process, depending on the selected spraying scheme. If droplets impact the substrates when already dry, a highly nonuniform deposition is obtained, while if too wet, the control of the evaporation may be difficult, leading to unpredicted rearrangement of the solute or coffee stain effects (Deegan et al., 1997; de Gans and Schubert, 2004; Lim et al., 2009). Therefore, an optimal trade-off has to be found, whilst controlling the substrate wettability. Control of the temperature substrate is typically adopted to obtain uniform depositions.

Suitably controlled spraying allows large area coatings of functional materials and it is compatible with roll-to-roll printing lines. The absence of jet collimation requires the use of shadow masking to achieve patterning and it makes less efficient the use of materials, with increased waste compared to ink or aerosol jet printing (Figure 3.4).

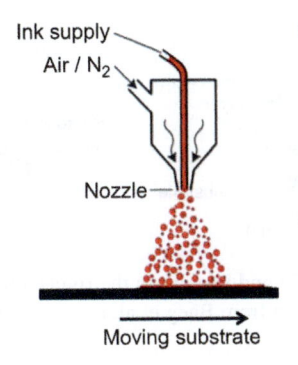

Figure 3.4 Schematic representation of air-assisted spraying apparatus.
Source: Adapted with permission from Søndergaard et al. (2013). Copyright 2013, Wiley.

3.3 Device structure and operation mechanisms

Photodetectors can be roughly divided into three subcategories: photodiodes, photoconductors, and phototransistors. We will focus on the first two, referring the reader interested in phototransistors to Baeg et al. (2013).

Photodiodes and *photoconductors* are two-terminal devices where the photoactive medium is contacted by two metal contacts (Figure 3.5). The most common topology is the vertical (sandwich-like) one. Usually, the photoactive layer has a thickness in the range of 100–200 nm, because given the fairly high absorption coefficient α, in the range of 10^5/cm, this is enough to absorb most of the incoming light. While many donor materials have been proposed, the choice of acceptors is somewhat more limited. Especially in the case of solution-processed devices, a soluble derivative of fullerene C_{60} phenyl-C_{61}-butyric acid methyl ester ($PC_{61}BM$) is a very common choice: it has good carrier mobility, high solubility, and low-lying LUMO. In addition, its peculiar spherical symmetry is favorable to intimate nanoscale intermixing with donor polymers (Verilhac, 2013). $PC_{61}BM$ has very low absorption in the visible: to circumvent this, higher analogs like [6,6]-phenyl-C_{71}-butyric acid methyl esterin ($PC_{71}BM$), which are characterized by a stronger optical absorption in the visible, have been proposed (Wienk et al., 2003). The ongoing research on nonfullerene acceptors has produced very interesting results, with the demonstration of long-range exciton diffusion (Firdaus et al., 2020) and quantum conversion efficiencies exceeding 16% in organic solar cells (Sun et al., 2019); such acceptors are likely to have a strong impact on photodetectors as well. The interested reader can find more details in the review by Hout et al. (2018).

As to the choice of metals, there are a number of constraints to be satisfied: at least one of the two needs to be transparent; in the case of solution processing, the bottom metal needs to be environmentally stable; to effectively collect holes and electrons, a high and a low work function metals are respectively needed. A very commonly adopted architecture employs indium tin oxide (ITO) as the transparent bottom anode and evaporated low work function metals (such as Al, Al:Ca) as the

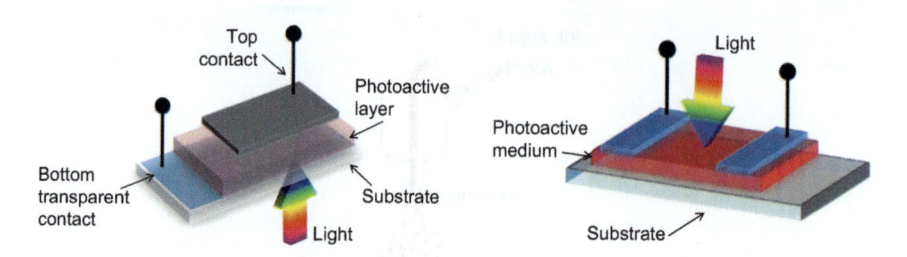

Figure 3.5 Vertical (on the left) and lateral (on the right) topology of OPDs.
Source: Adapted with permission from Baeg et al. (2013). Copyright 2013, Wiley.

top cathode. Indeed, since the ITO work function (about 4.7 eV) is not well suited to hole collection in many cases, an additional interlayer of PEDOT:PSS (work function of about 5.0 eV) is added; the thickness (few tens of nm) together with PEDOT:PSS electronic structure ensures a virtually negligible light absorption in the visible range. In some cases, an *inverted* structure is adopted, with a top anode and a bottom cathode (Arredondo et al., 2013). The high reactivity of low work function metal cathodes requires special care (such as etching of native aluminum oxide in the case of the aluminum cathode (Baierl et al., 2012)), or processing in an inert atmosphere (Baierl et al., 2010). As a nonmetal-based alternative, it was proposed to decrease PEDOT:PSS work function by 0.7–0.8 eV by means of a thin insulating interlayer based on polymers bearing polar groups, such as ethoxylated polyethyleneimine (PEIE) (Saracco et al., 2013; Cesarini et al., 2018).

In addition, there is ongoing research aimed at producing ITO-free devices, because of the following critical aspects of ITO: indium is not abundant; the ITO/PEDOT:PSS interface is chemically unstable; ITO is brittle and not well suited to flexible substrates; it is not transparent in the blue region of the spectrum (Arredondo et al., 2013).

The *lateral topology*, where contacts and active material lie in the same plane, is less popular: on the one hand, it offers direct access to the photoactive medium thus enabling direct investigations on working devices (Agostinelli et al., 2007b; Borel et al., 2014; Danielson et al., 2014), and since contacts are not along the optical path they are not required to be transparent. On the other hand, it is typically characterized by larger interelectrode spacings (from few to tens of μm, to be compared to the few hundreds of nm commonly adopted in vertical topology) and hence has lower operation bandwidth (BW).

3.3.1 Photodiodes operation mechanism

In photodiodes, metal-semiconductor contacts have to be noninjecting; an absorbed photon can result at best in one e/h pair collected at the electrodes [external quantum efficiency (EQE) ≤ 100%]. The behavior of a photodiode changes according to the fate of photogenerated carriers.

Let us assume that carriers can be collected before recombining. This means that their lifetime τ must exceed the time required to reach the electrodes, viz. $\tau > L/\mu E$, where L is the interelectrode spacing, E is the electric field, and μ is the charge mobility. We also introduce the net volume photogeneration rate G, which takes into account photon absorption and dissociation efficiency. If holes and electrons have comparable mobilities, then they are extracted at the same rate from cathode and anode and charge neutrality of the active layer is ensured. The photocurrent is $J = qGL$, where q is the elementary charge: the photodiode is working at its highest efficiency, as from each photogenerated pair an e^-/h^+ couple is collected in the outer circuit. The device BW is related to the reciprocal of the carrier transit time, hence it is proportional to $\mu V/L^2$. If mobilities are unbalanced a loss of photoactive area occurs (Agostinelli et al., 2007b). Let us assume that electron mobility (μ_n) is far larger than hole mobility (μ_p). Photogenerated holes will tend to accumulate in the device and a positive net space charge will be established. This charge reshapes the electric field to enhance (slow down) the collection of holes (electrons) to attain a regime condition: the externally applied voltage V drops almost entirely close to the anode across a region of length L', while in the remnant $L - L'$ part of the device no net photogeneration nearly occurs due to recombination. The net positive charge due to photogenerated holes is electrostatically upper bound, hence the photocurrent is space charge limited (SCL) (Lampert and Mark, 1970) and it can be shown to obey the following equation: $J \propto q\ (\varepsilon\mu_p/q)^{0.25} G^{0.75} V^{0.5}$ (Goodman and Rose, 1971; Mihailetchi et al., 2005).

Another possible scenario occurs when carriers recombine before collection. No clear consensus has been reached yet regarding the most effective recombination mechanism for excess carriers. Besides Langevin-type bimolecular recombination, modeling the annihilation of e^-/h^+ pairs driven by their mutual Coulombic attraction, trap-mediated monomolecular mechanisms are gaining consensus (Tzabari and Tessler, 2011; Kirchartz et al., 2011; Street, 2011; Lakhwani et al., 2014). We limit the discussion to a simplified case, assuming that electrons (holes) are characterized by a lifetime τ_n (τ_p) due to the dominant recombination mechanism.

To fix ideas we consider $\mu_n\tau_n \gg \mu_p\tau_p$ (the latter quantities are termed mobility-lifetime products). Analogously to the SCL case, the externally applied voltage drops almost entirely across a region close to the anode, sweeping holes to their collecting contact before recombination takes place. The extent of this region L' is recombination-limited and can be found by equating the hole transit time across L' with the hole lifetime. It can be shown that the current obeys the following expression: $J = qGL' = qG(\mu_p\tau_p)^{0.5}V^{0.5}$, hence it is linear on the photogeneration rate but sublinear on the applied voltage. Since an unbalanced charge region exists close to the anode, at very high G the device might enter into the SCL regime. It is worth noting that SCL and recombination-limited regime can be discriminated by looking at the photocurrent dependence on G. If the mobility-lifetime products are the same for holes and electrons then the photocurrent can be written as $J = qG\mu\tau V/L$ for low voltages and saturates to qGL for larger voltages, and no sublinear dependence on V occurs (Goodman and Rose, 1971).

This simplified description does not consider that carrier mobility is usually charge-density and electric-field dependent, and that the lifetime as well might depend on the recombination center occupancy, hence the photocurrent functional dependence on V and G could be more complicated.

3.3.2 Photoconductors

In photoconductors, contacts are injected for at least one kind of carrier. In addition, unbalanced carrier mobilities are needed. To illustrate the working principle, we consider a simplified case, and to fix ideas we assume holes to be deeply trapped and substantially immobile, and contacts to be ohmic for electrons (the interested reader can find more cases discussed in Bube (1960)). Contact ohmicity ensures electrical neutrality, which means that for every collected electron another one will be injected until electron/hole recombination takes place. According to the ratio between the recombination time τ and the electron transit time, which is termed *photoconductive gain*, an efficiency higher than 100% is possible. In this regime, the photocurrent follows the expression $J = qGL[\tau/(L^2/\mu V)]$. The device BW is proportional to the reciprocal of τ and is independent of the carrier transit time, hence a tradeoff exists between the photoconductive gain and the BW. It is worth noting that the recombination time can be a function of incident light intensity: at high light intensities more trapping sites are occupied, τ and hence the photoconductive gain gets lower but the device gets faster (Iacchetti et al., 2012b).

The requisite of unbalanced carrier mobilities are easily met in organic semiconductors. If the active medium is a blend, the ratio between donor and acceptor can be purposely tuned to avoid percolative paths for one of the carriers (Li et al., 2015a,b). A remarkable EQE as high as 37,500% has been recently achieved by exploiting a 100:1 BHJ of P3HT and $PC_{61}BM$ (Li et al., 2015b). As to contact ohmicity, it has been shown that it is possible to achieve such a regime in response to a light stimulus while leaving contacts non-injecting in the dark. To this extent, the accumulation of trapped carriers close to their collecting electrode is exploited: the electric field is locally enhanced thus easing the injection of the opposite carrier (Li et al., 2015a,b; Hammond et al., 2014; Melancon and Zivanovic, 2014). One of the most efficient realizations reports a gain of 10^5 in lateral devices based on the BHJ of poly dithienobenzodithiophene-co-diketopyrrolopyrrolebithiophene (PDPDBD) and $PC_{71}BM$. This impressive value has been obtained at a fairly low light intensity of 20 nW/cm^2; the gain decreases with impinging light intensity: in the tens of $\mu W/cm^2$ intensity range, it lies in the range of many tens (Park et al., 2014a,b). Some notable realizations are reported in Table 3.1. The interested reader can profit from the recent review of Miao and Zhang (Miao and Zhang, 2018).

3.3.3 Reduction of dark currents

Dark currents play a key role in determining the overall performance of a photodetector, as they increase *recombination losses* of photogenerated carriers and are a source of *electronic noise* determining the minimum detectable amplitude of the impinging optical signal. A commonly adopted figure of merit is the noise equivalent power (NEP), defined as the signal optical power yielding a signal to noise ratio (SNR) equal to 1. It can be written as:

$$NEP = \frac{S_{noise} \bullet \sqrt{\Delta f}}{R}$$

Table 3.1 Organic and hybrid photoconductors.

Active materials and device structurea	Spectral window (nm)	QE at V_{BIAS} (light source, intensity)	I_{dark} (V_{BIAS})	Speed	References
Me-PTCb, vertical (500 nm thick film)	350–700	~10,000 (IQE) at 15 V, (600 nm, 0.16 mW/cm^2) (−50°C)	50 mA/cm^2 (16 V)	—	Hiramoto et al. (1994)
Arylamino-PPV, vertical (6.7 μm thick film)	450–520	~20 (IQE) at 100 V (500 nm, 3 μW/cm^2)	—	—	Daubler et al. (1999)
NTCDAc, vertical (167 μm thick single crystal)	300–450	~200 (IQE) at 670 V (380 nm, 30 μW/cm^2)	30 μA/cm^2 (534 V)	Rise time 10%–90% 500 ms	Hiramoto et al. (2002)
C$_{60}$/BCP, vertical 80 nm/8 nm	350–700	~50 (EQE) at 4 V	—	—	Huang and Yang (2007)
MEH-PPV:PbSe, vertical	400–600	~100 (EQE) at 50 V (550 nm, 3 μW/cm^2)	0.1 mA/cm^2 (50 V)	Rise time of few ms	Campbell and Crone (2007)
P3HT:PC$_{61}$BM:CdTe, vertical	350–750	~80 (EQE) at 4.5 V (350 nm, -)	~0.2 mA/cm^2 (2.5 V)	—	Chen et al. (2008)
P3HT:PC$_{61}$BM:Ir125d, vertical	300–1000	~70 (EQE) at 1.5 V (500 nm, -)	—	—	Chen et al. (2010)
Rubrene:DPAe, lateral ($L = $ ~60 μm)	350–600	~82 (EQE) at ~600 V (430 nm, -)	—	—	Hernandez-Sosa et al. (2011)
CuPc:C$_{60}$ active layersf, NTCDA/C$_{60}$ hole BLs; vertical	400–800	~500 (IQE) at 4 V (690 nm, 40 μW/cm^2)	1 mA/cm^2 (4 V)	f_{3dB} 1 kHz	Hammond and Xue (2010)
Graphene/P3HT:PC$_{61}$BM, lateral $L = 12$ μm	325	~10^3 (IQE) (325 nm, ~30 μW/cm^2)	—	—	Tan et al. (2014)

(*Continued*)

Table 3.1 (Continued)

Active materials and device structure[a]	Spectral window (nm)	QE at V_{BIAS} (light source, intensity)	I_{dark} (V_{BIAS})	Speed	References
P3HT:ZnO, vertical	300–600	$\sim 3.4 \times 10^3$ (EQE) at 9 V (360 nm, 1.25 μW/cm^2)	100 nA/cm^2 (9 V)	f_{3dB} 9.4 kHz (−9 V, 1 μW/cm^2)	Guo et al. (2012)
P3HT: PC$_{61}$BM: Ir125: Q-switch1g, vertical	400–1200	~ 55 (EQE) at 3.7 V (560 nm, -)	~ 1 mA/cm^2 (3 V)	—	Chuang et al. (2012)
PbS:P3HT:PC$_{61}$BM and ZnO/ vertical	350–1000	16 (EQE) at 4 V (350 nm, -)	10 μA/cm^2 (4 V)	160 μs rise time (3 V, 0.1 mW/cm^2)	Dong et al. (2014)
P3HT:PC$_{61}$BM with semicontinuous Au interlayer, vertical	400–650	~ 15 (EQE) at 2 V (400 nm, —)	0.1 mA/cm^2 (2 V)	—	Melancon and Zivanovic (2014)
PDPBD:PC$_{71}$BM, lateral $L = 50$ μm	760	$\sim 10^5$ (EQE) (20 nW/cm^2, 760 nm)	~ 2 μA/cm (100 V)	—	Park et al. (2014a,b)
P3HT:PC$_{61}$BM 100:1, vertical	350–450, 625	375 (EQE) at 19 V (8.87 μW/cm^2, 625 nm)	<0.1 mA/cm^2 (19 V)	—	Li et al. (2015b)

[a] In case of bi- or multilayers active materials are separated by a slash, whereas in case of BHJ active materials are separated by a colon.
[b] N-methyl-3,4,9,10-perylenetetracarboxyl-diimide (Me-PTC).
[c] Naphthalene tetracarboxylic anhydride (NTCDA).
[d] 4,5-Benzoindotricarbocyanine (Ir125).
[e] 9,10-Diphenylanthracene (DPA).
[f] Copper(II) phthalocyanine (CuPc).
[g] 8-[5-[6,7-Dihydro-6-methyl-2,4-diphenyl-5H-1-benzopyran-8-yl] − 2,4-pentadienylidene] − 5,6,7,8-tetrahydro-6-methyl-2,4-diphenyl-1-benzopyrylium perchlorate (Qswitch1).

Source: Adapted with permission from Li et al. (2015b). Copyright American Chemical Society, 2015.

where S_{noise} is the noise spectral density in $A \cdot Hz^{-0.5}$ and Δf is the BW in Hz. The reciprocal of the NEP is referred to as the *detectivity D* of the device. D depends on the detector active area (A) and on the signal BW Δf. Normalization of D with respect to Δf and A is often applied to enable a straight comparison of different devices:

$$D^* = \frac{\sqrt{\Delta f \bullet A}}{NEP}$$

D^* is typically referred to as the *specific detectivity* of the detector and it is commonly expressed in Jones, that is, $cm \cdot Hz^{0.5}/W$.

In addition, dark currents adversely affect the device power consumption and may compromise the dynamic range of the front-end electronics adopted for signal readout.

Shot noise from the dark currents is commonly *assumed* [with some notable exceptions where it is actually measured (Yao et al., 2007; Armin et al., 2014, 2015; Lyons et al., 2014)] to be the dominant contribution (Gong et al., 2009). In this framework the expression of D^* becomes:

$$D^* = \frac{R \bullet \sqrt{A}}{\sqrt{2qI_{dark}}} = \frac{R}{\sqrt{2qJ_{dark}}}$$

with J_{dark} the dark current density and $2qJ_{dark}$ the shot noise power spectral density (S_{noise}).

A possible source of dark current is represented by film defects, pinholes, ITO spikes (Armin et al., 2014). To address these issues, the film thickness can be increased from a few to many hundreds of nanometers or even microns. For instance, dark currents below 1 nA/cm^2 have been reported (Ng et al., 2008) for photodiodes based on 4 µm thick BHJ based on poly(phenylene-vinylene) derivatives like poly[2-methoxy-5-(2-ethylhexyloxy) − 1,4-phenylene-vinylene] (MEH-PPV) and PC$_{61}$BM. With such large interelectrode spacing, it can be difficult to ensure an effective and balanced transport of charges, so a compromise between EQE and dark current has to be reached (Ramuz et al., 2008). Recently, thanks to the adoption of high mobility polymeric donor poly[N − 9-heptadecanyl-2,7-carbazole-alt-5,5-(4′,7′-di-2-thienyl-2′,1′,3′-benzothidizole)] (PCDTBT) in conjunction with PC$_{71}$BM acceptor, state of the art photodetectors were demonstrated: with an active layer thickness of 700 nm the dark current was in the range of few nA/cm^2 at 1 V and EQE was in excess of 60% (Armin et al., 2014).

Even with a defect-free active layer, dark injection of carriers can occur from the contacts (given the relatively large energy bandgap E_g of organic materials, *bulk thermal generation* can be typically neglected). To suppress contact charge injection, suitable engineering of the metal-semiconductor interfaces aimed at increasing the Schottky barrier is necessary. In the case of BHJ systems, where such energy barriers are typically small and ill-defined due to both the donor and the acceptor species being possibly present at both metal/organic interface (Figure 3.6A), large improvements can be in principle achieved by controlling donor/acceptor vertical phase

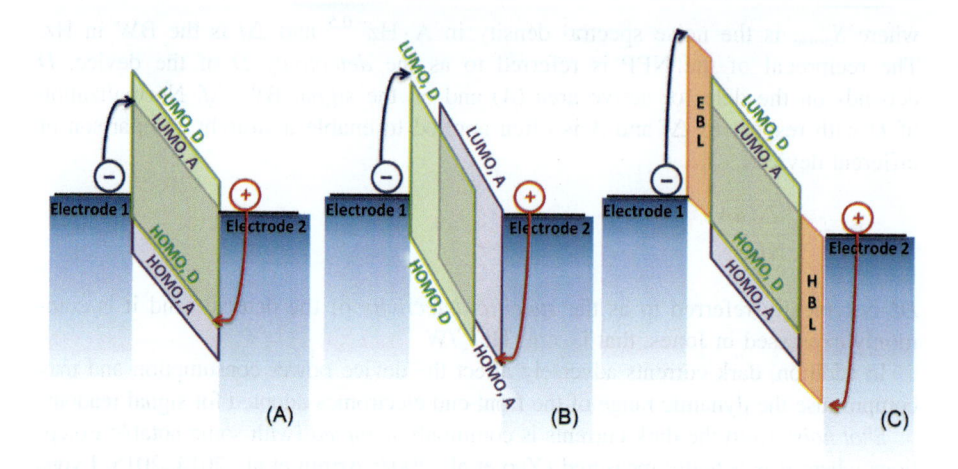

Figure 3.6 Sketch of charge injection phenomena from the metal contacts into the active material in the case of (A) a generic D/A BHJ, (B) a vertically segregated D/A BHJ, and (C) a device provided with dedicated injection BLs.
Source: Adapted with permission from Baeg et al. (2013). Copyright 2013, Wiley.

segregation (Figure 3.6B) (Agostinelli et al., 2008). Alternatively, extra layers interposed between the metal contact and the active material can be exploited (Figure 3.6C). If suitably engineered (Zaus et al., 2007) to suppress charge injection, while not hindering photogenerated carrier collection and not absorbing or scattering light, BLs should in principle allow the quantum efficiency of the device to be preserved. While multilayered structures are relatively easy to be developed in the case of small molecules by means of thermal evaporation (Xue and Forrest, 2004), the same is not true in the case of solution processing, due to the need of finding orthogonal solvents for successive depositions. Making underlying layers insoluble (e.g., by photo-crosslinking in presence of a suitable photo-crosslinking agent) is often required and can be a critical step because crosslinking can affect the optoelectronic properties of a material. For instance, a solution-processed squaraine:$PC_{61}BM$ BHJ photodetector was demonstrated (Binda et al., 2011) where, thanks to an electron BL based on MEH-PPV, dark currents are reduced by a factor of 30 and specific detectivity of 3.4×10^{12} cm·$Hz^{1/2}$/W at 700 nm is obtained. In this case, the layer which is crosslinked is the very BL, because being on the bottom contact it needs to withstand the deposition of the photoactive layer. If the top contact needs a BL, then crosslinking of the photoactive layer has to be undertaken, and this can be a highly critical step affecting EQE due to chemical reactions involved in crosslinking (Keivanidis et al., 2009).

Dark current reduction by means of BLs plays an even more strategic role when it comes to BHJ systems embedding low bandgap compounds, the Schottky barriers becoming correspondingly smaller. Combining a thick active layer with an evaporated C_{60} hole BL, impressive performances were obtained for a device based on the BHJ of PCDTBT and $PC_{71}BM$. In fact, with a dark current below 1 nA/cm^2, and a detectivity of 1×10^{13} Jones these devices favorably compare with commercial inorganic photodiodes (Armin et al., 2014).

3.4 Photoactive materials and detectors for different spectral regions

Here we will review photodiodes and photoconductors for different spectral regions, ranging from UV to NIR light. Instead of EQE, in some cases, we report responsivity (R), which can be expressed as $R = \text{EQE} \cdot \lambda q/hc$, where λ is the wavelength of interest, h the Planck constant, c the speed of light, q the electron charge. There is a rising interest also for adopting organic photodiodes in the field of radiation detection, mainly direct and indirect X-ray detection and alpha particle detection. The interested reader is referred to recent reviews and papers specific to the field (Chen et al., 2011; Binda et al., 2010; Fraboni et al., 2012; Iacchetti et al., 2012a; Chen et al., 2021; Basiricò et al., 2020).

3.4.1 Visible and NIR photodetectors

Organic semiconductors find natural and profitable exploitation as active materials in devices operating in the visible because of the typical size of their optical energy bandgap (Figure 3.7). The first examples were demonstrated between 1990 and 2000 (Sariciftci et al., 1993; Halls et al., 1995; Yu et al., 1999; Yu and Heeger, 1995). The vast majority of realizations exploit the D/A heterojunction concept. As far as devices

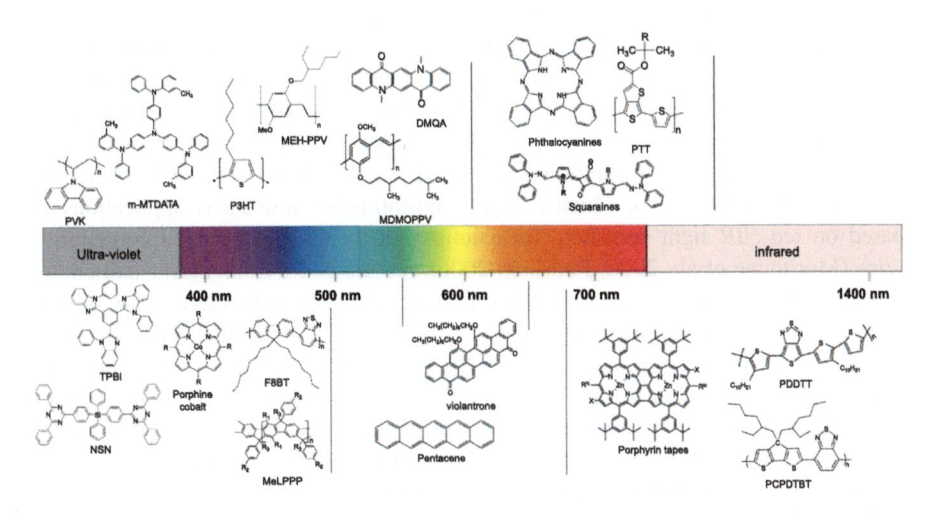

Figure 3.7 Some of the most widely adopted π-conjugated molecules from the UV to the NIR range of the spectrum.
Source: Reproduced with permission from Baeg et al. (2013). Copyright 2013, Wiley.

based on small molecules are concerned, molecular beam deposition techniques enable a relatively high degree of control over the active material nanostructure and morphology. By means of these techniques the fabrication of complex multilayered D/A architectures, resulting in extremely fast and efficient photodiodes, was possible, with EQE as high as 75% and BW up to 430 MHz (Peumans et al., 2000).

As far as *solution-processed devices* (Verilhac, 2013), the heterojunction morphology at the nanoscale and the crystallinity degree of D and A phases is far less controllable and critically depend on the specific mixing tendency of D/A couples and processing and postprocessing conditions.

In the case of *polymers*, molecular weight and polidispersity affect the solution rheology. As to the deposition process, the D/A ratio, the solution concentration, the presence of additives, the solvent or solvent mixtures and the solution drying time may affect the morphology of the thin film. Post-deposition treatments such as thermal or vapor annealing can be used as well to modify the film morphology, the degree of intermixing, and of crystallinity of D and A phases.

A very well-known D/A couple is given by MEH-PPV or poly(2-methoxy-5-$(3' - 7'$-dimethyloctyloxy) $-$ 1,4-phenylenevinylene) (MDMO-PPV) and $PC_{61}BM$: EQE around 60% is achieved for blue-green light detection up to about 550 nm (Ng et al., 2008; Niemeyer et al., 2007; Zhang et al., 2002; van Duren et al., 2004). One of the most studied D/A couples is given by P3HT and $PC_{61}BM$, which is characterized by broad absorption and relatively high carrier mobility, good coverage of the range between 400 and 600 nm with peak EQE above 70%, as reported by many groups (Ramuz et al., 2008; Tedde et al., 2009; Nalwa et al., 2010). Even though it is widely studied and adopted, and its synthesis can be conveniently scaled up to produce large quantities, P3HT has many drawbacks: it is poorly soluble in nonchlorinated solvents and it is subject to oxygen doping in ambient air, two aspects that complicate its processing. These issues have fostered the research of alternative donors to improve material solubility and purity, carrier mobility, and environmental stability (Facchetti, 2011). In addition, profiting from the ongoing material research in the field of organic photovoltaic cells, donor polymers with tuned absorption profiles and/or broadband absorption have been adopted. The development of the latter requires extending the absorption spectrum towards the red and NIR region by decreasing the optical gap, which is highly appealing because many applications are based on red-NIR light detection, for example, in the fields of optical communications (Morimune et al., 2006; Polishuk, 2006), remote control, environmental control or biomedicine (Pais et al., 2008). Various strategies have been successfully devised aimed at increasing the delocalization of π-electrons by extending the conjugation length, finally resulting in *panchromatic OPDs* with the broadband spectral operation, sometimes extended even from 700 to 1500 nm. Some applications, such as optical remote sensing for the reconstruction of morphological profiles (Zhang, 2002), environmental control (de Michele et al., 2012), or laboratory instrumentation for optical spectroscopy, can benefit from panchromatic OPDs.

In 2007, a photodetector with responsivity extended up to 900 nm (EQE $\sim 40\%$ at 800 nm) based on the BHJ of a suitably designed ester group modified polythieno [3,4-*b*]thiophene (PTT) polymer with $PC_{61}BM$ was demonstrated (Yao et al., 2007).

A similar approach based on the low bandgap poly(5,7-bis(4-decanyl-2-thienyl)-thieno(3,4-b)diathiazole-thiophene-2,5) (PDDTT) polymer again blended with $PC_{61}BM$ led to a detector with photoresponsivity ranging from 300 to 1450 nm (EQE ~30% at 900 nm) (Gong et al., 2009). Good photoresponse up to about 850 nm was also demonstrated for the low bandgap copolymer poly[2,6-(4,4-bis-(2-ethylhexyl) − 4H-cyclopenta[2,1-b;3,4-b']dithiophene)-alt-4,7-(2,1,3-benzothiadiazole)] (PCPDTBT) in blend with $PC_{71}BM$, a widely exploited D/A couple in photovoltaics (Soci et al., 2007; Peet et al., 2007). The development of NIR operating OPDs, is pursued by acting on the HOMO and LUMO levels, with the aim of reducing the energy gap. However, it has to be reminded that a sufficient energy offset between the donor and acceptor LUMOs has to be guaranteed for the charge photogeneration process to be effective. Recently, a synthetic strategy (based on electron-rich segments consisting of pyrrole combined with thiophene or selenophene, alternating with electron-deficient diketopyrrolepyrrole units) has been proposed that allows for reducing the gap by tailoring the HOMO level while leaving the LUMO level practically unchanged (Hendriks et al., 2014). Also thiadiazoloquinoxaline-based, push−pull type conjugated polymers have been employed as donors in BHJ, yielding devices capable of detection up to 1500 nm with EQE as high as 20% around 1200 nm (Verstraeten et al., 2020).

Concerning *small molecules, phthalocyanines* can be employed for red photon harvesting thanks to their typical absorption peak in the 600−700 nm range. While the unsubstituted form is substantially insoluble (Wang et al., 2011) and has to be deposited by thermal evaporation, substituted soluble forms exist as well and have enabled very good performance in the NIR. Campbell and Crone (2009) reported a photoconductor with a gain of 10 at 1000 nm based on soluble octabutoxy tin naphthalocyanine dichloride ($OSnNcCl_2$). *Squaraine compounds* are another interesting class of low bandgap molecules that are characterized by chemical and photochemical robustness and have an absorption band typically located between 600 and 800 nm. In blend with $PC_{61}BM$ they led to the development of a photodiode with peak EQE of about 15% in the 700−800 nm range (Binda et al., 2009, 2011). Square planar metal bis-*dithiolenes* (Garreau-de Bonneval et al., 2010) can be tuned as effective light harvesters between 700 and 1600 nm. Responsivity limitations encountered in the first realizations (Aragoni et al., 2004, 2007; Agostinelli et al., 2007a), have been partially solved by exploiting more favorable solid-state structures (Pintus et al., 2020). Another strategy exploits a peculiar metal/semiconductor/insulator/metal sandwich structure, which results in responsivities to periodic light signals larger than 0.2 A/W (up to 1.2 A/W upon blending with carbon black) with a BW up to 1 MHz (Dalgleish et al., 2012).

In *porphyrin-based compounds*, π-conjugation can be extended to address long wavelengths absorption. Recently, solution-processed tape-like porphyrin arrays have been exploited (Zimmerman et al., 2010, 2011) to achieve 10% EQE at 1400 nm in bilayer detectors with thermally evaporated C_{60} as the acceptor, showing a BW wider than 50 MHz.

In addition, extended photoresponse in the NIR can be achieved by exploiting *hybrid systems* where organic small molecules and polymers are used in combination with nanostructured organic or inorganic materials. The high polydispersity of semiconducting carbon nanotube (CNT) diameters enables broad spectral coverage from 400 to 1600 nm to be achieved. Excitons generated on CNTs are dissociated at the interface with a C_{60} layer (Arnold et al., 2009). Low bandgap semiconductors based on inorganic nanocrystals (NCs), like PbSe or PbS, have been exploited as well (Im et al., 2012a; Cho et al., 2007; Konstantatos et al., 2006; Xue et al., 2011; Ruiz-Hitzky et al., 2011) to detect light at wavelengths longer than 1 μm. Recently, a NIR lateral photoconductor based on PbS NCs-$PC_{61}BM$ D/A bilayer was presented (Osedach et al., 2010) where by fine engineering the NCs-organic hetero-interface the device performance is tuned. Hybrid inorganic/organic ternary blends have been recently proposed where small bandgap NCs, like PbS NCs, are used in combination with organic D/A couples, for example, P3HT:$PC_{61}BM$, exploiting a double CT of both the photoexcited hole and electron from the NC respectively towards the donor and the acceptor phase in the organic blend, profiting from the better charge transport of organic phases with respect to NCs network (Rauch et al., 2009; Jarzab et al., 2011).

The reader is referred to Table 3.2 for a collection of features and performance related to many organic and hybrid realizations for visible and NIR photodetectors.

3.4.2 Wavelength selective OPDs

The facile spectral tailoring of organic semiconductors has enabled the development of high-performance narrowband OPDs[1] (Jansen-van Vuuren et al., 2016; Pecunia, 2019a). In contrast to devices with a panchromatic photoresponse (Section 3.4.1), narrowband photodetectors can detect light within a narrow wavelength window in the UV, visible, or NIR range of the electromagnetic spectrum. This functionality is relevant to a plethora of applications, including colorimetry, digital photography, machine vision (e.g., for industrial inspection, security applications, and robot driving), lab-on-chip for biomedicine and chemical analysis, and optical communications (e.g., visible light communications), to mention but a few (Jansen-van Vuuren et al., 2016; Pecunia, 2019a). Narrowband OPDs have been a rather active area of research, leading to a wide range of device configurations and performance levels. In this section, we cover the salient properties of narrowband OPDs, while the interested reader is referred to a recent monograph on the subject (Pecunia, 2019a) for a more detailed treatment.

Aside from the manufacturing and application-related benefits associated with all types of OPDs (Section 3.2.2), the appeal of organic semiconductors for narrowband photodetection relates to the wide tunability of their absorption properties *via* facile chemical tailoring (Chochos and Choulis, 2011; Roncali, 2007) (see, for instance, Figure 3.8A). Crucially, chemical tailoring can deliver organic semiconductors selectively absorbing light within wavelength ranges (absorption bands)

---

[1] Narrowband photodetectors are also known as *spectrally-selective or wavelength-selective* photodetectors. If they are responsive within the visible range, they are also referred to as *color-selective* photodetectors.

Table 3.2 Visible-NIR photodetectors based on organic semiconductors or hybrid materials.

Method	Active material device structure[a]	Spectral window	Responsivity at V_{BIAS} (light source, intensity)	I_{dark} (V_{BIAS})	BW (t_R)[b]	Reference
Vacuum	CuPc/A1c multilayer alternated sandwich-like	525–725 nm	390 mA/W at −10 V (650 nm, <3 mW/cm^2)	NA	430 MHz (720 ps)	Peumans et al. (2000)
	Pentacene/C$_{60}$ multilayer alternated sandwich-like	450–690 nm	524 mA/W at −3.5 V (660 nm, 1.54 mW/cm^2)	0.2 mA/cm^2 (−0.5 V)	NA	Lee et al. (2009)
	CuPc/A2d bilayer sandwich-like	450–750 nm	30 mA/W at −3 V (525 nm, NA) 20 mA/W at −3 V (625 nm, NA)	1 mA/cm^2 (−7 V)	70 MHz (~40 ns)	Morimune et al. (2006)
	Pentacene/C$_{60}$ bilayer sandwich-like	500–690 nm	112 mA/W at −10 V (580 nm, NA) 56 mA/W at −10 V (690 nm, NA)	NA	80 MHz	Tsai et al. (2009)
	CoTPP/Alq3 bilayer sandwich-like	~400–460 nm	27 mA/W at >10 V (70 V) (425 nm, 50 mW/cm^2)	NA	NA	Seo et al. (2007)
	NN'QA/Alq3 sandwich-like	450–550 nm	25 mA/W at ~10 V (500 nm, 50 mW/cm^2)	NA	NA	
	ZnPc/Alq3 bilayer sandwich-like	550–750 nm	22 mA/W at ~10 V (600 nm, 50 mW/cm^2)	NA	>2 kHz	
	CuPc (a)/F16CuPc (b)e Three layers: (a)/(a):(b)/(b) sandwich-like	550–700 nm 750–850 nm	~60 mA/W at −9 V (808 nm, 3 mW/cm^2)	1 μA/cm^2 (−10 V)	14 MHz (80 ns)	Wang et al. (2011)

(*Continued*)

Table 3.2 (Continued)

Method	Active material device structure[a]	Spectral window	Responsivity at V_{BIAS} (light source, intensity)	I_{dark} (V_{BIAS})	BW (f_R)[b]	Reference
Solution/ vacuum	OSnNcCl$_2$ sandwich-like	400–600 nm	Photoconductive gain ~15 at −8 V (1064 nm, NA)	NA	<3 kHz	Campbell and Crone (2009)
	Zn-metallated porhpyrin/C$_{60}$ bilayer sandwich-like	800–1100 nm <1350 nm	~70 mA/W at 0 V (1345 nm, ~1 mW/cm^2)	NA	56 MHz (~2 ns)	Zimmerman et al. (2010)
	Zn-metallated porhpyrin:PC$_{61}$BM: pyridine BHJ/C$_{60}$ bilayer sandwich-like	<1600 nm	119 mA/W at 0 V (1400 nm, NA)	NA	NA	Zimmerman et al. (2011)
	MDMO-PPV:CNT/C$_{60}$ sandwich-like	400–1450 nm	23 mA/W at −0.7 V (1155 nm, NA)	NA	31 MHz (~7 ns)	Arnold et al. (2009)
	PCDTBT:PC$_{71}$BM BHJ/C$_{60}$ sandwich-like	300–650 nm	0.28 A/W at −1 V (350–600 nm, NA)	<1 nA/cm^2	1 MHz	Armin et al. (2014)
Solution	P3HT:PC$_{61}$BM BHJ sandwich-like	400–600 nm	339 mA/W at 0 V (600 nm, NA)	NA	NA	Nalwa et al. (2010)
	P3HT:PC$_{61}$BM BHJ sandwich-like	400–650 nm	100 mA/W at −1 V (468 nm, 40 mW/cm^2)	<10 nA/cm^2 (−1 V)	62 kHz	Ramuz et al. (2008)
	P3HT:PC$_{61}$BM BHJ sandwich-like	350–625 nm	177 mA/W at −1 V (550 nm, NA)	~100 nA/cm^2 (−1 V)	500 kHz	Baierl et al. (2010)
	P3HT:PC$_{61}$BM BHJ sandwich-like	400–600 nm	390 mA/W at −5 V (540 nm, NA)	65 nA/cm^2 (−5 V)	100 kHz	Tedde et al. (2009)
	P3HT:PC$_{61}$BM BHJ sandwich-like	400–600 nm	240 mA/W at −5 V (532 nm, 191 μW/cm^2)	~μA/cm^2 (−5 V)	1 MHz (~40 ns)	Punke et al. (2008)

MEH-PPV:PC$_{61}$BM BHJ sandwich-like	350–550 nm	140 mA/W at −4 V (488 nm, ~1 μW/cm^2)	<1 nA/cm^2 (−4 V)	(~ms)	Ng et al. (2008)
Polyfluorene F8T2:PC$_{61}$BMf BHJ sandwich-like	~400–500 nm	~670 mA/W at −10 V (460 nm, 9 mW/cm^2)	~1 mA/cm^2 (−10 V)	~50 MHz	Hamasaki et al. (2009)
F8BT:PDIg BHJ sandwich-like	400–600 nm	~40 mA/W at −0.5 V (530 nm, NA)	~8 nA/cm^2 (−0.5 V) with EBL	NA	Keivanidis et al. (2010)
P3HT:F8TBTh BHJ sandwich-like	400–600 nm	~100 mA/W at −0.5 V (500 nm, NA)	4 nA/cm^2 (−0.5 V) with EBL	NA	Keivanidis et al. (2009)
PDDTT:PC$_{61}$BM BHJ sandwich-like	~300–1400 nm	170 mA/W at −0.5 V (800 nm, 0.22 mW/cm^2)	<nA/cm^2 (−1 V) with EBL + HBL	NA	Gong et al. (2009)
PCPDTBT:PC$_{71}$BM BHJ sandwich-like	~300–900 nm	~300 mA/W at 0 V (800 nm, 80 mW/cm^2)	NA	NA	Peet et al. (2007)
PTT:PC$_{61}$BM BHJ sandwich-like	~300–950 nm	267 mA/W at −5 V (800 nm, NA)	~100 μA/cm^2 (−1 V)	4 MHz	Yao et al. (2007)
Squaraine:PC$_{61}$BM BHJ sandwich-like	650–850 nm	85 mA/W at −1 V (700 nm, ~10 mW/cm^2)	2 nA/cm^2 (−1 V) with EBL	1 MHz	Binda et al. (2011)

(*Continued*)

Table 3.2 (Continued)

Method	Active material device structure[a]	Spectral window	Responsivity at V_{BIAS} (light source, intensity)	I_{dark} (V_{BIAS})	BW (t_R)[b]	Reference
Solution/ hybrid	CdSe QDs/spiro-TPD bilayer lateral	350–600 nm	Photoconductive	~100 mA/cm^2 (~100 V)	NA	Osedach et al. (2009)
	PbS QDs:P3HT:PC$_{61}$BM BHJ sandwich-like	400–1450 nm	410 mA/W at −300 V (400 nm, NA) ~61 mA/W at −300 V (590 nm, NA) 160 mA/W at −5 V (1220 nm, NA)	~4 μA/cm^2 (−5 V)	2.5 kHz	Rauch et al. (2009)

[a] In case of bi- or multilayers active materials are separated by a slash, whereas in case of BHJ active materials are separated by a colon.
[b] t_R is the response time.
[c] A1: 3,4,9,10-perylenetetracarboxylic bis-benzimidazole (PTCBI).
[d] A2: N,N-bis (2, 5-di-tert-butylphenyl) 3, 4, 9, 10-perylenedicarboximide (BPPC).
[e] Copper(II) 1,2,3,4,8,9,10,11,15,16,17,18,22,23,24,25-hexadecafluoro-29H,31H-phthalocyanine (F$_{16}$CuPc).
[f] Poly(9,9-dioctylfluorene-alt-bithiophene) (F8T2).
[g] Perylene diimide (PDI).
[h] Poly[(9,9-dioctyluorene) − 2,7-diyl-alt-(4,7-bis(3-hexylthien-5-yl) − 2,1, 3-benzothiadiazole) − 2′,2″-diyl] (F8TBT).
Source: Adapted with permission from (Baeg et al., 2013). Copyright 2013, Wiley.

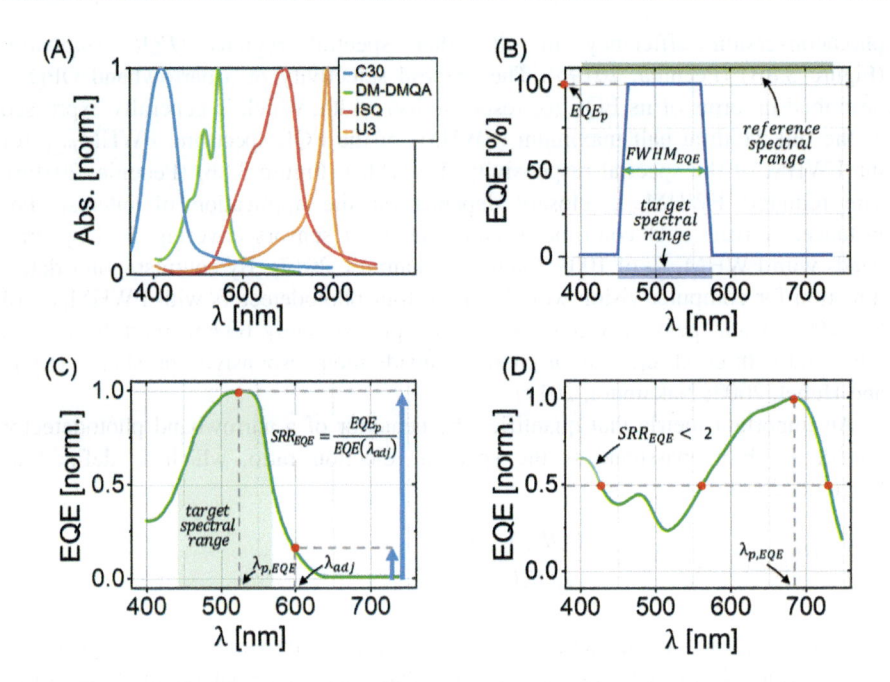

Figure 3.8 (A) Thin-film absorbance of representative organic semiconductors with narrowband absorption character: C30 (Sakai et al., 2012); DM-DMQA (Lim et al., 2015); ISQ (Li et al., 2016); and U3 (Osedach et al., 2012). (B) Salient features of the EQE spectrum of an ideal narrowband photodetector. (C) Spectral rejection ratio. (D) Example of a photodetector with color sensitivity (in the red and far-red ranges, specifically) but failing to meet the minimum spectral rejection criterion.
Source: Adapted from V. Pecunia, J. Phys. Mater., 2, 042001, 2019, DOI: 10.1088/2515-7639/ab336a, under the terms of a CC BY 3.0 License (https://creativecommons.org/licenses/by/3.0/legalcode). Copyright 2019 V. Pecunia.

that are considerably narrow, thereby enabling photodetectors with inherently narrowband responses (see below). Therefore, organic semiconductors provide a unique opportunity to deliver narrowband photodetectors beyond the limitations of mainstream semiconductors such as Si and III–V compounds, which have a broadband absorption character over fixed spectral ranges. Indeed, the use of such mainstream semiconductors for narrowband photodetection requires the concurrent use of input filters to selectively block incoming photons outside the target spectral range, which is problematic in terms of complexity and performance (Aihara et al., 2009; Park et al., 2014a; Takada et al., 2006).

Before we discuss the various strategies adopted for the realization of narrowband OPDs, it is useful to introduce some key properties and performance metrics relevant to such devices. Firstly, the spectral selectivity of a narrowband photodetector is specified in terms of the wavelength range within which it presents efficient photoconversion (*EQE passband*), which should ideally match the *target spectral range* relevant to the application of interest (Figure 3.8B) (Pecunia, 2019b). Concurrently, a narrowband photodetector should have negligible photoconversion efficiency in all other spectral regions (*EQE stopband*) (Figure 3.8B) (Pecunia, 2019b). The spectral selectivity of a narrowband OPD is described in terms of its EQE (or responsivity) width, which is generally expressed as the full-width at half maximum (FWHM) of the EQE spectrum, $FWHM_{EQE}$ (or the FWHM of the spectral responsivity, $FWHM_R$) (Figure 3.8b) (Pecunia, 2019b). The required $FWHM_{EQE}$ closely depends on the application of interest. For instance, conventional colorimetry can cope with sensors covering 3–4 spectral bands with $FWHM_{EQE}$ of 100–200 nm (Nakamura, 2017). By contrast, color determination for computer vision would require four photodetectors with $FWHM_{EQE}$ of 80–100 nm and peak wavelengths spaced approximately 60 nm apart from each other and with good rejection of photons outside their responsivity passbands (Lyon and Hubel, 2002; Nakamura, 2017).

An important metric that quantifies the behavior of a narrowband photodetector outside its EQE passband is the spectral rejection ratio, which is defined as (Pecunia, 2019b):

$$SRR_{EQE}(\lambda_{p,EQE}, \lambda_{adj}) := \frac{EQE(\lambda_{p,EQE})}{EQE(\lambda_{adj})}$$

Here, $\lambda_{p,EQE}$ is the wavelength at which the EQE is maximum (*peak EQE wavelength*), while λ_{adj} is a wavelength of particular significance outside the target spectral range. A large SRR_{EQE} is highly desired at wavelengths λ_{adj} whose rejection is particularly critical for the application of interest (Figure 3.8C). It is important to note that a narrowband photodetector requires at a minimum an $SRR_{EQE} > 2$ (*minimum spectral rejection criterion*) for all wavelengths λ_{adj} in the intended EQE stopband, otherwise its $FWHM_{EQE}$ would be ill-defined and its spectral selectivity utterly compromised (Figure 3.8d) (Pecunia, 2019a).

Due to their synthetic, spectral, and processing flexibility, organic semiconductors allow the realization of narrowband photodetectors through different approaches (narrowband strategies). The bulk of narrowband OPD research has pursued narrowband-absorption-type (NBA) photodetection (Pecunia, 2019a). As the name suggests, this strategy directly builds on the narrowband absorption nature of the organic semiconductors employed. Specifically, an NBA photodetector consists of an organic semiconductor layer with narrowband absorbance placed between two electrodes (Figure 3.9A). Consequently, only photons within its absorption range (ideally overlapping with the target spectral range) lead to an appreciable photocurrent, while all other photons are ideally rejected. In other words, the EQE spectrum of an NBA photodetector closely matches the absorbance spectrum of the photoactive layer (Figure 3.9A). This narrowband strategy is highly attractive due to its simplicity in terms of device architecture, fabrication process, and narrowband functionality, as the narrowband response is "hardwired" in the organic semiconductors used. Indeed, NBA OPDs are the only nonfiltered OPDs delivering narrowband functionality, given that they do not require any layer in the device stack exclusively dedicated to the suppression of the photoresponse outside the target spectral range. In light of the above, NBA photodetection allows efficient photoconversion, typically achieving the highest EQE values of all narrowband approaches (Pecunia, 2019a). Further, the

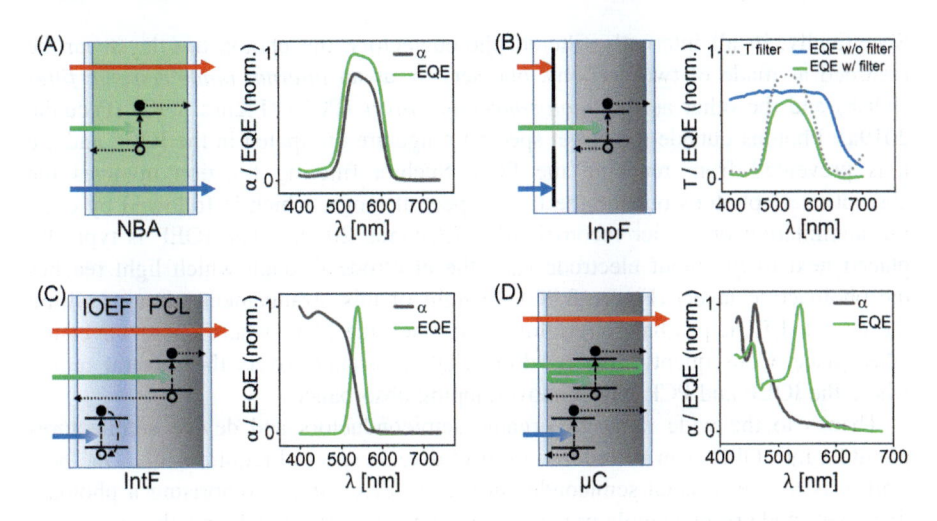

Figure 3.9 Narrowband approaches for the realization of narrowband OPDs: (A) NBA; (B) InpF; (C) IntF; and (D) μC. For each approach, the schematic of a representative green selective photodetector is presented alongside a typical spectral response. In (a), (c), and (d), the normalized absorption coefficient of the photoactive layer is also shown (assuming an IntF-1 or CCN configuration in (c)). The plot in (b) also presents the transmittance of the input optical filter.
Source: the schematics of all configurations are reproduced from V. Pecunia, *J. Phys. Mater.*, 2, 042001, 2019, DOI: 10.1088/2515–7639/ab336a, under the terms of a CC BY 3.0 License (https://creativecommons.org/licenses/by/3.0/legalcode). Copyright 2019 V. Pecunia.

spectral width of NBA OPDs directly relates to the width of the absorption bands of the organic semiconductors in the photoactive layer (Xia et al., 2020). Typical $FWHM_{EQE}$ values are in the range of ∼100 nm or larger due to the spectral broadening associated with aggregation effects in the solid state (Jansen-Van Vuuren et al., 2013). Nonetheless, numerous materials strategies have also been developed leading to NBA photodetectors with $FWHM_{EQE}$ <100 nm and even down to ∼20 nm (Bulliard et al., 2016; Li et al., 2016; Liess et al., 2019). Additionally, it has been shown that spectral broadening can be prevented altogether using a device architecture in which light absorption takes place only in isolated molecules, leading to $FWHM_{EQE}$ values down to 50 nm (Gao et al., 2017).

Departing from the filterless NBA approach, a number of filtered strategies have also been demonstrated. These strategies involve the (explicit or implicit) adoption of a filtering layer that suppresses the photoresponse of the organic semiconductor (s) used outside the target spectral range. The most basic approach is input optical filtering (InpF), which combines a panchromatic OPD with a narrowband optical filter placed atop the semitransparent input electrode (Figure 3.9B). Given its limitations (see discussion above), however, this narrowband strategy has captured little attention in OPD research.

Apart from input filtering, several internally filtered strategies have also emerged in recent years, which build on the unique optoelectronic properties of organic semiconductors to carry out the filtering function within the photoactive layer itself.

Specifically, in all internally filtered photodetectors, the photoactive layer can be regarded as made of two sections, one serving as an *internal optoelectronic filter* (IOEF) and the other as the *photoconversion layer* (PCL) (Figure 3.9C) (Pecunia, 2019a). Photons outside the target spectral range are dissipated in the IOEF and are thus prevented from reaching the PCL. Such a filtering function involves the absorption of photons outside the target spectral range, which is followed by exciton annihilation or carrier recombination (Pecunia, 2019a). The IOEF is typically placed next to the input electrode (i.e., the electrode through which light reaches the photoactive layer) (Figure 3.9C). In light of this arrangement, photons within the intended EQE passband are allowed to reach the PCL, where photoconversion takes place. Consequently, the resulting EQE passband covers the spectral region where the IOEF and PCL have nonoverlapping absorbance.

Thanks to the wide range of organic semiconductors and device architectures possible, IntF OPDs can be tailored to meet diverse spectral requirements. One possibility is to use a metal-semiconductor-metal device stack comprising a photoactive layer made of one single organic semiconductor with broadband absorption and unipolar transport. This configuration is referred to as *single-component internal filtering* (IntF-1) (Pecunia, 2019a). If the photoactive layer is sufficiently thick, its bulk will act as the IOEF via exciton annihilation or carrier recombination. Concurrently, the back interface will function as the PCL, leading to a narrowband photoresponse around the absorption onset of the semiconductor (Figure 3.9C). In recent years, this approach has found wide use in implementations with photocurrent gain, which have delivered apparent EQE values > 1000% and $FWHM_{EQE}$ of ~ 20–30 nm (Miao et al., 2018, 2017; Wang et al., 2018, 2017).

Alternatively, the internally filtered approach can be realized using two distinct organic semiconductors for the IOEF and PCL, with the IOEF being sufficiently thick to dissipate the incoming photons within its absorption range (Pecunia, 2019b). Therefore, the spectral selectivity of the resultant OPDs depends on the mismatch of the absorbance spectra of the two semiconductors employed for the IOEF and PCL (Kudo and Moriizumi, 1981; Sim et al., 2018; Xie et al., 2020).

An additional internally filtered configuration is *charge carrier narrowing* (CCN). This configuration is based on a photoactive layer made of a thick (typically >1 μm) broadband absorber with ambipolar charge transport (e.g., a donor-acceptor BHJ) (Pecunia, 2019a). In such a case, the region of the photoactive layer near the input electrode functions as the IOEF via carrier recombination, thereby suppressing the photoresponse to strongly absorbed photons (Armin et al., 2015). On the other hand, the bulk of the photoactive layer behaves as the PCL, allowing weakly absorbed photons to produce an appreciable photocurrent (Armin et al., 2015). Therefore, this approach delivers a narrowband photoresponse that peaks near the absorption onset of the photoactive layer (Figure 3.9C) (Arca et al., 2013; Zhong et al., 2017). Due to its inherent loss mechanism, CCN OPDs operating in primary photocurrent mode typically deliver peak EQEs < 10%, meanwhile achieving $FWHM_{EQE}$ in the range of 20–100 nm (Armin et al., 2015; Kim et al., 2018; Zhong et al., 2017).

A final approach to narrowband OPDs involves *microcavity resonance* (μC). μC-based OPDs rely on the enhancement of the optical field in the photoactive layer within a narrow wavelength range. In terms of device architecture, this approach requires the photoactive layer to be sandwiched between two partly or totally reflective electrodes, with the input one nonetheless having a finite transmittance to allow light in-coupling (Figure 3.9D). The light that is not strongly absorbed in the photoactive layer may be reflected back and forth several times between the two electrodes, enabling constructive interference if a resonance condition is met. In turn, such resonance leads to a peak in the photoresponse (Figure 3.9D). Importantly, however, the light that is strongly absorbed in the photoactive layer also leads to an appreciable photocurrent (Figure 3.9D). Therefore, microcavity resonance may enable truly narrowband photodetection only if combined with an input filter that suppresses the photoresponse associated with strongly absorbed photons (Siegmund et al., 2017). In such a case, this approach is attractive for ultranarrowband photodetection, as it can deliver $FWHM_{EQE}$ values <50 nm and it also allows the tuning of the peak EQE wavelength by varying the thickness of the photoactive layer (Tang et al., 2017). This configuration has attracted a considerable amount of attention for the realization of ultranarrowband OPDs selective in the NIR range, capitalizing on the weak subbandgap absorption band associated with the intermolecular charge-transfer states of BHJs (Siegmund et al., 2017; Tang et al., 2017; Ullbrich et al., 2017).

3.4.3 UV detectors

Even though UV detectors are usually considered a niche field, there is a strong interest in scientific, commercial, civil, and military applications, such as chemical and biological sensing, smoke and fire detection, missile warning, combustion monitoring, ozone sensing to cite but a few (Monroy et al., 2003; Razeghi and Rogalski, 1996). As far as inorganic semiconductors, GaN, SiC, or diamond (Sandvik et al., 2005; Liao et al., 2006; Carrano et al., 1997) are more suited than Si or GaAs, as they allow to achieve responsivities in the 100−200 mA/W range with a superior UV radiation hardness. Nevertheless, these technologies employ costly processes and hence organic-based UV detectors offer the opportunity for process simplification and widespread integration of cheap detectors, even though their radiation hardness has still to be properly assessed.

The most widely adopted structure is the vertical one. Most UV detectors are matched to the so-called UV-A range (310−420 nm) (Monroy et al., 2003), or near-UV (NUV) range which more or less corresponds. The deep-UV (DUV) range from 350 to 190 nm (Razeghi and Rogalski, 1996) is much less covered.

For various applications *spectral selectivity* is very important and in particular, visible blindness is a key property. This is very relevant in the case of DUV applications, like solar astronomy, intersatellite communications, missile detection, heat sensing, and sterilization monitors (Wu et al., 2010b). The relevant figure of merit is the UV to visible rejection ratio ($R_{UV/Vis}$). In the NUV range, detectors with an $R_{UV/Vis}$ of approximately 10^3 were demonstrated (Lin et al., 2005, 2008; Dai and Zhang, 2010). Selectivity to the DUV radiation has been demonstrated so far only with the adoption of filters placed on top of a semitransparent metallic cathode, ITO anode being not transparent to such high energy photons. Examples are N,N'-diphenyl-N,N'-bis(3-methyl-phenyl)(1,1'-biphenyl) − 4,4'-diamine (TPD)

(Wu et al., 2010b) and poly(vinyl alcohol) films blended with organic dyes (Li et al., 2011), where $R_{UV/Vis} \approx 10^3$ were also obtained.

Small molecules are quite commonly used as active materials because the absorption of high-energy photons requires a limited degree of π electron delocalization. As a consequence, vacuum thermal evaporation is the favored deposition method. In the case of donors, large EG triarylamines derivatives were extensively adopted (Lin et al., 2005; Wu et al., 2010a, 2011; Wang et al., 2010), and among these, 4,4′,4″-tri-(2-methylphenylphenylamino) triphenylaine (m-MTDATA) is the most common choice combining relatively high hole mobility of 3×10^{-5} cm^2/Vs (Zhang et al., 2009; Yan et al., 2009; Shirota et al., 1994), with a very low EA of 1.9 eV and an IP of 5.1 eV (Su et al., 2008; Giebeler et al., 1998). Other diamine derivatives were adopted as well, such as a starburst amine (though not yielding a visible-blind detector) (Li et al., 2005), TPD (Ray and Narasimhan, 2007), and N,N'-bis(naphthalene-1-yl)-N,N'-bis(phenyl)benzidine (NPB) (Dai and Zhang, 2010). While the choice of donor materials is quite limited, many more acceptor materials have been employed. Various 8-hydroxy-quinoline based complexes were tested varying the coordinating metal. Tris(8-hydroxyquinolinato)aluminum Alq$_3$ blended with TPD acting as a donor (Ray and Narasimhan, 2007) showed a limited responsivity R of approximately 30 mA/W at 360 nm (EQE $\approx$ 10%), due to the IP and EA of Alq$_3$ (5.5 and 2.4 eV, respectively) (Su et al., 2008) being not well matched with TPD levels (Cui et al., 2010). Tris(8-hydroxyquinoline) gallium (Gaq$_3$) blended with MTDATA (Su et al., 2008) showed a responsivity as high as 338 mA/W at 365 nm, with an applied bias voltage of 8 V, indicating a photoconductive behavior. To reduce absorption in the visible affecting both m-MTDATA:Alq$_3$ and m-MTDATA:Gaq$_3$ complexes in the 400–450 nm range, the coordinating metal atom was changed in favor of various rare earth complexes: a 1:1 blend of m-MTDATA and tris-(8-hydroxyquinoline) gadolinium (Gdq) yielded a detector with a 365 nm responsivity of 230 mA/W at 7.5 V (EQE $\approx$ 78%), with very suppressed photocurrent above 400 nm (Wang et al., 2010). Other successfully adopted acceptors are: bis(2-methyl-8-quinolinato) − 4-phenylphenolate aluminum (BAlq) which is characterized by high electron mobility and a good level matching with m-MTDATA (Cui et al., 2010; Wu et al., 2010a), 1,3,5-tris (N-phenyl-benzimidazol-2-yl) benzene (TPBi) which has relatively high electron mobility with a peak absorbance around 315 nm (Yan et al., 2009; Zhang et al., 2009), 1,3,4-oxadiazole-containing oligoaryls, also covalently linked to tryarildiamine compounds (Lin et al., 2005), and 2-(4-tertbutylphenyl) − 5-(4-biphenylyl) − 1,3,4-oxadiazole (PBD) (Dai and Zhang, 2010). Wide EG acceptors, such as 4,7-diphenyl-1,10-phenanthroline-(bathophenanthroline) (Bphen) (Wu et al., 2010b) or a silane-containing triazine derivative (NSN) (Li et al., 2011), allowed the development of DUV detectors in combination with the filters mentioned above and with proper donors. As an example, solution-processed poly(9,9-dihexylfluorene-2,7-diyl) (PFH) and poly(N-vinylcarbazole) (PVK) spin-coated on top of a vacuum deposited NSN acceptor, achieved responsivities comparable to all vacuum-deposited devices (10–100 mA/V in the 250–350 nm range) (Li et al., 2011).

The parameters describing the performances of many organic and hybrid UV detectors can be found in Table 3.3. The responsivity ranges from a few tens of

Table 3.3 UV photodetectors based on organic semiconductors or hybrid materials.

Method	Active material device structure[a]	Spectral window	Responsivity at V_{BIAS} (light source, intensity)	I_{light}/I_{dark} (V_{BIAS})	t_R[b]	Reference
Vapor	D2:A2c,d (BHJ)	300–410 nm	220 mA/W at −8 V (380 nm, 0.072 mW/cm^2)	~10^2 (−2 V)	<400 ns	Lin et al. (2005)
	m-MTDATA/Cu (DPEphos) ((Bphen))BF4e,f (bilayer)	300–420 nm	62 mA/W at 0 V (365 nm, 1.7 mW/cm^2)	NA	NA	Kong et al. (2006)
	TPD:Alq$_3$ (BHJ)	~200–450 nm	30 mA/W at −15 V (360 nm, 1.4 mW/cm^2)	~10^4 (−15 V)	~1 ms	Ray and Narasimhan (2007)
	m-MTDATA: Gaq$_3$ (BHJ)	~320–420 nm	338 mA/W at −8 V (365 nm, 1.2 mW/cm^2)	675 (−8 V)	NA	Su et al. (2008)
	m-MTDATA/TPBi (bilayer)	250–400 nm	75.2 mA/W at −6.5 V (365 nm, 1.0 mW/cm^2)	3405 (−6.5 V)	NA	Yan et al. (2009)
	m-MTDATA/TPBi/Bphenf (trilayer)	NA	79.9 mA/W at 0 V (365 nm, 2.0 mW/cm^2)	NA	NA	Zhang et al. (2010)
	m-MTDATA/TPBi/Bphenf (tandem)	NA	27.3 mA/W at 0 V (365 nm, 2.0 mW/cm^2)	NA	1.5 ms	Zhang et al. (2010)
	m-MTDATA/m-MTDATA: Bphen/Bphen/Cs2CO3:Bphenf (with filter)	250–400 nm	309 mA/W at −8 V (280 nm, 0.428 mW/cm^2)	510 (−8 V)	NA	Wu et al. (2010b)
	PCATA/Alq$_3$ (bilayer)	~300–450 nm	47.1 mA/W at 0 V (365 nm, 1.02 mW/cm^2)	NA	NA	Li et al. (2005)
	PCATA/TPBi (bilayer)	NA	64.7 mA/W at 0 V (365 nm, 1.02 mW/cm^2)	NA	NA	Li et al. (2005)

(*Continued*)

Table 3.3 (Continued)

Method	Active material device structure a	Spectral window	Responsivity at V_{BIAS} (light source, intensity)	I_{light}/I_{dark} (V_{BIAS})	t_R b	Reference
	NPB/PBD (bilayer)	300–420 nm	4.5 A/W at +3 V (350 nm, 60 μW/cm^2)	~24000 (+3 V)	NA	Dai and Zhang (2010)
	mMTDATA/TPBi/BCP (trilayer)	300–400 nm	100 mA/W at 0 V (365 nm, 0.426 mW/cm^2)	NA	NA	Zhang et al. (2009)
	m-MTDATA:TPBi/BCP (BHJ)	300–400 nm	135 mA/W at −4 V (365 nm, 0.426 mW/cm^2)	NA	NA	Zhang et al. (2009)
	m-MTDATA:Gdq (BHJ)	300–400 nm	230 mA/W at −7.5 V (365 nm, 1.2 mW/cm^2)	260 (−7.5 V)	NA	Wang et al. (2010)
	m-MTDATA/NSN (bilayer)	300–410 nm	334 mA/W at −12 V (365 nm, 1.0 mW/cm^2)	~300 (−12 V)	NA	Li et al. (2011)
	PFP/NSN (bilayer)	300–375 nm	245 mA/W at −12 V (340 nm, 0.5 mW/cm^2)	10^2	NA	Li et al. (2011)
	With NUV filter	200–300 nm	19.87 mA/W at −9 V (270 nm, NA)	NA	NA	Chen et al. (2006)
	m-MTDATA/CuPc (bilayer)	300–400 and 550–800 nm	32.1 mA/W at 0 V (365 nm, 1.7 mW/cm^2)	NA	NA	Si et al. (2007)
	m-MTDATA/(OXDPybm) Gd (DBM)$_3$ (bilayer)	300–400 nm	55 mA/W at 0 V (365 nm, 1.7 mW/cm^2)	NA	NA	Cui et al. (2010)
	m-MTDATA/m-MTDATA:BAlq/BAlq (BHJ)	300–400 nm	248 mA/W at −14 V (365 nm, 0.691 mW/cm^2)	NA	NA	Wu et al. (2010a)
		300–400 nm	514 mA/W at −7 V (365 nm, 1.2 mW/cm^2)	~10^3 (−7 V)	NA	

Solution and vacuum	PVK/NSN (bilayer)	300–370 nm	~400 mA/W at −12 V (340 nm, 0.5 mW/cm²)	~10	NA	Li et al. (2011)
	With NUV filter	200–300 nm	38.76 mA/W at −9 V (275 nm, NA)			
	PFH/NSN (bilayer)	300–425 nm	696 mA/W at −12 V (365 nm, 1.0 mW/cm²)	498 (−12 V)	NA	Li et al. (2011)

[a] In case of bi- or multilayers active materials are separated by a slash, whereas in case of BHJ active materials are separated by a colon.
[b] t_R is the device response time.
[c] D2: spirobifluorene-cored triaryldiamine.
[d] A2: spirobifluorene-cored oxadiazole-containing hexaaryl.
[e] DPEphos: 6,7-dicyanodipyrido [2,2-d:2′,3′-f] quinoxaline.
[f] Bphen: bathophenanthroline.
Source: Reproduced with permission from Baeg et al. (2013). Copyright 2013, Wiley.

mA/W to several hundreds of mA/W at wavelengths usually located around 350–360 nm with a light intensity that is often approximately 1 mW/cm^2. For the most responsive devices EQE exceeds 100%, clearly indicating a photoconducting behavior, hence concerns may arise regarding the device time response being ruled by charge recombination. Available data is limited: the fastest devices are characterized by a response time below 400 ns (Lin et al., 2005) or even of a few tens of ns (Wu et al., 2011), while usually a milliseconds regime is achieved.

A challenging aspect is related to the fact that many UV applications require devices capable of working in harsh environments (Monroy et al., 2003): it is then important to discuss whether organic-based UV detectors can qualify for such stringent applications. To date, most of the reports do not deal with device stability, and other reports contain only partial tests up to a maximum of 1000 min of operation in the air under UV light (Wu et al., 2010a,b; Wang et al., 2010). Preliminary data seem to suggest that organic UV detectors are capable of operating for up to hundreds of hours (Lin et al., 2005; Li et al., 2005). Further studies under controlled test conditions are therefore needed, though proper encapsulation will be required for harsh environmental conditions.

3.5 All-printed organic photodetectors fabricated by means of scalable solution-based processes

The use of solution-based processes to develop OPDs is very appealing because it promises a cost-effective deployment of light-sensing functionality on an ample set of substrates, rigid and flexible, planar and curved, and also back-end integration with existing technologies to realize hybrid systems. Since one of the main peculiarities of OPDs is also their compatibility with large areas, scalability of the processes is very important, also to sustain the development of products starting from laboratory demonstrators. Steps in this direction have been made and we revise in this section some of the most notable examples of OPDs fabricated by means of scalable processes.

3.5.1 Inkjet-printed organic photodetectors

Inkjet has been used in several OPD demonstrations to pattern the active material only or some of the layers of the photoactive stack. A very interesting approach is to pattern the complete stack by inkjet, as it would guarantee the scalability of the whole process. The challenge is to deposit multiple layers from the solution, one on top of the other, guaranteeing the electronic functionality of the underneath layer that should not be affected by the incoming wet droplets. In this section, we, therefore, report on OPDs fabricated mainly by inkjet printing, including examples of detectors where the entire functional stack is inkjet printed.

A fully inkjet printed detector for high illuminance detection was demonstrated by Lavery et al. (2011). High illuminance ($>$100 klux) detectors are useful to

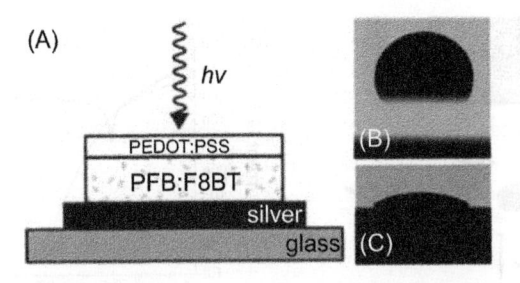

Figure 3.10 (A) Schematic cross-section of the all-printed PFB:F8BT photosensor. A droplet of (B) pristine PEDOT:PSS and (C) and PEDOT:PSS with a Zonyl fluorosurfactant additive on top of a PFB:F8BT film.
Source: Reproduced with permission from Lavery et al. (2011). Copyright 2011, Elsevier B. V.

monitor the environmental conditions of operators who may be exposed to life-threatening circumstances, such as explosions. Detection and recording of light emitted from explosions can provide useful medical indications on the operator's conditions. The detector demonstrated by Lavery et al. (Figure 3.10) comprises three inkjet-printed layers: (1) a bottom electrode of sintered silver nanoparticles; (2) a 1 μm thick photoactive layer composed of a blend of poly(9,9'-dioctylfluorine-co-bis-N,N'-(4-butylphenyl)-bis-N,N'-phenyl-1,4.phenylenediamine) (PFB) and poly(9,9'-dioctylfluorene-co-benzothiadiazole) (F8BT); (3) a semitransparent top-electrode made of PEDOT:PSS polymer conductor, to which a fluorosurfactant was added to wet the hydrophobic PFB:F8BT surface. The photosensor shows a peak EQE of 5.9% at 400 nm in the photovoltaic regime, that is, with 0 V bias, and a dark current density of 1 nA/cm^2 at 1 V reverse bias, yielding an estimated $D^* = 1 \times 10^{12}$ Jones. The detector shows a linear response as a function of light illuminance from 100 to 400 klux.

The low dark current achieved in the printed photosensor is mainly related to the thick photoactive layer, much thicker than necessary to absorb all incoming photons. Such thickness helps as well to obtain a good process yield, limiting short-circuits and avoiding unacceptably high leakage currents between the two printed electrodes, which is one of the problems to be solved when realizing a fully printed detector because of the possible formation of pinholes in the layers.

Inkjet printed OPDs with uniform and thinner photoactive layers were demonstrated by Azzellino et al. (2013) where, in a similar architecture to the previous case, a 120 nm thick P3HT:PC$_{61}$BM layer was adopted. Such optimized thickness allows obtaining an EQE exceeding 80% at 525 nm with a reverse bias of 0.9 V in an inkjet printed OPD on PEN foils. The photoresponse is well matched to the photoemission of a thallium doped CsI scintillator (CsI:Tl), characterized by a broad emission with a maximum at 550 nm. The printed OPD is therefore suitable for the development of X-ray imagers (Section 3.6). Despite the limited thickness of the printed blend, the yield obtained in laboratory scale experiments was above 85%, indicating that in the majority of cases the uniformity of the printed blend was sufficiently good to avoid

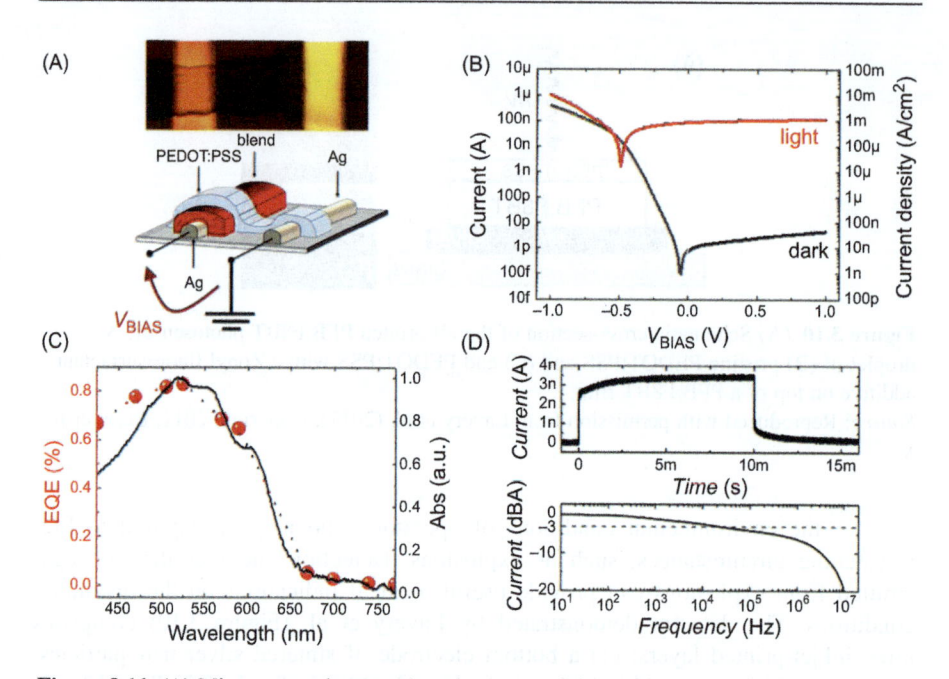

Figure 3.11 (A) Microscope photograph and scheme of the architecture of an inkjet printed OPD. (B) Photodetector dark current (black line) and photocurrent (red line, incident power density 3 mW/cm^2). (C) EQE spectrum (red circles) measured at 0.9 V bias under an incident power density of 10 mW/cm^2 and normalized active layer absorbance (black solid line). (D) Device response to a 10 ms light pulse and BW measured at 570 nm at an incident power density of 0.1 mW/cm^2 (1 V applied bias).
Source: Adapted with permission from Azzellino et al. (2013). Copyright 2013, Wiley.

short circuits between the printed electrodes. Critical to the realization of an optimized detector was the modification of the bottom silver layer thanks to the deposition of PEIE, which is effective in reducing the electrode work function and in obtaining a good rectifying diode. At 1 V reverse bias the diode shows dark currents as low as tens of nA/cm^2 (Figure 3.11), resulting in an estimated D^* well exceeding 10^{12} Jones. The working mechanism of the inkjet printed diode is consistent with a trap-dominated regime, where trapping and release time controls the photocurrent dynamics. Accordingly, while EQE is only weakly dependent on the light intensity, the response time strongly increases when the light intensity decreases, ranging from 100 μs to 100 ms with power density from 1 mW/cm^2 down to 4 μW/cm^2. In the frequency domain, a slow decrease in photocurrent occurs above 20 kHz, while a significant fall takes place only at 1 MHz.

Pace et al. (2014) further demonstrated a fully inkjet printed OPD where the donor, narrow bandgap small molecule 7,7'-(4,4-bis(2-ethylhexyl) − 4H-silolo[3,2-b:4,5-b']dithiophene-2,6-diyl)bis(6-fluoro-4-(5'-hexyl-[2,2'-bithiophen] − 5-yl) benzo[c][1,2,5] thiadiazole) was adopted in the photoactive ternary blend. Small molecules are desirable for easier scalability of their synthesis with respect to polymers.

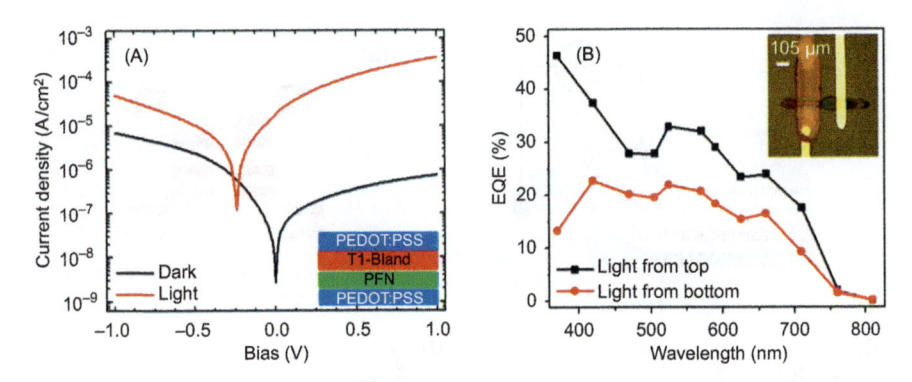

Figure 3.12 (A) Dark current (black line) and photocurrent (red line, incident power density 3 mW/cm^2) of a semitransparent OPD containing the P3HT/small molecule/PC$_{61}$BM ternary blend as active material. (B) EQE spectra of the printed OPD, acquired at 1 V bias, with light impinging from the bottom (red) and the top side (black). Inset: optical microscope image of all-organic, all-printed semitransparent OPD.
Source: Reproduced with permission from Pace et al. (2014). Copyright 2014, Wiley.

From a different point of view, the addition of small molecules in blends allows to extend and tune the spectral responsivity of the detector. Besides the PC$_{61}$BM acceptor, P3HT was added to improve the printability of the formulation on PEN substrates. To fabricate the fully printed detector, a thin poly((9,9-bis(3'-(N,N-dimethylamino)propyl) − 2,7-fluorene)-(alt-2,7-(9,9-dioctylfluorene)) (PFN) was printed on top of an Ag electrode in order to reduce its work function; the ternary blend was printed on top of PFN and the device was completed with a top PEDOT:PSS layer. Light sensitivity was extended up to 750 nm, where P3HT only based devices are blind. It was also demonstrated that the use of the PFN interlayer can decouple the optoelectronic performances of the printed OPD from the specific bottom electrode adopted: it is, for example, possible to replace the bottom opaque Ag electrode with a PEDOT:PSS one, thus demonstrating a semitransparent detector, where the signal can be detected from both sides (Figure 3.12).

3.5.2 Organic photodetectors fabricated by spray-coating

Spray-coating is a very effective technique for large-area coatings and may pave the way for very cost-effective processes for large-area detectors. In this direction, Tedde et al. (2009) demonstrated OPDs with all the organic layers coated thanks to a pneumatic spraying valve working at an air pressure of 4 bar (Figure 3.13). In the stack, which comprised a commercial ITO electrode and an evaporated Ca/Ag top cathode, both a P3HT:PC$_{61}$BM blend and a PEDOT:PSS electrode were sprayed. The device was finally encapsulated with a solvent-free epoxy resin and a glass slide. This allowed achieving a shelf-life exceeding 1 year in ambient conditions. One very interesting aspect of the sprayed device is that despite the higher surface roughness characterizing the spray-coated films, and the thicker photoactive layer (450 nm),

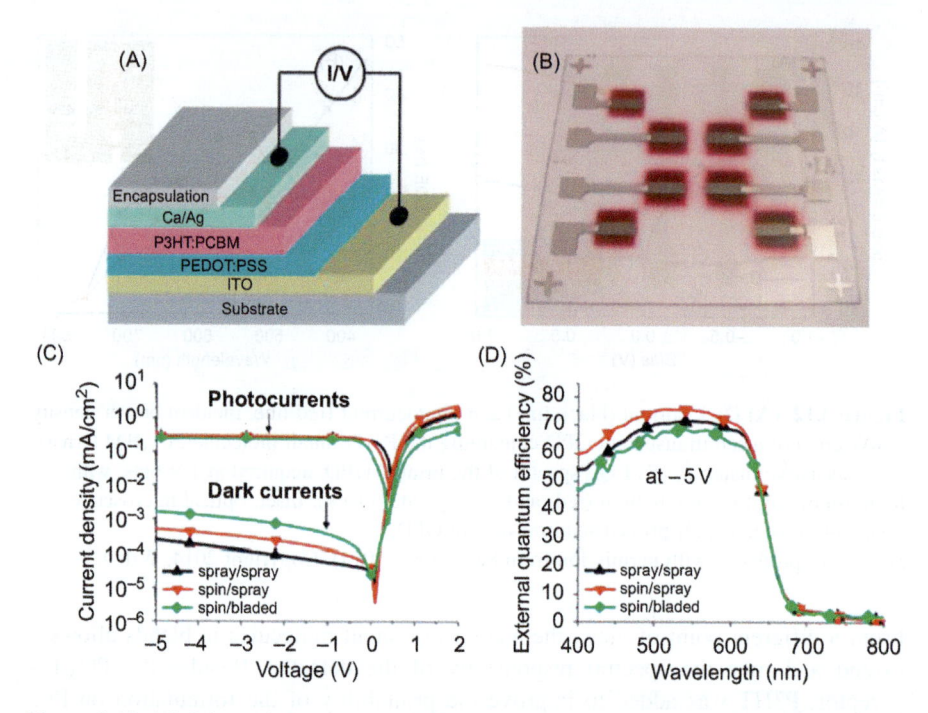

Figure 3.13 (A) Schematic layout of an OPD with spray-coated PEDOT:PSS and photoactive layer, including a glass encapsulation. (B) Photograph of eight spray-coated, 4 mm^2 OPDs. (C) J–V curves in dark and light conditions (780 μW/cm^2 at 532 nm) and (D) EQE spectra acquired at 5 V reverse bias of spray-coated OPD (black), compared to an OPD where only the P3HT:PC$_{61}$BM is spray-coated and the PEDOT:PSS is spin-coated (red) and an OPD where P3HT:PC$_{61}$BM is deposited by doctor blading and the PEDOT:PSS is spin-coated.
Source: Adapted with permission from Tedde et al. (2009). Copyright 2009, American Chemical Society.

EQE is not degraded with respect to films obtained by spin coating and doctor blade techniques (Søndergaard et al., 2013), while dark currents are reduced thanks to reduced film defectivity. Spray-coated detectors achieved a peak EQE of 71% at 550 nm at 5 V reverse bias, with a dark current density of 200–300 nA/cm^2, leading to an estimated specific detectivity of 8.8×10^{11} Jones. The spray-coated device reaches an interesting −3 dB cut-off frequency of 100 kHz.

ITO-free, fully spray-coated OPDs were demonstrated by replacing the ITO with a spray-coated transparent electrode based either on CNTs or on PEDOT:PSS (Falco et al., 2014; Schmidt et al., 2014). This allowed the fabrication of spray-coated P3HT:PC$_{61}$BM OPDs on flexible PET substrates (Fig. 3.14). The bottom electrode was composed of CNTs spray coated on PET, topped with a layer of sprayed PEDOT:PSS, while the top electrode was again sprayed PEDOT:PSS. EQE higher than 60% from 400 to 600 nm can be obtained, with dark currents lower than 100 nA/cm^2. The achieved fabrication yield exceeded 90%. Recently, spray-coated photodiodes based on a proprietary polymer have been reported to show

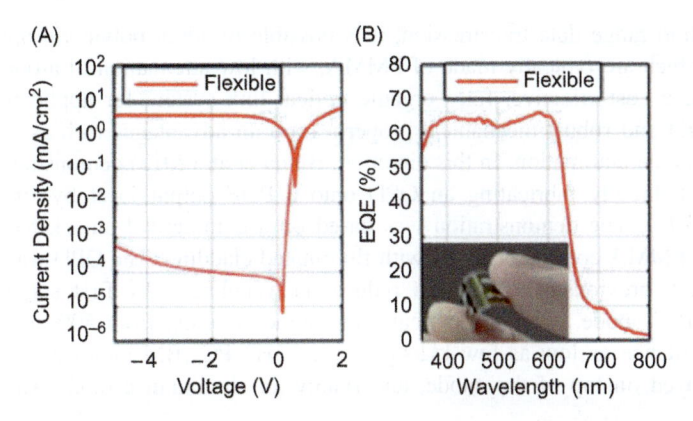

Figure 3.14 (A) Dark (dashed lines) and illuminated (solid lines) $J-V$ characteristics of OPDs obtained by spray-coating PEDOT:PSS and photoactive blend on a PET substrate. (B) EQE spectrum of the same sample. Inset: photograph of a spray-coated flexible photodiode. *Source*: Reproduced with permission from Falco et al. (2014). Copyright 2014, American Chemical Society.

performances comparable to commercially available state-of-the-art solid-state photodetectors: dark current densities in the tens of 34 pA /cm^2 range, the maximum responsivity of about 0.44 A/W at 660 nm, a BW of 50 kHz (Biele et al., 2019).

3.6 Applications of organic photodetectors

Thanks to the great flexibility in terms of fabrication processes offered by OPDs, and in general by organic optoelectronic devices, it is possible to devise and develop novel sensing applications, including wearables, distributed systems, and conformable large-area sensors. We revise in this section some of the most interesting and notable demonstrations. The interested reader can find more examples in recent reviews on this subject (Ren et al., 2021; Zhao et al., 2021; Chow and Someya, 2020).

3.6.1 Integrated transceivers

OPDs can be adopted as receivers in optical data transmission systems. Coupled with other polymeric elements and OLEDs, flexible integrated components can be fabricated. For example, an OPD can be directly fabricated on a polymeric waveguide to realize an integrated receiver. Taneda and coworkers (Ohmori et al., 2004) reported an organically integrated receiver optimized for maximum sensitivity in the red (640 nm) with a cut-off frequency of 5 MHz at 4 V, where the OPD was composed of vacuum deposited multilayers of donor and acceptor phthalocyanine derivatives. A complete electro-optical system was then realized by connecting the OPD integrated on the waveguide with a transmitter, fabricated by also integrating an OLED on a polymeric waveguide, through an optical fiber.

For short-range data transmission, it is possible to adopt polymeric optical fibers (POF), which are typically made of PMMA, with low attenuation of around 500 nm, to realize a cost-effective, fully organic optical link system for applications where lightweight and robust mechanical properties are an advantage, such as for automotive and home automation. In this direction, Binda et al. (2013) demonstrated the possibility of directly fabricating an OPD onto a POF output facet by spray coating (Figure 3.15). The demonstration was based on a commercial unjacketed 0.98 mm-diameter PMMA core-based POF, with fluorinated cladding. The OPD was fabricated on a facet, previously polished to reduce the roughness, by first spray-coating a PEDOT:PSS anode, obtaining optical transparency ranging from 50% to 85% and a sheet resistance as low as few kΩ/square. A P3HT:PC$_{61}$BM photoactive blend was then sprayed on top of the anode, and finally an aluminum cathode was thermally

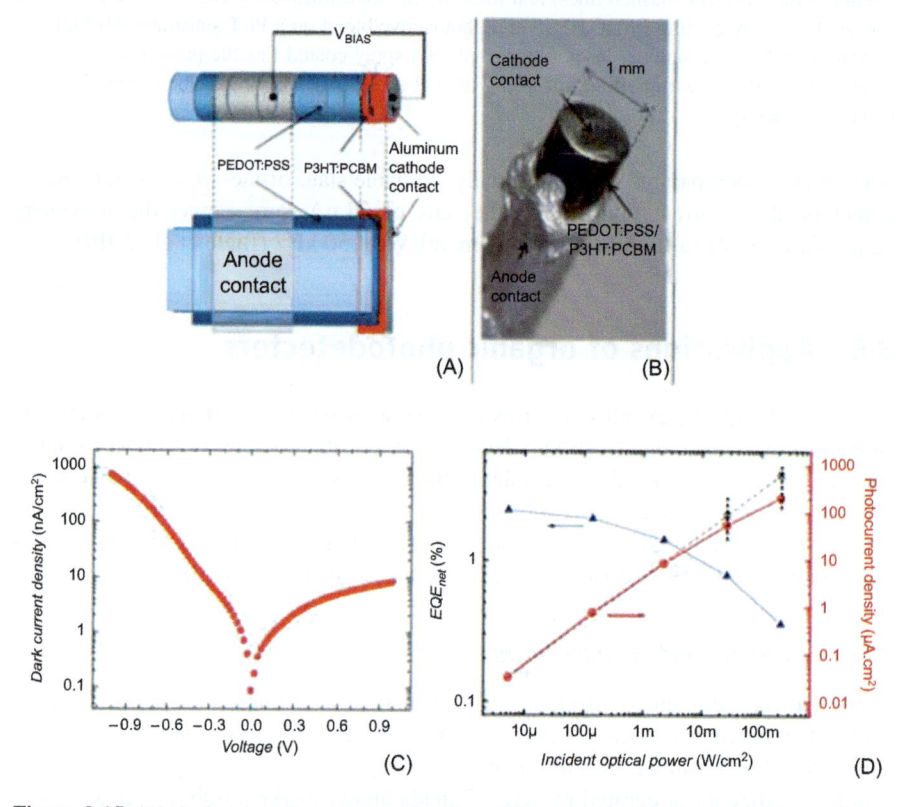

Figure 3.15 (A) A sketch and a photograph of the OPD integrated at the output facet of a POF. (B) The contact to the PEDOT:PSS anode is realized with silver paste. (C) Dark current of the on-fiber OPD. (D) EQE net of anode absorption and photocurrent density of the POF/OPD system under continuous light irradiation at 660 nm and with 1 V reverse bias applied. The dashed line refers to an ideal photocurrent density for a constant net EQE with light intensity.
Source: Reproduced with permission from Binda et al. (2013). Copyright 2013, Wiley.

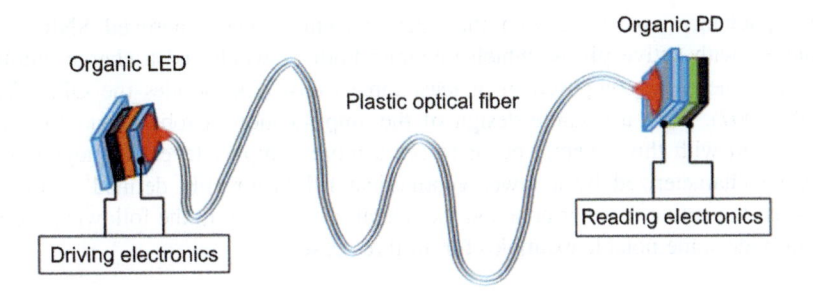

Figure 3.16 Exemplary scheme of an optical data link, including an OLED transmitter, a POF, and an OPD receiver.
Source: Adapted with permission from Baeg et al. (2013). Copyright 2013, Wiley.

evaporated to finalize the device. Such integrated OPD showed a rectification ratio of 100 in dark conditions at 1 V bias, with a low reverse dark current of approximately 10 nA/cm^2 at 1 V, an EQE = 2% at 660 nm under a light power density up to 100 μW/cm^2 and a −3 dB cut-off frequency of 15 kHz.

An optical data link composed only of organic elements has been proposed (Punke et al., 2008), including (1) a vacuum evaporated OLED with peak emission at 520 nm, coupled to (2) a PMMA POF, the latter coupled to an OPD based on a solution-processed P3HT:PC$_{61}$BM photoactive layer (an exemplary scheme is shown in Figure 3.16). Thanks to the 90%−10% fall time of the OPD, corresponding to 40 ns and compatible with the Sony/Philips Digital Interface Format standard (S/PDIF), a S/PDIF digitized audio signal can be transmitted at a bit rate of 2.8224 Mbit/s through the organic optical data link system.

3.6.2 Imaging applications

OPDs can represent a disruptive technology for large-area digital imaging (Natali, 2018), enabling lightweight, flexible, mechanically robust, and even conformable imagers, with applications in healthcare, industrial diagnostic, security, and gaming.

Photosensing arrays can be fabricated according to two main schemes, passive or active. A passive array is extremely simple, and it is composed of detectors that are connected through cross-bars. The pixel is composed of a single OPD only. This scheme is, therefore, the easiest to be developed, but it introduces limitations in the read-out and suffers from cross-talk of neighboring pixels because it is impossible to completely isolate a pixel during its read-out. Therefore, only relatively small arrays can be fabricated based on this approach. For large-area imagers, active solutions are required, where the pixel is more complex and is either passive or active. A passive pixel (Figure 3.18B) is composed of an OPD and a selector, typically a transistor, which allows for connection or isolates every single OPD during readout. In an array composed of passive pixels, the signal is conditioned by charge amplifies external to the array. It is therefore mandatory to carefully design the pixel and the array to guarantee a sufficient SNR, which degrades with the array dimension mainly because of the capacitance associated with the interconnection lines. Improved SNR can be achieved with active pixels, which integrate both a switch and a charge amplifier, and it is therefore composed of at least three transistors, besides the OPD (Tedde et al., 2007). With a suitable design of the amplification, a robust read-out can be guaranteed with this scheme, but it forces a more complex level of integration and may be characterized by a lower geometrical fill factor (FF, defined as the ratio between the active detector area and the overall pixel area). In the following sections, we provide some notable examples for all three cases.

3.6.2.1 Passive organic imager

The simplicity of the passive array was exploited to demonstrate a hemispherical focal plane detector array (Figure 3.17), fabricated on a plastic substrate with a curvature radius of 1 cm, adopted to mimic the human eye (Xu et al., 2008). As in a human eye, this sort of artificial eye requires a single focusing element owing to the nearly spherical shape forming a hemispherical focal plane. Metallic cross-bars were cold-welded on the curved surface by means of a transfer technique employing elastomeric stamps. The bottom bars, made of gold, served as anodes, while the silver top ones served as cathodes in a bilayer OPD composed of thermally evaporated phthalocyanine donor and fullerene acceptor layers. The spectral responsivity of the OPD was 65 mA/W, leading to $D^* = 5 \times 10^{10}$ Jones. The 90% − 10% fall time was less than 100 ns. Up to 100×100 pixels, with a lateral size of 40 μm, all correctly working, could be demonstrated.

Recently, a passive, 32×32 array (1×1 mm pixels) developed entirely by coating techniques (blade coating, spray coating, and screen printing) and capable of color recognition thanks to two color filters, has been presented (Deckman et al., 2018). Combining ink-jet and aerosol jet printing, an array comprised of 256 pixels

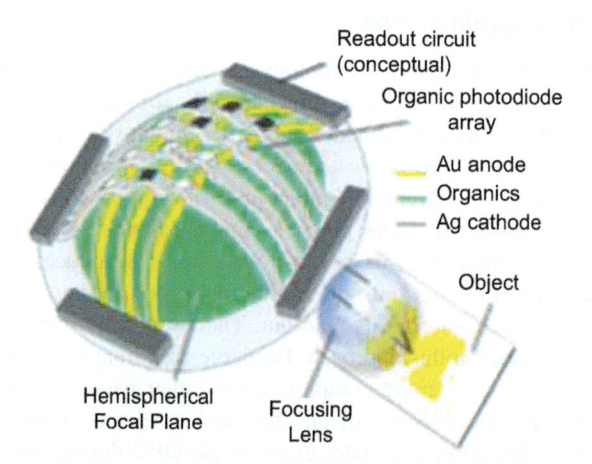

Figure 3.17 Schematic drawing of the hemispherical focal plane array based on an OPD passive array and its possible integration in an imaging system.
Source: Reproduced with permission from Xu et al. (2008). Copyrights 2008, Elsevier B.V.

(250 × 300 μm) exploiting a self-alignment process has also been reported (Eckstein et al., 2018).

3.6.2.2 Active organic imagers integrating passive pixels

Passive pixels can be fabricated both by integrating a transistor and an OPD in a coplanar geometry or by vertically stacking the OPD on top of the transistor. In the coplanar structure, the stacking complexity is simplified, both elements must be completely patterned, and the geometrical FF is reduced to allow the space for the transistor. Examples of integrated coplanar passive pixels, patterned through vacuum processes, were proposed by Forrest and coworkers starting in 2010 (Renshaw et al., 2010; Tong and Forrest, 2011).

Vertically stacked organic pixels (Jeong et al., 2010) improve the geometrical FF, while possibly complicating the fabrication because of the increased number of functional layers to be processed one on top of the other. Typically, this can be circumvented by decoupling the two processes: to this end between the OPD and the switch, a passivation layer is inserted through which a suitable via through has to be introduced. In a typical architecture, the transistor array is the first element to be fabricated, to create a selector backplane, which shows only the interconnection pads on top. In this way, the fabrication of the photoactive elements is largely simplified, because the patterning of the detectors is no longer required and the OPDs can be formed with two blanket depositions of the photoactive blend and the common counter electrode.

The previous approach enables the development of hybrid imagers, where the backplane is fabricated with an inorganic technology, and only the top detectors are organic. Wavelength selective passive pixels composed of stacked green (peak photoresponsivity at 540 nm) and red sensitive (peak photoresponsivity at 700 nm), vacuum deposited OPDs based on small molecules were demonstrated. Coupled with ZnO transistors (Aihara et al., 2009), a passive pixel capable of discriminating colors was realized. The approach was adopted to fabricate a 600 μm pitch, 47 × 30 pixels image sensor, with a TV frame rate. Extension of this concept may lead to alternative prism-less/filter-less color cameras that do not require a prism and/or a filter for color selectivity.

Hybrid imagers were also demonstrated building on more established thin-film technologies for the backplane, such for example amorphous silicon (a-Si:H) thin film transistors (TFTs). A flexible sensor array for the visible range of the spectrum was demonstrated by Street and coworkers in 2008 (Ng et al., 2008) starting from a flexible a-Si:H backplane, fabricated at temperatures below 150°C on a polyethylene naphthalate (PEN) substrate using inkjet-printed masks (Figure 3.18A−C). The sensor was composed of a 4 μm thick, unpatterned film of a blend of MEH-PPV and $PC_{61}BM$ deposited from solution on top of the backplane, followed by spin-coating of an ITO nanoparticles dispersion for the top semitransparent anode, realizing pixels with a geometrical FF of 0.76. The sensor was overall composed of 180 × 180 pixels, with 75 dots per inch resolution. Dark currents below 1 nA/cm^2 at a reverse bias of 4 V were achieved. The EQE of the integrated detector was

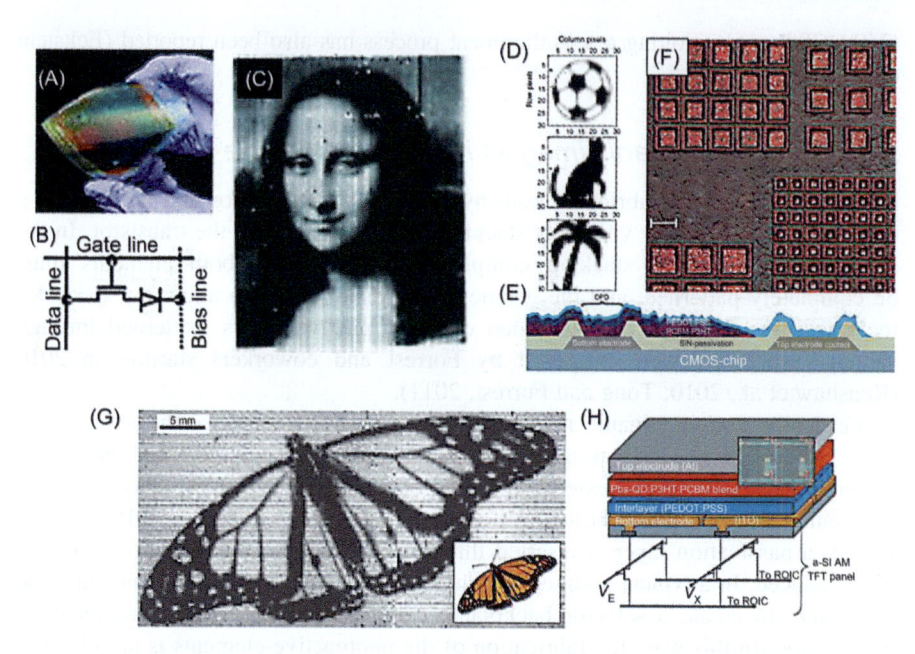

Figure 3.18 Hybrid imager based on solution-processed OPDs and a-Si:H backplane, fabricated at low temperature on a flexible PEN substrate: (A) photograph, (B) circuit diagram of an individual sensor passive pixel and (C) an ~5 × 3.8 cm image obtained with the flexible imager. CMOS-hybrid imager based on spray-coated photoactive material and a CMOS backplane: (D) examples of monochrome images acquired with a 30 × 30 array; © sketch of the cross-section of the imager architecture, comprising a CMOS chip, a SiN insulating layer, through which openings are created for the bottom Al electrode of the OPD; (F) top-view of the imager, with different pixel sizes (scale bar 50 μm). Hybrid NIR imager based on PbS quantum-dot sensitized OPDs and an a-Si:H backplane: (G) 1310 nm image of a butterfly; and (H) schematic view of the imager: the pixel area is defined by the patterned ITO bottom contact connected to its corresponding switch, while the photoactive blend is not patterned. *Source*: Adapted with permission (a)–(c) from Ng et al. (2008). Copyrights 2008, American Institute of Physics; (d)–(f) from Baierl et al. (2012). Copyrights 2012, Macmillan Publishers Limited; (g) and (h) from Rauch et al. (2009). Copyrights 2009, Macmillan Publishers Limited.

35% at 488 nm, with a NEP of 30 pW/cm^2 with an integration time of 100 ms. The sensor response time allows image detection with a frame rate of 500 Hz.

A similar hybrid approach was exploited to develop a *NIR imager*, where a non-patterned, solution-processed P3HT:PC$_{61}$BM photoactive layer was made sensitive to the NIR, up to 1800 nm, by sensitization with PbS quantum dots (Figure 3.18G,H) (Rauch et al., 2009). The PbS NCs act as antennas, absorbing the light and transferring a hole to P3HT and an electron to PC$_{61}$BM. The ternary, quantum-dot sensitized OPD showed an EQE as high as 51% at 1220 nm, attributed to the first excitonic transition in the PbS dots, at a reverse bias of 8 V. A specific detectivity D^* of

2.3×10^9 cm·Hz$^{1/2}$/W was obtained at a modulation frequency of 170 Hz. The pixel was found to have a -3 dB cut-off frequency of 2.5 kHz, limited by the parasitic capacitance of the TFT and compatible with conventional imaging applications at a few tens of Hz, such as in cell phone imagers. The demonstrative active array was composed of 256×256 pixels with 154 μm pitch and a geometrical FF of 83.3%.

Spray-coated organic photoactive blends were also combined with a CMOS backplane, replacing silicon photodiodes to obtain, with a cost-effective approach, a hybrid-CMOS imager (Figure 3.18D–F) composed of pixels with different pad sizes, from 50 μm down to 15 μm, arranged in arrays where the largest was composed of 30×30 pixels, suitable for acquiring low-resolution images (Baierl et al., 2012). CMOS imagers offer the advantage of improved electronic performances with respect to a-Si:H owing to larger electron mobility enabling much higher frame rates. The hybrid approach allows for the fabrication of a cost-effective CMOS-based imager with optimal geometrical FF, up to 100%, additionally characterized by tunability of the spectral responsivity, from the visible to the NIR region. The CMOS chip was fabricated in a commercial 0.35 μm CMOS-technology. For the visible range, a P3HT:PC$_{61}$BM layer was spray-coated on the CMOS backplane, and then coated with a transparent PEDOT:PSS electrode. The NIR sensitivity, up to 850 nm, was achieved by adding a squaraine dye to P3HT:PC$_{61}$BM, thus realizing an all-organic ternary blend. Dark currents were as low as 7 and 1.2 nA/cm^2 at 1.5 V reverse bias for the visible and NIR OPD, respectively. A peak EQE greater than 40% was achieved at 420 nm at 1.5 V reverse bias, with an almost constant EQE over the visible range. Typical rise and fall times of the OPDs are in the order of 10–15 μs, corresponding to a BW of 10–16 kHz, sufficient for imaging. Integration of faster OPDs (Baierl et al., 2010; Tedde et al., 2009) in the CMSO-hybrid imager may open applications in industrial high-speed imaging, requiring a BW of approximately 100 kHz.

An obvious and largely pursued alternative to hybrid imagers employing passive pixels is the fabrication of fully organic sensors, by employing organic transistors as selectors. A proof-of-concept 4×4 arrays were proposed (Nausieda et al., 2008) where coplanar organic passive pixels were fabricated by combining lithography and inkjet. The OPD had a lateral topology, allowing easy access to the light, and was based on a tytanil phthalocyanine. For the transistor, a bottom-contact pentacene device was employed. Such proof-of-concept showed a very limited responsivity of 6×10^{-5} A/W at 530 nm, in any case enough to image simple patterns, such as a "T," following a pixel by pixel precalibration. Earlier Someya et al. (2005) had proposed an all-organic passive pixels array with 36 dpi resolution on PEN, to be demonstratively adopted as an image scanner. The OPDs, composed of a blend of phthalocyanine and perylene derivatives, and the pentacene transistors, were deposited by thermal vacuum evaporation on two different sheets. The two sheets were laminated to obtain the imager. In the proposed scanner, OPDs were shielded from direct light and exposed only to reflected light. The scanning strategy was based on the possibility to distinguish black from white parts on paper thanks to their different reflectivity and did not require optics or mechanical parts.

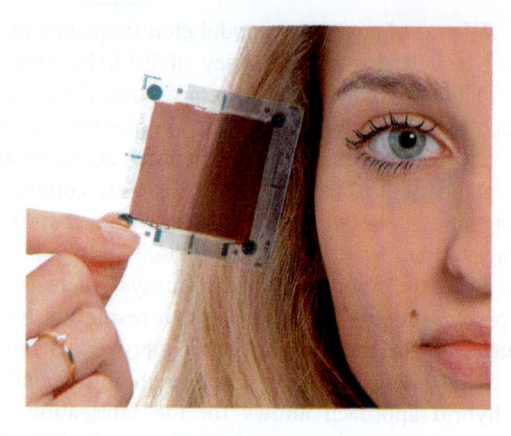

Figure 3.19 First conformable organic active image sensor, comprising 8930 passive pixels over an active area of 4×4 cm^2, with a 375 μm pitch.
Source: Image courtesy of ISORG. Copyright ISORG/Marvelpix.

These pioneering examples of all-organic imagers saw an advanced implementation in the first conformable image sensor (Figure 3.19) demonstrated by the company Isorg (http://www.isorg.fr), active in organic photodetectors, in collaboration with FlexEnable (http://www.flexenable.com), contributing the organic backplane. The flexible sensor has $94 \times 95 = 8930$ pixels resolution over an active area of 4×4 cm^2, with a 375 μm pitch (175 μm pixel size with 200 μm spacing).

An organic large-area active matrix can also be adopted to develop digital X-ray imagers when coupled with suitable scintillators, leading to very interesting biomedical and industrial diagnostic applications. Since X-rays cannot be focused on cost-effective and off-the-shelf solutions, the use of existing silicon imagers, either CMOS or CCD, while possible, is limited because of economic and handling reasons: large-area X-ray imagers imply heavy and bulky devices. Organic imagers would be lightweight, mechanically robust, and even shaped according to nonplanar geometries. Gelinck et al. (2013) reported an X-ray imager composed of an organic active matrix made of passive pixels, where all layers were solution processed but the electrodes, coupled to a CsI:Tl scintillator, and demonstrated X-ray imaging under conditions typical of biomedical applications. The organic backplane (Fig. 3.20) was fabricated on a very thin 25 μm PEN foil, adequately laminated onto a rigid carrier, and composed of solution-processed pentacene precursors transistors. The TFTs are isolated from the photoactive layer with a 1.7 μm thick SU-8 layer, and the opening of such interlayer on top of the Au electrode pad determines the area of the photodiode. The OPDs are fabricated by inkjet printing a PEDOT:PSS interlayer and subsequently by depositing an unpatterned P3HT:PC$_{61}$BM blend. A vacuum evaporated (0.5 nm) LiF/(2 nm) Al/(10 nm) Ag semitransparent electrode serves as the top cathode. The array, formed by 32 rows and 32 columns of pixels with a size of either 1×1 mm^2 or 200×200 μm^2, is then encapsulated with a multilayer organic/inorganic barrier. The organic imager provided performances in line with previous results on OPDs, allowing to resolve light intensity

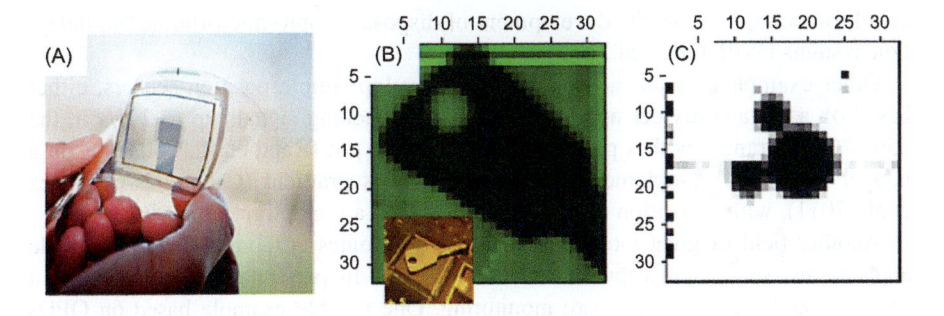

Figure 3.20 (A) Photograph of the organic backplane on PEN foil, plus fan out circuitry. (B) Recorded photo image of the 25 ppi array, where a key was placed on the array. (C) X-ray image of three circular lead objects recorded with the 25 ppi array with a dose of 1.2 mGy/s. *Source*: Reproduced with permission from Gelinck et al. (2013). Copyrights 2013, Elsevier B.V.

differences as low as approximately 1 μW/cm^2. When coupled with the scintillator, with light emission centered around 550 nm, images could be recorded using a typical radiography X-ray spectrum with a mean energy of 40 keV and dose rate levels down to 0.27 mGy/s, within the range normally used in medical applications. Robustness of the imager to prolonged 0.5 Gy X-ray exposure, in terms of the stability of the dark currents, was observed.

3.6.2.3 Active organic pixels

The active pixel architecture has been proposed in the literature following a hybrid approach, by coupling an OPD with a-Si:H TFTs (Tedde et al., 2007). In order to improve the geometrical FF, which may be limited in an active pixel to allow the space for the transistors, a stacked architecture was adopted also in this case, where the TFT backplane is first fabricated and then stacked with a P3HT:PC$_{61}$BM OPD array. This allowed obtaining a FF close to 100%. The bottom gold electrode of the OPD is connected to the gate of the amplifying transistor, the transconductance of which directly affects the charge amplification gain. It was demonstrated that the smallest detectable signal with such an active pixel can be as low as 1 μW/cm^2 at 522 nm, compared to a value of 6 μW/cm^2 obtained with a passive pixel fabricated with the same technology, thanks to the charge amplification integrated into the pixel.

3.6.3 Other applications

The ease of integration and the great tunability and flexibility of OPDs make them suitable for a large number of applications, which cannot be exhaustively covered here. Among the most interesting ones, we point to the integration of OPD in lab-on-chips, adding functionality that is currently missing as an integrated component (Kraker et al., 2008). The integration of OLEDs and OPDs in micro-fluidic systems could for example allow the development of disposable opto-microfluidic bio-detection systems (Williams et al., 2014).

Other examples regard the possibility to develop simple position sensors, either based on a voltage-divider approach, capable of tracking signals up to 2 m/s in the 550–700 nm range, with a position accuracy of 20 μm (Rand et al., 2003), or on a bilayer D/A with complementary lateral thickness gradient (Cabanillas-Gonzalez et al., 2011), with a position accuracy of 600 μm in a nonoptimized device.

Another field of great interest for organic electronics is represented by wearable applications, such as distributed wearable sensors. In particular, wearable medical sensors could improve healthcare monitoring. One notable example based on OPDs is represented by all-organic sensors for pulse oximetry (Fig. 3.21), for measuring pulse rate and arterial blood oxygenation. The device can be in principle easily worn in any part of the body and replace bulky and rigid oximeters which can be only attached at finger tips or ear lobes. An organic oximeter was realized by combining, through solution-based processes comprising spin-coating and printing, a green (532 nm) and a red (626 nm) OLED with an OPD covering the emitters' spectral range. The OPD was fabricated with a highly efficient and stable blend of

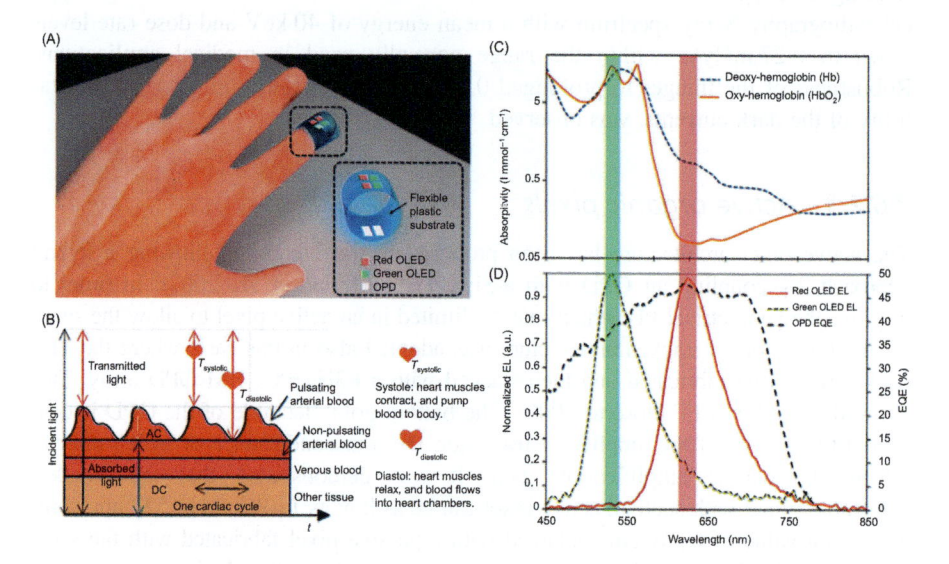

Figure 3.21 Schematic drawing of an all-organic pulse oximeter composed of two OLEDs and two OPDs. (A) The pulse oximetry sensor is composed of two OLED arrays and two OPDs. (B) Schematic illustration of a model for the pulse oximeter's light transmission path through pulsating arterial blood, nonpulsating arterial blood, venous blood, and other tissues. (C) Absorptivity of oxygenated (orange solid line) and deoxygenated (blue dashed line) hemoglobin in arterial blood. The wavelengths corresponding to the peak OLED electroluminescence (EL) spectra are highlighted. (D) OPD EQE (black dashed line) at short circuit, and EL spectra of red (red solid line) and green (green dashed line) OLEDs.
Source: Reproduced with permission from Lochner et al. (2014). Copyright 2014, Macmillan Publishers Limited.

poly({4,8-bis[(2-ethylhexyl)oxy] benzo[1,2-b:4,5-b$_0$]dithiophene-2,6-diyl}{3-fluoro-2-[(2-ethylhexyl)carbonyl] thieno [3,4−b] thiophenediyl} (PTB7) mixed with PC$_{71}$BM. A transparent PEDOT:PSS electrode and the active layer of the OPD have been printed on a plastic foil thanks to a surface tension-assisted blade-coating technique (Pierre et al., 2014), while an aluminum cathode was thermally evaporated. The OPD showed an EQE of 38% and 47% in short-circuit conditions, at 532 and 626 nm, respectively. The OPD had a −3 dB cut-off frequency of 10 kHz, much higher than required for oximetry. The organic sensor, coupled to conventional electronics operating at 1 kHz, is reported to accurately measure pulse rate and oxygenation with errors of 1% and 2%, respectively.

3.7 Conclusions

We have presented the most exemplary realizations of OPDs and recent progress in the field. Thanks to the vast library of materials and properties—constantly expanding and improving—and the ease of processing, this technology has the strong potential of being cost-effectively adopted for various exploitation cases. Furthermore, thanks to the maturity reached by other organic devices such as transistors, sensors, and memories, there is the opportunity to realize complex and integrated organic optoelectronic systems addressing a variety of applications such as short-range data transmission, digital imaging, and sensing.

At the single-device level, the current development status of organic photodiodes realized by means of laboratory techniques is very promising. In fact, EQE greater than 80% in the visible or in the NIR can be reached, with a spectrum that can be made panchromatic or wavelength selective. With the insertion of suitable charge BLs, dark currents can be reduced below 1 nA/cm^2 so that specific detectivities D^* close to 10^{13} Jones can be obtained. These figures merit well compare with commercial inorganic photodiodes, although it must be noted that the simultaneous optimization of EQE and detectivity at such levels is demonstrated in a limited number of publications only. Regarding the response time, there is a large dependence on device structure, materials, and deposition technique: the operation frequency of solution-processed BHJs can go beyond 1 MHz in various cases; for vacuum evaporated bilayers it is usually comprised between 10 and 100 MHz, and only in one case, thanks to a multilayer structure, it was possible to operate at frequencies in excess of 100 MHz.

To fully exploit the potential of OPDs it is mandatory to shift from laboratory-suited deposition techniques to scalable ones, enabling large-area addressing at low cost. A few examples of all-printed OPDs, realized through inkjet-printing and spray-coating, show state-of-the-art results in terms of EQE, whereas the dark current is still in the many tens/hundreds of nA/cm^2 range, likely due to printed layers nonuniformities and/or architecture restrictions imposed by the deposition techniques, which for example limit the choice of possible charge BLs.

In addition to single devices, more complex systems integrating OPDs with other devices can be found, imagers being the most strongly pursued. The integration of

OPDs and transistors often adopts hybrid approaches, either profiting from CMOS backplanes or transparent conductive oxides transistors, whereas the first examples of almost completely solution-processed organic matrices appeared only recently. The integration of OPDs and TFTs over large areas by means of scalable printing techniques has been recently demonstrated.

As a final note, we highlight the following criticalities, which have to be effectively addressed in the next future. The large variety of materials and processes that can be employed is a resource, but at the same time, the lack of a standard is hampering the development and adoption of simulation tools with effective predictive power. In addition, generically speaking, organic-based devices cannot be operated in ambient air without proper encapsulation. Even though very effective barriers are commercially available, a suitable trade-off between their effectiveness, their cost, the device lifetime, and the device's mechanical properties has to be found.

If the actual limitations will be effectively addressed and overcome, it is easy to foresee a strong penetration of lightweight, conformable, mechanically-robust, large-area, and distributed photosensors in several applications which will have an impact on our daily life, through interactive surfaces and personal health monitoring, and on innovative industrial diagnostic apparatus.

3.8 Note to the second edition

A thorough revision of the text has been performed and references to the most recent literature have been added. In addition, Section 3.4.2 has been completely rewritten by V. Pecunia, focusing on recent wavelength-selective detectors.

References

Agostinelli, T., Caironi, M., Natali, D., Sampietro, M., Arca, M., Devillanova, F.A., 2007a. Trapping effects on the frequency response of dithiolene-based planar photodetectors. Synth. Met. 157 (2007), 984–987.

Agostinelli, T., Caironi, M., Natali, D., Sampietro, M., Biagioni, P., Finazzi, M., 2007b. Space charge effects on the active region of a planar organic photodetector. J. Appl. Phys. 101 (2007).

Agostinelli, T., Campoy-Quiles, M., Blakesley, J.C., Speller, R., Bradley, D.D.C., Nelson, J., 2008. A polymer/fullerene based photodetector with extremely low dark current for x-ray medical imaging applications. Appl. Phys. Lett. 93 (2008).

Aihara, S., Seo, H., Namba, M., Watabe, T., Ohtake, H., Kubota, M., 2009. Stacked image sensor with green- and red-sensitive organic photoconductive films applying zinc oxide thin-film transistors to a signal readout circuit. IEEE Trans. Electron. Devices 56 (2009), 2570–2576.

Angione, M.D., Pilolli, R., Cotrone, S., Magliulo, M., Mallardi, A., Palazzo, G., 2011. Carbon based materials for electronic bio-sensing. Mater. Today 14 (2011), 424–433.

Aragoni, M.C., Arca, M., Caironi, M., Denotti, C., Devillanova, F.A., Grigiotti, E., 2004. Monoreduced [M(R,R′timdt)2]- dithiolenes (M = Ni, Pd, Pt; R,R′timdt = disubstituted imidazolidine-2,4,5-trithione): solid state photoconducting properties in the third optical fiber window. Chem. Commun. 10 (2004), 1882–1883.

Aragoni, M.C., Arca, M., Devillanova, F.A., Isaia, F., Lippolis, V., Mancini, A., 2007. First example of a near-IR photodetector based on neutral [M(R-dmet)2] bis(1,2-dithiolene) metal complexes. Inorg. Chem. Commun. 10 (2007), 191–194.

Arca, F., Sramek, M., Tedde, S.F., Lugli, P., Hayden, O., 2013. Near-infrared organic photodiodes. IEEE J. Quantum Electron 49 (2013), 1016–1025.

Arkhipov, V.I., Heremans, P., Bassler, H., 2003. Why is exciton dissociation so efficient at the interface between a conjugated polymer and an electron acceptor? Appl. Phys. Lett. 82 (2003), 4605–4607.

Armin, A., Hambsch, M., Kim, I.K., Burn, P.L., Meredith, P., Namdas, E.B., 2014. Thick junction broadband organic photodiodes. Laser Photonics Rev. 8 (2014), 924–932.

Armin, A., Jansen-van Vuuren, R.D., Kopidakis, N., Burn, P.L., Meredith, P., 2015. Narrowband light detection via internal quantum efficiency manipulation of organic photodiodes. Nat. Commun. 6 (2015), 6343.

Arnold, M.S., Zimmerman, J.D., Renshaw, C.K., Xu, X., Lunt, R.R., Austin, C.M., 2009. Broad spectral response using carbon nanotube/organic semiconductor/C60 photodetectors. Nano. Lett. 9 (2009), 3354–3358.

Arredondo, B., de Dios, C., Vergaz, R., Criado, A.R., Romero, B., Zimmermann, B., 2013. Performance of ITO-free inverted organic bulk heterojunction photodetectors: comparison with standard device architecture. Org. Electron. 14 (2013), 2484–2490.

Azzellino, G., Grimoldi, A., Binda, M., Caironi, M., Natali, D., Sampietro, M., 2013. Fully inkjet-printed organic photodetectors with high quantum yield. Adv. Mater. 25 (2013), 6829–6833.

Baeg, K.-J., Binda, M., Natali, D., Caironi, M., Noh, Y.-Y., 2013. Organic light detectors: photodiodes and phototransistors. Adv. Mater. (Deerfield Beach, Fla.) 25 (2013), 4267–4295.

Baierl, D., Fabel, B., Gabos, P., Pancheri, L., Lugli, P., Scarpa, G., 2010. Solution-processable inverted organic photodetectors using oxygen plasma treatment. Org. Electron. 11 (2010), 1199–1206.

Baierl, D., Pancheri, L., Schmidt, M., Stoppa, D., Dalla Betta, G.-F., Scarpa, G., 2012. A hybrid CMOS-imager with a solution-processable polymer as photoactive layer. Nat. Commun. 3 (2012), 1175.

Basiricò, L., Ciavatti, A., Fratelli, I., Dreossi, D., Tromba, G., Lai, S., et al., 2020. Medical applications of tissue-equivalent, organic-based flexible direct x-ray detectors'. Front. Phys. 8 (2020), 1–11.

Biele, M., Montenegro Benavides, C., Hürdler, J., Tedde, S.F., Brabec, C.J., Schmidt, O., 2019. Spray-coated organic photodetectors and image sensors with silicon-like performance'. Adv. Mat. Tech. 4 (2019), 1800158.

Binda, M., Agostinelli, T., Caironi, M., Natali, D., Sampietro, M., Beverina, L., 2009. Fast and air stable near-infrared organic detector based on squaraine dyes. Org. Electron. 10 (2009), 1314–1319.

Binda, M., Iacchetti, A., Natali, D., Beverina, L., Sassi, M., Sampietro, M., 2011. High detectivity squaraine-based near infrared photodetector with nA/cm[sup 2] dark current. Appl. Phys. Lett. 98 (2011).

Binda, M., Natali, D., Iacchetti, A., Sampietro, M., 2013. Integration of an organic photodetector onto a plastic optical fiber by means of spray coating technique. Adv. Mater. 25 (2013), 4335–4339.

Binda, M., Natali, D., Sampietro, M., Agostinelli, T., Beverina, L., 2010. Organic based photodetectors: suitability for X- and Γ-rays sensing application. Nucl. Instrum. Methods Phys. Res. Sect. A 624 (2010), 443–448.

Borel, T., Wang, Q., Aziz, H., 2014. Factors governing photo- and dark currents in lateral organic photo-detectors. Org. Electron. 15 (2014), 1096–1104.

Brady, M.A., Su, G.M., Chabinyc, M.L., 2011. Recent progress in the morphology of bulk heterojunction photovoltaics. Soft Matter 7 (2011), 11065–11077.

Braun, S., Salaneck, W.R., Fahlman, M., 2009. Energy-level alignment at organic/metal and organic/organic interfaces. Adv. Mater. 21 (2009), 1450–1472.

Briand, D., Oprea, A., Courbat, J., Bârsan, N., 2011. Making environmental sensors on plastic foil. Mater. Today. 14 (2011), 416–423.

Bube, R.H., 1960. Photoconductivity of Solids. John Wiley & Sons, Inc, New York, p. NY1960.

Bulliard, X., Jin, Y.W., Lee, G.H., Yun, S., Leem, D., Ro, T., et al., 2016. Dipolar donor−acceptor molecules in the cyanine limit for high efficiency green-light-selective organic photodiodes. J. Mater. Chem. C. 4 (2016), 1117–1125.

Cabanillas-Gonzalez, J., Pena-Rodriguez, O., Lopez, I.S., Schmidt, M., Alonso, M.I., Goni, A.R., 2011. Organic position sensitive photodetectors based on lateral donor-acceptor concentration gradients. Appl. Phys. Lett. 99 (2011).

Caironi, M., Gili, E., Sirringhaus, H., Klauk, H., 2012. Ink−jet printing of downscaled organic electronic devicesorganic electronics II: more materials and applications. In: Klauk, H. (Ed.), Organic Electronics II: More Materials and Applications. Wiley-VCH Verlag GmbH & Co. KGaA, Weinheim, Germany, p. 2012.

Campbell, I.H., Crone, B.K., 2007. Bulk photoconductive gain in poly(phenylene vinylene) based diodes. J. Appl. Phys. 2007, 101.

Campbell, I.H., Crone, B.K., 2009. A near infrared organic photodiode with gain at low bias voltage. Appl. Phys. Lett. 95 (2009), 263302–263303.

Carrano, J.C., Li, T., Grudowski, P.A., Eiting, C.J., Dupuis, R.D., Campbell, J.C., 1997. High quantum efficiency metal-semiconductor-metal ultraviolet photodetectors fabricated on single-crystal GaN epitaxial layers. Electron. Lett. 33 (1997), 1980–1981.

Cesarini, M., Brigante, B., Caironi, M., Natali, D., 2018. Reproducible, high performance fully printed photodiodes on flexible substrates through the use of a polyethylenimine interlayer. ACS Appl. Mater. Interfaces 10 (2018), 32380.

Chen, F.C., Chien, S.C., Cious, G.L., 2010. Highly sensitive, low-voltage, organic photomultiple photodetectors exhibiting broadband response. Appl. Phys. Lett. 2010, 97.

Chen, Q., Hajagos, T., Pei, Q., 2011. Conjugated polymers for radiation detection. Annu. Rep. Sec. "C" (Phys. Chem.) 107 (2011), 298.

Chen, L.L., Li, W.L., Wei, H.Z., Chu, B., Li, B., 2006. Organic ultraviolet photovoltaic diodes based on copper phthalocyanine as an electron acceptor. Sol. Energy Mater. Sol. Cells. 90 (2006), 1788–1796.

Chen, H.Y., Lo, M.K.F., Yang, G.W., Monbouquette, H.G., Yang, Y., 2008. Nanoparticle-assisted high photoconductive gain in composites of polymer and fullerene. Nat. Nanotechnol. 3 (2008), 543–547.

Chen, M., Wang, C., Hu, W., 2021. Organic photoelectric materials for x-ray and gamma ray detection: mechanism, material preparation and application. J. Mater. Chem. C. 9 (2021), 4709.

Chochos, C.L., Choulis, S.A., 2011. How the structural deviations on the backbone of conjugated polymers influence their optoelectronic properties and photovoltaic performance. Prog. Polym. Sci. 36 (2011), 1326–1414.

Chow, P.C.Y., Someya, T., 2020. Organic photodetectors for next-generation wearable electronics. Adv. Mat. 32 (2020), 902045.

Cho, N., Roy Choudhury, K., Thapa, R.B., Sahoo, Y., Ohulchanskyy, T., Cartwright, A.N., 2007. Efficient photodetection at IR wavelengths by incorporation of PbSe−carbon-nanotube conjugates in a polymeric nanocomposite. Adv. Mater. 19 (2007), 232–236.

Cho, B., Song, S., Ji, Y., Kim, T.-W., Lee, T., 2011. Organic resistive memory devices: performance enhancement, integration, and advanced architectures. Adv. Funct. Mater. 21 (2011), 2806–2829.

Chuang, S.T., Chien, S.C., Chen, F.C., 2012. Extended spectral response in organic photomultiple photodetectors using multiple near-infrared dopants. Appl. Phys. Lett. 2012, 100.

Clarke, T.M., Durrant, J.R., 2010. Charge photogeneration in organic solar cells. Chem. Rev. 110 (2010), 6736–6767.

Cui, Y., Liu, L., Liu, C., Wang, Q., Li, W., Che, G., 2010. High response organic ultraviolet photodetector based on bis (2-methyl-8-quinolinato) − 4-phenylphenolate aluminum. Synth. Met. 160 (2010), 373–375.

Dai, Q., Zhang, X.Q., 2010. High-response ultraviolet photodetector based on N,N'-bis (naphthalen-1-yl)-N,N'-bis(phenyl)benzidine and 2-(4-tertbutylphenyl) − 5-(4-biphenylyl) − 1,3,4-oxadiazole. Opt. Express. 18 (2010), 11821–11826.

Dalgleish, S., Matsushita, M.M., Hu, L., Li, B., Yoshikawa, H., Awaga, K., 2012. Utilizing photocurrent transients for dithiolene-based photodetection: stepwise improvements at communications relevant wavelengths. J. Am. Chem. Soc. 134 (2012), 12742–12750.

Danielson, E., Ooi, Z., Liang, K., Morris, J., Lombardo, C., Dodabalapur, A., 2014. Analysis of bulk heterojunction material parameters using lateral device structures. J. Photonics Energy. 4 (2014).

Daubler, T.K., Neher, D., Rost, H., Horhold, H.H., 1999. Efficient bulk photogeneration of charge carriers and photoconductivity gain in arylamino-PPV polymer sandwich cells. Phys. Rev. B. 59 (1999), 1964–1972.

Deckman, I., Lechêne, P.B., Pierre, A., Arias, A.C., 2018. All-printed full-color pixel organic photodiode array with a single active layer. Org. Electron. 56 (2018), 139–145.

Deegan, R.D., Bakajin, O., Dupont, T.F., Huber, G., Nagel, S.R., Witten, T.A., 1997. Capillary flow as the cause of ring stains from dried liquid drops. Nature. 389 (1997), 827–829.

Deibel, C., Strobel, T., Dyakonov, V., 2010. Role of the charge transfer state in organic donor−acceptor solar cells. Adv. Mater. 22 (2010), 4097–4111.

de Gans, B.-J., Schubert, U.S., 2004. Inkjet printing of well-defined polymer dots and arrays. Langmuir 20 (2004), 7789–7793.

de Michele, M., Leprince, S., Thiébot, J., Raucoules, D., Binet, R., 2012. Direct measurement of ocean waves velocity field from a single SPOT-5 dataset. Remote. Sens. Environ. 119 (2012), 266–271.

Dong, R., Bi, C., Dong, Q.F., Guo, F.W., Yuan, Y.B., Fang, Y.J., 2014. An ultraviolet-to-NIR broad spectral nanocomposite photodetector with gain. Adv. Opt. Mater. 2 (2014), 549–554.

Dong, H., Zhu, H., Meng, Q., Gong, X., Hu, W., 2012. Organic photoresponse materials and devices. Chem. Soc. Rev. 41 (2012), 1754–1808.

Eckstein, R., Eckstein, N., Strobel, T., Rödlmeier, K., Glaser, U., Lemmer, G., et al., 2018. Fully Digitally Printed Image Sensor Based on Organic Photodiodes. Adv. Opt. Mater. 6 (2018), 1701108.

Elmqvist, R. (1951). Measuring instrument of the recording type. US Patent 2,566,443.

Facchetti, A., 2011. π-conjugated polymers for organic electronics and photovoltaic cell applications. Chem. Mater. 23 (2011), 733–758.

Falco, A., Cina, L., Scarpa, G., Lugli, P., Abdellah, A., 2014. Fully-sprayed and flexible organic photodiodes with transparent carbon nanotube electrodes. ACS Appl. Mater. Interfaces 6 (2014), 10593–10601.

Firdaus, Y., Le Corre, V.M., Karuthedath, S., et al., 2020. Long-range exciton diffusion in molecular non-fullerene acceptors. Nat. Commun. 11 (2020), 5220.

Fraboni, B., Ciavatti, A., Merlo, F., Pasquini, L., Cavallini, A., Quaranta, A., 2012. Organic semiconducting single crystals as next generation of low-cost, room-temperature electrical x-ray detectors. Adv. Mater. 24 (2012), 2289–2293.

Gao, L., Ge, C., Li, W., Jia, C., Zeng, K., Pan, W., et al., 2017. Flexible filter-free narrowband photodetector with high gain and customized responsive spectrum. Adv. Funct. Mater. 27 (2017), 1702360.

Gao, F., Inganas, O., 2014. Charge generation in polymer-fullerene bulk-heterojunction solar cells. Phys. Chem. Chem. Phys. 16 (2014), 20291–20304.

Garreau-de Bonneval, B., Moineau-Chane Ching, K.I., Alary, F., Bui, T.T., Valade, L., 2010. Neutral d8 metal bis-dithiolene complexes: synthesis, electronic properties and applications. Coord. Chem. Rev. 254 (2010), 1457–1467.

Gelinck, G.H., Kumar, A., Moet, D., van der Steen, J.-L., Shafique, U., Malinowski, P.E., 2013. X-ray imager using solution processed organic transistor arrays and bulk heterojunction photodiodes on thin, flexible plastic substrate. Org. Electron. 14 (2013), 2602–2609.

Giebeler, C., Antoniadis, H., Bradley, D.D.C., Shirota, Y., 1998. Space-charge-limited charge injection from indium tin oxide into a starburst amine and its implications for organic light-emitting diodes. Appl. Phys. Lett. 72 (1998), 2448–2450.

Gili, E., Caironi, M., Sirringhaus, H., Klauk, H., 2012. Inkjet printing of downscaled organic electronic devicesorganic electronics II: more materials and applications. In: Klauk, H. (Ed.), Organic Electronics II: More Materials and Applications. Wiley-VCH, Weinheim, p. 2012.

Gili, E., Caironi, M., Sirringhaus, H., Wiederrecht, G., 2009. Picoliter printing handbook of nanofabrication. In: Wiederrecht, G. (Ed.), Handbook of Nanofabrication. Elsevier, Amsterdam, Netherlands, p. 2009.

Gong, X., Tong, M., Xia, Y., Cai, W., Moon, J.S., Cao, Y., 2009. High-detectivity polymer photodetectors with spectral response from 300 nm to 1450 nm. Science. 325 (2009), 1665–1667.

Goodman, A.M., Rose, A., 1971. Double extraction of uniformly generated electron-hole pairs from insulators with noninjecting contacts. J. Appl. Phys. 42 (1971), 2823–2830.

Grancini, G., Maiuri, M., Fazzi, D., Petrozza, A., Egelhaaf, H.J., Brida, D., 2013. Hot exciton dissociation in polymer solar cells. Nat. Mater. 12 (2013), 29–33.

Guo, F.W., Yang, B., Yuan, Y.B., Xiao, Z.G., Dong, Q.F., Bi, Y., 2012. A nanocomposite ultraviolet photodetector based on interfacial trap-controlled charge injection. Nat. Nanotechnol. 7 (2012), 798–802.

Halls, J.J.M., Walsh, C.A., Greenham, N.C., Marseglia, E.A., Friend, R.H., Moratti, S.C., 1995. Efficient photodiodes from interpenetrating polymer networks. Nature. 376 (1995), 498–500.

Hall, D.B., Underhill, P., Torkelson, J.M., 1998. Spin coating of thin and ultrathin polymer films. Polym. Eng. Sci. 38 (1998), 2039–2045.

Hamasaki, T., Morimune, T., Kajii, H., Minakata, S., Tsuruoka, R., Nagamachi, T., 2009. Fabrication and characteristics of polyfluorene based organic photodetectors using fullerene derivatives. Thin Solid. Films 518 (2009), 548–550.

Hammond, W.T., Mudrick, J.P., Xue, J., 2014. Balancing high gain and bandwidth in multilayer organic photodetectors with tailored carrier blocking layers. J. Appl. Phys. 116 (2014), 214501.

Hammond, W.T., Xue, J.G., 2010. Organic heterojunction photodiodes exhibiting low voltage, imaging-speed photocurrent gain. Appl. Phys. Lett. 97 (2010).

Hansell, W. 1950. Jet sprayer actuated by supersonic waves. US Patent 2,512,743.
Hedley, G.J., Ruseckas, A., Samuel, I.D.W., 2017. Light harvesting for organic photovoltaics. Chem. Rev. 117 (2017), 796–837.
Hendriks, K.H., Li, W., Wienk, M.M., Janssen, R.A., 2017. Small-bandgap semiconducting polymers with high near-infrared photoresponse. J. Am. Chem. Soc. 136 (2014), 12130–12136.
Hernandez-Sosa, G., Coates, N.E., Valouch, S., Moses, D., 2011. High photoconductive responsivity in solution-processed polycrystalline organic composite films. Adv. Funct. Mater. 21 (2011), 927–931.
Hiramoto, M., Imahigashi, T., Yokoyama, M., 1994. Photocurrent multiplication in organic pigment films. Appl. Phys. Lett. 64 (1994), 187–189.
Hiramoto, M., Miki, A., Yoshida, M., Yokoyama, M., 2002. Photocurrent multiplication in organic single crystals. Appl. Phys. Lett. 81 (2002), 1500–1502.
Hong, K., Kim, Y.H., Kim, S.H., Xie, W., Xu, W.D., Kim, C.H., 2014. Aerosol jet printed, sub-2 V complementary circuits constructed from P- and N-type electrolyte gated transistors. Adv. Mater. 26 (2014), 7032–7037.
Hou, J., Inganäs, O., Friend, R., Gao, F., 2018. Organic solar cells based on non-fullerene acceptors. Nat. Mater. 17 (2018), 119–128.
Huang, J.S., Yang, Y., 2007. Origin of photomultiplication in C-60 based devices. Appl. Phys. Lett. 91 (2007).
Iacchetti, A., Binda, M., Natali, D., Giussani, M., Beverina, L., Fiorini, C., 2012a. Multilayer organic squaraine-based photodiode for indirect x-ray detection. IEEE. Trans. Nucl. Sci. 59 (2012), 1862–1867.
Iacchetti, A., Natali, D., Binda, M., Beverina, L., Sampietro, M., 2012b. Hopping photoconductivity in an exponential density of states. Appl. Phys. Lett. 2012, 101.
Im, S.H., Chang, J.A., Kim, S.W., Kim, S.-W., Seok, S.I., 2012a. Near-infrared photodetection based on PbS colloidal quantum dots/organic hole conductor. Org. Electron. 11 (2010), 696–699.
Jacobs, I.E., Moulé, A.J., 2017. Controlling molecular doping in organic semiconductors. Adv. Mater. 2017, 1703063.
Jansen-van Vuuren, R.D., Armin, A., Pandey, A.K., Burn, P.L., Meredith, P., 2016. Organic photodiodes: the future of full color detection and image sensing. Adv. Mater. 2016, 4766–4802.
Jansen-van Vuuren, R.D., Pivrikas, A., Pandey, A.K., Burn, P.L., 2013. Colour selective organic photodetectors utilizing ketocyanine-cored dendrimers. J. Mater. Chem. C. 1 (2013), 3532.
Jarzab, D., Szendrei, K., Yarema, M., Pichler, S., Heiss, W., Loi, M.A., 2011. Charge-separation dynamics in inorganic–organic ternary blends for efficient infrared photodiodes. Adv. Funct. Mater. 21 (2011), 1988–1992.
Jeong, S.W., Jeong, J.W., Chang, S., Kang, S.Y., Cho, K.I., Ju, B.K., 2010. The vertically stacked organic sensor-transistor on a flexible substrate. Appl. Phys. Lett. 97 (2010).
Jung, S., Sou, A., Banger, K., Ko, D.H., Chow, P.C.Y., McNeill, C.R., 2014. All-inkjet-printed, all-air-processed solar cells. Adv. Energy Mater. 4 (2014).
Kawase, T., Shimoda, T., Newsome, C., Sirringhaus, H., Friend, R.H., 2003. Inkjet printing of polymer thin film transistors. Thin Solid. Films 438 (2003), 279–287.
Keivanidis, P.E., Ho, P.K.H., Friend, R.H., Greenham, N.C., 2010. The dependence of device dark current on the active-layer morphology of solution-processed organic photodetectors. Adv. Funct. Mater. 20 (2010), 3895–3903.

Keivanidis, P.E., Khong, S.-H., Ho, P.K.H., Greenham, N.C., Friend, R.H., 2009. All-solution based device engineering of multilayer polymeric photodiodes: minimizing dark current. Appl. Phys. Lett. 94 (2009).

Kim, S.K., Park, S., Son, H.J., Chung, D.S., 2018. Synthetic approach to achieve a thin-film red-selective polymer photodiode: difluorobenzothiadiazole-based donor-acceptor polymer with enhanced space charge carriers. Macromolecules 51 (2018), 8241−8247.

Kippelen, B., Bredas, J.-L., 2009. Organic photovoltaics. Energy Environ. Sci. 2 (2009).

Kirchartz, T., Pieters, B.E., Kirkpatrick, J., Rau, U., Nelson, J., 2011. Recombination via tail states in polythiophene:fullerene solar cells. Phys. Rev. B 83 (2011), 115209.

Klauk, H., 2006. Organic electronics: materials, manufacturing, and applications. In: Klauk, H. (Ed.), Organic Electronics: Materials, Manufacturing, and Applications. Wiley-VCH Verlag GmbH & Co. KGaA, Weinheim, FRG2006.

Kong, Z., Li, W., Che, G., Chu, B., Bi, D., Han, L., 2006. Highly sensitive organic ultraviolet optical sensor based on phosphorescent Cu (I) complex. Appl. Phys. Lett. 89 (2006), 161112−161113.

Konstantatos, G., Howard, I., Fischer, A., Hoogland, S., Clifford, J., Klem, E., 2006. Ultrasensitive solution-cast quantum dot photodetectors. Nature. 442 (2006), 180−183.

Kraker, E., Haase, A., Lamprecht, B., Jakopic, G., Konrad, C., Kostler, S., 2008. Integrated organic electronic based optochemical sensors using polarization filters. Appl. Phys. Lett. 92 (2008), 033302−033303.

Krebs, F.C., 2009. Fabrication and processing of polymer solar cells: a review of printing and coating techniques. Sol. Energy Mater. Sol. Cells. 93 (2009), 394−412.

Kudo, K., Moriizumi, T., 1981. Spectrum-controllable color sensors using organic dyes. Appl. Phys. Lett. 39 (1981), 609−611.

Lakhwani, G., Rao, A., Friend, R.H., 2014. Bimolecular recombination in organic photovoltaics. Annu. Rev. Phys. Chem. 65 (2014), 557−581.

Lampert, M.A., Mark, P., 1970. Current Injection in Solids. Academic Press, New York, p. NY1970.

Lavery, L.L., Whiting, G.L., Arias, A.C., 2011. All ink-jet printed polyfluorene photosensor for high illuminance detection. Org. Electron. 12 (2011), 682−685.

Lee, J., Jadhav, P., Baldo, M.A., 2009. High efficiency organic multilayer photodetectors based on singlet exciton fission. Appl. Phys. Lett. 95 (2009), 033301−033303.

Liao, M., Koide, Y., Alvarez, J., 2006. Photovoltaic Schottky ultraviolet detectors fabricated on boron-doped homoepitaxial diamond layer. Appl. Phys. Lett. 88, 2006.

Liess, A., Arjona-Esteban, A., Kudzus, A., Albert, J., Krause, A., Lv, A., et al., 2019. Ultranarrow bandwidth organic photodiodes by exchange narrowing in merocyanine H- and J-aggregate excitonic systems. Adv. Funct. Mater. 29 (2019), 1805058.

Lim, S.J., Leem, D.S., Park, K.B., Kim, K.S., Sul, S., Na, K., et al., 2015. Organic-on-silicon complementary metal-oxide-semiconductor colour image sensors. Sci. Rep. 5 (2015), 7708.

Lim, J.A., Lee, H.S., Lee, W.H., Cho, K., 2009. Control of the morphology and structural development of solution-processed functionalized acenes for high-performance organic transistors. Adv. Funct. Mater. 19 (2009), 1515−1525.

Lin, Y.-Y., Chen, C.-W., Yen, W.-C., Su, W.-F., Ku, C.-H., Wu, J.-J., 2008. Near-ultraviolet photodetector based on hybrid polymer/zinc oxide nanorods by low-temperature solution processes. Appl. Phys. Lett. 92 (2008), 233301−233303.

Lin, H.W., Ku, S.Y., Su, H.C., Huang, C.W., Lin, Y.T., Wong, K.T., 2005. Highly efficient visible-blind organic ultraviolet photodetectors. Adv. Mater. 17 (2005), 2489−2493.

Lin, Y., Li, Y., Zhan, X., 2012. Small molecule semiconductors for high-efficiency organic photovoltaics. Chem. Soc. Rev. 41 (2012), 4245–4272.
Li, J., Lee, C.-S., Lee, S., 2005. Efficient UV-sensitive organic photovoltaic devices using a starburst amine as electron donor. J. Mater. Chem. 15 (2005), 3268.
Li, W., Li, D., Dong, G., Duan, L., Sun, J., Zhang, D., et al., 2016. High-stability organic red-light photodetector for narrowband applications. Laser Photon. Rev. 10 (2016), 473–480.
Li, J., Naiini, M.M., Vaziri, S., Lemme, M.C., Östling, M., 2014. Inkjet printing of MoS2. Adv. Funct. Mater. 24 (2014), 6524–6531.
Li, H.-G., Wu, G., Chen, H.-Z., Wang, M., 2011. Spectral response tuning and realization of quasi-solar-blind detection in organic ultraviolet photodetectors. Org. Electron. 12 (2011), 70–77.
Li, J., Ye, F., Vaziri, S., Muhammed, M., Lemme, M.C., Ostling, M., 2013. Efficient inkjet printing of graphene. Adv. Mater. 25 (2013), 3985–3992.
Li, L., Zhang, F., Wang, J., An, Q., Sun, Q., Wang, W., 2015a. Achieving EQE of 16,700% in P3HT:PC71BM based photodetectors by trap-assisted photomultiplication. Sci. Rep. 5 (2015), 9181.
Li, L., Zhang, F., Wang, W., An, Q., Wang, J., Sun, Q., 2015b. Trap-assisted photomultiplication polymer photodetectors obtaining an external quantum efficiency of 37500. ACS Appl. Mater. Interfaces. 7 (2015), 5890–5897.
Lochner, C.M., Khan, Y., Pierre, A., Arias, A.C., 2014. All-organic optoelectronic sensor for pulse oximetry. Nat. Commun. 5, 5745.
Lussem, B., Riede, M., Leo, K., 2013. Doping of organic semiconductors. Phys. Status Solidi A Appl. Mater. Sci. 210 (2013), 9–43.
Lyons, D.M., Armin, A., Stolterfoht, M., Nagiri, R.C.R., Jansen-van Vuuren, R.D., Pal, B.N., 2014. Narrow band green organic photodiodes for imaging. Org. Electron. 15 (2014), 2903–2911.
Lyon, R.F., Hubel, P.M. 2002. Eyeing the camera: into the next century. In: Proceedings of the Tenth Color Imaging Conference: Color Science and Engineering Systems, Technologies, Applications. Scottsdale, Arizona, USA, pp. 349–355.
Mandal, S., Dell'Erba, G., Luzio, A., Bucella, S.G., Perinot, A., Calloni, A., 2015. Fully-printed, all-polymer, bendable and highly transparent complementary logic circuits. Org. Electron. 20 (2015), 132–141.
Melancon, J.M., Zivanovic, S.R., 2014. Broadband gain in poly(3-hexylthiophene):phenyl-C-61-butyric-acid-methyl-ester photodetectors enabled by a semicontinuous gold interlayer. Appl. Phys. Lett. 105 (2014).
Mette, A., Richter, P.L., Glunz, S.W., Willeke, G. 2006. Novel metal jet printing technique for the front side metallization of highly efficient industrial silicon solar cells. In: Proceedings of the Twenty-first European Photovoltaic Solar Energy Conference. Dresden, Germany, 4–8 September 2006. München: WIP-Renewable Energies, ISBN: 3-936338-20-5, pp.1174–1177.
Miao, J., Zhang, F., 2018. Recent progress on photomultiplication type organic photodetectors. Laser Photonics Rev. 13 (2), 1800204.
Miao, J., Zhang, F., Du, M., Wang, W., Fang, Y., 2017. Photomultiplication type narrowband organic photodetectors working at forward and reverse bias. Phys. Chem. Chem. Phys. 19 (2017), 14424–14430x.
Miao, J., Zhang, F., Du, M., Wang, W., Fang, Y., 2018. Photomultiplication type organic photodetectors with broadband and narrowband response ability. Adv. Opt. Mater. 6 (2018), 1800001.

Mihailetchi, V.D., Wildeman, J., Blom, P.W.M., 2005. Space-charge limited photocurrent. Phys. Rev. Lett. 94 (2005), 126602.

Minemawari, H., Yamada, T., Matsui, H., Tsutsumi, J.Y., Haas, S., Chiba, R., 2011. Inkjet printing of single-crystal films. Nature. 475 (2011), 364−367.

Monroy, E., Omnès, F., Calle, F., 2003. Wide-bandgap semiconductor ultraviolet photodetectors. Semicond. Sci. Technol. 18 (2003), R33.

Morimune, T., Kajii, H., Ohmori, Y., 2006. Photoresponse properties of a high-speed organic photodetector based on copper−phthalocyanine under red light illumination. IEEE Photonics Technol. Lett. 18 (2006), 2662−2664.

Morteani, A.C., Sreearunothai, P., Herz, L.M., Friend, R.H., Silva, C., 2004. Exciton regeneration at polymeric semiconductor heterojunctions. Phys. Rev. Lett. 92 (2004), 247402.

Najafov, H., Lyu, B., Biaggio, I., Podzorov, V., 2008. Investigating the origin of the high photoconductivity of rubrene single crystals. Phys. Rev. B. 77 (2008), 125202.

Nakamura, J. (Ed.), 2017. Image Sensors and Signal Processing for Digital Still Cameras, 2017. CRC Press, Boca Raton, FL.

Nalwa, K.S., Cai, Y., Thoeming, A.L., Shinar, J., Shinar, R., Chaudhary, S., 2010. Polythiophene-fullerene based photodetectors: tuning of spectral response and application in photoluminescence based (bio)chemical sensors. Adv. Mater. 22 (2010), 4157−4161.

Natali, D., Caironi, M., 2012. Charge injection in solution processed organic field-effect transistors: physics, models and characterization methods. Adv. Mater. 24 (2012), 1357.

Natali, D., 2020. Fundamentals of organic electronic devices. In: Cosseddu, P., Caironi, M. (Eds.), Organic Flexible Electronics, Woodhead Publishing Series in Electronic and Optical Materials. .

Natali, D., 2018. Organic imagers. In: Kevin, K., Basiricò, L., Iniewski, K. (Eds.), Sensors for Diagnostics and Monitoring, 2018. CRC Press.

Nausieda, I., Ryu, K., Kymissis, I., Akinwande, A.I., Bulovic, V., Sodini, C.G., 2008. An organic active-matrix imager. IEEE Trans. Electron. Devices 55, 527−532. 2008.

Nelson, J., 2011. Polymer:fullerene bulk heterojunction solar cells. Mater. Today. 14 (2011), 462−470.

Ng, T.N., Wong, W.S., Chabinyc, M.L., Sambandan, S., Street, R.A., 2008. Flexible image sensor array with bulk heterojunction organic photodiode. Appl. Phys. Lett. 92 (2008).

Niemeyer, A.C., Campbell, I.H., So, F., Crone, B.K., 2007. High quantum efficiency polymer photoconductors using interdigitated electrodes. Appl. Phys. Lett. 91 (2007).

Noh, Y.-Y., Zhao, N., Caironi, M., Sirringhaus, H., 2007. Downscaling of self-aligned, all-printed polymer thin-film transistors. Nat. Nanotechnol. 2 (2007), 784−789.

Ohmori, Y., Kajii, H., Kaneko, M., Yoshino, K., Ozaki, M., Fujii, A., 2004. Realization of polymeric optical integrated devices utilizing organic light-emitting diodes and photodetectors fabricated on a polymeric waveguide. IEEE J. Sel. Top. Quantum Electron. 10 (2004), 70−78.

Olivier, Y., Niedzialek, D., Lemaur, V., Pisula, W., Müllen, K., Koldemir, U., 2014. 25th anniversary article: high-mobility hole and electron transport conjugated polymers: how structure defines function. Adv. Mater. 26 (2014), 2119−2136.

Osedach, T.P., Geyer, S.M., Ho, J.C., Arango, A.C., Bawendi, M.G., Bulovic, V., 2009. Lateral heterojunction photodetector consisting of molecular organic and colloidal quantum dot thin films. Appl. Phys. Lett. 94 (2009).

Osedach, T.P., Iacchetti, A., Lunt, R.R., Andrew, T.L., Brown, P.R., Akselrod, G.M., et al., 2012. Near-infrared photodetector consisting of J-aggregating cyanine dye and metal oxide thin films. Appl. Phys. Lett. 101 (2012), 113303.

Osedach, T.P., Zhao, N., Geyer, S.M., Chang, L.-Y., Wanger, D.D., Arango, A.C., 2010. Interfacial recombination for fast operation of a planar organic/QD infrared photodetector. Adv. Mater. 22 (2010), 5250–5254.

Pace, G., Grimoldi, A., Natali, D., Sampietro, M., Coughlin, J.E., Bazan, G.C., 2014. All-organic and fully-printed semitransparent photodetectors based on narrow bandgap conjugated molecules. Adv. Mater. 26 (2014), 6773–6777.

Pais, A., Banerjee, A., Klotzkin, D., Papautsky, I., 2008. High-sensitivity, disposable lab-on-a-chip with thin-film organic electronics for fluorescence detection. Lab. Chip 8 (2008).

Park, H., Dan, Y., Seo, K., Yu, Y.J., Duane, P.K., Wober, M., et al., 2014a. Filter-free image sensor pixels comprising silicon nanowires with selective color absorption. Nano Lett. 14 (2014), 1804–1809.

Park, S., Lim, B.T., Kim, B., Son, H.J., Chung, D.S., 2014b. High mobility polymer based on a π-extended benzodithiophene and its application for fast switching transistor and high gain photoconductor. Sci. Rep. 4, 2014.

Pecunia, 2019a. Organic Narrowband Photodetectors: Materials, Devices and Applications. IOP Publishing, Bristol.

Pecunia, 2019b. Efficiency and spectral performance of narrowband organic and perovskite photodetectors: a cross-sectional review. J. Phys. Mater. 2 (2019), 042001.

Peet, J., Kim, J.Y., Coates, N.E., Ma, W.L., Moses, D., Heeger, A.J., 2007. Efficiency enhancement in low-bandgap polymer solar cells by processing with alkane dithiols. Nat. Mater. 6 (2007), 497–500.

Peumans, P., Bulovic, V., Forrest, S.R., 2000. Efficient photon harvesting at high optical intensities in ultrathin organic double-heterostructure photovoltaic diodes. Appl. Phys. Lett. 76 (2000), 2650–2652.

Pierre, A., Sadeghi, M., Payne, M.M., Facchetti, A., Anthony, J.E., Arias, A.C., 2014. All-printed flexible organic transistors enabled by surface tension-guided blade coating. Adv. Mater. 26 (2014), 5722–5727.

Pintus, A., Ambrosio, L., Aragoni, M.C., Binda, M., Coles, S.J., Hursthouse, M.B., et al., 2020. Photoconducting devices with response in the visible-near-infrared region based on neutral ni complexes of aryl-1,2-dithiolene ligands. Inorg. Chem. 59 (9), 6410–6421.

Polishuk, P., 2006. , Plastic optical fibers branch out. IEEE Commun. Mag. 44 (2006), 140–148.

Punke, M., Valouch, S., Kettlitz, S.W., Gerken, M., Lemmer, U., 2008. Optical data link employing organic light-emitting diodes and organic photodiodes as optoelectronic components. J. Lightwave Technol. 26 (2008), 816–823.

Ramuz, M., Bürgi, L., Winnewisser, C., Seitz, P., 2008. High sensitivity organic photodiodes with low dark currents and increased lifetimes. Org. Electron. 9 (2008), 369–376.

Rand, B.P., Xue, J., Lange, M., Forrest, S.R., 2003. Thin-film organic position sensitive detectors. IEEE Photonics Technol. Lett. 15 (2003), 1279–1281.

Rauch, T., Boberl, M., Tedde, S.F., Furst, J., Kovalenko, M.V., Hesser, G., 2009. Near-infrared imaging with quantum-dot-sensitized organic photodiodes. Nat. Photon. 3 (2009), 332–336.

Rayleigh, L., 1878. On the instability of jets. Proc. Lond. Math. Soc. 10 (1878), 4–13.

Ray, D., Narasimhan, K.L., 2007. High response organic visible-blind ultraviolet detector. Appl. Phys. Lett. 91 (2007).

Razeghi, M., Rogalski, A., 1996. Semiconductor Ultrav. detectors. J. Appl. Phys. 79 (1996), 7433–7473.

Renshaw, C.K., Xu, X., Forrest, S.R., 2010. A monolithically integrated organic photodetector and thin film transistor. Org. Electron. 11 (2010), 175–178.

Ren, H., Chen, J.-D., Li, Y.-Q., Tang, Jian-Xin, 2021. Recent progress in organic photodetectors and their applications. Adv. Sci. 8 (2021), 2002418.

Roncali, J., 2007. Molecular engineering of the band gap of π-conjugated systems: facing technological applications. Macromol. Rapid Commun. 28 (2007), 1761−1775.

Ruderer, M.A., Muller-Buschbaum, P., 2011. Morphology of polymer-based bulk heterojunction films for organic photovoltaics. Soft Matter 7 (2011), 5482−5493.

Ruiz-Hitzky, E., Darder, M., Fernandes, F.M., Zatile, E., Palomares, F.J., Aranda, P., 2011. Supported graphene from natural resources: easy preparation and applications. Adv. Mater. 23 (2011), 5250−5255.

Sakai, T., Seo, H., Aihara, S., Kubota, M., Egami, N., Mori, K., et al., 2012. Doping effect of silole derivative in coumarin 30 photoconductive film. Mol. Cryst. Liq. Cryst. 568, 74−81.

Salzmann, et al., 2016. Molecular electrical doping of organic semiconductors: fundamental mechanisms and emerging dopant design rules. Acc. Chem. Res. 2016, 370−378.

Sandvik, P., Brown, D., Fedison, J., Matocha, K., Kretchmer, J., 2005. Dual-SiC photodiode devices simultaneous two-band detection. J. Electrochem. Soc. 152 (2005), G199−G202.

Saracco, E., Bouthinon, B., Verilhac, J.M., Celle, C., Chevalier, N., Mariolle, D., 2013. Work function tuning for high-performance solution-processed organic photodetectors with inverted structure. Adv. Mater. 25 (2013), 6534−6538.

Sariciftci, N.S., Braun, D., Zhang, C., Srdanov, V.I., Heeger, A.J., Stucky, G., 1993. Semiconducting polymer-buckminsterfullerene heterojunctions: diodes, photodiodes, and photovoltaic cells. Appl. Phys. Lett. 62 (1993), 585−587.

Savart, F., 1833. Memoire sur la constitution des veines liquides lancees par des orifices circulaires en mince paroi. Annales de. Chimie et. de Phys. 53 (1833), 337−386.

Schmidt, G.C., Bellmann, M., Kempa, H., Hambsch, M., Reuter, K., Stanel, M., 2011. Mass-printed integrated circuits with enhanced performance using novel materials and concepts. MRS Online Proc. Libr. 1285 (2011).

Schmidt, M., Falco, A., Loch, M., Lugli, P., Scarpa, G., 2014. Spray coated indium-tin-oxide-free organic photodiodes with PEDOT:PSS anodes. AIP Adv. 4 (2014).

Seo, H., Aihara, S., Watabe, T., Ohtake, H., Kubota, M., Egami, N., 2007. Color sensors with three vertically stacked organic photodetectors. Jpn. J. Appl. Phys. 46 (2007), L1240.

Shirota, Y., Kuwabara, Y., Inada, H., Wakimoto, T., Nakada, H., Yonemoto, Y., 1994. Multilayered organic electroluminescent device using a novel starburst molecule, 4,4[script'],4[script']-tris(3-methylphenylphenylamino)triphenylamine, as a hole transport material. Appl. Phys. Lett. 65 (1994), 807−809.

Siegmund, B., Mischok, A., Benduhn, J., Zeika, O., Ullbrich, S., Nehm, F., et al., 2017. Organic narrowband near-infrared photodetectors based on intermolecular charge-transfer absorption. Nat. Commun. 8, 15421.

Sim, K.M., Yoon, S., Cho, J., Jang, M.S., Chung, D.S., 2018. Facile tuning the detection spectrum of organic thin film photodiode via selective exciton activation. ACS Appl. Mater. Interfaces 10 (2018), 8405−8410.

Singh, M., Haverinen, H.M., Dhagat, P., Jabbour, G.E., 2010. Inkjet printing-process and its applications. Adv. Mater. 22 (2010), 673−685.

Sirringhaus, H., 2009. Materials and applications for solution-processed organic field-effect transistors. Proc. IEEE. 97 (2009), 1570−1579.

Sirringhaus, H., Kawase, T., Friend, R.H., Shimoda, T., Inbasekaran, M., Wu, W., 2000. High-resolution inkjet printing of all-polymer transistor circuits. Science. 290 (2000), 2123−2126.

Si, Z., Li, B., Wang, L., Yue, S., Li, W., 2007. OPV devices based on functionalized lanthanide complexes for application in UV—light detection. Sol. Energy Mater. Sol. Cells. 91 (2007), 1168—1171.

Soci, C., Hwang, I.W., Moses, D., Zhu, Z., Waller, D., Gaudiana, R., 2007. Photoconductivity of a low-bandgap conjugated polymer. Adv. Funct. Mater. 17 (2007), 632—636.

Someya, T., Kato, Y., Shingo, I., Noguchi, Y., Sekitani, T., Kawaguchi, H., 2005. Integration of organic FETs with organic photodiodes for a large area, flexible, and lightweight sheet image scanners. IEEE Trans. Electron. Devices 52 (2005), 2502—2511.

Street, R.A., 2011. Localized state distribution and its effect on recombination in organic solar cells. Phys. Rev. B. 84 (2011), 075208.

Sun, H., Liu, T., Yu, J., Lau, T.-K., Zhang, G., Zhang, Y., et al., 2019. Energy Environ. Sci. 12 (2019), 3328—3337.

Su, Z., Li, W., Chu, B., Li, T., Zhu, J., Zhang, G., 2008. High response organic ultraviolet photodetector based on blend of 4,4',4-tri-(2-methylphenyl phenylamino) triphenylaine and tris-(8-hydroxyquinoline) gallium. Appl. Phys. Lett. 93, 2008.

Søndergaard, R.R., Hösel, M., Krebs, F.C., 2013. Roll-to-roll fabrication of large area functional organic materials. J. Polym. Sci. Part. B: Polym. Phys. 51 (2013), 16—34.

Takada, S., Ihama, M., Inuiya, M. 2006. CMOS image sensor with organic photoconductive layer having narrow absorption band and proposal of stack type solid-state image sensors. In: Blouke, M.M. (Ed.), Proceedings Volume 6068, Sensors, Cameras, and Systems for Scientific/Industrial Applications VII. San Jose, California, United States, p. 60680 A, 2006.

Tang, Z., Ma, Z., Sánchez-Díaz, A., Ullbrich, S., Liu, Y., Siegmund, B., et al., 2017. Polymer:fullerene bimolecular crystals for near-infrared spectroscopic photodetectors. Adv. Mater. 29 (2017), 1702184.

Tan, W.C., Shih, W.H., Chen, Y.F., 2014. A highly sensitive graphene-organic hybrid photodetector with a piezoelectric substrate. Adv. Funct. Mater. 24 (2014), 6818—6825.

Tedde, S.F., Kern, J., Sterzl, T., Furst, J., Lugli, P., Hayden, O., 2009. Fully spray coated organic photodiodes. Nano. Lett. 9 (2009), 980—983.

Tedde, S., Zaus, E.S., Furst, J., Henseler, D., Lugli, P., 2007. Active pixel concept combined with organic photodiode for imaging devices. IEEE Electron. Device Lett. 28 (2007), 893—895.

Tenent, R.C., Barnes, T.M., Bergeson, J.D., Ferguson, A.J., To, B., Gedvilas, L.M., 2009. Ultrasmooth, large-area, high-uniformity, conductive transparent single-walled-carbon-nanotube films for photovoltaics produced by ultrasonic spraying. Adv. Mater. 21 (2009), 3210.

Thejo Kalyani, N., Dhoble, S.J., 2012. Organic light emitting diodes: energy saving lighting technology—a review. Renew. Sustain. Energy Rev. 16 (2012), 2696—2723.

Tong, X., Forrest, S.R., 2011. An integrated organic passive pixel sensor. Org. Electron. 12 (2011), 1822—1825.

Torrisi, F., Hasan, T., Wu, W., Sun, Z., Lombardo, A., Kulmala, T.S., 2012. Inkjet-printed graphene electronics. ACS Nano 6 (2012), 2992—3006.

Tsai, W.-W., Chao, Y.-C., Chen, E.-C., Zan, H.-W., Meng, H.-F., Hsu, C.-S., 2009. Increasing organic vertical carrier mobility for the application of high speed bilayered organic photodetector. Appl. Phys. Lett. 95 (2009).

Tzabari, L., Tessler, N., 2011. Shockley—Read—Hall recombination in P3HT:PCBM solar cells as observed under ultralow light intensities. J. Appl. Phys. 109 (2011), 064501—064505.

Ullbrich, S., Siegmund, B., Mischok, A., Hofacker, A., Benduhn, J., Spoltore, D., et al., 2017. Fast organic near-infrared photodetectors based on charge-transfer absorption. J. Phys. Chem. Lett. 8 (2017), 5621−5625.

van Duren, J.K.J., Yang, X., Loos, J., Bulle-Lieuwma, C.W.T., Sieval, A.B., Hummelen, J.C., 2004. Relating the morphology of poly(p-phenylene vinylene)/methanofullerene blends to solar-cell performance. Adv. Funct. Mater. 14 (2004), 425−434.

Venkateshvaran, D., Nikolka, M., Sadhanala, A., Lemaur, V., Zelazny, M., Kepa, M., 2014. A2014pproaching disorder-free transport in high-mobility conjugated polymers. Nature. 515, 384−388.

Verilhac, J.-M., 2013. Recent developments of solution-processed organic photodetectors. Eur. Phys. J. Appl. Phys 63 (2013), 14405.

Verlaak, S., Heremans, P., 2007. Molecular microelectrostatic view on electronic states near pentacene grain boundaries. Phys. Rev. B. 75 (2007), 115127.

Verstraeten, F., Gielen, S., Verstappen, P., Raymakers, J., Penxten, H., Lutsen, L., et al., 2020. J. Mater. Chem. C. 8 (2020), 10098−10103.

Wang, W., Du, M., Zhang, M., Miao, J., Fang, Y., Zhang, F., 2018. Organic photodetectors with gain and broadband/narrowband response under top/bottom illumination conditions. Adv. Opt. Mater. 6 (2018), 1800249.

Wang, J.B., Li, W.L., Chu, B., Chen, L.L., Zhang, G., Su, Z.S., 2010. Visible-blind ultraviolet photo-detector using tris-(8-hydroxyquinoline) rare earth as acceptors and the effects of the bulk and interfacial exciplex emissions on the photo-responsivity. Org. Electron. 11 (2010), 1301−1306.

Wang, J.B., Li, W.L., Chu, B., Lee, C.S., Su, Z.S., Zhang, G., 2011. High speed responsive near infrared photodetector focusing on 808 nm radiation using hexadecafluoro−copper−phthalocyanine as the acceptor. Org. Electron. 12 (2011), 34−38.

Wang, W., Zhang, F., Du, M., Li, L., Zhang, M., Wang, K., et al., 2017. highly narrowband photomultiplication type organic photodetectors. Nano Lett. 17 (2017), 1995−2002.

Weng, B., Shepherd, R.L., Crowley, K., Killard, A.J., Wallace, G.G., 2010. Printing conducting polymers. Analyst. 135 (2010), 2779−2789.

Wienk, M.M., Kroon, J.M., Verhees, W.J.H., Knol, J., Hummelen, J.C., van Hal, P.A., 2003. Efficient methano[70]fullerene/MDMO-PPV bulk heterojunction photovoltaic cells. Angew. Chem. Int. (Ed.) 42 (2003), 3371−3375.

Williams, G., Backhouse, C., Aziz, H., 2014. Integration of organic light emitting diodes and organic photodetectors for lab-on-a-chip bio-detection systems. Electronics. 3 (2014), 43−75.

Wudl, F., 2014. Spiers memorial lecture organic electronics: an organic materials perspective. Faraday Discuss. 174 (2014), 9−20.

Wu, S.-H., Li, W.-L., Chu, B., Lee, C.S., Su, Z.-S., Wang, J.-B., 2010a. Visible-blind ultraviolet sensitive photodiode with high responsivity and long term stability. Appl. Phys. Lett. 97 (2010), 023306−023313.

Wu, S.-H., Li, W.-L., Chu, B., Lee, C.S., Su, Z.-S., Wang, J.-B., 2010b. High response deep ultraviolet organic photodetector with spectrum peak focused on 280 nm. Appl. Phys. Lett. 96 (2010), 093302−093303.

Wu, S.-H., Li, W.-L., Chu, B., Su, Z.-S., Zhang, F., Lee, C.S., 2011. High performance small molecule photodetector with broad spectral response range from 200 to 900 nm. Appl. Phys. Lett. 99 (2011).

Xia, K., Li, Y., Wang, Y., Portilla, L., Pecunia, V., 2020. Narrowband-absorption-type organic photodetectors for the far-red range based on fullerene-free bulk heterojunctions. Adv. Opt. Mater. 2020, 1902056.

Xie, R., Zhang, K., Yin, Q., Hu, Z., Yu, G., Huang, F., et al., 2020. Self-filtering narrowband high performance organic photodetectors enabled by manipulating localized Frenkel exciton dissociation. Nat. Commun. 11 (2020), 2871.

Xue, J., Forrest, S.R., 2004. Carrier transport in multilayer organic photodetectors: I. Effects of layer structure on dark current and photoresponse. J. Appl. Phys. 95 (2004), 1859−1868.

Xue, D.-J., Wang, J.-J., Wang, Y.-Q., Xin, S., Guo, Y.-G., Wan, L.-J., 2011. Facile synthesis of germanium nanocrystals and their application in organic−inorganic hybrid photodetectors. Adv. Mater. 23 (2011), 3704−3707.

Xu, X., Davanco, M., Qi, X., Forrest, S.R., 2008. Direct transfer patterning on three dimensionally deformed surfaces at micrometer resolutions and its application to hemispherical focal plane detector arrays. Org. Electron. 9 (2008), 1122−1127.

Yan, F., Liu, H., Li, W., Chu, B., Su, Z., Zhang, G., 2009. Double wavelength ultraviolet light sensitive organic photodetector. Appl. Phys. Lett. 95 (2009).

Yao, Y., Liang, Y., Shrotriya, V., Xiao, S., Yu, L., Yang, Y., 2007. Plastic near-infrared photodetectors utilizing low band gap polymer. Adv. Mater. 19 (2007), 3979−3983.

Yu, G., Cao, Y., Wang, J., McElvain, J., Heeger, A.J., 1999. High sensitivity polymer photosensors for image sensing applications. Synth. Met. 102 (1999), 904−907.

Yu, G., Heeger, A.J., 1995. Charge separation and photovoltaic conversion in polymer composites with internal donor/acceptor heterojunctions. J. Appl. Phys. 78 (1995), 4510−4515.

Zaus, E.S., Tedde, S., Furst, J., Henseler, D., Dohler, G.H., 2007. Dynamic and steady state current response to light excitation of multilayered organic photodiodes. J. Appl. Phys 101 (2007), 044501−044507.

Zhang, Y. 2002. Problems in the fusion of commercial high-resolution satellite images as well as Landsat 7 images and Initial solutions. In: ISPRS, CIG, SDH Joint International Symposium on "GeoSpatial Theory, Processing and Applications," July 8−12 2002 Ottawa, Canada.

Zhang, F., Johansson, M., Andersson, M.R., Hummelen, J.C., Inganäs, O., 2002. Polymer photovoltaic cells with conducting polymer anodes. Adv. Mater. 14 (2002), 662−665.

Zhang, G., Li, W., Chu, B., Su, Z., Yang, D., Yan, F., 2009. Highly efficient photovoltaic diode based organic ultraviolet photodetector and the strong electroluminescence resulting from pure exciplex emission. Org. Electron. 10 (2009), 352−356.

Zhang, G., Li, W., Chu, B., Yan, F., Zhu, J., Chen, Y., 2010. Very high open-circuit voltage ultraviolet photovoltaic diode with its application in optical encoder field. Appl. Phys. Lett. 96 (2010), 073301−073303.

Zhao, D., Liu, T., Park, J.G., Zhang, M., Chen, J.-M., Wang, B., 2012. Conductivity enhancement of aerosol-jet printed electronics by using silver nanoparticles ink with carbon nanotubes. Microelectron. Eng. 96 (2012), 71−75.

Zhao, Z., Xu, C., Niu, L., Zhang, X., Zhang, F., 2021. Recent Progress on Broadband Organic Photodetectors and their Applications. Laser Photonics Rev. 2020, 2000262.

Zhong, Y., Sisto, T.J., Zhang, B., Miyata, K., Zhu, X.-Y., Steigerwald, M.L., et al., 2017. Helical Nanoribbons for Ultra-Narrowband Photodetectors. J. Am. Chem. Soc. 139 (2017), 5644−5647.

Zimmerman, J.D., Diev, V.V., Hanson, K., Lunt, R.R., Yu, E.K., Thompson, M.E., 2010. Porphyrin-Tape/C60 organic photodetectors with 6.5% external quantum efficiency in the near infrared. Adv. Mater. 22 (2010), 2780−2783.

Zimmerman, J.D., Yu, E.K., Diev, V.V., Hanson, K., Thompson, M.E., Forrest, S.R., 2011. Use of additives in porphyrin-tape/C60 near-infrared photodetectors. Org. Electron. 12 (2011), 869−873.

Nanowires for photodetection

Badriyah Alhalaili[1,*], Elif Peksu[2,3,*], Lisa N. Mcphillips[2], Matthew M. Ombaba[2], M. Saif Islam[2] and Hakan Karaagac[3]
[1]Nanotechnology and Advanced Materials Program, Kuwait Institute for Scientific Research, Kuwait City, Kuwait, [2]Integrated Nanodevices and Nanosystems Research, Electrical and Computer Engineering, University of California, Davis, CA, United States, [3]Physics Department, Istanbul Technical University, Istanbul, Turkey

4.1 Introduction

During the last two decades, significant progress has been made in developments and advances in the "bottom-up" synthesis of one-dimensional nanowires (NWs) (1D-NWs) for scalable, miniature, and energy-efficient electronics, photonics, magnetics, and electromechanical systems to transform computing and communication (Cui et al., 2001; Quitoriano and Kamins, 2008; Do et al., 2006; Jiang et al., 2007; Björk et al., 2002; Wallentin et al., 2010; Lieber, 2001; Kong et al., 1999; Huang et al., 2001b; Wang et al., 2006; Chaudhry and Islam, 2008; Fan et al., 2008; Logeeswaran et al., 2008). Simultaneously, the current state-of-the-art silicon CMOS technology has already been scaled down to nanometer feature sizes and is approaching the physical lower limit of beneficial scaling. These trends motivate a search for new technologies that may allow widespread and cost-effective integration of NWs in devices and circuits for electronic as well as optoelectronic applications.

There have been many publications on the field of NWs that have mainly focused on the various material growth aspects, mechanisms, and techniques to control NW crystal growth (Mieszawska et al., 2007; Ruda and Shik, 2005) with the recent exception of an in-depth review paper specifically dedicated to the device physics of NW photodetectors (PDs) by (Soci et al., 2010). In this chapter, we have identified a wide range of literature in the field and aim to provide the reader with the overall concepts and general aspects that are important as a starting point for a discussion on NW-based PDs for communication and sensing systems. We present some exciting recent developments in NW-based detectors that have the potential to play a critical role in the development and transformation of future communication and sensing networks.

Direct integration of an assortment of semiconductor NWs on single crystal surfaces offers attractive opportunities in several areas of high-performance communication

---

* Both authors contributed equally to this work.

electronics, optoelectronics, sensing, energy conversion, and imaging. Beyond these applications, the integration of a range of nanostructures on amorphous surfaces offers unlimited capabilities for multifunctional materials and device integration. The possibility of low-cost electronics and photonics based on such an approach would dwarf silicon photonics and other competing technologies. The key constraints in growing planar epitaxial thin film of a semiconductor on another single crystal substrate are lattice and thermal expansion coefficient mismatches, material incompatibilities, and differences in the crystal structure. These limitations can now be circumvented by growing semiconductor nanoheterostructures that can accommodate large mismatches due to their small crystallite dimensions.

With recent rapid progress in the growth of various inorganic NWs, mostly using the vapor-liquid-solid (VLS) growth method developed by Wagner and others in the mid-1960s (Wagner and Ellis, 1964; Givargizov and Sheftal, 1971; Givargizov, 1975; Thelander et al., 2003), there have been several studies on the growth of NW heterostructures such as InAs/InP (3.1% lattice mismatch) (Borgström et al., 2001), GaAs/GaP (3.1% lattice mismatch) (Gudiksen et al., 2002), InP/Si (8% lattice mismatch) (Mårtensson et al., 2004; Yi et al., 2006), Ge/GaAs (0.1% mismatch) (Kim et al., 2007), GaAs/Si ($\sim$4% lattice mismatch) (Kobayashi et al., 2009), and Si/Ge (4% lattice mismatch) (Kamins et al., 2004). These results indicate that the lattice misfit can be effectively accommodated in NW heterostructures. However, nanoheteroepitaxy is still in the early stage for the effective growth of ternary and quaternary materials that helped in making revolutionary advances in traditional optoelectronic devices in the last two decades.

The ability to grow dissimilar materials with large lattice mismatches on a single substrate removes the integration-related constraints of incorporating photonics with electronics, and therefore an associated challenge of on-chip electronic-to-optical and optical-to-electronic conversions for communication and information processing using primarily electrons and the majority of the information transfer using photons. Semiconductor nanostructures with direct bandgap grown on silicon can offer high gain for designing laser diodes as well as superior absorption characteristics needed for high-speed photodetection on a silicon surface.

We reported the heteroepitaxial growth of highly aligned InP NWs on silicon substrates and introduced a new method of synthesizing III−V NWs on a *nonsingle crystalline surface* that directly relaxes any lattice mismatching conditions. We also demonstrated a device for high-speed photodetection based on InP NWs grown in the form of nanobridges between prefabricated electrodes made of hydrogenated microcrystalline silicon (Logeeswaran et al., 2008). The device was fabricated on an amorphous glass surface and was measured to have a bandwidth above 30 GHz. We reported the integration of bridged NWs between a pair of vertically oriented nonsingle crystal surfaces etched into rib optical waveguides to design NW integrated waveguide PDs on amorphous surfaces (Logeeswaran et al., 2011a,b). These capabilities offer opportunities for ultra-fast low-cost free-space optical interconnect for future computers and servers.

Heteroepitaxial has grown optically active NWs such as III−V and II−VI for integration with the mainstream Si technology may enable low-cost, and highly integrated ultra-fast devices due to their high carrier mobilities and optical absorption coefficients. This opens opportunities for a wide variety of applications including intrachip, interchip, and free-space communications. The capabilities of generating and detecting photons by direct bandgap materials on Si substrate, which is known for its low efficiency in electron-to-photon conversion, will bring about a myriad of challenges along with revolutionary opportunities that will impact a large sector of computing, communication, and sensing industry.

4.2 Nanowires photodetector fabrication themes

NW integration modalities for optoelectronic devices have largely relied on expensive sequential interfacing procedures that at times rely on serendipity. The most common method is the deposition of blanket electrode pads atop ensembles of randomly oriented NWs atop a given substrate (Thelander et al., 2006; Huang et al., 2010; Lee et al., 2007a). Without any NW alignment, typically a statistically low number of devices are obtained. These are then tracked microscopically and analyzed. Development of massively parallel, barrier-free, low-noise and preferably "in situ" electrical connections to semiconductor NWs would therefore find widespread adaption in many other applications outside the photodetection realm. Photoactive single semiconductor NWs can only be fashioned into realistic photodetection devices upon robustly being anchored into mechanical support using low resistance electrical contacts fabricated via methods that either involve *direct growth* or transfer *printing (or pick-and-place) onto* secondary receiver substrates in several configurations as reviewed in one of our recent papers as summarized in Figure 4.1 (Logeeswaran et al., 2011a,b).

4.2.1 Direct nanowire integration

Where catalyst deposition is permissible, semiconductor NWs PD fabrication can be most efficiently integrated to device sites of interest if grown directly therein, creating a mechanically and electrically stable monolithic ensemble. The success of this technique hinges on the success of the definition of the catalytic sites using techniques such as nanoimprint and e-beam lithography. Catalytic nanoparticles can also nucleate from thin films of the catalytic material upon annealing even though catalytic diameter control is difficult while catalyst nanoparticle migration and coalescence lead to NW position uncertainties (Kayes et al., 2007).

We pioneered a technique for making one-directional silicon NW connections in 2004 (Saif Islam et al., 2005), which was coined "bridging" growth (Sharma et al., 2005; Tabib-Azar et al., 2005; Saif Islam et al., 2005). NW growth initiated from a catalytic site atop an electrode and continued until or even beyond impingement of the opposite electrode, forming a crystalline bridge without having any of the original catalysts as part of the bridging process as illustrated in Figure 4.3. A variety of both Group IV (Islam et al., 2004), III–V (Yi et al., 2006) and II–VI (Lee et al., 2006)

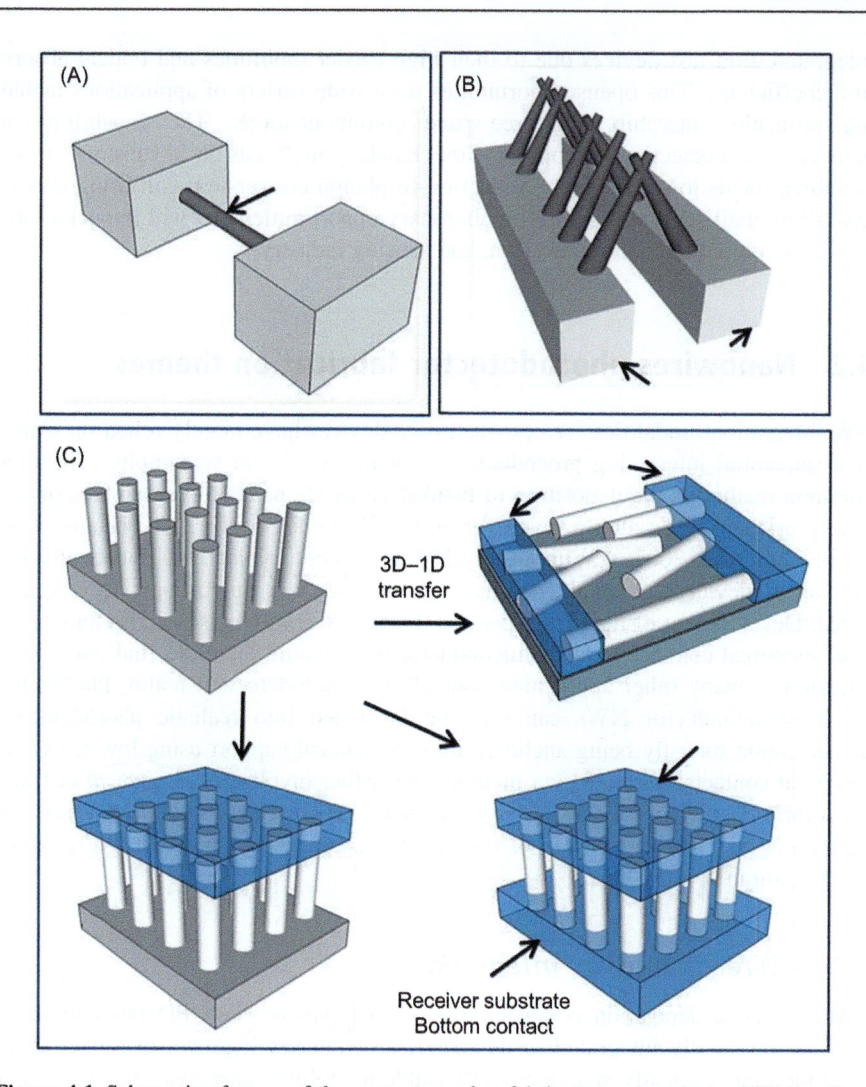

Figure 4.1 Schematic of some of the avenues used to fabricate semiconductor NW-based devices for photonic application. (A) NWs can be directly grown between two electrodes using both topdown and bottomup approaches. (B) NW bridges can form on isolated electrode contacts as a result of a serendipitous fusing phenomenon during their growth. (C) Vertically oriented NWs atop the mother wafer can be fashioned into a PD by a lateral 3D–1D contact printing onto receiver substrates followed by deposition of electrical contacts or by a 3D–3D print transfer mechanism onto thermally pliable matrices followed by top contact deposition. Or whilst still, atop the mother wafer, isolation and top contact layers can be deposited onto the array to create a device.

semiconductor NWs have been bridged using this technique. Our recent advancement of this technique has led to the development of gate-all-round phototransistors for electro-optical OR gate circuits and frequency doubler applications amongst other devices as depicted in Figure 4.2 (Oh et al., 2014; Yong Oh and Saif Islam, 2014). While deposition of the catalyst onto the desired region is highly challenging, this technique presents new opportunities for utilization of the more often unexploited third dimension of a wafer for photodetection device applications.

The semiconductor NW bridge is a versatile architecture that allows for subsequent postgrowth NW manipulations with no thermal restrictions. Processes such as thermal oxide growth, high-temperature doping, and annealing can be swiftly performed without compromising the mechanical integrity of the suspended NW. By virtue of its suspension, a bridged semiconductor NW is ideal for conformal growth and deposition of gate channels, creating well-defined and isolated core-shell devices. Recently, we ventured into the fabrication of gated devices based on this architecture (Oh et al., 2014). Photoresponsive field-effect-transistor analogs were realized upon the creation of Gate-all-round transistors using conventional CMOS compatible operations (Yong Oh and Saif Islam, 2014) as illustrated in Figure 4.3. The off currents of these devices were found to increase by over three orders of magnitude upon illumination with a wide spectral light source. The output current linearly increased with an increase in the illumination intensity. Due to the photosensitivity of these devices, digital and analog, electro-optical gate, and illumination frequency doubler devices were realized.

Oriented NWs have also been demonstrated by Dzbanovsky et al., and Englander et al., by using an electric field between two silicon surfaces (Dzbanovsky et al., 2005; Englander et al., 2005). An advantage of this technique is the ability to grow NW between any kind of electrodes and control the orientation. However, creating an electric field and sustaining silicon plasma between electrodes could present unique industrial challenges. Eventually, for large-scale integrated circuits or integrated sensor applications, the "bridging" process has the added advantage to meet industry requirements of low cost and high throughput. The in

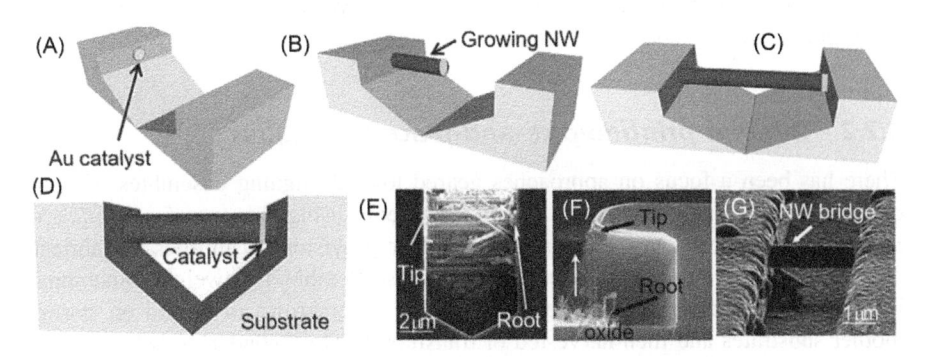

Figure 4.2 (A) A schematic illustrating lithography-aided deposit of a gold catalyst onto the side of the side wall of a trench. (B) Catalyst grown NW produces oriented NW directional growth. (C) Continued NW growth causes the NW to grow onto the opposing sidewall. (D) As the NW grows, the catalyst disperses away from the NW connection as the NW crystallizes to the opposing sidewall. (E) and (F) Lateral and vertical bridging of several Si NWs. (G) A single lateral Si NW bridge.

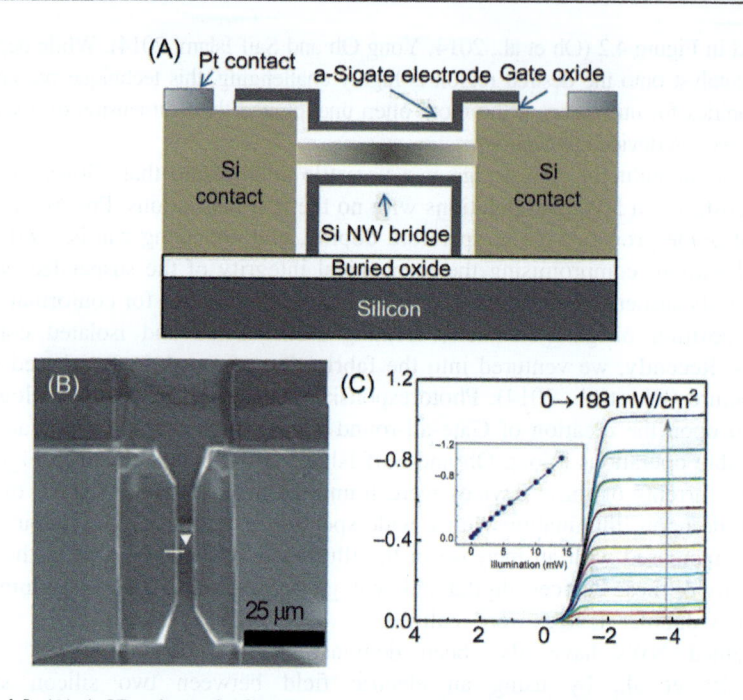

Figure 4.3 (A) A 2D schematic diagram of a suspended Si NW bridge Gate-all-around phototransistor and (B) an SEM micrograph of the device. The length and thickness of the nanobridges were 5 μm and 180 nm, respectively. (C) Output characteristics (I_S–V_{DS}) of the device with photo-gate configuration. The inset graph depicts a linear correlation between the applied wide spectral illumination and the collected photocurrent of the detector.

situ fabricated NW devices offer a massively parallel, self-assembling technique that allows controllable interconnections of the NW devices between electrodes using only relatively coarse lithography.

4.2.2 *Transfer-printing/pick-and-place techniques*

There has been a focus on approaches geared toward aligning ensembles of semiconductor NWs into large-scale complex systems using the *transfer-printing* or *pick-and-place* techniques more so due to their provision of abilities to fabricate devices on plastic substrates that are lower in cost, flexible, lightweight, biocompatible and possess optical transparency. NWs are either grown or etched on starting mother substrates and then harvested or transferred onto secondary substrates using the dry transfer (Yao et al., 2013; McAlpine et al., 2007; Chang and Hong, 2009; Jong-Hyun et al., 2007; Yuan et al., 2009; Javey et al., 2007; Bower et al., 2008), wet transfer (Huang et al., 2001a; Dorn et al., 2009; Kang et al., 2010) or contact printing (Baca et al., 2007; Baca et al., 2008; Kim and Rogers, 2008; Lee et al., 2007b; Yoon et al., 2008; Hines et al., 2007; Sun and Rogers, 2004; Kim et al., 2009; Qiang et al., 2008; Tunnell et al., 2008) techniques. These techniques have challenges and, therefore, advantages are highly localized. Additionally, most of the devices fabricated end up being single NW devices with stochastic characteristics.

A noteworthy approach by Lieber and coworkers entails a solution-based hierarchical organization of NWs to develop building blocks for nanosystems (Whang et al., 2004; Jin et al., 2004). The NWs are ingeniously aligned with controlled nanometer to micrometer scale separation using the Langmuir-Blodgett (LB) technique to facilitate the transfer of pregrown NWs to planar substrates in a layer-by-layer process, forming parallel and crossed NW structures over centimeter length scales. A subsequent patterning technique was used to connect the arrays to electrical contacts of controlled dimensions and pitch using photolithography as illustrated in Figure 4.4.

Lieber and his coworkers have further reported on a type of dry contact printing technique that provides for massively aligning NWs of dissimilar lengths on a substrate before contact deposition (Yao et al., 2013). Javey and his coworkers used a similar technique to transfer compositionally graded CdSxSe NW parallel arrays with tunable photodetection capabilities (Takahashi et al., 2012).

The above-described transfer methodologies render devices with laterally aligned semiconductor NWs. Such an orientation of semiconductor NWs robs them of a significant surface area for photon detection since approximately half their surface area is in contact with the substrates. This, however, can be addressed by transfer printing semiconductor NWs in their vertical orientation.

4.2.3 Transfer printing of horizontally oriented semiconductor nanowires

Emergent themes on the transfer of semiconductor NWs from their formative wafers into arbitrary substrates with retention of their horizontal orientation are powerful tools for maximizing PD sensitivity due to the increased surface area. Lewis recently described a protocol for cutting vertically oriented Si NWs from their formative wafer upon integration with a PDMS film using a blade (Plass et al., 2009). Photonic devices such as solar cells were fabricated thereof. We have further pioneered an even better approach that entails imprinting semiconductor NWs into

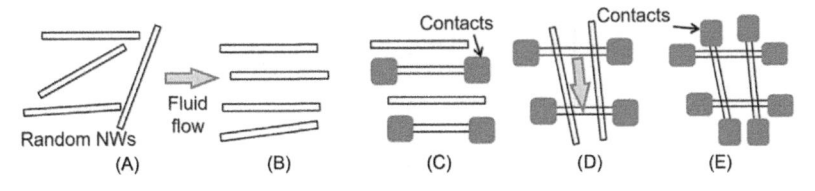

Figure 4.4 A schematic of LB-assisted technique for producing connections to NWs. (A) Randomly dispersed NWs on a substrate. (B) The NWs are moved and oriented in a single direction by flowing fluid. (C) To fabricate a device, contacts are deposited on either NWs end. (D) A subsequent secondary set of NWs is deposited orthogonal to the patterned NWs by the flowing fluid. (E) After a subsequent patterning, contact deposition is effected to yield the device.

thermally pliable polymer films and then instantaneously fracturing them from their mother substrate with retention of their orientation (Logeeswaran et al., 2010). Additional in-house built tools for precision control of this process have been custom-built (Triplett et al., 2014). Upon subtle modification, they can accommodate roll-to-roll transfer printing of semiconductor NWs and thus allow for high-speed industrial manufacturability. Moreover, this technique allows for the reuse of the mother substrate for the sequential fabrication of additional devices. A single wafer therefore could yield a large area of NW nanocarpets integrated onto arbitrary substrates. The precision tools we developed allow for other 1-D micro and nanostructures to be transferred (Triplett et al., 2014). Coupled with the versatility of choice of receiver substrates, flexible plastic films with photonic NW arrays can be easily fabricated inexpensively. Other transfer techniques similar to this have been reported by (Weisse et al., 2011) while we continue to develop other related protocols as illustrated in Figure 4.5 (Ombaba et al., 2014). Crozier et al., have subsequently exploited the quantum confinement effect of Si NWs to fabricate multispectral PD cameras (Seo et al., 2011; Park and Crozier, 2013; Park et al., 2012). They harvested pixelated SiNW arrays with different diameters and pitches embedded in PDMS from their mother wafer onto a readout circuit (Park and Crozier, 2013).

To date, truly wide spectral NW PDs have not been realized even though several studies have borne materials that are photoactive in different regions of the spectral

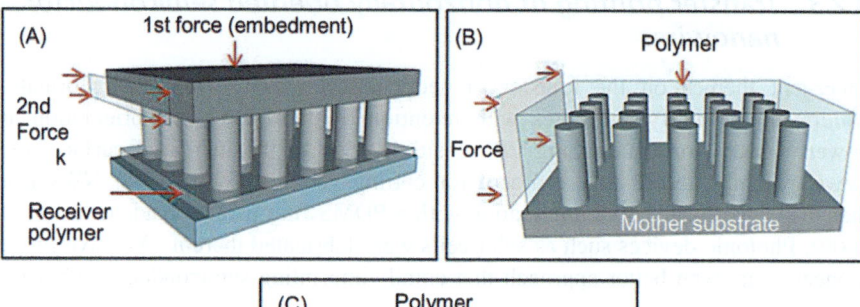

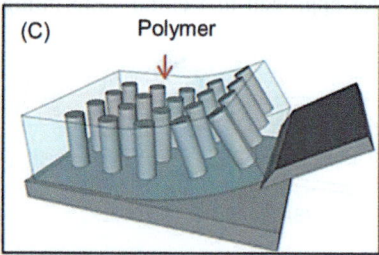

Figure 4.5 (A) Fracture transfer of semiconductor arrays onto secondary substrates by embedment onto a polymer film (Logeeswaran et al., 2010; Triplett et al., 2014). (B) A similar approach wherein the entire array is embedded in a transfer matrix with high impact resistance (Ombaba et al., 2014). (C) An alternative approach of detach semiconductor arrays from their mother wafer by cutting them with a blade.

band. A wide multispectral PD would need to be responsive to photons over a large spectral band rather than isolated regions within the said band. No amount of processing of a single photoactive material such as Crozier and Javey's approaches can bear such a detector (Park and Crozier, 2013; Takahashi et al., 2012). As such, combining several well-known photoactive materials into a single device and exploiting their collective photoactivity of the ensemble as a whole, would suffice.

4.3 Recent device demonstrations

There have been several genres of semiconductor NW PDs reported so far. This section will look at some of the notable contributions that have been made in the recent past.

4.3.1 Demonstrations of direct-growth photodetectors

So far, a considerable amount of literature has been published on NW-based PDs using the direct growth approach. Tsakalakos et al. reported the growth of randomly oriented Si NWs on a metal foil. We used a similar approach to fabricate an InP ultra-fast photoconductor on silicon dioxide with a response above 30 GHz (Logeeswaran et al., 2008). Designed by integrating it with a coplanar waveguide transmission line, the architecture facilitates DC and high-frequency measurements as shown in Figure 4.6.

Using a locked fiber laser with a wavelength of 780 nm as a pulsing source with a total output power of approximately 90 μW, and pulse width of 1 ps, and a repetition rate of 20 MHz, the top surface of the device was excited via free space coupling onto the active region of the photoconductor. The obtained FWHM from the oscilloscope was 16 ps and considering the 11.2 ps FWHM response for the 40-GHz oscilloscope and the laser pulse width of 1 ps, the device temporal response was estimated to be 11.4 ps at 780 nm based on Eq. (4.1) of Rush et al. (1990) which is given by Eq. (4.1).

$$\tau_{meas} = \sqrt{\tau_{actual}^2 + \tau_{scope}^2 + \tau_{optical}^2} \qquad (4.1)$$

where τ_{meas}, τ_{actual}, τ_{scope}, and $\tau_{optical}$ are the measured, actual, oscilloscope, and optical pulse widths in time, respectively. Since we have neglected the microwave components and laser timing jitter that contribute to the measured pulse width, the response estimate is truly conservative.

The direct growth approach was also employed for various optoelectronic devices. For example, (Huang et al., 2009) embedded Si NWs into a polymer (poly (3-hexylthiophene)) for enhancing the performance of hybrid photovoltaic devices. Wei et al. (2009) synthesized vertical InAs NWs on Si (111) substrates that were sensitive to visible and IR photodetection. InP NWs/polymer hybrid photodiodes

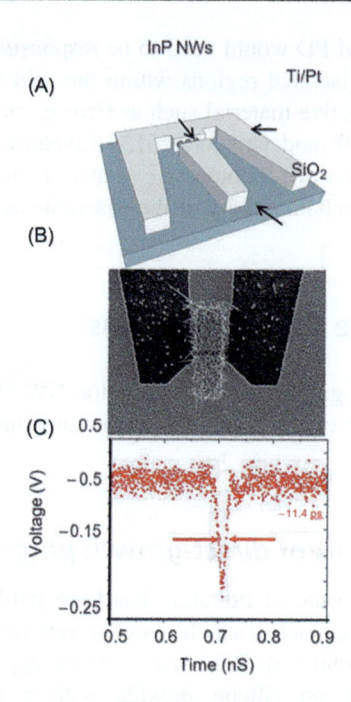

Figure 4.6 (A) A schematic of the high-speed InP NW PD. (B) An SEM micrograph image of the device the InP NW device. The random growth of the NWs on the μC-Si:H electrode is ascribed to the lack of a long-range translational crystallographic symmetry between the growth of the substrate and InP. (C) A pulse response ($\sim$11.4 ps) of the InP NW photoconductor (0.7 μm gap) was triggered by ultra-short 780 nm laser pulses of width 1 ps.

were fabricated (Novotny et al., 2008) as well as NIR PD based on indium nitride (InN) nanorod/poly (3-hexylthiophene) hybrids (Lai et al., 2010). Additionally, (nc-CdSe) NWs for detecting light as photoconductors (Kung et al., 2010), Ge NW PDs (Yu et al., 2006), a random network of silicon NWs (Huang et al., 2007), and nanostructured amorphous-silicon (a-Si:H)/polymer hybrid photocells (Gowrishankar et al., 2008) were also reported.

4.3.2 Waveguide coupled photodetectors

Greco and coworkers pioneered the fabrication of a monolithic waveguide-integrated photoconductor with Si-NWs on an amorphous substrate (Sonia et al., 2009; Grego et al., 2011). This could potentially bear new building blocks for a self-contained CMOS-compatible photonic chip for light guiding and detection of high-speed optical interconnects (Wu et al., 2009; Haurylau et al., 2006; Hosako and Hiromoto, 2004; Hu et al., 2009; Chang et al., 2007). The waveguide was composed of a SiOxNy core and an amorphous SiO$_2$ cladding.

Device photoresponsivity was measured by probing the two electrodes bridged by the Si NWs while coupling the waveguide with a 2 mW laser input of wavelength 780 nm. Devices with NWs of different diameters were prepared to elicit dimensional effects on device photodetection performance. Photoresponsivity was found to correlate to NW density irrespective of NW diameter, attributed to high bridging connections of the denser NWs or lower optical loss via optical scattering. Such devices could therefore benefit from improved NW growth density and bridging modalities. The measured responsivity of the device with denser NWs was about 0.03 A/W at 5 V bias for a waveguide illumination of 300 μW as depicted in Figure 4.7.

Further progress in this device can be realized based on the understanding of how passive waveguides and NWs interact with photons at the optical index transition, and how propagating optical modes from a micron-size waveguide switch to nanoscale semiconductor wires, and the impact of misorientation and nonuniformity

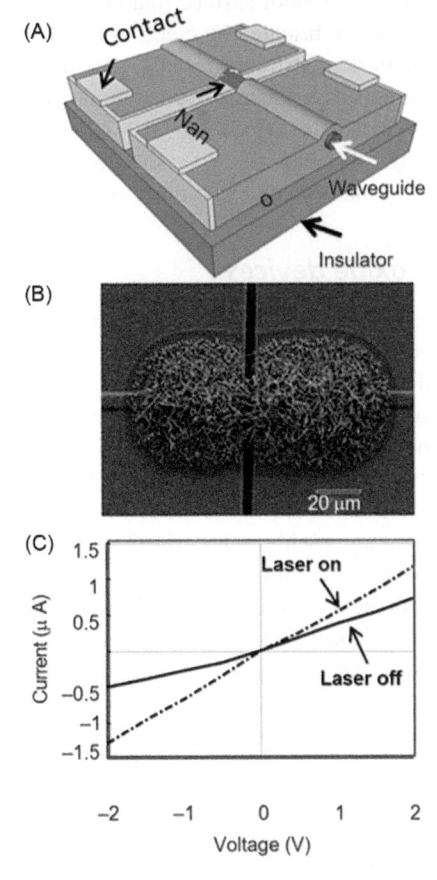

Figure 4.7 (A) Schematic of the waveguide PD where InP NWs are the core photoactive material. (B) Top view micrograph of the bridged NWs integrated waveguide device grown at 680°C. (C) Current-voltage characterization of waveguide integrated NW photoconductors across a 7 μm trench when illuminated with a 2 mW laser input of wavelength 780 nm.

in the distribution of the NWs. Photon propagation in a waveguide carries a specific polarization and as such NWs oriented along the direction of the waveguide will demonstrate varying coupling efficiency for different polarizations.

4.3.3 Plasmonic photodetectors

An exciting new research area in nanoelectronics is represented by NW plasmonics. There have been notable efforts by researchers geared towards creating nanoscale photonic circuits based on plasmonics. Controlling surface plasmons and guiding them through a specific path could lead to the optical equivalent of electronic circuits. Several studies of optical waveguide miniaturization using plasmonics have been undertaken (Maier et al., 2003a,b). Silver NWs have unique properties that make them particularly attractive for nanoscale confinement and guiding of light to nanoscale objects due to their smooth surface that contributes to lower propagation loss than metallic waveguides fabricated by electron-beam lithography (Pyayt et al., 2008). Falk and coworkers demonstrated an NW plasmonic PD that consists of a silver NW (Ag NW) that acts as a plasmonic waveguide and a crossing Ge NW that is connected to two metal pads (Falk et al., 2009). Current pick-and-place approaches to transferring NWs can allow for the coupling of individual semiconductor NWs to a single Ag nanorod after the detector fabrication.

4.3.4 Photoactive oxide devices

In recent years, one-dimensional (1D) semiconductors, including NWs, nanobelts, nanotubes, and nanorods, have received significant attention due to their unique physical, chemical, and mechanical properties that can be used for the development of many high-performance electronic and optoelectronic devices at the nanoscale, such as solar cells, diodes, biosensors, transistors, displays, PDs, magnetoresistive sensors, and light emitting diodes (LEDs) (Florica et al., 2015; Florica et al., 2014; Zhang et al., 2016; Enculescu et al., 2007; Costas et al., 2019; Chen et al., 2020; Barrigón et al., 2020). Among a wide range of 1D nanostructured materials, metal-oxide NWs, including TiO_2, ZnO, SnO_2, and Ga_2O_3, have gained a particular interest in the development of next-generation optoelectronic devices because of their peculiar properties, such as their wide band gap, simple fabrication process, high optical transparency, good mechanical stress tolerance, strong adsorption ability, large area electrical uniformity, superconductivity, scalability, high surface reaction activity and radiation resistance, high chemical stability and excellent carrier mobility (Devan et al., 2012; Yu et al., 2016). Metal-oxide NWs have recently been largely employed as PDs, which use the photoelectric effect to convert the light signal into an electrical signal, for light detection ranging from ultraviolet (UV) to near-infrared (NIR) frequencies (Ouyang et al., 2019). In particular, the UV PDs, detecting light in the wavelength range of 100–400 nm effectively, have attracted much research interest in the past decade for a wide range of applications, such as flame detection, space exploration, air purification, ozone sensing, leak detection, missile plume detection, UV communication, medical imaging, power grid safety monitoring, defense warning system, industrial production, environmental testing and healthcare (Xu et al., 2019b; Zou et al., 2018; Chen et al., 2015a). Due to the unique properties of NWs, such as large surface-to-volume ratio, isolated optical absorption, direct carrier transport paths, high mobility, efficient light trapping, light polarization sensitivity, and high photoconductivity gain, the integration of metal oxide NWs into UV PDs is expected to greatly boost the light detection performance of NWs-based devices over the PDs constructed with bulk films or crystals (Soci et al., 2007; Soci et al., 2010; Logeeswaran et al., 2011a,b). Until now, various types of metal oxide NWs, such as ZnO, CuO, Ga_2O_3, TiO_2, and SnO_2, have been successfully synthesized via many fabrication methods and widely used for light detection from UV to NIR regions via different photodetector device configurations, such as Schottky barrier, photoconductive, p-n junction and p-i-n junction types (Zou et al., 2018; Tian et al., 2015; Monroy et al., 2001; Liu et al., 2010), the schematic structure of which are given in Figure 4.8. Therefore, in this section, we provide a brief overview of recent research conducted on several common metal-oxide NWs-based PDs.

4.3.4.1 ZnO nanowires photodetectors

Zinc oxide (ZnO) is considered one of the most ideal metal-oxide semiconductor materials for optoelectronic applications such as solar cells, gas sensors, and field emission devices, especially for UV PDs due to its wide band gap of 3.37 eV at room temperature, a large exciton binding energy of 60 meV, high chemical

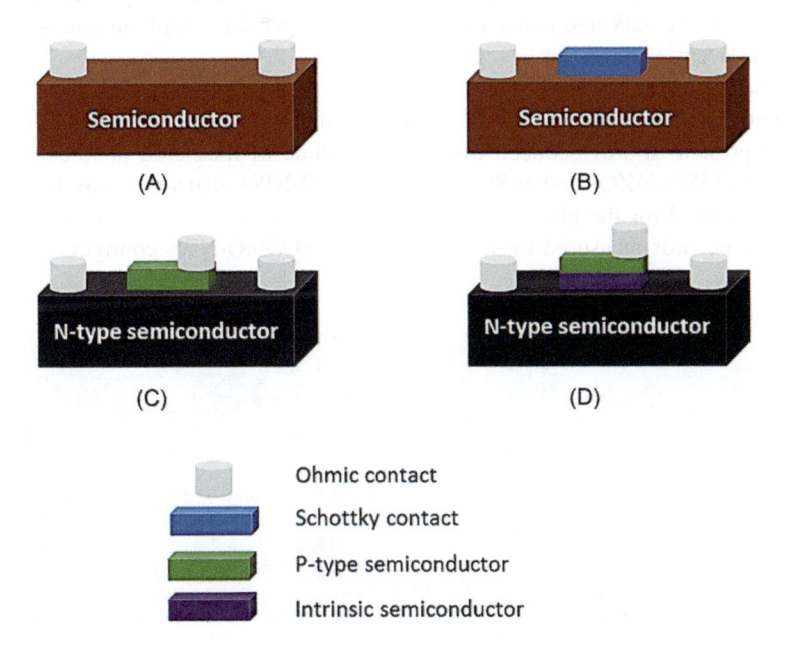

Figure 4.8 Schematic device structure of (A) photodetector, (B) Schottky barrier, (C) p-n junction, and (D) p-i-n junction type photodetectors (Monroy et al., 2001; Liu et al., 2010).

stability and ease of manufacturing (Pearton et al., 2003; Soci et al., 2007; Lu et al., 2018; Yin et al., 2018; Zou et al., 2018). Compared to ZnO film, ZnO NWs have been reported to have much greater photo-electronic sensitivity when exposed to UV illumination owing to their large surface area relative to volume ratio, and the presence of deep level surface trap states, and excellent carrier transport channel (Zou et al., 2018). The defects that occur during the synthesis of some materials limit their potential applications. In this regard, ZnO is considered to be an important material for a wide range of applications because it can be synthesized with comparatively low defect concentrations and fairly simple approaches.

In 2013, we studied 3D micro/nano p-n heterojunctions between ZnO NWs and Si micropillar/walls with the aim of fabricating high-performance PDs (Karaagac et al., 2013). Figure 4.9A and B show SEM images of ZnO NWs grown on p-Si microwalls (MWs) and n + -Si (highly-doped) micropillars (MPs), respectively. First, high aspect ratio vertically oriented p- and n-silicon MPLs and MWLs were synthesized using the deep reactive ion etching process. Following the deposition of AZO thin film using the hydrothermal technique, vertically oriented dense ZnO NWs arrays were grown onto the AZO precoated p- and n-type Si MPLs and MWLs. This resulted in p-n heterojunctions as tested for UV and visible light sensing.

The current (I)–voltage (V) characteristics of the 3D PDs having the p-Si-MWL/AZO/ZnO-NWs configuration were determined in the dark and under light illumination of both UV and white light at room temperature. The $I-V$ characteristics of planar-p-Si/AZO, p-Si-MWL/AZO, and p-Si-MWL/AZO/ZnO-NWs heterojunctions measured in the dark and under white light (a halogen lamp with an intensity lower than 5 mW/cm^2) are shown in Figure 4.10A.

A clear standard diode behavior was observed in all structures both in the dark and under white light illumination. This verified that p-n heterojunctions were indeed present. It was deduced that the magnitude of measured photocurrents for both p-Si-MWL/AZO, and p-Si-MWL/AZO/ZnO-NWs structures was larger than those measured for the planar-p-Si/AZO. In addition, it was observed that a higher photocurrent was measured for the p-Si-MWL/AZO/ZnO-NWs configuration, when

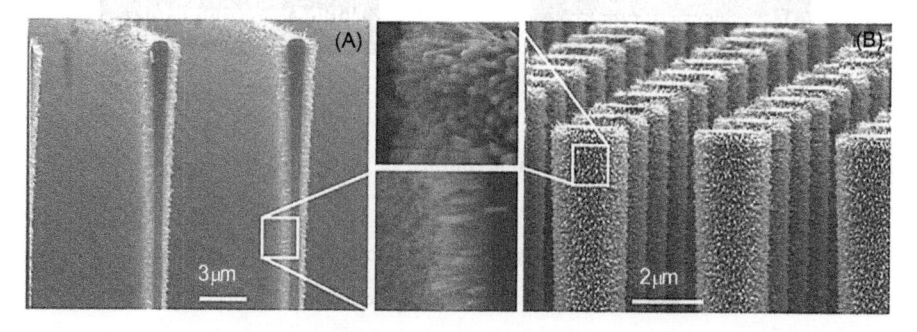

Figure 4.9 Scanning electron microscopy images showing ZnO NWs grown on (A) p-Si MWLs and (B) n + -Si MPLs using the hydrothermal technique (Karaagac et al., 2013).

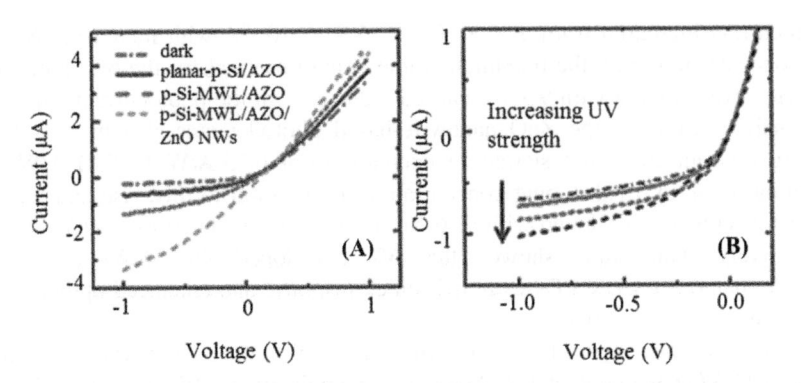

Figure 4.10 (A) I–V characteristics of p-Si wall/AZO/ZnO NWs heterojunction measured in the dark and under a white light with an intensity lower than 5 mW/cm^2. (B) The I–V characteristics of the PD with the p-Si wall/AZO/ZnO NWs configuration, were recorded in the dark and under different UV light intensities (Karaagac et al., 2013).

compared to the p-Si-MWL/AZO, which highlighted the importance of a combination of micro/nanostructures to improve the performance of conventional planer PDs by either reducing light reflection deduced from the reflectance measurements or enhancing free-charge collection efficiency. Figure 4.10B shows the rectification under both conditions (dark and UV illumination), which is indicative of the formation of a p-n heterojunction between p-Si MWLs and ZnO NWs. It is clear that the reverse current increased gradually with increasing UV light intensity. The observed increase in reverse current is indicative of the photogeneration of electron-hole pairs in ZnO NWs and their collection through the built-in electric field formed between the p-Si walls and ZnO NWs. Since more electron-hole pairs are photogenerated as the UV light intensity increases ($UV_3 > UV_2 > UV_1$), an increase in photocurrent was observed explicitly. Based on these observations, we suggested that the constructed 3D p-n heterojunction could be used to sensitively detect UV light.

It is well known that doping of ZnO with appropriate metal ions can enhance the sensing, electrical, structural, optical, and morphological properties of the material (Peksu and Karaagac, 2018; Yang et al., 2017; Al Farsi et al., 2021). In 2020, Raj and coworkers fabricated Nd-doped ZnO NWs with different Nd concentrations using chemical bath deposition for photodetector applications. A photoconductor-type UV photodetector was constructed using silver (Ag) as metal electrodes. It was determined that the crystal quality and surface area increased by doping ZnO NWs with 6% Nd. Especially the increase in the surface area provides high absorption, which is expected to increase the photodetection performance of the PDs. When the I-V characteristics were examined in the dark and under UV illumination (between -5 V and 5 V), the measured dark current in the dark for the Nd-doped NWs was found to be slightly higher than obtained for the device constructed with pristine NWs, which was attributed to the presence of excess electrons in the doped ZnO NWs. As exposed to the UV light, however, both sensors exhibited a significant increase in current as compared to the dark current. It was also revealed that the photocurrent increased from 5.37×10^{-7} to 3.77×10^{-6} A with increasing Nd doping concentration, with the maximum value being observed for the 6% doped sample. However, doping with 9% Nd caused a decrease in the photocurrent value. The responsivity value of the ZnO nanowire-based photodetector also increased with doping concentration and started to decrease from 0.73 A/W to 0.41 A/W after reaching up % to 6 Nd doping concentration. In addition to this, rise and fall time for the ZnO nanowire doped with % 6 Nd have been calculated to be 0.3 and 1.8 s, respectively. This study showed that 6% Nd doped ZnO NWs are a very suitable material for UV PDs with their quick response and enhanced optoelectronic properties (Raj et al., 2020).

In another study, hydrothermally grown and well-aligned ZnO NWs were doped with Al (1.57 at %) content for photodetector applications. When Hsu et al., compared the responsivities of ZnO and Al-doped ZnO NWs, they found that the responsivity of AZO NWs was twofold greater than that of the ZnO NWs in the UV region. The spectral responsivity was carried out under 360 nm UV illumination. The max responsivity of AZO NWs was calculated to be 3.61 A/W, while the max responsivity of ZnO NWs was calculated to be 1.824 A/W at a bias of 5 V. Additionally, quantum efficiency was estimated to be 84.9% for the AZO NWs (Hsu et al., 2017).

Te doped ZnO NWs photodetector has been studied by Khosravi-Nejad et al., ZnO NWs were synthesized through physical vapor deposition. In this study, the ZnO NW photodetector exhibited a fast dynamic response and showed a response time of 57 s (Khosravi-Nejad et al., 2019).

In 2018, L. Yin et al. (2018) successfully demonstrated the fabrication of a high-performance transparent UV photodetector based on ZnO NAs. ZnO NWs were synthesized on an FTO-glass substrate (interdigital patterned electrode) by chemical bath deposition at low temperatures. The performance of the photodetector was determined under UV light of 0.1 mW/cm^2 (365 nm) at different voltages ranging from 0 to +5 V. The responsivity, response time, and decay time of the fabricated UV photodetector were found to be 113 A/W, 23 s, and 73 s, respectively. The detected high responsivity was attributed to the large surface-to-volume ratio of the ZnO NAs as they provide the direct electrical pathways for the photo-generated carriers.

In a study conducted by Yang et al., the ZnO NWs were assembled on Pt electrodes with a photoresist to enhance the photo response and recovery speed of UV PDs. Prior to photoresist coating, it was observed that the photocurrent under UV light increased abruptly, then decreased after a few seconds. The reason for this rapid increment is the photogenerated e-h pairs that occur with UV illumination. Then the decrement in the rate of increase can be attributed to the oxygen molecules' desorption from the surface of the ZnO NWs following the UV light illumination. Once UV light is turned off, the photocurrent decreased sharply at first and could not reach its initial value even after seconds. Since the absorption rate of the oxygen molecules is quite low to initiate the photocurrent, the photocurrent rate may not return to its original level rapidly. According to the carrier transport mechanism model shown in Figure 4.11 (Dogar et al., 2016), oxygen molecules from the

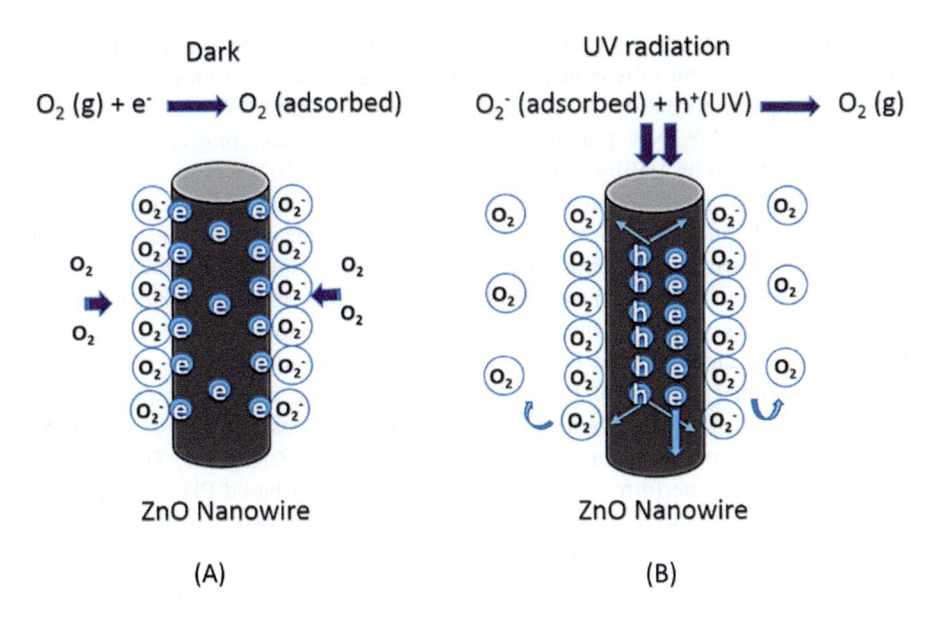

Figure 4.11 (A) Adsorption of O_2 in the dark condition and (B) desorption of O_2^- in the presence of UV illumination (Dogar et al., 2016).

atmosphere are adsorbed by the n-type ZnO NWs and capture free electrons from ZnO NWs to form O_2^- ion, which subsequently triggers the formation of a depletion region near the surface of ZnO NWs. When the light is turned on, e-h pairs are created and some O_2^- ions lose their electrons and migrate from the nanowire surface. If the photoresist is coated on the NWs, however, very few oxygen molecules are adsorbed. Furthermore, because of the presence of the photoresist film, oxygen molecules that lose electrons would be unable to escape from the NW surface when the light is turned on. As a result, when the UV light is switched off, the photocurrent immediately returns to its initial level. Following the photoresist coating, it was revealed that when exposed to UV radiation, the photocurrent increases 20 times in 3 s and returns to its initial level in 20 s. These findings indicated that photoresist coating had a substantial impact on detector photoresponse and recovery time (Yang et al., 2016).

Z. Zhan et al. reported a high-speed UV photodetector based on ZnO NWs grown via catalyst-free chemical vapor deposition (CVD) method. The UV PDs were fabricated on both entangled ZnO NAs and single ZnO NW. For the case of entangled ZnO NAs-based PD, Ag paste was applied to both sides of ZnO NWs as ohmic electrodes. A dielectrophoretic process was used to integrate a single ZnO NW into a UV photodetector. Following the dispersion of ZnO NWs into isopropanol, a droplet of this solution was dropped over the interdigital microelectrodes. ZnO NWs bridging was then achieved by applying a sinusoidal voltage (5 V at 103 kHz) supplied by a function generator for a 60 s time duration. The results demonstrated that high-speed UV PDs based on entangled ZnO NW arrays synthesized on SiO$_2$ using a catalyst-free CVD process could be manufactured. The observed improvement in response speed of photodetector based on entangled ZnO NAs as compared to the one constructed with a single ZnO NW was attributed to the height of the NW-NW junction established between interconnected ZnO NWs that could be tuned by UV light illumination (Zhan et al., 2017).

In 2021, Noh et al., fabricated a vertically aligned and closely packed ZnO NWs-based UV photodetector to reveal the changes inside the nano-material when repeatedly exposed to an intense pulse light (IPL). ZnO NWs were grown on SiO2 substrate using a hydrothermal technique. In order to improve the electrical and optical properties of the synthesized NWs, they were subjected to IPL exposure treatment. Both untreated and IPL-treated ZnO NWs-based PDs were tested under UV illumination (365 nm) of 0.85 mW/cm^2 at 1 V. The results showed that the electrical output of the IPL-treated ZnO NWs with a processing time of just 6.25 s was 3.31 times better than that recorded for the untreated ZnO NWs. The observed improvement in the performance of IPL-treated ZnO NWs-based PD was attributed to the controlled surface/bulk defect ratio following the post-IPL treatment, which triggered the formation of a narrow conduction channel in the dark and widened conduction channel under UV light illumination due to the improved crystallinity via the IPL-treatment (Noh et al., 2021).

In addition to the known photodetector types such as pn-junction (or p-i-n) and Schottky barrier, self-powered PDs with photoelectrochemical structure have attracted considerable attention in recent years due to their advantages such as high sensitivity, fast response and its ability to provide high performance at zero bias voltage. Liquid electrolytes are used for photoelectrochemical type PDs and unfortunately, liquid electrolytes can damage the device due to stability problems. In 2019, Xie et al. fabricated a photoelectrochemical-type UV photodetector composed of ZnO NWs embedded in a liquid crystal instead of using liquid electrolytes to overcome this problem. ZnO nanowire arrays (NAs) were grown on fluorine-doped-tin-oxide (FTO) precoated glass substrates via the hydrothermal method. In the constructed device structure, the synthesized ZnO NAs and 20 nm Pt thin film deposited on FTO served as active photoanode and counter electrodes, respectively. A special mixture of electrolytes (LC-embedded) melted at 60°C was then injected into the space between the photoanode and counter electrode before assembly. The performance of the detector was tested under UV light provided by a 365 nm LED and the generated photocurrent was recorded by an electrochemical workstation. From the I-V characteristics recorded in the dark condition, it was realized that the device was exhibiting a typical Schottky-barrier type diode behavior with a 0.4 V forward turn-on voltage. Under illumination with UV light of 2 mW/cm^2 (365 nm), the detector exhibited a perfect photovoltaic effect, with open-circuit voltage (V_{oc}) and short circuit current (I_{sc}) of 0.32 V and 18 μA, respectively, suggesting that the detector could be operated at both photodiode mode (at reverse bias) and photovoltaic mode (at zero bias). The constructed detector had a high responsivity and a fast photoresponse, which was ascribed to the light trapping feature of the electrolyte injected into the gap between the ZnO NAs photoanode and Pt counter electrode. The responsivity and IPCE values at the wavelength of 365 nm were calculated to be 0.05 A/W and greater than 17%, respectively. It was also found that the measured photoresponsivity is two orders of magnitude higher than that achieved with a self-powered UV photodetector constructed with ZnO nanorod arrays (Shen et al., 2015; Boruah et al., 2016). Moreover, the obtained IPCE value was found to be quite higher (3 times) than that achieved with a hybrid (organic-inorganic) ZnO nanorods-based photodetector (Game et al., 2014). The rise time and decay time constants were calculated from the transient photoresponse under chopped illumination of UV light of 365 nm and found to be 0.15 and 0.05 s, indicating the fabrication of a long-term stable UV detector with a rapid response property. This study is an important step toward the realization of next-generation high-performance, inexpensive, and high-speed self-powered UV PDs (Xie et al., 2019).

In 2016, a self-powered UV-Visible photodetector with ZnO/Cu2O nanowire-electrolyte device architecture was successfully manufactured by Bai et al., Zinc-oxide NAs were synthesized on an FTO-coated glass substrate via hydrothermal technique. The grown ZnO NAs were then decorated with branched Cu_2O NWs using chemical bath deposition, which was followed by a thermal reduction process. A 50-nm thick Pt film deposited on the FTO layer was assigned as the counter electrode. A thermoplastic hot-melt sealing foil was used to adhere the ZnO/Cu_2O photoanode to Pt/FTO counter electrode. The gap between the electrodes was then filled with a Na_2SO_4 aqueous solution, functioning as an electrolyte. It was found that the PDs at zero bias exhibited an excellent performance not only in the UV region but also in the visible region. The on/off ratios for UV light of 6 mW/cm^2 (355 nm) and visible light of 25 mW/cm^2 ($>$425 nm) was calculated to be 525 and 1945, respectively. Furthermore, the rise and decay times were determined to be 0.14 s and 0.36 s, respectively, indicating that photoelectrochemical class PDs have a high potential in high-speed UV-visible light detecting applications (Bai and Zhang, 2016).

A study on the fabrication of a fully wide band gap type-II ZnO/ZnS core-shell structured UV-Visible photodetector was reported by Rai et al., ZnO NAs were grown on SiO_2 passivated indium-tin-oxide (ITO) precoated glass substrate using thermal evaporation of Zn powder. To construct the ZnO-NAs/ZnS core-shell structure with a total active area of 30 mm^2, the synthesized ZnO NAs on the ITO were then coated with a thin layer of ZnS using a pulsed laser deposition system. A 30-nm thick Ag film was then deposited on top of the NAs as the top electrode. For a 0.4 kgf compressive load (at 1.5 V bias), the manufactured photodetector exhibited 2.5, 0.54, and 0.13 A/W responsivities under UV light of 1.32 mW/cm^2, blue light of 3 mW/cm^2 and green light of 3.2 mW/cm^2, respectively, suggesting at least one order of magnitude higher responsivity than that observed under no load conditions. The observed improvement in responsivity under an externally applied load was attributed to the piezo-phototronic effect, which had a significant effect on bending the valance band and conduction band edges to higher energies at the ZnO-ZnS interface. The upward band bending eliminated a preexisting barrier that was preventing an effective charge carrier separation in ZnO and ZnS. The elimination of this barrier, therefore, resulted in an increase in the generated photocurrent at the same bias and illumination compared to the one noted with no strain (Rai et al., 2015).

So far, many approaches, such as annealing and doping, have previously been employed to improve the efficiency of ZnO nanowire-based PDs. Despite the fact that these approaches led to success, the expected improvements in performance could not be achieved. Jiang and colleagues improved the efficiency of ZnO nanowire-based UV PDs using spin-coated nanodiamonds, known to be a significant member of the carbon family. Nanodiamond is an excellent light collector in UV PDs due to its wide band gap. ZnO NWs were synthesized on the SiO_2/Si substrate using the CVD method. The nanodiamond solution prepared from commercial nanodiamond powder was coated on the ZnO NWs with a spin coater. The I-V characteristics of the device were carried out under 365 nm UV illumination (14 mW/cm^2). Under a bias of 5 V, the dark current and the light current of nanodiamond/ZnO nanowire heterostructure UV photodetector were measured to be 2.8 μA and 29.5 μA, respectively. However, after 20 min of annealing in a vacuum at 600°C, the light current value increased 11 times (327.8 A), which was 16.5 times higher than that obtained with a pure ZnO NWs-based photodetector. The I–V curve demonstrated that the Schottky contact was established between the NWs and silver electrodes. The Schottky barrier formed in the contacts both accelerated the separation of photogenerated electron-hole pairs and reduced the rate of electron-hole recombination, which contributes to the enhancement of photocurrent and response speed of the manufactured photodetector. In addition, a high photocurrent responsivity of 0.59 A/W and a high photocurrent intensity of 819.75 A/cm^2 was achieved with such a device configuration (Jiang et al., 2018).

4.3.4.2 Ga_2O_3 nanowires photodetectors

Gallium oxide (Ga_2O_3) is a semiconductor with a wide bandgap of 4.8eV-making making it a suitable material for harsh environment applications. Compared to Si, SiC, and GaN, it has great potential to perform in high power and optoelectronics applications due to its high breakdown voltage and low on-resistance. Hence, Ga_2O_3 has received comparatively greater recognition among wide bandgap materials for diverse applications such as high power, high voltage, high-temperature devices, solar-blind UV detectors, and deep UV sensors. Wide bandgap-based sensors are an exceptional choice for extreme temperatures above 600°C and have been demonstrated to operate at even 1000°C (Maier et al., 2012; Tanaka et al., 2015).

Although Ga_2O_3 is highly desirable for use in semiconductor devices, the cost of Ga_2O_3 wafers is still prohibitive as shown in Figure 4.12. Furthermore, methods for the top-down fabrication of Ga_2O_3 nanostructures are still immature. Hence, a wide variety of processes to grow Ga_2O_3 NWs have been explored, including thermal oxidation (Patil-Chaudhari et al., 2017; Bayam et al., 2015), VLS growth, molecular beam epitaxy (Ghose et al., 2016), laser ablation (Feng et al., 2015) and CVD (Pallister et al., 2015) and the hydrothermal method (Zhao et al., 2008; Reddy et al., 2015).

Recently, the interest of researchers has been focusing on the fabrication of low-dimensional monoclinic crystal structure β-Ga_2O_3 NWs. Compared to bulk structures, β-Ga_2O_3 NWs exhibit a higher surface-to-volume ratio, leading to more surface states at the interface and therefore increasing the sensitivity of detection.

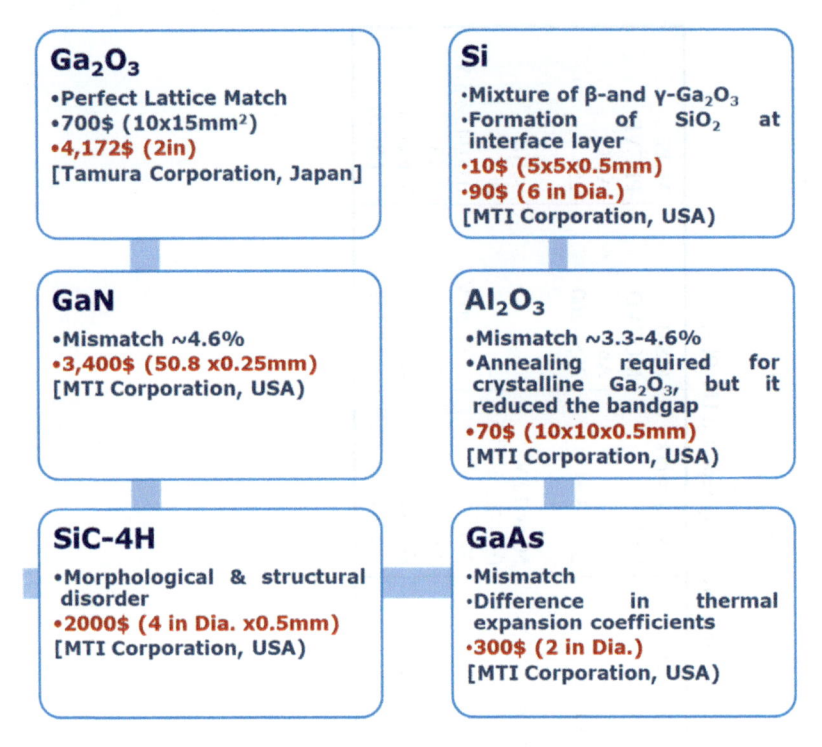

Ga$_2$O$_3$
- Perfect Lattice Match
- 700$ (10x15mm²)
- 4,172$ (2in)
[Tamura Corporation, Japan]

Si
- Mixture of β-and γ-Ga$_2$O$_3$
- Formation of SiO$_2$ at interface layer
- 10$ (5x5x0.5mm)
- 90$ (6 in Dia.)
[MTI Corporation, USA]

GaN
- Mismatch ~4.6%
- 3,400$ (50.8 x0.25mm)
[MTI Corporation, USA]

Al$_2$O$_3$
- Mismatch ~3.3-4.6%
- Annealing required for crystalline Ga$_2$O$_3$, but it reduced the bandgap
- 70$ (10x10x0.5mm)
[MTI Corporation, USA]

SiC-4H
- Morphological & structural disorder
- 2000$ (4 in Dia. x0.5mm)
[MTI Corporation, USA]

GaAs
- Mismatch
- Difference in thermal expansion coefficients
- 300$ (2 in Dia.)
[MTI Corporation, USA]

Figure 4.12 The price and features of Ga$_2$O$_3$ substrate compared to other substrates to grow which used to grow Ga$_2$O$_3$.

Ga$_2$O$_3$ NWs have been used in numerous applications such as PDs (Lopez et al., 2014; Alhalaili et al., 2020c), and gas sensors (Wu et al., 2014), and transistors (Yu et al., 2010). Various types of photoconductors have been made with Ga$_2$O$_3$ including MSM photodiodes, Schottky diodes, p-n junctions, p-i-n junctions, and avalanche photodiodes. These types and materials have been discussed and reported in detail in ref. (Alhalaili et al., 2018).

Ga$_2$O$_3$ NWs have shown better performance in the applications of nanosensors (Zhang et al., 2015). These Ga$_2$O$_3$ NWs also have the potential to increase light absorption and confinement to improve their photosensitivity (Dai et al., 2014). Studies have also focused on the synthesis and fabrication of Ga$_2$O$_3$-based UV PDs (Table 4.1), and research has increased to investigate the electrical properties of UV PDs. It has been shown that the presence of Ag nanoparticles can improve the electrical properties of the β-Ga$_2$O$_3$ NWs (Alhalaili et al., 2020b,c). The listed parameters are crucial for evaluating the performance of UV PDs:

4.3.4.3 Growth process of Ga$_2$O$_3$ nanowires

Thermal oxidation is a process performed at high temperatures within a furnace. This technique is simple, inexpensive, and leads to faster growth. In

Table 4.1 Summary of β-Ga$_2$O$_3$ device performance of the present device and other previously reported UV PDs.

Device structure	MSM	MSM	GR/oxide/GR	NW network	MSM	MOS
Fabrication method	CVD	MBE	LMBE	CVD	MOCVD	PECVD
Electrode	Cr/Au	Ti/Al		Au	Au	Au/Cr
Light intensity (nm)	280–450 nm	255	254	290–340	255–260	255
I$_{Photo}$/I$_{Dark}$	38.3 @10 V	5.58 × 10^4	82.88	50 @ 10 V	-	<10^3
Td [s]	0.33s–2.2 s	4.00		0.2–7.1		4
R [A/W]		54.9@ 10 V	9.66 @10 V	377	17@ 20 V	0.00343@5 V
Year	2020	2017	2017	2016	2015	2013
Reference	Alhalaili et al. (2020c)	Qian et al. (2017)	Ai et al. (2017)	Du et al. (2016)	Hu et al. (2015)	Wu et al. (2013)

addition, it generally will trigger the self-assembly growth of NWs (Mao et al., 2017). However, it has some disadvantages such as the potential for contamination from the furnace, possible difficulty in catalyst selection due to high-temperature constraints, and limitations in controlling the film deformation by thermal stress due to expansion mismatch. Furthermore, it increases the number of defects and dislocations within the substrate. However, using a catalyst that can withstand higher temperatures beyond the Ag catalyst melting point could be a beneficial choice to increase the rate of the growth mechanism spontaneously.

Ga_2O_3 NWs were grown at 1000°C under the presence of an Ag catalyst in different substrates such as p-Si substrate doped with boron, 250 nm SiO_2 on n-Si, 250 nm Si_3N_4 on p-Si, quartz, n-Si substrates (Alhalaili et al., 2020a) and GaAs (Alhalaili et al., 2020b). The detailed growth process and the intense characterization performed on the oxidation of Ga on quartz substrate (with and without Ag) are explained in detail in ref (Alhalaili et al., 2020c). Figure 4.13 summarizes the fabrication process of the UV photodetector. In brief, the suberates will be sputtered by 5 nm Ag and positioned to face the Ga pool into the furnace. Then, the substrates are heated up to 1000°C for 60 min, leading to the growth of the NWs (Figure 4.13B). Finally, a shadow mask was used with gold metal contacts on the surface of Ga_2O_3 NWs (Figure 4.13C).

The growth mechanism was improved by the Ag catalyst. All the morphological, structural, optical, and electrical characterizations are presented in detail in ref. (Alhalaili et al., 2020c). The Ag catalyst leads to the formation of a homogeneous coating and a dense growth of Ga_2O_3 NWs. In addition, the catalyst assists in reducing the diameters of the Ga_2O_3 NWs and increasing their density. In the case of Ag-free growth, it has been shown that the diameters of NWs range from 150–270 nm (Figure 4.14A). When Ag was applied, the diameters were reduced to 120–160 nm (Figure 4.14B) (Alhalaili et al., 2020c). Consequently, using Ag as a catalyst improves the synthesis of Ga_2O_3 NWs by producing a majority of NWs with a thinner diameter and longer length.

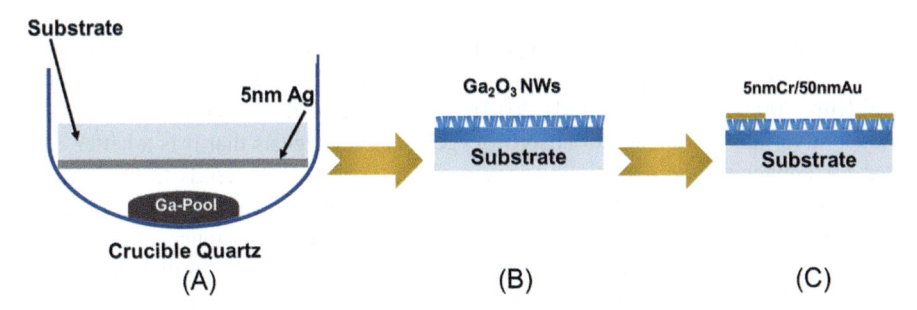

Figure 4.13 Experimental Process of the UV detector fabrication steps. (A) substrate fabrication process. (B) Ga_2O_3 nanowires growth. (C) metal contacts deposition.

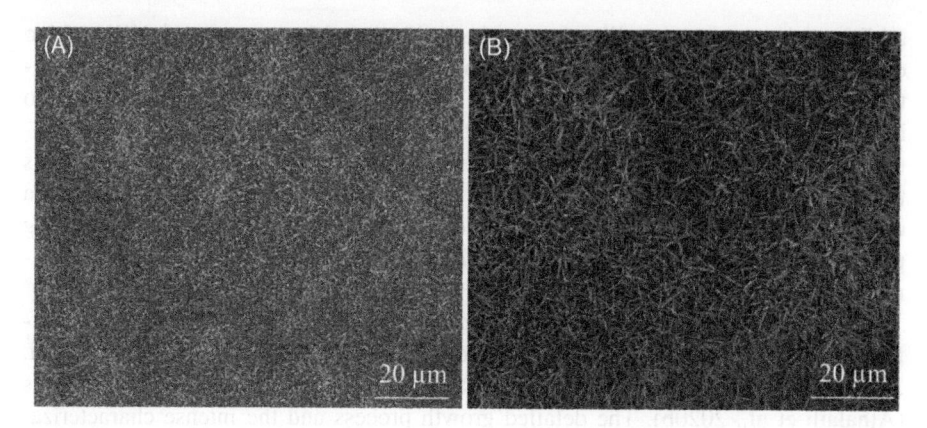

Figure 4.14 Ga_2O_3 nanowires grown at 1000°C. (A) Ag-free quartz. (B) 5 nm Ag deposited on quartz.

4.3.4.4 Silver catalyst

Silver has been used as a catalyst to enhance growth and improve its properties (Nguyen et al., 2015; Yoshida et al., 2018; Kanika Arora et al., 2018). Ag thin films with their simple fabrication could be a potential alternative to nanoparticles as catalysts for Ga_2O_3 nanowire growth (Alhalaili et al., 2019). Silver is an effective catalyst to enhance oxygen absorption. Due to the combination of high oxygen diffusivity and solubility at temperatures above the melting point of Ag ($T = 961.8$°C), more accumulation of atmospheric O_2 leads to effective transport into Ga. Since, Ga has a higher affinity for O_2 molecules than Ag, resulting in more O_2 diffusion into the Ag (Alcock and Jacob, 1977). Consequently, as Ag nanoparticles are incorporated into Ga molecules, there are more O_2 molecules to react with Ga, leading to a dense growth of Ga_2O_3 NWs. However, other potential metal catalysts are unsuitable due to either a lack of solubility (such as in Au) or diffusivity (such as in Fe). Thus, some parameters that affect the growth process such as the diffusion and solubility of oxygen in silver are compared to that of other materials.

1. Oxygen diffusivity

The diffusion coefficient of oxygen in a silver thin film is relatively high when compared to other catalysts such as Au (Zhou et al., 2016; Yasui et al., 2008; Kim and Kim, 2005), Fe (Kumar et al., 2014), or Pt (Song et al., 2017). Due to its higher value, silver has a tendency to absorb oxygen which suggests that it is a better catalyst for enhancing the density and length of Ga_2O_3 NWs. Au tends to have higher diffusion of oxygen than Ag; however, it has been shown that the concentration of oxygen at interstitial sites in Ag is larger than that in Au. For instance, at a temperature of 527°C, the oxygen concentration in Ag ($1.34 \times 10°$) is higher than in Au (4.87×10^{-7}) by seven orders of magnitude (Zhou et al., 2016). In fact, it has even been shown that while oxygen easily diffuses through copper and silver, this is not the case for gold (Bergwerff, 2003). In Pt films, diffusion of oxygen is lower because it occurs at the interface while in Ag films, the transport mechanism for oxygen is bulk diffusion (Moghadam and Stevenson, 1986). The higher oxygen diffusivity of silver is even more apparent with platinum, which has a solubility on the order of 10^{-9} (Velho and Bartlett, 1972) compared to silver which has solubility on the order of 10^5. Table 4.2. shows the reported values for the diffusion coefficients of oxygen at different temperatures in various materials that have been used as catalysts for Ga_2O_3 nanowire growth. Using silver, we were able to achieve higher oxygen diffusion at a higher temperature which resulted in enhanced nanowire growth.

2. Solubility

The second parameter that makes silver an effective catalyst for Ga_2O_3 nanowire growth is its solubility of oxygen. Although the solubility of gases dissolved in liquids usually decreases with increasing temperature, in a few cases, solubility actually increases with increased temperature (Fabbro and Paolo, 2004; Holmes, 1996). In fact, silver is one of the few cases (Holmes, 1996). The solubility of oxygen in a molten silver thin film can be increased by increasing its temperature above its melting point of 961.8°C (which is lower compared to other metals such as Pt (1768°C), Ni (1455°C) and Au (1064°C)). This makes it relatively easy to use silver as it does not require the high temperatures that other catalysts require. Various studies have demonstrated the low solubility of oxygen in Au (Zhou et al., 2016; Toole and Johnson, 1933; Jones et al., 2013). In Ni, the solubility decreases with increasing temperature (Park and Altstetter, 1987; Brown and Alcock, 1969; Seybolt and Fullman, 1954). For Fe and Cu, the low solubility of oxygen has also been noted (Meijering, 1955). Based on the solubility results summarized in Table 4.2, oxygen has a solubility of about 0.05 at. % in silver at 890°C (Ramanarayanan and Rapp, 1972), which is almost twice that reported for iron with

Table 4.2 Summary of previous studies that focus on the oxygen diffusivity and solubility of different metals Ag, Ni, and Fe at different temperatures.

Metal	T (°C)	Diffusivity (cm^2/s)	Ref.	Solubility (at. %)	Ref.
Ag	763	1.58×10^5	Ramanarayanan and Rapp (1972)	0.0281	Ramanarayanan and Rapp (1972)
	804	2.37×10^5		0.0306	
	812	2.39×10^5		0.0339	
	845	2.44×10^5		0.0443	
	890	3.25×10^5		0.0519	
	935	3.76×10^5		0.0629	
Ni	800	2×10^{-8}	Konrad et al. (2012)	0.0001858	Konrad et al. (2012)
	1000	5×10^{-7}		0.0005836	
	1100	2×10^{-10}		0.0009128	
Fe	700	No data exists for comparison		0.008	
	900			0.03	Seybolt (1959)

a solubility of 0.03 at. % at 900°C (Meijering, 1955). Thus, silver is a more attractive candidate as a catalyst.

Table 4.3 summarizes the solubility of oxygen molecules in Ag and Ga at different temperatures. Gallium has a higher tendency to absorb oxygen than silver. However, silver assists in enhancing absorption and hence results in a higher density of nanowire growth. The solubility of oxygen in liquid gallium increases with increasing temperature and follows the Arrhenius equation (Zinkevich and Aldinger, 2004). In general, different mechanisms could explain the oxygen molecule's incorporation at the surface of the Ag and the Ag/Ga_2O_3 interface. However, there is a lack of experimental and research work that addresses the chemical and physical interactions between Ag, Ga, and O.

4.3.4.5 SnO_2 nanowires-based photodetectors

Because of its extraordinary optical and electronic properties, tin dioxide (SnO_2) is one of the most extensively studied metal oxide semiconductors that have been employed for a wide range of applications, including solar cells, LEDs, LCDs, transparent conductive electrodes, sensors and batteries (Park et al., 2017; Lee et al., 2016; Singh et al., 2017; Zhao et al., 2017). Specifically, because of its high transparency, it has been widely employed as flat touch panels, LCDs, and plasma display panels (Ramarajan et al., 2020). It is a well-known n-type semiconductor with a direct band gap of around 3.6 eV and high excitonic binding energy (180 meV) (Dias et al., 2020). SnO_2 NWs have been one of the most popular oxide nanostructures over the last decade due to their excellent properties and potential applications (Chen and Lou, 2013). In particular, they have gained much importance in light detection in recent years.

For instance, a thin-sized SnO_2 nanowire-based UV photodetector has been reported recently by Hu et al., SnO_2 NWs were fabricated using the VLS method. The resulting device structure demonstrated excellent light selectivity and stability. When the device was illuminated with UV light (320 nm), a significant increase in photocurrent was detected. The photocurrent value was measured to be 2.1 μA under a bias of 1 V. SnO_2 nanowire UV photodetector exhibited ultrahigh external quantum efficiency with the value of 1.32×10^7 (Hu et al., 2011).

Table 4.3 Summary of previously reported solubility of oxygen in Ag and Ga at different temperatures.

Metal	T (°C)	Solubility (at. %)	Ref.
Ag	763	0.0281	Rapp and Ramanarayanan (1972)
	804	0.0306	
	935	0.0629	
Ga	775	0.15×10^2	Alcock and Jacob (1977)
	800	$0.23-0.25 \times 10^2$	
	900	$0.85-0.97 \times 10^2$	
	1000	3.08×10^2	

In 2020, De Araujo et al. reported the fabrication of SnO_2 nanowire network photodetector in metal-semiconductor-metal architecture. In this study, SnO_2 NWs were synthesized using the VLS mechanism. The photodetector performance of SnO_2 NWs was investigated under three different illumination consist of UV light (9.5 mW/cm^2), white light (2 mW/cm^2), and sunlight (78.6 mW/cm^2). Under illumination, the current is expected to increase as photoionized charges will be activated and will change the depletion region. In addition, the barrier height at the metal-semiconductor interface will be adjusted. When I-V characteristics of the device were examined, it was observed that the device exhibited different responses to different sources, and the current increased with illumination. Compared with the white light source, the device showed high sensitivity in the UV region under both UV lamps and sunlight. Although the UV part of the solar spectrum is small, it demonstrates this sensitivity owing to its wide wavelength that creates sunlight. Since white light has an energy between 1.65 and 3.1 V, this energy will be insufficient to excite the electrons in the valence band. Therefore, photocurrent is only due to the excitation of levels within the energy band under white light. The maximum photocurrent will only be achieved with UV illumination, where the energy is greater than or equal to the energy band gap of SnO_2. Under UV exposure, electrons can easily overcome potential barriers and contribute significantly to current. In this study, the photoresponse (on/off ratio) of the SnO_2 nanowire network UV photodetector was calculated to be 10^4 for all illumination sources, and response times were summarized in Table 4.4 (De Araújo et al., 2020).

In another study, a high current on/off ratio of the order of 10^5 was reported for SnO_2 NWs with carbon nanotubes by Kim et al., However, the response and fall time remained relatively low compared to the high photoresponse of the device reported by (Kim et al., 2013).

In 2019, Yan et al., fabricated high-performance solar-blind SnO_2 nanowire PDs. Their device demonstrated a high current on/off ratio of 2.99 $\times$ 10^5 under 275 nm UV illumination. In addition, a large photo responsivity value of 4.3 $\times$ 10^4 was obtained with an excellent external quantum efficiency value of 1.94 $\times$ 10^5. The rise and fall times were calculated to be 60 and 100 Ms, respectively (Yan et al., 2019).

Lupan et al., successfully fabricated SnO_2 NWs using a flame-transport-synthesis process and functionalized the surface with Zn_2SnO_4 dots to build a UV

Table 4.4 The rise and fall times of SnO_2 nanowire-based UV photodetector for different illumination sources (De Araújo et al., 2020).

Sources	Rise time (s)		Fall time (s)
	Min	Max	
UV lamp	0.3	1.2	< 0.8
Sunlight	0.8	2.8	< 0.8
White light	56.3	280	90

photodetector. The results showed that a UV photodetector based on SnO_2:Zn_2SnO_4 NWs had an outstanding photoconductive performance and an ultra-fast reaction time when exposed to 375 nm UV light. The UV response (I_{uv}/I_{dark}) ratio for SnO_2:Zn_2SnO_4 NWs was calculated to be around 11, while it was approximately 4 for the pristine SnO_2 NWs-based detector. Furthermore, both the rise and fall times were found to be the same and estimated to be 0.1 s (Lupan et al., 2018).

In 2019, Li et al. demonstrated stretchable SnO_2-CdS interlaced-nanowire film UV PDs. SnO_2-CdS interlaced NW photodetector was built by a multiple lithographic filtration method as an attempt to combine the high spectral responsivity of SnO_2 NW PDs for UV light with the fast response speed of CdS NW PDs. The SnO_2 and CdS NWs were synthesized by CVD and solvothermal techniques, respectively. The photoresponse characteristics of the constructed devices were analyzed in detail and their performances were compared with those built via pure SnO_2 NWs and CdS NWs. The findings showed that the SnO_2-CdS interlaced device had a smaller dark current and a faster response speed compared to the one constructed with the pure SnO_2 NWs. Moreover, the devices exhibited better spectral responsivities compared with those fabricated with pure CdS NWs. It was also realized that the interlaced PDs had perfect electrical stability and stretching cyclability, linked to the waved wrinkles produced on the surface of the NWs/PDMS layer during the pre-stretching cycles. The findings of this study can be considered a critical step toward the manufacture and widespread use of next-generation wearable and stretchable PDs (Li et al., 2019).

In a recent study, a self-powered Schottky type Au/SnO_2 structured UV photodetector based on SnO_2 NWs was demonstrated by Chetri and Dhar (2019). The glancing angle deposition (GLAD) technique was used for the controlled growth of SnO_2 NW arrays (NAs) on n-type Si <100> wafer. The Schottky barrier-type UV detector was manufactured by evaporating a 30 nm thick gold thin film with a 1.5 mm circular diameter on top of SnO_2 NW arrays. Due to the existence of good Schottky contact between Au and SnO_2 NWs, the photodetector exhibited promising responsivity and fast response. Under the dark condition, the rectification ratio (at 5 V) was found to be 5.86, verifying the formation of a good Schottky contact for the Au/SnO_2-NWs/Si device. Under UV illumination, a dramatic increase in rectification ratio (482%) was measured, which was linked to the existence of a high-quality Schottky junction. The detector exhibited a maximum photoresponse of 0.36 mA/W under 370 nm UV light exposure at zero bias, which is almost as good as commercially available SiC UV detectors. The rise time and decay time parameters were also determined and found to be 0.72 s and 1.72 s, respectively, which are significantly better than those reported in the literature for a number of self-powered PDs (Fu et al., 2018).

In another study, a high-performing CuO/SnO_2 p-n structured visible-blind UV photodetector was reported by Xie et al., The responsivity for the nanoheterojunction photodetector was found to be 10.3 A/W. The photocurrent-to-dark current ratio was also determined, which was calculated to be around 592 at 290 nm (Xie et al., 2015).

Recently, Zheng et al. used a near-field electrospinning method to grow ZnO-SnO_2 NW arrays for flexible UV photodetector applications. The manufactured photodetector based on ZnO-SnO$_2$ heterojunction NW arrays exhibited excellent performance, such as high on/off current ratios (up to 1000) and good stability/reproducibility, under UV light of 300 nm (Zheng et al., 2018).

4.3.4.6 TiO$_2$ nanowires-based photodetectors

Titanium dioxide (TiO$_2$) is a versatile functional oxide-semiconductor with a large band gap ranging from 3.0 eV to 3.6 eV (3.0 eV for rutile, 3.2 eV for anatase, and 3.6 eV for brookite phases). Due to its excellent chemical stability, nontoxicity, thermal stability, eco-friendliness, and abundance in the earth's crust, it is among the most widely studied materials over the years (Zou et al., 2018). Because of their excellent optical, chemical, and physical properties, TiO$_2$ NWs have been widely used for a number of electronic and optoelectronic device applications, including dye-sensitized solar cells, supercapacitors, light detection, lithium batteries, data storage systems, medical engineering, and gas sensors (Yu et al., 2021; Khanna et al., 2021; Tang et al., 2016; Kim et al., 2015; Bjursten et al., 2010; Natarajan et al., 1998). Specifically, TiO$_2$ NWs based PDs have attracted great attention in the last few years, which are expected to combine the distinctive characteristics of the sensing material, like hole trapping mechanism via O$_2$ adsorption and effective UV light absorption capability, with special features of NWs such as high mobility, effective charge collection, quantum confinement and large surface-to-volume ratio (Tsai et al., 2012; Yu and Park, 2010; Guller et al., 2018).

Flexible PDs based on bendable substrates create great potential for next-generation smart electronic devices. Wang et al. (2017) reported the fabrication of a transparent and flexible UV photodetector based on TiO$_2$ NWs with high performance. For the manufacture of transparent and flexible UV PDs, TiO$_2$ NWs were grown on a mica surface by the electrospinning method. Then, interdigitated platinum (Pt) was deposited on the surface via a mask using the vacuum evaporation method. According to the XRD results, it was seen that the TiO$_2$ samples had an anatase phase when heated at 500°C and a rutile crystal phase when heated at 800°C. It was reported that TiO$_2$ samples in this study had a mixture of anatase (83%) and rutile (17%) phases in the annealing range of 500°C–700°C. Due to its wide surface affinity, anatase TiO$_2$ is known to have higher photocatalytic activity than rutile TiO$_2$ (Yang et al., 2009). Although TiO$_2$ NWs showed a low dark current at pA level even when applied with 10 V external bias, they exhibited three times greater I_{uv}/I_{dark} ratio under 254 nm UV light. After annealing at 550°C, the photocurrent of the device reached up to 47 nA and 38 nA under 254 nm and 365 nm UV illumination at a 10 V bias. In addition, the UV detectors showed excellent mechanical flexibility and durability. The results confirmed the fabrication of highly flexible and transparent PDs with quick response times, which is an important step toward the realization of wearable optoelectronic device applications.

In 2019, Deb and a coworker synthesized TiO$_2$ nanowire on graphene oxide thin film using the glancing angle deposition technique. Because of its high mobility and wide bandgap, graphene oxide has a lot of promise in photocatalytic applications. In this study, TiO$_2$ nanowire on graphene oxide thin film was used to fabricate the fast response UV photodetector with low noise. In comparison to bare TiO_2 nanowire, the responsivity, sensitivity, and photocurrent were enhanced and the dark current was reduced in Au/TiO_2 NW/ graphene oxide thin film structure. High photocurrent under UV illumination was largely due to interface trap charges. The dark current was measured to be 0.0254 μA at the bias of 4 V. The responsivity and sensitivity were found to be 48.3 mA/W and 82.7, respectively (under 380 nm). Furthermore, it was observed a strong photon absorption in the UV region. A high rise time value of 48 Ms and recovery time of 40 Ms owing to efficient charge separation between the TiO_2 NWs and graphene oxide thin film was reported (Deb and Dhar, 2019).

In another study, TiO_2 NWs were synthesized via an inkjet printing technique by Chen et al., and TiO_2 NWs were used to fabricate a highly transparent photodetector. The device demonstrated a low dark current value of 10^{-12} A and a high response with an on/off current ratio of around 2000 (Chen et al., 2015b).

In a recent study, the manufacture of a highly photosensitive low-noise UV photodetector with n-TiO_2/In_2O_2 coaxial heterostructure was reported by Pooja et al., As-grown and annealed n-type TiO_2/In_2O_3 nanowire heterojunction UV PDs were investigated. Vertically aligned TiO_2/In_2O_3 NAs were successfully synthesized on TiO_2 thin film precoated p-Si substrate using an electron-beam evaporator assisted with GLAD. To construct the coaxial TiO_2/In_2O_3 structure, In_2O_3 NWs were grown on the top portion of TiO_2 NWs using the same growth process. Silver (Ag) dot contacts with a diameter of 1 mm and the ITO conductive layer were assigned as top electrode and back electrode, respectively. The schematic architecture of the fabricated TiO_2/In_2O_3 coaxial heterojunction photodetector is shown in Figure 4.15. To investigate the effect of the postannealing process on the performance of the coaxial UV heterojunction PDs, both as-grown and annealed devices were built, structural, electrical, and optical properties of which were studied in detail. The results revealed that the device annealed at 600°C was exhibiting an excellent photo responsivity of 325 A/W under UV light of 330 nm at -5 V bias, which was nearly

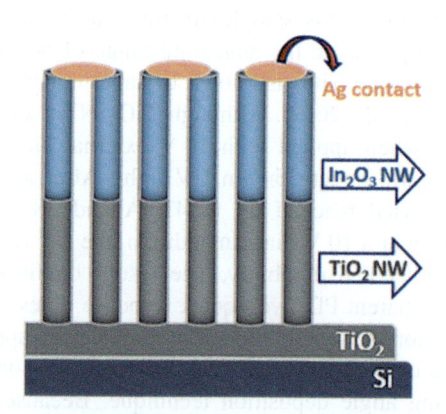

Figure 4.15 The schematic architecture of the fabricated TiO_2/In_2O_3 coaxial heterojunction photodetector (Pooja and Chinnamuthu, 2021).

15,000 times higher than that obtained for the as-grown UV photodetector. The observed drastic improvement in photosensitivity was attributed to the enhancement in crystallinity and reduction of interface trap states following the annealing process conducted at the high annealing temperature. In addition, under UV illumination of 280 nm, the device annealed at 600°C exhibited a high responsivity of 325 A/W and quantum efficiency of 1,20,000 suggesting the suitability of the annealed TiO_2/In_2O_3 coaxial heterojunction devices for the realization of low-cost, efficient and high-performance UV PDs (Pooja and Chinnamuthu, 2021).

4.3.4.7 Other metal oxide nanowires-based photodetectors

There are several alternative metal oxide NWs used for UV photodetection. With its 2.8 eV band gap and high photoresponse capability, WO_3 (tungsten trioxides) is an excellent candidate for the realization of high-performance PDs (Cai et al., 2015). Recently, one-dimensional WO_3 NWs grown on carbon papers by CVD technique have been reported by Li et al., The WO_3 NWs-based photodetector exhibited high responsivity at 375 nm UV light. The I-V characteristics of the device were investigated under UV light illumination (375 nm) and in dark conditions. Because of its high sensitivity to UV light, the device generated a higher current under UV illumination than the one produced in the dark condition (Li et al., 2011).

Moudgil et al. fabricated the high-speed efficient UV photodetector based on 500 nm width WO_3 NWs. The WO_3 NWs were synthesized through the physical deposition technique, and the photodetector device structure was produced using the conventional complementary metal-oxide-semiconductor. The manufactured photodetector structure demonstrated high responsivity of 47.3 A/W and gain of 172 under 340 nm UV light (at a 2 V bias). It was determined that the photocurrent was four times greater than the dark current. Rise and fall time measurements of the photodetector were reported to be 112 µs and 84 µs, respectively (Moudgil et al., 2018).

A fabrication of ultrafast with high-performance UV/IR PDs based on CuO NWs was reported by Ate et al., CuO NWs were synthesized from the oxidation of Cu foils using the thermal oxidation method. CuO NWs photodetector with platinum (Pt) contact electrodes exhibited a high sensitivity to 390 nm UV illumination at 5 and 10 V external bias. In addition, fast response and recovery time of 0.05 s were achieved for CuO NWs photodetector (Ate et al., 2014).

In_2O_3 is another crucial transparent conducting metal oxide semiconductor for light detection. It is a well-known fact that In_2O_3 only reacts to UV illumination due to its wide energy band gap. Since it has a high dark current value, a slower response speed is to be predicted. The combination of a broadband material and a narrow band gap semiconductor is supposed to enhance PDs' optoelectronic properties by adjusting the energy band gap structure and allowing for efficient electron-hole separation and transportation at the interface (Zheng et al., 2015). Li et al. showed in 2017 that coating In_2O_3 NWs with a thin layer of CuO improves the photosensitivity of the photodetector. Coating the In_2O_3 NWs with CuO having a

Table 4.5 The performance comparison of metal oxide nanowire-based UV photodetectors (Zou et al., 2018).

Metal oxide nanowires	Photocurrent (A)	Responsivity (A/W)	Response time (s) Rise time fall time		EQE/ Gain (%)	Reference
VO_2	–	7.07×10^3	0.126		2.4×10^{10}	Wu and Chang (2014)
$InGaO_3$ (ZnO)	4.71×10^{-7}	5.3×10^4	0.4	2.5	1.9×10^9	Lou et al. (2015)
Zn_2GeO_4	–	5.11×10^3	0.01	0.013	2.45×10^8	Zhou et al. (2016)
In_2O_3	2.17×10^{-2}	4.8×10^6	3	13	1.46×10^9	Meng et al. (2017)

band gap of 1.2–1.35 eV resulted in increased UV-Vis light absorption capability in the UV-visible regions. The results showed that illumination at 300 nm resulted in an increase in conductivity since more effective electron-hole pairs were generated, separated, and transported. Furthermore, a fivefold increase in response time was recorded following the CuO coating (Li et al., 2017). Table 4.5 summarizes the rest of the studies on metal oxide nanowire-based UV PDs (Zou et al., 2018).

4.4 Device design challenges

There are serious challenges that impede the design, fabrication, and operation of reliable NW PDs. Key amongst them is the development of efficient and massively parallel protocols for creating mechanically stable ohmic contacts to the NWs for their deployment. Others include modalities that could maximize or enhance photon trapping capabilities. Other important aspects such as controlling their doping levels, dimensions, limitation of surface defects, optimum charge mobility, and impedance matching for high-speed operations are core challenges that cripple mass deployment and adoption of NW-based devices. This section will look into some of these challenges.

4.4.1 Impediments on contact formation

Difficulties in the formation of reproducible low resistivity electrical contacts are a barrier to the wide-scale integration of functional NWs in devices and systems. Unlike the research-based approach of sequentially connecting electrodes to individual NWs for device physics studies, a massively parallel and manufacturable interfacing technique is crucial for the reproducible fabrication of dense and low-cost nanodevice arrays with reliability that matches existing IC processing techniques.

Our direct growth of semiconductor NWs directly forming electrically integrated bridges between two contacts is a powerful tool that bears reproducible ohmic contacts with formidable mechanical integrity. Both group IV and III–V NWs were bridged between Si electrodes (Yi et al., 2006; Islam et al., 2004). Based on our linear current-voltage measurements and a constructed model, we calculated the specific contact resistance to be in the range of 4×10^{-6} Ω-cm^2 for bridged Si NWs, a value that is more than two orders of magnitudes lower than that of other approaches to evaporating metals (Chaudhry et al., 2007) (Figure 4.16).

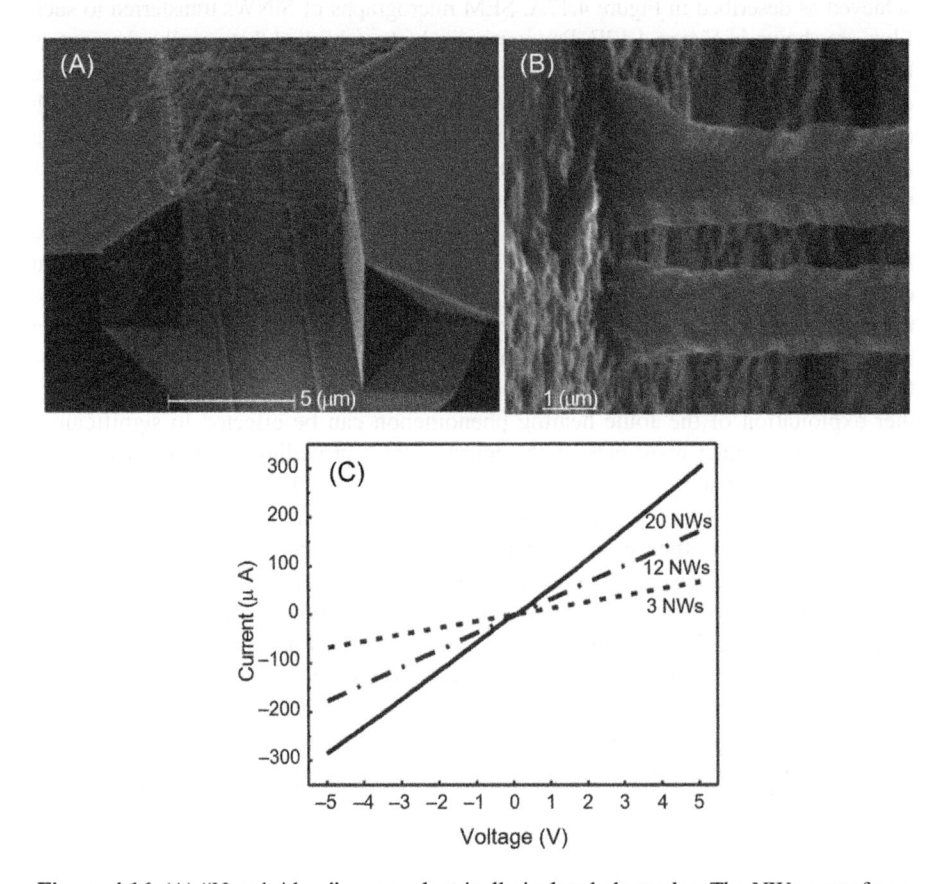

Figure 4.16 (A) "Nanobridges" across electrically isolated electrodes. The NWs grow from left to right in these SEM views. Au-nucleated NWs grown mostly perpendicular to the (111) sidewalls. (B) An SEM of actual contact of the NW to the Si electrode. (C) Highly linear current-voltage ($I-V$) characteristics of bridged Si NWs.

4.4.2 Hybrid contacts for print transferred nanowires

Alongside the print transfer process, modalities of electrical interfacing of harvested structures provide means of harnessing photons collected by the semiconductor NWs. Efforts toward fabricating ohmic contacts based on composites made of metallic nanoparticles (AgNPs) inside organic matrices have been developed. The synergetic approach adopted mirrors that employed in the flip-chip packaging industry (Ombaba et al., 2013; Ombaba et al., 2013). Here, composite films of select polymers that contain spatially distributed AgNPs are prepared. We have recently reported on how such films can be prepared from heterogenous ternary mixtures via spin coating (Ombaba et al., 2013). By controlling the spin speed and solution properties of the ternary mixture, the good spatial distribution of the metallic particle within the ensuing solid film can be realized as shown in Figure 4.17D. During the transfer process, a simultaneous anchoring of the semiconductor NW to the film and contact formation to the AgNPs is achieved as described in Figure 4.17A. SEM micrographs of SiNWs transferred to such films are shown in Figure 4.17E. Previously we had established that usually, the contact between AgNP and the underlying gold (Au) film has ohmic characteristics (Ombaba et al., 2013). This was achieved by assembling an AgNP array atop the Au film. With an electrical osmium probe atop the film while another touches the film, the ensuing I–V characteristics indicate that the resistivity across the AgNP film was more or less similar to that of the Au film, as shown in Figure 4.17C.

The choice of AgNPs is mitigated by their propensity to form deformable and flexible electrodes within a specific bending radius of curvature (Logeeswaran et al., 2012a). Additionally, the used AgNPs are not solid all the way through since they are merely agglomerates of Ag nanorods of about 20 nm diameters that are soft and pliable upon application of pressure from incoming semiconductor NWs. Once the semiconductor NWs are into contact with the AgNPs (Figure 4.17E), further exploitation of the Joule heating phenomenon can be effected to significantly decrease the contact resistance at the interfaces by providing localized heating in the ambient (Ombaba et al., 2013; Logeeswaran et al., 2012a; Logeeswaran et al., 2012b). We carried out a study to ascertain this by arraigning AgNPs on an Au-coated substrate and then probing them inside a custom-modified SEM instrument equipped with electrical probes capable of inserting known pressure and current while the sample is being imaged by the microscope. Initially, the current magnitude across the AgNPs was in the nano ampere range (Figure 4.17B). Upon joule heating, charge transport across the NP attained ohmic characteristics as evidenced by the increase in current to the milliampere range. Additionally, good ohmic contacts on two terminal semiconductor NWs array devices have been realized using precision controlling of the array transfer and integration process using the developed in-house tools (Triplett et al., 2014).

4.4.3 Control of nanowires doping

There are existing challenges associated with NW doping even though both p-type and n-type NWs have been widely reported. In the "bottom-up" growth technique,

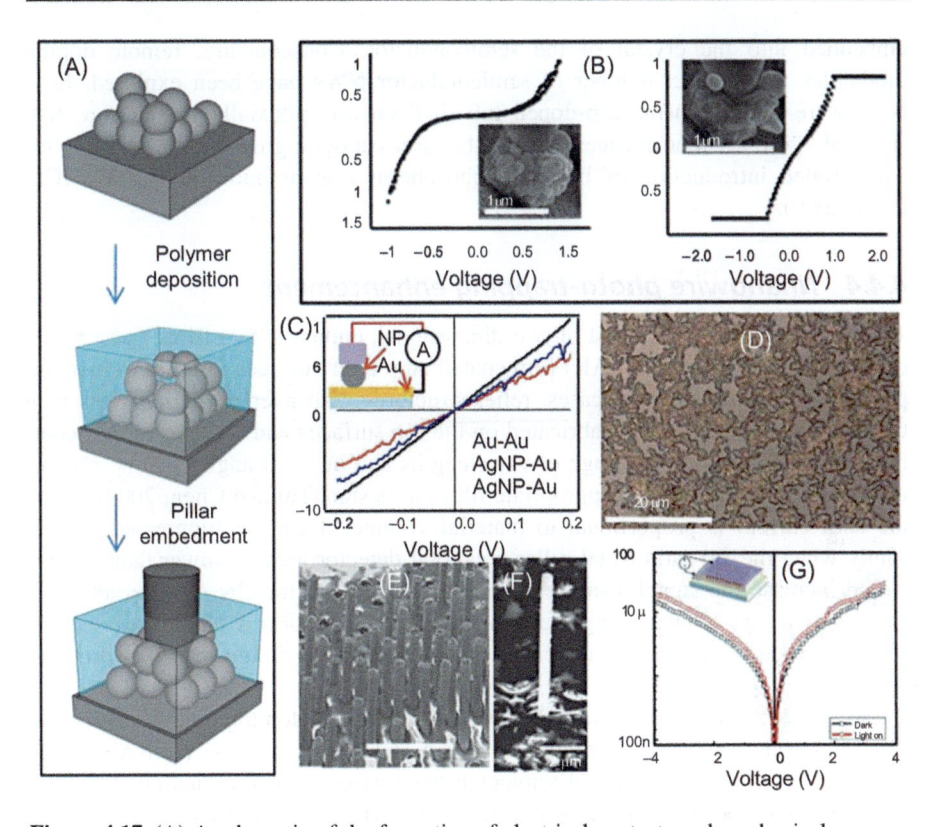

Figure 4.17 (A) A schematic of the formation of electrical contacts and mechanical attachment of a semiconductor micro- and nanostructure into a heterogeneous polymer/nanoparticle composite film. (B) Formation of ohmic AgNP electrical interface via joule heating. (C) Characterization of electrical charge transport across a silver nanoparticle film using a microprobe. (D) Electrically conductive heterogeneous polymer/silver nanoparticle films prepared via spin coating are used as receiver substrates on which semiconductor nano- and microstructures are transferred into. (E) Semiconductor MPs are embedded into the heterogeneous film described in (D). (F) A single semiconductor NW is embedded onto a poly(aniline)/poly(methyl methacrylate) composite. (G) Electrical characterization of a two-terminal device of transfer printed semiconductor NWs onto polymeric, electrically conductive substrates.

doping occurs via either the vapor-solid or VLS mechanism. The former allows for direct atomistic deposition of the dopant material on the NW surface while the latter entails the incorporation of the dopant into the catalyst such that as growth progresses, dopant atoms at the interface between the catalyst and the NW diffuse and hinge themselves into the growing NW. These two methods have different rates of diffusion (Perea et al., 2009). The flow rate of doping gases contributes to different doping concentrations and it remains dependent on other physical parameters of a CVD chamber. Using the VLS mechanism to grow NWs does not allow for determining how impurity atoms interact with the metal particle and how they are embedded into the crystal of the semiconductor. Consequently, remote doping modalities for a select number of semiconductor NWs have been explored. InAs NWs were P-doped using a p-doped InP shell grown epitaxially on the core NW (Li et al., 2007). Other avenues such as the laser catalytic growth method allow for a controlled introduction of boron or phosphorus dopant into silicon NWs (Cui et al., 2000).

4.4.4 Nnanowire photo-trapping enhancement

Traditional PD devices consist of two-dimensional, epitaxial, thin films and a single or multilayer antireflection (AR) film coated atop their surfaces in order to increase photon absorption. In some cases, relief structures with average periods less than the incident wavelength are fabricated on the top surfaces and serve as an AR coating. Photons with a broad range of wavelengths and incident angles can be accommodated efficiently with an appropriate fill ratio design (Hu and Chen, 2007). Since the dark current is proportional to material volume, it can be suppressed significantly when the fill ratio (total pillar area/total detector area) is lower than 1. This would increase the signal-to-noise ratio of the detectors, thereby offering an interesting advantage of increasing the temperature of operation if challenges such as surface leakage and traps associated with the pillar design can be properly addressed.

Appropriate nanotexturing can also lead to efficient photon trapping of even ultrathin semiconductor surfaces, leading to thinner and lighter devices. We have carried out the effect of nanotexturing of semiconductor surface with NWs and holes in view of determining the appropriate photon trapping surfaces as shown in Figure 4.18B. This comprehensive study will be disseminated elsewhere. Meanwhile, traditional approaches such as the decoration of Semiconductor NWs with various nanoparticles or secondary growth of smaller semiconductor NWs atop a photoactive NW continue to elicit considerable interest (Kelzenberg et al., 2010).

A dense network of NWs enables multiple reflections and scatterings which enhance incident photon trapping, a feature not present in conventional PDs. Lo and coworkers showed that the fraction of incident photon energy overlapping with the NW volume can be significantly higher than the fill ratio due to the high confinement of energy in a relatively high refractive index medium of NWs (Zhang et al., 2008). Bayandir and his coworkers have recently described macroscopic photoactive devices made of ultra-long Se NW ensembles aligned parallel to each other (Ozgur et al., 2012). A subsequent study describes how photon harvesting of a photoactive NW can be significantly enhanced by laying it atop an NW film of smaller diameter that acts as a photon collection enhancer as depicted in Figure 4.18A (Khudiyev and Bayindir, 2014).

The metal-catalyzed VLS method is a widely used approach to growing NWs. Nanoparticles of metal-semiconductor eutectic alloy are formed on the NW surfaces under specific growth conditions. These eutectic alloys on the surface of NWs can be used as catalysts to enable one-step growth of branched NWs as shown in Figure 4.19. Based on the same principle demonstrated by Kelzenberg et al. (2010),

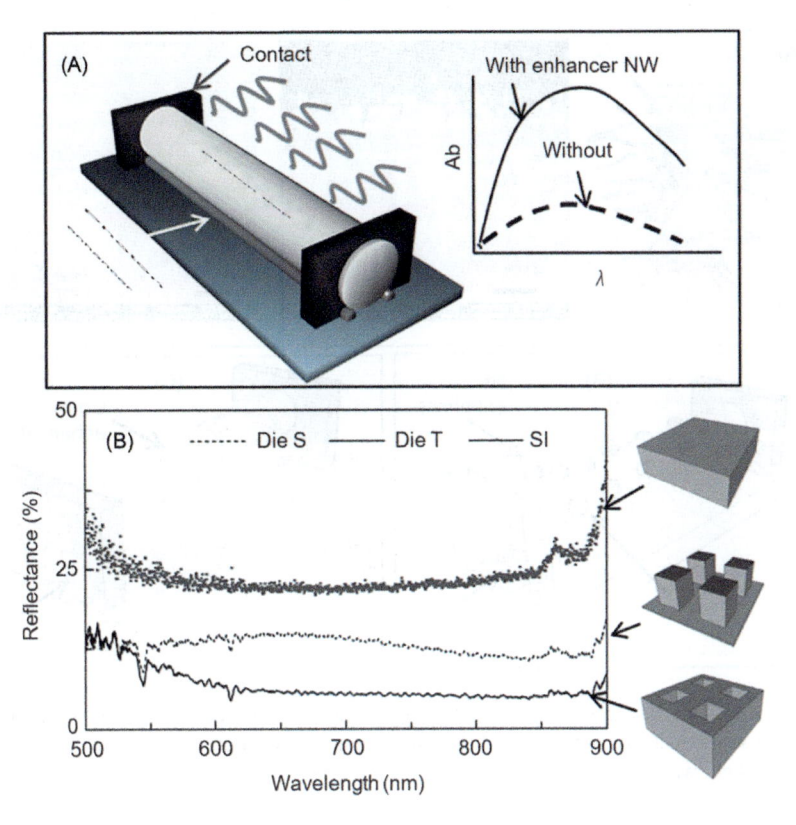

Figure 4.18 (A) Optical reflectance of a bare surface and InP NW coated surface. An NW coated (100)-oriented GaAs surface shows greatly reduced reflectance over the spectrum ranging from 400 to 1150 nm (Logeeswaran et al., 2008). (B) Optical transmission data at normal incidence for the transferred devices in the shape of MPs onto the respective polymer-coated carrier substrates. The transmission curve of each sample is normalized to its individual reference without the device (MPs). The devices on each sample absorb a fraction of incident photons, reducing the total transmission (Logeeswaran et al., 2010).

the branches can be used for enhanced light scattering in NW PDs. In order to prevent photo-absorption from scattering branches, dimensions need to be selected below the quantum confinement regime. We recently grew such structures on Si NWs for field ionization applications as depicted in the SEM micrograph in Figure 4.19A and we are in the process of harnessing them for photonic applications (Banan Sadeghian and Saif Islam, 2011). Their diameter and density can be controlled thermally as well as by controlling the catalytic amount on the surface.

Introducing a light scattering media on the surface of a photoactive surface has a propensity of increasing photon collection abilities. Materials such as Al_2O_3 nanoparticles scatter the light to enhance photo-NW interactions even when the fill ratio is very low (Kelzenberg et al., 2010). The wide bandgap of the Al_2O_3 suppresses absorption in the NPs and increases photon absorption in the semiconductor NW.

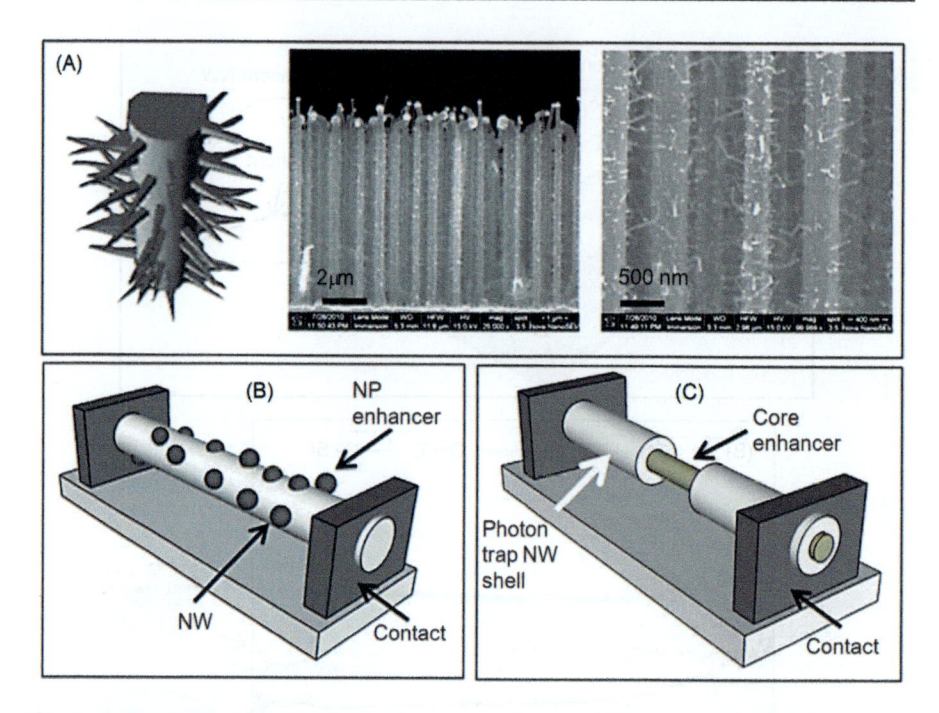

Figure 4.19 (A) A schematic of secondary regrowth nanowhiskers on a primary photon trapping semiconductor NW. (B) and (C) The ensuing SEM micrographs show these nanowhiskers which are enjoined onto the primary NW.

Localized plasmonic effects on the core of photoactive NW shells can improve their photoabsorption abilities. LI and company carried out a theoretical study on Ag NW cores inside a Si shell using a finite element method. The device's photocurrent could be modulated by varying AgNW thickness (Zhan et al., 2012; Mann and Garnett, 2013).

We demonstrated that the NW-coated surface contributes to suppressing the reflection of wide spectral photons using a reference GaAs substrate and a second GaAs substrate with an InP NW coating. On (100)-GaAs, InP NWs grow along two equivalent (111) directions of the substrate and show a dramatically reduced reflectance over the spectral range from 400 to 1150 nm even without using any AR coating. In the device that was designed with InP NWs on n-Si:H electrodes, as described in Section 4.3, InP NWs grow with a high degree of randomness in their orientations and can offer an intrinsic capability for significantly reduced reflection loss. The optical absorption and transmission measurements of the sample are testaments to the enhanced absorption and reduced reflection from a highly oriented micropillar-based structure with a pillar size of approximately 10 μm and diameter of approximately 1 μm that is transferred subsequently to a thin film.

A wide variety of organic materials have also been harnessed as photon enhancement agents. Baek et al., have described inorganic/organic field-effect transistors made of multiple shells of porphyrin, SiO_2, and Si NW cores. The device's photoswitching capabilities were ascribed to the electric field effect due to charge redistribution within the porphyrin film (Baek et al., 2015). Graphene and AuNPs have been credited for photo-absorption and surface plasmon enhancement in Si and InAs array PDs (Luo et al., 2014; Miao et al., 2015).

4.4.5 Additional challenges

There are more additional challenges befitting the mass deployment of NW-based PDs that were not dealt with within this chapter. The issue of how to eliminate the negative effects of surface traps on semiconductor NWs has been a subject of many studies and therefore not dealt with here. Semiconductor NWs especially those prepared via bottomup techniques have stochastic variations from each other due to their defects, which are credited for charge trapping and delayed carrier mobilities. While greater control of NW growth and the use of high purity chemical precursors alleviates some of these defects, it is virtually impossible to prepare identical NWs by bottomup approaches due to other physical and thermodynamics that occur during NW growth. We have recently presented an in-depth discussion of these challenges amongst others in a review paper, which could be a resourceful reference for further reading (Logeeswaran et al., 2011a,b).

4.5 Towards development of integrated multispectral nanowires photodetectors

PDs capable of undertaking, simultaneous detection of multiple wavelengths would allow for enhanced identification of samples of choice with higher certainty. When a photodetecting system is set to sense separate spectral bands at one go, it enables the accurate determination of several unique signatures of an analyte in a certain setting instantaneously. Such a system greatly increases the sensitivity of a PD beyond what single spectral devices are capable of. The notion of obtaining a multispectral photon signal of an object allows for the design of more intelligent systems that speed up decision-making for the end user. Such a system when employed in applications such as the military could allow for informed and appropriate splitsecond decisions in life or death situations.

Current multispectral imaging devices rely on cumbersome detecting processes that either disperse the optical signal across multiple infrared sensors or use a filter wheel to spectrally discriminate images focused on a single PD. These systems include beam splitters, lenses, and bandpass filters placed in the optical path to focus images onto separate sensors responding to different bands. This can be overcome by having a multispectral integrated PD in which each sensing NW pixel is able to read multiple spectral bands in parallel. Since no known single semiconductor is responsive to the entire spectrum, a PD that combines several semiconductors will have to be designed.

The emergent themes on the transfer of semiconductor micro and nanostructures from their formative wafers into arbitrary substrates offer insights into how multiple arrays of nano/microstructures fabricated from different semiconductor materials capable of sensing a particular spectral wavelength ranges could be serially incorporated onto a single substrate then postprocessed into a single multispectral PD (Logeeswaran et al., 2010; Islam and Logeeswaran, 2010; Triplett et al., 2014). With few modifications to the tools that we have developed, a roll-to-roll processing capability may be incorporated, as illustrated in Figure 4.20. Although we have already demonstrated the ability to print transfer NW arrays of a single semiconductor, we do recognize that there will be physical and mechanical challenges that will need to be addressed in order to achieve this. These are tasks that we are currently pursuing and the preliminary processes that we have developed show great promise.

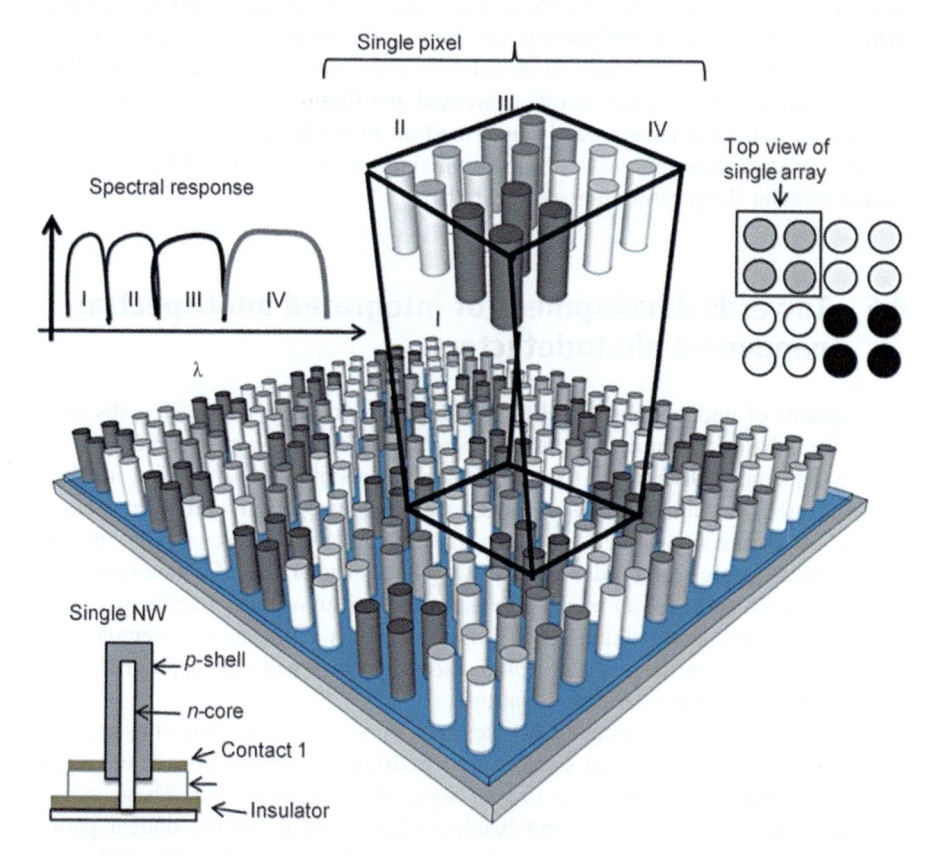

Figure 4.20 Conceptual multispectral pixelated PD composed of four arrays made of different semiconductor materials that individually detect part of the overall spectral bandwidth. A single pixel will have four NW arrays integrated via the print transfer method. Contact isolation on NWs with core-shell junctions can easily be achieved using conventional methods.

4.6 Conclusions and perspectives

The ability to design PDs using high-quality semiconductor and oxide NWs based on bottomup and self-assembly processes and integrate them onto various substrates can lead to cost-effective, superior, and novel functionalities. In this review, we provided a brief description of different approaches to realizing an integrated NW device for light detection. We show that both synthesis in-place and transfer printing on an arbitrary substrate offers high flexibility in the design and fabrication of NW-based PDs, which can be used in a wide range of technologies. However, for the NW integration process to be economically viable and attractive for CMOS applications, it must offer the possibility of direct growth and integration of non-Si-based NW devices onto large Si wafers. Currently, issues such as lattice constant, material and thermal mismatch, lack of good control over atomic structures, assembly into functional devices, difficulty in forming ternary and quaternary NW alloys, surface states, persistent photocurrent, contact resistance and noise, controlled doping for sharp homo- and hetero-junctions, nano-to-micro impedance matching, catalyst-induced contamination, etc., makes it difficult to grow and design NW devices on Si or any other substrates, which is a setback in the application of NW PDs at the industrial level.

A successful NW detection and sensing device would require a meticulous control of surface states, defects, traps, orientation, polarization, and light coupling as well as good control of generating sharp axial or radial junctions in the NWs. There are obvious challenges in implementing the NW PDs. Most conventional semiconductor PDs operate at essentially 100% quantum efficiency, whereas NW PDs' efficiencies are much lower in value. The development of feasible techniques is now crucial to match the efficiency of NW PDs to that of classical counterparts and a combination of plasmonic techniques and NW heteroepitaxy may offer the potential to realize substrate-independent NW PDs with high efficiency and bandwidth.

The basic devices and laboratory demonstrations of the key elements of technology exist for NW PDs and no physical breakthrough is required to implement the device in most systems. Substantial technological work remains, however. Controlling the physical properties such as diameter, doping uniformity, orientations, identical surface, and contacts are required to meet the strict requirements for practical systems with future generations of silicon photonics and CMOS. In addition, work will be required in developing impedance matching techniques between NW PDs and conventional high-speed circuits. The technology for integrating III–V and II–VI NWs with silicon integrated circuits is still at an early stage, though there have been considered key demonstrations in the last few years. It is likely that the first introductions of NW PDs will use hybrid approaches, such as transfer printing or pick-and-place, such hybrid approaches require no modifications to the current process for fabricating silicon integrated circuits except to add processes to fabricated NW PDs. Longer-term approaches will probably focus on monolithic integration via nanoheteroepitaxy. As the technology matures, techniques used in conventional devices such as resonant cavity enhanced PDs, edge illuminated waveguides, and evanescently coupled distributed PDs and avalanche PDs can be implemented in NW-based counterparts.

A recent trend in employing 1D inverse-NWs or nanoholes for designing high-speed PDs with high efficiency based on a light-trapping mechanism. Such light-trapping can cause slow propagating modes to enhance photon-material interactions and dramatically increase the light absorption capabilities of semiconductors that are weak in absorption characteristics. Using this technology, researchers developed extremely fast Si and Ge-based PDs with a high quantum efficiency ($\sim 10\times$) (Gao et al., 2017; Zang et al., 2017; Cansizoglu et al., 2018b), which opens opportunities for integrating amplifiers, equalizers, and other CMOS circuit elements with photodiodes, overcoming the challenges of cost, the complexity of integration, and mass production of receivers for rapidly growing data centers with less complexity and low-parasitic (Bartolo-Perez et al., 2021). The nature of inverse or negative structures such as holes, inverted pyramids, or cones can keep the surface material unfragmented and offer a solution to the first challenge by designing contacts that can be fabricated with CMOS-compatible conventional processes. On the other hand, the challenges of additional surface area and surface damage along with higher surface resistance are added issues of this approach. Therefore, CMOS-compatible surface passivation techniques need to be developed to fully tap into the potential of micro-/nanostructures in Si PDs (Mayet et al., 2018). Even though there are challenges that need to be addressed to benefit the full potential of inverse-NWs PDs with the advantages of monolithic integration and reduced cost of the transceivers that utilized such PDs, they promise cost-effective solutions for the high demand for connectivity in modern data centers that are an integral part of modern society(Cansizoglu et al., 2018a,c).

Acknowledgments

The authors are truly grateful for financial support from the DoD (ARO Research Grant # W911NF-14-4-0341) and National Science Foundation (Grant CMMI-1235592).

References

Ai, M.L., Guo, D.Y., Qu, Y.Y., Cui, W., Wu, Z.P., Li, P.G., et al., 2017. Fast-response solar-blind ultraviolet photodetector with a graphene/beta-Ga2O3/graphene hybrid structure. J. Alloy. Compd. 692, 634–638. Available from: https://doi.org/10.1016/j.jallcom.2016.09.087.

Al Farsi, B., Souier, T.M., Al Marzouqi, F., Al Maashani, M., Bououdina, M., Widatallah, H.M., et al., 2021. Structural and optical properties of visible active photocatalytic Al doped ZnO nanostructured thin films prepared by dip coating. Optical Mater. 113, 110868. Available from: https://doi.org/10.1016/j.optmat.2021.110868.

Alcock, C.B., Jacob, K.T., 1977. Solubility and activity of oxygen in liquid gallium and gallium-copper alloys. J. Less-Common Met. 53 (2), 211–222. Available from: https://doi.org/10.1016/0022-5088(77)90106-0.

Alhalaili, B., Mao, H., Islam, S. 2018. Ga2O3 nanowire synthesis and device applications.

Alhalaili, B., Bunk, V., Islam, 2019. Dynamics contributions to the growth mechanism of Ga2O3 thin film and NWs enabled by Ag catalyst. Nanomaterials 9, 1272. Available from: https://doi.org/10.3390/nano9091272.

Alhalaili, B., Vidu, R., Mao, H., Islam, M.S., 2020a. Comparative study of growth morphologies of Ga2O3 nanowires on different substrates. Nanomaterials 10, 1920. Available from: https://doi.org/10.3390/nano10101920.

Alhalaili, B., Mao, H., Dryden, D., Cansizoglu, H., Bunk, R., Vidu, R., et al., 2020b. Influence of silver as a catalyst on the growth of β- Ga2O3 nanowires on GaAs. Materials 13, 5377. Available from: https://doi.org/10.3390/ma13235377.

Alhalaili, B., Bunk, R., Mao, H., Cansizoglu, H., Vidu, R., Woodall, J., et al., 2020c. Gallium oxide nanowires for UV detection with enhanced growth and material properties. Sci. Rep. 10, 21434. Available from: https://doi.org/10.1038/s41598-020-78326-x.

Ate, A., Zhu, H., X. Q, H., Cai, X.W., Tang, Z., 2014. Ultrahigh responsivity UV/IR photodetectors based on pure CuO nanowires. AIP Conf. Proc. 1586 (1), 92–96. Available from: https://doi.org/10.1063/1.4866737.

Baca, A.J., Meitl, M.A., Ko, H.C., Mack, S., Kim, H.S., Dong, J., et al., 2007. Printable single-crystal silicon micro/nanoscale ribbons, platelets and bars generated from bulk wafers. Adv. Funct. Mater. 17 (16), 3051–3062. Available from: https://doi.org/10.1002/adfm.200601161.

Baca, A.J., Ahn, J.-H., Sun, Y., Meitl, M.A., Menard, E., Kim, H.-S., et al., 2008. Semiconductor wires and ribbons for high- performance flexible electronics. Angew. Chem. Int. (Ed.) 47 (30), 5524–5542. Available from: https://doi.org/10.1002/anie.200703238.

Baek, E., Pregl, S., Shaygan, M., Römhildt, L., Weber, W.M., Mikolajick, T., et al., 2015. Optoelectronic switching of nanowire-based hybrid organic/oxide/semiconductor field-effect transistors. Nano Res. 8 (4), 1229–1240. Available from: https://doi.org/10.1007/s12274-014-0608-7.

Bai, Z., Zhang, Y., 2016. Self-powered UV–visible photodetectors based on ZnO/Cu2O nanowire/electrolyte heterojunctions. J. Alloy. Compd. 675, 325–330. Available from: https://doi.org/10.1016/j.jallcom.2016.03.051.

Banan Sadeghian, R., Saif Islam, M., 2011. Ultralow-voltage field-ionization discharge on whiskered silicon nanowires for gas-sensing applications. Nat. Mater. 10 (2), 135–140. Available from: https://doi.org/10.1038/nmat2944.

Barrigón, E., Hrachowina, L., Borgström, M.T., 2020. Light current-voltage measurements of single, as-grown, nanowire solar cells standing vertically on a substrate. Nano Energy 78, 105191. Available from: https://doi.org/10.1016/j.nanoen.2020.105191.

Bartolo-Perez, C., Qarony, W., Ghandiparsi, S., Mayet, A.S., Ahamed, A., Cansizoglu, H., et al., 2021. Maximizing absorption in photon-trapping ultrafast silicon photodetectors. Adv. Photonics Res. 2 (6), 2000190. Available from: https://doi.org/10.1002/adpr.202000190.

Bayam, Y., Logeeswaran, V.J., Katzenmeyer, A.M., Sadeghian, R.B., Chacon, R.J., Wong, M.C., et al., 2015. Synthesis of Ga2O3 nanorods with ultra-sharp tips for high-performance field emission devices. Sci. Adv. Mater. 7 (2), 211–218. Available from: https://doi.org/10.1166/sam.2015.2160.

Bergwerff, J., 2003. The interaction of O2 with silver. A survey on the molecular processes in the passage of O2 through a silver membrane. Supervised by Bibi Dauvillier Dr A.J. van Dillen. Available from: https://anorg.chem.uu.nl/PDF/Bergwerff_silver%20literature.pdf.

Björk, M., Ohlsson, J., Thelander, C., Persson, A., Deppert, K., Wallenberg, R., et al., 2002. Nanowire resonant tunneling diodes. Appl. Phys. Lett. 81, 4458–4460. Available from: https://doi.org/10.1063/1.1527995.

Bjursten, L.M., Rasmusson, L., Oh, S., Smith, G.C., Brammer, K.S., Jin, S., 2010. Titanium dioxide nanotubes enhance bone bonding in vivo. J. Biomed. Mater. Res. A 92 (3), 1218–1224. Available from: https://doi.org/10.1002/jbm.a.32463.

Borgström, M., Bryllert, T., Sass, T., Gustafson, B., Wernersson, L.E., Seifert, W., et al., 2001. High peak-to-valley ratios observed in InAs/InP resonant tunneling quantum dot stacks. Appl. Phys. Lett. 78, 3232–3234. Available from: https://doi.org/10.1063/1.1374235.

Boruah, B.D., Mukherjee, A., Misra, A., 2016. Sandwiched assembly of ZnO nanowires between graphene layers for a self-powered and fast responsive ultraviolet photodetector. Nanotechnology 27 (9), 095205. Available from: https://doi.org/10.1088/0957-4484/27/9/095205.

Bower, C.A., Menard, E., Garrou, P.E., 2008. Transfer printing: an approach for massively parallel assembly of microscale devices. In: 2008 58th Electronic Components and Technology Conference, 27–30 May 2008.

Brown, C.B., Alcock, P.B., 1969. Physicochemical factors in the dissolution of thoria in solid nickel. 3 (1), 111–120. https://doi.org/10.1179/msc.1969.3.1.116.

Cai, Z.-X., Li, H.-Y., Yang, X.-N., Guo, X., 2015. NO sensing by single crystalline WO3 nanowires. Sens. Actuators B: Chem. 219, 346–353. Available from: https://doi.org/10.1016/j.snb.2015.05.036.

Cansizoglu, H., Elrefaie, A.F., Bartolo-Perez, C., Yamada, T., Gao, Y., Mayet, A.S., et al., 2018a. A new paradigm in high-speed and high-efficiency silicon photodiodes for communication—Part II: device and VLSI integration challenges for low-dimensional structures. IEEE Trans. Electron. Devices 65 (2), 382–391. Available from: https://doi.org/10.1109/TED.2017.2779500.

Cansizoglu, H., Bartolo-Perez, C., Gao, Y., Devine, E.P., Ghandiparsi, S., Polat, K.G., et al., 2018b. Surface-illuminated photon-trapping high-speed Ge-on-Si photodiodes with improved efficiency up to 1700 nm. Photonics Res. 6 (7), 734–742. Available from: https://doi.org/10.1364/PRJ.6.000734.

Cansizoglu, H., Ponizovskaya Devine, E., Gao, Y., Ghandiparsi, S., Yamada, T., Elrefaie, A.F., et al., 2018c. A new paradigm in high-speed and high-efficiency silicon photodiodes for communication—Part I: enhancing photon–material interactions via low-dimensional structures. IEEE Trans. Electron. Devices 65 (2), 372–381. Available from: https://doi.org/10.1109/TED.2017.2779145.

Chang, Y.K., Hong, F.C., 2009. The fabrication of ZnO nanowire field-effect transistors by roll-transfer printing. Nanotechnology 20 (19), 195302. Available from: https://doi.org/10.1088/0957-4484/20/19/195302.

Chang, D.E., Sørensen, A.S., Hemmer, P.R., Lukin, M.D., 2007. Strong coupling of single emitters to surface plasmons. Phys. Rev. B 76 (3), 035420. Available from: https://doi.org/10.1103/PhysRevB.76.035420.

Chaudhry, A., Islam, M.S., 2008. Examining the anomalous electrical characteristics observed in InN nanowires. J. Nanosci. Nanotechnol. 8 (1), 222–227. Available from: https://doi.org/10.1166/jnn.2008.n18.

Chaudhry, A., Vishwanath Ramamurthi, E.F., Saif Islam, M., 2007. Ultra-low contact resistance of epitaxially interfaced bridged silicon nanowires. Nano Lett. 7 (6), 1536–1541. Available from: https://doi.org/10.1021/nl070325e.

Chen, J.S., Lou, X.W., 2013. SnO2-Based nanomaterials: synthesis and application in Lithium-Ion batteries. Small 9 (11), 1877–1893. Available from: https://doi.org/10.1002/smll.201202601.

Chen, H., Liu, K., Hu, L., Ahmed, A.A.-G., Fang, X., 2015a. New concept ultraviolet photodetectors. Mater. Today 18 (9), 493–502. Available from: https://doi.org/10.1016/j.mattod.2015.06.001.

Chen, S.-P., Retamal, J.R.D., Lien, D.-H., He, J.-H., Liao, Y.-C., 2015b. Inkjet-printed transparent nanowire thin film features for UV photodetectors. RSC Adv. 5 (87), 70707–70712. Available from: https://doi.org/10.1039/C5RA12617G.

Chen, M., Mu, L., Wang, S., Cao, X., Liang, S., Wang, Y., et al., 2020. A single silicon nanowire-based ratiometric biosensor for Ca2 + at various locations in a neuron. ACS Chem. Neurosci. 11 (9), 1283–1290. Available from: https://doi.org/10.1021/acschemneuro.0c00041.

Chetri, P., Dhar, J.C., 2019. Self-powered UV detection using SnO2 nanowire arrays with Au Schottky contact. Mater. Sci. Semicond. Process. 100, 123–129. Available from: https://doi.org/10.1016/j.mssp.2019.05.003.

Costas, A., Florica, C., Preda, N., Apostol, N., Kuncser, A., Nitescu, A., et al., 2019. Radial heterojunction based on single ZnO-CuxO core-shell nanowire for photodetector applications. Sci. Rep. 9 (1), 5553. Available from: https://doi.org/10.1038/s41598-019-42060-w.

Cui, Y., Duan, X., Hu, J., Lieber, C.M., 2000. Doping and electrical transport in silicon nanowires. J. Phys. Chem. B 104 (22), 5213–5216. Available from: https://doi.org/10.1021/jp0009305.

Cui, Y., Wei, Q., Park, H., Lieber, C.M., 2001. Nanowire nanosensors for highly sensitive and selective detection of biological and chemical species. Science 293 (5533), 1289. Available from: https://doi.org/10.1126/science.1062711.

Dai, X., Zhang, S., Wang, Z.L., Adamo, G., Liu, H., Huang, Y.Z., et al., 2014. GaAs/AlGaAs nanowire photodetector. Nano Lett. 14 (5), 2688–2693. Available from: https://doi.org/10.1021/nl5006004.

De Araújo, E.P., Arantes, A.N., Costa, I.M., Chiquito, A.J., 2020. Reliable tin dioxide based nanowire networks as ultraviolet solar radiation sensors. Sens. Actuators A: Phys. 302, 111825. Available from: https://doi.org/10.1016/j.sna.2019.111825.

Deb, P., Dhar, J.C., 2019. Fast response UV photodetection using TiO2 nanowire/graphene oxide thin-film heterostructure. IEEE Photonics Technol. Lett. 31 (8), 571–574. Available from: https://doi.org/10.1109/LPT.2019.2900283.

Devan, R.S., Patil, R.A., Lin, J.-H., Ma, Y.-R., 2012. One-dimensional metal-oxide nanostructures: recent developments in synthesis, characterization, and applications. Adv. Funct. Mater. 22 (16), 3326–3370. Available from: https://doi.org/10.1002/adfm.201201008.

Dias, J.S., Batista, F.R.M., Bacani, R., Triboni, E.R., 2020. Structural characterization of SnO nanoparticles synthesized by the hydrothermal and microwave routes. Sci. Rep. 10 (1), 9446. Available from: https://doi.org/10.1038/s41598-020-66043-4.

Do, M.-Y., Lee, S.-H., Seo, S.-H., Shin, J.-K., Choi, P., Park, J.-H., et al. 2006. Silicon-on-insulator complementary metal oxide semiconductor image sensor using a nanowire metal oxide semiconductor field-effect transistor-structure photodetector.

Dogar, S., Khan, W., Kim, S.-D., 2016. Ultraviolet photoresponse of ZnO nanostructured AlGaN/GaN HEMTs. Mater. Sci. Semicond. Process. 44, 71–77. Available from: https://doi.org/10.1016/j.mssp.2016.01.004.

Dorn, A., Allen, P.M., Bawendi, M.G., 2009. Electrically controlling and monitoring InP nanowire growth from solution. ACS Nano 3 (10), 3260–3265. Available from: https://doi.org/10.1021/nn900820h.

Du, J.Y., Xing, J., Ge, C., Liu, H., Liu, P.Y., Hao, H.Y., et al., 2016. Highly sensitive and ultrafast deep UV photodetector based on a beta-Ga2O3 nanowire network grown by CVD. J. Phys. D-Appl. Phys. 49 (42). Available from: https://doi.org/10.1088/0022-3727/49/42/425105Artn 425105.

Dzbanovsky, N.N., Dvorkin, V.V., Pirogov, V.G., Suetin, N.V., 2005. The aligned Si nanowires growth using MW plasma enhanced CVD. Microelectron. J. 36 (7), 634–638. Available from: https://doi.org/10.1016/j.mejo.2005.04.035.

Enculescu, I., Toimil-Molares, M.E., Zet, C., Daub, M., Westerberg, L., Neumann, R., et al., 2007. Current perpendicular to plane single-nanowire GMR sensor. Appl. Phys. A 86 (1), 43–47. Available from: https://doi.org/10.1007/s00339-006-3738-2.

Englander, O., Christensen, D., Kim, J., Lin, L., Morris, S.J.S., 2005. Electric-field assisted growth and self-assembly of intrinsic silicon nanowires. Nano Lett. 5 (4), 705–708. Available from: https://doi.org/10.1021/nl050109a.

Fabbro, M.R. Paolo. June 2004. The Difficulty of Working with Silver Alloys with a High Oxygen Content. Legor Group SpA.

Falk, A.L., Koppens, F.H.L., Yu, C.L., Kang, K., de Leon Snapp, N., Akimov, A.V., et al., 2009. Near-field electrical detection of optical plasmons and single-plasmon sources. Nat. Phys. 5 (7), 475–479. Available from: https://doi.org/10.1038/nphys1284.

Fan, Z., Ho, J.C., Jacobson, Z.A., Yerushalmi, R., Alley, R.L., Razavi, H., et al., 2008. Wafer-scale assembly of highly ordered semiconductor nanowire arrays by contact printing. Nano Lett. 8 (1), 20–25. Available from: https://doi.org/10.1021/nl071626r.

Feng, Q., Li, F.G., Dai, B., Jia, Z.T., Xie, W.L., Xu, T., et al., 2015. The properties of gallium oxide thin film grown by pulsed laser deposition. Appl. Surf. Sci. 359, 847–852. Available from: https://doi.org/10.1016/j.apsusc.2015.10.177.

Florica, C., Matei, E., Costas, A., Molares, M.E.T., Enculescu, I., 2014. Field effect transistor with electrodeposited ZnO nanowire channel. Electrochim. Acta 137, 290–297. Available from: https://doi.org/10.1016/j.electacta.2014.05.124.

Florica, C., Costas, A., Boni, A.G., Negrea, R., Ion, L., Preda, N., et al., 2015. Electrical properties of single CuO nanowires for device fabrication: diodes and field effect transistors. Appl. Phys. Lett. 106 (22), 223501. Available from: https://doi.org/10.1063/1.4921914.

Fu, Q.-M., He, D.-C., Yao, Z.-C., Peng, J.-L., Zhao, H.-Y., Tao, H., et al., 2018. Self-powered ultraviolet photodetector based on ZnO nanorod arrays decorated with sea anemone-like CuO nanostructures. Mater. Lett. 222, 74–77. Available from: https://doi.org/10.1016/j.matlet.2018.03.147.

Game, O., Singh, U., Kumari, T., Banpurkar, A., Ogale, S., 2014. ZnO(N)–Spiro-MeOTAD hybrid photodiode: an efficient self-powered fast-response UV (visible) photosensor. Nanoscale 6 (1), 503–513. Available from: https://doi.org/10.1039/C3NR04727J.

Gao, Y., Cansizoglu, H., Polat, K.G., Ghandiparsi, S., Kaya, A., Mamtaz, H., et al., 2017. Photon-trapping microstructures enable high-speed high-efficiency silicon photodiodes. Nat. Photonics 11, 301–308.

Ghose, S., Rahman, M.S., Rojas-Ramirez, J.S., Caro, M., Droopad, R., Arias, A., et al., 2016. Structural and optical properties of beta-Ga2O3 thin films grown by plasma-assisted molecular beam epitaxy. J. Vac. Sci. Technol. B 34 (2), . Available from: https://doi.org/10.1116/1.4942045Artn 021109.

Givargizov, E.I., 1975. Fundamental aspects of VLS growth. J. Cryst. Growth 31, 20–30. Available from: https://doi.org/10.1016/0022-0248(75)90105-0.

Givargizov, E.I., Sheftal, N.N., 1971. Morphology of silicon whiskers grown by the VLS-technique. J. Cryst. Growth 9, 326–329. Available from: https://doi.org/10.1016/0022-0248(71)90250-8.

Gowrishankar, V., Scully, S.R., Chan, A.T., McGehee, M.D., Wang, Q., Branz, H.M., 2008. Exciton harvesting, charge transfer, and charge-carrier transport in amorphous-silicon nanopillar/polymer hybrid solar cells. J. Appl. Phys. 103 (6), 064511. Available from: https://doi.org/10.1063/1.2896583.

Grego, S., Gilchrist, K.H., Kim, J.-Y., Kwon, M.-K., Islam, M.S., 2011. Nanowire-based devices combining light guiding and photodetection. Appl. Phys. A 105 (2), 311–316. Available from: https://doi.org/10.1007/s00339-011-6623-6.

Gudiksen, M.S., Lauhon, L.J., Wang, J., Smith, D.C., Lieber, C.M., 2002. Growth of nanowire superlattice structures for nanoscale photonics and electronics. Nature 415 (6872), 617–620. Available from: https://doi.org/10.1038/415617a.

Guller, O., Peksu, E., Karaagac, H., 2018. Synthesis of TiO2 nanorods for Schottky-Type UV-photodetectors and third-generation solar cells. Phys. Status Solidi (a) 215 (4), 1700404. Available from: https://doi.org/10.1002/pssa.201700404.

Haurylau, M., Chen, G., Chen, H., Zhang, J., Nelson, N.A., Albonesi, D.H., et al., 2006. On-Chip optical interconnect roadmap: challenges and critical directions. IEEE J. Sel. Top. Quantum Electron. 12 (6), 1699–1705. Available from: https://doi.org/10.1109/JSTQE.2006.880615.

Hines, D.R., Ballarotto, V.W., Williams, E.D., Shao, Y., Solin, S.A., 2007. Transfer printing methods for the fabrication of flexible organic electronics. J. Appl. Phys. 101 (2), 024503. Available from: https://doi.org/10.1063/1.2403836.

Holmes, L.H., 1996. The solubility of gases in liquids. J. Chem. Educ. 73 (2), 143. Available from: https://doi.org/10.1021/ed073p143. 143.

Hosako, I., Hiromoto, N., 2004. A novel wave-guide Ge:Ga photoconductor. In: Infrared and Millimeter Waves, Conference Digest of the 2004 Joint 29th International Conference on 2004 and 12th International Conference on Terahertz Electronics, 2004, September 27–October 1, 2004.

Hsu, C.-L., Hsu, D.-X., Hsueh, T.-J., Chang, S.-P., Chang, S.-J., 2017. Transparent gas senor and photodetector based on Al doped ZnO nanowires synthesized on glass substrate. Ceram. Int. 43 (7), 5434–5440. Available from: https://doi.org/10.1016/j.ceramint.2017.01.035.

Hu, L., Chen, G., 2007. Analysis of optical absorption in silicon nanowire arrays for photovoltaic applications. Nano Lett. 7 (11), 3249–3252. Available from: https://doi.org/10.1021/nl071018b.

Hu, X., Holzwarth, C., Masciarelli, D., Dauler, E., Berggren, K., 2009. Efficiently coupling light to superconducting nanowire single-photon detectors. IEEE 19. Available from: https://doi.org/10.1109/TASC.2009.2018035.

Hu, L., Yan, J., Liao, M., Wu, L., Fang, X., 2011. Ultrahigh external quantum efficiency from thin SnO2 nanowire ultraviolet photodetectors. Small 7 (8), 1012–1017. Available from: https://doi.org/10.1002/smll.201002379.

Hu, G.C., Shan, C.X., Zhang, N., Jiang, M.M., Wang, S.P., Shen, D.Z., 2015. High gain Ga2O3 solar-blind photodetectors realized via a carrier multiplication process. Opt. Express 23 (10), 13554–13561. Available from: https://doi.org/10.1364/Oe.23.013554.

Huang, Y., Duan, X., Wei, Q., Lieber, C.M., 2001a. Directed assembly of one-dimensional nanostructures into functional networks. Science 291 (5504), 630. Available from: https://doi.org/10.1126/science.291.5504.630.

Huang, M.H., Mao, S., Feick, H., Yan, H., Wu, Y., Kind, H., et al., 2001b. Room-temperature ultraviolet nanowire nanolasers. Science 292 (5523), 1897. Available from: https://doi.org/10.1126/science.1060367.

Huang, B.-R., Hsu, J.-F., Huang, C.-S., Shih, Y.-T., Lu, K.-S., 2007. Silicon nanowire networks for the application of field effect phototransistor. Mater. Sci. Eng.: C. 27 (5), 1197−1200. Available from: https://doi.org/10.1016/j.msec.2006.08.015.

Huang, J.-S., Hsiao, C.-Y., Syu, S.-J., Chao, J., Lin, C.-F., 2009. Well-aligned single-crystalline silicon nanowire hybrid solar cells on glass. Sol. Energy Mater. Sol. Cell 93, 621−624.

Huang, C.-C., Brian, D.P., John Jr., F.C., 2010. Directed integration of ZnO nanobridge sensors using photolithographically patterned carbonized photoresist. Nanotechnology 21 (19), 195307. Available from: https://doi.org/10.1088/0957-4484/21/19/195307.

Islam, M.Saif, Logeeswaran, V.J., 2010. Nanoscale materials and devices for future communication networks. Commun. Magazine, IEEE 48, 112−120. Available from: https://doi.org/10.1109/MCOM.2010.5473872.

Islam, M.Saif, Sharma, S., Kamins, T.I., Williams, R.S., 2004. Ultrahigh-density silicon nanobridges formed between two vertical silicon surfaces. Nanotechnology 15 (5), L5−L8. Available from: https://doi.org/10.1088/0957-4484/15/5/l01.

Javey, A., Nam, Friedman, R.S., Yan, H., Lieber, C.M., 2007. Layer-by-layer assembly of nanowires for three-dimensional, multifunctional electronics. Nano Lett. 7 (3), 773−777. Available from: https://doi.org/10.1021/nl0630561.

Jiang, X., Xiong, Q., Nam, S., Qian, F., Li, Y., Lieber, C.M., 2007. InAs/InP radial nanowire heterostructures as high electron mobility devices. Nano Lett. 7 (10), 3214−3218. Available from: https://doi.org/10.1021/nl072024a.

Jiang, H., Li, L., Feng, S., Lu, W., 2018. Nanodiamond enhanced ZnO nanowire based UV photodetector with a high photoresponse performance. Phys. E: Low-Dimens. Syst. Nanostruct. 104, 314−319. Available from: https://doi.org/10.1016/j.physe.2018.07.041.

Jin, S., Whang, D., McAlpine, M.C., Friedman, R.S., Wu, Y., Lieber, C.M., 2004. Scalable interconnection and integration of nanowire devices without registration. Nano Lett. 4 (5), 915−919. Available from: https://doi.org/10.1021/nl049659j.

Jones, T.E., Piccinin, S., Stampf, C., 2013. Relativity and the nobility of gold. Mater. Chem. Phys. 141, 14−17.

Jong-Hyun, A., Kim, H.-S., Menard, E., Lee, K.J., Zhu, Z., Kim, D.-H., et al., 2007. Bendable integrated circuits on plastic substrates by use of printed ribbons of single-crystalline silicon. Appl. Phys. Lett. 90 (21), 213501. Available from: https://doi.org/10.1063/1.2742294.

Kamins, T.I., Li, X., Stanley Williams, R., Liu, X., 2004. Growth and structure of chemically vapor deposited ge nanowires on Si substrates. Nano Lett. 4 (3), 503−506. Available from: https://doi.org/10.1021/nl035166n.

Kang, M.-G., Ahn, J.-H., Lee, J., Hwang, D.-H., Kim, H.-T., Rieh, J.-S., et al., 2010. Microwave characterization of a field effect transistor with dielectrophoretically-aligned single silicon nanowire. Jpn J. Appl. Phys. 49 (6), 06GG12. Available from: https://doi.org/10.1143/jjap.49.06gg12.

Kanika, Arora, Kumar, V., Kumar, M., 2018. Silver plasmonic density tuned polarity switching and anomalous behaviour of high performance self-powered β-gallium oxide solar-blind photodetector. Appl. Phys. Lett. .

Karaagac, H., Logeeswaran, V., Saif Islam, M., 2013. Fabrication of 3D-silicon micro-pillars/walls decorated with aluminum-ZnO/ZnO nanowires for optoelectric devices. Phys. Status Solidi A 210. Available from: https://doi.org/10.1002/pssa.201329135.

Kayes, B., Filler, M., Putnam, M.C., Kelzenberg, M., Lewis, N., Atwater, H., 2007. Growth of vertically aligned Si wire arrays over large areas (>1 cm^2) with Au and Cu catalysts. Appl. Phys. Lett. 91, 103110.

Kelzenberg, M.D., Boettcher, S.W., Petykiewicz, J.A., Turner-Evans, D.B., Putnam, M.C., et al., 2010. Enhanced absorption and carrier collection in Si wire arrays for photovoltaic applications. Nat. Mater. 9 (3), 239−244. Available from: https://doi.org/10.1038/nmat2635.

Khanna, S., Marathey, P., Paneliya, S., Chaudhari, R., Vora, J., 2021. Fabrication of rutile − TiO2 nanowire on shape memory alloy: a potential material for energy storage application. Mater. Today: Proc. . Available from: https://doi.org/10.1016/j.matpr.2021.01.012.

Khosravi-Nejad, F., Teimouri, M., Marandi, S.J., Shariati, M., 2019. The highly crystalline tellurium doped ZnO nanowires photodetector. J. Cryst. Growth 522, 214−220. Available from: https://doi.org/10.1016/j.jcrysgro.2019.06.020.

Khudiyev, T., Bayindir, M., 2014. Superenhancers: novel opportunities for nanowire optoelectronics. Sci. Rep. 4 (1), 7505. Available from: https://doi.org/10.1038/srep07505.

Kim, H.W., Kim, N.H., 2005. Synthesis of beta-Ga2O3 nanowires by an MOCVD approach. Appl. Phys. A:Mater. Sci. Process. 81 (4), 763−765. Available from: https://doi.org/10.1007/s00339-004-2982-6.

Kim, D.-H., Rogers, J.A., 2008. Stretchable electronics: materials strategies and devices. Adv. Mater. 20 (24), 4887−4892. Available from: https://doi.org/10.1002/adma.200801788.

Kim, Y., Song, M.S., Kim, Y., Jung, J.H., Gao, Q., Tan, H., et al., 2007. Epitaxial Germanium nanowires on GaAs grown by chemical vapor deposition. J. Korean Phys. Soc. 51. Available from: https://doi.org/10.3938/jkps.51.120.

Kim, T.-H., Carlson, A., Ahn, J.-H., Won, S.M., Wang, S., Huang, Y., et al., 2009. Kinetically controlled, adhesiveless transfer printing using microstructured stamps. Appl. Phys. Lett. 94 (11), 113502. Available from: https://doi.org/10.1063/1.3099052.

Kim, D., Shin, G., Yoon, J., Jang, D., Lee, S.-J., Zi, G., et al., 2013. High performance stretchable UV sensor arrays of SnO2 nanowires. Nanotechnology 24 (31), 315502. Available from: https://doi.org/10.1088/0957-4484/24/31/315502.

Kim, W.T., Kim, I.H., Choi, W.Y., 2015. Fabrication of TiO2 nanotube arrays and their application to a gas sensor. J. Nanosci. Nanotechnol. 15 (10), 8161−8165. Available from: https://doi.org/10.1166/jnn.2015.11278.

Kobayashi, N.P., Mathai, S., Li, X., Logeeswaran, V.J., Islam, M.S., Lohn, A., et al., 2009. Ensembles of indium phosphide nanowires: physical properties and functional devices integrated on non-single crystal platforms. Appl. Phys. A 95 (4), 1005−1013. Available from: https://doi.org/10.1007/s00339-009-5110-9.

Kong, J., Zhou, C., Morpurgo, A., Soh, H.T., Quate, C.F., Marcus, C., et al., 1999. Synthesis, integration, and electrical properties of individual single-walled carbon nanotubes. Appl. Phys. A 69 (3), 305−308. Available from: https://doi.org/10.1007/s003390051005.

Konrad, C.H., Volkl, R., Glatzel, U., 2012. Determination of oxygen diffusion along nickel/zirconia phase boundaries by internal oxidation. Oxid. Met. 77 (3-4), 149−165. Available from: https://doi.org/10.1007/s11085-011-9278-y.

Kumar, S., Sarau, G., Tessarek, C., Bashouti, M.Y., Hahnel, A., Christiansen, S., et al., 2014. Study of iron-catalysed growth of beta-Ga2O3 nanowires and their detailed characterization using TEM, Raman and cathodoluminescence techniques. J. Phys. D-Appl. Phys. 47 (43), . Available from: https://doi.org/10.1088/0022-3727/47/43/435101Artn 435101.

Kung, S.-C., van der Veer, W.E., Yang, F., Donavan, K.C., Penner, R.M., 2010. 20 μs photocurrent response from lithographically patterned nanocrystalline cadmium selenide nanowires. Nano Lett. 10 (4), 1481−1485. Available from: https://doi.org/10.1021/nl100483v.

Lai, W.-J., Li, S.-S., Lin, C.-C., Kuo, C.-C., Chen, C.-W., Chen, K.-H., et al., 2010. Near infrared photodetector based on polymer and indium nitride nanorod organic/inorganic hybrids. Scr. Mater. 63 (6), 653−656. Available from: https://doi.org/10.1016/j.scriptamat.2010.05.035.

Lee, J.S., Saif Islam, M., Kim, S., 2006. Direct formation of catalyst-Free ZnO nanobridge devices on an etched Si substrate using a thermal evaporation method. Nano Lett. 6 (7), 1487−1490. Available from: https://doi.org/10.1021/nl060883d.

Lee, J.S., Islam, M.S., Kim, S., 2007a. Photoresponses of ZnO nanobridge devices fabricated using a single-step thermal evaporation method. Sens. Actuators B: Chem. 126 (1), 73−77. Available from: https://doi.org/10.1016/j.snb.2006.10.042.

Lee, K.-N., Jung, S.-W., Kim, W.-H., Lee, M.-H., Shin, K.-S., Seong, W.-K., 2007b. Well controlled assembly of silicon nanowires by nanowire transfer method. Nanotechnology 18 (44), 445302. Available from: https://doi.org/10.1088/0957-4484/18/44/445302.

Lee, J., Kim, N.-H., Park, Y.S., 2016. Characteristics of SnO2:Sb films as transparent conductive electrodes of flexible inverted organic solar cells. J. Nanosci. Nanotechnol. 16 (5), 4973−4977. Available from: https://doi.org/10.1166/jnn.2016.12173.

Li, H.Y., Wunnicke, O., Borgström, M.T., Immink, W.G.G., van Weert, M.H.M., Verheijen, M.A., et al., 2007. Remote p-doping of InAs nanowires. Nano Lett. 7 (5), 1144−1148. Available from: https://doi.org/10.1021/nl0627487.

Li, L., Zhang, Y., Fang, X., Zhai, T., Liao, M., Sun, X., et al., 2011. WO3 nanowires on carbon papers: electronic transport, improved ultraviolet-light photodetectors and excellent field emitters. J. Mater. Chem. 21 (18), 6525−6530. Available from: https://doi.org/10.1039/C0JM04557H.

Li, X., Xiong, X., Zhang, Q., 2017. Performance-enhancing ultraviolet photodetectors established on individual In2O3nanowires via coating a CuO layer. Mater. Res. Express 4 (4), 045018. Available from: https://doi.org/10.1088/2053-1591/aa6c39.

Li, L., Lou, Z., Chen, H., Shi, R., Shen, G., 2019. Stretchable SnO2-CdS interlaced-nanowire film ultraviolet photodetectors. Sci. China Mater. 62 (8), 1139−1150. Available from: https://doi.org/10.1007/s40843-019-9416-7.

Lieber, C.M., 2001. The incredible shrinking circuit. Sci. Am. 285 (3), 58−64. Available from: https://doi.org/10.1038/scientificamerican0901-58.

Liu, K., Sakurai, M., Aono, M., 2010. ZnO-based ultraviolet photodetectors. Sens. (Basel) 10 (9), 8604−8634. Available from: https://doi.org/10.3390/s100908604.

Logeeswaran, V.J., Sarkar, A., Islam, M.S., Kobayashi, N.P., Straznicky, J., Li, X., et al., 2008. A 14-ps full width at half maximum high-speed photoconductor fabricated with intersecting InP nanowires on an amorphous surface. Appl. Phys. A 91 (1), 1−5. Available from: https://doi.org/10.1007/s00339-007-4394-x.

Logeeswaran, V., Katzenmeyer, A., Islam, M., 2010. Harvesting and transferring vertical pillar arrays of single-crystal semiconductor devices to arbitrary substrates. IEEE Trans. Electron. Devices 57, 1856−1864.

Logeeswaran, Vj, Oh, J., Nayak, A., Katzenmeyer, A., Gilchrist, K., Grego, S., et al., 2011a. A perspective on nanowire photodetectors: current status, future challenges, and opportunities. Sel. Top. Quantum Electron. IEEE J. of 17, 1002−1032. Available from: https://doi.org/10.1109/JSTQE.2010.2093508.

Logeeswaran, V.J., Oh, J., Nayak, A.P., Katzenmeyer, A.M., Gilchrist, K.H., Grego, S., et al., 2011b. A perspective on nanowire photodetectors: current status, future challenges, and opportunities. IEEE J. Sel. Top. Quantum Electron. 17 (4), 1002−1032. Available from: https://doi.org/10.1109/JSTQE.2010.2093508.

Logeeswaran, V.J., Katzenmeyer, A.M., Triplett, M., Ombaba, M., Islam, M.S., 2012a. Interfacing Ag nanoparticles with 1D semiconductor micro/nanostructures via joule heating for transfer printing nanodevices at room ambient. MRS Online Proc. Library 1429 (1), 1−6. Available from: https://doi.org/10.1557/opl.2012.1531.

Logeeswaran, V.J., Triplett, M., Lam, D., Yengel, E., Grewal, H., Ombaba, M., et al., 2012b. Electrical contact characteristics between silicon micropillars and Ag nanoparticles with controlled mechanical load. MRS Online Proc. Library 1429 (1), 20−24. Available from: https://doi.org/10.1557/opl.2012.1532.

Lopez, I., Castaldini, A., Cavallini, A., Nogales, E., Mendez, B., Piqueras, J., 2014. Beta-Ga2O3 nanowires for an ultraviolet light selective frequency photodetector. J. Phys. D-Appl. Phys. 47 (41), . Available from: https://doi.org/10.1088/0022-3727/47/41/415101 Artn 415101.

Lou, Z., Li, L., Shen, G., 2015. InGaO3(ZnO) superlattice nanowires for high-performance ultraviolet photodetectors. Adv. Electron. Mater. 1 (7), 1500054. Available from: https://doi.org/10.1002/aelm.201500054.

Lu, N., Zhou, C., Wang, Y., Elquist, A., Ghods, A., Ferguson, I., et al. 2018. A review of earth abundant ZnO-based materials for thermoelectric and photovoltaic applications.

Luo, L.-B., Zeng, L.-H., Xie, C., Yu, Y.-Q., Liang, F.-X., Wu, C.-Y., et al., 2014. Light trapping and surface plasmon enhanced high-performance NIR photodetector. Sci. Rep. 4 (1), 3914. Available from: https://doi.org/10.1038/srep03914.

Lupan, O., Wolff, N., Postica, V., Braniste, T., Paulowicz, I., Hrkac, V., et al., 2018. Properties of a single SnO2:Zn2SnO4 − Functionalized nanowire based nanosensor. Ceram. Int. 44 (5), 4859−4867. Available from: https://doi.org/10.1016/j.ceramint.2017.12.075.

Maier, S.A., Kik, P.G., Harry, A.A., 2003a. Optical pulse propagation in metal nanoparticle chain waveguides. Phys. Rev. B 67 (20), 205402. Available from: https://doi.org/10.1103/PhysRevB.67.205402.

Maier, S.A., Brongersma, M.L., Kik, P.G., Meltzer, S., Requicha, A.A.G., Koel, B.E., et al., 2003b. Plasmonics—a route to nanoscale optical devices (Advanced Materials, 2001, 13, 1501. Adv. Mater. 15 (7-8), 562. Available from: https://doi.org/10.1002/adma.200390134. 562.

Maier, D., Alomari, M., Grandjean, N., Carlin, J.F., Diforte-Poisson, M.A., Dua, C., et al., 2012. InAlN/GaN HEMTs for operation in the 1000 degrees C regime: a first experiment. IEEE Electron. Device Lett. 33 (7), 985−987. Available from: https://doi.org/10.1109/Led.2012.2196972.

Mann, S.A., Garnett, E.C., 2013. Extreme light absorption in thin semiconductor films wrapped around metal nanowires. Nano Lett. 13 (7), 3173−3178. Available from: https://doi.org/10.1021/nl401179h.

Mao, H., Alhalaili, B., Kaya, A., Dryden, D.M., Woodall, J.M., Saif Islam, M., 2017. Oxidation of GaAs substrates to enable β-Ga2O3 films for sensors and optoelectronic devices, SPIE Optical Engineering + Applications, vol. 10381. SPIE.

Mårtensson, T., Carlberg, P., Borgström, M., Montelius, L., Seifert, W., Samuelson, L., 2004. Nanowire arrays defined by nanoimprint lithography. Nano Lett. 4 (4), 699−702. Available from: https://doi.org/10.1021/nl035100s.

Mayet, A.S., Cansizoglu, H., Gao, Y., Ghandiparsi, S., Kaya, A., Bartolo-Perez, C., et al., 2018. Surface passivation of silicon photonic devices with high surface-to-volume-ratio nanostructures. J. Opt. Soc. Am. B 35 (5), 1059–1065. Available from: https://doi.org/10.1364/JOSAB.35.001059.

McAlpine, M.C., Ahmad, H., Wang, D., Heath, J.R., 2007. Highly ordered nanowire arrays on plastic substrates for ultrasensitive flexible chemical sensors. Nat. Mater. 6 (5), 379–384. Available from: https://doi.org/10.1038/nmat1891.

Meijering, J.L., 1955. On the diffusion of oxygen through solid iron. Acta Metal. 3 (2), 157–162. Available from: https://doi.org/10.1016/0001-6160(55)90085-7.

Meng, M., Wu, X., Ji, X., Gan, Z., Liu, L., Shen, J., et al., 2017. Ultrahigh quantum efficiency photodetector and ultrafast reversible surface wettability transition of square In2O3 nanowires. Nano Res. 10 (8), 2772–2781. Available from: https://doi.org/10.1007/s12274-017-1481-y.

Miao, J., Hu, W., Guo, N., Lu, Z., Liu, X., Liao, L., et al., 2015. High-responsivity Graphene/InAs NAnowire heterojunction near-infrared photodetectors with distinct photocurrent on/off ratios. Small 11 (8), 936–942. Available from: https://doi.org/10.1002/smll.201402312.

Mieszawska, A.J., Jalilian, R., Sumanasekera, G.U., Zamborini, F.P., 2007. The synthesis and fabrication of one-dimensional nanoscale heterojunctions. Small 3 (5), 722–756. Available from: https://doi.org/10.1002/smll.200600727.

Moghadam, F.K., Stevenson, D.A., 1986. Oxygen diffusion and solubility studies in Ag and Pt using Ac impedance spectroscopy. J. Electrochem. Soc. 133 (7), 1329–1332. Available from: https://doi.org/10.1149/1.2108864.

Monroy, E., Calle, F., Pau, J.L., Muñoz, E., Omnès, F., Beaumont, B., et al., 2001. AlGaN-based UV photodetectors. J. Cryst. Growth 230 (3), 537–543. Available from: https://doi.org/10.1016/S0022-0248(01)01305-7.

Moudgil, A., Dhyani, V., Das, S., 2018. High speed efficient ultraviolet photodetector based on 500 nm width multiple WO3 nanowires. Appl. Phys. Lett. 113 (10), 101101. Available from: https://doi.org/10.1063/1.5045249.

Natarajan, C., Setoguchi, K., Nogami, G., 1998. Preparation of a nanocrystalline titanium dioxide negative electrode for the rechargeable lithium ion battery. Electrochim. Acta 43 (21), 3371–3374. Available from: https://doi.org/10.1016/S0013-4686(97)10140-2.

Nguyen, T.D., Kim, E.T., Dao, K.A., 2015. Ag nanoparticle catalyst based on Ga2O3/GaAs semiconductor nanowire growth by VLS method. J. Mater. Sci.Mater. Electron. 26 (11), 8747–8752. Available from: https://doi.org/10.1007/s10854-015-3552-8.

Noh, Y., Jeong, H., Lee, D., 2021. Enhanced ultraviolet photodetector using zinc oxide nanowires with intense pulsed light post-treatment. J. Alloy. Compd. 871, 159537. Available from: https://doi.org/10.1016/j.jallcom.2021.159537.

Novotny, C.J., Edward, T.Y., Paul, K.L.Y., 2008. InP Nanowire/polymer hybrid photodiode. Nano Lett. 8 (3), 775–779. Available from: https://doi.org/10.1021/nl072372c.

Oh, J.Y., Park, J.-T., Jang, H.-J., Cho, W.-J., Islam, M.S., 2014. 3D-transistor array based on horizontally suspended silicon nano-bridges grown via a bottom-up technique. Adv. Mater. 26 (12), 1929–1934. Available from: https://doi.org/10.1002/adma.201304245.

Ombaba, M.M., Logeeswaran, V.J., Islam, M.S., 2013. Electrically conducting film of silver sub-micron particles as mechanical and electrical interfaces for transfer printed micro- and nano-pillar devices. Appl. Phys. A 111 (1), 251–259. Available from: https://doi.org/10.1007/s00339-012-7516-z.

Ombaba, M., Hasegawa, T., Lu, L., Yasuda, Y., Nishida, M., Koh, S., et al., 2013. Hierarchical silver nanoparticle micro-clustering in poly(methyl methacrylate) matrix in spin-coatable electrically conductive thermoplastics. Sci. Adv. Mater. 5, 1546–1555. Available from: https://doi.org/10.1166/sam.2013.1672.

Ombaba, M.M., Jayaraman, L.V., Islam, M.S., 2014. Precision stress localization during mechanical harvesting of vertically oriented semiconductor micro- and nanostructure arrays. Appl. Phys. Lett. 104 (24), 243109. Available from: https://doi.org/10.1063/1.4884200.
Ouyang, W., Teng, F., He, J.-H., Fang, X., 2019. Enhancing the photoelectric performance of photodetectors based on metal oxide semiconductors by charge-carrier engineering. Adv. Funct. Mater. 29 (9), 1807672. Available from: https://doi.org/10.1002/adfm.201807672.
Ozgur, E., Aktas, O., Kanik, M., Yaman, M., Bayindir, M., 2012. Macroscopic assembly of indefinitely long and parallel nanowires into large area photodetection circuitry. Nano Lett. 12 (5), 2483−2487. Available from: https://doi.org/10.1021/nl300597c.
Pallister, P.J., Buttera, S.C., Barry, S.T., 2015. Self-seeding gallium oxide nanowire growth by pulsed chemical vapor deposition. Phys. Status Solidi A-Appl. Mater. Sci. 212 (7), 1514−1518. Available from: https://doi.org/10.1002/pssa.201532275.
Park, J.W., Altstetter, C.J., 1987. The diffusion and solubility of oxygen in solid nickel. Metall. Trans. A-Phys. Metall. Mater. Sci. 18 (1), 43−50. Available from: https://doi.org/10.1007/Bf02646220.
Park, H., Crozier, K.B., 2013. Multispectral imaging with vertical silicon nanowires. Sci. Rep. 3 (1), 2460. Available from: https://doi.org/10.1038/srep02460.
Park, H., Seo, K., Crozier, K.B., 2012. Adding colors to polydimethylsiloxane by embedding vertical silicon nanowires. Appl. Phys. Lett. 101 (19), 193107. Available from: https://doi.org/10.1063/1.4766944.
Park, H., Alhammadi, S., Bouras, K., Schmerber, G., Ferblantier, G., Dinia, A., et al., 2017. Nd-Doped SnO2 and ZnO for application in Cu(InGa)Se2 solar cells. Sci. Adv. Mater. 9 (12), 2114−2120. Available from: https://doi.org/10.1166/sam.2017.3207.
Patil-Chaudhari, D., Ombaba, M., Oh, J.Y., Mao, H., Montgomery, K.H., Lange, A., et al., 2017. Solar blind photodetectors enabled by nanotextured beta-Ga2O3 films grown via oxidation of GaAs substrates. IEEE Photon. J. 9 (2), . Available from: https://doi.org/10.1109/Jphot.2017.2688463Artn 2300207.
Pearton, S.J., Norton, D.P., Ip, K., Heo, Y.W., Steiner, T., 2003. Recent progress in processing and properties of ZnO. Superlattices Microstruct. 34 (1), 3−32. Available from: https://doi.org/10.1016/S0749-6036(03)00093-4.
Peksu, E., Karaagac, H., 2018. Doping and annealing effects on structural, electrical and optical properties of tin-doped zinc-oxide thin films. J. Alloy. Compd. 764, 616−625. Available from: https://doi.org/10.1016/j.jallcom.2018.06.101.
Perea, D.E., Hemesath, E.R., Schwalbach, E.J., Lensch-Falk, J.L., Voorhees, P.W., Lauhon, L.J., 2009. Direct measurement of dopant distribution in an individual vapour−liquid−solid nanowire. Nat. Nanotechnol. 4 (5), 315−319. Available from: https://doi.org/10.1038/nnano.2009.51.
Plass, K.E., Filler, M.A., Spurgeon, J.M., Kayes, B.M., Maldonado, S., Brunschwig, B.S., et al., 2009. Flexible polymer-embedded Si wire arrays. Adv. Mater. 21 (3), 325−328. Available from: https://doi.org/10.1002/adma.200802006.
Pooja, P., Chinnamuthu, P., 2021. Annealed n-TiO2/In2O3 nanowire metal-insulator-semiconductor for highly photosensitive low-noise ultraviolet photodetector. J. Alloy. Compd. 854, 157229. Available from: https://doi.org/10.1016/j.jallcom.2020.157229.
Pyayt, A.L., Wiley, B., Xia, Y., Chen, A., Dalton, L., 2008. Integration of photonic and silver nanowire plasmonic waveguides. Nat. Nanotechnol. 3 (11), 660−665. Available from: https://doi.org/10.1038/nnano.2008.281.

Qian, L.X., Wang, Y., Wu, Z.H., Sheng, T., Liu, X.Z., 2017. beta-Ga2O3 solar-blind deep-ultraviolet photodetector based on annealed sapphire substrate. Vacuum 140, 106−110. Available from: https://doi.org/10.1016/j.vacuum.2016.07.039.

Qiang, Z., Yang, H., Chen, L., Pang, H., Ma, Z., Zhou, W., 2008. Fano filters based on transferred silicon nanomembranes on plastic substrates. Appl. Phys. Lett. 93 (6), 061106. Available from: https://doi.org/10.1063/1.2971199.

Quitoriano, N.J., Kamins, T.I., 2008. Integratable nanowire transistors. Nano Lett. 8 (12), 4410−4414. Available from: https://doi.org/10.1021/nl802292h.

Rai, S.C., Wang, K., Ding, Y., Marmon, J.K., Bhatt, M., Zhang, Y., et al., 2015. Piezo-phototronic effect enhanced UV/Visible photodetector based on fully wide band gap Type-II ZnO/ZnS core/shell nanowire array. ACS Nano 9 (6), 6419−6427. Available from: https://doi.org/10.1021/acsnano.5b02081.

Raj, I.L.P., Valanarasu, S., Hariprasad, K., Ponraj, J.S., Chidhambaram, N., Ganesh, V., et al., 2020. Enhancement of optoelectronic parameters of Nd-doped ZnO nanowires for photodetector applications. Opt. Mater. 109, 110396. Available from: https://doi.org/10.1016/j.optmat.2020.110396.

Ramanarayanan, T.A., Rapp, R.A., 1972. The diffusivity and solubility of oxygen in liquid tin and solid silver and the diffusivity. Metall. Trans. 3 (12), 3239−3246. Available from: https://doi.org/10.1007/bf02661339.

Ramarajan, R., Kovendhan, M., Thangaraju, K., Paul Joseph, D., Ramesh Babu, R., Elumalai, V., 2020. Enhanced optical transparency and electrical conductivity of Ba and Sb co-doped SnO2 thin films. J. Alloy. Compd. 823, 153709. Available from: https://doi.org/10.1016/j.jallcom.2020.153709.

Rapp, R.A., Ramanarayanan, T.A., 1972. The diffusivity and solubility of oxygen in liquid Tin and solid silver and the diffusivity of oxygen in solid nickel ". Metall. Trans. 3 (12), 3239−3246.

Reddy, L.S., Ko, Y.H., Yu, J.S., 2015. Hydrothermal synthesis and photocatalytic property of beta-Ga2O3 nanorods. Nanoscale Res. Lett. 10, . Available from: https://doi.org/10.1186/s11671-015-1070-5ARTN 364.

Ruda, H.E., Shik, A., 2005. Polarization-sensitive optical phenomena in semiconducting and metallic nanowires. Phys. Rev. B 72 (11), 115308. Available from: https://doi.org/10.1103/PhysRevB.72.115308.

Rush, K., Draving, S., Kerley, J., 1990. Characterizing high-speed oscilloscopes. IEEE Spectr. 27 (9), 38−39. Available from: https://doi.org/10.1109/6.58452.

Saif Islam, M., Sharma, S., Kamins, T.I., Williams, R.S., 2005. A novel interconnection technique for manufacturing nanowire devices. Appl. Phys. A 80 (6), 1133−1140. Available from: https://doi.org/10.1007/s00339-004-3177-x.

Seo, K., Wober, M., Steinvurzel, P., Schonbrun, E., Dan, Y., Ellenbogen, T., et al., 2011. Multicolored vertical silicon nanowires. Nano Lett. 11 (4), 1851−1856. Available from: https://doi.org/10.1021/nl200201b.

Seybolt, A.U., 1959. Solubility of oxygen in alpha iron—a revision. Trans. Am. Inst. Min. Metall. Eng. 215 (2), 298−301.

Seybolt, A.U., Fullman, R.L., 1954. A rationalization of the oxygen solid solubility in some transition metals. Trans. Am. Inst. Min. Metall. Eng. 200 (5), 548−549.

Sharma, S., Kamins, T., Islam, M., Williams, R., Marshall, A., 2005. Structural characteristics and connection mechanism of gold-catalyzed bridging silicon nanowires. J. Cryst. Growth 280, 562−568.

Shen, Y., Yan, X., Bai, Z., Zheng, X., Sun, Y., Liu, Y., et al., 2015. A self-powered ultraviolet photodetector based on solution-processed p-NiO/n-ZnO nanorod array

heterojunction. RSC Adv. 5 (8), 5976−5981. Available from: https://doi.org/10.1039/C4RA12535E.
Singh, K., Malakar, R., Narzary, R., Kakoty, P., Mondal, B., 2017. Hydrogen sensing properties of pure and composites of ZnO and SnO2 particles: understanding sensing mechanism. Sens. Lett. 15 (9), 771−778. Available from: https://doi.org/10.1166/sl.2017.3870.
Soci, C., Zhang, A., Xiang, B., Dayeh, S.A., Aplin, D.P.R., Park, J., et al., 2007. ZnO nanowire UV photodetectors with high internal gain. Nano Lett. 7 (4), 1003−1009. Available from: https://doi.org/10.1021/nl070111x.
Soci, C., Zhang, A., Bao, X.Y., Kim, H., Lo, Y., Wang, D., 2010. Nanowire photodetectors. J. Nanosci. Nanotechnol. 10 (3), 1430−1449. Available from: https://doi.org/10.1166/jnn.2010.2157.
Song, P.Y., Wu, Z.Y., Shen, X.Y., Kang, J.Y., Fang, Z.L., Zhang, T.Y., 2017. Self-consistent growth of single-crystalline ((2)over-bar01)beta-Ga2O3 nanowires using a flexible GaN seed nanocrystal. Crystengcomm 19 (4), 625−631. Available from: https://doi.org/10.1039/c6ce02319c.
Sonia, G., Kristin, H.G., Kim, J.-Y., M.-Ki, K., Islam, M.S., 2009. Waveguide-integrated nanowire photoconductors on a non-single crystal surface. Proc.SPIE. .
Sun, Y., Rogers, J.A., 2004. Fabricating semiconductor nano/microwires and transfer printing ordered arrays of them onto plastic substrates. Nano Lett. 4 (10), 1953−1959. Available from: https://doi.org/10.1021/nl048835l.
Tabib-Azar, M., Nassirou, M., Wang, R., Sharma, S., Kamins, T.I., Saif Islam, M., et al., 2005. Mechanical properties of self-welded silicon nanobridges. Appl. Phys. Lett. 87 (11), 113102. Available from: https://doi.org/10.1063/1.2042549.
Takahashi, T., Nichols, P., Takei, K., Ford, A.C., Jamshidi, A., Wu, M.C., et al., 2012. Contact printing of compositionally graded CdSxSe1 − xnanowire parallel arrays for tunable photodetectors. Nanotechnology 23 (4), 045201. Available from: https://doi.org/10.1088/0957-4484/23/4/045201.
Tanaka, A., Chen, R.J., Jungjohann, K.L., Dayeh, S.A., 2015. Strong geometrical effects in submillimeter selective area growth and light extraction of gan light emitting diodes on sapphire. Sci. Rep. 5, . Available from: https://doi.org/10.1038/srep17314ARTN 17314.
Tang, Y., Wang, C., Hu, Y., Huang, L., Fu, J., Yang, W., 2016. Preparation of anatase TiO2 nanorods with high aspect ratio for high-performance dye-sensitized solar cells. Superlattices Microstruct. 89, 1−6. Available from: https://doi.org/10.1016/j.spmi.2015.11.003.
Thelander, C., Björk, M., Persson, A.I., Ohlsson, J., Sass, T., Wallenberg, R., et al., 2003. Heterostructures incorporated in one-dimensional semiconductor materials and devices. Inst. Phys. Conf. Ser. 171, 253−260.
Thelander, C., Agarwal, P., Brongersma, S., Eymery, J., Feiner, L.F., Forchel, A., et al., 2006. Nanowire-based one-dimensional electronics. Mater. Today 9 (10), 28−35. Available from: https://doi.org/10.1016/S1369-7021(06)71651-0.
Tian, W., Lu, H., Li, L., 2015. Nanoscale ultraviolet photodetectors based on onedimensional metal oxide nanostructures. Nano Res. 8 (2), 382−405. Available from: https://doi.org/10.1007/s12274-014-0661-2.
Toole, F.J., Johnson, F.M.G., 1933. The solubility of oxygen in gold and in certain silver-gold alloys. J. Phys. Chem. 37 (3), 331−346. Available from: https://doi.org/10.1021/j150345a005.
Triplett, M., Nishimura, H., Ombaba, M., Logeeswarren, V.J., Yee, M., Polat, K.G., et al., 2014. High-precision transfer-printing and integration of vertically oriented semiconductor arrays for flexible device fabrication. Nano Res. 7 (7), 998−1006. Available from: https://doi.org/10.1007/s12274-014-0462-7.

Tsai, T.-Y., Chang, S.-J., Weng, W.-Y., Hsu, C.L., Wang, S.-H., Chiu, C.-J., et al., 2012. A visible-blind TiO2Nanowire photodetector. J. Electrochem. Soc. 159 (4), J132−J135. Available from: https://doi.org/10.1149/2.008205jes.

Tunnell, A.J., Ballarotto, V.W., Hines, D.R., Williams, E.D., 2008. Vertical integration on plastic substrates using transfer printing. Appl. Phys. Lett. 93 (19), 193113. Available from: https://doi.org/10.1063/1.3026744.

Velho, L.R., Bartlett, R.W., 1972. Diffusivity and solubility of oxygen in platinum and Pt-Ni Alloys. Metall. Trans. 3 (1), . Available from: https://doi.org/10.1007/Bf0268058665-&.

Wagner, R.S., Ellis, W.C., 1964. Vapor-liquid-solid mechanism of single crystal growth. Appl. Phys. Lett. 4 (5), 89−90. Available from: https://doi.org/10.1063/1.1753975.

Wallentin, J., Persson, J.M., Wagner, J.B., Samuelson, L., Deppert, K., Borgström, M.T., 2010. High-performance single nanowire tunnel diodes. Nano Lett. 10 (3), 974−979. Available from: https://doi.org/10.1021/nl903941b.

Wang, G.T., Talin, A.A., Werder, D.J., Creighton, J.R., Lai, E., Anderson, R.J., et al., 2006. Highly aligned, template-free growth and characterization of vertical GaN nanowires on sapphire by metal−organic chemical vapour deposition. Nanotechnology 17 (23), 5773−5780. Available from: https://doi.org/10.1088/0957-4484/17/23/011.

Wang, Y., Cheng, J., Shahid, M., Zhang, M., Pan, W., 2017. A high-performance TiO2 nanowire UV detector assembled by electrospinning. RSC Adv. 7 (42), 26220−26225. Available from: https://doi.org/10.1039/C7RA03072J.

Wei, W., Bao, X.-Y., Soci, C., Ding, Y., Wang, Z.-L., Wang, D., 2009. Direct heteroepitaxy of vertical InAs nanowires on Si substrates for broad band photovoltaics and photodetection. Nano Lett. 9 (8), 2926−2934. Available from: https://doi.org/10.1021/nl901270n.

Weisse, J.M., Kim, D.R., Lee, C.H., Zheng, X., 2011. Vertical transfer of uniform silicon nanowire arrays via crack formation. Nano Lett. 11 (3), 1300−1305. Available from: https://doi.org/10.1021/nl104362e.

Whang, D., Jin, S., Lieber, C.M., 2004. Large-scale hierarchical organization of nanowires for functional nanosystems. Jpn. J. Appl. Phys. 43 (7B), 4465−4470. Available from: https://doi.org/10.1143/jjap.43.4465.

Wu, J.Ming, Chang, W.E., 2014. Ultrahigh responsivity and external quantum efficiency of an ultraviolet-light photodetector based on a single VO2 microwire. ACS Appl. Mater. Interfaces 6 (16), 14286−14292. Available from: https://doi.org/10.1021/am503598g.

Wu, F., Logeeswaran Vj, M.S., Islam, D., Horsley, R., Walmsley, S., Mathai, D., et al., 2009. Integrated receiver architectures for board-to-board free-space optical interconnects. Appl. Phys. A 95, 1079−1088. Available from: https://doi.org/10.1007/s00339-009-5114-5.

Wu, Y.L., Chang, S.J., Weng, W.Y., Liu, C.H., Tsai, T.Y., Hsu, C.L., et al., 2013. Ga2O3 nanowire photodetector prepared on SiO2/Si template. IEEE Sens. J. 13 (6), 2368−2373. Available from: https://doi.org/10.1109/Jsen.2013.2247996.

Wu, Y.L., Luan, Q.P., Chang, S.J., Jiao, Z.Y., Weng, W.Y., Lin, Y.H., et al., 2014. Highly sensitive beta-Ga2O3 nanowire nanowires isopropyl alcohol sensor. IEEE Sens. J. 14 (2), 401−405. Available from: https://doi.org/10.1109/Jsen.2013.2283885.

Xie, T., Hasan, M.R., Qiu, B., Arinze, E.S., Nguyen, N.V., Motayed, A., et al., 2015. Highperforming visible-blind photodetectors based on SnO2/CuO nanoheterojunctions. Appl. Phys. Lett. 107 (24), 241108. Available from: https://doi.org/10.1063/1.4938129.

Xie, Y., Li, H., Zhang, D., Zhang, L., 2019. High-performance quasi-solid-state photoelectrochemical-type ultraviolet photodetector based on ZnO nanowire arrays. Vacuum 164, 58−61. Available from: https://doi.org/10.1016/j.vacuum.2019.03.003.

Xu, J., Zheng, W., Huang, F., 2019. Gallium oxide solar-blind ultraviolet photodetectors: a review. J. Mater. Chem. C. 7 (29), 8753−8770. Available from: https://doi.org/10.1039/C9TC02055A.

Yan, J., Chen, Y., Wang, X., Fu, Y., Wang, J., Sun, J., et al., 2019. High-performance solar-blind SnO2 nanowire photodetectors assembled using optical tweezers. Nanoscale 11 (5), 2162−2169. Available from: https://doi.org/10.1039/C8NR07382A.

Yang, D., Liu, H., Zheng, Z., Yuan, Y., Zhao, J.-cai, Waclawik, E.R., et al., 2009. An efficient photocatalyst structure: TiO2(B) nanofibers with a shell of anatase nanocrystals. J. Am. Chem. Soc. 131 (49), 17885−17893. Available from: https://doi.org/10.1021/ja906774k.

Yang, X., Chen, Z., Xie, X., Xu, X., Xiong, W., Li, W., et al., 2016. Enhanced response speed of ZnO nanowire photodetector by coating with photoresist. J. Nanomater. 2016, 1367095. Available from: https://doi.org/10.1155/2016/1367095.

Yang, J., Jiang, Y., Li, L., Gao, M., 2017. Structural, morphological, optical and electrical properties of Ga-doped ZnO transparent conducting thin films. Appl. Surf. Sci. 421, 446−452. Available from: https://doi.org/10.1016/j.apsusc.2016.10.079.

Yao, J., Yan, H., Lieber, C.M., 2013. A nanoscale combing technique for the large-scale assembly of highly aligned nanowires. Nat. Nanotechnol. 8 (5), 329−335. Available from: https://doi.org/10.1038/nnano.2013.55.

Yasui, K., Kohiki, S., Shimooka, H., Shishido, T., 2008. Effects of Au catalyst on growth of beta-Ga2O3 nanostructure at alpha-Al2O3 (0001) surface. Solid. State Sci. 10 (12), 1860−1863. Available from: https://doi.org/10.1016/j.solidstatesciences.2008.04.005.

Yi, S., Girolami, G., Amano, J., Islam, M.S., Sharma, S., Kamins, T., et al., 2006. InP nanobridges epitaxially formed between two vertical Si surfaces by metal-catalyzed chemical vapor deposition. Appl. Phys. Lett. 89, 133121. Available from: https://doi.org/10.1063/1.2357890. 133121.

Yin, L., Ding, H., Yuan, Z., Huang, W., Shuai, C., Xiong, Z., et al., 2018. A simple and transparent well-aligned ZnO nanowire array ultraviolet photodetector with high responsivity. Optical Mater. 80, 149−153. Available from: https://doi.org/10.1016/j.optmat.2018.04.047.

Yong Oh, J., Saif Islam, M., 2014. Nanobridge gate-all-around phototransistors for electro-optical OR gate circuit and frequency doubler applications. Appl. Phys. Lett. 104 (2), 022110. Available from: https://doi.org/10.1063/1.4862328.

Yoon, J., Baca, A.J., Park, S.-I., Elvikis, P., Geddes, J.B., Li, L., et al., 2008. Ultrathin silicon solar microcells for semitransparent, mechanically flexible and microconcentrator module designs. Nat. Mater. 7 (11), 907−915. Available from: https://doi.org/10.1038/nmat2287.

Yoshida, T., Yamamoto, N., Mizutani, T., Yamamoto, M., Ogawa, S., Yagi, S., et al., 2018. Synthesis of Ag nanoparticles prepared by a solution plasma method and application as a cocatalyst for photocatalytic reduction of carbon dioxide with water. Catal. Today 303, 320−326. Available from: https://doi.org/10.1016/j.cattod.2017.08.047.

Yu, C., Park, J., 2010. Thermal annealing synthesis of titanium-dioxide nanowire−nanoparticle hetero-structures. J. Solid. State Chem. 183 (10), 2268−2273. Available from: https://doi.org/10.1016/j.jssc.2010.08.022.

Yu, B., Sun, X.H., Calebotta, G.A., Dholakia, G.R., Meyyappan, M., 2006. One-dimensional germanium nanowires for future electronics. J. Clust. Sci. 17 (4), 579−597. Available from: https://doi.org/10.1007/s10876-006-0081-x.

Yu, J.W., Wu, Y.R., Huang, J.J., Peng, L.H., 2010. 75GHz Ga2O3/GaN Single nanowire metal-oxide-semiconductor field-effect transistors. In: 2010 IEEE Compound Semiconductor Integrated Circuit Symposium (Csics).

Yu, X., Marks, T.J., Facchetti, A., 2016. Metal oxides for optoelectronic applications. Nat. Mater. 15 (4), 383−396. Available from: https://doi.org/10.1038/nmat4599.

Yu, Z., Sun, T., Liu, B., Zhang, L., Chen, H., Fan, X., et al., 2021. Self-rectifying and forming-free nonvolatile memory behavior in single-crystal TiO2 nanowire memory device. J. Alloy. Compd. 858, 157749. Available from: https://doi.org/10.1016/j.jallcom.2020.157749.

Yuan, H.-C., Shin, J., Qin, G., Sun, L., Bhattacharya, P., Lagally, M.G., et al., 2009. Flexible photodetectors on plastic substrates by use of printing transferred single-crystal germanium membranes. Appl. Phys. Lett. 94 (1), 013102. Available from: https://doi.org/10.1063/1.3062938.

Zang, K., Jiang, X., Huo, Y., Ding, X., Morea, M., Chen, X., et al., 2017. Silicon single-photon avalanche diodes with nano-structured light trapping. Nat. Commun. 8 (1), 628. Available from: https://doi.org/10.1038/s41467-017-00733-y.

Zhan, Y., Zhao, J., Zhou, C., Alemayehu, M., Li, Y., Li, Y., 2012. Enhanced photon absorption of single nanowire α-Si solar cells modulated by silver core. Opt. Express 20 (10), 11506−11516. Available from: https://doi.org/10.1364/OE.20.011506.

Zhan, Z., Xu, L., An, J., Du, H., Weng, Z., Lu, W., 2017. Direct catalyst-free chemical vapor deposition of ZnO nanowire array UV photodetectors with enhanced photoresponse speed. Adv. Eng. Mater. 19 (8), 1700101. Available from: https://doi.org/10.1002/adem.201700101.

Zhang, A., You, S., Soci, C., Liu, Y., Wang, D., Lo, Y.-H., 2008. Silicon nanowire detectors showing phototransistive gain. Appl. Phys. Lett. 93 (12), 121110. Available from: https://doi.org/10.1063/1.2990639.

Zhang, Y.Y., Wu, J., Aagesen, M., Liu, H.Y., 2015. III-V nanowires and nanowire optoelectronic devices. J. Phys. D:Appl. Phys. 48 (46), . Available from: https://doi.org/10.1088/0022-3727/48/46/463001Artn 463001.

Zhang, G., Li, Z., Yuan, X., Wang, F., Fu, L., Zhuang, Z., et al., 2016. Single nanowire green InGaN/GaN light emitting diodes. Nanotechnology 27 (43), 435205. Available from: https://doi.org/10.1088/0957-4484/27/43/435205.

Zhao, Y.Y., Frost, R.L., Yang, J., Martens, W.N., 2008. Size and morphology control of gallium oxide hydroxide GaO(OH), nano- to micro-sized particles by soft-chemistry route without surfactant. J. Phys. Chem. C. 112 (10), 3568−3579. Available from: https://doi.org/10.1021/jp710545p.

Zhao, Z., Wang, B., Ma, J., Zhan, W., Wang, L., Guo, Y., et al., 2017. Catalytic combustion of methane over Pd/SnO2 catalysts. Chin. J. Catal. 38 (8), 1322−1329. Available from: https://doi.org/10.1016/S1872-2067(17)62864-X.

Zheng, Z., Gan, L., Li, H., Ma, Y., Bando, Y., Golberg, D., et al., 2015. a fully transparent and flexible ultraviolet−visible photodetector based on controlled electrospun ZnO-CdO heterojunction nanofiber arrays. Adv. Funct. Mater. 25 (37), 5885−5894. Available from: https://doi.org/10.1002/adfm.201502499.

Zheng, L., Yang, X., Chen, H., Liang, Z., 2018. Flexible ultraviolet photodetectors based on ZnO−SnO2 heterojunction nanowire arrays. J. Semicond. 39 (2), 024002. Available from: https://doi.org/10.1088/1674-4926/39/2/024002.

Zhou, X., Zhang, Q., Gan, L., Li, X., Li, H., Zhang, Y., et al., 2016. High—performance solar-blind deep ultraviolet photodetector based on individual single-crystalline

Zn2GeO4 nanowire. Adv. Funct. Mater. 26 (5), 704−712. Available from: https://doi.org/10.1002/adfm.201504135.

Zhou, Z.Y., Ma, Y.M., Han, Q.F., Liu, Y.L., 2016. Solubility, permeation, and capturing of impurity oxygen in Au/Ag: a comparative investigation from first-principles. Comput. Mater. Sci. 114, 79−85. Available from: https://doi.org/10.1016/j.commatsci.2015.11.023.

Zinkevich, M., Aldinger, F., 2004. Thermodynamic assessment of the gallium-oxygen system. J. Am. Ceram. Soc. 87 (4), 683−691. Available from: https://doi.org/10.1111/j.1551-2916.2004.00683.x.

Zou, Y., Zhang, Y., Hu, Y., Gu, H., 2018. Ultraviolet detectors based on wide bandgap semiconductor nanowire: a review. Sens. (Basel) 18 (7). Available from: https://doi.org/10.3390/s18072072.

High-speed InAs quantum dot photodetectors for data/telecom

Adriano Cola[1], Gabriella Leo[2], Annalisa Convertino[3], Anna Persano[1], Fabio Quaranta[1], Marc Currie[4] and Bahram Nabet[5]

[1]IMM-CNR, Institute for Microelectronics and Microsystems, Unit of Lecce, National Research Council, Lecce, Italy, [2]Institute for the Study of Nanostructured Materials (ISMN)-CNR, Rome, Italy, [3]Institute for Microelectronics and Microsystems (IMM)-CNR, Rome Unit, Rome, Italy, [4]Optical Sciences Division, Naval Research Laboratory, Washington, DC, United States, [5]Electrical and Computer Engineering Department, Drexel University, Philadelphia, PA, United States

5.1 Introduction

The pervasive need for connected devices in the Internet of Things, the high throughput demanded by the 5th generation (5G) communications network, and the explosion of demands for data processing by applications such as autonomous vehicles in hyper-scale data-centers call for high-performance photonics components, such as lasers, modulators, switches, routers, and detectors, operating at the wavelength range 1.3–1.55 μm, the O- and C-Bands. This corresponds to the window of minimum optical loss for silica optical fiber, pivotal in optical communication and data systems. In particular, high-speed photodetectors (PDs), efficiently converting optical into an electrical signal are strategic components for the next generation of fast (100 Gbit/s and above) communication systems and data interconnects, strongly affecting the overall performance of the photonics integrated systems.

To date fast, low-power, low-cost PDs for communication systems based on Ge (Thompson et al., 2016; Benedikovic et al., 2019; Michel et al., 2010) and III−V InGaAs alloys as absorbing material (http://www.osioptoelectronics.com/Libraries/Datasheets/FCI-InGaAs-36C.sflb.ashx; Chen et al., 2021) are available on the market. However, the demand for increasingly better-performing optoelectronic and photonic devices, which are easy to integrate with the existing platforms, has continuously motivated the research in the material science and technology of a variety of semiconductors. In the last 30 years, advancement in epitaxial semiconductor growth and chemical route to nanostructured material preparation has led to the development of high-quality low dimensional materials such as semiconductor quantum wells (QWs) (Physics and Applications, 2012), quantum wires (Levine, 1996), quantum dots (QDs) (Bimberg, 2018; Ren et al., 2019; Wu et al., 2015), quantum dashes (Qdashes) (Khan et al., 2014; Wan et al., 2020), colloidal nanocrystals (Gao et al., 2016; Xu et al., 2020a,b) or, more recently, 2D materials [e.g., graphene (Mueller et al., 2010; Pospischil et al., 2013; Ma et al., 2019; Ding et al., 2020)] and MXenes (Naguib et al., 2011; Montazeri et al., 2019; Kim et al., 2021).

Unlike the traditional bulk semiconductors, low-dimensional semiconductors show energy quantization of the electronic levels in the confinement directions allowing to tailor the semiconductor's electronic and optical properties by changing the size, and therefore the confined spatial dimension (2D, 1D, 0D). In particular, 0-dimensional semiconductors (0D), the so-called QDs, have attracted great interest for applications in optoelectronics and photonics devices due to their unique bandgap tunability, discrete energy level with quasi-δ-like density of states, and 3-D confinement of carriers (Bimberg, 2018). In addition, QDs provide many parameters for tuning the device operating spectral range, such as QD size and shape, strain, and alloy composition.

In the early 1990s defect-free, nanometer-sized self-assembled Stranski-Krastanov InAs QDs had been intensively studied (Leonard et al., 1993) for their potential to replace epitaxial semiconductors and QWs in infrared (IR) devices by exploiting QDs intraband transitions (Campbell and Madhukar, 2007). Building on such results, the development of high performances PDs in the telecommunication spectral range based on InAs QDs has been pursued by researchers and industry as a sound alternative to the higher cost/lower size InP-based devices exploiting the technological maturity of the GaAs platform (Persano et al., 2010; Golovynskyi et al., 2018). From such a perspective, the outstanding results achieved in the fabrication of highly performing lasers using self-assembled InAs QDs active layer on GaAs substrate [(Bimberg, 2018; Wu et al., 2015) and references therein] have driven the scientific and technology research to achieve high detectivity, high responsivity, low dark current and high-speed InAs QDs based PDs operating at 1.3 μm.

More recently, heterogeneous integration of III−V semiconductor photonics components with silicon technology has been pursued as a promising approach to achieving low-cost, high-performing photonic integrated circuits (PICs). Integrating III−V semiconductors on Si overcomes the lack of laser sources on the Si platform and permits high-speed Si-based PDs in the communication spectral band. The very promising results obtained with 1.3 μm InAs QD lasers integrated on Si boosted the research activity of many groups to explore new technology strategies and design for efficient and high-speed InAs QD PDs on Si. PD structures based on InAs directly grown on Si (Wan et al., 2017; Inoue et al., 2018; Chen et al., 2020) and grown on III−V substrates wafer bonded to Si (Tossoun et al., 2019) have been reported in the most recent literature. This represents a significant stride toward PICs fully compatible with complementary metal-oxide semiconductor (CMOS) foundries.

At 1.55 μm, commercial PDs still rely on InP or Ge platform as thin films (Piels and Bowers, 2016). However, the optimization of III−V QDs growth on InP substrates and the advancement in 1.55 μm QD laser performance (Wan et al., 2019) along with their monolithic integration on Si platforms (Shi et al., 2017) is prompting research towards InAs QDs and Qdash based PDs at 1.55 μm (Wan et al., 2020; Umezawa et al., 2014).

This chapter aims to provide an in-depth view of the state of the art with recent advances in III−V QD growth and processing technologies, as well as PDs design and performance for the 1.3−1.55 μm spectral range. Section 5.2 reviews the most recent advances and trends in the growth of the III−V QDs and QD structures for application in the communication spectral range. Section 5.3 reports on the main electrical and photoelectrical properties of InAs QDs. Section 5.4 describes the basic configurations of PDs used for optical communications. Section 5.5 provides an overview of InAs QD PDs performance, compared with the main figures-of-merit of the present tele- and data-communication PDs. Finally, Section 5.6 discusses the future of InAs QD PDs.

5.2 InAs quantum dots: epitaxial growth advances

The classical approach to hetero-epitaxial growth focuses mainly on the general paradigm of lattice-matched or nearly lattice-matched growth: in order to obtain a coherent hetero-structure, that is defect free, the lattice constant of both the growing material and substrate has to be very similar (i.e. lattice mismatch $\leq 1\%$). Lattice mismatched systems can grow coherently only below a certain critical thickness, beyond which dislocations and defects will affect the crystal structure. By looking at the bandgap/wavelength and lattice constant diagram for typical III−V semiconductors and their alloys in Figure 5.1 it is clear that only the In_xGa_{1-x} As alloy with

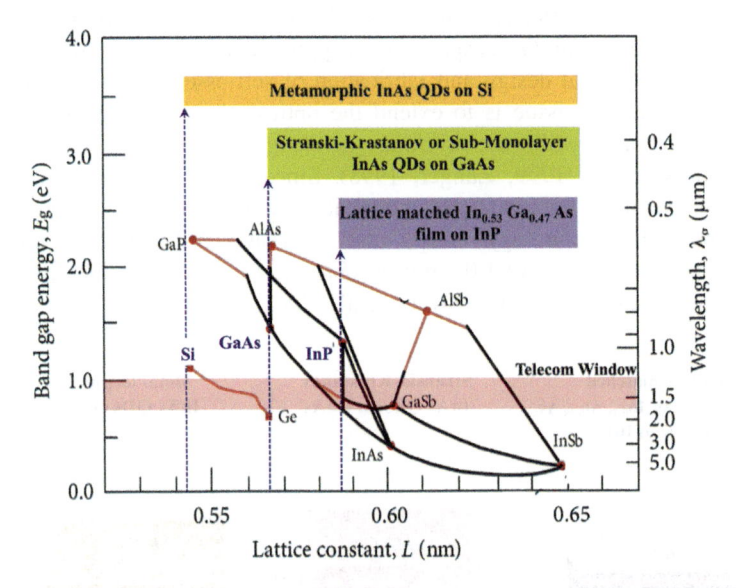

Figure 5.1 Plot of bandgap energy/wavelength versus lattice constant for III−V, Si, and Ge. The black lines indicate direct bandgaps. The vertical blue arrows point to the lattice constants of common commercial substrates for optoelectronic devices (InP, GaAs, and Si); for each substrate, the corresponding growth mode for obtaining In(Ga) As-based structures optically active in the telecom band are indicated: lattice-matched InGaAs films on InP, InAs QDs via Stranski-Krastanov or submonolayer growth modes on GaAs, and metamorphic InAs QDs on Si. The horizontal pink bar shows the wavelength range of the telecommunication window, 1.3−1.55 μm.

$x \approx 0.5$ ($E_g \approx 0.73$ eV) enables the achievement of light absorption in the range 1.3–1.55 μm and a complete or nearly lattice match ($\leq 1\%$) with the InP substrate. Figure 5.2 schematically shows $In_{0.53}Ga_{0.47}As/In_{0.52}Al_{0.48}$ As QWs lattice matched to InP substrate.

In the early 1990s, the self-assembly growth mechanism of Stranski–Krastanov (S–K), allowed hetero-epitaxial growth of material on lattice-mismatched substrates, fabricating high density of self-assembled dislocation-free, nanometric islands of semiconductors, achieving quantum confinement of carriers, the so-called QDs. In S-K epitaxial growth, after reaching a critical thickness of deposited materials, the strain accumulated in the grown material due to the large mismatch (2%–10%) with the substrate turns the epitaxial growth into 3D islands nucleation on a thin layer known as wetting layer (WL) (see Figure 5.2 for sketch of InAs QDs on GaAs). The WL for the lattice-mismatched InAs/GaAs system (about 7%) is about 1.7 ML thick (around 0.56 nm) (Leonard et al., 1993, 1994).

This new route of epitaxial growth, obtained by molecular beam epitaxy or metal-organic chemical vapor deposition, created the unprecedented opportunity to fabricate self-assembled InAs QDs on a GaAs substrate, which can be optically active in the 1.3–1.55 μm wavelength region (Passaseo et al., 2001a; Convertino et al., 2004; Ustinov et al., 1999). Figure 5.3 shows the typical atomic force microscopy image of a single layer of InAs QDs on GaAs substrate (Figure 5.3A) and the scanning transmission electron microscope image of the cross-section of 10 InAs QD layers on GaAs with GaAs spacer and capping layer (Figure 5.3B).

However, the actual design and fabrication of self-assembled InAs QDs remain challenging. The first issue is to extend the optical absorption of the QDs to the 1.3–1.55 μm range, while the working wavelength of the InAs QDs is ~1 μm (Grassi Alessi et al., 1999; Zunger, 1998). Since the early 2000s, many efforts focused on extending the InAs QDs optical activity in the telecom windows by controlling the QD morphology (shape and size) and In-Ga intermixing effects. Emission wavelengths above 1.0 μm were obtained by optimizing the growth conditions such as deposition rate (Joyce et al., 2001; Gerard et al., 1995), arsenic

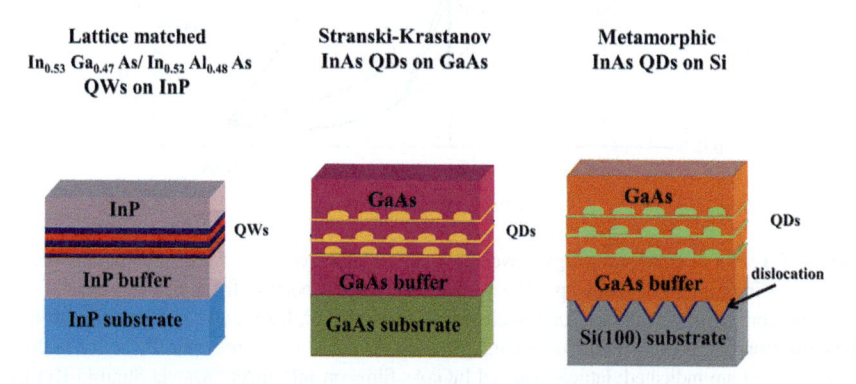

Figure 5.2 Schematic diagram of three different growth modes of III–V semiconductor nanostructures on InP, GaAs, and Si substrates.

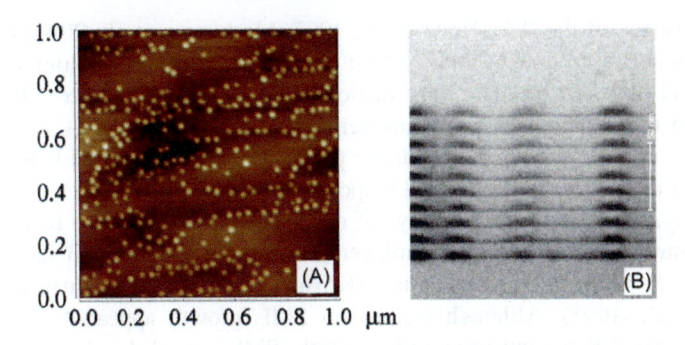

Figure 5.3 (A) Atomic force microscopy (AFM) plan view image of a single InAs QDs on GaAs substrate. The average QD diameter and height turn out to be 21.9 ± 5.2 and 2.2 ± 0.7 nm. The error reported is the standard deviation of the QD size distributions (Persano et al., 2007); (B) Scanning transmission electron microscope (STEM) image through a stack of 10 InAs QD layers, with 10 nm thick GaAs spacer/capping layers (Martí et al., 2007).
Source: (A) Reproduced with permission from Persano, A., Cola, A., Taurino, A, Catalano, M., Lomascolo, M., Convertino, A., et al. 2007. Electronic structure of double stacked InAs/GaAs quantum dots: experiment and theory, J. Appl. Phys. 102, 094314; (B) Reproduced with permission from Martí, A., López, N., Antolín, E., Cánovas, E., Luque, A., Stanley, C.R., et al. 2007. Emitter degradation in quantum dot intermediate band solar cells. Appl. Phys. Lett. 90, 233510.

pressure (Chu et al., 1999a; Riel et al., 2002), substrate temperature (Chu et al., 1999a) and introducing growth interruptions (Convertino et al., 2004). Another effective way to achieve the long-wavelength telecom window is to cap the QDs with InGaAs strain-reducing layer, instead of GaAs (Tatebayashi et al., 2001; Chang et al., 2003). More recently, the use of quaternary III–V alloys in the capping layer of InAs QDs has reduced the noise in the current density and improved the thermal stability of the carriers in the QDs-based PDs (Tongbram et al., 2015; Chatterjee et al., 2019).

The second important issue is the low quantum efficiency (Pan et al., 1998; Liu et al., 2001; Chu et al., 2001) of QD-based structures due to the relatively small absorption volume of a single layer of QDs, being the dot density in the range $10^{10}-10^{11}$ cm^{-2} (see Figure 5.3A). Although several improvements have been made over the years by exploring different material systems, including multiple QD layers (Umezawa et al., 2014; Ledentsov et al., 1996; Lao et al., 2013; Passaseo et al., 2001b) (see Figure 5.3B), QD in a well (DWELL) (Persano et al., 2010; Barve et al., 2010; Wang et al., 2016), QD incorporated in periodic current blocking layers (usually composed of either AlGaAs or InGaAs) (Wang et al., 2001a; Ling et al., 2008; Murata et al., 2020), the low quantum efficiency remains a significant limitation for the exploitation of InAs QDs as active material in optoelectronic devices. We should also consider that the self-assembling mechanism is an intrinsically uneven phenomenon due to unavoidable slight variations in the dot size, shape, and composition. Ultimately, the electronic states of the QDs can be affected by the presence of the WL (Heitz et al., 1997; Markussen et al., 2006; Berg et al., 2001; Gomis-Bresco et al., 2008), whose composition and morphology (i.e. indium content and thickness) remain very difficult to control and engineer. These drawbacks lead to an inhomogeneous broadening of transition energies within the dot ensemble with deleterious effects on the performance of QD-based PDs, especially in terms of dark current and spectral response.

Recently, dense and ordered arrays of nanoscaled and crystalline InAs islands on GaAs substrate obtained by submonolayer (SML) growth have been shown to overcome several of the SK growth issues (Krishna et al., 1999; Mikhrin et al., 2000; Tsyrlin et al., 1995). Although studies on SML growth appeared quite simultaneously to the S-K growth ones, only recently SML growth has been considered a promising growth strategy for the fabrication of optoelectronic devices (Xu et al., 2004; Germann et al., 2008; Ledentsov et al., 2007; Lai et al., 2007; Lenz et al., 2010; Kim et al., 2013, 2018; Das et al., 2020). The SML growth involves the deposition of a fraction of a monolayer of material (e.g., In(Ga)As) on the host matrix (GaAs). Then, a thin spacer layer (a few angstroms to nanometers of GaAs or InGaAs or InAlGaAs) is deposited before the next deposition of the InAs SML. The related local increase in lattice constant provides favorable sites for the nucleation of InAs also in the subsequent layer, leading in the ideal case to a vertical alignment of the SML InAs material. SML QDs do not develop a WL and are characterized by ultrahigh dot density ($\geq 10^{12}$ cm^{-2}) (Xu et al., 2003; Lenz et al., 2011), excellent size uniformity, and high crystal quality, leading to higher absorption efficiency (Germann et al., 2008; Lenz et al., 2010), controlled dark current and improved sharper photoresponse (Kim et al., 2013, 2018). Very recently Das (Das et al., 2020) has proposed the mixing of the S-K and SML approaches by realizing vertically coupled heterostructures of combined S-K and SML QD layers showing improvements in the optical and photoelectrical properties from the near-IR up to about 8 μm (Dongre et al., 2020).

The need to further improve the 1.55 μm optoelectronic devices, turned research towards the growth of S-K InAs QDs on InP as an active component of high performances lasers and PDs. The small lattice mismatch and the resulting complex strain distribution in InAs/InP structures resulted in the formation of a new class of self-assembled nanostructures, the Qdashes (Sauerwalda et al., 2005). These elongated dot-like nanostructures, active between ~ 1.5 and ~ 2.0 μm wavelength range, have been used as laser sources (Wan et al., 2019; Khan et al., 2019), optical amplifiers (Lelarge et al., 2007) and low dark current photodiodes (Wan et al., 2020).

Lastly, the direct growth of III−V heterostructures on Si wafers is a long-standing goal and the integration of optoelectronics based on III−V semiconductors with the well-established and low-cost Si manufacturing technologies remains a formidable challenge. The first attempts to grow GaAs thin films on Si substrates date the back to early 1980s (Fischer et al., 1985; Wang, 1984). The monolithic growth of III−Vs on Si is affected by several drawbacks due to the large material dissimilarity between group III−V and group IV materials, including polar versus nonpolar surfaces, lattice mismatch ($>10\%$ between InAs and Si), and different coefficients of thermal expansion (CTE). These differences tend to produce antiphase boundaries (APBs), threading dislocations (TDs), and thermal cracks, respectively. All these defects generate nonradiative recombination centers and to a certain extent can be treated as leakage paths for electronic devices causing a significant reduction in electron mobility and compromising device performance and lifetime (Alcotte et al., 2016; Akiyama et al., 1984; Norman et al., 2018). However, high densities of QDs in the device active region can largely limit the undesirable effects of crystal defects thanks to the efficient confinement of photo-excited carriers by individual QDs, making the III−V QD-based PDs more advantageous than the ones based on the III−V QW.

To date, several approaches have been explored for the monolithic growth of III−V semiconductors on Si based on the idea of growing fully relaxed III−V materials on a Si substrate by accommodating crystal defects in engineered regions away from the active areas. This metamorphic growth approach in principle accommodates an unlimited degree of lattice mismatch (Norman et al., 2018). Also, in metamorphic growth, there is no limit to the thickness of the relaxed epilayer. Within the metamorphic approach, three different strategies of confining the defects in the device region have emerged: abrupt interfaces, buffer layers, and structured interfaces (Richardson and Lee, 2016).

An abrupt interface is an atomically abrupt transition from the substrate material to the thin-film layer that is overgrown. A typical abrupt interface is GaAs on a Si substrate. A reduction of the TDs and crack defect density can be obtained by optimizing the growth conditions, annealing treatments of the deposited layer, or inserting strained superlattice structures as dislocation filters (Park et al., 2020). In 2015, monolithic growth of InAs QD structures on Si substrates for use as both photodiodes and avalanche photodiodes (APD) operating at a wavelength of 1.3 μm was demonstrated (Sandall et al., 2012).

The second approach is the insertion of an intermediate buffer layer of other materials whose lattice constant and coefficient of thermal expansion are matched with the substrate, for example, Si. For GaAs/Si heteroepitaxy, a wide variety of buffers using Ge, GaAsP, and InGaP has been explored. Among these materials, Ge has been most widely used because of its complete miscibility with Si, well-developed Ge-on-Si growth technology, and nearly the same lattice constant and CTE matching between GaAs and Ge (Lee et al., 2013; Tsuji et al., 2004; Groenert et al., 2003; Carlin and Ringel, 2000). Moreover, the compositionally graded Ge/Ge$_x$Si$_{1-x}$ structure offers efficient strain relaxation in the buffer layer, and therefore a final Ge cap layer serves as a virtual substrate for GaAs growth. Although this strategy can be very effective in producing very low dislocation densities and successful demonstrations of III−V QD laser and photodiode integration on Si have been achieved (Lee et al., 2012; Wan et al., 2015), the compositionally graded buffers often require several micrometers of thickness and come with significant surface roughening which is detrimental for the QD nucleation.

The third main strategy for metamorphic growth is to use structured or patterned interfaces for manipulating dissimilarities in structure between the substrate and the thin film. The techniques for creating these interfaces include selective area growth, where growth occurs on a pedestal or rib such that the dislocations can terminate at a free sidewall before the device region is grown. For example, growth on

V-grooved Si(100) substrates (Wan et al., 2017) whose surface can localize most of the APBs generated at the GaAs/Si hetero-interface in the V-grooves, has been pursued to achieve high-quality GaAs-on-V-grooves-Si (GoVS) (001) substrates (see Figure 5.2). The use of GoVS substrate is very promising for effective monolithic fabrication of III–V QDs photodiodes on Si. These engineered GoVS substrates offer the further advantage of trapping and reducing TDs, as Wan et al. (Wan et al., 2017, 2015) reported for InAs/GaAs QDs on GoVS with an emission wavelength of 1.3 μm at room temperature. In 2020 Chen (Chen et al., 2020) obtained InAs QD waveguide (WG) APDs monolithically grown on GoVS.

5.3 Electrical and photoelectrical properties of quantum dots

The striking feature of QDs-based PDs is that, upon absorption of light, free carriers are generated by electronic transitions between the QD discrete levels, which result from the 3D quantum confinement of the dot in the embedding matrix. The discrete band structure of semiconductor InAs QDs allows wide tuning of the operating wavelength of QDs PDs from about 1 μm up to about 20 μm by exploiting interband or intraband transitions, as shown in Figure 5.4.

QDs PDs designed for mid (MWIR, 3–8 μm) and long (LWIR, 8–14 μm) wavelength IR range usually rely on unipolar n-type/intrinsic/n-type device structures, the QDs being embedded in the intrinsic region (Stiff-Roberts, 2009). From the occupied QD electronic ground state, intraband transitions to excited states are induced

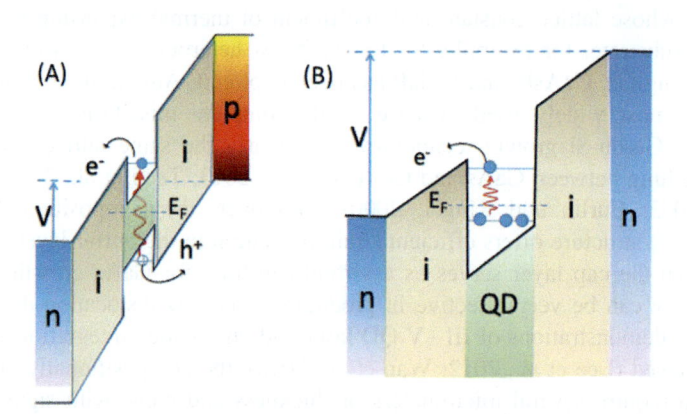

Figure 5.4 Sketch of biased QD photodetector structure (A) interband transitions in a typical p-type/intrinsic/n-type (PIN) configuration for optical communications; (B) intraband transitions in a typical n-type/intrinsic/n-type (NIN) configuration for infrared wavelengths detection. Note that in (A) both carriers contribute to photocurrent, while in (B) only one carrier type (electrons) does. The Fermi level, E_F, in the QD is also drawn to highlight the difference in the QD occupation states in the two PDs.

by IR photons (Figure 5.4B). It has been shown that a suitable QD doping enables fine control of its electron occupation which results in an optimized photoresponse (Lin et al., 2004; Attaluri et al., 2006). Following photon absorption, tunneling and thermal emission of the excited electrons outside the QDs are the main mechanisms that give rise to the photocurrent (PC). Another possibility is the bound-to-continuum transition, in which an electron in the ground- or excited state is photoexcited into the continuum of energy levels out of the QD (Stiff-Roberts, 2009).

On the other hand, QDs based PDs, operating in the 1.3–1.55 μm detection window, rely on interband optical transitions from QD valence band states into the empty conduction band electronic states, as sketched in Figure 5.4A in a typical p-type/intrinsic/n-type (PIN) configuration for optical communications. The photo-generated electrons and holes drift towards their collecting electrode having escaped the potential well by tunneling and/or thermal emission. However, irrespective of the optical transitions that produce free carriers, trapping or recombination occurs during the transport in the matrix and across the photodetector structure. Hence, understanding the absorption and carrier transport mechanisms, including the limiting processes, is crucial for an effective design to enhance QD PD's performance.

Since carrier transport mechanisms perpendicular and parallel to QD plane(s) are different, in the following sections the properties of both vertical and planar device structures will be discussed separately.

5.3.1 Electrical properties of quantum dots in vertical device structures

Vertical structures offer the best configuration to investigate the electronic properties of the QDs, allowing for a uniform and fine control of the charge occupation and the electric field surrounding the dots. Photocurrent Spectroscopy (PS) is largely used to study QD systems as it directly addresses the optical absorption and carrier transport mechanisms. The competition between optical recombination and separation/drift of carriers, photogenerated in the QDs, has been often highlighted by the complementarity of photoluminescence (PL) and PC signals dominant at low/high electric fields, respectively (Chang et al., 2001a; Brunkov et al., 2002; Fry et al., 2000b; Findeis et al., 2001). Energy values corresponding to the interband transition between the electron/hole ground states ($e_0 - h_0$), and often the first ($e_1 - h_1$) and the second ($e_2 - h_2$) excited states have been obtained from PC spectra of a single layer of InAs QDs in both Schottky (Brunkov et al., 2002; Persano et al., 2005) and PIN vertical structures (Fry et al., 2000b; Chang et al., 2000). In a study on ten monolayers of $In_{0.35}Ga_{0.65}As$ QDs, Passmore et al. (2008) identified PC peaks corresponding to several interband transitions $hh_i \rightarrow e_i$, from heavy hole states hh_i (but possibly also some transitions from light hole states) to electron states e_i, with i ranging from 1 to 6. In addition, a continuum signal associated to QDs discrete transitions, increasing with the energy of the incident photons, is commonly observed in PC signals as shown in Figure 5.5 (Brunkov et al., 2002; Persano et al., 2005; Chang et al., 2000; Fry et al., 2000a). In agreement with the

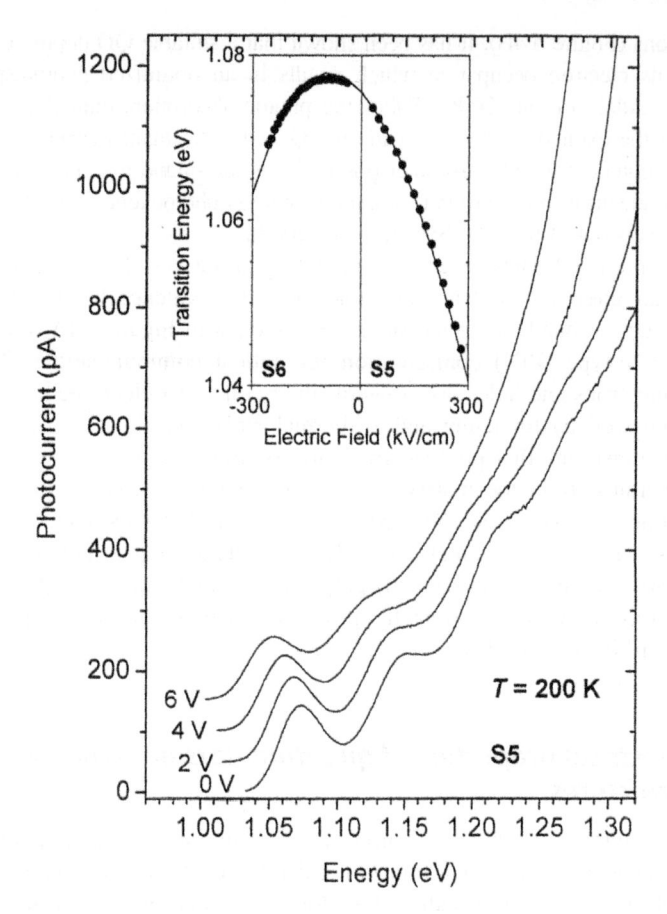

Figure 5.5 Photocurrent spectra as a function of applied bias at a temperature of 200K. A strong Stark shift to the lower energy of interband transitions is seen. The inset shows the ground state transition energies for PIN ($E > 0$) and NIP ($E < 0$) samples as a function of the electric field, together with a parabolic fit.
Source: Reproduced with permission from Fry, P.V., Skolnick, M.S., Mowbray, D.J., Itskevich, I.E., Finley, J.J., Wilson L.R., et al. 2001. Electronic properties of InAs/GaAs self-assembled quantum dots studied by photocurrent spectroscopy. Phys. E 9, 106–113.

calculations published by Vasanelli et al. (2002), Mowbray and Skolnick (2005) showed that such signal arises from "crossed" transitions between the valence band of the WL and the first confined state of the dots. Further evidence for interaction between discrete dot states and the WL has been presented by Karrai et al. (2004).

The response to polarized light is an important aspect of the QD optical absorption. PDs with high polarization sensitivity are in great demand in advanced optical communication (Tan and Mohseni, 2017). Polarization sensitivity implies a different PC response to in-plane polarized light (Transverse Electric TE) and light with polarization in the growth direction (Transverse Magnetic). Fry et al. (2000a) measured strong TE polarized PC spectra, either under normal and parallel light incidence with respect to the QD plane, ascribed to heavy hole absorption. It has also been pointed out in the literature that polarization sensitivity in self-assembled QDs depends on the biaxial strain, dot shape, numbers of stacking layers, and vertical coupling among dots (see for example Saito et al., 2008; Chu et al., 1999b).

The carrier escape mechanism is at the core of the PD performance since it enables optical absorption to be turned into the measured signal, that is the PC. Insights on carrier escape from the QDs can be found by studying PC spectra as a function of the temperature and applied electric fields. Thermionic emission and tunneling are the main competing escape mechanisms in QD-based structures. Tunneling dominates at low temperatures, while at higher temperatures, carrier thermal activation prevails (thermionic emission) and increases until all carriers are emitted with the consequent PC saturation. Figure 5.6 shows how the applied voltage can

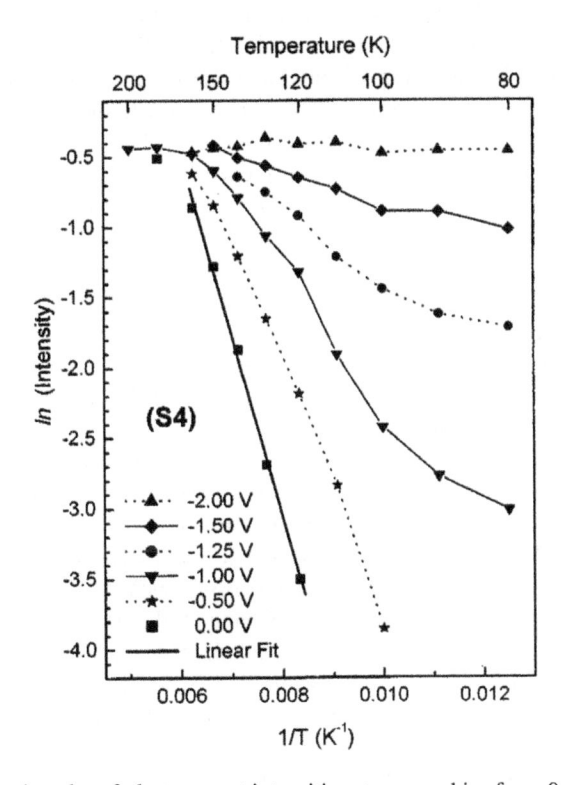

Figure 5.6 Arrhenius plot of photocurrent intensities at reverse bias from 0 to 2 V. From the slope of the 0 V results, where thermal escape dominates, the activation energy of 130 meV is deduced. With increasing bias the slopes decrease as tunneling becomes increasingly important until at -2 V the thermal escape becomes temperature independent.
Source: Reproduced with permission from Fry, P.V., Itskevich, I.E., Parnell, S.R., Finley, J. J., Wilson, L.R., Schumacher, K.L., et al. 2000a. Photocurrent spectroscopy of InAs/GaAs self-assembled quantum dots, Phys. Rev. B 62, 16784–16791.

switch the transition from a full thermally activated behavior (for $V=0$) to a full temperature-independent behavior, that is dominated by the tunneling ($V = -2V$) (Fry et al., 2000a). Thermally activated PC indicates thermionic emission from the localized states. The escape activation energy can be extracted from the slope of the corresponding Arrhenius plot. Interestingly, the deduced activation energies for QDs are typically quite small, 2–3 times smaller than the corresponding exciton localization energies (Chang et al., 2000, 2001a; Brunkov et al., 2002; Fry et al., 2000a), which is indicative that indirect emission processes take place.

The mechanisms for electron and hole emission from InAs/GaAs QDs have also been investigated by space-charge techniques, such as admittance and deep-level transient spectroscopy (DLTS) technique (see for example Wang et al., 2000; Chang et al., 2001b; Schulz et al., 2004; Kapteyn et al., 2000; Kaniewska et al., 2010). DLTS allows us to characterize the deep levels and the band offset of semiconductor heterostructures. QDs embedded in an n-type or p-type matrix allow separate investigation of capture/emission mechanisms of electrons and holes, respectively (Wang et al., 2000; Geller et al., 2006a). The control of the QD charge occupation occurs by applying a suitable external bias. The applied bias also changes the electric field experienced by carriers inside the QDs hence enhancing tunneling escape. Also, it has been shown (Chang et al., 2001b; Schulz et al., 2004) that the charge state of the dot strongly influences the emission process.

As for PC measurements, DLTS measurements at low temperature demonstrate pure tunneling conditions, which specifically addresses only the effect of the electric field. Under such conditions (see Figure 5.7) Geller et al. (Geller et al., 2006a,b) measured tunneling barrier height and calculated ground state localization energies, for both electrons and holes, in good agreement with theoretical predictions. In addition, at room temperature and zero fields, the escape time for thermal excitation from the electron/hole ground states was estimated at 200 and 0.5 ns for holes and electrons, respectively (Geller et al., 2006b).

DLTS studies of QD charge emission allowed Engström et al. (2003) and Schulz et al. (2004) to distinguish the s1 and s2 energies associated with the single- and two-electron state in the s-shell of the QDs. Similar to the PC, the thermal activation energies extracted by DLTS studies are often smaller than localization energies. The main reason is attributed to indirect emission processes such as thermal excitation to a bound dot level and subsequent tunneling (Chang et al., 2001b, 2002; Kapteyn et al., 1999, 2000) or phonon-assisted tunneling (Schulz et al., 2004). In particular, Chang et al. (2001b) showed that electrons escape is a two-step process involving excited states. By performing DLTS measurements Kapteyn et al. (1999) were able to confirm the two-step process for the electron escape at a temperature above 40K, while direct tunneling occurred at lower temperatures.

In addition, and closely related to carrier escape mechanisms, the electron and hole contribution to PC has been intensively debated since the early 2000s. According to Fry et al. (2001), both in the case of tunneling and thermionic emission, carrier escape occurs from the same state in the dot, probably the electron/hole ground state. They pointed out the major contributions of electrons when tunneling is the dominant mechanism (i.e. low temperature/high electric field),

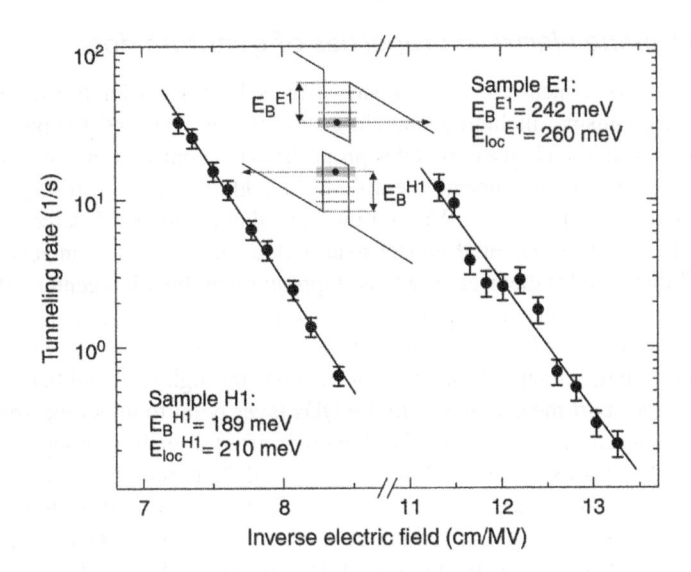

Figure 5.7 Tunneling rate for the electron (sample E1) and hole (sample H1) as a function of the inverse of the electric field. The inset displays schematically the tunneling process through the GaAs barrier E_B. The barrier is almost equal to the entire localization energy E_{loc}, the difference being due to the Poole-Frenkel effect.
Source: Reproduced with permission from Geller, M., Stock, E., Sellin, R.L., Bimberg, D. 2006a. Direct observation of tunneling emission to determine localization energies in self-organized In(Ga)As quantum dots, Phys. E 32, 171–174.

since ground state tunneling times are expected to be in the range of 1–10 ns for electrons, and many orders of magnitude greater for heavy holes due to 5x larger hole mass. Brunkov et al. (2002) also demonstrated that the temperature dependence of the PC signal from the QDs is governed by the relatively fast thermal escape of electrons. Unlike these authors, Chang et al. (2000) have reported that, for lens-shaped dots, the temperature dependence of the PC signal from InAs QDs is controlled by thermally assisted hole tunneling via the WL, implying that holes are less confined than electrons. Chang et al. (2002) somehow concluded the dispute by showing that both the electron/hole localization energies of the dots and the applied field will determine the carrier emission rate. This question is relevant for the QD PDs speed, which will be limited by the carrier exhibiting the lower emission rate (Wan et al., 2017).

It is worth noting that the asymmetry in electron/hole escape rates raises the issue of charge storage in the QDs. Charge storage in the QDs can induce band bending (Wang et al., 2001b), photoconductive effects (Vakulenko et al., 2011), and eventually persistent PC (Chang et al., 2002). Such a charge storing effect has been exploited in an ad-hoc designed structure showing that InAs QDs may be selectively charged and emptied with long data retention times at low temperatures (Finley et al., 1998; Pettersson et al., 2001; Marent et al., 2006) which can be of potential interest for optical memories.

5.3.2 In-plane electrical properties of quantum dots

A multiplicity of concurrent mechanisms affects lateral carrier transport. Several transport mechanisms, including tunneling, thermal escape, and hopping, strongly related to the single and ensemble QDs properties (i.e. composition, size, size distribution, density), can be indeed identified. In addition, carrier transport can be affected by the presence of the WL and the traps due to structural defects as well as by the vertical energy potential barrier near and at the electrical contacts. It should be noticed that as QDs themselves act as trapping/recombination centers, the charge occupancy can influence their trapping properties. This has been pointed out by Wang et al. (2001b), who attributed hysteresis loops in the current-voltage characteristic to the charging and discharging of the QDs through the band bending effect.

In-plane transport measurements in the QDs layer point to direct electronic transport through an ensemble of QDs via phonon-assisted tunneling (hopping) between localized QD states (Song et al., 2001). Negative differential photoconductance has been observed at 4K in high density, size homogenous QDs and attributed to resonant tunneling (Song et al., 2001; Jung et al., 1999). Moreover, Heller et al. (1998) have observed PL lines broadening at high lateral electric fields, attributed to field-induced tunneling out of the dot.

In the planar configuration, the WL can play a role as charge transport media but its electrical properties have not been examined in depth. Carrier mobility in such a thin layer is expected to be low due to fluctuations in thickness, composition, and strain (Chu et al., 2001). Accordingly, Mukherjee et al. (2019) considered photocarrier losses due to the WL and remarked the absence of the WL as one of the advantages of high-density SML InAs QDs.

Carrier traps close to QDs, besides affecting the photoresponse, could facilitate trap-assisted tunneling and could provide an additional current path for the dark current when dealing with planar structures. The presence of carrier traps in PDs containing In(Ga)As QD has been widely investigated by using thermally stimulated current (TSC) (see the table in Ref. Golovynskyi et al., 2017) and by DLTS (Kaniewska et al., 2010; Sobolev et al., 2000; Chen and Wang, 2007; Hamdaoui et al., 2011). It should be noted that such electrical techniques rely on the thermal emission of trapped carriers and experienced the QD as a "giant" trap. Hence neither TSC nor DLTS can easily distinguish the QD contribution from those of conventional traps (Sobolev et al., 2000). Golovynskyi et al. (2017, 2020, 2019) have extensively investigated the role of defects in lateral transport in metamorphic InAs/In$_x$Ga$_{1-x}$As QD structures. The traps occur as extended defects located in the InGaAs embedding layer, or close to the QD (Golovynskyi et al., 2017) and their density increases with the In content. Moreover, charges captured near interface defects around QDs can induce strong built-in electric fields around the QDs (Vakulenko et al., 2011). It has been also reported that the charge trapped by defects located in the QD surrounding could induce the Coulomb screening/narrowing of the WL conductivity channel (Golovynskyi et al., 2019). The QDs in-plane carrier transport is limited by all the above issues, yet high responsivities (10–20 A/W) have been reported in the literature under low excitation intensity (10^{-2}–10^{-1} μW/cm^2) in both S-K and metamorphic InAs QD PDs (Golovynskyi et al., 2020, 2016).

Interestingly, while the trap/recombination effect of InAs QDs in planar carrier transport is detrimental in PDs, it has been favorably exploited in lateral photoconductive THz detectors (Kumagai et al., 2021) and emitters (Gorodetsky et al., 2019). In both cases, the role of QDs is to shorten the carrier lifetime to a few picoseconds (ps), similar to what happens in GaAs grown at low temperature, where the lifetime is reduced because of high trap density.

5.3.3 Stark effect in quantum dots

QD-based structures under high electric fields show a significant quantum confined Stark effect (QCSE) (Miller et al., 2004), resulting in a quadratic behavior of the interband transition energy with the applied electric field (see inset of Figure 5.5). The observed strong Stark red-shift of the QDs photoresponse peak wavelength (see Figure 5.5, Fry et al., 2001) has a deep impact on PDs, where high electric fields can be applied. QCSE has thus been largely exploited in terms of bias tunability of the spectral response in IR PDs (Krishna, 2005a; Hwang et al., 2020) or to develop optical modulators based on InAs QDs (Malins et al., 2007; Sandall et al., 2013). In the telecommunication spectral range, improvement of the InAs QDs PDs responsivity at 1.3 μm related to the Stark red-shift of the cut-off wavelength has been reported (Inoue et al., 2018; Chen et al., 2020; Tossoun et al., 2019; Sandall et al., 2012).

The presence of a linear term in the quadratic energy dependence, as observed in Figure 5.5, points out the presence of a permanent dipole, corresponding to electron (hole) attraction to the base (apex) of the dots, ascribed to a graded $In_{1-x}Ga_x$ As dot composition and to the dot shape (Findeis et al., 2001; Fry et al., 2000c,d; Oulton et al., 2002). It should be noted that less commonly, but consistently with theoretical calculations (Barker and O'Reilly, 2000), a permanent dipole with reversed sign has been found (Jin et al., 2004) for ideal pyramidal InAs QDs.

QCSE in the QD plane(s) is predicted to be larger than that along the vertical direction because of the typical elongated QD shape (Gotoh et al., 2000). However, a weak Stark effect has been observed in planar device structure (Heller et al., 1998; Wolst et al., 2002) as the Stark redshift is compensated by the decrease of the excitonic binding energy (blueshift) with increasing field (Wolst et al., 2002). Further investigations suggest the absence of a permanent dipole moment in the lateral QD plane (Vogel et al., 2007).

5.4 Photodetectors for optical communication

InAs QDs-based structures allow large flexibility in PD fabrication while retaining the advantages of localized optical absorption. Three basic PD configurations are usually employed in InAs QD-based PDs for the 1.3–1.55 μm optical telecommunication spectral region, as shown in Figure 5.8 and described below.

(A) PIN photodetector   **(B) Avalanche photodetector**   **(C) MSM photodetector**

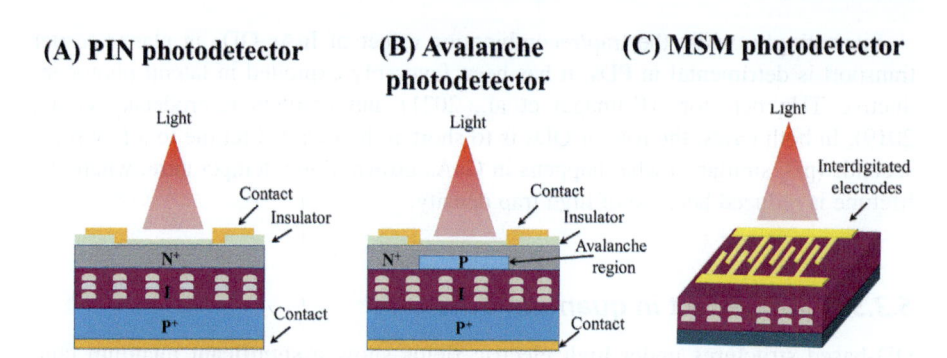

Figure 5.8 Sketch of the PIN (A), APD (B), and MSM (C) photodetector configurations under normal light incidence.

1. The vertical PIN photodetector based on QDs is sketched in Figure 5.8A. In such a structure p-type layer, the intrinsic layer, and the n-type layer are vertically stacked. Light is absorbed by the QDs in the intrinsic layer generating the PC. Absorption in the doped layer should be minimized because photogenerated carriers would move under the slow process of diffusion.
2. The APD is a vertical PIN PD (see Figure 5.8B) exploiting impact ionization to generate more electron-hole pairs. This amplifies the PC generated after light absorption. An APD usually has four regions: n^+ region, p (carrier multiplication) region, intrinsic layer (where absorption occurs by the QDs), and p^+ region. In QD-based PDs, the absorption region is usually separated from the carrier multiplication region, so that only one carrier moves (by drift or tunneling) to an extremely high field region where it is subjected to the avalanche effect. Compared to the PIN PD, the APD PD has a higher sensitivity and is regarded as the better candidate for long-haul communications although it has a more complex epitaxial wafer structure than PIN PDs and higher noise due to the avalanche multiplication process.
3. The planar metal-semiconductor-metal (MSM) photodetector is sketched in Figure 5.8C. In this configuration, the device contains two Schottky contacts as interdigitated electrodes, which collect the charge photogenerated in the underneath QD plane(s). MSM PD configuration advantages are small intrinsic RC constants and short carrier transit time. In addition, MSM structures can significantly reduce fabrication complexity. Dark currents depend on the specific Schottky barrier height between the metal and semiconductor. This is an issue for Ge, due to poor Schottky contacts (Michel et al., 2010), while this is not an issue with the wide-gap III–V layers on top of QDs (Persano et al., 2010).

PDs configurations are further categorized according to the geometry of light incidence. In PDs designed for normally incident light, the light impinges perpendicular to the device, from the top or bottom of the device. This is the simplest light coupling scheme in telecommunication PDs due to ease of fabrication and low polarization-dependent loss. However, it suffers from an inherent drawback due to the bandwidth-efficiency trade-off. This trade-off results from the opposing requirements of photo-absorption layer thickness: thick for high efficiency but thin for high bandwidth (short transit time). Furthermore, normal incidence PDs must be a reasonable size for fiber coupling, yet a larger area increases the capacitance and dark current, and therefore also limits the bandwidth and sensitivity.

Resonant-Cavity-Enhanced (RCE) PDs, embedding the photo-absorbing material into light reflectors to enhance the cavity resonance, is one of the strategies to overcome the trade-off between bandwidth and efficiency in normal incidence detectors (El-Batawy et al., 2016). Although the greater complexity in the design, process, and integration of this kind of PD due to the fabrication of high reflectivity mirror and multiple layers, RCE detectors are useful in optical telecommunication as their increased wavelength selectivity makes RCE PDs especially useful for wavelength division multiplexing systems.

One of the most promising approaches for overcoming the bandwidth-efficiency trade-off affecting normal incidence detectors is the lateral (or edge) optical excitation. Indeed, compared to normal incidence, Umezawa et al. (2014) reported under lateral (or edge) light incidence a factor of ten increase in responsivity in InAs QD-based PDs. The optical WG is a well-established basic element of any optical circuit, with huge flexibility in fabrication, which perfectly fits the needs of lateral and confined optical excitation of any kind of PD. Hence, lateral coupled WG-integrated PDs, where the light signal is delivered to the device by an in-plane optical WG, permit the bandwidth and efficiency to be determined almost independently. The efficiency in these devices is no longer dependent on the photo-absorption layer thickness, but the WG length. Moreover, WG PDs can be monolithically integrated with other photonic components (Benedikovic et al., 2019; Wan et al., 2020; Tossoun et al., 2019; Schuler et al., 2018).

5.5 InAs quantum dot-based photodetectors: trends and performance

QDs-based PDs meeting the requirements of optical communications systems such as high detectivity, high speed, large bandwidth, and low dark currents are extensively reported in the literature. In particular, recent research on QD PDs has shown their potential in terms of speed, probably the most crucial figure of merit for modern telecommunications. In Table 5.1, the principal figures of merit of currently best-performing PDs for optical communications based on InAs QDs (Wan et al., 2017; Inoue et al., 2018; Chen et al., 2020; Tossoun et al., 2019; Umezawa et al., 2014; Huang et al., 2019) and Qdashes (Wan et al., 2020) are summarized and compared to the main commercial competitors as Ge (Benedikovic et al., 2019; Fard et al., 2016; Vivien et al., 2012; Kang et al., 2009; Salamin et al., 2018) and InGaAs (http://www.osioptoelectronics.com/Libraries/Datasheets/FCI-InGaAs-36C.sflb.ashx; Kinsey et al., 2001) thin film-based PDs. The Table also includes the performance of 1.3–1.55 μm PDs based on new and promising materials, for example, graphene (Mueller et al., 2010; Ding et al., 2020; Schuler et al., 2018; Schall et al., 2018; Song, 2016), as well as PDs based on innovative structures such as plasmonics (Salamin et al., 2018) and membranes (Baumgartner et al., 2021).

The optoelectronic performances of PD for telecommunication applications are often observed by exciting with sine- and/or square-wave modulated light over a

Table 5.1 Performance comparison for state-of-art 1.3–1.55 μm photodetectors based on InAs QDs and other material systems.

Material/substrate	λ (μm)	Responsivity[a] (A/W)	I_{Dark} or J_{Dark}[a]	3 dB Bandwidth (GHz)	PD structure	Year	References
InGaAs/InP	1.3–1.55	0.85/0.80	0.5 nA @5 V	9	FCI InGaAs 36 C (Commercial)	—	http://www.osioptoelectronics.com/Libraries/Datasheets/FCI-InGaAs-36C.sflb.ashx
InAs QD/Si (Monolithic)	1.3	≈0.2 @6 V	0.8 nA @-1V	1.5 @-4V	PIN WG	2017	Wan et al. (2017)
InAs QDs/InP	1.55	0.48[c]	<1 nA @-20V	50 (estimated)	PIN, 20 QD layers	2014	Umezawa et al. (2014)
InAs QD/GaP/Si (Monolithic)	1.3	0.08	1.3×10^{-4} A/cm^2 @-3V	5.5 @-5V	PIN WG	2018	Inoue et al. (2018)
InAs QDs/Si (Monolithic)	1.3	0.26	3.5×10^{-7} A/cm^2 @-1V	—	PIN WG	2019	Huang et al. (2019)
Graphene/SOI	1.55	6.1×10^{-3} @0.4 V	$>10^{-3}$ A @0.4 V	16	MSM (asymmetric)	2010	Mueller et al. (2010)
Graphene/SOI	1.55	0.18 @0.5 V	—	≥128	MSM WG	2018	Schall et al. (2018)
Graphene/Si	1.55	0.17 @0.4 V	—	18	MSM WG	2019	Schuler et al. (2018)
Graphene/SOI	1.54	0.25	—	110	MSM (asymmetric) WG plasmonic	2020	Ding et al. (2020)
InAlGaAs QWs/Si (Heterogeneous)	1.26	0.4	0.2 nA @4 V	65	Lateral PIN membrane WG	2021	Song (2016)
InAs QDash/InP	1.55	0.26 @-3V	5 pA @-1V	4.1	PIN WG	2020	Wan et al. (2020)
Ge/Si	1.31	IQE[d] = 0.36 @10 V	20 mA @10 V	>100 GHz	MSM WG plasmonic	2018	Salamin et al. (2018)

Material	Wavelength (μm)	Responsivity (A/W)	Dark current	Bandwidth	Type	Year	Reference
Ge/Si	1.55	1.09 @-2V	5.5 A/cm^2 @-2V	42.5 @-2V	PIN WG	2016	Fard et al. (2016)
Ge/Si	1.55	0.8	80 A/cm^2 @-1V	120 @-2V	Lateral PIN WG	2012	Vivien et al. (2012)
Ge/Si	1.55	1.19	100 nA @-1V	9 @-1V	Lateral PIN WG	2019	Benedikovic et al. (2019)
InGaAs	1.55	EQE = 0.16	10^{-2} A/cm^2 @V = $0.9V_{BD}$[e]	28 @gain = 6 GBW[b] = 320 GHz	APD SACM WG	2001	Kinsey et al. (2001)
InAs QD/Si (Monolithic)	1.31	0.234 @-5V	6.7×10^{-5} A/cm^2 @-5V	2.26 @-6V gain$_{max}$ = 198	APD PIN WG	2020	Chen et al. (2020)
InAs QD/Si (Heterogeneous)	1.31	0.34 @-9V	1×10^{-6} A/cm^2 @-1V	8 @gain = 30 GBW = 240 GHz	APD PIN WG	2019	Tossoun et al. (2019)
Ge/Si	1.3	0.55	19 μA/cm^2 @-5V	11.5 @gain = 20 GBW = 340 GHz	APD SACM	2008	Kang et al. (2009)

[a]For APDs, responsivity and dark current densities refer to low voltages (no gain).
[b]GBW = maximum gain bandwidth product (for APDs).
[c]edge illumination.
[d]IQE = Internal Quantum Efficiency.
[e]V_{BD}: breakdown voltage.

wide range of modulation frequencies. However, better insight into the physical mechanisms giving rise to the observed performance as well as limitations of QD PDs can be achieved by investigating the optical impulse photoresponse in the time domain. Persano et al. (2010) used response to short pulses of light to investigate the fast dynamics of carriers generated in a single layer of InAs QDs, identifying the importance of the QD environment where the transport of the photogenerated carriers occurs. By fabricating planar MSM detectors on top of GaAs epitaxial structures containing a single InAs QD layer in an InGaAs well embedded in GaAs matrix, Persano et al. (2010) reported the PC response to ps optical pulses quasi-resonant with the QD localized states. Transients exhibit fast and slow components with an order of magnitude difference in decay time (see Figure 5.9), with the former almost doubling the second in terms of integrated charge. This is explained by the large difference in mobility of the electrons and holes in the GaAs host matrix, resulting in a much slower collection time for holes. After deconvolving the excitation pulse (see inset of Figure 5.9) and the microwave measurement system, the fast component has a pulse width of about 10 ps, equivalent to a bandwidth of 16 GHz. In agreement with the expected optical absorption, External Quantum Efficiency (EQE) was in the range $10^{-5}-10^{-4}$, with values varying with the specific QD structure. In particular, the EQE of the QD layer deposited on the InGaAs strain reducing QW was found to be larger than the EQE of InAs QD on the GaAs buffer layer. This result strongly suggests that the InGaAs QW underlying the InAs WL provides an efficient channel for the lateral transport of the carriers escaping from

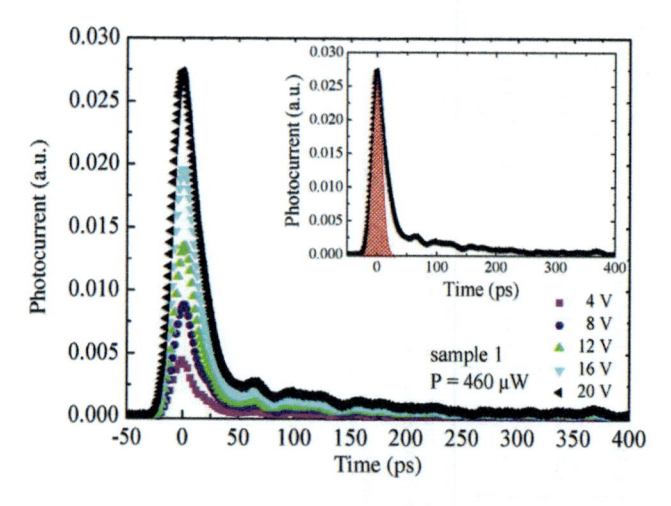

Figure 5.9 Time response of planar MSM QD PD under different applied voltages. The inset shows the optical excitation pulse (*dashed line*) quasiresonant with the QD $e_o - h_o$ ground state interband transition.
Source: Reproduced with permission from Persano, A., Nabet, B., Currie, M, Convertino, A., Leo, G., Cola, A. 2010. Single-layer InAs quantum dots for high-performance planar photodetectors near 1.3 µm, IEEE Trans. Electron. Devices 57, 1237–1242.

QDs hence avoiding the uncontrolled, defective WL. Indeed, most of the recent works for optical communication InAs QD PDs rely on DWELL design (Ren et al., 2019; Wu et al., 2015; Kim et al., 2018). Advantages have been demonstrated for IR PDs, exploiting intersubband transitions, which include dark current reduction and better control of operating wavelength, demonstrating high potential for IR imaging (Krishna, 2005b).

As mentioned in Section 5.2 of this chapter, the low absorption volume of single or multilayer QDs does not provide sufficient responsivity for telecommunication applications. Strategies to improve efficiency include incorporating gain and/or increasing the optical absorption, as described in basic PD structures in Section 5.4. Increasing gain refers to the carrier multiplication process, such as those already exploited in standard APDs. As a general strategy, submicron scaling of the multiplication region can reduce noise and improve performance. For APDs in the optical communication window, separate absorption, charge, and multiplications (SACM) APDs have been proposed (Campbell, 2016) and successfully implemented in the case of Ge/Si and InGaAs. However, the high gain does not necessarily translate into better detectivity due to the fundamental excess noise associated with the avalanche process or dark current fluctuations. In particular, Ge/Si PDs, APDs and PINs, suffer from dark currents, higher than their III−V counterparts, due to the high density of lattice mismatch defects (Michel et al., 2010).

Besides carrier multiplication, the optical absorption of PDs can be enhanced by placing the absorbing QDs in a resonant cavity, or WG, or taking advantage of plasmonic enhancement. Incorporating the PD absorber in a resonant cavity is a widely employed design in PDs. Many recent studies embedding InAs QDs in RCEs have been published (El-Batawy et al., 2016; Kim et al., 2006). However, with growing interest in Si integration, the complexity of RCE has been replaced by integrating a WG into the PD structure. Currently, structures embedding both the APD and the WG are implemented in PDs based on InAs QDs, and high gain-bandwidth products and low dark current have been demonstrated (Chen et al., 2020; Tossoun et al., 2019; Umezawa et al., 2014).

The optical WG is a well-established basic element of any optical circuit, with huge flexibility in fabrication, which perfectly fits the needs of lateral and confined optical excitation of any kind of PD. The integration of WG with PD is essential in photonics integrated circuits. For the past ten years, the integration of InAs QDs as a gain material on Si has proven to be a promising platform for CMOS-compatible, uncooled on-chip lasers (Khan et al., 2019; Wang et al., 2011). Progress in InAs QDs laser performance at 1.3 μm combined with the advances in the growth of III−V materials on Si substrates has boosted the monolithic and heterogeneous (e.g., wafer-bonding) integration of InAs QD-based PDs on Si (see Section 5.2). It is worth noting that both approaches offer the possibility of fabricating lasers and PDs from the same epitaxial structure and following the same processing steps (Chen et al., 2020; Tossoun et al., 2019). InAs-based QD PDs and APDs monolithically grown on Si substrates operating at a wavelength of 1.3 μm, have been already reported in 2012 by Sandall et al. (Sandall et al., 2012). More recently high performance InAs QD PDs monolithically grown on GoVS have been obtained

(Wan et al., 2017; Chen et al., 2020; Huang et al., 2019). Sharing the same material platform that previously demonstrated micro-ring lasers, Wan et al. (Wan et al., 2017) developed QD-on-Si WG PDs operating with an internal responsivity of 0.9 A/W at 1.25 μm, the low dark current of 4.8 nA at −3 V (density 4.8×10^{-4} A/cm^2) and high-speed performance:1.5 GHz bandwidth at 4 V and 1.3 μm. Moreover, they also identified hole trapping in the QD as the main cause for the limitation of the device speed. As a striking feature, Wan et al. (2017) also presented proof of principle demonstration for on-chip photodetection via free-space coupling (see Figure 5.10), the first step to integrating QD-based PDs with lasers on Si substrates (Bowers et al., 2019).

For monolithic InAs QD APDs on Si, very low dark current densities (6.7×10^{-5} A/cm^2 at −5 V) have been recently obtained (Chen et al., 2020), with a 1.3 nA dark current at 99% of the breakdown voltage, avalanche gain up to 198, and open-eye diagrams up to 8 Gbit/s at 1.31 μm optical modulation. The APD structure was grown by the GoVS approach and integrated with a WG for lateral optical excitation. As an alternative to using APDs, fabricating PIN QD PDs in WG configuration produced a low dark current density of 3.5×10^{-7} A/cm^2 (at −1 V and 300K) with a responsivity of 0.26 A/W at 1.3 μm with −1 V bias (Huang et al., 2019).

However, monolithic integration of InAs QDs onto Si presently requires the growth of thick buffer layers to mitigate TDs in the active device layers. This buffer layer limits the performance and design flexibility in the integration of active devices. Moreover, the thick buffer layers prevent efficient, evanescent coupling

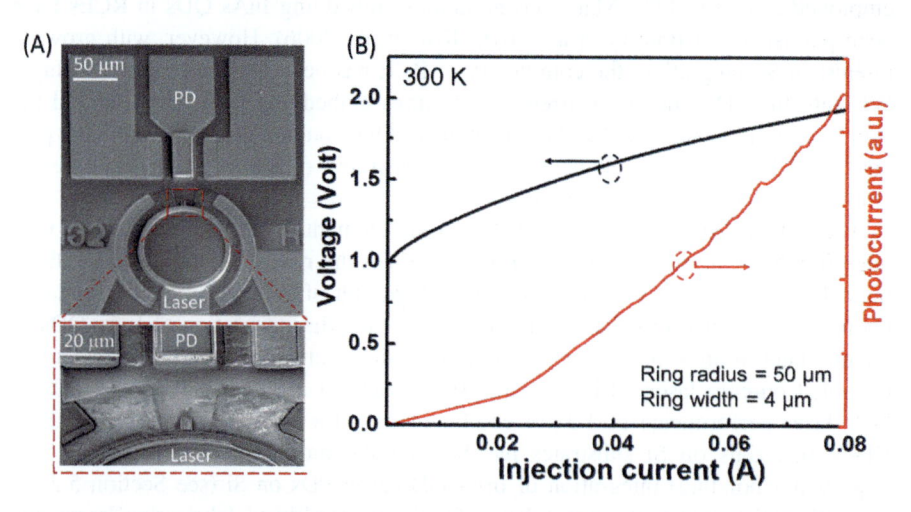

Figure 5.10 (A) Top-view of InAs QD-on-Silicon micro-ring laser-PD system through free-space coupling; (B) measured laser voltage and detector photocurrent as a function of the injection current for the micro-ring laser (50-μm outer-ring radius and 4 μm mesa width). *Source*: Reproduced with permission from Wan, Y., Zhang, Z., Chao, R., Norman, J., Jung, D., Shnag, C. et al. 2017. Monolithically integrated InAs/InGaAs quantum dot photodetectors on silicon substrates, Opt. Express 25, 27715–27723 @ The Optical Society.

between Si WGs and the active devices (Tossoun et al., 2019). These limitations can be alleviated through heterogenous integration by wafer-bonding. Heterogeneous integration is a low-cost/high-performance alternative route currently explored by research (Tossoun et al., 2019) and industry (Kurczveil et al., 2020). Uncooled InAs QD/GaAs QD-based APD with low dark current (density as low as $10-6$ A/cm^2), 15 GHz bandwidth at 1.3 μm, 240 GHz gain-bandwidth product, and reasonably responsivity at the telecommunication wavelengths has been demonstrated by bonding the III−V structure directly to the SOI substrate (Tossoun et al., 2019). Such a device exploits the same epitaxial layers and processing steps as an on-chip laser, which significantly simplifies the processing for a fully integrated transceiver on Si (Kurczveil et al., 2020).

As the advancement in InAs QD-based lasers has promoted the progress of InAs QD-based PDs at 1.3 μm wavelength, similar technological evolution has occurred at 1.55 μm for InAs Qdash devices. The elongated InAs structures grown on InP have allowed the development of lasers operating at 1.55 μm (Wan et al., 2019; Wang et al., 2011; Bowers et al., 2019; Kurczveil et al., 2020; Nötzel et al., 2006) and Qdash lasers are being integrated on Si (Xue et al., 2020). The Qdash laser progress inspired the recent development of InAs Qdash WG PDs with a record low dark current of 5 pA, a responsivity of 0.26 A/W at 1.55 μm, and open eye diagrams up to 10 Gbit/sec (Wan et al., 2020). Hence the Qdash is a promising alternative to InGaAs and Ge in the long wavelength range of optical telecommunications.

5.6 InAs quantum dots photodetectors: the way ahead

The InAs QDs unique features combined with device engineering strategies meet the stringent performance requirements of PDs operating in the optical telecom/datacom windows, that is high responsivity, high speed, large bandwidth, high quantum efficiency, low dark currents, low cost, and small footprint. In addition, integration of InAs QD-based PDs on the Si platform has been successful in view of photonics integration. Indeed, InAs QD-based lasers at 1.3 μm on Si are now a mature technology which implies that integrated PDs are being established in Silicon Photonics technology. This shows a path for QDs to become the platform for the first integrated optical circuits with an on-chip optical source working at the 1.3 μm optical communication wavelength.

It is worth reminding that a narrow active region is crucial for achieving a short carrier transit time, hence fast photoresponse. Narrow active regions can be obtained by resorting to plasmonics. In fact, the ability to confine light below the diffraction limit, enabling light-matter interaction on a deep subwavelength scale of plasmonic PDs (Ding et al., 2020; Dorodnyy et al., 2018; Muehlbrandt et al., 2016) allows shrinking of device size while ensuring the high speed required by photonics devices (up to 100 GHz) (Salamin et al., 2018). This size reduction brings the technology one step closer to a fusion of optical and electronic components at the same size scale. Plasmonic enhancements in PDs for optical telecommunication have been recently fabricated in many different materials, see for example 100-GHz PD at 1.3 μm in Ge (Salamin et al., 2018) and 110-GHz PD at 1.55 μm in graphene (Ding et al., 2020; Dorodnyy et al., 2018). InAs QD PDs enhanced by plasmonics have been also demonstrated for the IR (Lee et al., 2009), with notable results in imaging (Hwang et al., 2020; Lee et al., 2011), but not yet reported in the 1.3–1.55 μm range. It is thus only a matter of time until plasmonic enhancements are incorporated in InAs QD PDs for optical communications.

References

Akiyama, M., Kawarada, Y., Kaminishi, K., 1984. Growth of single domain GaAs layer on (100)-oriented Si substrate by MOCVD. Jpn. J. Appl. Phys. 23, L843.
Alcotte, R., Martin, M., Moeyaert, J., Cipro, R., David, S., Bassani, F., et al., 2016. Epitaxial growth of antiphase boundary free GaAs layer on 300 mm Si(001) substrate by metalorganic chemical vapour deposition with high mobility. APL. Mater. 4, 46101.
Attaluri, R.S., Annamalai, S., Posani, K.T., Stinz, A., Krishna, S., 2006. Effects of Si doping on normal incidence InAs/In0.15Ga0.85As dots-in-well quantum dot infrared photodetectors. J. Appl. Phys. 99, 083105.
Barker, J.A., O'Reilly, E.P., 2000. Theoretical analysis of electron-hole alignment in InAs-GaAs quantum dots. Phys. Rev. B 61, 13840–13851.
Barve, A.V., Rotter, T., Sharma, Y., Lee, S.J., Noh, S.K., Krishna, S., 2010. Systematic study of different transitions in high operating temperature quantum-dots-in-a-well photodetectors. Appl. Phys. Lett. 97, 061105.
Baumgartner, T., Caimi, D., Sousa, M., Hopstaken, N., Salamin, Y., Baeuerle, B., et al., 2021. High-speed CMOS-compatible III-V on Silicon membrane photodetectors. Opt. Express 29, 509–516.
Benedikovic, D., Virot, L., Aubin, G., Szelag, B., Karakus, B., Hartamann, J.-M., et al., 2019. 25 Gbps low-voltage hetero-structured silicon-germanium waveguide pin photodetectors for monolithic on-chip nanophotonic architectures. Photonics Res. 7, 437–444.
Berg, W., Bischoff, S., Magnusdottir, I., Mork, J., 2001. Ultrafast gain recovery and modulation limitations in self-assembled quantum-dot devices. IEEE Photonics Technol. Lett. 13, 541–543.
Bimberg, D., 2018. Semiconductor nanostructures for flying q-bits and green photonics. Nanophotonics 7, 1245–1257.
Bowers J.E. et al. 2019. Monolithic integrated quantum dot photonic integrated circuits, WIPO Patent WO2019/227935, Published 28 November 2019.
Brunkov, P.N., Patane, A., Levine, A., Eaves, L., Main, P.C., Musikhin, Y.G., et al., 2002. Photocurrent and capacitance spectroscopy of Schottky barrier structures incorporating InAs/GaAs quantum dots. Phys. Rev. B 65, 085326.
Campbell, J.C., 2016. Recent advances in avalanche photodiodes. J. Ligthwave Technol. 34 (2), 278–285.
Campbell, J.C., Madhukar, A., 2007. Quantum-dot infrared photodetectors. Proc. IEEE 95 (9), 1815–1827.
Carlin, J.A., Ringel, S.A., 2000. Impact of GaAs buffer thickness on electronic quality of GaAs grown on graded Ge/GeSi/Si substrates. Appl. Phys. Lett. 76, 1884.

Chang, W.-H., Hsu, T.M., Huang, C.C., Hsu, S.L., Lai, C.Y., Yeh, N.T., et al., 2000. Photocurrent studies of the carrier escape process from InAs self-assembled quantum dots. Phys. Rev. B 62, 6959–6962.

Chang, W.-H., Hsu, T.M., Huang, C.C., Hsu, S.L., Lai, C.Y., Yeh, N.T., et al., 2001a. A carrier escape study from InAs self-assembled quantum dots in photocurrent measurements. Phys. Stat. Sol. B 224, 85–88.

Chang, W.-H., Chen, W.Y., Cheng, M.C., Lai, C.Y., Hsu, T.M., Yeh, N.-T., et al., 2001b. Charging of embedded InAs self-assembled quantum dots by space-charge techniques. Phys. Rev. B 64, 125315.

Chang, W.-H., Chen, W.Y., Hsu, T.M., Yeh, N.-T., Chyi, J.-I., 2002. Hole emission processes in InAs/GaAs self-assembled quantum dots. Phys. Rev. B 66, 195337.

Chang, F.Y., Wu, C.C., Lin, H.H., 2003. Effect of InGaAs capping layer on the properties of InAs/InGaAs quantum dots and lasers. Appl. Phys. Lett. 82, 4477.

Chatterjee, A., Panda, D., Patwari, J., Tongbram, B., Chakrabarti, S., Pa, S.K., 2019. Strain relaxation in InAs quantum dots through capping layer variation and its impact on the ultrafast carrier dynamics. Semicond. Sci. Technol. 34, 095017.

Chen, J.F., Wang, J.S., 2007. Electron emission properties of relaxation-induced traps in InAs/GaAs quantum dots and the effect of electronic band structure. J. Appl. Phys. 102, 043705.

Chen, B., Wan, Y., Xie, Z., Huang, J., Zhang, N., Shang, C., et al., 2020. Low dark current high gain InAs quantum dot avalanche photodiodes monolithically grown on Si. ACS Photonics 7, 528–533.

Chen, B., Chen, Y., Deng, Z., 2021. Recent advances in high-speed photodetectors for eSWIR/MWIR/LWIR applications. Photonics 8 (1), 14.

Chu, L., Arzberger, M., Bohm, G., Abstreiter, G., 1999a. Influence of growth conditions on the photoluminescence of self-assembled InAs/GaAs quantum dots. J. Appl. Phys. 85, 2355.

Chu, L., Arzberger, M., Zrenner, A., Böhm, G., Abstreiter, G., 1999b. Polarization dependent photocurrent spectroscopy of InAs/GaAs quantum dots. App. Phys. Lett. 75, 2247–2249.

Chu, L., Zrenner, A., Bichler, M., Abstreiter, G., 2001. Quantum-dot infrared photodetector with lateral carrier transport. Appl. Phys. Lett. 79, 2249–2251.

Convertino, A., Cerri, L., Leo, G., Viticoli, S., 2004. Growth interruption to tune the emission of InAs quantum dots embedded in InGaAs matrix in the long wavelength region. J. Cryst. Growth 261, 458–465.

Das, D., Saha, J., Panda, D., Tongbram, B., Raut, P.P., Ramavath, R., et al., 2020. Vertically coupled hybrid InAs sub-monolayer on InAs Stranski–Krastanov quantum dot heterostructure: toward next generation broadband IR detection. IEEE Trans. Nanotechnol. 19, 76–83.

Ding, Y., Cheng, Z., Zhu, X., Yvid, Y., Domg, J., Galili, M., et al., 2020. Ultra-compact integrated graphene plasmonic photodetector with bandwidth above 110GHz. Nanophotonics 9, 317–325.

Dongre, S., Panda, D., Alam Gazi, S., Das, D., Kumar, R., Pandey, N., et al., Submonolayer quantum dots in P-i-P configuration: study on effects of monolayer coverage and stacking variations. In: Proceedings Volume 11291, Quantum Dots, Nanostructures, and Quantum Materials: Growth, Characterization and Modeling XVII; 112910Q (2020).

Dorodnyy, A., Salamin, Y., Ma, P., Vukajlovic Plestina, J., Lassaline, N., Mikulik, D., et al., 2018. Plasmonics photodetectors. IEEE J. Sel. Top. Quantum Electron. 24 (6), 1–13.

El-Batawy, Y., Mohammedy, F.M., Deen, M.J., 2016. Resonant cavity enhanced photodetectors: theory, design and modelling in photodetectors. In: Nabet, B. (Ed.), Photodetectors. Woodhead Publishing, pp. 415–470.

Engström, O., Malmkvist, M., Fu, Y., Olafsson, H.O., Sveinbjörnsson, E.O., 2003. Thermal emission of electrons from selected s-shell configurations in InAs/GaAs quantum dots. Appl. Phys. Lett. 83, 3578–3580.

Fard, M.M.P., Cowan, G., Liboiron-Ladouceur, G., 2016. Responsivity optimization of a high-speed germanium-on-silicon photodetector. Opt. Express 24 (24), 27738–27752.

Findeis, F., Baier, M., Beham, E., Zrenner, A., Abstreiter, G., 2001. Photocurrent and photoluminescence of a single self-assembled quantum dot in electric fields. Appl. Phys. Lett. 78, 2958–2960.

Finley, J.J., Skalitz, M., Arzberger, M., Zrenner, A., Böhm, G., Abstreiter, G., 1998. Electrical detection of optically induced charge storage in self-assembled InAs quantum dots. Appl. Phys. Lett. 73, 2618–2620.

Fischer, R., Masselink, W.T., Klem, J., Henderson, T., McGlinn, T.C., Klein, M.V., et al., 1985. Growth and properties of GaAs/AlGaAs on nonpolar substrates using molecular beam epitaxy. J. Appl. Phys. 58, 374.

Fry, P.V., Itskevich, I.E., Parnell, S.R., Finley, J.J., Wilson, L.R., Schumacher, K.L., et al., 2000a. Photocurrent spectroscopy of InAs/GaAs self-assembled quantum dots. Phys. Rev. B 62, 16784–16791.

Fry, P.W., Finley, J.J., Wilson, L.R., Lemaitre, A., Mowbray, D.J., Stolnick, M.S., et al., 2000b. Electric-field-dependent carrier capture and escape in self-assembled InAs/GaAs quantum dots. Appl. Phys. Lett. 77, 4344–4346.

Fry, P.W., Itskevich, I.E., Mowbray, D.J., Skolnick, M.S., Finley, J.J., Barker, J.A., et al., 2000c. Inverted electron-hole alignment in InAs-GaAs self-assembled quantum dots. Phys. Rev. Lett. 84, 733–736.

Fry, P.W., Itskevich, I.E., Mowbray, D.J., Skolnick, M.S., Barker, J., O'Reilly, E.P., et al., 2000d. Quantum confined stark effect and permanent dipole moment of InAs/GaAs self-assembled quantum dots. Phys. Stat. Sol. a 178, 269–275.

Fry, P.V., Skolnick, M.S., Mowbray, D.J., Itskevich, I.E., Finley, J.J., Wilson, L.R., et al., 2001. Electronic properties of InAs/GaAs self-assembled quantum dots studied by photocurrent spectroscopy. Phys. E 9, 106–113.

Gao, J., Nguyen, S.C., Bronstein, N.D., Alivisatos, A.P., 2016. Solution processed, high-speed and high-quantum-efficiency quantum dot infrared photodetectors. ACS Photonics 3, 1217–1222.

Geller, M., Stock, E., Sellin, R.L., Bimberg, D., 2006a. Direct observation of tunneling emission to determine localization energies in self-organized In(Ga)As quantum dots. Phys. E 32, 171–174.

Geller, M., Stock, E., Kapteyn, C., Sellin, R.L., Bimberg, D., 2006b. Tunneling emission from self-organized In(Ga)As/GaAs quantum dots observed via time-resolved capacitance measurements. Phys. Rev. B 73, 205331.

Gerard, J.M., Genin, J.B., Lefebvre, J., Moison, J.M., Lebouche, N., Barthe, F., 1995. Optical investigation of the self-organized growth of InAs/GaAs quantum boxes. J. Cryst. Growth 150, 351–356.

Germann, T.D., Strittmatter, A., Pohl, J., Pohl, U.W., Bimberg, D., Rautiainen, J., et al., 2008. High-power semiconductor disk laser based on InAs/GaAs submonolayer quantum dots. Appl. Phys. Lett. 92, 101123.

Golovynskyi, S.L., Dacenko, O.I., Kondratenko, S.V., Lavoryk, S.R., Mazur, Yu.I., Wang, Zh.M., et al., 2016. Intensity-dependent nonlinearity of the lateral photoconductivity in InGaAs/GaAs dot-chain structures. J. Appl. Phys. 119, 184303.

Golovynskyi, S., Datsenko, O., Seravalli, L., Kozak, O., Trevisi, G., Frigeri, P., et al., 2017. Deep levels in metamorphic InAs/InGaAs quantum dot structures with different composition of the embedding layers. Semicon. Sci. Technol. 32, 125001.

Golovynskyi, S., Datsenko, O.I., Seravalli, L., Trevisi, G., Frigeri, P., Babichuk, I.S., et al., 2018. Interband photoconductivity of metamorphic InAs/InGaAs quantum dots in the 1.3−1.55μm window. Nanoscale Res. Lett. 13, 103.

Golovynskyi, S., Datsenko, O.I., Seravalli, L., Trevisi, G., Frigeri, P., Babichuk, I.S., et al., 2019. Nanotechnology 30, 305701.

Golovynskyi, S., Datsenko, O.I., Seravalli, L., Kondratenko, S.V., Trevisi, G., Frigeri, P., et al., 2020. Near-infrared lateral photoresponse in InGaAs/GaAs quantum dot. Semicon. Sci. Technol. 35, 055029.

Gomis-Bresco, J., Dommers, S., Temnov, V.V., Woggon, U., Laemmlin, M., Bimberg, D., et al., 2008. Impact of Coulomb scattering on the ultrafast gain recovery in InGaAs quantum dots. Phys. Rev. Lett. 101, 256803.

Gorodetsky, A., Bazieva, N., Rafailov, E.U., 2019. Pump dependent carrier lifetimes in InAs/ GaAs quantum dot photoconductive terahertz antenna structures. J. Appl. Phys. 125, 151606.

Gotoh, H., Kamada, H., Ando, H., Temmyo, J., 2000. Lateral electric-field effects on excitonic photoemissions in InGaAs quantum disks. Appl. Phys. Lett. 76, 867−869.

Grassi Alessi, M., Capizzi, M., Bhatti, A.S., Frova, A., Martelli, F., Frigeri, P., et al., 1999. Optical properties of InAs quantum dots: common trends. Phys. Rev. B 59, 7620−7623.

Groenert, M.E., Leitz, C.W., Pitera, A.J., Yang, V., 2003. Monolithic integration of room-temperature cw GaAs/AlGaAs lasers on Si substrates via relaxed graded GeSi buffer layers. J. Appl. Phys. 93, 362.

Hamdaoui, N., Ajjel, R., Salem, B., Gendry, M., Maaref, H., 2011. Coulomb charging effect of electrons in InAs/InAlAs quantum dots studied by capacitance techniques. Phys. B 406, 3531−3533.

Heitz, R., Veit, M., Ledentsov, N.N., Hoffmann, A., Bimberg, D., Ustinov, V.M., et al., 1997. Energy relaxation by multiphonon processes in InAs/GaAs quantum dots. Phys. Rev. B 56, 10435.

Heller, H., Bockelmann, U., Abstreiter, G., 1998. Electric-field effects on excitons in quantum dots. Phys. Rev. B 75, 6270−6275.

Huang, J., Wan, Y., Jung, D., Norman, J., Shang, C., Li, Q., et al., 2019. Defect characterization of InAs/InGaAs quantum dot p-i-n photodetector grown on GaAs-on-V-grooved-Si substrate. ACS Photonics 6 (5), 1100−1105.

Hwang, J., Ku, Z., Jeon, J., Kim, Y., Kim, J.O., Kim, D.-K., et al., 2020. Plasmonic-layered InAs/InGaAs quantum-dots-in-a-well pixel detector for spectral-shaping and photocurrent enhancement. Nanomaterials 10 (1−14), 1827.

Inoue, D., Wan, Y., Jung, D., Norman, J., Shang, C., Nishiyama, N., et al., 2018. Low-dark current 10 Gbit/s operation of InAs/InGaAs quantum dot p-i-n photodiode grown on-axis (001) GaP/Si. App. Phys. Lett. 113 (1−3), 093506.

Jin, P., Li, C.M., Zhang, Z.Y., Liu, F.Q., Chen, Y.H., Ye, X.L., et al., 2004. Quantum-confined Stark effect and built-in dipole moment in self-assembled InAs/GaAs quantum dots. Appl. Phys. Lett. 85, 2791−2793.

Joyce, P.B., Kryzewski, T.J., Bell, G.R., Jones, T.S., Malik, S., Childs, D., et al., 2001. Growth rate effects on the size, composition and optical properties of InAs/GaAs quantum dots grown by molecular beam epitaxy. J. Cryst. Growth 227, 1000−1004.

Jung, S.K., Hwang, S.W., Choi, B.H., Kim, S.I., Park, J.H., Kim, Y., et al., 1999. Direct electronic transport through an ensemble of InAs self-assembled quantum dots. Appl. Phys. Lett. 74, 714−716.

Kang, Y., Liu, H.-D., Morse, M., Paniccia, M.J., Zadka, M., Litski, S., et al., 2009. Monolithic germanium/silicon avalanche photodiodes with 340 GHz gain−bandwidth product. Nat. Photonics 3, 59−63.

Kaniewska, M., Engström, O., Kaczmarczyk, M., 2010. Classification of energy levels in quantum dot structures by depleted layer spectroscopy. J. Electron. Mater. 39, 766–772.

Kapteyn, C.M.A., Heinrichsdorff, F., Stier, O., Heitz, R., Grundmann, M., Zakharov, N.D., et al., 1999. Electron escape from InAs quantum dots. Phys. Rev. B 60, 14265–14268.

Kapteyn, C.M.A., Lion, M., Heitz, R., Bimberg, D., Brunkov, P.N., Volovik, B.V., et al., 2000. Hole and electron emission from InAs quantum dots. Appl. Phys. Lett. 76, 1573–1575.

Karrai, K., Warburton, R.J., Schulhauser, C., Högele, A., Urbaszek, B., McGhee, E.J., et al., 2004. Hybridization of electronic states in quantum dots through photon emission. Nature 427, 135–138.

Khan, M.Z.M., Ng, T.K., Ooi, B.S., 2014. Self-assembled InAs/InP quantum dots and quantum dashes: materials structures and devices. Prog. Quantum Electron. 28, 237–313.

Khan, M.Z.M., Alkhazraji, E.A., Khan, M.T.A., Ng, T.K., Ooib, B.S., 2019. Chapter 5 - InAs/InP quantum-dash lasers. In: Tong, C. (Ed.), Chennupati Jagadish in Micro and Nano Technologies, Nanoscale Semiconductor Lasers. Elsevier, pp. 109–138. ISBN 9780128141625.

Kim, D.H., Roh, C.H., Song, H.J., Choi, Y.-S., Hahn, C.-K., Kim, H., et al., 2006. Design and fabrication of InAs/GaAs quantum-dot resonant-cavity avalanche photodetectors. Curr. Appl. Phys. 6 (Suppl. 1), e172–e175.

Kim, J.O., Sengupta, S., Barve, A.V., Sharma, Y.D., Adhikary, S., Lee, S.J., et al., 2013. Multi-stack InAs/InGaAs sub-monolayer quantum dots infrared photodetectors. Appl. Phys. Lett. 102, 011131.

Kim, Y., Kim, J., Lee, S., Noh, S.K., 2018. Submonolayer quantum dots for optoelectronic devices. J. Korean Phys. Soc. 73, 833–840.

Kim, H., Nugraha, M.I., Guan, X., Wang, Z., Hota, M.K., Xu, X., et al., 2021. All-solution-processed quantum dot electrical double-layer transistors enhanced by surface charges of Ti3C2Tx MXene contacts. ACS Nano 15 (3), 5221–5229.

Kinsey, G.S., Campbell, J.C., Dentai, A.G., 2001. Waveguide avalanche photodiode operating at 1.55 μm with a gain-bandwidth product of 320 GHz. IEEE Photonics Technol. Lett. 13 (8), 842–844.

Krishna, S., 2005a. Quantum dots-in-a-well infrared photodetectors. Infrared Phys. Technol. 47, 153–163.

Krishna, S., 2005b. Quantum dots-in-a-well infrared photodetectors. J. Phys. D: Appl. Phys. 38, 2142–2150.

Krishna, S., Zhu, D., Xu, J., Linder, K.K., Qasaimeh, O., Bhattacharya, P., 1999. Structural and luminescence characteristics of cycled submonolayer InAs/GaAs quantum dots with room-temperature emission at 1.3 μm. J. Appl. Phys. 86, 6135–6138.

Kumagai, N., Lu, X., Minami, Y., Kitada, T., Isu, T., 2021. Mobility and activation energy of lateral photocurrent of InAs quantum dot layers with ultrafast carrier relaxation. Phys. E 126, 114478.

Kurczveil G. et al. Quantum-dot based avalanche photodiodes on silicon. US Patent US2020/0243701, Published 30-07-2020.

Lai, F.-I., Yang, H., Lin, G., Hsu, I.-C., Liu, J.-N., Maleev, N., et al., 2007. High-power single-mode submonolayer quantum-dot photonic crystal vertical-cavity surface-emitting lasers. IEEE J. Sel. Top. Quantum Electron. 13, 1318–1323.

Lao, Y.F., Wolde, S., Unil Perera, A.G., Zhang, Y.H., Wang, T.M., Liu, H.C., et al., 2013. InAs/GaAs p-type quantum dot infrared photodetector with higher efficiency. Appl. Phys. Lett. 103 (1–3), 241115.

Ledentsov, N.N., Shchukin, V.A., Grundmann, M., Kirstaedter, N., Böhrer, J., Schmidt, O., et al., 1996. Direct formation of vertically coupled quantum dots in Stranski-Krastanow growth. Phys. Rev. B 54, 8743–8750.

Ledentsov, N.N., Bimberg, D., Hopfer, F., Mutig, A., Shchukin, V.A., Savel'ev, A.V., et al., 2007. Submonolayer quantum dots for high-speed surface emitting lasers. Nanoscale Res. Lett. 2, 417.

Lee, S.C., Krishna, S., Brueck, S.R.J., 2009. Quantum dot infrared photodetector enhanced by surface plasma wave excitation. Opt. Express 17 (25), 23160–23168.

Lee, S.J., Ku, Z., Barve, A., Montoya, J., Jang, W.-Y., Brueck, S.R.J., et al., 2011. A monolithically integrated plasmonic infrared quantum dot camera. Nat. Commun. 2 (286). Available from: https://doi.org/10.1038/ncomms1283.

Lee, A., Jiang, Q., Tang, M., Seeds, A., Liu, H., 2012. Continuous-wave InAs/GaAs quantum-dot laser diodes monolithically grown on Si substrate with low threshold current densities. Opt. Express 20, 22181–22187.

Lee, A.D., Jiang, Q., Tang, M., Zhang, Y., Seeds, A.J., Liu, H., et al., 2013. InAs/GaAs quantum-dot lasers monolithically grown on Si, Ge, and Ge-on-Si substrates. IEEE J. Sel. Top. Quantum Electron. 19, 1901107.

Lelarge, F., Dagens, B., Renaudier, J., Brenot, R., Accard, A., van Dijk, F., et al., 2007. Recent advances on InAs/InP quantum dash based semiconductor lasers and optical amplifiers operating at 1.55 μm. IEEE J. Sel. Top. Quantum Electron. 13, 111–124.

Lenz, A., Eisele, H., Becker, J., Ivanova, L., Lenz, E., Luckert, F., et al., 2010. Atomic structure of buried InAs sub-monolayer depositions in GaAs. Appl. Phys. Express 3, 105602.

Lenz, A., Eisele, H., Becker, J., Schulze, J.-H., Germann, T.D., Luckert, F., et al., 2011. Atomic structure and optical properties of submonolayer InAs depositions in GaAs. J. Vac. Sci. Technol. B 29, 04D104.

Leonard, D., Krishnamurthy, M., Reaves, C.M., Denbaars, S.P., Petroff, P.M., 1993. Direct formation of quantum-sized dots from uniform coherent islands of InGaAs on GaAs surfaces. Appl. Phys. Lett. 63, 3203–3205.

Leonard, D., Pond, K., Petroff, P.M., 1994. Critical layer thickness for self-assembled InAs islands on GaAs. Phys. Rev. B 50, 11 687.

Levine, B.F., 1996. In: Bhattacharya, P. (Ed.), III–V Quantum Wells and Superlattices. INSPEC, London.

Lin, S.-Y., Tsai, Y.-J., Lee, S.-C., 2004. Effect of silicon dopant on the performance of InAs/GaAs quantum-dot infrared photodetectors. Jap. J. Appl. Phys. 43, L167–L169.

Ling, H.S., Wang, S.Y., Lee, C.P., Lo, M.C., 2008. High quantum efficiency dots-in-a-well quantum dot infrared photodetectors with AlGaAs confinement enhancing layer. Appl. Phys. Lett. 92, 193506.

Liu, H.C., Gao, M., McCaffrey, J., Wasilewski, Z.R., Fafard, S., 2001. Quantum dot infrared photodetectors. Appl. Phys. Lett. 78, 79–81.

Ma, P., Salamin, Y., Bauerle, B., Josten, A., Heni, W., Emboras, A., et al., 2019. Plasmonic enhanced graphene photodetector featuring 100 Gbit/s date reception, high responsivity and compact size. ACS Photonics 6, 154–161.

Malins, D.B., Gomez-Iglesias, A., Rafailov, E.U., Sibbett, W., Miller, A., 2007. Electroabsorption and Electrorefraction in an InAs quantum-dot waveguide modulator. IEEE Photonics Technol. Lett. 19, 1118–1120.

Marent, A., Geller, M., Bimberg, D., Vasi'ev, A.P., Semenova, E.S., Zhukov, A.E., et al., 2006. Carrier storage time of milliseconds at room temperature in self-organized quantum dots. Appl. Phys. Lett. 89, 072103.

Markussen, T., Kristensen, P., Tromborg, B., Berg, T.W., Mork, J., 2006. Influence of wetting-layer wave functions on phonon-mediated carrier capture into self-assembled quantum dots. Phys. Rev. B 74, 195342.

Martí, A., López, N., Antolín, E., Cánovas, E., Luque, A., Stanley, C.R., et al., 2007. Emitter degradation in quantum dot intermediate band solar cells. Appl. Phys. Lett. 90, 233510.

Michel, J., Liu, J., Kimerling, L.C., 2010. High-performance Ge-on-Si photodetectors. Nat. Photonics 4, 527–534.

Mikhrin, S.S., Zhukov, A.E., Kovsh, A.R., Maleev, N.A., Ustinov, V.M., Shernyakov, Y.M., et al., 2000. 0.94 μm diode lasers based on Stranski-Krastanov and sub-monolayer quantum dots. Semicond. Sci. Technol. 15, 1061.

Miller, D.A.B., Chemla, D.S., Damen, T.C., Gossard, A.C., Wiegmann, W., Wood, T.H., et al., 2004. Band-edge electroabsorption in quantum well structures: the quantum-confined stark effect. Phys. Rev. Lett. 53, 2173–2176.

Montazeri, K., Currie, M., Verger, L., Dianat, P., Barsoum, M.W., Nabet, B., 2019. Beyond Gold: spin-coated Ti3C2-based MXene photodetectors. Adv. Mater. 31 (43), 1903271.

Mowbray, D.J., Skolnick, M.S., 2005. New physics and devices based on self-assembled semiconductor quantum dots. J. Phys. D: Appl. Phys. 38, 2059–2076.

Muehlbrandt, S., Melikyan, A., Harter, T., Köhnle, K., Muslija, A., Vincze, P., et al., 2016. Silicon-plasmonic internal-photoemission detector for 40 Gbit/s data reception. Optica 3, 741–747.

Mueller, T., Xia, F., Avouris, P., 2010. Graphene photodetectors for high-speed optical communications. Nat. Photonics 4, 297–301.

Mukherjee, S., Mukherjee, S., Pradhan, A., Maitra, T., Sengupta, S., Chakrabarti, S., et al., 2019. Carrier transport and recombination dynamics of InAs/GaAS sub-monolayer quantum dot near infrared photodetector. J. Phys. D: Appl. Phys. 52, 1–9.

Murata, T., Asahi, S., Sanguinetti, S., Kita, T., 2020. Infrared photodetector sensitized by InAs quantum dots embedded near an Al0.3Ga0.7As/GaAs heterointerface. Sci. Rep. 10, 11628.

Naguib, M., Kurtoglu, M., Presser, V., Lu, J., Niu, J., Heon, M., et al., 2011. Two-Dimensional nanocrystals produced by exfoliation of Ti3AlC2. Adv. Mater. 23 (37), 4248–4253.

Norman, J.C., Jung, D., Wan, Y., Bowers, J.E., 2018. Perspective: the future of quantum dot photonic integrated circuits. APL. Photonics 3, 30901.

Nötzel, R., Anantathanasarn, S., Van Veldhoven, R.P.J., Van Ottens, F.W.M., Eijkemans, T.J., Trampert, A., et al., 2006. Self-assembled InAs/InP quantum dots for telecom applications in the 1.55 μm wavelength range: wavelength tuning, stacking, polarization control, and lasing. Jap. J. Appl. Phys. 45 (8B), 6544–6549.

Oulton, R., Finley, J.J., Ashmore, A.D., Gregory, I.S., Mowbray, D.J., Skolnick, M.S., et al., 2002. Manipulation of the homogeneous linewidth of an individual In(Ga)As quantum dot. Phys. Rev. B 66 (1–4), 045313.

Pan, D., Towe, E., Kennerly, S., 1998. Normal-incidence intersubband (In, Ga)As/GaAs quantum dot infrared photodetectors. Appl. Phys. Lett. 73, 1937–1939.

Park, J.-S., Tang, M., Chen, S., Liu, H., 2020. Heteroepitaxial growth of III-V semiconductors on silicon. Crystals 10, 1163.

Passaseo, A., Rinaldi, R., Longo, M., Antonaci, S., Convertino, A., Cingolani, R., et al., 2001a. Structural study of InGaAs/GaAs quantum dots grown by metalorganic chemical vapor deposition for optoelectronic applications at 1.3 μm. J. Appl. Phys. 89, 4341–4348.

Passaseo, A., Maruccio, G., De Vittorio, M., Rinaldi, R., Cingolani, R., 2001b. Wavelength control from 1.25 to 1.4 μm in $In_xGa_{1-x}As$ quantum dot structures grown by metal organic chemical vapor deposition. Appl. Phys. Lett. 78, 1382–1384.

Passmore, B.S., Wu, J., Manasreh, M.O., Kunets, V.P., Lytvyn, P.M., Salamo, G.J., 2008. Room temperature near-infrared photoresponse based on interband transitions in In0.35Ga0.65As multiple quantum dot photodetector. IEEE Electron. Dev. Lett. 29, 224–227.

Persano, A., Cola, A., Vasanelli, L., Convertino, A., Leo, G., Cerri, L., et al., 2005. Photoelectrical properties of 1.3 μm emitting InAs quantum dots in InGaAs Matrix. Acta Phys. Polonica A 107, 381−387.

Persano, A., Cola, A., Taurino, A., Catalano, M., Lomascolo, M., Convertino, A., et al., 2007. Electronic structure of double stacked InAs/GaAs quantum .0 and theory. J. Appl. Phys. 102, 094314.

Persano, A., Nabet, B., Currie, M., Convertino, A., Leo, G., Cola, A., 2010. Single-layer InAs quantum dots for high-performance planar photodetectors near 1.3 μm. IEEE Trans. Electron. Devices 57, 1237−1242.

Pettersson, H., Bååth, L., Carlsson, N., Seifert, W., Samuelson, L., 2001. Case study of an InAs quantum dot memory: optical storing and deletion of charge. App. Phys. Lett. 79, 78−81.

Physics and Applications of Quantum Wells and Superlattices. 2012. Edited by E.E. Mendez, K. Von Klizing, NATO ASI Series B, 170.

Piels, M., Bowers, J.E., 2016. Photodetectors for silicon photonic integrated circuits. In: Nabet, B. (Ed.), Photodetectors. Woodhead Publishing, pp. 3−20.

Pospischil, A., Humer, M., Furchi, A.M., Bachmann, D., Guider, R., Fromherz, T., et al., 2013. CMOS-compatible graphene photodetector covering all optical communication bands. Nat. Photonics 7, 892−896.

Ren, A., Yuan, L., Xu, H., Wu, J., Wang, Z., 2019. Recent progress of III−V quantum dot infrared photodetectors on silicon. J. Mat. Chem. C. 7, 14441−14453.

Richardson, C.J.K., Lee, M.L., 2016. Metamorphic epitaxial materials. MRS Bull. 41, 193.

Riel, B.J., Hinzer, K., Moisa, S., Fraser, J., Finnie, P., Piercy, P., et al., 2002. InAs/GaAs(1 0 0) self-assembled quantum dots: arsenic pressure and capping effects. J. Cryst. Growth 236, 145−154.

Saito, T., Ebe, H., Arakawa, Y., Kakitsuka, T., Sugawara, M., 2008. Optical polarization in columnar InAs/GaAs quantum dots: 8-band k·p calculations. Phys. Rev. B 195318.

Salamin, Y., Ma, P., Baeuerle, B., Emboras, A., Fedoryshyn, Y., Heni, W., et al., 2018. 100GHz plasmonic photodetectors. ACS Photonics 5, 3291−3297.

Sandall, I., Ng, J.S., David, J.P.R., Tan, C.H., Wang, T., Liu, H., 2012. 1300 nm wavelength InAs quantum dot photodetector grown on silicon. Opt. Express 20, 10446.

Sandall, I.C., Ng, J.S., David, J.P.R., Liu, H., Tan, C.H., 2013. Evaluation of InAs quantum dots on Si as optical modulator. Semicond. Sci. Technol. 28 (5p), 094002.

Sauerwalda, A., Kümmell, T., Bacher, G., 2005. Size control of InAs quantum dashes. Appl. Phys. Lett. 86, 253112.

Schall, D., Pallecchi, E., Ducournau, G., Avramovic, V., Otto, M., Neumaier, D. Record high bandwidth integrated graphene photodetectors for communication beyond 180 Gb/s, Opt. Fiber Commun. Conf. 2018, OSA Technical Digest, sM2I.

Schuler, S., Schall, D., Neumaier, D., Schwarz, B., Watanabe, K., Taniguchi, T., et al., 2018. Graphene photodetector integrated on a photonic crystal defect waveguide. ACS Photonics 5, 4758−4763.

Schulz, S., Schnüll, S., Heyn, Ch, Hansen, W., 2004. Charge-state dependence of InAs quantum-dot emission energies. Phys. Rev. B 69, 15317.

Shi, B., Zhu, S., Li, Q., Wan, Y., Hu, E.L., Lau, K.M., 2017. Continuous-wave optically pumped 1.55 μm InAs/InAlGaAs quantum dot microdisk lasers epitaxially grown on silicon. ACS Photonics 4 (2), 204−210.

Sobolev, M.M., Kochnev, I.V., Lantratov, V.M., Bert, N.A., Cherkashin, N.A., Ledentsov, N.N., et al., 2000. Thermal annealing of defects in InGaAs/GaAs heterostructures with three-dimensional islands. Semiconductors 34, 195−204.

Song, Y.-W., 2016. Carbon nanotube and graphene photonic devices. In: Nabet, B. (Ed.), Photodetectors. Woodhead Publishing, pp. 47–85.

Song, H.Z., Akahane, K., Lan, S., Xu, H.Z., Okada, Y., Kawabe, M., 2001. In-plane photocurrent of self-assembled InxGa1-xAs/GaAs(311)B quantum dot arrays. Phys. Rev. B 64, 085303.

Stiff-Roberts, A.D., 2009. Quantum-dot infrared photodetector: a review. J. Nanophoton 3 (1), 1–19. 031607.

Tan, C.L., Mohseni, H., 2017. Emerging technologies for high performance infrared detectors. Nanophotonics 7, 169–197.

Tatebayashi, J., Nishioka, M., Arakawa, Y., 2001. Over 1.5 μm light emission from InAs quantum dots embedded in InGaAs strain-reducing layer grown by metalorganic chemical vapor deposition. Appl. Phys. Lett. 78, 3469.

Thompson, D., Zilkie, A., Bowers, J.E., Komljenovic, T., Reed, G.T., Vivien, L., et al., 2016. Roadmap on silicon photonics. J. Opt. 18 (7), 073003.

Tongbram, B., Shetty, S., Ghadi, H., Adhikary, S., Chakrabarti, S., 2015. Enhancement of device performance by using quaternary capping over ternary capping in strain-coupled InAs/GaAs quantum dot infrared photodetectors. Appl. Phys. A 118, 511–517.

Tossoun, B., Kurczveil, G., Zhang, C., Descos, A., Huang, Z., Beling, A., et al., 2019. Indium arsenide quantum dot waveguide photodiodes heterogeneously integrated on silicon. Optica 6 (10), 1277–1281.

Tsuji, T., Yonezu, H., Ohshima, N., 2004. Selective epitaxial growth of GaAs on Si with strained short-period superlattices by molecular beam epitaxy under atomic hydrogen irradiation. J. Vac. Sci. Technol. B 22, 1428.

Tsyrlin, G.É., Golubok, A.O., Tipisev, S.Y., Ledentsov, N.N., 1995. InAs/GaAs quantum dots obtained by submonolayer migration-enhanced epitaxy. Semiconductors 29, 884–886.

Umezawa, T., Akahane, K., Yamamoto, N., Kanno, A., Kawanishi, T., 2014. Highly sensitive photodetector using ultra-high-density 1.5-μm quantum dots for advanced optical fiber communications. IEEE J. Sel. Top. Quantum Electron. 20, 147–153.

Ustinov, V.M., Maleev, N.A., Zhukov, A.E., Kovsh, A.R., Egorov, A.Y., Lunev, A.V., et al., 1999. InAs/InGaAs quantum dot structures on GaAs substrates emitting at 1.3 μm. Appl. Phys. Lett. 74, 2815–2817.

Vakulenko, O.V., Golovynskyi, S.L., Kondratenko, S.V., 2011. Effect of carrier capture by deep levels on lateral photoconductivity of InGaAs/GaAs quantum dot structures. J. Appl. Phys. 110, 043717.

Vasanelli, A., Ferreira, R., Bastard, G., 2002. Continuous absorption background and decoherence in quantum dots. Phys. Rev. Lett. 89, 216804.

Vivien, L., Polzer, A., Marris-Morini, D., Osmond, J., Hartmann, J.M., Crozat, P., et al., 2012. Zero-bias 40Gbit/s germanium waveguide photodetector on silicon. Opt. Express 20 (2), 1096–1101.

Vogel, M.M., Ulrich, S.M., Hafenbrak, R., Michler, P., Wang, L., Rastelli, A., et al., 2007. Influence of lateral electric fields on multiexcitonic transitions and fine structure of single quantum dots. Appl. Phys. Lett. 91, 051904.

Wan, Y., Li, Q., Geng, Y., Shi, B., Lau, K.M., 2015. InAs/GaAs quantum dots on GaAs-on-V-grooved-Si substrate with high optical quality in the 1.3 μm band. Appl. Phys. Lett. 107, 081106.

Wan, Y., Zhang, Z., Chao, R., Norman, J., Jung, D., Shnag, C., et al., 2017. Monolithically integrated InAs/InGaAs quantum dot photodetectors on silicon substrates. Opt. Express 25, 27715–27723.

Wan, Y., Jung, D.N., Shang, C., Collins, N., MacFarlane, I., Norman, J., et al., 2019. Low-threshold continuous-wave operation of electrically pumped 1.55 μm InAs quantum dash microring lasers. ACS Photonics 6, 279—285.

Wan, Y., Shnag, C., Huang, J., Xie, Z., Jain, A., Norman, J., et al., 2020. Low dark current 1.55 micrometer InAs quantum dash waveguide photodiodes. ACS Nano 14, 3519—3527.

Wang, W.I., 1984. Molecular beam epitaxial growth and material properties of GaAs and AlGaAs on Si (100). J. Appl. Phys. 44, 1149.

Wang, H.L., Feng, S.L., Zhu, H.J., Ning, D., Chen, F., Wang, X.D., 2000. Electronic characteristics of InAs self-assembled quantum dots. Phys. E 7, 383—387.

Wang, S., Lin, S., Wu, H., Lee, C., 2001a. High performance InAs/GaAs quantum dot infrared photodetectors with AlGaAs current blocking layer. Infrared Phys. Technol. 42, 473—477.

Wang, T.H., Li, H.W., Zhou, J.M., 2001b. Charging effect in InAs self-assembled quantum dots. Appl. Phys. Lett. 79, 1537—1539.

Wang, T., Liu, H., Lee, A., Pozzi, F., Seeds, A., 2011. 1.3-μm InAs/GaAs quantum-dot lasers monolithically grown on Si substrates. Opt. Express 19, 11381—11386.

Wang, C., Ke, S., Hu, W., Yang, J., Yang, Y., 2016. Review of quantum dot-in-a-well infrared photodetectors and prospect of new structures. J. Nanosci. Nanotechnol. 16, 8046—8054.

Wolst, O., Schardt, M., Kahl, M., Malzer, S., Döhler, G.H., 2002. A combined investigation of lateral and vertical Stark effect in InAs self-assembled quantum dots in waveguide structures. Phys. E 13, 283—288.

Wu, J., Chen, S., Seeds, A., Liu, H., 2015. Quantum dot optoelectronics devices: lasers, photodetectors and solar cells. J. Phys. D: Appl. Phys 48, 363001—363030.

Xu, Z., Birkedal, D., Hvam, J.M., Zhao, Z., Liu, Y., Yang, K., et al., 2003. Structure and optical anisotropy of vertically correlated submonolayer InAs/GaAs quantum dots. Appl. Phys. Lett. 82, 3859—3861.

Xu, Z., Birkedal, D., Juhl, M., Hvam, J.M., 2004. Submonolayer InGaAs/GaAs quantum dot lasers with high modal gain and zero-linewidth enhancement factor. Appl. Phys. Lett. 85, 3259—3261.

Xu, Q., Meng, L., Sinha, K., Chowdhury, F.I., Hu, J., Wang, X., 2020a. Ultrafast colloidal quantum dot infrared photodiode. ACS Photonics 7, 1297—1303.

Xu, K., Zhou, W., Ning, Z., 2020b. Integrated structure and device engineering for high performance and scalable quantum dot infrared photodetectors. Small 16, 2003397—2003410.

Xue, Y., Lue, W., Zhu, S., Lin, L., Shi, B., Lau, K.M., 2020. 1.55 μm electrically pumped continuous wave lasing of quantum dash lasers grown on silicon. Opt. Express 28 (12), 18172—18179.

Zunger, A., 1998. Electronic-structure theory of semiconductor quantum dots. MRS Bull. 23, 35.

Advances in chip-integrated silicon-germanium photodetectors

Daniel Benedikovič
Department of Multimedia and Information-Communication Technologies, University of Žilina, Žilina, Slovakia

6.1 Introduction

In recent years, the research on integrated photonics has evolved rapidly and many promising results were demonstrated. Recent achievements in integrated photonics have propelled advances in optoelectronics, health, and biomedicine, or information, communication, and quantum technologies, to name a few. Integrated photonics thus paves the way to light-speed, more compact, and high-performance components (Atabaki et al., 2018; Sabella, 2020; Rodt and Reitzenstein, 2021). These components are based on semiconductor chips, with efficiencies that can be orders of magnitude beyond table-sized or bulk systems. Semiconductor chips offer a high integration density of complex photonic functions, while advantaged by the maturity and know-how of complementary metal-oxide-semiconductor (CMOS) processes and toolsets developed by the microelectronic industry. Moreover, recent open-access foundry models form a vital ecosystem that supports a complete chain of steps, ranging from device design, fabrication, testing, and assembly to device packaging (Rahim et al., 2019, 2018). The overreaching goal of semiconductor photonics is the chip-scale integration of different functionalities. The aim here is to address the information-communication bottleneck of metalized connections as those start to approach their fundamental limits (Alduino and Paniccia, 2007; Beausoleil et al., 2008). Indeed, the bottleneck of electrical interconnects mirrors a major challenge in many rising applications. Issues are constrained by interlink shrinkage, high energy consumption (power and heat dissipation), large latency and low throughput, limited bandwidth and speed, higher production costs, and restricted space for dense integration. To step over these hurdles, optical technologies, where lightwave signals are used to transmit information data instead of electrical ones (Alduino and Paniccia, 2007), are arguably the path to the future. Here, information data can be processed at the speed of light, utilizing a much larger bandwidth with less noise. Optical technologies, at an early stage represented by optical fibers only, have revolutionized global communications, the rise of the Internet era, and most recently, allowing for unprecedented data traffic growth with streaming services and instant connection via social networks (Shi et al., 2020). Fast optical solutions have progressively supplanted copper-based wires in different links, ranging from long-haul and metro systems down to access networks. Indeed, this trend continues to evolve towards short-range architectures, that is, communication concepts as diverse as board-to-board, chip-to-chip, and intra- or interchip (Alduino and Paniccia, 2007; Cheng et al., 2018).

A special focus is next-generation systems based on data-intensive applications. Examples are intra- and interdata centers (Cheng et al., 2018), 5G/6G cellular networks and communications (Sabella, 2020; Doerr and Chen, 2018; Bernabé et al., 2021), exascale computers (Bernabé et al., 2021; Liang et al., 2020a), or high-capacity storages, clouds, and servers (Liang et al., 2020a; Benner et al., 2010; Filer et al., 2019), among a wealth of others. In this context, the drive for such applications has leveraged semiconductor photonics, which becomes indispensable for optical transceivers implemented on a chip (Hu et al., 2020a; Timurdogan et al., 2020). Historically, integrated circuits have been enabled by using crystalline inorganic materials from groups III, IV, and V of an atomic column. These elemental semiconductors are widely used in modern fields of electronics, computation, and optoelectronics (Ieong et al., 2004; del Alamo, 2011). The integrated group-IV photonics, having a wide material base with Silicon (Si), Germanium (Ge), Silicon Nitride (SiN), or Germanium-Tin (GeSn), is particularly appealing as it aims to deliver solutions at near-infrared (near-IR) wavebands—spectral range harnessed by commercial fiber-optic communications. Besides that, group-IV materials can cover practically attractive visible spectral range (Blumenthal et al., 2018) or can extend operation towards promising short-wave-, mid-wave-, and long-wave-infrared wavelengths (Marris-Morini et al., 2018). Recent progress in group-IV photonics has proved promises to cointegrate electronics and photonics within the same circuit (Atabaki et al., 2018; Sun et al., 2015), aided by well-mastered fabrication at reduced complexity and production cost (Rahim et al., 2018). The optoelectronic properties of group-IV materials can be exploited for the multifunctional and scalable monolithic platform. Such a platform includes low-loss elements for light guiding, routing, and manipulating (Halir et al., 2018; Wilmart et al., 2021) as well as elements with gain-based functions to enable light emission/amplification, modulation, and photodetection on one chip (Fadaly et al., 2020; Rahim et al., 2021; Benedikovic et al., 2021). Group-IV photonic platform thus holds prospects to tackling the direct bandgap deficiency intrinsically presented in this platform, whereas enabling chip realization on a wafer-scale leveraging mature Si-foundry environment. This opens up bountiful opportunities for cost-effective and high-volume applications, and in turn, the group-IV platform helps to expand photonic frontiers from laboratories to industrial production (Network transformation with world-class optics; 100GBaud + Silicon Photonics Solutions Drive Optical Network Evolution). To date, a rich set of fundamental devices that form photonic circuits has been developed (Halir et al., 2018; Wilmart et al., 2021; Fadaly et al., 2020; Rahim et al., 2021; Benedikovic et al., 2021; Network transformation with world-class optics; 100GBaud + Silicon Photonics Solutions Drive Optical Network Evolution; Benedikovic et al., 2019a; Alonso-Ramos et al., 2019; Sylvain et al., 2018; Durán-Valdeiglesias et al., 2019; Benedikovic et al., 2017; Otterstrom et al., 2018; Armand Pilon et al., 2019; Berciano et al., 2018; Boeuf et al., 2017; Michel et al., 2010). This includes low-loss, wideband, and well-performing passives (waveguides, couplers, splitters, filters, resonators, and multiplexers) (Halir et al., 2018; Wilmart et al., 2021; Benedikovic et al., 2019a; Alonso-Ramos et al., 2019; Sylvain et al., 2018; Durán-Valdeiglesias et al., 2019; Benedikovic et al., 2017) as well as a great number of actives such as light sources (Fadaly et al., 2020; Otterstrom et al., 2018; Armand Pilon et al., 2019), optical modulators (Rahim et al., 2021; Berciano et al., 2018; Boeuf et al., 2017), and photodetectors (Benedikovic et al., 2021; Michel et al., 2010).

High-speed transceivers in optical communication links, for both short-distance and long-haul hierarchies, call upon high-performance photodetectors within their receiver subassembly system. The optical photodetector is one of the key building elements in the optoelectrical link chain, conventionally situated at the end of a link. Photodetectors perform a signal conversion from optical into the electrical domain, and thus are key for photo-receivers and further electronic processing. For optical communications, discrete photodetectors are typically made out of III/V materials due to their direct and well-tailored bandgaps, governing their superior optoelectronic characteristics over the near-IR wavelengths (Mauthe et al., 2020; Shen et al., 2017). First reports on Ge photodetectors, as devices for optical communications, dated back to the early 50s and late 70s of the 20th century (Shive, 1953; Melchior and Lynch, 1966). Since then, Ge became a close-to-optimal material for photodetectors in the near-IR range. Now, it is widely accepted that Ge photodetectors are very good alternatives to III/V counterparts, yielding mature and reliable performances. Currently, many Si-foundries over the globe afford Ge photodiodes grown directly on Si templates or realized on silicon-on-insulator (SOI) platforms (Rahim et al., 2019, 2018; Siew et al., 2021), where their optoelectrical performances are well-balanced with compliant CMOS manufacturing and overall production cost. Thus, a realization of complex photonic receivers harnessing group-IV photodetectors can provide a promising path toward future optoelectronic applications. The key performance metrics of group-IV photodetectors include (1) high photo-responsivity close to the quantum efficiency limit for low-voltage-operated devices, or even beyond this limit in case of avalanche photodiodes; (2) fast frequency response and high-speed operation at levels of tens of GHz and Gbps to facilitate optical communication scaling and future capacity needs; (3) low-noise performance, including low dark currents, small dark-current densities, and low levels of an avalanche noise to guarantee high signal-to-noise ratio for low bit-error-rate (BER) thresholds; (4) device drivers and operation compatible with CMOS circuitry; (5) reliable sensitivity to low-intensity optical signals modulated at high speeds for diverse modulation formats; (6) compact footprint to favor dense device integration, and (7) low energy consumption to improve optical power budgets of next-generation chip-scale optical links.

In this Chapter, we review recent advances in group-IV photodetectors. In Section 2, we start with a short categorization of materials for on-chip photodetection, including conventional and surging material classes. Section 3 provides an overview of different approaches and techniques to grow Ge on Si substrates as well as refers to the photodetector integration in existing Si waveguide platforms, CMOS foundries, and research and development (R&D) lines. Contemporary state-of-the-art Si-Ge photodetectors are then surveyed in Section 4, overviewing relevant designs and recent achievements. Section 5 concludes the Chapter with prospects for Si-Ge photodetectors.

6.2 Photodetection material systems: standard semiconductors and beyond

The key mechanism, on which light detection relies, is the transformation of incident photons into electrical signals. A converted signal typically takes the form of either voltage or current pulse. This allows subsequent signal storage, image reconstruction, or information processing by available electronics. Conveniently, optical photodetectors are square-law active devices, responding to the optical power (or intensity) rather than the amplitude of the electromagnetic field. The foremost task in the optical photodetector is to use material with a high absorption coefficient for the spectral range to be detected. Then, depending on the material system in use (and detector structure as well), different mechanisms are at the center of the optical-to-electrical conversion. Indeed, due to the difference in the fundamental physics behind the optoelectrical conversion (Singh, 1996; Romagnoli et al., 2018; Manzeli et al., 2017; Koppens et al., 2014; Akinwande et al., 2019), there are essential performance tradeoffs, different integration approaches, and target applications.

Rapidly advancing on-chip optoelectronics is advantaged by crystalline semiconductors due to their unique electronic and photonic characteristics (Ieong et al., 2004; del Alamo, 2011; Singh, 1996; Romagnoli et al., 2018; Manzeli et al., 2017; Koppens et al., 2014; Akinwande et al., 2019). Crystalline semiconductors, particularly elemental semiconductors from the III, IV, and V atomic columns, have many compelling properties that are not shared by other materials. For that reason, they are still broadly employed in modern computation and communication and are heavily used in photonics. Efficient conversion of absorbed photons into free electron-hole pairs hinges upon material electronic bandgap (and its energy), enabling electron scattering from a fully occupied valence band into the empty conduction band (Singh, 1996). The photon energy has to be greater or at least equal to the bandgap of the semiconductor material. Figure 6.1 shows an absorption coefficient for a few well-adopted semiconductors used in recent optical photodetectors as a function of the wavelength and the bandgap energy. Table 6.1 sums up the essential properties of Si and Ge, respectively.

6.2.1 Silicon photodetectors

Crystalline semiconductors, in general, and Si in particular, have revolutionized many aspects of the world. Si is now feeding a multibillion-dollar industry that touches our daily life. Si has an indirect bandgap with an energy level of 1.12 eV (Singh, 1996). This has two major consequences: Si does not allow photon emission and weakly favors photon detection of near-IR radiation at wavelengths above

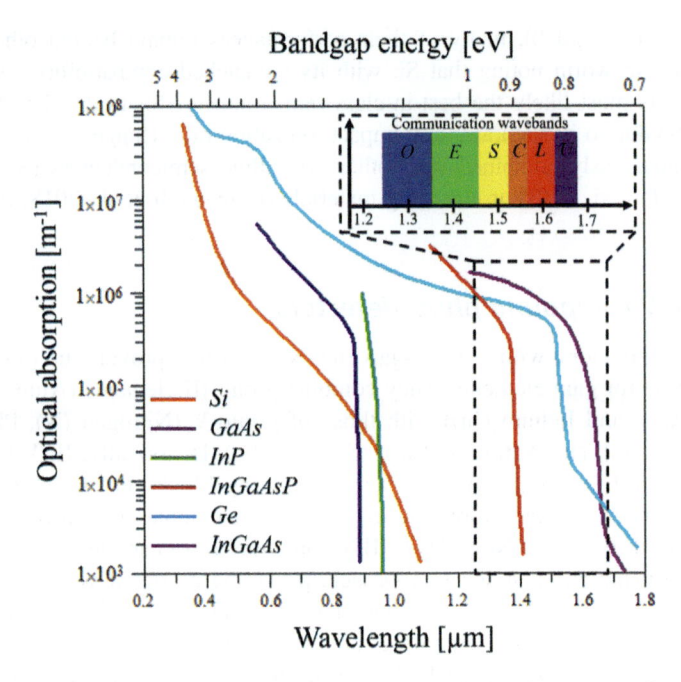

Figure 6.1 Optical absorption versus the wavelength and the bandgap energy for several crystalline semiconductors used in optical photodetectors.

Table 6.1 Properties of group-IV photonic materials.

Group-IV material	Bandgap energy (eV)@ 300 K	Cut-off wavelength [μm]@ 300 K	Lattice constant [Å]@ 300 K	Ionization ratio [-] @ 300 K	Transparency window [μm]
Si	1.12	1.1	5.4310	$k < 1$	Up to 8
Ge	0.66	1.8	5.6579	$k \sim 1$	Up to 14

1.1 μm. In opposition, these properties make Si an excellent material for passive photonics, affording wide wavelength transparency up to ~8 μm. Si photodiodes succeeded in the visible wavelengths (Shi and Nihtianov, 2012) and their operation is adequate at standard communication bands (1.26–1.675 μm) unless high detection speeds are needed (Casalino et al., 2010; Vivien and Pavesi, 2013). The intrinsic deficiency of Si absorption can be compensated by designs, exploiting two-photon absorption (Tanabe et al., 2010), structured opto-cavities (Ackert et al., 2011), or poly-Si overlayers (Preston et al., 2011). Furthermore, defects-based photo-generation (Ackert et al., 2015; Desiatov et al., 2014) and an internal photo-emission in Schottky diodes can be used (Goykhman et al., 2016). Even though the research on native Si photodiodes moves continuously forward (Tasker et al., 2021; Gherabli et al., 2020;

Frydendahl et al., 2020), many of their performances remain behind other alternatives. It is also worth noting that Si, with its unmatched compatibility with CMOS electronics, is most likely the best-in-class semiconductor for avalanche photodetectors (APDs) due to its naturally low impact ionization ratio (Singh, 1996). Therefore, it is advantageously combined with other crystalline semiconductors (Singh, 1996; Vivien and Pavesi, 2013) or emerging materials (Romagnoli et al., 2018; Akinwande et al., 2019).

6.2.2 III/V compound photodetectors

III/V semiconductors, with a direct-gap energy transition, provide the most evolved material base for gain elements. They combine group-III elements (Aluminum (Al), Gallium (Ga), and Indium (In)) with those of group-V (Nitrogen (N), Phosphorus (P), Arsenic (As), and Antimony (Sb)) (Singh, 1996). Historically, III/V technology generates a full set of passive and active building blocks in one chip, uniquely offering native lasers and amplifiers as well as fast modulators and photodetectors (Bowers and Liu, 2017). Nowadays, III/V foundries can guarantee seamless device prototyping using an open-access approach, providing mastered skills ranging from design through manufacturing to characterization and packaging (Helkey et al., 2019; Hoefler et al., 2019; Lemaitre et al., 2019). The appeal for III/V materials in light detection stems from their high optical absorption and high carrier velocities in the near-IR spectrum, yielding good responsivities and frequency responses of tens of GHz (Mauthe et al., 2020; Shen et al., 2017). However, intrinsic noise properties of native III/V materials are not as good as those of Si or other III/V-founded material systems (Woodson et al., 2016; Yi et al., 2019), but are substantially better compared to Ge. Moreover, a compound form of III-V's enables them to tailor their bandgaps as well as favors a good lattice matching between them. Photodetectors with binary, ternary, or quaternary compound alloys such as gallium arsenide (GaAs), indium phosphide (InP), indium gallium arsenide (InGaAs), or indium gallium arsenide phosphide (InGaAsP) are gold-standards in optical communications. Most recently, III/V materials are brought to Si platforms through hybrid or heterogeneous integration (Lemaitre et al., 2019; Roelkens et al., 2010; Ramirez et al., 2020; Takenaka et al., 2017). For example, in optical communication wavebands, SOI-to-III/V hybrid integration, chip-to-chip bonding, selective hetero-epitaxy growth, or transfer printing are common and well-utilized approaches. Indeed, such strategies provide singular advantages over monolithic solutions; however, a range of challenges still preserves (Benedikovic et al., 2021).

6.2.3 Germanium photodetectors

Promises to use Ge in electronics and photonics have been recognized for a long time (Wada and Kimerling, 2015). Indeed, Ge provides desired advantages and complementarities to Si, but unlike Si, Ge is a strong candidate for reliable and fast photodetection at near-IR wavelengths (Benedikovic et al., 2021; Michel et al., 2010), among other vital functionalities in Ge lasers (Fadaly et al., 2020; Armand

Pilon et al., 2019) and electro-absorption modulators (Rahim et al., 2021). Besides that, Ge also has a broad transparency window, attracting applications in short-wave-, mid-wave-, and long-wave-infrared ranges (Marris-Morini et al., 2018; Alonso-Ramos et al., 2016; Li et al., 2019; Liu et al., 2018). Ge has a quasidirect bandgap with an energy of 0.8 eV, which is only slightly higher than the dominant indirect bandgap energy (0.66 eV). Thus, Ge is a very good material for active photonic devices, offering high optical absorption in optical communication wavelengths. In general, Ge is a key material to form an intrinsic diode region given its advantageous optoelectronic properties. This includes high optical absorption in a broad spectrum ($\sim$0.4 to 1.8 μm) and uniquely high mobility of photo-carriers (also appreciated feature in microelectronics, especially in advanced CMOS transistors) (Benedikovic et al., 2021; Singh, 1996). From a technological perspective, mastered selective area epitaxy enables growing Ge layers on Si with low defects and high crystalline quality (Hartmann et al., 2004). Moreover, CMOS-friendly fabrication and accessible foundry model support monolithic integration (Rahim et al., 2018), where a low-loss Si platform can meet advanced optoelectrical efficiency and speed. On the other hand, a large 4.2% lattice mismatch between Ge and Si and high intrinsic noise (as a consequence of Ge low bandgap energy) are known difficulties that stay present (Singh, 1996; Vivien and Pavesi, 2013).

6.2.4 Germanium-tin photodetectors

Besides pure Ge and its benefits for electronics and photonics, alloying Ge with Tin (Sn) is rather attractive for photonic-to-electronic integration on Si chips. Sn is a semimetal element that also belongs to column IV of the periodic table. The semimetal nature of Sn yields 0 eV bandgap, while, as mentioned earlier, Ge is a quasidirect bandgap semiconductor with an energy level of 0.8 eV. Alloyed GeSn layers on Si and/or Ge appear promising for photodetectors for several reasons (Soref, 2014). GeSn structures are harnessed to realize improved diode designs at conventional communication wavebands, leveraging foundry-enabled processes and affordable 200 and 300 mm SOI wafers. Moreover, GeSn gained considerable attention as this alloyed group-IV material enables to extend photodetector operation range well beyond the Ge cut-off ($\sim$1.6 μm). By modifying the Sn content in the unstrained and compressively strained GeSn layer, the direct bandgap of the material can be advantageously tailored and tuned. This improves the control over the photodetection spectral range and helps to enhance the photodetector performance. For the most part, GeSn photodetectors are realized as surface-illuminated devices (Xu et al., 2019; Tsai et al., 2020; Tran et al., 2019; Ghosh et al., 2020), while reliable waveguide-integrated designs are being steadily developed and research moves forward.

6.2.5 New materials beyond semiconductors

Although semiconductors will shape many applications in the future, there are emerging material classes leaping integrated photonics right now. In recent years, we have seen a growing interest and extensive research effort to integrate novel materials on photonic waveguide platforms (Romagnoli et al., 2018; Manzeli et al., 2017; Koppens et al., 2014; Akinwande et al., 2019). These new materials can provide impressive opportunities for optical photodetectors and their optoelectrical performances, which are otherwise challenging to obtain using standard approaches. Moreover, bringing novel materials into group-IV photonic platforms preserves the manufacturing and processing advantages offered by this technology. Broadly, surging approaches can be categorized as follows: (1) plasmonics (Goykhman et al., 2016; Salamin et al., 2018; Ma et al., 2019; Dorodnyy et al., 2018) and a variety of (2) atomically thick two-dimensional (2D) materials, mostly covered by graphene (Romagnoli et al., 2018; Akinwande et al., 2019; Goykhman et al., 2016; Ma et al., 2019; Kim et al., 2011) and transition-metal dichalcogenides (Manzeli et al., 2017; Koppens et al., 2014; Akinwande et al., 2019). A new degree of freedom and attractiveness of plasmonics and novel 2D materials stems from their tailorable optical and electrical properties and strongly enhanced light-matter interaction. Moreover, ultra-short plasmonic structures and one- or few-layer-thick 2D materials show great prospects to develop ultra-short devices with compact footprints, which, in turn, enables high integration density. The implementation of efficient plasmonics and 2D-material-based photodetectors by integrating them with CMOS-compatible (Si, SiN, or Ge) waveguides is most appealing. Such integration can be, in principle, easier compared to complex semiconductor-founded schemes as they are not limited to the matching of crystal lattices or contamination problems (Michel et al., 2010; Lemaitre et al., 2019; Ramirez et al., 2020; Takenaka et al., 2017). Instead, fabless chip production (Rahim et al., 2018) coupled with novel micro-transfer printing, advanced local layer deposition methods, or local growth techniques appears promising for future development (Romagnoli et al., 2018; Koppens et al., 2014). However, available fabrication facilities and processing technologies are not yet mature enough to deliver automatic process flow as is the case of standard group-IV or III/V production lines (Rahim et al., 2018; Helkey et al., 2019; Hoefler et al., 2019). Integrating novel materials with photonic platforms has the potential to enhance the performance of photodetectors. Early-stage results and proof-of-concept demonstrations showed promises in terms of improved responsivity, spectral coverage, voltage operation, noise levels, and improved detection speeds (Salamin et al., 2018; Dorodnyy et al., 2018; Kim et al., 2011; Bie et al., 2017; Ma et al., 2018; Flöry et al., 2020).

6.3 Processing methods and integration opportunities

Group-IV materials are widely used in electronics and photonics. As outlined earlier, growing Ge on Si is difficult due to the 4.2% mismatch in lattice constants. This lattice mismatch is revealed via high-density threading defects and dislocations. The threading dislocations are responsible for mid-gap generation-recombination centers. This, however, deteriorates the optoelectrical performance of Si-Ge photodetectors by increasing the dark current and reducing the mobility of photo carriers.

6.3.1 Growth methods: overview and recent trends

Earlier attempts that elaborated on Ge films growth go back to the 1980s and were primarily focused on stand-alone devices (Luryi et al., 1984). In the beginning, Ge films were only realized thanks to graded SiGe buffers. This was the only solution available to address threading dislocations. Optimized graded SiGe buffers provided low defect densities and low dark currents. Initially, thick buffer-layered technology with layer grading towards low Ge content (of about 20%–25%) was interesting mainly for field-effect transistors (Fitzgerald et al., 1991). However, this approach was not well-suited for full Ge photodetectors, targeting 100% Ge content. From an integration perspective, ultra-thick graded buffers (of the order of several microns) are not ideal for several reasons. Recent planar waveguide processing within Si CMOS foundries favors efficient connection between waveguides and devices, which is difficult to obtain with large graded buffers. Such structures also prevent the use of buried oxides to isolate carriers or to form dielectric bottom mirrors at the substrate backside (Rahim et al., 2019, 2018; Luryi et al., 1984; Fitzgerald et al., 1991; Currie et al., 1998). Alternatively, direct Ge growth on Si can be used to alleviate a 4.2% lattice mismatch between two materials by utilizing a two-step growth process (Colace et al., 1998; Luan et al., 1999; Hartmann et al., 2005, 2009). This approach yields high crystalline quality and suits better for device integration compared to thick graded buffers. At first, a thin Ge seed layer is grown at low temperatures (300°C–400°C), followed by faster growth at high temperatures (600°C–700°C), forming μm-thick active layers. To improve threading dislocations in the final Ge film, postprocess thermal annealing at temperatures of about 900°C can be employed. A high-quality Ge layer can also be grown using sub-μm SiGe graded structures coupled with the two-step growth method, plasma-enhanced chemical vapor deposition, or selective Ge epitaxy inside etched cavities (Huang et al., 2004; Osmond et al., 2009; Reboud et al., 2017). In addition, Ge epitaxially grown on Si forms a tensile strain of about 0.2% in the top Ge layer (Michel et al., 2010; Vivien and Pavesi, 2013). This not only improves the absorption in the C-band (1.53–1.565 μm) compared to the bulk Ge, but also shifts the bandgap energy edge towards L- and U- wavebands (1.565–1.625 μm and 1.625–1.675 μm), and thus extending the Ge spectral coverage.

6.3.2 Detector-waveguide integration

First Si-Ge photodetectors were realized as discrete normal-incidence components (Samavedam et al., 1998; Colace et al., 2000). Here, the diode junction was vertical and the path of photon absorption is oriented in the same way as the path for carrier collection. Then, larger devices yield higher responsivities, however, with a slower response, lower operation speeds, and typically higher noise. Although many vertically-illuminated Si-Ge photodiodes were reported (Kim et al., 2013, 2018a;

Song et al., 2021), their performance is still hindered by deleterious responsivity-bandwidth trade-offs. In contrast, waveguide integration of Si-Ge photodetectors is crucial for modern photonic circuits and essentially improves their performances in terms of responsivity, bandwidth, and dark current (Benedikovic et al., 2021; Michel et al., 2010; Vivien and Pavesi, 2013). Photodetector integration in the Si waveguide platform de-couples the light path (along the mode traveling direction) from the carrier collection (across the device cross-section). This way, a photon absorption path is not shared by a direction, where the electric field extracts generated photo-carriers. The intrinsic device region can be as long as needed to entirely absorb the injected power. Then, the device cross-section can be engineered to reduce the photo-carrier transit time, and thus enhanced the photodetector bandwidth. This way, the responsivity (or equivalently quantum efficiency) and bandwidth can be tailored separately to obtain optimal optoelectrical performance. Thus, waveguide-integrated Si-Ge photodetectors provide a range of advantages over their normal-incidence alternatives. The free-space photodiodes require a comparatively thicker active area to absorb fiber-coupled or free-space light inputs. They are then larger, which yields higher noise (due to the larger dark current), lower bandwidth (due to the higher capacity), slower detection speeds, and a limited by-product of responsivity and bandwidth. Conversely, waveguide-integrated photodetectors tackle these shortcomings to the full extent. They have compact footprints and lower device noise while providing no significant responsivity-bandwidth trade-offs, allowing higher detection speeds, and improving optical power sensitivities.

6.3.3 On-chip schemes for waveguide coupling

Low-loss coupling between the photodetector and the waveguide is crucial for the device's performance. Besides the normal-incidence (vertical illumination) coupling, injecting the light into integrated Ge photodiodes can be performed through evanescent or butt-coupling schemes. Coupling techniques for waveguide-integrated photodetectors are schematically shown in Figure 6.2.

6.3.3.1 Evanescent light coupling

The evanescent coupling gradually transfers optical power between an input waveguide (of a lower refractive index) to a Ge photodetector (with a higher material index). In evanescently-coupled photodetectors, we can have vertically-coupled or laterally-coupled schemes. In the former case, this can be implemented as a top-to-bottom (Ahn et al., 2007, 2011) or bottom-to-top (Yin et al., 2007; Masini et al., 2008; Hu et al., 2020b) configuration, with the injection waveguide situated either on top of (Figure 6.2A) or beneath the Ge photodetector (Figure 6.2B), respectively. The top-to-bottom approach provides an added degree of freedom by selecting a material of the injection waveguide that is different from Si. This can be additional overlayers with amorphous Si, SiN, or Silicon Oxi-Nitride (SiON). Indeed, these materials are CMOS compatible and can be used in back-end-of-line (BEOL) processing, thus supporting multilevel integration with enhanced functions (Wilmart

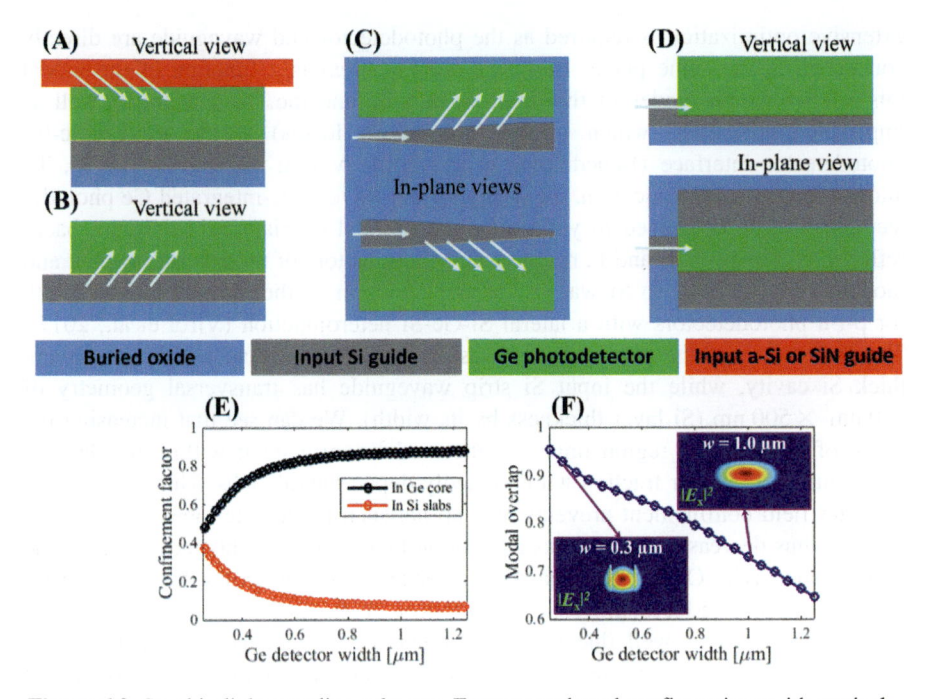

Figure 6.2 On-chip light coupling schemes. Evanescent-based configurations with vertical (A) top-to-bottom or (B) bottom-to-top arrangements and (C) lateral coupling. (D) Butt-waveguide-coupling configuration. (E) Modal confinement and (F) modal overlap as a function of the Ge photodetector width. Note: for (E) and (F> p-i-n photodetector with lateral Si-Ge-Si junction was considered (Virot et al., 2017).

et al., 2020). Figure 6.2C shows a laterally-coupled scheme. Here, the injection waveguide is positioned near the Ge photodetector and is gradually tapered to narrower or wider waveguides (Byrd et al., 2017; Zuo et al., 2019; Hu et al., 2021). Factors that affect the coupling loss are geometrical and material parameters (waveguide and detector dimensions as well as index contrast between them), light polarization, and operating wavelength. Recently, evanescently-coupled photodetectors are appealing for microwave photonics and analog photonic links that require an improved power-handling capability (Tzu et al., 2019).

6.3.3.2 Direct butt-waveguide-coupling

Butt-coupling, as shown in Figure 6.2D, is a low-loss light coupling scheme, where optical power is directly transferred from the input waveguide into the optical photodetector. From an optical point of view, butt-coupled photodiodes have much better modal confinement within the light absorbing region. In turn, ultra-compact structures with abrupt absorption profile, high responsivity, low noise, and fast frequency response can be realized. Butt-coupling schemes are easier to design and fabricate since the photodetector is an extended part of a waveguide. Typically, no extensive optimization is required as the photodetector and waveguide are directly connected on the same plane. Here, the coupling loss is a function of the modal mismatch between modes of the input waveguide and the Ge detector as well as impedance mismatch (which governs the return losses) at the waveguide-to-photodetector interface (Benedikovic et al., 2021; Vivien and Pavesi, 2013). To allow a low-loss light injection, the design of the waveguide-integrated Ge photodetector has to be optimized to yield maximum modal overlap and minimum back-reflections. Figure 6.2E and F, respectively, show factors of modal confinement and modal overlap with strip Si waveguide as a function of the intrinsic region width for p-i-n photodetectors with a lateral Si-Ge-Si heterojunction (Virot et al., 2017). The active area of the photodetector was 260 nm thick, sitting within a ~60 nm thick Si cavity, while the input Si strip waveguide has transversal geometry of 220 nm × 500 nm (Si layer thickness by its width). We can see that increasing the width of the intrinsic region improves the modal confinement within the Ge core and reduces the power fraction within heavily doped lateral slabs. This is important as better field confinement prevents overlap between mode side lobes and lossy Si regions, thus decreasing deleterious absorption losses. Indeed, this may improve the photo-responsivity On the other hand, a larger cross-section slows down the carrier transit time and reduces the bandwidth (Virot et al., 2017). Second, the modal overlap decreases as well due to the enlarged mode size disparity. This is also accompanied by a larger impedance mismatch (and thus higher spurious optical back-reflections), which in turn, deteriorates the overall coupling loss. Better coupling calls upon tapered waveguide inputs to fit the intrinsic diode width. This can be realized by using additional top layers made out of Ge (Chen et al., 2008) or additional Si overlayers (Chen et al., 2016).

6.3.4 Si-complementary metal-oxide-semiconductor and Si-foundry integration

Nowadays, seamless integration of Ge into Si CMOS circuits and well-developed fabrication routines are largely possible thanks to strong research and rapid technological expansion. Baseline process issues, overload in the thermal budget, potential line contamination, or large device footprints were concerns that are solved now. Up-to-date, integrated photonics offers solutions sufficiently good to bridge the gap between research laboratories and commercial production (Network transformation with world-class optics; 100GBaud + Silicon Photonics Solutions Drive Optical Network Evolution). In turn, this rich environment can support a myriad of applications, where the datacom and telecom market are envisioned to dominate (Bernabé et al., 2021; Hu et al., 2020a; Timurdogan et al., 2020; Network transformation with world-class optics; Wilmart et al., 2020). Besides that, optical sensing in biomedicine, health monitoring, photonic computation, or Light Detection and Ranging are other rising areas. Moreover, an open environment and healthy infrastructure open up a space for companies and start-ups to make their leap from prototyping to the final product. Available technology harnesses a modular approach with mature front-end-of-line and BEOL processing, where SOI is the primary integration platform. SOI is advantaged by access to 200 mm or 300 mm wafers, which are much larger than III/V or lithium niobate wafers. The such platform provides many low-loss passive and high-speed active photonic devices, albeit lasers are excluded for now. This is supported by multiproject wafer (MPW) shuttles, custom-oriented services, or volume production (Rahim et al., 2019, 2018). MPW processes include various modules (library of photonic components) to satisfy different application needs. Moreover, the SiN platform is introduced into the SOI environment (Wilmart et al., 2020; Fahrenkopf et al., 2019). Although the SiN platform has no active role, it offers ultra-low losses and much better fabrication tolerances. Multilayer SiN-on-SOI platforms facilitate broadband, polarization-insensitive, and low-loss access to the photonic chips via inverted-taper couplers (Papes et al., 2016; Guerber et al., 2018). Besides that, SiN can be advantageously utilized for light coupling into optical photodetectors (Masini et al., 2008). Photonic-electronic integration is possible (Atabaki et al., 2018) with Si CMOS foundries, yielding complete transceiver designs (transmitter + receiver), together with additional driving electronic components (Zilkie et al., 2019; Giewont et al., 2019).

6.4 Contemporary advances and state-of-the-art silicon-germanium photodetectors

The profound kinds of planar waveguide-integrated Si-Ge photodetectors are metal-semiconductor-metal (MSM) diodes and PIN structures. They are schematically shown in Figure 6.3A–C, respectively.

6.4.1 Metal-semiconductor-metal diodes

A planar MSM photodiode, shown in Figure 6.3A, consists of back-to-back Schottky contacts, that is, connected metallic electrodes, positioned on top of an undoped semiconductor. MSM photodiodes are thus a type of Schottky barrier device, and for operation, electrical voltage is applied to the interdigitated metal contacts. When the incident light enters the active layer between electrodes, free electron-hole pairs (photo-carriers) are generated (Sarto and Van Zeghbroeck, 1997; Vivien et al., 2008; Assefa et al., 2010a; Cervantes-González et al., 2012; Pan et al., 2012; Dushaq et al., 2017). MSM devices have a simple design and fabrication, which is well-suited for monolithic integration with other optoelectronic and photonic devices. Moreover, in MSM structures, there is no need for semiconductor doping. In turn, the impact of parasitic capacitance on the bandwidth is lower compared to other photodiodes. The optoelectrical bandwidth (and overall device speed) is virtually limited by the transit time of photo carriers only. MSM' are attractive as they operate at low voltages, facilitating compatibility with CMOS circuitry. Over years, improved responsivities and fast light detection were demonstrated using MSM devices. However, they still have excessively large dark currents at low

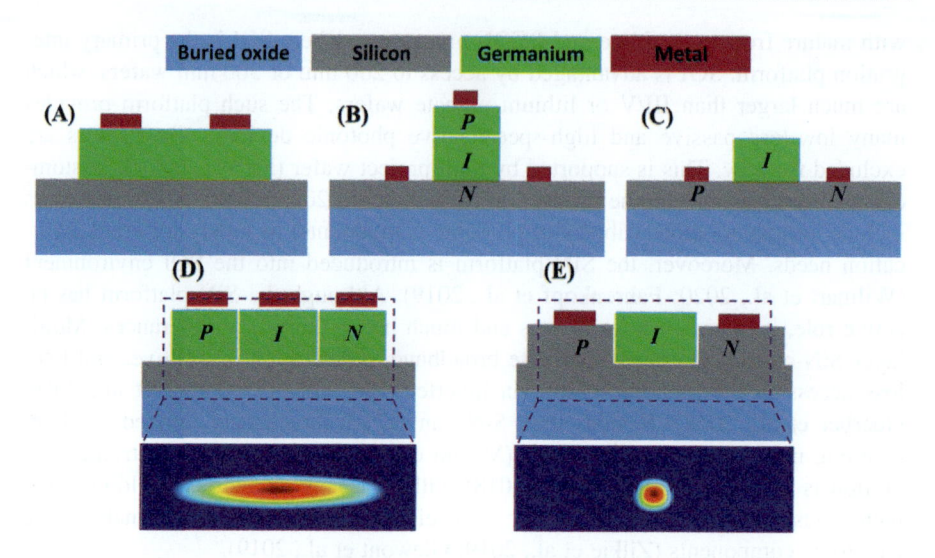

Figure 6.3 Types of waveguide-integrated Si-Ge photodetectors. (A) MSM structure, PIN diode with (B) vertical and (C) lateral arrangement. Lateral PIN diode architecture with (D) homo-junction (full-Ge) and (E) hetero-junction (Si-Ge) configuration. Insets of (D) and (Ee): optical field profiles of fundamental quasi-transverse electrical modes excited in homo- and hetero-junction waveguide photodetectors.

voltages (Vivien et al., 2008; Assefa et al., 2010a; Cervantes-González et al., 2012; Pan et al., 2012), which prevents their wider use. Typically, high dark currents are mainly attributed to the small Schottky barrier height for any metal on Ge, threading dislocations in the Ge layer, and the contact quality between Ge and metal. Consequently, MSM photodetectors have low signal-to-noise ratios, limited sensitivity, and dissipate large amounts of energy. On the other hand, improved design and fabrication of MSM detectors were recently reported, suppressing the dark current by more than four orders of magnitude (Dushaq et al., 2017).

6.4.2 PIN diodes

The most common type of photodetectors used in integrated photonics is PIN structure. For now, many outstanding demonstrations were reported, yielding compact devices, low-voltage operation, improved power-handling capability, and superior optoelectrical performances. PIN photodiode comprises an active material zone (intrinsic light absorbing region) that is sandwiched between two heavily doped (*p*-type and *n*-type) slab regions. PIN photodiodes are widely implemented in either vertical or lateral arrangement, schematically shown in Figure 6.3B and C, respectively, following either homo-junction (full-Ge, Figure 6.3D) or hetero-junction (Si-Ge, Figure 6.3E) configuration.

For vertical and lateral designs, the intrinsic diode thickness and diode width can be tailored by ion implantations and through the Ge epitaxial growth with in situ doping, respectively (Virot et al., 2013). Vertical PIN photodiodes can provide some advantages over lateral ones. The prime benefit of vertical PIN photodiodes is better power-handling capacity. In a vertical diode arrangement, the photodiode width can be as large as needed. This helps to reduce the optical power density within the intrinsic device region and thus enlarging the output power saturation. On the other hand, a wider intrinsic diode zone increases the junction capacitance, which, however, restricts the frequency response via a larger resistance-capacitance delay (RC-delay). In contrast, lateral PIN photodetectors are preferred due to the simpler design as the Ge thickness is not a design parameter, which simplifies process control. Local tailoring of Ge epi-layers is not commonly available in open-access Si-foundries. Lateral PIN photodetectors are then less challenging for integration, manufacturing, and planar wafer processing. Lateral PINs are also easier to fabricate than vertical PIN counterparts since the metallic contacts on doped regions are performed on the same level. This then requires a single-etch step only. In opposition, two different etch-depth levels are indeed required to define a layout for vertical PIN photodiode (Benedikovic et al., 2021; Michel et al., 2010; Virot et al., 2013).

Homo-junction photodetectors, shown in Figure 6.3D, are composed of Ge only, which is epitaxially grown on Si (Virot et al., 2013; DeRose et al., 2011; Vivien et al., 2012; Li et al., 2012; Nam et al., 2015; de Cea et al., 2021; Liu et al., 2017). Such single-material devices have poor optical mode confinement. As p-type and n-type Ge side regions are formed by doping, there is only a slight change in the Ge refractive index compared to the undoped intrinsic region, resulting in a weak index contrast. The optical mode in the homo-junction photodiode spreads over the entire structure and largely overlaps the P, I, and N regions (see inset of Figure 6.3D). However, this may introduce an optical recombination of generated photo carriers. In turn, this impairs the photo-current, enlarging the recombination loss, and deteriorating the device's photo-responsivity. Moreover, from a technological and device processing perspective, full-Ge photodiodes require heavily doped Ge and metal-contact formation directly on Ge. This increases the process complexity as both fabrication steps are less evolved compared to the full-Si processing schemes. This also affects optoelectrical performances of photodetectors by lowering responsivity-bandwidth product and enlarging dark currents. The singular advantage of full-Ge photodetectors is that they favor fast signal detection at a zero-voltage due to the missing discontinuities in the bandgap energy (Vivien et al., 2012; de Cea et al., 2021; Liu et al., 2017). The electrical field generated within the depletion zone (at 0 V) is sufficiently strong to effectively extract the majority of carriers. Although there are existing concerns with higher noise and lower responsivities, homo-junction photodetectors can become an appealing solution for high-speed and energy-aware photonic applications.

PIN photodiodes with hetero-junction arrangements were demonstrated as vital alternatives to homo-junction devices. Technologically, hetero-junction schemes leverage mature Si processes (ion implantations/metal-vias directly performed in/on Si) (Benedikovic et al., 2021; Virot et al., 2017; Chen et al., 2016). In the available fabrication flow, the same procedures are harnessed to dope and contact optical modulators and optical photodetectors. This dramatically simplifies wafer-scale manufacturing and reduces overall production expenses. Doped Si slabs are defined either on pure Si templates (Cui and Zhou, 2017; Zhou and Sun, 2018) or within deep-etched Si cavities (Virot et al., 2017; Chen et al., 2016; Zhang et al., 2014; Lischke et al., 2015; Chen et al., 2015a; Benedikovic et al., 2019b, 2020a). Then, the intrinsic region situated between p-type and n-type slabs is created by the selective epitaxial growth of Ge. Optically, diodes with hetero-structured Si-Ge-Si PIN junction offer improved control over the modal confinement. The modal confinement in hetero-junction devices can be readily engineered, which is not the case in homo-junction configurations. Optical modes excited in the photodetector's intrinsic region are strongly confined (see Figure 6.3E), which in turn, reduces overlap between evanescent tails of the mode and heavily doped side regions. Larger modal confinement yields smaller overlaps with heavily doped Si slabs. This improves modal absorption and minimizes absorption losses of photo carriers, and in the end, yields better responsivities (Virot et al., 2017; Chen et al., 2016). Hetero-junction devices typically provide insufficient optoelectrical performances at zero voltages. Indeed, hetero-junction Si-Ge-Si devices have a strong bandgap energy barrier that results in weak in-built electrical fields in the junction. Then, the additional external voltage of about 0.5 V (Virot et al., 2017), 1 V (Chen et al., 2016), or even higher (Zhang et al., 2014) are applied to improve collection efficiency. In particular, Zhang et al. reported Si-contacted photodetector with a responsivity of 1.14 A/W and a bandwidth of 20 GHz, supporting 40 Gbps operation (Zhang et al., 2014). Improved vertical (Chen et al., 2015a) and lateral (Chen et al., 2016) hetero-structured Si-Ge-Si designs were demonstrated by Chen et al., yielding impressive optoelectrical performance at mainstream optical communication wavebands. Such PINs provide fast signal detection at -1V bias up to 28 (Chen et al., 2015a) and 56 Gbps (Chen et al., 2016), respectively. Benedikovic et al. presented a -0.5-V-operated lateral Si-Ge-Si diodes with a near-unity quantum efficiency (responsivity of 1.2 A/W at 1.55 µm wavelength) (Benedikovic et al., 2019b, 2020a). Moreover, such designs yield high-speed detection with power sensitivities of -14 and -11 dBm for established 10 and 25 Gbps bit rates (at an error level of 10^{-9}) (Benedikovic et al., 2019b). These fundamental building blocks were proved reliable for use in 80 and 100 Gbps data reception systems (Chen et al., 2017; Shi et al., 2021) or in complex 90 Gbps optical receivers (Lambrecht et al., 2019; Li et al., 2021).

Optical PIN photodetectors that handle high powers are essential for microwave and analog photonic links (Marpaung et al., 2019; Beling et al., 2016). Typically, normal-incidence photodiodes were used to that end as they have larger active areas, which is indeed better for high saturated output powers (Benedikovic et al., 2021; Li et al., 2020). Most recently, there is a strong quest to improve the power-handling capacity of chip-integrated Si-Ge photodetectors. Butt-coupled PIN photodetectors have an abrupt absorption profile. This configuration offers the realization of devices with compact footprints, high-speed, and sensitive operation at low driving voltages. However, in these devices, the light is coupled from an input waveguide into the photodetector at once using Si/Ge interface. This typically produces strong modal interferences between excited photodetector modes with locally high optical intensities. Additionally, most of the input light is absorbed within the first few μm's of the photodetector only. At high operating powers, both problems lead to local saturation effects and photo-carrier screening. This, however, lowers the photo-responsivity and reduces the bandwidth. Photodetectors with high output power call upon structures with dilute absorption profiles (Hu et al., 2020b; Byrd et al., 2017; Zuo et al., 2019; Hu et al., 2021). Byrd et al. proposed a neat mode-evolution Si waveguide coupler that laterally transfers the power to the photodetector (Byrd et al., 2017). Uniform light injection via tapered coupler reduces photo-current saturation at high powers. This scheme generated net-light-current up to ~ 16 mA and 40-times improved the bandwidth compared to the butt-coupling alternative. Moreover, an ultra-low dark current of only 1.16 nA and a high 40 (A/W GHZ) responsivity-bandwidth product was obtained. Si-Ge photodetectors with an engineered light field manipulation were demonstrated by Zuo et al. (Zuo et al., 2019). Here, a judiciously designed input multimode waveguide taper is leveraged to enable inhomogeneous light distribution into the photodetector. With a high-power injection, generated photo-current starts to saturate at a level of ~ 27 mA. In addition, high-speed detection at a 10 Gbps signal rate was comprehensively demonstrated in a 5 to 11 mA range. Simultaneous high-power and high-speed operations were also obtained with an optimized Si-Ge photodetector, having 4-directional light inputs (Hu et al., 2020b). Full-Ge lateral PIN photodetector with 4-input optical ports facilitates uniform distribution of the optical field in the light-absorbing region. At 1 V reverse bias, the homo-junction diode has a dark-current of 4 nA, photo-responsivity of 1.23 A/W, and bandwidth of 3.5 GHz. At 8 V reverse bias, PIN photodetector preserves high-speed operation at 20 Gbps within a photo-current range of 0.5 to 7 mA. Most recently, an evanescently-coupled Si-Ge photodetector with double-sided SiN waveguides was demonstrated to favor uniform light injection into the Ge absorber (Hu et al., 2021). The such design offers unprecedented optoelectrical characteristics. Specifically, internal device photo-responsivity was 0.52 A/W with a saturation level up to 25 mA. In addition, operating at low voltages (at 3 V reverse bias in particular), the photodetector has a bandwidth of 60 and 36 GHz for output photo-currents of 4 and 12 mA, respectively. Eye diagram inspection at 1 mA showed clear eye openings for NRZ signals modulated at data rates from 70 to 100 Gbps and for multilevel 4-PAM signaling with bit rates between 100 and 150 Gbps. Last, but not least, 60 Gbps signal detection was elaborated in the 5–20 mA photo-current range. Table 6.2 sums up the optoelectrical performances of the state-of-the-art low-voltage and high-power Si-Ge photodetectors operating at mainstream optical communication wavelengths. Examples of high-power photodetectors are shown in Figure 6.4.

6.4.3 Avalanche photodiodes

Avalanche photodiodes (APDs) are key devices for optical communications, enabling the detection of high-speed optical signals with low intensities. While conventional photodiodes (both MSMs and PINs) typically operate with unity gain, APDs are attractive as they offer an additional internal photo gain that is larger

Table 6.2 State-of-the-art low-voltage and high-power Si-Ge photodetectors operated at communication wavebands.

References	Diode	λ_o [μm]	V_o [V]	I_d [A]	R_p [A/W]	Δf [GHz]	TBR [Gbps]
Chen et al. (2008)	MSM	1.55	1	100n	0.4	40	–
Vivien et al. (2008)	MSM	1.55	1	92–155μ	1	4.2–9	40@1.5 V
Assefa et al. (2010a)	MSM	1.31	1	90μ	0.42	35@2 V	–
Cervantes-González et al. (2012)	MSM	1.55	1	1–2μ	0.33	2@2 V	40@1 V
Pan et al. (2012)	MSM	1.31	1	120μ	0.5	40	–
Dushaq et al. (2017)	MSM	1.55	1	76n	0.8	–	–
Ahn et al. (2007)	V-PIN	1.55	0.1	60n	1.08	7.2	40@5 V
Yin et al. (2007)	V-PIN	1.55	2	169–267n	0.89–1.16	29–31	10@0.3 V
Masini et al. (2008)	V-PIN	1.55	1	3μ	0.85	20	50–90@8 V
Hu et al. (2020b)	L-PIN	1.55	1	4n	1.23	3.5	–
Byrd et al. (2017)	V-PIN	1.55	1	1.6n	1	40	10@3 V
Zuo et al. (2019)	V-PIN	1.55	3	1.26–1.41μ	0.8–0.88	20.4–20.8	70–150@3 V
Hu et al. (2021)	V-PIN	1.55	3	7–84n	0.4–0.52	60	–
Virot et al. (2017)	L-PIN	1.55	1	207n	1.1	5–37	50–56@3 V
Chen et al. (2016)	L-PIN	1.31/1.55	1	2.5n	0.72–0.98	45–50	45–90
Virot et al. (2013)	L-PIN	1.55	1	32n	0.52–0.78	45	–
DeRose et al. (2011)	V-PIN	1.53	1	3n	0.8	120	40@0 V
Vivien et al. (2012)	L-PIN	1.55	1	4n	0.8	14–19	–
Li et al. (2012)	L-PIN	1.55	0.5	500n	0.8	14	5–15@0 V
de Cea et al. (2021)	μ-ring	1.18/1.27	0	0	0.043/0.35	13–27	40@0 V
Liu et al. (2017)	V-PIN	1.31/1.55	0	5.2n@1 V	0.27–0.59	31.7	–
Cui and Zhou (2017)	V-PIN	1.55	1	7.7n	0.75	17@4 V	40@4 V
Zhou and Sun (2018)	V-PIN	1.55	1	160n	0.75	20	–
Zhang et al. (2014)	L-PIN	1.55	4	120n	1.14	70	28@1 V
Lischke et al. (2015)	L-PIN	1.55	1	100n	1	50	10–40@1 V
Chen et al. (2015a)	V-PIN	1.55	1	11n	0.4–0.6	7	10–40@1 V
Benedikovic et al. (2019b)	L-PIN	1.55	1	100n	1.20	7–35	–
Benedikovic et al. (2020a)	L-PIN	1.55	1	100n	0.17–1.16	4.14@3 V	–
Li et al. (2020)	NIPIN	1.31/1.55	1	0.45–1.31μ	0.31–0.52		

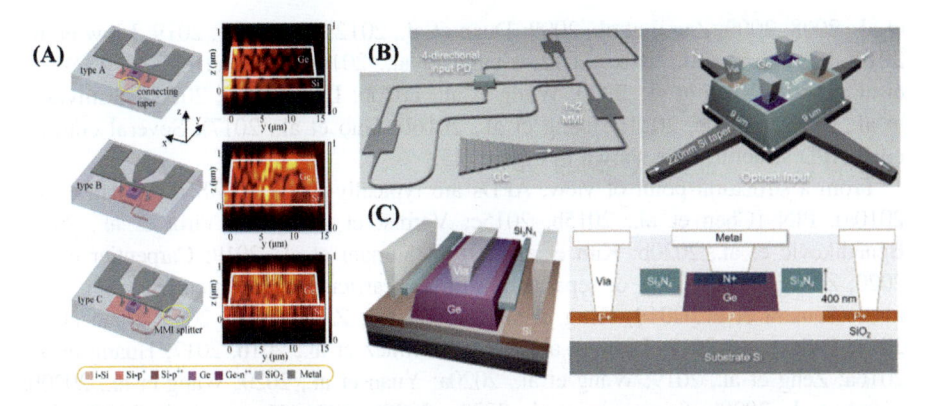

Figure 6.4 Waveguide-integrated Si-Ge photodetectors with high-power-handling capacity. (A) Vertical PIN photodetector with engineered light field manipulation. (B) Lateral PIN photodetector with four-directional light injections. (C) Vertical PIN photodetector with double-stage laterally-coupled SiN input waveguides.
Source: Adapted with permission from part (A) Zuo et al. (2019)—The Optical Society; (B) Hu et al. (2020b)—The Optical Society; (C) Hu et al. (2021)—Chinese Laser Press.

than one. The multiplication gain is generated through the photo-carrier-initiated impact ionization process, taking place under sufficiently high electric fields. If the generated electrical field is strong enough, the impact ionization effect dominates and drives the generation of multiple photo-carriers, that is, several electron-hole pairs are newly created for one absorbed photon. As a result, the generated signal (multiplied photo-current) is amplified. This, in turn, improves the device's photo-responsivity beyond the limit imposed by quantum efficiency, boosts signal-to-noise ratio and sensitivity as well as lowers the thermal noise. Reduced thermal noise can even suppress the need for amplification circuits with additional electronic gain elements, thus favoring the development of receiver-less integrated photonic circuits for future low-energy consumption optical interconnects (Assefa et al., 2010b). However, full-Ge APDs are noisy devices, which is a result of Ge low bandgap energy. For this reason and for a long time, conventional full-Ge APDs were assumed to be unreliable for use in optical communication systems. Ionization coefficients of photo-carriers (electrons (α) and holes (β)) in Ge have similar magnitude, and thus effective ionization parameter is close to one ($k = \beta/\alpha$ ~ 0.9). In contrast, ionization coefficients in Si are quite different. Then, the impact ionization ratio of Si is very low ($k \sim 0.01$), indeed, one of the lowest of established crystalline semiconductors. The k-factor is a function of an applied electric field and directly affects the APD performances, especially in terms of gain and bandwidth, noise, and sensitivity. This way, combing low-noise multiplication properties of Si and high optical absorption of Ge is recognized to be advantageous for the development of waveguide-integrated APDs (Assefa et al., 2010a; Chen et al., 2015b, 2015c; Verbist et al., 2017; Virot et al., 2014; Benedikovic et al., 2020b; Kim et al., 2018b; Samani et al., 2019; Carpentier et al., 2020; Zhang et al., 2020; Kang et al., 2008, 2009; Zaoui et al., 2009; Duan et al., 2012; Park et al., 2019; Liow et al., 2014; Martinez et al., 2016, 2017; Huang et al., 2016a; Zeng et al., 2019; Wang et al., 2020a; Yuan et al., 2020; Wang et al., 2020b; Liang et al., 2020b; Srinivasan et al., 2020a, 2020b, 2021; Huang et al., 2016b; Guo et al., 2017). Several cutting-edge APD structures are shown in Figure 6.5.

From a structural point of view, APDs are typically made as MSM (Assefa et al., 2010a), PIN (Chen et al., 2015b, 2015c; Verbist et al., 2017; Virot et al., 2014; Benedikovic et al., 2020b; Kim et al., 2018b; Samani et al., 2019; Carpentier et al., 2020; Zhang et al., 2020), or separate absorption carrier multiplication (SACM) structures (Samani et al., 2019; Kang et al., 2008, 2009; Zaoui et al., 2009; Duan et al., 2012; Park et al., 2019; Liow et al., 2014; Martinez et al., 2016, 2017; Huang et al., 2016a; Zeng et al., 2019; Wang et al., 2020a; Yuan et al., 2020; Wang et al., 2020b; Liang et al., 2020b; Srinivasan et al., 2020a, 2020b, 2021; Huang et al., 2016b; Guo et al., 2017). Performance characteristics of the state-of-the-art APDs are summed up in Table 6.3. Although internal avalanche multiplication in Ge generates a lot of noise, it is easier to design, integrate, and fabricate such a device in both MSM and PIN configurations compared to the SACM alternative. IBM demonstrated a 10 Gbps MSM-based Si-Ge APD, achieving a gain-bandwidth product of 300 GHz at a reverse bias of 1.5 V only. Indeed, these photodetectors would be highly attractive for CMOS-compatible photonic-electronic circuits. However, its limited opto-electronic performance (high noise, low sensitivity, and large power consumption) is so dominant that jeopardizes future implementations.

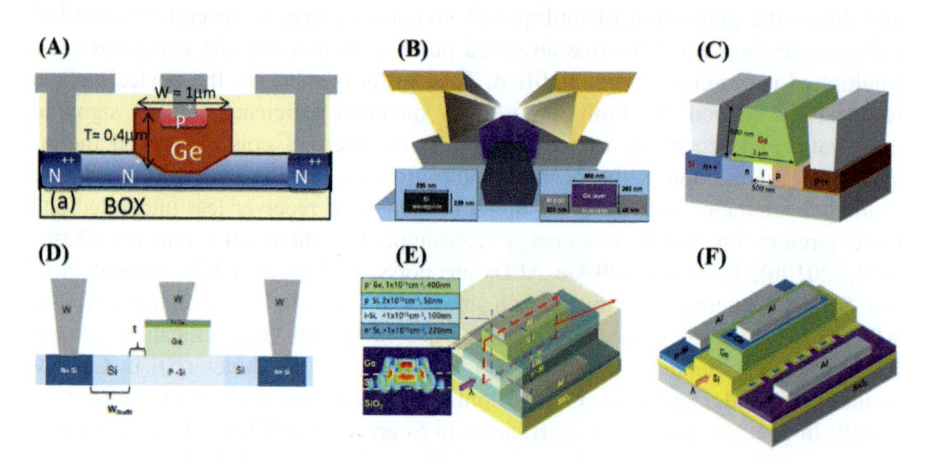

Figure 6.5 Si-Ge avalanche photodetectors. (A) 10 Gbps vertical PIN APD. (B) 40 Gbps lateral PIN APD. (C) 160 Gbps full-Ge vertical PIN APD. (D) Linear mode 13 Gbps SACM APD. (E) 25 Gbps two-stage vertical SACM APD. (F) 25 Gbps three-terminal vertical SACM APD.
Source: Adapted with permission from part (A) Chen et al. (2015b)—The Optical Society; (B) Benedikovic et al. (2020b)—The Optical Society; (C) Zhang et al. (2020)—The Optical Society; (D) Martinez et al. (2016)—The Optical Society; (E) Huang et al. (2016a)—The Optical Society; (F) Zeng et al. (2019)—The Optical Society.

Table 6.3 State-of-the-art Si-Ge avalanche photodetectors operated at communication wavebands.

References	Type	Modulation	λ_o [μm]	V_b [V]	GBP [GHz]	k	TBR [Gbps]	P_m [dBm]@BER
Assefa et al. (2010a)	MSM	OOK	1.31/1.5	3.5	300	0.20	10	$-13.9@10^{-9}$
Chen et al. (2015b)	PIN	OOK	1.55	6.2	100	0.50	10	$-24.4@10^{-9}$
Chen et al. (2015c)	PIN	OOK	1.31	5	140	0.20	25	$-14.8@10^{-9}$
Verbist et al. (2017)	PIN	4-PAM	1.31	5	106	–	32–50	$-6 @3.8 \times 10^{-3}$
Virot et al. (2014)	PIN	OOK	1.55	7	190	0.40	10	$-26@10^{-7}$
Benedikovic et al. (2020b)	PIN	OOK	1.55	11	210	0.25	40	$-11.2@10^{-9}$
Kim et al. (2018b)	PIN	OOK	1.31/1.55	>30	460	–	40	$-13.9@10^{-12}$
Samani et al. (2019)	PINSACM	4-PAMOOK	1.31	4>15	–	–	11225	$-8@2.0 \times 10^{-4}$ $-12@2.0 \times 10^{-4}$
Carpentier et al. (2020)	PIN	OOK	1.55	4	400	–	28	$-9@10^{-12}$
Zhang et al. (2020)	PIN	4-PAM16-QAM	1.55	>12	528	0.15–0.25	64160	–
Kang et al. (2009)	SACM	OOK	1.30	>25	340	0.09	10	$-28@10^{-12}$
Zaoui et al. (2009)	SACM	–	1.31	>25	840	–	–	–
Duan et al. (2012)	SACM	–	1.55	>25	310	–	–	–
Park et al. (2019)	SACM	OOK	1.31	12	150	–	50	$-16@10^{-4}$
Liow et al. (2014)	SACM	OOK	1.30	20	115	–	25	$-30.5@10^{-10}$
Martinez et al. (2016)	SACM	OOK	1.55	31	432	–	13	$-18.3@10^{-12}$
Huang et al. (2016a)	SACM	OOK	1.55	10	276	–	25	$-16@10^{-12}$
Zeng et al. (2019)	SACM	OOK	1.55	6	280	0.05	25	$-11.4@10^{-4}$
Wang et al. (2020a)	SACM	4-PAM	1.55	10	500	–	64	$-13@10^{-4}$
Srinivasan et al. (2020b)	SACM	OOK	1.31/1.55	>12	297	–	25–50	–
Srinivasan et al. (2021)	SACM	OOK	1.31	>12	238	<0.25	56	$-18.6@10^{-4}$
Huang et al. (2016b)	SACM	OOK	1.31	>20	240	–	25	$-22.5@10^{-12}$

Notes: λ_o, operating wavelength; BER, Bit-error-rate threshold; GBP, diode gain-bandwidth product; k, effective ionization ratio; P_m, optical power sensitivity; TBR, transmission bit rate; V_b, diode breakdown voltage.

On the other hand, simple PIN structures, biased at higher voltages to trigger an avalanche multiplication, are an appealing solution for APDs. Similar to MSM structures, PIN-based APDs favor facile design and manufacturing in Si-foundries (Rahim et al., 2019, 2018) and may operate at low voltages, typically at sub10V level (Chen et al., 2015b, 2015c; Verbist et al., 2017; Virot et al., 2014; Benedikovic et al., 2020b; Kim et al., 2018b; Samani et al., 2019; Carpentier et al., 2020; Zhang et al., 2020). PIN-based APDs also provide attractive and tailorable optoelectrical performances, yielding high-speed and low-noise operation, coupled with sensitive signal detection. A strong advantage of PIN APDs, compared to other approaches, is the straightforward control of PIN geometry. In other words, optimizing the intrinsic diode region through its judicious thinning (Chen et al., 2015b, 2015c) or narrowing (Virot et al., 2014; Benedikovic et al., 2020b) enables avalanching the diode at reduced voltages as well as helps to lower the multiplication noise via dead space effect. In particular, vertical Si-Ge PIN APD optical receivers operated at low-voltages (sub6V (Chen et al., 2015b) and sub-5V (Chen et al., 2015c)) with 10 (Chen et al., 2015b) and 25 Gbps (Chen et al., 2015c) data rates were demonstrated by Chen et al., providing a credible light detection at commercial communication wavebands (Chen et al., 2015b, 2015c). Moreover, such devices were also used in short-link photo-receivers to detect a spectrally efficient 4-level pulse amplitude modulation (4-PAM) (Verbist et al., 2017). Moreover, compact homo-junction (Virot et al., 2014) and hetero-junction (Benedikovic et al., 2020b) lateral PIN APDs were proposed to yield fast signal detection for 10 and 40 Gbps bit rates, alongside high gain-bandwidth products (up to 210 GHz) and low excess noise (providing effective ionization coefficients in a 0.25 to 0.4 range). Photo-receiver with normal-incidence vertical APD was demonstrated by Kim et al. (Kim et al., 2018b). leveraging the negative photo-conductance effect. This APD affords a gain-bandwidth product of 460 GHz, high-speed operation up to 50 GHz, and low sensitivities for 40 Gbps data rates. However, practical use is mostly hindered by very large operating voltages (>26 V) and vertical-illuminated coupling. Judiciously designed vertical PIN-APD was proposed for 25 and 28 Gbps error-free operation at O- and C-band, respectively, for conventional on-off keying (OOK) signaling (Samani et al., 2019; Carpentier et al., 2020). The same structures also showed high gain-bandwidth products (>400 GHz) and great promises to detect 112 Gbps signals under a higher-order 4-PAM modulation scheme (Samani et al., 2019). Most recently, full-Ge vertical PIN-APD was demonstrated by Zhang et al. (Zhang et al., 2020). providing a low-noise structure with a high gain-bandwidth product (>500 GHz). This enables an ultra-fast signal detection for 64 Gbps 4-PAM and 160 Gbps 16-Quadrature amplitude modulation (16-QAM) signals.

SACM APDs at near-IR wavelengths are advantaged by combing high optical absorption of Ge and low-noise properties of Si that are exploited for carrier multiplication. Compared to MSMs and PINs, SACM-based APDs bring an additional degree of freedom to flexibly engineer the output optoelectrical performances. Higher gain-bandwidth products and lower operating voltages are typically obtained with SACM APDs, and thus they are promising for transmission systems operating beyond 25 Gbps per single wavelength. Typically, SACM APD structures can be realized in a form of surface-illuminated (Kang et al., 2008, 2009; Zaoui et al., 2009; Duan et al., 2012; Park et al., 2019) or waveguide-coupled (Samani et al., 2019; Liow et al., 2014; Martinez et al., 2016, 2017; Huang et al., 2016a; Zeng et al., 2019; Wang et al., 2020a; Yuan et al., 2020; Wang et al., 2020b; Liang et al., 2020b; Srinivasan et al., 2020a, 2020b) device architectures, using either vertical or lateral diode configuration. For the most part, they are found in a vertical diode arrangement (Samani et al., 2019; Kang et al., 2009; Duan et al., 2012; Park et al., 2019; Liow et al., 2014; Martinez et al., 2016, 2017; Huang et al., 2016a; Zeng et al., 2019) instead of lateral-based designs (Srinivasan et al., 2020a, 2020b, 2021). However, vertical SACM diode designs require more complex fabrication processes, including the development of dedicated procedures for Ge epi-layers and/or metal-via contacts made directly on the absorbing Ge region. Consequently, this may impair the overall APD performance by reducing the nominal (unity-gain) responsivity, enlarging the dark-current tunneling, and decreasing the bandwidth. These shortcomings can be challenged by lateral SACM designs (Srinivasan et al., 2020a, 2020b, 2021). Lateral arrangements afford more flexibilities with minimized complexity for epitaxy and metal-contact processing. In addition, conventional SACM designs typically have only one contact electrode to control generated electric fields in both distinctly separated regions (light absorbing and carrier multiplication zone) (Huang et al., 2016a; Zeng et al., 2019). In addition, SACM is a type of APD structure that requires fine optimization of the charge layer thickness/width and a careful selection of its doping profile. It is then of critical importance to control the generation of the electric field. This field should reach a level that initiates impact ionization in the Si charge zone, while at the same time, the electrical field is high enough in the Ge layer to extract the generated photo-carriers and has a threshold level that avoids impact ionization in Ge. Beyond this threshold, the dark current increases excessively, which limits the overall device performance (Benedikovic et al., 2021; Michel et al., 2010). Up to date, many outstanding SACM APDs were reported. More specifically, 10 Gbps APD operation was initially demonstrated by Kang et al. using a normal-incidence SACM device (Kang et al., 2009). Improved SACM design, with over 400 GHz gain-bandwidth product and 13 Gbps operation, was reported by Martinez et al. (Martinez et al., 2016). The same SACM APD was leveraged to demonstrate a single-photon detection (Martinez et al., 2017). However, these SACM designs are driven with very high bias supplies (>20 V), which is not practical for future low-energy-consumption interconnect scenarios. More recent SACM APDs reduce the operating voltage while improving detection speeds. In particular, Hewlett Packard Laboratories demonstrated two-stage (Huang et al., 2016a) and optimized three-terminal (Zeng et al., 2019) APD receivers operating at sub10V reverse voltages and 25 Gbps speeds, alongside with high gain-bandwidth products (up to 280 GHz) and credible sensitivities down to -12 and -16 dBm, both obtained for conventional OOK modulation (Huang et al., 2016a; Zeng et al., 2019). Moreover, improved APD designs with distributed Bragg waveguide mirrors (Wang et al., 2020a) and better temperature stability (Yuan et al., 2020) were demonstrated for 64 Gbps 4-PAM signaling. Additionally, such APDs were also analyzed for use in low-voltage, high-speed, and energy-efficient photonics links (Wang et al., 2020b).

Analysis revealed reliable performance for 50 Gbps OOK signal detection, while detection of 50 to 100 Gbps signals was readily possible with a 4-PAM modulation format. Besides that, low energy consumption in the 5 to 25 pJ/bit range was predicted for on-chip signal detection up to 100 Gbps. SACM APD also proved to be a great choice to develop green-energy dense wavelength division multiplexing (DWDM) systems for next-generation chip-scale interconnects (Liang et al., 2020b). Lateral Si-contacted SACM APDs were demonstrated by Srinivasan, et al. (Srinivasan et al., 2020a, 2020b, 2021). showing great potential for mass-scale fabrication within Si-foundries. Full APD-based receivers using hybrid BiCMOS-Si technology were reported, operating at speeds up to 56 Gbps and yielding power consumption of 140 mW only (Srinivasan et al., 2020b, 2021). SACM Si-Ge APDs operating at 25 Gbps (Huang et al., 2016b) were reported for next-generation 100G Passive Optical Network (100G-PON) as well (Guo et al., 2017).

6.5 Conclusion

We reviewed contemporary progress made in the field of Si–Ge photodetectors at commercial optical communication wavelengths. Over the decades of research, chip-integrated Si–Ge photodetectors obtained remarkable performances, leveraging advances in design strategies, material processing and fabrication, and seamless Si CMOS integration. Recent Si–Ge photodetectors are mainly advantaged by the well-mastered Si chip industry, providing a rich environment for innovations and breakthroughs. This encompasses reasonably priced and widely accessible SOI platforms, high-quality Ge epitaxy, and open-access foundry models. Si and Ge elemental semiconductors are well-complemented to each other, which speeds up the development of reliable receivers, aiming for a fast, sensitive, and energy-aware operation. A myriad of low-voltage or high-power PIN devices, as well as APD structures, are now available for the integrated photonics community, yielding optoelectrical performances that start to be unmatched and, in many situations, superior to the existing III/V-based counterparts. Most recently, enhancing the detector detection speeds, lowering its external driving voltage as well as obtaining much better power sensitivities are ongoing research tasks that will drive future works and development. All of this will make Si-Ge photodetectors critically important building blocks in chip-scale systems to support surging applications in integrated photonics. This mainly includes short-reach optical interconnects for telecom and datacom up to progressive areas such as sensing and monitoring as well as biomedicine and health sciences.

Funding

This work received a funding support by the Slovak Grant Agency VEGA 1/0113/22 and Slovak Research and Development Agency under the project APVV-21-0217.

References

100GBaud + Silicon Photonics Solutions Drive Optical Network Evolution. Available from <https://acacia-inc.com/acacia-resources/100gbaud-silicon-photonics-solutions-drive-optical-network-evolution/>. [accessed 14 Jun 2021].

Ackert, J.J., Thomson, D.J., Shen, L., Peacock, A.C., Jessop, P.E., Reed, G.T., et al., 2015. High-speed detection at two micrometres with monolithic silicon photodiodes. Nat. Photonics 9, 393–396.

Ackert, J.J., Fiorentino, M., Logan, D.F., Beausoleil, R., Jessop, P.E., Knights, A.P., 2011. Silicon-on-insulator microring resonator defect-based photodetector with 3.5-GHz bandwidth. J. Nanophotonics 5, 059507.

Ahn, D., Hong, C.-Y., Liu, J., Giziewicz, W., Beals, M., Kimerling, L.C., et al., 2007. High performance, waveguide integrated Ge photodetectors. Opt. Express 15, 3916–3921.

Ahn, D., Kimerling, L.C., Michel, J., 2011. Efficient evanescent wave coupling conditions for waveguide-integrated thin-film Si/Ge photodetectors on silicon-on-insulator/germanium-on-insulator substrates. J. Appl. Phys. 110, 083115.

Akinwande, D., Huyghebaert, C., Wang, C.-H., Serna, M.I., Goossens, S., Li, L.-J., et al., 2019. Graphene and two-dimensional materials for silicon technology. Nature 573, 507–518.

Alduino, A., Paniccia, M., 2007. Wiring electronics with light. Nat. Photonics 1, 153–155.

Alonso-Ramos, C., Nedeljkovic, M., Benedikovic, D., Soler Penadés, J., Littlejohns, C.G., Khokhar, A.Z., et al., 2016. Germanium-on-silicon mid-infrared grating couplers with low-reflectivity inverse taper excitation. Opt. Lett. 41, 4324–4327.

Alonso-Ramos, C., Le Roux, X., Zhang, J., Benedikovic, D., Vakarin, V., Durán-Valdeiglesias, E., et al., 2019. Diffraction-less propagation beyond the sub-wavelength regime: a new type of nanophotonic waveguide. Sci. Rep. 9, 5347.

Armand Pilon, F.T., Lyasota, A., Niquet, Y.-M., Reboud, V., Calvo, V., Pauc, N., et al., 2019. Lasing in strained germanium microbridges. Nat. Commun. 10, 2724.

Assefa, S., Xia, F., Vlasov, Y.A., 2010a. Reinventing germanium avalanche photodetector for nanophotonic on-chip optical interconnects. Nature 464, 80–84.

Assefa, S., Xia, F., Bedell, S.W., Zhang, Y., Topuria, T., Rice, P.M., et al., 2010a. CMOS-integrated high-speed MSM germanium waveguide photodetector. Opt. Express 18, 4986–4999.

Assefa, S., Xia, F., Green, W.M.J., Schow, C.L., Rylyakov, A.V., Vlasov, Y.A., 2010b. CMOS-integrated optical receivers for on-chip interconnects. IEEE J. Sel. Top. Quantum Electron. 16, 1376–1385.

Atabaki, A.H., Moazeni, S., Pavanello, F., Gevorgyan, H., Notaros, J., Alloati, L., et al., 2018. Integrating photonics with silicon nanoelectronics for next generation of systems on a chip. Nature 556, 349–354.

Beausoleil, R.G., Kuekes, P.J., Snider, G.S., Wang, S.-Y., Williams, R.S., 2008. Nanoelectronic and nanophotonic interconnect. Proc. IEEE 96, 230–247.

Beling, A., Xie, X., Campbell, J.C., 2016. High-power, high-linearity photodiodes. Optica 3, 328–338.

Benedikovic, D., Alonso-Ramos, C., Guerber, S., Le Roux, X., Cheben, P., Dupré, C., et al., 2019a. Sub-decibel silicon grating couplers based on L-shaped waveguides and engineered subwavelength metamaterials. Opt. Express 27, 26239–26250.

Benedikovic, D., Virot, L., Aubin, G., et al., 2019b. 25 Gbps low-voltage hetero-structured silicon-germanium waveguide pin photodetectors for monolithic on-chip nanophotonic architectures. Photon. Res. 7, 437–444.

Benedikovic, D., Virot, L., Aubin, G., et al., 2020b. 40 Gbps heterostructure germanium avalanche photo receiver on a silicon chip. Optica 7, 775–783.

Benedikovic, D., Virot, L., Aubin, G., et al., 2020a. Comprehensive study on chip-integrated germanium pin photodetectors for energy-efficient silicon interconnects. IEEE J. Quantum Electron. 56, 8400409.

Benedikovic, D., Virot, L., Aubin, G., Hartmann, J.-M., Amar, F., Le Roux, X., et al., 2021. Silicon–germanium receivers for short-wave-infrared optoelectronics and communications. Nanophotonics 10, 1059–1079.

Benedikovic, D., Berciano, M., Alonso-Ramos, C., Le Roux, X., Cassan, E., Marris-Morini, D., et al., 2017. Dispersion control of silicon nanophotonic waveguides using subwavelength grating metamaterials in near- and mid-IR wavelengths. Opt. Express 25, 19468–19478.

Benner, A.F., Kuchta, D.M., Pepeljugoski, P.K., Budd, R.A., Hougham, G., Fasano, B.V., et al., 2010. Optics for high-performance servers and supercomputers. In: Optical Fiber Communication Conference, Optical Society of America, p. OTuH1.

Berciano, M., Marcaud, G., Damas, P., Le Roux, X., Crozat, P., Alonso Ramos, C., et al., 2018. Fast linear electro-optic effect in a centrosymmetric semiconductor. Commun. Phys. 1, 64.

Bernabé, S., Wilmart, Q., Hasharoni, K., Hassan, K., Thonnart, Y., Tissier, P., et al., 2021. Silicon photonics for terabit/s communication in data centers and exascale computers. Solid-State Electron. 179, 107928.

Bie, Y.-Q., Grosso, G., Heuck, M., Furchi, M.M., Cao, Y., Zheng, J., et al., 2017. A MoTe2-based light-emitting diode and photodetector for silicon photonic integrated circuits. Nat. Nanotechnol. 12, 1124–1129.

Blumenthal, D.J., Heideman, R., Geuzebroek, D., Leinse, A., Roeloffzen, C., 2018. Silicon Nitride in Silicon Photonics. Proc. IEEE 106, 2209–2231.

Boeuf, F., Han, J.-H., Takagi, S., Takenaka, M., 2017. Benchmarking Si, SiGe, and III–V/Si hybrid SIS optical modulators for datacenter applications. J. Light. Technol. 35, 4047–4055.

Bowers, J.E., Liu, A.Y., 2017. A comparison of four approaches to photonic integration. In: Optical Fiber Communication Conference, Optical Society of America, p. M2B.4.

Byrd, M.J., Timurdogan, E., Su, Z., Poulton, C.V., Fahrenkopf, N.M., Leake, G., et al., 2017. Mode-evolution-based coupler for high saturation power Ge-on-Si photodetectors. Opt. Lett. 42, 851–854.

Carpentier, O., Samani,A., Jacques, M., et al., 2020. High gain-bandwidth waveguide coupled silicon germanium avalanche photodiode. In: Conference on Lasers and Electro-Optics, Optical Society of America, p. STh4O.3.

Casalino, M., Coppola, G., Iodice, M., Rendina, I., Sirleto, L., 2010. Near-infrared sub-bandgap all-silicon photodetectors: state of the art and perspectives. Sensors (Basel) 10, 10571–10600.

Cervantes-González, J.C., Ahn, D., Zheng, X., Banerjee, S.K., Jacome, A.T., Campbell, J.C., et al., 2012. Germanium metal-semiconductor-metal photodetectors evanescently coupled with upper-level silicon oxynitride dielectric waveguides. Appl. Phys. Lett. 101, 261109.

Chen, H.T., Verbist, J., Verheyen, P., et al., 2015c. 25-Gb/s 1310-nm optical receiver based on a sub-5-V waveguide-coupled germanium avalanche photodiode. IEEE Photonics J. 7, 7902909.

Chen, H.T., Verbist, J., Verheyen, P., et al., 2015b. High sensitivity 10Gb/s Si photonic receiver based on a low-voltage waveguide-coupled Ge avalanche photodetector. Opt. Express 23, 815–822.

Chen, H.T., Verheyen, P., De Heyn, P., et al., 2015a. High-responsivity low-voltage 28-Gb/s Ge p-i-n photodetector with silicon contacts. J. Light. Technol. 33, 820–824.

Chen, H., Galili, M., Verheyen, P., et al., 2017. 100-Gbps RZ data reception in 67-GHz Si-contacted germanium waveguide p-i-n photodetectors. J. Light. Technol. 35, 722–726.

Chen, H., Verheyen, P., De Heyn, P., Lepage, G., De Coster, J., Balakrishnan, S., et al., 2016. − 1 V bias 67 GHz bandwidth Si-contacted germanium waveguide p-i-n photodetector for optical links at 56 Gbps and beyond. Opt. Express 24, 4622–4631.

Chen, L., Dong, P., Lipson, M., 2008. High performance germanium photodetectors integrated on submicron silicon waveguides by low temperature wafer bonding. Opt. Express 16, 11513–11518.

Cheng, Q., Bahadori, M., Glick, M., Rumley, S., Bergman, K., 2018. Recent advances in optical technologies for data centers: a review. Optica 5, 1354–1370.

Colace, L., Masini, G., Galluzzi, F., Assanto, G., 1998. Metal–semiconductor–metal near-infrared light detector based on epitaxial Ge/Si. Appl. Phys. Lett. 72, 3175–3177.

Colace, L., Masini, G., Assanto, G., Luan, H.-C., Wada, K., Kimerling, L.C., 2000. Efficient high-speed near-infrared Ge photodetectors integrated on Si substrates. Appl. Phys. Lett. 76, 1231.

Cui, J., Zhou, Z., 2017. High-performance Ge-on-Si photodetector with optimized DBR location. Opt. Lett. 42, 5141–5144.

Currie, M.T., Samavedam, S.B., Langdo, T.A., Leitz, C.W., Fitzgerald, E.A., 1998. Controlling threading dislocation densities in Ge on Si using graded SiGe layers and chemical-mechanical polishing. Appl. Phys. Lett. 72, 1718–1720.

de Cea, M., Van Orden, D., Fini, J., Wade, M., Ram, R.J., 2021. High-speed, zero-biased silicon-germanium photodetector. APL. Photonics 6, 041302.

del Alamo, J.A., 2011. Nanometre-scale electronics with III−V compound semiconductors. Nature 479, 317–323.

DeRose, C.T., Trotter, D.C., Zortman, W.A., et al., 2011. Ultra compact 45 GHz CMOS compatible germanium waveguide photodiode with low dark current. Opt. Express 19, 24897–24904.

Desiatov, B., Goykhman, I., Shappir, J., Levy, U., 2014. Defect-assisted sub-bandgap avalanche photodetection in interleaved carrier-depletion silicon waveguide for telecom band. Appl. Phys. Lett. 104, 091105.

Doerr, C., Chen, L., 2018. Silicon photonics in optical coherent systems. Proc. IEEE 106, 2291–2301.

Dorodnyy, A., Salamin, Y., Ma, P., Vukajlovic Plestina, J., Lassaline, N., Mikulik, D., et al., 2018. Plasmonic photodetectors. IEEE J. Sel. Top. Quantum Electron. 24, 4600313.

Duan, N., Liow, T.-Y., Lim, A.E.-J., Ding, L., Lo, G.Q., 2012. 310 GHz gain-bandwidth product Ge/Si avalanche photodetector for 1550 nm light detection. Opt. Express 20, 11031–11036.

Durán-Valdeiglesias, E., Guerber, S., Oser, D., Le Roux, X., Benedikovic, D., Pérez-Galacho, D., et al., 2019. Dual-polarization silicon nitride Bragg filters with low thermal sensitivity. Opt. Lett. 44, 4578–4581.

Dushaq, G., Nayfeh, A., Rasras, M., 2017. Metal-germanium-metal photodetector grown on silicon using low temperature RF-PECVD. Opt. Express 25, 32110–32119.

Fadaly, E.M.T., Dijkstra, A., Suckert, J.R., Ziss, D., van Tilburg, M.A.J., Mao, C., et al., 2020. Direct-bandgap emission from hexagonal Ge and SiGe alloys. Nature 580, 205–209.

Fahrenkopf, N.M., McDonough, C., Leake, G.L., Su, Z., Timurdogan, E., Coolbaugh, D.D., 2019. The AIM photonics MPW: a highly accessible cutting edge technology for rapid prototyping of photonic integrated circuits. IEEE J. Sel. Top. Quantum Electron. 25, 8201406.

Filer, M., Gaudette, J., Yin, Y., Billor, D., Bakhtiari, Z., Cox, J.L., 2019. Low-margin optical networking at cloud scale. J. Opt. Commun. Netw. 11, C94–C105.

Fitzgerald, E.A., Xie, Y.-H., Green, M.L., Brasen, D., Kortan, A.R., Michel, J., et al., 1991. Totally relaxed GeSi layers with low threading dislocation densities grown on Si substrates. Appl. Phys. Lett. 59, 811–814.

Flöry, N., Ma, P., Salamin, Y., Emboras, A., Taniguchi, T., Watanabe, K., et al., 2020. Waveguide-integrated van der Waals heterostructure photodetector at telecom wavelengths with high speed and high responsivity. Nat. Nanotechnol. 15, 118–124.

Frydendahl, C., Grajower, M., Bar-David, J., Zektzer, R., Mazurski, N., Shappir, J., et al., 2020. Giant enhancement of silicon plasmonic shortwave infrared photodetection using nanoscale self-organized metallic films. Optica 7, 371–379.

Gherabli, R., Grajower, M., Shappir, J., Mazurski, N., Wofsy, M., Inbar, N., et al., 2020. Role of surface passivation in integrated sub-bandgap silicon photodetection. Opt. Lett. 45, 2128–2131.

Ghosh, S., Lin, K.-C., Tsai, C.-H., Kumar, H., Chen, Q., Zhang, L., et al., 2020. Metal-semiconductor-metal gesn photodetectors on silicon for short-wave infrared applications. Micromachines 11, 795.

Giewont, K., Nummy, K.A., Anderson, F.A., et al., 2019. 300-mm monolithic silicon photonics foundry technology. IEEE J. Sel. Top. Quantum Electron. 25, 8200611.

Goykhman, I., Sassi, U., Desiatov, B., Mazurski, N., Milana, S., de Fazio, D., et al., 2016. On-Chip integrated, silicon–graphene plasmonic schottky photodetector with high responsivity and Avalanche photogain. Nano Lett. 16, 3005–3013.

Guerber, S., Alonso-Ramos, C., Benedikovic, D., Pérez-Galacho, D., Le Roux, X., Vulliet, N., et al., 2018. Integrated SiN on SOI dual photonic devices for advanced datacom solutions. In: Proceedings Volume 10686, Silicon Photonics: From Fundamental Research to Manufacturing, Strasbourg, France.

Guo, Y., Yin, Y., Song, Y., et al., 2017. Demonstration of 25Gbit/s per channel NRZ transmission with 35 dB power budget using 25G Ge/Si APD for next generation 100G-PON. In: Optical Fiber Communication Conference, Optical Society of America, p. M3H.6.

Halir, R., Ortega-Moñux, A., Benedikovic, D., Mashanovich, G.Z., Wangüemert-Pérez, J.G., Schmid, J.H., et al., 2018. Subwavelength-grating metamaterial structures for silicon photonic devices. Proc. IEEE 106, 2144–2157.

Hartmann, J.M., Abbadie, A., Papon, A.M., et al., 2004. Reduced pressure–chemical vapor deposition of Ge thick layers on Si(001) for 1.3–1.55-μm photodetection. J. Appl. Phys. 95, 5905–5913.

Hartmann, J.M., Abbadie, A., Cherkashin, N., Grampeix, H., Clavelier, L., 2009. Epitaxial growth of Ge thick layers on nominal and 6° off Si(0 0 1); Ge surface passivation by Si. Semicond. Sci. Technol. 24, 055002.

Hartmann, J.M., Damlencourt, J.-F., Bogumilowicz, Y., Holliger, P., Rolland, G., Billon, T., 2005. Reduced pressure-chemical vapor deposition of intrinsic and doped Ge layers on Si(0 0 1) for microelectronics and optoelectronics purposes. J. Cryst. Growth 274, 90–99.

Helkey, R., Saleh, A.A.M., Buckwalter, J., Bowers, J.E., 2019. High-performance photonic integrated circuits on silicon. IEEE J. Sel. Top. Quantum Electron. 25, 8300215.

Hoefler, G.E., Zhou, Y., Anagnosti, M., Bhardwaj, A., Abolghasem, P., James, A., et al., 2019. Foundry development of system-on-chip inp-based photonic integrated circuits. IEEE J. Sel. Top. Quantum Electron. 25, 6100317.

Hu, X., Wu, D., Zhang, H., Li, W., Chen, D., Wang, L., et al., 2020b. High-speed lateral PIN germanium photodetector with 4-directional light input. Opt. Express 28, 38343–38354.

Hu, X., Wu, D., Zhang, H., Li, W., Chen, D., Wang, L., et al., 2021. High-speed and high-power germanium photodetector with a lateral silicon nitride waveguide. Photon. Res. 9, 749–756.

Hu, Z., Shao, S., Xiao, Z., Zhu, X., Wu, Y., Feng, J., et al., 2020a. 100 Gb/s PAM-4 Silicon Photonics Transceiver for Intra-Datacenter on a 200-mm Wafer. In: Asia Communications and Photonics Conference/International Conference on Information Photonics and Optical Communications 2020, Optical Society of America, p. S3D.3.

Huang, M., Cai, P., Li, S., et al., 2016b. Breakthrough of 25Gb/s germanium on silicon avalanche photodiode. In: Optical Fiber Communication Conference, Optical Society of America, p. Tu2D.2.

Huang, Z., Li, C., Liang, D., et al., 2016a. 25 Gbps low-voltage waveguide Si–Ge avalanche photodiode. Optica 3, 793–798.

Huang, Z., Oh, J., Campbell, J.C., 2004. Back-side-illuminated high-speed Ge photodetector fabricated on Si substrate using thin SiGe buffer layers. Appl. Phys. Lett. 85, 3286–3288.

Ieong, M., Doris, B., Kedzierski, J., Rim, K., Yang, M., 2004. Silicon device scaling to the sub-10-nm regime. Science 306, 2057–2060.

Kang, Y., Liu, H.-D., Morse, M., et al., 2009. Monolithic germanium/silicon avalanche photodiodes with 340 GHz gain–bandwidth product. Nat. Photonics 3, 59–63.

Kang, Y., Zadka, M., Litski, S., Sarid, G., Morse, M., Paniccia, M.J., et al., 2008. Epitaxially-grown Ge/Si avalanche photodiodes for 1.3μm light detection. Opt. Express 16, 9365–9371.

Kim, G., Kim, S., Kim, S.A., Oh, J.H., Jang, K.-S., 2018a. NDR-effect vertical-illumination-type Ge-on-Si avalanche photodetector. Opt. Lett. 43, 5583–5586.

Kim, G., Kim, S., Kim, S.A., Oh, J.H., Jang, K.-S., 2018b. NDR-effect vertical-illumination-type Ge-on-Si avalanche photodetector. Opt. Lett. 43, 5583–5586.

Kim, I.G., Jang, K.-S., Joo, J., Kim, S., Kim, S., Choi, K.-S., et al., 2013. High-performance photoreceivers based on vertical-illumination type Ge-on-Si photodetectors operating up to 43 Gb/s at $\lambda \sim 1550$nm. Opt. Express 21, 30716–30723.

Kim, K., Choi, J.-Y., Kim, T., Cho, S.-H., Chung, H.-J., 2011. A role for graphene in silicon-based semiconductor devices. Nature 479, 338–344.

Koppens, F.H.L., Mueller, T., Avouris, P., Ferrari, A.C., Vitiello, M.S., Polini, M., 2014. Photodetectors based on graphene, other two-dimensional materials and hybrid systems. Nat. Nanotechnol. 9, 780–793.

Lambrecht, J., Ramon, H., Moeneclaey, B., et al., 2019. 90-Gb/s NRZ optical receiver in silicon using a fully differential transimpedance amplifier. J. Light. Technol. 37, 1964–1973.

Lemaitre, F., Fortin, C., Lagay, N., Binet, G., Pustakhod, D., Decobert, J., et al., 2019. Foundry photonic process extension with bandgap tuning using selective area growth. IEEE J. Sel. Top. Quantum Electron. 25, 6100708.

Li, G., Luo, Y., Zheng, X., et al., 2012. Improving CMOS-compatible germanium photodetectors. Opt. Express 20, 26345–26350.

Li, T., Nedeljkovic, M., Hattasan, N., Cao, W., Qu, Z., Littlejohns, C.G., et al., 2019. Ge-on-Si modulators operating at mid-infrared wavelengths up to 8μm. Photon. Res. 7, 828–836.

Li, W., Zhang, H., Hu, X., Lu, D., Chen, D., Chen, S., et al., 2021. 100 Gbit/s co-designed optical receiver with hybrid integration. Opt. Express 29, 14304–14313.

Li, X., Peng, L., Liu, Z., et al., 2020. High-power back-to-back dual-absorption germanium photodetector. Opt. Lett. 45, 1358–1361.

Liang, D., Kurczveil, G., Huang, Z., Wang, B., Descos, A., Srinivasan, S., et al., 2020a. Integrated green DWDM photonics for next-gen high-performance computing. In: Optical Fiber Communication Conference, Optical Society of America, p. Th1E.2.

Liang, D., Kurczveil, G., Huang, Z., et al., 2020b. Integrated green DWDM photonics for next-gen high-performance computing. In: Optical Fiber Communication Conference, Optical Society of America, p. Th1E.2.

Liow, T.-Y., Duan, N., Lim, A.E.-J., Tu, X., Yu, M., Lo, G.-Q., 2014. Waveguide Ge/Si avalanche photodetector with a unique low-height-profile device structure. In: Optical Fiber Communication Conference, Optical Society of America, p. M2G.6.

Lischke, S., Knoll, D., Mai, C., et al., 2015. High bandwidth, high responsivity waveguide-coupled germanium p-i-n photodiode. Opt. Express 23, 27213–27220.

Liu, Q., Ramirez, J.M., Vakarin, V., Le Roux, X., Frigerio, J., Ballabio, A., et al., 2018. On-chip Bragg grating waveguides and Fabry-Perot resonators for long-wave infrared operation up to 8.4 μm. Opt. Express 26, 34366–34372.

Liu, Z., Yang, F., Wu, W., Cong, H., Zheng, J., Li, C., et al., 2017. 48 GHz high-performance Ge-on-SOI photodetector with zero-bias 40 Gbps grown by selective epitaxial growth. J. Light. Technol. 35, 5306–5310.

Luan, H.-C., Lim, D.R., Lee, K.K., Chen, K.M., Sandland, J.G., Wada, K., et al., 1999. High-quality Ge epilayers on Si with low threading-dislocation densities. Appl. Phys. Lett. 75, 2909–2911.

Luryi, S., Kastalsky, A., Bean, J., 1984. New infrared detector on a silicon chip. IEEE Trans. Electron. Devices 31, 1135–1139.

Ma, P., Flöry, N., Salamin, Y., Baeuerle, B., Emboras, A., Josten, A., et al., 2018. Fast MoTe2 waveguide photodetector with high sensitivity at telecommunication wavelengths. ACS Photonics 5, 1846–1852.

Ma, P., Salamin, Y., Baeuerle, B., Josten, A., Heni, W., Emboras, A., et al., 2019. Plasmonically enhanced graphene photodetector featuring 100 Gbit/s data reception, high responsivity, and compact size. ACS Photonics 6, 154–161.

Manzeli, S., Ovchinnikov, D., Pasquier, D., Yazyev, O.V., Kis, A., 2017. 2D transition metal dichalcogenides. Nat. Rev. Mater. 2, 17033.

Marpaung, D., Yao, J., Capmany, J., 2019. Integrated microwave photonics. Nat. Photonics 13, 80–90.

Marris-Morini, D., Vakarin, V., Ramirez, J.M., Liu, Q., Ballabio, A., Frigerio, J., et al., 2018. Germanium-based integrated photonics from near- to mid-infrared applications. Nanophotonics 7, 1781–1793.

Martinez, N.J.D., Derose, C.T., Brock, R.W., et al., 2016. High performance waveguide-coupled Ge-on-Si linear mode avalanche photodiodes. Opt. Express 24, 19072–19081.

Martinez, N.J.D., Gehl, M., Derose, C.T., Starbuck, A.L., Pomerene, A.T., Lentine, A.L., et al., 2017. Single photon detection in a waveguide-coupled Ge-on-Si lateral avalanche photodiode. Opt. Express 25, 16130–16139.

Masini, G., Sahni, S., Capellini, G., Witzens, J., Gunn, C., 2008. High-speed near infrared optical receivers based on Ge waveguide photodetectors integrated in a CMOS process. Adv. Opt. Technol. 2008, 196572.

Mauthe, S., Baumgartner, Y., Sousa, M., Ding, Q., Rossell, M.D., Schenk, A., et al., 2020. High-speed III-V nanowire photodetector monolithically integrated on Si. Nat. Commun. 11, 4565.

Melchior, H., Lynch, W.T., 1966. Signal and noise response of high speed germanium avalanche photodiodes. IEEE Trans. Electron. Devices 13, 829–838.

Michel, J., Liu, J., Kimerling, L.C., 2010. High-performance Ge-on-Si photodetectors. Nat. Photonics 4, 527–534.

Nam, J.H., Afshinmanesh, F., Nam, D., et al., 2015. Monolithic integration of germanium-on-insulator p-i-n photodetector on silicon. Opt. Express 23, 15816–15823.

Network transformation with world-class optics. <https://www.cisco.com/c/en/us/products/interfaces-modules/transceiver-modules/index.html>. [accessed 14 Jun 2021].

Osmond, J., Isella, G., Chrastina, D., Kaufmann, R., Acciarri, M., von Känel, H., 2009. Ultralow dark current Ge/Si(100) photodiodes with low thermal budget. Appl. Phys. Lett. 94, 201106.

Otterstrom, N.T., Behunin, R.O., Kittlaus, E.A., Wang, Z., Rakich, P.T., 2018. A silicon Brillouin laser. Science 360, 1113–1116.

Pan, H., Assefa, S., Green, W.M.J., Kuchta, D.M., Schow, C.L., Rylyakov, A.V., et al., 2012. High-speed receiver based on waveguide germanium photodetector wire-bonded to 90nm SOI CMOS amplifier. Opt. Express 20, 18145–18155.

Papes, M., Cheben, P., Benedikovic, D., Schmid, J.H., Pond, J., Halir, R., et al., 2016. Fiber-chip edge coupler with large mode size for silicon photonic wire waveguides. Opt. Express 24, 5026–5038.

Park, S., Malinge, Y., Dosunmu, O., et al., 2019. 50-Gbps receiver subsystem using Ge/Si avalanche photodiode and integrated bypass capacitor. In: Optical Fiber Communication Conference, Optical Society of America, p. M3A.3.

Preston, K., Lee, Y.H.D., Zhang, M., Lipson, M., 2011. Waveguide-integrated telecom-wavelength photodiode in deposited silicon. Opt. Lett. 36, 52–54.

Rahim, A., Hermans, A., Wohlfeil, B., Petousi, D., Kuyken, B., Van Thourhout, D., et al., 2021. Taking silicon photonics modulators to a higher performance level: state-of-the-art and a review of new technologies. Adv. Photonics 3, 024003.

Rahim, A., Goyvaerts, J., Szelag, B., Fedeli, J.-M., Absil, P., Aalto, T., et al., 2019. Open-access silicon photonics platforms in Europe. IEEE J. Sel. Top. Quantum Electron. 25, 1–18.

Rahim, A., Spuesens, T., Baets, R., Bogaerts, W., 2018. Open-access silicon photonics: current status and emerging initiatives. Proc. IEEE 106, 2313–2330.

Ramirez, J.M., Elfaiki, H., Verolet, T., Besancon, C., Gallet, A., Néel, D., et al., 2020. III-V-on-silicon integration: from hybrid devices to heterogeneous photonic integrated circuits. IEEE J. Sel. Top. Quantum Electron. 26, 6100213.

Reboud, V., Gassenq, A., Hartmann, J.M., Widiez, J., Virot, L., Aubin, J., et al., 2017. Germanium based photonic components toward a full silicon/germanium photonic platform. Prog. Cryst. Growth Charact. Mater. 63, 1–24.

Rodt, S., Reitzenstein, S., 2021. Integrated nanophotonics for the development of fully functional qunatum circuits based on on-demand single-photon emitters. APL Photonics 6, 010901.

Roelkens, G., Liu, L., Liang, D., Jones, R., Fang, A., Koch, B., et al., 2010. III–V/silicon photonics for on-chip and intra-chip optical interconnects. Laser Photonics Rev. 4, 751–779.

Romagnoli, M., Sorianello, V., Midrio, M., Koppens, F.H.L., Huyghebaert, C., Neumaier, D., et al., 2018. Graphene-based integrated photonics for next-generation datacom and telecom. Nat. Rev. Mater. 3, 392–414.

Sabella, R., 2020. Silicon photonics for 5G and future networks. IEEE J. Sel. Top. Quantum Electron. 26, 8301611.

Salamin, Y., Ma, P., Baeuerle, B., Emboras, A., Fedoryshyn, Y., Heni, W., et al., 2018. 100 GHz plasmonic photodetector. ACS Photonics 8, 3291–3297.

Samani, A., Carpentier, O., El-Fiky, E., et al., 2019. Highly sensitive, 112 Gb/s O-band waveguide coupled silicon-germanium avalanche photodetectors. In: Optical Fiber Communication Conference, Optical Society of America, p. Th3B.1.

Samavedam, S.B., Currie, M.T., Langdo, T.A., Fitzgerald, E.A., 1998. High-quality germanium photodiodes integrated on silicon substrates using optimized relaxed graded buffers. Appl. Phys. Lett. 73, 2125.

Sarto, A.W., Van Zeghbroeck, B.J., 1997. Photocurrents in a metal-semiconductor-metal photodetector. IEEE J. Quantum Electron. 33, 2188–2194.

Shen, L., Pun, E.Y.B., Ho, J.C., 2017. Recent developments in III–V semiconducting nanowires for high-performance photodetectors. Mater. Chem. Front. 1, 630–645.

Shi, L., Nihtianov, S., 2012. Comparative study of silicon-based ultraviolet photodetectors. IEEE Sens. J. 12, 2453–2459.

Shi, W., Tian, Y., Gervais, A., 2020. Scaling capacity of fiber-optic transmission systems via silicon photonics. Nanophotonics 9, 4629–4663.

Shi, Y., Zhou, D., Yu, Y., Zhang, X., 2021. 80 GHz germanium waveguide photodiode enabled by parasitic parameter engineering. Photon. Res. 9, 605–609.

Shive, J.N., 1953. The properties of germanium phototransistors. J. Opt. Soc. Am. 43, 239–244.

Siew, S.Y., Li, B., Gao, F., Zheng, H.Y., Zhang, W., Guo, P., et al., 2021. Review of silicon photonics technology and platform development. J. Light. Technol., Early Access. .

Singh, J., 1996. Optoelectronics: An Introduction to Materials and Devices. The McGraw-Hill Companies.

Song, J., Yuan, S., Cui, C., Wang, Y., Li, Z., Wang, A.X., et al., 2021. High-efficiency and high-speed germanium photodetector enabled by multiresonant photonic crystal. Nanophotonics 10, 1081–1087.

Soref, R., 2014. Silicon-based silicon–germanium–tin heterostructure photonics. Philos. Trans. R. Soc. A 372, 20130113.

Srinivasan, S.A., Lambrecht, J., Guermandi, D., Lardenois, S., Berciano, M., Absil, P., et al., 2021. 56 Gb/s NRZ O-band hybrid BiCMOS-silicon photonics receiver using Ge/si avalanche photodiode. J. Light. Technol. 39, 1409–1415.

Srinivasan, S.A., Lambrecht, J., Berciano, M., et al., 2020a. Highly sensitive 56 Gbps NRZ O-band BiCMOS-silicon photonics receiver using a Ge/Si avalanche photodiode. In: Optical Fiber Communication Conference, Optical Society of America, p. W4G.7.

Srinivasan, S.A., Berciano, M., De Heyn, P., Lardenois, S., Pantouvaki, M., Van Campenhout, J., 2020b. 27 GHz silicon-contacted waveguide-coupled Ge/Si avalanche photodiode. J. Light. Technol. 38, 3044–3050.

Sun, C., Wade, M.T., Lee, Y., Orcutt, J.S., Alloatti, L., Georgas, M.S., et al., 2015. Single-chip microprocessor that communicates directly using light. Nature 528, 534–538.

Sylvain, G., Alonso-Ramos, C., Benedikovic, D., Duran-Valdeiglesias, E., Le Roux, X., Vulliet, N., et al., 2018. Broadband polarization beam splitter on a silicon nitride platform for O-Band operation. IEEE Photon. Technol. Lett. 30, 1679–1682.

Takenaka, M., Kim, Y., Han, J., Kang, J., Ikku, Y., Cheng, Y., et al., 2017. Heterogeneous CMOS photonics based on SiGe/Ge and III–V semiconductors integrated on Si platform. IEEE J. Sel. Top. Quantum Electron. 23, 8200713.

Tanabe, T., Sumikura, H., Taniyama, H., Shinya, A., Notomi, M., 2010. All-silicon sub-Gb/s telecom detector with low dark current and high quantum efficiency on chip. Appl. Phys. Lett. 96, 101103.

Tasker, J.F., Frazer, J., Ferranti, G., Allen, E.J., Brunel, L.F., Tanzilli, S., et al., 2021. Silicon photonics interfaced with integrated electronics for 9 GHz measurement of squeezed light. Nat. Photonics 15, 11–15.

Timurdogan, E., Su, Z., Shiue, R.-J., Byrd, M.J., Poulton, C.V., Jabon, K., et al., 2020. 400G silicon photonics integrated circuit transceiver chipsets for CPO, OBO, and pluggable modules. In: Optical Fiber Communication Conference, Optical Society of America, p. T3H.2.

Tran, H., Pham, T., Margetis, J., Zhou, Y., Dou, W., Grant, P.C., et al., 2019. "Si-based GeSn photodetectors toward mid-infrared imaging applications. ACS Photonics 6, 2807–2815.

Tsai, C.-H., Huang, B.-J., Soref, R.A., Sun, G., Cheng, H.H., Chang, G.-E., 2020. GeSn resonant-cavity-enhanced photodetectors for efficient photodetection at the 2 μm wavelength band. Opt. Lett. 45, 1463–1466.

Tzu, T.-C., Su, K., Costanzo, R., Ayoub, D., Bowers, S.M., Beling, A., 2019. Foundry-enabled high-power photodetectors for microwave photonics. IEEE J. Sel. Top. Quantum Electron. 25, 3800111.

Verbist, J., Lambrecht, J., Moeneclaey, B., et al., 2017. 40-Gb/s PAM-4 transmission over a 40 km amplifier-less link using a sub-5V Ge APD. IEEE Photon. Technol. Lett. 29, 2238–2241.

Virot, L., Benedikovic, D., Szelag, B., Alonso-Ramos, C., Karakus, B., Hartmann, J.-M., et al., 2017. Integrated waveguide PIN photodiodes exploiting lateral Si/Ge/Si heterojunction. Opt. Express 25, 19487–19496.

Virot, L., Vivien, L., Fédéli, J.-M., et al., 2013. High-performance waveguide-integrated germanium PIN photodiodes for optical communication applications. Photon. Res. 1, 140–147.

Virot, L., Crozat, P., Fédéli, J.-M., et al., 2014. Germanium avalanche receiver for low power interconnects. Nat. Commun. 5, 4957.

Vivien, L., Pavesi, L., 2013. Handbook of Silicon Photonics. CRC Press.

Vivien, L., Polzer, A., Marris-Morini, D., et al., 2012. Zero-bias 40Gbit/s germanium waveguide photodetector on silicon. Opt. Express 20, 1096–1101.

Vivien, L., Marris-Morini, D., Fédéli, J.-M., Rouvière, M., Damlencourt, J.-F., El Melhaoui, L., et al., 2008. Metal-semiconductor-metal Ge photodetectors integrated in silicon waveguides. Appl. Phys. Lett. 92, 151114.

Wada, K., Kimerling, L.C., 2015. Photonics and Electronics with Germanium. Wiley-VCH.

Wang, B., Huang, Z., Sorin, W.V., et al., 2020b. A low-voltage Si-Ge avalanche photodiode for high-speed and energy efficient silicon photonic links. J. Light. Technol. 38, 3156–3163.

Wang, B., Huang, Z., Yuan, Y., et al., 2020a. 64 Gb/s low-voltage waveguide SiGe avalanche photodiodes with distributed Bragg reflectors. Photonics Res. 8, 1118–1123.

Wilmart, Q., Brision, S., Hartmann, J.-M., Myko, A., Ribaud, K., Petit-Etienne, C., et al., 2021. A complete Si photonics platform embedding ultra-low loss waveguides for O- and C-band. J. Light. Technol. 39, 532–538.

Wilmart, Q., Mang, T., Fowler, D., Brision, S., Ribaud, K., Malhouitre, S., et al., 2020. Advanced Si photonics platform for high-speed and energy-efficient optical transceivers for datacom. In: Proceedings Volume 11285, Silicon Photonics XV, 112850B, San Francisco, CA.

Woodson, M.E., Ren, M., Maddox, S.J., Chen, Y., Bank, S.R., Campbell, J.C., 2016. Low-noise AlInAsSb avalanche photodiode. Appl. Phys. Lett. 108, 081102.

Xu, S., Han, K., Huang, Y.-C., Hong L., K., Kang, Y., Masudy-Panah, S., et al., 2019. Integrating GeSn photodiode on a 200 mm Ge-on-insulator photonics platform with Ge

CMOS devices for advanced OEIC operating at 2 μm band. Opt. Express 27, 26924−26939.

Yi, X., Xie, S., Liang, B., Lim, L.W., Cheong, J.S., Debnath, M.C., et al., 2019. Extremely low excess noise and high sensitivity AlAs0.56Sb$_{0.44}$ avalanche photodiodes. Nat. Photonics 13, 683−686.

Yin, T., Cohen, R., Morse, M.M., Sarid, G., Chetrit, Y., Rubin, D., et al., 2007. 31GHz Ge n-i-p waveguide photodetectors on Silicon-on-Insulator substrate. Opt. Express 15, 13965−13971.

Yuan, Y., Huang, Z., Wang, B., et al., 2020. Superior temperature performance of Si-Ge waveguide avalanche photodiodes at 64Gbps PAM4 operation. In: Optical Fiber Communication Conference, Optical Society of America, p. M2A.2.

Zaoui, W.S., Chen, H.-W., Bowers, J.E., et al., 2009. Frequency response and bandwidth enhancement in Ge/Si avalanche photodiodes with over 840GHz gain-bandwidth-product. Opt. Express 17, 2641−12649.

Zeng, X., Huang, Z., Wang, B., Liang, D., Fiorentino, M., Beausoleil, R.G., 2019. Silicon−germanium avalanche photodiodes with direct control of electric field in charge multiplication region. Optica 6, 772−777.

Zhang, J., Kuo, B.P.-P., Radic, S., 2020. 64Gb/s PAM4 and 160Gb/s 16QAM modulation reception using a low-voltage Si-Ge waveguide-integrated APD. Opt. Express 28, 23266−23273.

Zhang, Y., Yang, S., Yang, Y., et al., 2014. A high-responsivity photodetector absent metal-germanium direct contact. Opt. Express 22, 11367−11375.

Zhou, H., Sun, Y., 2018. Size reduction of Ge-on-Si photodetectors via a photonic bandgap. Appl. Opt. 57, 2962−2966.

Zilkie, A.J., Srinivasan, P., Trita, A., et al., 2019. Multi-micron silicon photonics platform for highly manufacturable and versatile photonic integrated circuits. IEEE J. Sel. Top. Quantum Electron. 25, 8200713.

Zuo, Y., Yu, Y., Zhang, Y., Zhou, D., Zhang, X., 2019. Integrated high-power germanium photodetectors assisted by light field manipulation. Opt. Lett. 44, 3338−3341.

Ultraviolet detectors for harsh environments

7

Ruth A. Miller[1,2], Hongyun So[3], Thomas A. Heuser[4], Ananth Saran Yalamarthy[5], Peter F. Satterthwaite[6] and Debbie G. Senesky[1,7]

[1]Department of Aeronautics and Astronautics, Stanford University, Stanford, CA, United States, [2]NASA Ames Research Center, Mountain View, CA, United States, [3]Department of Mechanical Engineering, Hanyang University, Seoul, South Korea, [4]Department of Materials Science and Engineering, Stanford University, Stanford, CA, United States, [5]Department of Mechanical Engineering, Stanford University, Stanford, CA, United States, [6]Department of Electrical Engineering and Computer Science, Massachusetts Institute of Technology, Cambridge, MA, United States, [7]Department of Electrical Engineering, Stanford University, Stanford, CA, United States

7.1 Introduction

On the electromagnetic spectrum, the ultraviolet (UV) region spans wavelengths from 400 to 10 nm (corresponding to photon energies from 3 to 124 eV) and is typically divided into three spectral bands: UV-A (400−320 nm), UV-B (320−280 nm), and UV-C (280−10 nm). The Sun is the most significant natural UV source. The approximate solar radiation spectrum just outside Earth's atmosphere, calculated as black body radiation at 5800K using Plank's law, is shown in Figure 7.1. At the top of Earth's atmosphere, about 9% of the Sun's total radiation is in the UV regime (Diffey, 2002). Earth's ozone layer makes the planet habitable to humans by almost completely absorbing UV-C and significantly attenuating UV-B, which causes cataracts, burns, and skin cancer (Biesalski and Obermueller-Jevic, 2001). Artificial sources of UV radiation can be created by passing an electric current through a gas, typically mercury, causing energy to be released in the form of optical radiation. Further information on natural and artificial UV sources can be found in (Diffey, 2002).

There is a wide range of ground and space-based applications that require UV detection including optical communication (Xu and Sadler, 2008), flame detection (Oshima et al., 2009), combustion monitoring (Pau et al., 2006), chemical analysis (Golimowski and Golimowska, 1996), astronomy (Robichaud, 2003), etc. Many of these applications require UV instrumentation capable of operating in high-temperature harsh environments. While most sensors and electronics today are silicon (Si)-based due to well-established manufacturing processes, easy circuit integration, and low cost, Si has shown limited usefulness as a high-temperature UV detecting material platform (Lien et al., 2012; Vijayakumar et al., 2007). Si and

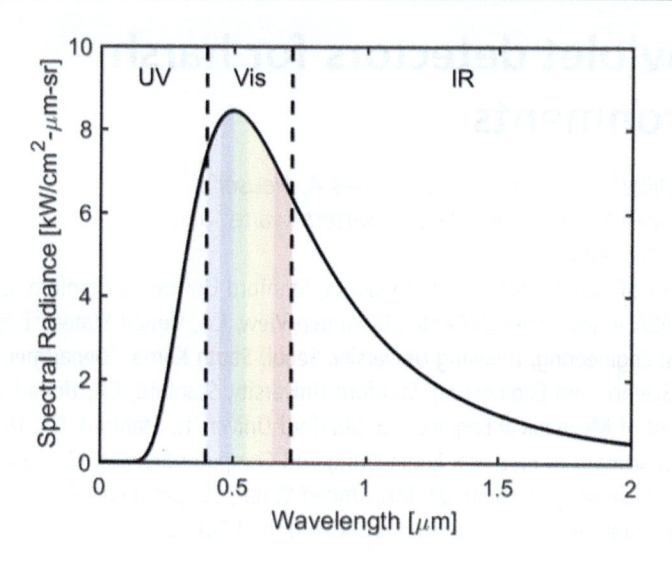

Figure 7.1 Black body radiation at 5800K approximating the solar radiation spectrum outside Earth's atmosphere.

other narrow bandgap semiconductors are not able to directly measure UV light due to their bandgaps corresponding to near-infrared or visible wavelengths. Since UV light is higher energy than the bandgap of Si and other narrow bandgap materials, part of the energy will be lost to heating, resulting in low quantum efficiency. To use these materials for UV photodetection, filtering devices that absorb UV light and reemit lower energy light need to be incorporated (Decoster and Harari, 2009), or detectors with wide, shallow charge collection zones are required since the absorption depth of UV energy in Si is very small (less than 10 nm for wavelengths between 100–300 nm) (Alaie et al., 2015; Shi and Nihtianov, 2012). Additionally, Si-based photodetectors have been shown to be limited to temperatures less than about 125°C (Lien et al., 2012; Vijayakumar et al., 2007). Si has a relatively high intrinsic carrier concentration at room temperature (10^{10} cm^{-3}) which increases exponentially with temperature to reach 10^{15} cm^{-3} at 300°C (Pierret, 1996). Additionally, the lightest doping concentrations of Si devices range from 10^{14} to 10^{17} cm^{-3} (Neudeck et al., 2002). Therefore, when operating at high temperature, the intrinsic carrier concentration will overwhelm the doping concentration and the device will no longer function as intended.

Instead of Si, wide bandgap semiconductors such as III-nitrides, silicon carbide (SiC), and zinc oxide (ZnO) are being explored as material platforms for UV detection. The bandgaps of these materials allow for the detection of UV light while remaining blind to visible light. Additionally, electronics based on wide bandgap materials have demonstrated high-temperature operation indicating their applicability to harsh environment sensing (Neudeck et al., 2002; Buttay et al., 2011). A comparison of the material properties of Si and wide bandgap semiconductors used for UV detection is listed in Table 7.1. Reviews of the latest progress in wide bandgap

Table 7.1 Material properties of semiconductors used to manufacture UV photodetectors.

	Si	AlN	Diamond	GaN	4H-SiC	6H-SiC	ZnO	β-Ga$_2$O$_3$	h-BN
Bandgap (eV)	1.12 Indirect	6.2 Direct	5.5 Indirect	3.4 Direct	3.2 Indirect	2.86 Indirect	3.37 Direct	4.9 Direct	~6 Indirect
Cut-off wavelength (nm)	1107	200	225	365	387	434	368	253	207
Melting point (°C)	1410	2200	3500	2500	2700	2700	1975	1780	~3000
Thermal conductivity (W/cm-K)	1.5	3.2	20	1.3	4.9	4.9	5.4	2.9	2.8
Electron mobility (cm^2/V-s)	1400	135	2200	1000	400	400	205	182	-
Hole mobility (cm^2/V-s)	600	14	1600	30	75	75	70	0.2	16
Dielectric constant	11.8	8.1	5.5	8.9	9.7	9.7	9.1	3.29	4
Breakdown field (10^5 V/cm)	3	20	100	26	24	24	35	80	~20

semiconductor UV photodetectors focusing on new device architectures and novel material combinations have been recently published (Alaie et al., 2015; Monroy et al., 2003; Sang et al., 2013; Razeghi and Rogalski, 1996; Munoz et al., 2001; Rivera et al., 2010; Monroy et al., 2001; Moustakas and Paiella, 2017; Omnes and Monroy, 2003; Liu et al., 2010; Chen et al., 2015; Konstantatos and Sargent, 2010; Peng et al., 2013).

To complement and expand on these recent reviews, this review presents a summary of the high-temperature characterization and operation of UV photodetectors using wide bandgap semiconductors as well as the challenges that persist in furthering this technology for use in a variety of applications.

7.2 Photodetector parameters

There are several types of semiconductor device architectures that can be used to create photodetectors including photoconductor, metal-semiconductor-metal (MSM), Schottky, p-n, p-i-n, and avalanche. A schematic of each device architecture is shown in Figure 7.2 and the theoretical framework that describes their operation can be found in (Sze and Ng, 2007; Sze et al., 1971; Bhattacharya, 1997). Regardless of the photodetector architecture, the basic principle of operation of all semiconductor photodetectors is the same; photons with energy greater than or equal to the bandgap of the semiconductor are absorbed and generated electron-hole pairs. Under an applied electric field, the electron-holes pairs separate and a photo-generated current can be measured. The main figures of merit that are used to quantify photodetector performance are photocurrent-to-dark current ratio (PDCR), normalized photocurrent-to-dark current ratio (NPDR), responsivity (R), and quantum efficiency (η), and response time.

Additionally, there exist a number of parameters that can help to quantify the behavior of a given device with regard to noise. Although these parameters are

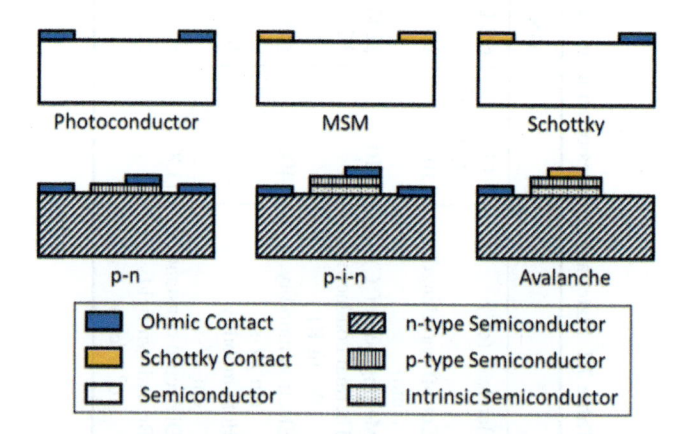

Figure 7.2 Different semiconductor photodetector device architectures.

important for fully quantifying photodetector performance, they are often not reported for new and developing device technologies. Some of the more significant parameters pertaining to noise include signal-to-noise ratio (SNR), noise-equivalent power (NEP), and specific detectivity (D^*).

PDCR is a measure of the photodetector sensitivity with respect to the dark (or leakage) current. PDCR is defined as

$$PDCR = \frac{I_{photo} - I_{dark}}{I_{dark}} \qquad (7.1)$$

where I_{photo} is the measured photocurrent and I_{dark} is the dark current Responsivity is another measure of photodetector sensitivity and is defined as the photo-generated current ($I_{photo} - I_{dark}$) per unit of incident optical power, that is

$$R = \frac{I_{photo}}{P_{opt}} \qquad (7.2)$$

where P_{opt} is the applied optical power. Responsivity is directly proportional to the external quantum efficiency (η) which is a measure of the number of photo-generated electron-hole pairs per incident photon. The external quantum efficiency is defined as

$$\eta = R\frac{hc}{q\lambda} \qquad (7.3)$$

where h is Plank's constant, c is the speed of light, q is the elementary charge, and λ is the wavelength. Photodetector response time (which is related to bandwidth) is quantified in terms of photocurrent 10% to 90% rise time and photocurrent 90% to 10% decay time. SNR is the ratio of the power of the desired signal to the power of noise in a given system. An SNR greater than 1:1 means that the signal is distinguishable from the noise, while an SNR equal to or less than 1:1 means that the signal cannot be distinguished. As it is a ratio of powers, SNR is unitless.

$$SNR = \frac{P_{signal}}{P_{noise}} \qquad (7.4)$$

NPDR is the ratio of photocurrent to dark current normalized by the incident optical power. It can therefore also be defined as the ratio of the responsivity R to the dark current.

$$NPDR = \frac{\left(\frac{I_{photo}}{I_{dark}}\right)}{P_{opt}} = \frac{\left(\frac{I_{photo}}{P_{opt}}\right)}{I_{dark}} = \frac{R}{I_{dark}} \qquad (7.5)$$

NEP is a measure of the sensitivity of a system. At an output bandwidth of one hertz, NEP is defined as the signal power that gives an SNR of 1:1 and is equivalent

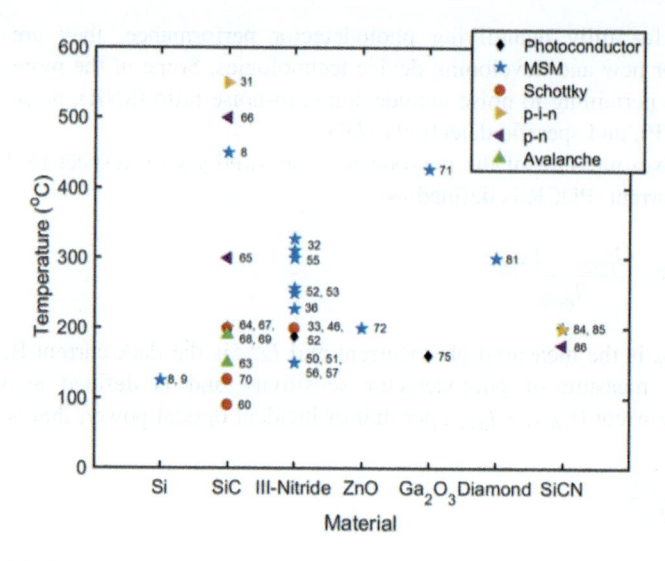

Figure 7.3 Maximum operational temperatures reported in the literature for Si photodetectors compared to photodetectors based on wide bandgap materials.

to the noise spectral density divided by the responsivity. The units of NEP are watts per square root hertz ($W/\sqrt{Hz}$).

$$NEP = \frac{N_0}{R} \tag{7.6}$$

D^* is a figure of merit used to quantify photodetector performance. It is equal to the reciprocal NEP normalized by the square root of the detector area multiplied by the frequency bandwidth and has units of centimeters multiplied by square root hertz per watt ($cm\times\sqrt{Hz}/W$).

$$D^* = \frac{\sqrt{Af}}{NEP} \tag{7.7}$$

A comparison of the maximum operating temperatures reported in the literature for Si, SiC, III-nitride, and other wide bandgap-based UV photodetectors is shown in Figure 7.3. To date, SiC-based photodetectors have demonstrated the highest operational temperatures, up to 550°C (Hou et al., 2016). The following sections detail the high-temperature photodetector response in terms of PDCR, R, η, and response time for each of these material platforms. Additionally, the development and continuing challenges of using wide bandgap semiconductors for high-temperature UV sensing are discussed.

7.3 III–nitride-based ultraviolet photodetectors

The III-nitrides, consisting of GaN, InN, AlN, and their ternary compounds, present some unique benefits for UV photodetection when compared to other wide bandgap materials. III-nitride semiconductors have direct bandgaps and thus higher photon absorption than their indirect counterparts (such as Si or SiC). Also, the region of the electromagnetic spectrum, between 200 nm (AlN, bandgap of 6.2 eV) and 650 nm (InN, bandgap of 1.9 eV), that III-nitride detectors are sensitive to can be selected by changing the ternary compound mole fraction (Alaie et al., 2015; Monroy et al., 2003). Lastly, heterojunctions can be formed from III-nitride semiconductors and the highly conductive two-dimensional electron gas (2DEG) formed at the material interface can be leveraged as a photo-sensing element (De Vittorio et al., 2004; So et al., 2016; So and Senesky, 2016a; Hou et al., 2017).

On the other hand, III-nitride materials have some limitations when used for UV sensing, the most significant of which is known as persistent photoconductivity (PPC). PPC is a phenomenon in which the photocurrent response remains after the illumination source has been removed as shown in Figure 7.4, resulting in photocurrent decay times on the order of hours to days (De Vittorio et al., 2004; Hou et al., 2017; Poti et al., 2004; Hirsch et al., 1997; Katz et al., 2004; Qiu and Pankove, 1997; Chen et al., 1997; Li et al., 1997). PPC has been attributed to excitons (De Vittorio et al., 2004), negatively charged surface states (Katz et al., 2004),

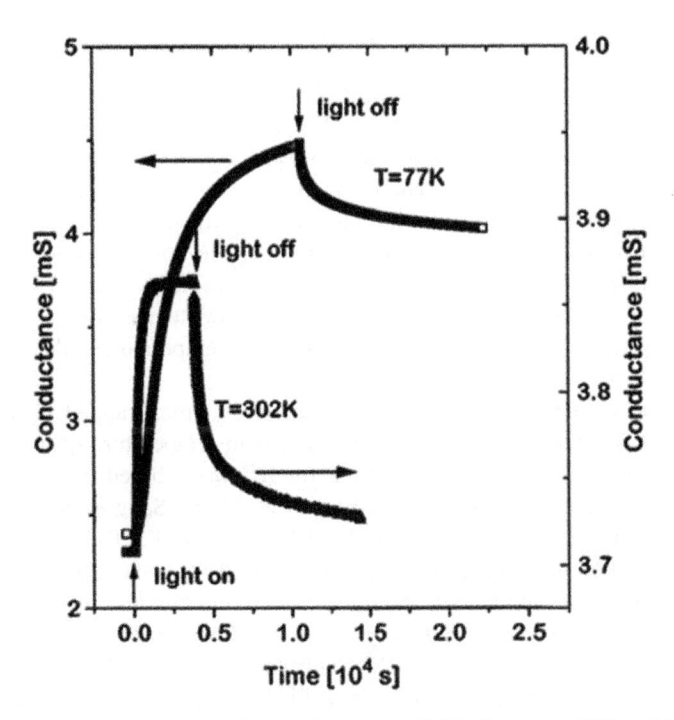

Figure 7.4 Buildup and decay transients of an n-type GaN substrate at 77K and 302K demonstrating decreased time constants at higher temperatures (Zhou et al., 2014a).
Source: Adopted with permission from Zhou, H., Zhu, J., Liu, Z., Yan, Z., Fan, X., Lin, J., et al. 2014a. High thermal conductivity of suspended few-layer hexagonal boron nitride sheets. Nano Res. 7 (8), 1232−1240.

metastable defects (Hirsch et al., 1997; Qiu and Pankove, 1997; Chen et al., 1997), gallium vacancies (Poti et al., 2004; Qiu and Pankove, 1997), nitrogen antisites (Chen et al., 1997), and deep-level defects (Qiu and Pankove, 1997; Li et al., 1997) trapping photo-generated carriers. The falling transient is often fit by a stretched exponential function of the form

$$I_{ppc}(t) = I_0 \exp\left[-\left(\frac{t}{\tau}\right)^\beta\right] \tag{7.8}$$

where I_0 is the photocurrent before the illumination source is removed, τ is the decay time constant, and β is the decaying exponential ($0 < \beta < 1$) (Poti et al., 2004; Qiu and Pankove, 1997; Chen et al., 1997; Li et al., 1997; Dang et al., 1998). Experimental results have shown that at elevated temperatures, thermal energy is able to release trapped photo-generated carriers and thus reduce PPC effects (De Vittorio et al., 2004; Hou et al., 2017; Hirsch et al., 1997). To account for this temperature dependence, the decay time constant has been modeled as

$$\tau = \tau_0 \exp\left[\frac{\Delta E}{kT}\right] \tag{7.9}$$

where ΔE is the carrier capture barrier, k is Boltzmann's constant, and T is the temperature (Hirsch et al., 1997; Chen et al., 1997; Li et al., 1996). The carrier capture barrier is thought to originate from the non-overlapping vibronic states of unfilled and filled defects (Lang and Logan, 1977). To be captured, charge carriers require additional energy to get into the vibronic states of filled defects. Reported values for ΔE in III-nitride photodetectors range from 132 to 360 meV (Hou et al., 2017; Hirsch et al., 1997; Chen et al., 1997; Li et al., 1997). This large energy barrier prevents the decay of photo-generated carriers at low temperatures. However, as trapped carriers gain thermal energy at elevated temperatures, the capture rate increases, and thus the decay time is reduced.

In addition to increasing the photocurrent decay time, trapped charge carriers also affect photodetector quantum efficiency. Room temperature quantum efficiencies greater than 100% are often reported for III-nitride-based photodetectors and attributed to a photo-gain mechanism (Katz et al., 2004; Sang et al., 2011; Munoz et al., 1997; Binet et al., 1996; Xie et al., 2011a; Liu et al., 2017). The photo-gain has been attributed to photo-generated holes trapped at negatively-charged surface states at the metal-semiconductor interface causing the Schottky barrier to lower or bend with respect to the dark Schottky barrier as shown in Figure 7.5 (Katz et al., 2004; Xie et al., 2011a; Liu et al., 2017). This change in the Schottky barrier results in an enhanced leakage current. Similar to PPC, thermal energy is able to release trapped photo-generated holes (Munoz et al., 1997; Xie et al., 2011a). Therefore, at elevated temperatures, the change in the Schottky barrier from dark to illuminated conditions is less pronounced resulting in minimal photo-gain. In addition to a reduced photo-gain at high temperature, increased dark (or leakage) current,

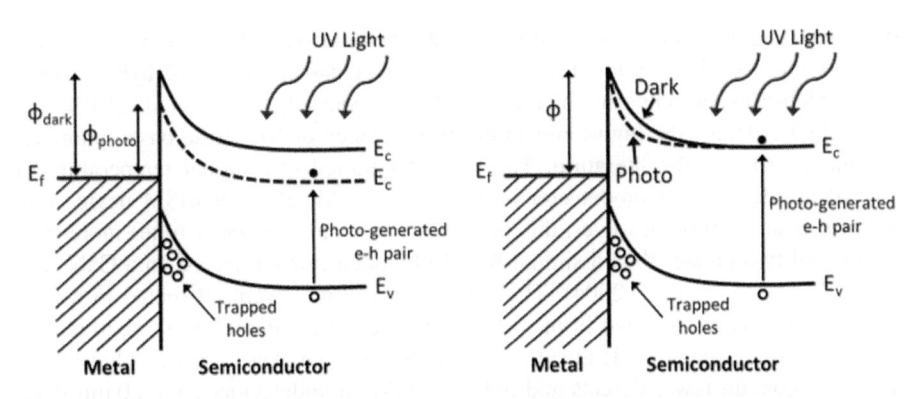

Figure 7.5 Trapped photo-generated holes lowering or bending the Schottky barrier with respect to the dark Schottky barrier.

enhanced carrier recombination, increased lattice scattering, and shifting of the bandgap to longer wavelengths have been cited as reasons for lower quantum efficiencies, responsivities, and PDCRs at elevated temperatures (De Vittorio et al., 2004; So et al., 2016; Chang et al., 2017; Xie et al., 2012). Increased responsivity at higher temperatures has also been reported and has been attributed to thermal ionization of trapped photo-generated carriers providing additional carriers (De Vittorio et al., 2004; Sang et al., 2011, 2012).

Amongst photodetectors with quantum efficiency >100%, a recent variant on the MSM architecture has shown excellent properties at both room temperature and high temperature (Satterthwaite et al., 2018). This device uses an array of interdigitated AlGaN/GaN 2DEG electrodes spaced by intrinsic GaN channels instead of the Schottky metal that is typical of MSM photodetector. This device was shown to have a record high NPDR of 6×10^{14} W^{-1} at room temperature (Satterthwaite et al., 2018). This device also shows a high responsivity (7800 A/W) and an UV-visible rejection ratio of 10^6 at room temperature, which are among the highest reported values for any GaN photodetector architecture (Satterthwaite et al., 2018). This performance arises from a gain mechanism where holes generated from incident photons accumulate at the AlGaN/GaN valence band interface, which leads to the lowering of the energy barrier for electrons entering the conduction band. Furthermore, photocurrent rise and fall times were measured to be 32 and 76 Ms for this device, defined here as the time it takes the photocurrent to go from 10% to 90% of its final value (Satterthwaite et al., 2018). These rise and fall times are long relative to MSM photodetectors with no internal gain (Monroy et al., 2003), however, they present a significant improvement relative to photoconductors with comparable gain to this device, which have rise and fall times of several seconds (Liu et al., 2017). The high temperatures properties of this 2DEG-IDT photodetector were investigated to evaluate its performance in extreme conditions (Alpert et al., 2019). The peak spectral responsivity of the photodetector fell by a factor of nearly 2000 from room temperature to 250°C, but the NPDR still remained above 109 W-1 even at the highest temperatures, which is higher than the values of several GaN photodetectors at room temperature in the published literature. This photodetector was used to successfully measure UV emission from a gaseous CH4/O2 hybrid rocket igniter plume owing to its good high-temperature responsivity (Alpert et al., 2019).

Table 7.2 details the high-temperature performance of III-nitride-based photodetectors reported in the literature. The highest reported operational temperature of any III-nitride-based photodetector is 327°C by an AlGaN/GaN MSM device (De Vittorio et al., 2004). However, this photodetector exhibited decay times more than a factor of two greater than a similar MSM photodetector fabricated on a GaN substrate (De Vittorio et al., 2004). These long decay times were attributed to a high defect density as determined from surface roughness measurements. Thus, to reduce PPC effects, high-quality III-nitride films are needed. Thicker GaN films have been shown to contain fewer defects and indeed, MSM photodetectors on a 4.0 μm thick GaN film showed improved performance compared to photodetectors fabricated on 1.5 μm thick GaN film (Carrano et al., 1998). In AlGaN-based devices, the band gap varies with the Al:Ga ratio, from a minimum of 3.4 eV (GaN) to a maximum of 6.2 eV (AlN) in accordance with Vegard's Law:

$$E(A_xB_{1-x}N) = xE(AN) + (1-x)E(BN) - bx(1-x) \tag{7.10}$$

where the bowing parameter b determines the deviation from linearity (Pelá et al., 2011). InGaN and InAlN, also ternary III-nitrides, behave similarly. Calculated bowing parameters for these material systems can be found in Table 7.3 (Pelá et al., 2011; Meyer and Vurgaftman, 2003).

Aside from growing higher quality, less defective III-nitride films, another proposed approach to mitigate PPC effects is to use in situ heating. Suspended AlGaN/GaN photodetectors utilizing the 2DEG as both a sensor and heater demonstrated a photocurrent decay time of 24 s which was more than three orders of magnitude less than the 39-h decay time of their solid state counterpart (Hou et al., 2017). The suspended photodetectors used joule heating of the 2DEG to accelerate the carrier capture rate once the UV source was removed which successfully mitigated PPC effects.

To increase quantum efficiency, PDCR, and responsivity at high temperatures, thermally stable contacts with low dark current are needed. To accomplish this, calcium fluoride (CaF2) was tested as an insulation layer in InGaN-based metal-insulator-semiconductor (MIS) Schottky photodiodes (Sang et al., 2011, 2012). Compared to the commonly used SiO2 insulation layer, CaF2 has an extremely wide bandgap (12 eV), high thermal conductivity, and is robust to radiation (Sang et al., 2012). The addition of the CaF2 insulation layer decreased the reverse leakage current by three orders of magnitude compared to Schottky photodiodes without the insulation layer (Sang et al., 2011, 2012). The MIS device demonstrated operation up to 250°C with PDCR values of 1317 at room temperature and 72 at 250°C (Sang et al., 2011, 2012). Whereas Schottky photodiodes without the insulation layer were only operational up to 200°C with lower PDCRs of 79 at room temperature and 2.2 at 200°C (Sang et al., 2011, 2012).

Table 7.2 High-temperature III-nitride-based UV photodetectors.

Material	Photodetector architecture	Electrode material	Temperature (°C)	PDCR	Responsivity (A/W)	Notes	References
GaN	MSM	Ti/Au	310		~8 (77°C) ~13 (310°C)	Thermal ionization of trapped photo-generated carriers increases responsivity.	De Vittorio et al. (2004)
GaN	MSM	Ni/Au	150		~0.1 (RT, 3 V) ~0.04 (150°C, 3 V)	Internal gain is attributed to trapped photo-generated holes reducing the Schottky barrier height.	Xie et al. (2011b)
GaN		Al Wirebond	250	2.63 (RT, 1 V) 0.27 (250°C, 1 V)		Direct wire-bonding. ZnO nanorods for antireflective coating.	So and Senesky (2016b)
GaN	MSM	Ni/Au	150			Dark current ~ 10^{-12} A at RT and 150°C. Photocurrent ~ 10^{-5} at RT and ~ 10^{-6} at 150°C.	Chang et al. (2017)
GaN	Photoconductor	Ti/Al or In	187			Very high gains (~10^4 at 187°C and 10^{-2} W/cm^2 optical power) attributed to a modulation mechanism of the conductive volume of the GaN layer.	Munoz et al. (1997)
GaN	MSM	Pt	257			Thicker GaN devices showed improved performance due to a reduction in defects and deep level traps.	Carrano et al. (1998)
GaN	MSM	Ti/Au	227			Studied trapping mechanisms responsible for persistent photoconductivity	Poti et al. (2004)
AlGaN	MSM	Ni/Au	150		0.14 (RT, 10 V) 0.11 (150°C, 10 V)	Ultra-low dark current (fA range at 150°C) achieved using a high temperature AlN buffer layer.	Xie et al. (2012)
AlGaN	MSM	Ni/Au	150			QE: 11.3% (RT, 2 V, front illumination), 21.1% (RT, 2 V, back-illumination)	Wang et al. (2013)

(*Continued*)

Table 7.2 (Continued)

Material	Photodetector architecture	Electrode material	Temperature (°C)	PDCR	Responsivity (A/W)	Notes	References
InGaN	MSM	Ti/Al/Ni/Au Ni/WC	250	1317 (RT, −1V) 72 (250°C, −1V)	3.3 (RT, −3 V) 5.6 (250°C, −3 V)	UV/visible light discrimination ratio of 10^5 at 250°C. Used CaF_2 insulation layer to reduce the dark current.	Sang et al. (2011, 2012)
InGaN	Schottky	Ti/Al/Ni/Au Ni/WC	200	79 (RT, −1V) 2.2 (200°C, −1V)			Sang et al. (2012)
AlGaN/GaN	MSM	Ti/Au	327		~1 (27°C) ~18 (327°C)	Thermal ionization of trapped photo-generated carriers increases responsivity.	De Vittorio et al. (2004)
AlGaN/GaN	Photoconductor	Ti/Al/Pt/Au	200	1.4 (RT, 1 V) 0.3 (200°C, 1 V)	0.00035 (RT and 200°C, 1 V)	V-grooved photodetector increases absorption of incident light resulting in higher sensitivity compared to the planar photodetector.	So et al. (2016)
AlGaN/GaN		Al Wirebond	100			Direct wire-bonding enabling rapid fabrication and packaging. Signal to noise ratio: 29.2 (RT) and 14.4 (100°C)	So and Senesky (2016a)
AlGaN/GaN		Ti/Al/Pt/Au	270 (Membrane Heating)	0.04 (RT, 30 V)		Photocurrent decay time: 39 h (1 V = membrane temperature 25°C) and 24 s (30 V = membrane temperature 270°C)	Hou et al. (2017)
AlN	MSM	Ti/Pt	300	60 (RT, 5 V) 3.5 (300°C, 5 V)	0.015 (RT, 5 V)	Rise time ~ 110 Ms Decay time ~ 80 Ms	Tsai et al. (2013)
AlGaN/GaN	2DEG-IDT	Ti/Al/Pt/Au	250		7800 (RT, 5 V) 3.9 (250°C, 5 V)	Rise time: 32 Ms, Decay time: 76 Ms	Satterthwaite et al. (2018)

Table 7.3 Calculated bowing parameters for III-nitride compound semiconductors.

	LDA	LDA-1/2	Refernces (Meyer and Vurgaftman, 2003).
$In_xGa_{1-x}N$	1.4	1.3	1.4
$Al_xGa_{1-x}N$	0.4	0.8	0.7
$Al_xIn_{1-x}N$	2.7	$3.4x + 1.2$	2.4

Another approach to increase quantum efficiency, PDCR, and responsivity at high temperatures is to increase UV light absorption. Three-dimensional AlGaN/GaN UV photodetectors using a "v-grooved" Si substrate demonstrated operation up to 200°C with improved PDCR over planar AlGaN/GaN UV photodetectors (~ 1.4 compared to ~ 0.9 at room temperature and ~ 0.3 compared to ~ 0.1 at 200°C) (So et al., 2016). The increased sensitivity was attributed to the increased absorption (via redirection of reflected light) of the incident UV light in the three-dimensional substrate. III-nitride photodetectors with ZnO nanorod arrays have also demonstrated enhanced sensitivity at high temperatures due to their antireflective properties (So and Senesky, 2016b). Increased photodetector sensitivity due to ZnO nanorod arrays were demonstrated with a direct wire bond photodetector where a GaN chip was directly wire-bonded to a carrier chip thus avoiding time-consuming and costly microfabrication steps such as photolithography, metal sputtering (or evaporation), and etching (or lift-off) (So and Senesky, 2016b). These rapidly fabricated/packaged photodetectors had a PDCR of 1.11 at room temperature and demonstrated operation up to 250°C with a PDCR of 0.11. With the addition of an antireflective ZnO nanorod array coating to enhance light trapping, the direct wire bond photodetector sensitivity improved to 2.63 and 0.27 at room temperature and 250°C, respectively (So and Senesky, 2016b).

Lastly, few reports of AlN photodetectors have been published. However, AlN is a promising material candidate for high-temperature deep-UV (< 200 nm) sensing applications. AlN has a band-gap of 6.2 eV which corresponds to a cut-off wavelength of 200 nm, making AlN photodetectors not only visible blind but also solar blind. MSM photodetectors fabricated on AlN thin films grown on a Si substrate demonstrated room temperature dark currents below 1 nA at applied bias voltages up to 200 V and room temperature PDCRs as high as 63 (Tsai et al., 2013). The AlN photodetectors demonstrated operation up to 300°C (PDCR of 3.5) and the room temperature response was fully recovered after thermal cycling up to 400°C (Tsai et al., 2013).

7.4 SIC-based ultraviolet photodetectors

SiC is another attractive wide bandgap semiconductor for UV detection at high temperatures. In particular, 4H-SiC has strong chemical bonds, high thermal conductivity ($4 - 4.9$ W/cm/K) (Saddow and Agarwal, 2004), and high electron saturation velocity (2.7×10^7 cm/s) (Joshi, 1995) enabling 4H-SiC based UV photodetectors to operate in high-temperature and high-radiation environments with fast response speed. Table 7.4 summarizes the type (structure) of the device, electrode metal, operation temperature, PDCR, and responsivity of SiC-based UV photodetectors. The average operation temperature is slightly higher than the operating temperature of GaN-based UV photodetectors while the overall values of responsivity are relatively smaller than those of GaN-based photodetectors. The highest reported operational temperature of SiC-based photodetectors is 550°C by p-i-n structure (Hou et al., 2016). From room temperature to 550°C, the photocurrent increased by 9 times at 365 nm and decreased by 2.6 times at 275 nm due to band-gap narrowing effect at high temperature as shown in Figure 7.6 (Hou et al., 2016; Mazzillo et al., 2014). This thermally-induced bandgap narrowing effect also induces an increase in the optical absorption coefficient (Cha et al., 2008; Brown et al., 1993), thus increasing the quantum efficiency of photodetectors as temperature increases. Based on this effect, the highest quantum efficiency of 53.4% and 63.6% were reported at room temperature and 150°C, respectively, by 4H-SiC avalanche photodiodes (Cha et al., 2008). For long-term reliability at high temperatures, a 4H-SiC photodetector using Schottky metal contacts was exposed to 200°C in the air for 100 h (Xu et al., 2015). After thermal storage, the responsivity and dark current level of the device remained unchanged, thus indicating the reliable operation of 4H-SiC UV photodetectors at high temperatures up to 200°C (Xu et al., 2015). The temperature-independent responsivity of SiC photodetectors was also studied since the photoresponse usually depends on temperature. By controlling the reverse-bias voltage from 0 to 150 V, 4H-SiC p-n photodiodes were shown to have temperature-independent responsivity under the 280 − 300 nm wavelength range as shown in Figure 7.7 (Watanabe et al., 2012). This phenomenon was explained by a combination of temperature-dependent optical absorption coefficient and surface recombination effects (Watanabe et al., 2012).

Compared to GaN-based photodetectors, a severe PPC effect was not observed in SiC-based photodetectors which enables photodiodes to have a fast response time. The rise/fall time of an MSM photodetector using a 7 μm p-type 4H-SiC epitaxial layer was reported as 594 μs/699 μs and 684 μs/786 μs at room temperature and 400°C, respectively (Lien et al., 2012). Because the bandwidth of 4H-SiC is decreased due to the decrease in hole/electron saturation velocity as temperature increases, the rise/fall time of the photodetector is slightly increased at high temperatures (Lien et al., 2012). Although many SiC-based photodetectors are fabricated on 4H-SiC epitaxial layer, 6H-SiC (Ueda et al., 2008), nanocrystalline SiC (Lien et al., 2011), and β-SiC (or 3C-SiC) (Hsieh et al., 2001) have also been used for fabrication of UV photodetectors. In particular, β-SiC on Si substrate was used to obtain a high gain of the optical sensor (Hsieh et al., 2001). To extend the operating temperature and improve the high-temperature performance, the porous silicon substrate was additionally adopted as the semiinsulating substrate, suppressing the leakage current and thus resulting in a low dark current level (Hsieh et al., 2001). The synthesis of ZnO nanorod arrays (i.e., antireflective coating) on SiC layer was reported as an alternative method to increase the operating temperature of the photodetectors (Lien et al., 2011).

Table 7.4 High-temperature SiC-based UV photodetectors.

Material	Photodetector architecture	Electrode material	Temperature (°C)	PDCR	Responsivity (A/W)	Notes	References
4H-SiC	p-i-n	Ni Ni/Ti/Al	550	45.4 (500°C) 7.3 (550°C)	0.12 (RT) 0.046 (550°C)	From RT to 550°C, photocurrent increased by 9 times at 365 nm and decreased by 2.6 times at 275 nm due to bandgap narrowing.	Hou et al. (2016)
4H-SiC	MSM	Cr/Pd	450	1.3×10^5 (RT) 0.62 (450°C)	0.305 (RT, 20 V)	Rise time: 594 μs (RT) and 684 μs (400°C) Fall time: 699 μs (RT) and 786 μs (400°C)	Lien et al. (2012)
4H-SiC	Schottky	Ti/AlSiCu Ti/Ni/Au	90		0.046 (RT)	QE: 19% (RT)	Xie et al. (2011b)
4H-SiC	Avalanche	Ni/Ti/Al/ Au	150		0.125 (RT)	Maximum QE: 53.4% (RT, 290 nm) and 63.3% (150°C, 295 nm)	Zhou et al. (2014b)
4H-SiC	Avalanche	Ni/Ti/Al/ Au	190		0.093 (RT)	QE: 41% (RT)	Bai et al. (2007)
4H-SiC	p-n	Ti/Al/Ni	300		Between 0.15 and 0.2 (RT and 300°C)	Achieved a temperature-independent photoresponse at targeted wavelengths by controlling the reverse bias voltage. QE: 50% (RT)	Watanabe et al. (2012)
4H-SiC	Schottky	Ni/Ti/Al/ Au Ni	200			After thermal storage at 200°C in the air for 100 h, the dark current increased lightly but remained less than 1 pA at 20 V.	Xu et al. (2015)
4H-SiC	Schottky	Cr	127			QE: 26% (−198°C), 27.4% (102°C)	Blank et al. (2005)
6H-SiC	n⁺-p	Ni Ti/Al	500			QE: 24.9% (RT) Photocurrent increased with increasing temperature.	Ueda et al. (2008)

(Continued)

Table 7.4 (Continued)

Material	Photodetector architecture	Electrode material	Temperature (°C)	PDCR	Responsivity (A/W)	Notes	References
Nanocrystalline SiC	MSM	Au	200	RT: 4.9, 13.3 (with ZnO Nanorods) 200°C: 4.9, 7.6 (with ZnO nanorods)		Used ZnO nanorod arrays as an antireflective coating.	Lien et al. (2011)
β-SiC on Porous Si	MSM	Al	200	30 (RT, 5 V) 15 (200°C, 5 V)		High resistivity and flexible porous silicon substrate result in low dark current and improved high-temperature performance.	Hsieh et al. (2001)
β-SiC on Si	MSM	Al	200	8 (RT, 5 V) 1(200°C, 5 V)			Hsieh et al. (2001)

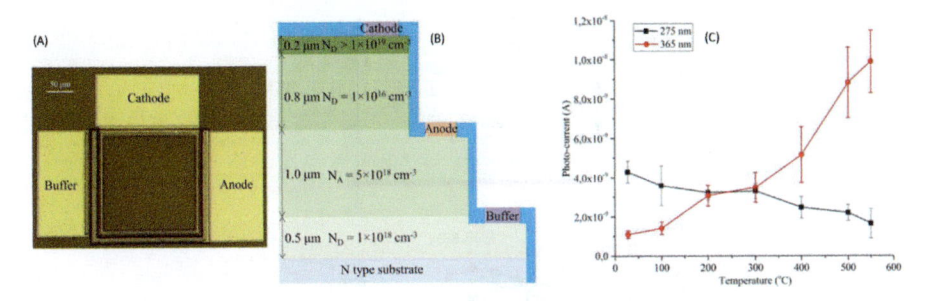

Figure 7.6 (A) Microscope image, (B) cross-section, and (C) photocurrent as a function of temperature for a SiC p-i-n photodetector. As temperature increases, the photocurrent generated by 275 nm light decreases while the photocurrent generated by 365 nm light increases indicating bandgap narrowing at high temperature affects the photodetector response (Chikoidze et al., 2017). *Source*: Adopted with permission from Chikoidze, E., Fellous, A., Perez-Tomas, A., Sauthier, G., Tchelidze, T., Ton-That, C., et al. 2017. P-type β-gallium oxide: A new perspective for power and optoelectronic devices, Mater. Today Phys. 3, 118−126 (Copyright 2016, IEEE).

7.5 Other types of ultraviolet photodetectors

In addition to GaN and SiC as discussed above, there are other less developed materials with bandgaps appropriate for UV photodetection that are also capable of operating at elevated temperatures including zinc oxide (ZnO), gallium oxide (Ga2O3), diamond, and boron nitride (BN) (Liu et al., 2010; Chen et al., 2015; Konstantatos and Sargent, 2010; Peng et al., 2013). To date, a Ga2O3 MSM photodetector with transparent indium zinc oxide (IZO) electrodes has the highest reported operating temperature (427°C) of all photodetectors based on these other materials (Wei et al., 2014). Table 7.5 details the reported operating temperatures of photodetectors based on these materials with the benefits and drawbacks of each material discussed further below.

ZnO, an II−VI semiconductor, is often compared to GaN. Both preferentially form in the wurtzite crystal structure, and they have direct band gaps with very similar energies (3.37 eV vs 3.4 eV for GaN, corresponding to a cut-off wavelength of 368 nm) (Li et al., 2014). This, combined with a high decomposition temperature ($\sim$1975°C), means that ZnO is very well suited for use in high-temperature UV photodetectors. Although ZnO has been studied extensively as a material for UV photodetection, there has been little investigation into the behavior of ZnO-based devices at elevated temperatures, with the highest reported testing (of an MSM device with Al electrodes) carried out at 200°C (Li et al., 2014). Operation at this temperature showed reduced photocurrent and responsivity, as well as increased dark current compared with samples tested at room temperature. These changes were attributed to band gap shrinkage and increased lattice scattering (Li et al., 2014).

β-Ga2O3, with a band gap of 4.9 eV (253 nm), is the most stable of the five polymorphs of Ga2O3. It has a melting point of $\sim$1780°C (α-Ga2O3 has a higher theoretical melting point of $\sim$1900°C but transforms into β-Ga2O3 at temperatures over

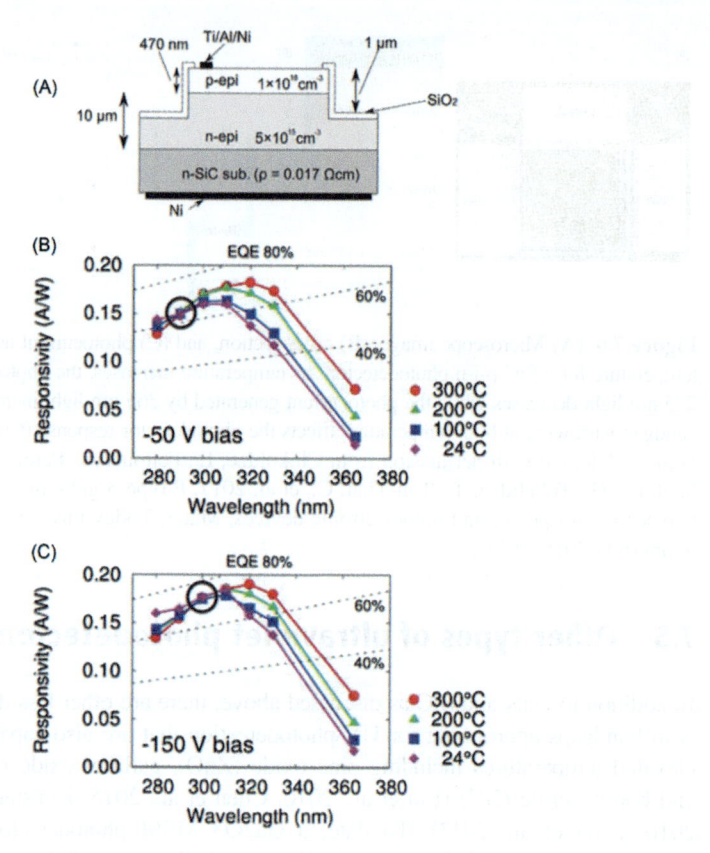

Figure 7.7 (A) cross-section, (B) responsivity spectrum as a function of temperature for a SiC p-i-n photodetector at 50V reverse bias, and (C) same as (b) at 150V reverse bias. As temperature increases, the photocurrent generated by 275 nm light decreases while the photocurrent generated by 365 nm light increases indicating bandgap narrowing at high temperature affects the photodetector response.
Source: Adopted with permission from Xie, F., Lu, H., Chen, D., Zhang, R., Zheng, Y., 2011b. GaN MSM photodetectors fabricated on bulk GaN with low dark-current and high UV/visible rejection ratio. Phys. Status Solidi C. 8 (7–8), 2473–2475. (Copyright 2012, IOP Publishing Ltd)

800°C) (Stepanov et al., 2016). Much of the interest in β-Ga2O3 stems from its potential as a transparent conducting oxide that allows for the transmission of long-wavelength UV light, a result of its large band gap (Hrong et al., 2017). A variety of β-Ga2O3 UV photodetectors (mostly Schottky diodes, and MSM devices) have been tested at elevated temperatures, and are particularly notable for their short decay times, extremely low dark currents, and solar-blind photoresponses (Wei et al., 2014; Zou et al., 2014; Alema, 2017; Ghose, 2017; Ai et al., 2017). β-Ga2O3 UV photodetectors have also demonstrated robustness through little variation in device behavior with changing atmospheric oxygen concentration and good resistance to permanent

Table 7.5 High-temperature UV photodetectors from other wide bandgap materials.

Material	Photodetector architecture	Electrode material	Temperature (°C)	PDCR	Responsivity (A/W)	Notes	References
ZnO	MSM	Al	200		9 (RT) 2 (200°C)	Responsivity decreases with increasing temperature due to bandgap shrinkage and lattice scattering.	Li et al. (2014)
β-Ga$_2$O$_3$	MSM	IZO	427	14 (RT, 10 V) 1.5 (427°C, 10 V)	0.00032 (RT, 10 V)		Wei et al. (2014)
Ga$_2$O$_3$ Nanobelt	Photoconductor	Au	160	72,463 (55°C, 20 V) 8892 (160°C, 20 V)	870 (55°C, 20 V) 650 (160°C, 20 V)		Zou et al. (2014)
Diamond	MSM	Ag	300			UV/visible (200–800 nm) discrimination of four orders of magnitude up to 300°C	Salvatori et al. (1999)
BNNS	MSM	Au	400		9 µA/W (RT, 0 V)	Photocurrent increased by a factor of four from RT to 400°C. Thermal noise increased by a factor of 3 from RT to 400°C.	Brown et al. (1993)
SiCBN	MSM	Al	200	6.5 (RT)			Vijayakumar et al. (2007)
SiCN	MSM	Au	200	2.3 (200°C, −5V)			Chang et al. (2003)
n-SiCN/i-SiCN/p-SiCN	n-i-p	Au	200	3180 (RT, −5V) 135.65 (200°C, −5V)	0.14 (RT, −5V)	QE: 67% (RT, −5V)	Juang et al. (2011)
n-SiCN/i-SiCN/p-Si	n-i-p	Ni	200	60 (RT, −5V) 4.7 (200°C, −5V)			Juang et al. (2011)
n-SiCN/p-SiCN	n-p	Ni Al	175	1940 (RT, −5V) 96.3 (175°C, −5V)			Chou et al. (2008)

degradation at elevated temperatures, with one device (MSM with IZO electrodes) showing full recovery of room-temperature behavior after testing at 427°C, the current high-temperature record for β-Ga2O3 UV photodetectors (Wei et al., 2014).

Diamond has a band gap of ∼ 5.5 eV (225 nm), is one of the hardest known materials, has a very high (pressure-dependent) theoretical sublimation temperature (over 3000°C), but readily transforms into graphite at temperatures above ∼1200°C. (Diamond is thermodynamically unstable with respect to graphite at all temperatures, but the transformation is extremely slow at lower temperatures) (Pierson). Pure diamond is often doped (particularly with boron) to improve its semiconducting properties (Liao et al., 2012). MSM photodetectors fabricated on 1 μm-thick polycrystalline CVD-deposited diamond films with Ag electrodes have been tested up to 300°C, displaying a responsivity of over 100 mA/W for the range from 225–350 nm (Salvatori et al., 1999).

Hexagonal boron nitride (h-BN) has a band gap of ∼6 eV (206 nm) corresponding to the deep UV and is particularly notable, even among wide-band-gap semiconductors, for its high decomposition temperature (2973°C). Due to its extremely wide bandgap, h-BN is solar-blind and therefore does not need solar rejection filters to operate in the deep UV (Rivera et al., 2017). H-BN can be synthesized in a variety of forms including nanorods, nanotubes, and nanosheets (BNNS) with 2D nanosheets of special interest due to their physical and structural similarities (in addition to thermal stability) to graphene (Rivera et al., 2018). Photodetectors based on BNNS with Au electrodes tested up to 400°C displayed a factor of four increase in photocurrent and a factor of three increase in thermal noise compared to devices tested at room temperature (Rivera et al., 2017, 2018). This combination of high photocurrent, solar-blind photoresponse, and high thermal stability means that BNNS is particularly well suited to use in high-temperature UV photodetectors.

7.6 Conclusions

Wide bandgap semiconductors have advanced high-temperature UV sensing beyond the capabilities of silicon due to their intrinsic thermal stability. Of the wide bandgap semiconductors used for high-temperature UV detection, the III-nitrides and SiC is the most developed and have demonstrated operating temperatures upwards of 300°C. However, further development of all of these material platforms is needed to enable high UV sensitivity with long-term reliable operation at elevated temperatures. For these materials to reach their full potential, material quality issues need to be addressed, high-temperature capable electrode technology needs to be furthered, and high-temperature packaging schemes need to be developed.

References

Ai, M., Guo, D., Qu, Y., Cui, W., Wu, Z., Li, P., et al., 2017. Fast-response solar-blind ultraviolet photodetector with a graphene/β-Ga2O3/graphene hybrid structure. J. Alloy. Compd. 692, 634–638.

Alaie, Z., Mohammad Nejad, S., Yousefi, M.H., 2015. Recent advances in ultraviolet photodetectors. Mater. Sci. Semicond. Process. 29, 16−55.

Alema, F., 2017. Vertical solar blind Schottky photodiode based on homoepitaxial Ga2O3 thin film. In: Proc. SPIE 10105, Oxide-based Materials and Devices VIII, p. 101051M.

Alpert, H.S. et al., 2019. Gallium nitride photodetector measurements of UV emission from a gaseous CH4/O2 hybrid rocket igniter plume. In: 2019 IEEE Aerospace Conference, Big Sky, MT, USA, pp. 1−8, https://doi.org/10.1109/AERO.2019.8741713.

Bai, X., Guo, X., Mcintosh, D.C., Liu, H.-D., Campbell, J.C., 2007. High detection sensitivity of ultraviolet 4H-SiC avalanche photodiodes. IEEE J. Quantum Electron. 43 (12), 1159−1163.

Bhattacharya, P., 1997. Semiconductor Optoelectronic Devices, second ed. Prentice Hall, Hoboken, NJ.

Biesalski, H.K., Obermueller-Jevic, U.C., 2001. UV light, beta-carotene and human skin-beneficial and potentially harmful effects. Arch. Biochem. Biophys. 389 (1), 1−6.

Binet, F., Duboz, J.Y., Rosencher, E., Scholz, F., Harle, V., 1996. Mechanisms of recombination in GaN photodetectors. Appl. Phys. Lett. 69 (9), 1202−1204.

Blank, T.V., Goldberg, Y.A., Kalinina, E.V., Konstantinov, O.V., Konstantinov, A.O., Hallen, A., 2005. Temperature dependence of the photoelectric conversion quantum efficiency of 4H-SiC Schottky UV photodetectors. Semicond. Sci. Technol. 20 (8), 710−715.

Brown, D.M., Downey, E.T., Ghezzo, M., Kretchmer, J.W., Saia, R.J., Liu, Y.S., et al., 1993. Silicon carbide UV photodiodes. IEEE Trans. Electron. Devices 40 (2), 325−333.

Buttay, C., Planson, D., Allard, B., Bergogne, D., Bevilacqua, P., Joubert, C., et al., 2011. State of the art of high temperature power electronics. Mater. Sci. Eng. B 176 (4), 283−288.

Carrano, J.C., Li, T., Grudowski, P.A., Eiting, C.J., Dupuis, R.D., Campbell, J.C., 1998. Current transport mechanisms in GaN-based metal-semiconductor-metal photodetectors. Appl. Phys. Lett. 72 (5), 542−544.

Cha, H.Y., Soloviev, S., Zelakiewicz, S., Waldrab, P., Sandvik, P.M., 2008. Temperature dependent characteristics of nonreach-through 4H-SiC separate absorption and multiplication APDs for UV detection. IEEE Sens. J. 8 (3), 233−237.

Chang, W.-R., Fang, Y.-K., Ting, S.-F., Tsair, Y.-S., Chang, C.-N., Lin, C.-Y., et al., 2003. The hetero-epitaxial SiCN/Si MSM photodetector for high-temperature deep-UV detecting applications. IEEE Electron. Device Lett. 24 (9), 565−567.

Chang, S., Chang, M., Yang, Y., 2017. Enhanced responsivity of GaN metal-semiconductor-metal (MSM) photodetectors on GaN substrate. IEEE Photon. J. 9 (2), 6801707.

Chen, H.M., Chen, Y.F., Lee, M.C., Feng, M.S., 1997. Persistent photoconductivity in n-type GaN. J. Appl. Phys. 82 (2), 899−901.

Chen, H., Liu, K., Hu, L., Al-Ghamdi, A.A., Fang, X., 2015. New concept ultraviolet photodetectors. Mater. Today 18 (9), 493−502.

Chikoidze, E., Fellous, A., Perez-Tomas, A., Sauthier, G., Tchelidze, T., Ton-That, C., et al., 2017. P-type β-gallium oxide: a new perspective for power and optoelectronic devices. Mater. Today Phys. 3, 118−126.

Chou, T.-H., Fang, Y.-K., Chiang, Y.-T., Lin, C.-I., Yang, C.-Y., 2008. A low cost n-SiCN/p-SiCN homojunction for high temperature and high gain ultraviolet detecting applications. Sens. Actuators A 147 (1), 60−63.

Dang, X.Z., Wang, C.D., Yu, E.T., Boutros, K.S., Redwing, J.M., 1998. Persistent photoconductivity and defect levels in n-type AlGaN/GaN heterostructures. Appl. Phys. Lett. 72 (21), 2745−2747.

De Vittorio, M., Poti, B., Todaro, M.T., Frassanito, M.C., Pomarico, A., Passaseo, A., et al., 2004. High temperature characterization of GaN-based photodetectors. Sens. Actuators A 113 (3), 329−333.

Decoster, D., Harari, J., 2009. Ultraviolet Photodetectors. Optoelectronic Sensors. John Wiley & Sons, Inc, Hoboken, NJ, pp. 181−222.

Diffey, B.L., 2002. Sources and measurement of ultraviolet radiation. Methods 28 (1), 4−13.

Ghose S., 2017. Growth and characterization of wide bandgap semiconductor oxide thin films. Ph.D. dissertation, Dept. of Materials Science, Engineering, and Commercialization, Texas State Univ., San Marcos, TX.

Golimowski, J., Golimowska, K., 1996. UV-photooxidation as pretreatment step in inorganic analysis of environmental samples. Analytica Chim. Acta 325 (3), 111−133.

Hirsch, M.T., Wolk, J.A., Walukiewicz, W., Haller, E.E., 1997. Persistent photoconductivity in n-type GaN. Appl. Phys. Lett. 71 (8), 1098−1100.

Hou, S., Hellstrom, P.-E., Zetterling, C.-M., Ostling, M., 2016. 550°C 4H-SiC p-i-n photodiode array with two-layer metallization. IEEE Electron. Device Lett. 37 (12), 1594−1596.

Hou, M., So, H., Suria, A.J., Yalamarthy, A.S., Senesky, D.G., 2017. Suppression of peresistent photoconductivity in AlGaN/GaN ultraviolet photodetectors using in situ heating. IEEE Electron. Device Lett. 38 (1), 56−59.

Hrong, R.-H., Zeng, Y.-Y., Wang, W.-K., Tsai, C.-L., Fu, Y.-K., Kuo, W.-H., 2017. Transparent electrode design for AlGaN deep-ultraviolet light-emitting diodes. Opt. Express 25 (25), 32206−32213.

Hsieh, W.T., Fang, Y.K., Wu, K.H., Lee, W.J., Ho, J.J., Ho, C.W., 2001. Using porous silicon as semi-insulating substrate for β-SiC high temperature optical-sensing devices. IEEE Trans. Electron. Devices 48 (4), 801−803.

Joshi, R.P., 1995. Monte Carlo calculations of the temperature- and field-dependent electron transport parameters for 4H-SiC. J. Appl. Phys. 78 (9), 5518-1521.

Juang, F.-R., Fang, Y.-K., Chiang, Y.-T., Chou, T.-H., Lin, C.-I., 2011. A high-performance n-i-p SiCN homojunction for low-cost and high-temperature ultraviolet detecting applications. IEEE Sens. J. 11 (1), 150−154.

Katz, O., Bahir, G., Salzman, J., 2004. Persistent photocurrent and surface trapping in GaN Schottky ultraviolet detectors. Appl. Phys. Lett. 84 (20), 4092−4094.

Konstantatos, G., Sargent, E.H., 2010. Nanostructured materials for photon detection. Nat. Nanotechnol. 5, 391−400.

Lang, D.V., Logan, R.A., 1977. Large-lattice-relaxation model for persistent photoconductivity in compound semiconductors. Phys. Rev. Lett. 39 (10), 635−639.

Li, J.Z., Lin, J.Y., Jiang, H.X., Salvador, A., Botchkarev, A., Morkoc, H., 1996. Nature of Mg impurities in GaN. Appl. Phys. Lett. 69 (10), 1474−1476.

Li, J.Z., Lin, J.Y., Jiang, H.X., Asif Khan, M., Chen, Q., 1997. Persistent photoconductivity in a two-dimensional electron gas system formed by an AlGaN/GaN heterostructure. J. Appl. Phys. 82 (3), 1227−1230.

Li, G., Zhang, J., Hou, X., 2014. Temperature dependence of performance of ZnO-based metal-semiconductor-metal ultraviolet photodetectors. Sens. Actuators A 209 (1), 149−153.

Liao, M., Sang, L., Teraji, T., Imura, M., Alvarez, J., Koide, Y., 2012. Comprehensive investigation of single crystal diamond deep-ultraviolet detectors. Jpn. J. Appl. Phys. 51 (9R), 090115.

Lien, W.-C., Tsai, D.-S., Chiu, S.-H., Senesky, D.G., Maboudian, R., Pisano, A.P., et al., 2011. Low-temperature, ion beam-assisted SiC thin films with antireflective ZnO nanorod arrays for high-temperature photodetection. IEEE Electron. Device Lett. 32 (11), 1564–1566.

Lien, W.-C., Tsai, D.-S., Lien, D.-H., Senesky, D.G., He, J.-H., Pisano, A.P., 2012. 4H-SiC metal-semiconductor-metal ultraviolet photodetectors in operation of 450°C. IEEE Electron. Device Lett. 33 (11), 1586–1588.

Liu, K., Sakurai, M., Aono, M., 2010. ZnO-based ultraviolet photodetectors. Sensors 10 (9), 8604–8634.

Liu, L., Yang, C., Patane, A., Yu, Z., Yan, F., Wang, K., et al., 2017. High-detectivity ultraviolet photodetectors based on laterally mesoporous GaN. Nanoscale 9 (24), 8142–8148.

Mazzillo, M., Sciuto, A., Marchese, S., 2014. Impact of the epilayer doping on the performance of thin metal film Ni2Si/4H-SiC Schottky photodiodes. J. Instrum. 9, P12001.

Meyer, J.N., Vurgaftman, I., 2003. Band parameters for nitrogen-containing semiconductors. J. Appl. Phys. 94 (6), 3675–3696.

Monroy, E., Calle, F., Pau, J.L., Munoz, E., Omnes, F., Beaumont, B., et al., 2001. AlGaN-based UV photodetectors. J. Cryst. Growth 230 (3–4), 537–543.

Monroy, E., Omnes, F., Calle, F., 2003. Wide-bandgap semiconductor ultraviolet photodetectors. Semicond. Sci. Technol. 18 (4), R33–R51.

Moustakas, T.D., Paiella, R., 2017. Optoelectronic device physics and technology of nitride semiconductors from the UV to the terahertz. Rep. Prog. Phys. 80, 106501.

Munoz, E., Monroy, E., Garrido, J.A., Izpura, I., Sanchez, F.J., Sanchez-Garcia, M.A., et al., 1997. Photoconductor gain mechanisms in GaN ultraviolet detectors. Appl. Phys. Lett. 71 (7), 870–872.

Munoz, E., Monroy, E., Pau, J.L., Calle, F., Omnes, F., Gibart, P., 2001. III nitrides and UV detection. J. Phys.: Condens. Matter 13, 7115–7137.

Neudeck, P.G., Okojie, R.O., Chen, L.-Y., 2002. High-temperature electronics—a role for wide bandgap semiconductors? Proc. IEEE 90 (6), 1065–1076.

Omnes, F., Monroy, E., 2003. GaN-based UV photodetectors. Nitride Semiconductors: Handbook on Materials and Devices. Wiley-VCH Verlag GMbH & Co., Weinheim, Germany, pp. 627–660.

Oshima, T., Okuno, T., Arai, N., Suzuki, N., Hino, H., Fujita, S., 2009. Flame detection by a b-Ga2O3-based sensor. Jpn. J. Appl. Phys. 48, 011605.

Pau, J.L., Anduaga, J., Rivera, C., Navarro, A., Alava, I., Redondo, M., et al., 2006. Optical sensors based on III-nitride photodetectors for flame sensing and combustion monitoring. Appl. Opt. 45 (28), 7498–7503.

Pelá, R.R., Caetano, C., Marques, M., Ferreira, L.G., Furthmüller, J., Teles, L.K., 2011. Accurate band gaps of AlGaN, InGaN, and AlInN alloys calculations based on LDA-1/2 approach. Appl. Phys. Lett. 98 (15), 151907.

Peng, L., Hu, L., Fang, X., 2013. Low-dimensional nanostructure ultraviolet photodetectors. Adv. Mater. 25 (37), 5321–5328.

Pierret, R.F., 1996. Semiconductor Device Fundamentals. Addison-Wesley, Boston, MA.

Pierson H.O., Structure and properties of diamond and diamond polytypes. In: Handbook of Carbon, Graphite, Diamonds and Fullerenes: Processing, Properties and Applications, first ed. Park Ridge, New Jersey.

Poti, B., Passaseo, A., Lomascolo, M., Cingolani, R., De Vittorio, M., 2004. Persistent photocurrent spectroscopy of GaN metal-semiconductor-metal photodetectors on long time scale. Appl. Phys. Lett. 85 (25), 6083–6085.

Qiu, C.H., Pankove, J.I., 1997. Deep levels and persistent photoconductivity in GaN thin films. Appl. Phys. Lett. 70 (15), 1983–1985.

Razeghi, M., Rogalski, A., 1996. Semiconductor ultraviolet detectors. J. Appl. Phys. 79 (10), 7433–7473.

Rivera, C., Pereiro, J., Navarro, A., Munoz, E., Brandt, O., Grahn, H.T., 2010. Advances in group-III-nitride photodetectors. Open. Electr. Electron. Eng. J. 4, 1–9.

Rivera, M., Velazquez, R., Aldalbahi, A., Zhou, A.F., Feng, P., 2017. High operating temperature and low power consumption boron nitride nanosheets based broadband UV photodetector. Sci. Rep. 7, 42973.

Rivera, M., Velazquez, R., Aldalbahi, A., Zhou, A.F., Feng, P.X., 2018. UV photodetector based on energy bandgap shifted hexagonal boron nitride nanosheets for high-temperature environments. J. Phys. D: Appl. Phys. 51 (4), 045102.

Robichaud, J.L. 2003. SiC optics for EUV, UV, and visible space missions. Proc. SPIE 4854, 39–49.

Saddow, S.E., Agarwal, A., 2004. Advances in Silicon Carbide Processing and Applications. Artech House, Norwood, MA.

Salvatori, S., Scotti, F., Conte, G., Rossi, M.C., 1999. Diamond-based UV photodetectors for high-temperature applications. Electron. Lett. 35 (20), 1768–1770.

Sang, L., Liao, M., Koide, Y., Sumiya, M., 2011. High-temperature ultraviolet detection based on InGaN Schottky photodiodes. Appl. Phys. Lett. 99 (3), 031115.

Sang, L.W., Liao, M.Y., Koide, Y., Sumiya, M., 2012. InGaN photodiodes using CaF2 insulator for high-temperature UV detection. Phys. Status Solidi C. 9 (3–4), 953–956.

Sang, L., Liao, M., Sumiya, M., 2013. A comprehensive review of semiconductor ultraviolet photodetectors: from thin film to one-dimensional nanostructures. Sensors 13 (8), 10482–10518.

Satterthwaite, P.F., Yalamarthy, A.S., Scandrette, N.A., Newaz, A.K.M., Senesky, D.G., 2018. High responsivity, low dark current ultraviolet photodetectors based on two-dimensional electron gas interdigitated transducers,". ACS Photon. 2018 5 (11), 4277–4282.

Shi, L., Nihtianov, S., 2012. Comparative study of silicon-based ultraviolet photodetectors. IEEE Sens. J. 12 (7), 2453–2459.

So, H., Senesky, D.G., 2016a. Rapid fabrication and packaging of AlGaN/GaN high-temperature ultraviolet photodetectors using direct wire bonding. J. Phys. D: Appl. Phys. 49 (28), 285109.

So, H., Senesky, D.G., 2016b. ZnO nanorod arrays and direct wire bonding on GaN surfaces for rapid fabrication of antireflective, high-temperature ultraviolet sensors. Appl. Surf. Sci. 387, 280–284.

So, H., Lim, J., Senesky, D.G., 2016. Continuous V-grooved AlGaN/GaN surfaces for high-temperature ultraviolet photodetectors. IEEE Sens. J. 16 (10), 3633–3639.

Stepanov, S.I., Nikolaev, V.I., Bougrov, V.E., Romanov, A.E., 2016. Gallium oxide: properties and applications - a review. Rev. Adv. Mater. Sci. 44, 63–86.

Sze, S.M., Ng, K.K., 2007. Physics of Semiconductor Devices, third ed. John Wiley & Sons, Inc, Hoboken, NJ.

Sze, S.M., Coleman Jr., D.J., Loya, A., 1971. Current transport in metal-semiconductor-metal (MSM) structures. Solid-State Electron. 14 (12), 1209–1218.

Tsai, D.-S., Lien, W.-C., Lien, D.-H., Chen, K.-M., Tsai, M.-L., Senesky, D.G., et al., 2013. Solar-blind photodetectors for harsh electronics. Sci. Rep. 3, 2628.

Ueda, Y., Akita, S., Nomura, Y., Nakayama, Y., Naito, H., 2008. Study of high temperature photocurrent properties of 6H-SiC UV sensor. Thin Solid. Films 517 (4), 1471–1473.

Vijayakumar, A., Todi, R.M., Sundaram, K.B., 2007. Amorphous-SiCBN-based metal-semiconductor-metal photodetector for high-temperature applications. IEEE Electron. Device Lett. 28 (8), 713–715.

Wang, G., Xie, F., Lu, H., Chen, D., Zhang, R., Zheng, Y., et al., 2013. Performance comparison of front- and back-illuminated AlGaN-based metal-semiconductor-metal solar-blind ultraviolet photodetectors. J. Vac. Sci. Technol. B 31 (1), 011202.

Watanabe, N., Kimoto, T., Suda, J., 2012. 4H-SiC pn photodiodes with temperature-independent photoresponse up to 300°C. Appl. Phys. Express 5, 094101.

Wei, T.-C., Tsai, D.-S., Ravadgar, P., Ke, J.-J., Tsai, M.-L., Lien, D.-H., et al., 2014. See-through Ga2O3 solar-blind photodetectors for use in harsh environments. IEEE J. Sel. Top. Quantum Electron. 20 (6), 3802006.

Xie, F., Lu, H., Xiu, X., Chen, D., Han, P., Zhang, R., et al., 2011a. Low dark current and internal gain mechanism of GaN MSM photodetectors fabricated on bulk GaN substrate. Solid-State Electron. 57 (1), 39−42.

Xie, F., Lu, H., Chen, D., Zhang, R., Zheng, Y., 2011b. GaN MSM photodetectors fabricated on bulk GaN with low dark-current and high UV/visible rejection ratio. Phys. Status Solidi C. 8 (7−8), 2473−2475.

Xie, F., Lu, H., Chen, D., Ji, X., Yan, F., Zhang, R., et al., 2012. Ultra-low dark current AlGaN-based solar-blind metal-semiconductor-metal photodetectors for high-temperature applications. IEEE Sens. J. 12 (6), 2086−2090.

Xu, Z., Sadler, B.M., 2008. Ultraviolet communications: potential and state-of-the-art. IEEE Commun. Mag. 46 (5), 67−73.

Xu, Y., Zhou, D., Lu, H., Chen, D., Ren, F., Zhang, R., et al., 2015. High-temperature and reliablity performance of 4H-SiC Schottky-barrier photodiodes for UV detection. J. Vac. Sci. Technol. B 33 (4), 040602.

Zhou, H., Zhu, J., Liu, Z., Yan, Z., Fan, X., Lin, J., et al., 2014a. High thermal conductivity of suspended few-layer hexagonal boron nitride sheets. Nano Res. 7 (8), 1232−1240.

Zhou, D., Liu, F., Lu, H., Chen, D., Ren, F., Zhang, R., et al., 2014b. High-temperature single photon detection performance of 4H-SiC avalanche photodiodes. IEEE Photon. Technol. Lett. 26 (11), 1136−1138.

Zou, R., Zhang, Z., Liu, Q., Hu, J., Sang, L., Liao, M., et al., 2014. High detectivity solar-blind high-temperature deep-ultraviolet photodetector based on multi-layered (100) facet-oriented β-Ga2O3 nanobelts. Small 10 (9), 1848−1856.

Low-temperature grown gallium arsenide (LT-GaAs) high-speed detectors

8

Marc Currie
Optical Sciences Division, Naval Research Laboratory, Washington, DC, United States

8.1 Introduction

Creating short carrier lifetimes in semiconductors has been achieved by noncrystalline material growth or by inducing damage into an otherwise crystalline material. Some of the first experimental evidence of picosecond photoresponse (and carrier lifetime) was demonstrated by amorphous growth of silicon in a photodetector (Auston et al., 1980). This remarkable demonstration reduced photodetector response times to 40 ps, and was explained by rapid relaxation of photoexcited carriers to localized states. Later experimental efforts employed (oxygen) ion implantation into silicon-on-sapphire to reach a 600 fs photoexcited carrier lifetime (Doany et al., 1987). The explanation was trapping of the photoexcited carriers at states induced by the oxygen implantation, and the trapping time decreased linearly with the implantation density.

Around the same time, buffer layers were being created for gallium arsenide (GaAs) devices. The GaAs substrate can conduct enough current to modulate the space-charge region in field-effect transistors. This can reduce the source-drain current (a process known as backgating). Growing a buffer layer of low-temperature-grown GaAs (LT-GaAs) isolates the substrate and eliminates backgating (Smith et al., 1988). This buffer layer has been successfully used in metal−semiconductor field-effect transistors (MESFETs) and is potentially useful in metal-insulator-semiconductor field-effect transistors (MISFETs), high-electron-mobility transistor (HEMTs), and other field-effect devices (Smith et al., 1988).

This produced investigations into the properties of these LT-GaAs buffer layers. Some of the early research found that arsenic antisites were predicted to produce deep level donors (Bachelet et al., 1983), as well as metastable states when optically excited (Dabrowski and Scheffler, 1989). Experimental efforts revealed arsenic precipitate formation whose size ranged from 2 to 10 nm (Melloch et al., 1990).

The drive for producing shorter photoexcited carrier lifetimes in GaAs would follow the earlier experiments performed using defects in silicon. This new method of low-temperature growth of GaAs soon established even shorter responses than in silicon, with 460 fs electrical pulsewidths (Warren et al., 1990), and <400 fs photoexcited carrier lifetimes (Gupta et al., 1991). Remarkably, this same LT-GaAs material exhibited a sub-bandgap photoresponse at 1300 nm explained by internal Schottky barriers from the arsenic precipitates, with a speed limited by test equipment to 50–100 GHz (Warren et al., 1991).

The rapid response times of LT-GaAs photodetectors led to production of pulsed THz radiation from these devices. Soon, LT-GaAs photoconductive switches (PCSs) were used to realize all-optoelectronic, 1-THz continuous-wave (cw) imaging systems (Siebert et al., 2002). The signal-to-noise ratio was greater than 100:1 and had a similar performance to that of THz pulsed systems.

Another recent interesting application has been in magnetic spin dynamics where LT-GaAs PCSs achieved high fields and short pulses (0.6 T, 3 ps pulses), which rivals those pulses from linear accelerators (Wang et al., 2008).

In addition to photodetectors, the electro-optic properties of LT-GaAs (as well as AlGaAs) showed initial improvement over standard GaAs. Enhancements of $2-6\times$ ($6-12\times$) in electroabsorption figure of merit for LT-GaAs (LT-AlGaAs) were explained as arising from excitonic states sensitive to local fields created by arsenic precipitates (Nolte et al., 1993).

Controlling the properties (e.g., carrier lifetime) in low-temperature-grown materials is essential in enabling devices for specific applications. While our understanding is progressing, many groups are still producing widely varying results. These previous studies are summarized here to provide a guide to understanding and comparison of techniques, measurements, and models for producing more consistent low-temperature-grown materials.

8.2 Attributes of low-temperature-grown photodetectors

8.2.1 Growth temperature

Low-temperature-grown semiconductor materials operate as very high speed photodetectors due to their very short excited carrier lifetime as compared to regular temperature grown materials. In GaAs the temperature range for normal growth temperatures are around 580°C–600°C (Look et al., 1994). Experiments with low-temperature growth have been anywhere from 180°C to 300°C. This is followed by intermediate growth temperatures from 350°C to 500°C (Nabet et al., 1994). While the samples are tested as-grown, annealing is often performed at around 600°C to alter the material properties (more on this later). Figure 8.1 shows the measured carrier lifetimes as a function of the growth temperature. The shaded symbols show lifetime after annealing, while the open symbols show the as-grown lifetime. All of the lifetime data are for electron lifetime except the data show by the asterisk which measured the hole lifetime (after annealing). The overall trend from this figure is that increasing growth temperature appears to increase the carrier lifetime. A more thorough discussion of carrier lifetime occurs later in this section.

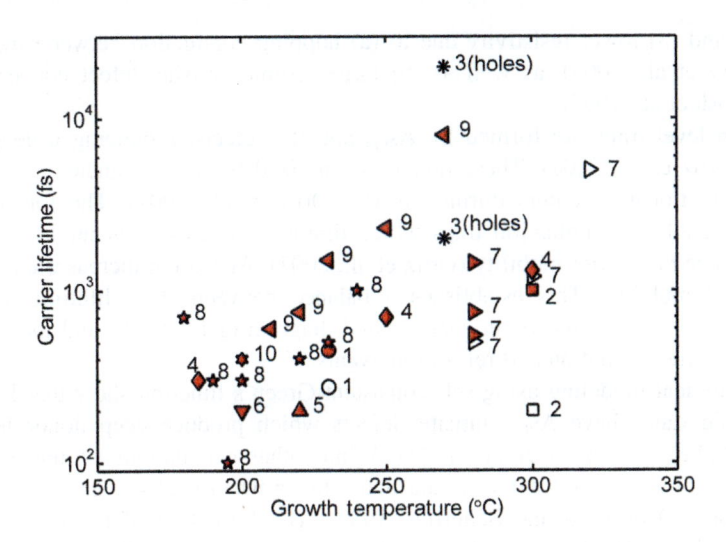

Figure 8.1 Measured carrier lifetimes as a function of LT-GaAs growth temperature. 1: (Pastor et al., 2013), 2: (Prabhu et al., 1997), 3: (Adomavičius et al., 2003), 4: (Loukakos et al., 2001), 5: (Zamdmer et al., 1999), 6: (Gregory et al., 2005), 7: (Stellmacher et al., 2000), 8: (McIntosh et al., 1997), 9: (Sosnowski et al., 1997), 10: (Gupta et al., 1991).

LT-GaAs has been shown to be crystalline but non-stoichiometric, with ~1% excess arsenic. Regardless of growth temperature, stoichiometric LT-GaAs can be fabricated by careful control of the Ga/As flux ratio (Fukushima et al., 2001). Furthermore, adding dopants can increase the As point defects, and thus create and modify carrier traps. In addition, the carrier trap density increases as the flux ratio moves from the stoichiometric ratio (as measured with a fixed dopant concentration) (Fukushima et al., 2001).

Arsenic defects and clustering occur as a result of low-temperature growth. GaAs grown by molecular beam epitaxy (MBE) with low substrate temperatures (250°C) was observed to contain arsenic precipitates. The density of these precipitates was found to vary with the substrate temperature (Melloch et al., 1990). Annealing was observed to convert defects into arsenic clusters (Loukakos et al., 2001). Carrier lifetimes were shown to be proportional to the arsenic cluster spacing and inversely proportional to cluster size (Loukakos et al., 2001; Melloch et al., 1995). This modification of the distribution of arsenic defects and clusters produced changes in the resistivity (McIntosh et al., 1997).

8.2.2 Antisite defects and carrier traps

In compound semiconductors (like GaAs), atoms can exchange sites in the lattice creating an antisite pair. This is a point defect in an otherwise perfectly ordered crystal lattice. Point defects and clusters form with low-temperature growth and cause GaAs to (1) create carrier traps, (2) lower carrier mobilities (Ortiz et al.,

2007), and (3) lower resistivity due to (a) hopping conduction between trap states (Gregory et al., 2003), as well as (b) large arsenic antisite defect concentrations (McIntosh et al., 1997).

Deep level traps are formed by As_{Ga} antisite defects, producing a deep donor band (Ortiz et al., 2007). These donors are ionized by V_{Ga} vacancies acceptors or by unintentional acceptors during growth (Ortiz et al., 2007). The ionized deep donors enable recombination traps which dramatically lower carrier lifetimes and also reduce the carrier mobility (Ortiz et al., 2007). Annealing increases carrier lifetime and mobility. This establishes a balance between short lifetime and high mobility. Figure 8.2 shows an energy band diagram of LT-GaAs and indicates the optical carrier excitation and relaxation events.

Theoretical modeling using self-consistent Green's function show that LT-GaAs electronic states have As_{Ga} antisite defects which produce deep donor levels of 0.27 ± 0.2 eV (Bachelet et al., 1983). While other calculations of the electronic structure show the EL2 defect is the main donor when GaAs is grown in As-rich conditions (Dabrowski and Scheffler, 1989). The formation of the EL2 defect is possibly due to (1) As_{Ga} antisite (arsenic atom occupying a gallium site), (2) As_{Ga} with vacancies, and/or (3) As_{Ga} antisite combined with an arsenic interstitial.

An experiment using deep-level, transient spectroscopy observed a trap at 0.65 eV below the conduction bandedge and identified it with the As_{Ga} antisite defect in relation to the EB4 deep level trap in GaAs (Look et al., 1994). The calculated capture cross section for this trap was 1.5×10^{-15} cm^2. At high concentrations (10^{20} cm^{-3}), deep level trap states that demonstrate EL2-like defects can dominate the observed photoexcited carrier dynamics. Carrier populations in the bottom of the conduction band relax to form populations in the trap states and these allow transitions to excited states higher in the conduction band. Benjamin et al. (1996)

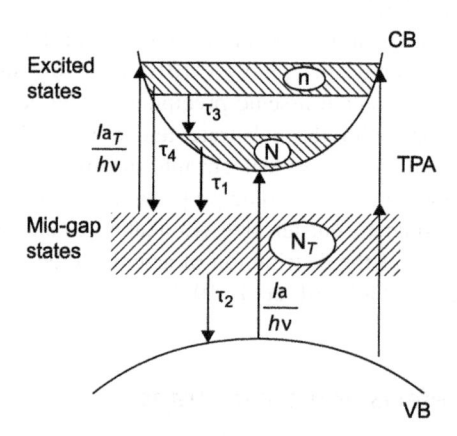

Figure 8.2 Energy-band diagram LT-GaAs with optical carrier excitation and relaxation events, where VB and CB are the valence and conduction bands, and α is the mid-gap state absorption coefficient.
Source: From Benjamin et al. (1996).

model these dynamics and describe the experimentally measured absorption dynamics of 150-fs pulses at 870 nm, which is at the bandedge of GaAs.

Recently, antisite defect concentration was determined using X-ray diffraction (Pastor et al., 2013). For LT-GaAs a second diffraction peak emerges near the peak for semi-insulating GaAs and is associated with the As_{Ga} antisites. After annealing the second peak due to the As_{Ga} antisites disappears leaving only the peak corresponding to stoichiometric GaAs (Pastor et al., 2013). Gregory et al. (2003) previously observed how the X-ray peak angle decreases with increasing annealing temperature, until it reaches that of stoichiometric GaAs. These are further observations that demonstrate the critical role annealing plays in the material properties of LT-GaAs.

8.2.3 Annealing

After low-temperature growth, annealing tends to increase the resistivity, which is a benefit for many photonic applications, with measured values as high as 10^6 Ω cm (McIntosh et al., 1997). However, annealing has also been shown to increase the carrier lifetime (although some controversy exists over the interpretation of this increase), and this increased lifetime reduces the performance of many photonic devices.

While the annealing process involves many subtleties, the following summarizes the basic changes which impact carrier lifetime. Before annealing, photoexcited carriers may become trapped rapidly (e.g., via point defects), however, carrier recombination may take substantially longer. In LT-GaAs this recombination lifetime decreases after annealing. In post annealed samples, the point defects disappear, and thus cannot provide the short carrier lifetimes (Lochtefeld et al., 1996). Carrier recombination exhibits a single exponential in annealed samples, demonstrating that, unlike trap states, arsenic precipitates are the reason for the short carrier lifetimes (Lochtefeld et al., 1996).

As-grown LT-GaAs shows additional light absorption below the GaAs absorption gap (Pastor et al., 2013). This sub-bandgap light absorption is attributed to the As_{Ga} antisite defects in samples before annealing. After annealing the sub-bandgap absorption disappears due to the reduction of As_{Ga} defects and the formation of As nanoinclusions in stoichiometric GaAs (Pastor et al., 2013). Annealing forms arsenic precipitates from point defects in the LT-GaAs. Longer annealing converted more of the point defects into arsenic clusters. Using transmission electron microscopy (TEM), the spacing and size of the arsenic clusters were determined (Loukakos et al., 2001).

Most experiments anneal samples at 600°C for about 10 min. To investigate the properties of LT-GaAs change at different anneal temperatures, samples were annealed at 600°C−900°C for 30 s, which is short but sufficient to form precipitates from excess As (Prabhu et al., 1997). This study showed that carrier lifetimes increased with increasing anneal temperature. Their interpretation was that the short (picosecond and subpicosecond) lifetime was related to the arsenic point defects, and that higher temperature annealing removes these defects and restores the typical 1-ns recombination times of bulk GaAs.

To maintain arsenic concentration while annealing, a 10-min rapid thermal anneal was performed in nitrogen while the LT-GaAs was in contact with semi-insulating GaAs wafer for surface passivation. Using this technique, investigation of annealing temperatures between 300°C and 600°C was performed. The results demonstrate that subpicosecond lifetimes are maintained up to nearly 550°C, during which time the resistance increases by nearly two orders of magnitude (Gregory et al., 2003), see Figure 8.3. The short lifetimes combined with high resistivity are important in many photodetection applications.

8.2.4 Mobility and resistivity

In early studies, the Hall mobility of annealed LT-GaAs was shown to be large (1000 cm^2/V s), while the mobility estimated from photocurrent measurements were nearly an order of magnitude lower (120–150 cm^2/V s) (Gupta et al., 1991). Using ultrafast optoelectronic measurements at 800-nm wavelength, the electron and hole mobilities in Be-doped LT-GaAs (280°C, annealed at 600°C) were 540 and 90 cm^2/V s, respectively (Eusèbe et al., 2005). This resulted in a fast response at low fluence, but at higher fluences trap saturation reduces the electron recombination time to that of the holes, since excited electrons wait for traps to free via hole capture (Eusèbe et al., 2005), see Figure 8.4.

For photodetectors, high mobility provides high efficiency, since the higher mobility allows more photoexcited carriers to get collected at the electrodes. The relatively low mobility of LT-GaAs can be enhanced by incorporating a thin

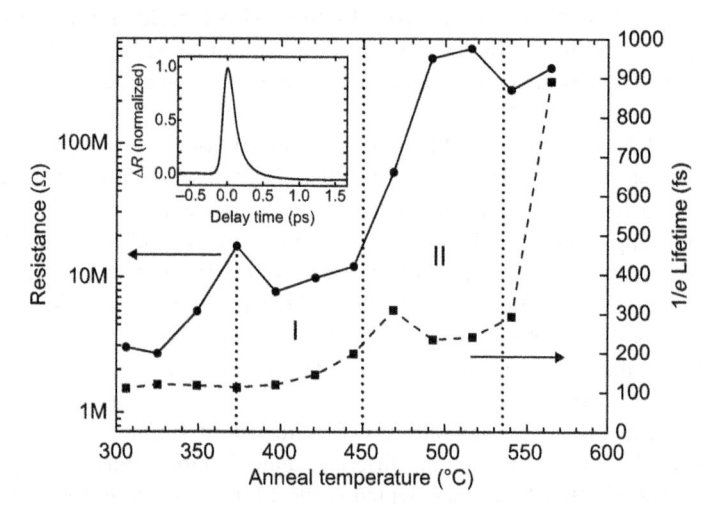

Figure 8.3 Measured resistance and lifetime for THz antenna fabricated on LT-GaAs. The inset shows the time-resolved, transient reflectivity used for determining the lifetime.
Source: From Gregory et al. (2003).

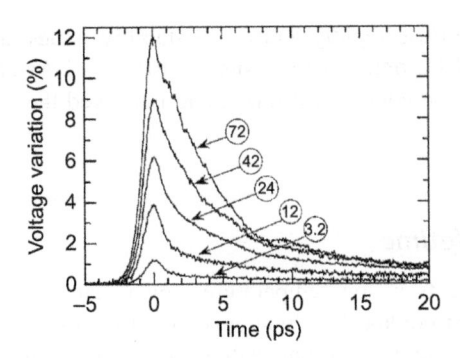

Figure 8.4 Optical pulse-energy dependent voltage transients measured from a LT-GaAs optoelectronic switch. Low fluence produce fast responses, but higher fluence saturates trap and reduces the electron recombination time to that of the holes.
Source: From Eusèbe et al. (2005).

channel layer with higher mobility above the LT-GaAs layer. This was shown to increase the carrier collection efficiency while maintaining the short lifetimes necessary for high-speed photodetector operation (Currie et al., 2011).

High resistivity is important in devices, for example, to enable large biases and reduce dark currents (Gregory et al., 2003). Annealing modifies the distribution of arsenic defects and clusters and increases the resistivity as high as $10^6 \, \Omega \, cm$ (McIntosh et al., 1997). As-grown samples have low resistivity due to (1) hopping conduction between trap states (Gregory et al., 2003), and (2) large arsenic antisite defect concentrations (McIntosh et al., 1997). Upon annealing arsenic precipitates form in LT-GaAs and behave like Schottky barriers. The annealed LT-GaAs can appear semi-insulating for high precipitate densities in which overlapping Schottky depletion regions could create the semi-insulating behavior (Warren et al., 1992, 1990).

The high resistivity of LT-GaAs has also been attributed to space-charge effects near a GaAs/LT-GaAs junction region. This space-charge results in inhomogeneous electric fields in the LT-GaAs, and can result in high-field regions that suppress hoping conductance, leading to resistivities $>10^8 \, \Omega \, cm$ even with low applied field (Kordos et al., 1997).

The contact resistance with metal electrodes is also an important parameter in photodetector performance. Creating ohmic contacts on GaAs is difficult due to high doping densities and the formation of native oxides on the surface. However, ohmic contacts to LT-GaAs using non-alloyed metals have achieved contact resistances of 10^{-3} and $10^{-7} \, \Omega \, cm^2$. Carrier transport models demonstrate that dense EL2-like deep donor bands in the LT-GaAs allow carriers bypass the Schottky barrier and move directly from the contact metal to the deep donor state producing room temperature contact resistances as low as $10^{-3} \, \Omega \, cm^2$ (Yamamoto et al., 1990). In addition, using methods that passivate the high space charge density layer on the surface resulted in even lower contact resistances of $10^{-7} \, \Omega \, cm^2$ (Patkar et al., 1995). To achieve this, Be doping $>5 \times 10^{19}$ was used. However, since interstitial Be can diffuse rapidly and cause stability issues, a post doping anneal drove the interstitial Be onto acceptor sites in the LT-GaAs and thereby increased the doping, lowered the contact resistance, and improved the stability (Patkar et al., 1995).

8.2.5 Carrier lifetime

The creation of defects in low-temperature-grown materials decreases the carrier lifetime, as discussed previously. Upon annealing this lifetime increases again, and for longer and higher temperature annealing this value approaches the lifetime for standard growth materials. Several studies, detailed below, use optical pump-probe techniques to observe the carrier lifetimes. These are mostly the electron lifetime, but hole traps and lifetimes were also investigated (Adomavičcius et al., 2003). To further understand the physical processes, the effects of high optical fluence, material doping, annealing, and electrical bias on carrier lifetime are also reviewed. Before beginning it is worthwhile to note one caveat on measured lifetimes: surface recombination can significantly reduce carrier lifetimes. To reduce this influence and concentrate on the material performance, cap layers have been implemented in more recent studies (Ortiz et al., 2007; Stellmacher et al., 2000).

Material growth temperatures are correlated to relaxation times (see Figure 8.1). In femtosecond optical reflectivity measurements, the shortest measured relaxation time (90 fs) occurred for growth at 195°C and annealed at 580°C for 10 min (McIntosh et al., 1997). Growth at temperatures above this produced a lower concentration of arsenic defects, while growth at temperatures below this produced strain, extended defects and polycrystalline material. Therefore, at higher temperatures the reduced defect concentration results in fewer traps and leads to a longer response, while the lower temperatures may show longer response times due to a variety of mechanisms, such as larger clusters, incorporation as As_4 or As_8, but are not well understood at this time (McIntosh et al., 1997).

While previous optical measurements were performed near 800 nm (1.55 eV), these photon energies are far (energetically speaking) from the bandedge of GaAs. In contrast, optical pump-probe measurements were performed at 860 nm (1.44 eV), a wavelength selected due to its proximity to the GaAs bandedge at 873 nm (1.42 eV) to minimize any response due to carrier cooling effects (Loukakos et al., 2001). Here, electron lifetimes were studied by observing the ultrafast response which results from carrier trapping on the arsenic clusters. For the TEM measurement and to isolate the low-temperature grown sample, an aluminum arsenides (AlAs) sacrificial layer was grown and the LT-GaAs layer was lifted off and bonded (via van der Waals interaction) to glass slides. Lifetimes of 300, 700, and 1300 fs were measured for growth temperatures of 185°C, 250°C, and 300°C, respectively. The carrier lifetimes were compared with the average spacing R and average size (alpha) of the arsenic precipitates. The lifetime dependence was proportional to the spacing cubed and was inversely proportional to the size ($\tau \sim R^3/\text{alpha}$) (Loukakos et al., 2001), see Figure 8.5. This expanded upon previous

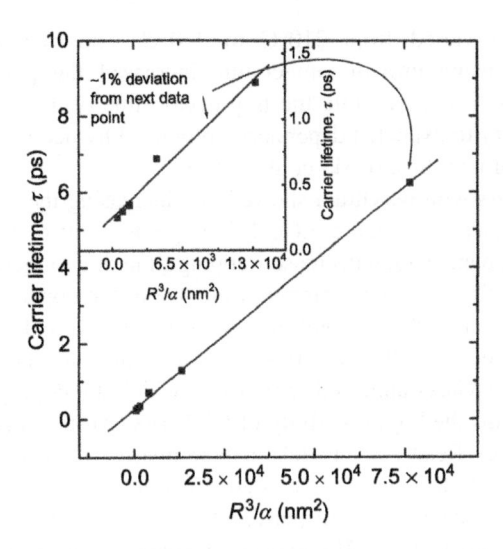

Figure 8.5 The measured electron lifetime of LT-GaAs is proportional to the spacing cubed and was inversely proportional to the size ($\tau \sim R^3/\alpha$).
Source: From Loukakos et al. (2001).

studies which did not take into account the size, and found the lifetime proportional to the square of the spacing between arsenic clusters (Melloch et al., 1995).

Deep donor as well as deep acceptor concentrations determine the carrier lifetimes in LT-GaAs. However, combinations of these concentrations do not produce unique lifetimes. Experimentally driven models for the carrier dynamics in LT-GaAs demonstrate that the same lifetime can occur for different donor and acceptor concentrations (Stellmacher et al., 2000). These models based upon Shockley-Read-Hall recombination recreate the experimentally observed lifetimes by varying only one model parameter: the number of acceptors (Stellmacher et al., 2000). While different donor and acceptor concentrations may provide similar lifetimes, other properties such as transport (e.g., mobility) and nonlinear optoelectronic effects (e.g., saturation) may not be similar. However, the model of Stellmacher et al. shows that two parameters can control the carrier dynamics, growth, and annealing temperatures. These control the concentrations of deep donors and acceptors and confirm that deep donors are the main recombination center, but that the contribution from the acceptor concentration is needed to accurately determine the carrier lifetime (Stellmacher et al., 2000).

To more fully characterize the material response, the large optical signal regime needs to be studied as well. For large photoexcited carrier densities (i.e., as the input fluence was increased) the relaxation time increased due to filling of the trap states at high fluence. Optical pump-probe studies used an 800-nm pump wavelength, but probed at 860 nm to remove carrier cooling effects from the measured decay time, and demonstrated almost linear decay when the photoexcited carrier density was much greater than the trap density (Sosnowski et al., 1997). Growth temperatures were varied from 210°C to 270°C. The modeled time response showed that increasing growth temperature increased the photoexcited electron decay time from 0.6 to 8 ps while the trap decay time decreased from 20 to 1 ps, the latter is contrary to expected dependence produced by decreased defects at higher growth temperatures (Sosnowski et al., 1997).

Doping LT-GaAs with beryllium showed that increasing the Be-dopant increased carrier trapping times (see Figure 8.6). This experiment also showed the complex interaction with dopants which decrease trapping time due to increased ionized antisites in opposition to an increase trapping time due to the dopant decreasing antisite density (Krotkus et al., 1999). In addition, multiple optical and optoelectronic techniques demonstrated a 520 fs electron trapping time in moderately Be-doped (3×10^{17} cm^{-3}) LT-GaAs annealed samples, in which the doping hinders As precipitates and maintains the high resistivity of LT-GaAs (Krotkus et al., 2002).

Using a Shockley-Read-Hall model for nonradiative recombination, Ortiz et al. (2007) found that the electron lifetime is governed by the ionized trap density, with the trap absorption causing a small perturbation to this lifetime. In this model the influence of the holes (from V_{Ga} vacancies or unintentional acceptors) varies depending on wavelength and pump power. For optimal performance high excess arsenic concentrations should be combined with short (~ 1 min) annealing at temperatures below 600°C (Ortiz et al., 2007).

In addition to the electron dynamics, hole trapping times were investigated by optically probing the intervalence band states (i.e., light hole/heavy hole populations) using 9-μm-wavelength optical pulses. LT-GaAs as-grown samples exhibited a 2 ps trapping time, and this time increased by an order of magnitude after annealing at 600°C for 10 min (Adomavičcius et al., 2003). In addition in the as-grown samples doping with silicon reduced this time to 1 ps, while doping with beryllium

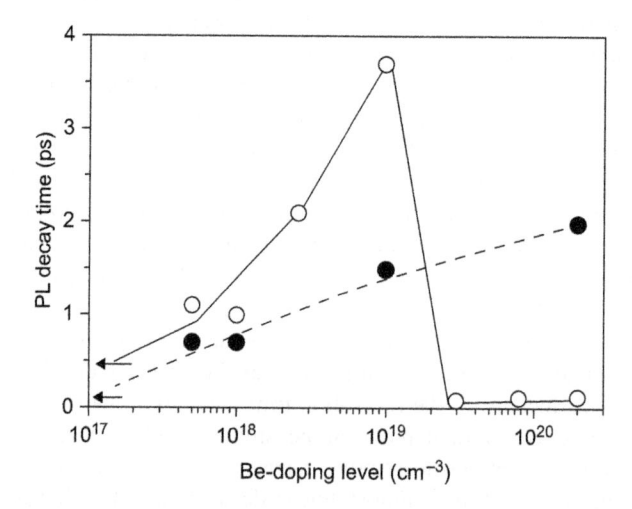

Figure 8.6 Be-doping concentration influences the carrier trapping time in LT-GaAs.
Source: From Krotkus et al. (1999).

increased this time to about 10 ps (Adomavicčius et al., 2003). After annealing the Be-doped sample showed an order of magnitude increase in hole trapping time (no results were reported on an annealed Si-doped sample). These results suggest that the mechanism for trapping is due nonequilibrium hole capture at neutral arsenic antisites (As_{Ga}) (Adomavicčius et al., 2003).

While as-grown samples showed short carrier lifetimes, annealing changes the characteristics of the low-temperature growth, as discussed previously. Carrier lifetime increases with annealing, due to decreased point defects (Prabhu et al., 1997). Carrier dynamics probed using THz pulses show shorter lifetimes than those probe using visible and near infrared (IR) light (Prabhu et al., 1997). The THz/far-IR wavelengths are low energy probes and are sensitive to the carrier conductivities. In GaAs, since the electron mobility is an order of magnitude greater than that of the holes, the probed response is dominated by the electron dynamics (Prabhu et al., 1997).

While annealing increases lifetime, there are tradeoffs for material performance (e.g., as THz emitter/detectors). The resistivity, which is important in devices for minimizing dark currents, also increases with annealing. Gregory et al. (2003) observe that these do not occur at the same rate, that is, the lifetime remains fairly constant up to anneal temperatures of 500°C (see Figure 8.3).

In contrast to short lifetimes in the unbiased low-temperature grown materials, long lifetimes can exist under electrical bias. This arises from trapped carriers in the space-charge region near the contacts, with screening times of 1–40 ns (Loata et al., 2007). This can adversely impact the performance in THz devices (especially cw, but also low energy pulsed radiation).

Increased response times have also been observed in LT-GaAs PCSs with increasing voltage bias. This increase in response time at high bias can be explained by a reduction in the carrier caption cross section due to carrier heating and Coulomb-barrier lowering. A similar effect has also been found in traps in SiO_2. In contrast with suggestions of space-charge current limiting, the photo-induced current scales linearly with optical intensity (suggesting photogenerated mechanisms), and the observed threshold voltage is $1000 \times$ greater than expected for an injection-limited (space-charge) effect (Zamdmer et al., 1999).

From reviewing the material properties related to carrier lifetime, it is important to realize the trade-off in material properties when considering an application. The important parameters are growth temperature, material doping, annealing temperature (as well as annealing time), electrical bias, optical fluence, and surface recombination (minimized by adding capping layers). All of these should be balanced in our application of LT-GaAs to photodetectors.

8.2.6 Optical absorption at mid-gap states

Absorption of photons whose energy is below the GaAs bandedge (1.42 eV) can occur due to mid-gap defect states, which are about 0.7 eV below the conduction band. This enables absorption in photodetectors made from LT-GaAs from the visible (with photon energies greater than the GaAs bandgap) to about 1600 nm

(0.775 eV). However, at intermediate wavelengths, between the energies of the GaAs bandedge and the mid-gap states, nonlinear behavior can occur. The observed nonlinear behavior broadens the optoelectronic time response by a factor of three with increasing optical power, as shown in Figure 8.7.

One explanation is that photons with energies between the bandedge and mid-gap states generate carriers with sufficient excess energy (~ 0.3 eV) that they can scatter out of the gamma valley and into the L valley. This hot-electron effect involving the L-valley can lead to longer carrier lifetimes and therefore broaden the time response. While this nonlinearity reduces the ultimate photodetector performance, these devices still have ~ 100-GHz electrical bandwidths (Shi et al., 2004).

The carrier dynamics follow four time regimes: <1, ~ 1, ~ 10, and ~ 100 ps. The shortest is for intraband hot-carrier relaxation, followed by free-carrier trapping in mid-gap states. The slow responses at ~ 10 and ~ 100 ps involve carrier recombination of the mid-gap states, which suggests individual trapping states for electrons as well as holes (Grenier and Whitaker, 1997).

Femtosecond transmission experiments as well as electro-optic sampling measurements probed the carrier relaxation dynamics for 1230 nm, sub-bandgap, excitation of LT-GaAs. The sample fabrication used epitaxial lift-off and transfer to glass substrates to reduce substrate effects. For excitation producing a carrier density $>3 \times 10^{17}$ cm^{-3}, intervalley scattering and hot-electron effects increase the carrier lifetime (Sun et al., 2003). In addition, electric-field dependent studies demonstrate and further confirm increased lifetime from intervalley scattering. Electro-optic sampling device measurements also confirm this effect (Sun et al., 2003).

These results show that LT-GaAs is useful not only to make fast GaAs photonic devices but also for increasing its optical detection bandwidth. LT-GaAs extends the optical detection wavelength range from that of GaAs detectors (500–870 nm) to nearly 1600 nm, a range covered by InGaAs detectors.

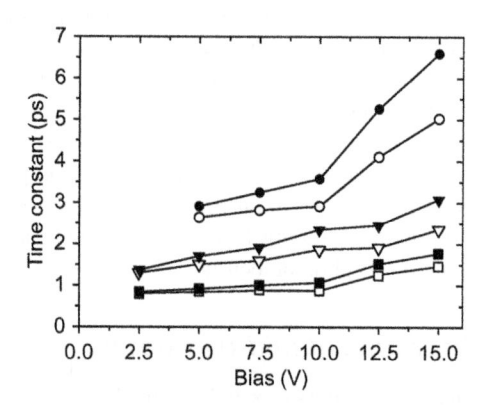

Figure 8.7 The optoelectronic time response of a LT-GaAs photodetector nonlinearly broadens by a factor of three with increasing optical power.
Source: From Sun et al. (2003).

8.3 Material systems

While this chapter has focused on LT-GaAs it is important to recognize that while this low-temperature growth technique for modifying the material properties began with GaAs it has spread to a variety of other materials such as InGaAs, GaAsSb, and InP. In addition, various growth techniques have been developed to emphasize particular material properties, as well as enable cost-effective commercial material and device fabrication.

To fabricate LT-GaAs MBE is widely used. However, metal organic chemical vapor deposition (MOCVD) low temperature growth (300°C–400°C) of GaAs on a semi-insulating GaAs substrate has achieved comparable characteristics to those of LT-GaAs fabricated via MBE (Boutros et al., 1995). In addition, photoassisted MOCVD growth using an argon laser selectively grew the LT-GaAs only in the laser irradiated areas, thus providing the potential for in situ, maskless integrated devices (Boutros et al., 1995).

InP grown at low temperatures has shown similar photoexcited carrier dynamics as those in LT-GaAs. Carrier trapping times of 500 and 1600 fs (as-grown) were measured for growth temperatures of 200°C and 300°C, respectively (Kostoulas et al., 1995).

Using InGaAs, less than 500 fs trapping times were achieved using 1060 nm excitation. The material also maintained a high resistance, 10,000 Ω cm (Baker et al., 2004). By incorporating multiple quantum well (MQW) of InGaAs/InAlAs, picosecond carrier lifetimes were demonstrated along with 1.5 ps optical switching at 1550 nm with low-energy (2 pJ) pulses (Takahashi et al., 1994, 1996). And as in LT-GaAs doping experiments using beryllium-doped InGaAs/InAlAs multiple quantum wells grown by MBE at low temperatures (250°C) demonstrated subpicosecond (~ 100 fs) carrier lifetimes with high sheet resistance ($10^5 - 10^7$ Ω/sq) and good (500–1800 cm^2/V s) carrier mobility (Kostakis et al., 2012). This material system shows potential for THz emitter and detector applications with photomixers in the 1500-nm region. For further application to THz photomixers, GaAsSb grown at low temperatures with an excess of Sb showed picosecond carrier lifetimes (Wallart et al., 2010).

Alternative methods to achieve short carrier lifetimes used material defects through ion bombardment or the growth of amorphous or polycrystalline material. To create defects some groups have tried Be doping using He-plasma-assisted epitaxy. Optical pulsed response at 1550 nm have demonstrated picosecond and subpicosecond carrier lifetimes in InGaAsP with heavy beryllium doping ($10^{18} - 10^{19}$ cm^{-3}) using He-plasma-assisted MBE (Kang et al., 1998; Qian et al., 1997a,b, 2000). This can be performed under normal growth conditions, therefore, unlike low-temperature-grown materials their properties are relatively unaltered by annealing (Qian et al., 2000). This can be beneficial to minimize processing steps, but, it also lacks another control for optimizing the performance.

8.4 Principle of operation for LT-GaAs photodetectors

Low-temperature-grown semiconductor materials achieve ultra-high-speed operation by creating defects and trapping carriers to produce shorter carrier lifetimes in contrast to conventionally grown materials. As discussed in the previous sections on material properties, with these shorter lifetimes comes a decrease in the mobility. While annealing changes the mobility and resistivity it also alters carrier lifetimes requiring optimization of material properties to achieve the most favorable balance.

Let us briefly review these parameters. Carrier trapping time was determined as a function of spacing and size of As precipitates (Loukakos et al., 2001). The carrier trapping and carrier lifetime were influenced by annealing. The material resistivity increased with annealing (good for device performance) more quickly than the carrier lifetime increased (which reduces device speed) (Gregory et al., 2003). High contact resistivity in LT-GaAs was attributed to space-charge regions (Kordos et al., 1997) and was found to be reduced by doping and subsequent annealing as well as by surface passivation (Patkar et al., 1995).

The epitaxial growth of LT-GaAs on top of a buffer layer (e.g., AlAs) was lifted off the substrate and deposited on other host substrates (Yablonovitch et al., 1987). Several studies used this technique to isolate the response of the LT-GaAs from (usually) GaAs substrates. This enabled material measurements (such as TEM) as well as improving device performance (dark current, resistivity, dielectric load, etc.) by transferring to alternative substrates (Heiliger et al., 1996; Keil et al., 1997; Loukakos et al., 2001; Mikulics et al., 2003). Using this summary will help understand the benefits and trade-offs in the device technologies presented in the next section.

8.5 Photodetector technologies

Enhanced operation over traditional GaAs photodetectors is achieved using LT-GaAs. There are four main photodetector types reviewed here: (1) PIN photodiode, (2) PCS, (3) metal-semiconductor-metal (MSM), and (4) waveguide detectors. Each style balances particular aspects of photodetection to achieve unique performance characteristics.

8.5.1 PIN photodiodes

For use in PIN photodetectors, the intrinsic (i) region is the active absorption region, and for increasing device speed this i region is fabricated with LT-GaAs. For operation between 1300 and 1500 nm, a 1-µm-thick LT-GaAs (grown at 225°C and annealed at 600°C) was used for the intrinsic layer. The LT-GaAs was found to have a 0.7 eV Schottky barrier due to As precipitates, thus, potentially enabling operation at wavelengths as long as 1700 nm. While the responsivity of this device was low (~1 mA/W), the dark current for the 100-μm-diameter photodetector was also low (~10 nA @ 10 V_{bias}). This potentially allowed operation at higher bias voltages to improve the collection efficiency. The photodetector speed was shown to be >2 GHz, and was limited by the measurement system rather than the photodiode (Warren et al., 1991).

In a more recent experiment at 1550 nm, a PIN photodetector was again created using LT-GaAs as a lightly-doped (n- = 10^{16} cm^{-3}) intrinsic (i) absorption layer for 1550 nm light (Butun et al., 2004). Since the sub-bandgap absorption efficiency in LT-GaAs is much lower than light above the bandgap, a resonant cavity was created using a Bragg mirror buried below the n-layer. This enhanced the device performance by 750% and demonstrated a measured quantum efficiency of 0.07% at 1548 nm, which is larger than previous PIN photodetectors but lower than their simulated predictions, as shown in Figure 8.8. This geometry produced an 11.2 GHz frequency response (3 dB) with pulse rise times of 12−16 ps and fall times of ~80 ps. An issue arose with excitation power of the 1550 nm, 1 ps source: increasing the excitation intensity increased the rise time from 12 to 16 ps (with a roughly constant fall time). This nonlinear response has been observed before in time-resolved, near-IR measurements of the material (Grenier and Whitaker, 1997; Shi et al., 2004). The authors suggest that both this nonlinear optical response with increasing power as well as their lower that expected quantum efficiency were due slowly emptying trap states in the LT-GaAs (Butun et al., 2004).

A similar PIN structure to Buntun et al. was fabricated with improved responsivity of 7.1 mA/W at 1563 nm. However, the resonant cavity structure only provides this enhancement over a narrow (4 nm) spectral region (Han et al., 2006).

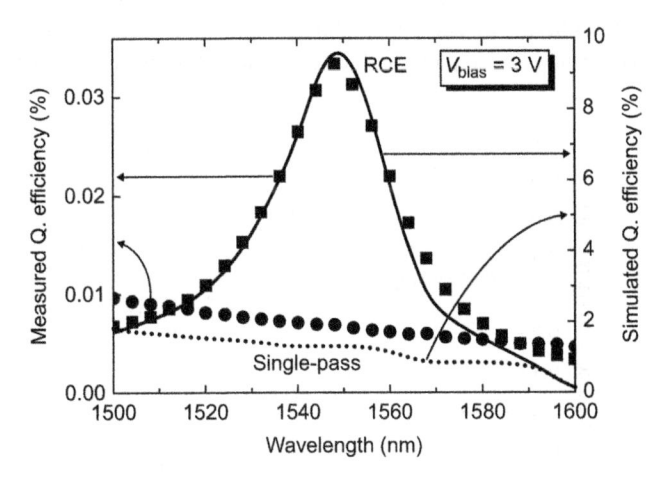

Figure 8.8 Quantum efficiency (vs wavelength) for LT-GaAs PIN photodetectors (circles). To increase the sub-gap detection efficiency a resonant-cavity enhanced (RCE) PIN detector was also fabricated (squares), and compared with models, solid and dotted curves are calculated values with and without the RCE.
Source: From Butun et al. (2004).

8.5.2 Photoconductive switches

One of the earliest optoelectronic signals generated using LT-GaAs was performed by creating the PCS from a 10 μm gap in a gold transmission line (Gupta et al., 1991), as shown in Figure 8.9. Optical pulses at 620 nm wavelength excited the gap and produced a 0.6 ps FWHM electrical transient (as measured by an electro-optic sampling system) with a 0.7 V peak response. However, the sample's dark resistance was low and produced poorer performance than expected (Gupta et al., 1991).

In other devices, epitaxial grown LT-GaAs of 0.5–1.5 μm thickness were lifted off and were attached to SiO_2/Si substrates by van der Waals bonding. Recessed wells were fabricated in the SiO_2/Si substrate to accommodate the thickness of the LT-GaAs so that continuous Ti/Au coplanar transmission lines could be evaporated. This processing enabled low dark currents and high breakdown voltages. In addition, this method offers the advantage of integrating these PCS in any position on wafer scale devices (Mikulics et al., 2003).

Additional work fabricated recessed electrical contacts into the LT-GaAs layer. This provided faster responses (1 ps FWHM) due to a 3× reduction in fall time and with a 30% increase in responsivity (0.12 A/W). This is attributed to the recessed contact geometry which provides a significant reduction in capacitance as well as more efficient carrier collection due to the enhanced electric-field profile (Mikulics et al., 2006).

A further sensitivity enhancement was achieved by fabricating a PCS using alloyed AuGe eutectic contacts. This changed the electric field within the LT-GaAs and demonstrated a 200% improvement in the device's responsivity as compared with non-alloyed metal contacts (Mikulics et al., 2008).

Epitaxial low-temperature growth (250°C) with lift-off and transfer to plastic (polyethylene terephthalate) substrates showed an 850 fs time response (FWHM). The PCS was formed by a patterned 20 μm gap CPS made of Ti/Au. The low

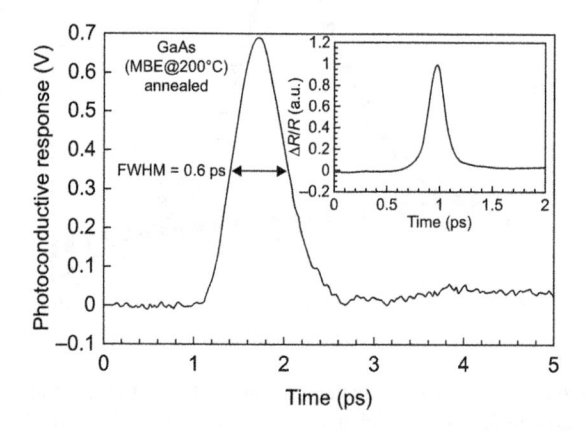

Figure 8.9 Ultrafast optoelectronic transient generated by LT-GaAs PCS, measured by electro-optic sampling. The time-resolved reflectance measurement is shown in the inset. *Source*: From Gupta et al. (1991).

(40 nA) dark current was probably due to the relatively wide (20 μm) electrode spacing. Since the plastic substrate is flexible, the sample was bent at an angle between 90 and 120 degrees over 100 times without performance degradation in the PCS (Mikulics et al., 2005). This demonstrates capabilities for hybrid, flexible optoelectronic systems.

Further work on the transfer of free-standing LT-GaAs films demonstrated faster responses, at 360 fs (FWHM), producing electrical bandwidths of 1.25 THz (3-dB) with quantum efficiencies as high as 0.07 at 810 nm (Zheng et al., 2003). This free-standing LT-GaAs did not need patterning, it was laid on top of a CPS transmission line. This further increases the applicability since these type of optoelectronic switches can be placed on top of any metalized circuit to inject subpicosecond (THz bandwidth) signals.

To extend the optical bandwidth, sub-bandgap photodetectors for wavelengths as long as 1500 nm were fabricated using 750 nm of LT-GaAs grown on 500 nm of undoped GaAs (Harmon et al., 1995). The tradeoff of carrier lifetime to transit time was engineered to produce a high internal gain, with the sub-bandgap absorption explained by photoemission from arsenic precipitates. The spectral response of the detector showed an order-of-magnitude decrease in responsivity when moving from above to below the bandgap (from 850 to 950 nm). Another order-of-magnitude decrease in responsivity was observed as the excitation was changed from 950 to 1300 nm, and yet another order-of-magnitude decrease from 1300 to 1500 nm, providing 0.1 and 0.01 A/W, respectively (Harmon et al., 1995). The time response (measured above the bandgap) yielded devices with electrical responses of ~ 10 GHz bandwidth (Harmon et al., 1995).

Additional sub-bandgap performance of LT-GaAs at 1550 nm demonstrated 450 fs transient pulses due to two-photon absorption (Erlig et al., 1999). The PCS was fabricated by attaching a coplanar transmission line to the LT-GaAs substrate and illuminating with two optical pulses with a varying time delay. This provided time-resolved data to complement the nonlinear intensity vs photocurrent data. While the responsivity is low (~ 0.5 mA/W), these are fast photodetectors (450 fs response) in the 1550 nm region with a 190 GHz 3 dB bandwidth (Erlig et al., 1999).

As demonstrated, PCS fabricated from LT-GaAs have shown some of the fastest photodetector response times. This comes at the potential expense of responsivity, or collection efficiency. The material resistivity, carrier mobility, and carrier lifetime are essential to the performance of these PCS photodetectors.

8.5.3 MSM photodetectors

MSM photodetectors differ from PCS technology in their metal-semiconductor contact. In general PCS rely on an ohmic contact with the semiconductor, allowing photoconductive gain to increase the photocurrent signal at the potential expense of photodetector speed. On the other hand, MSM photodetectors employ Schottky contacts with the semiconductor, which enable rapid collection of photoexcited carriers, at the expense of the photoconductive gain. Traps in LT-GaAs decrease the response time by trapping carriers at the expense of also reducing the photocurrent. Thus, with photoconductivity thwarted by Schottky contacts and carrier trapping, MSM techniques offer the ability to rapidly sweep out carriers by applying high electric fields in the absorption region.

An early experiment compared semi-insulating GaAs MSM photodetectors to those fabricated on LT-GaAs (grown at 210°C, annealed 600°C 60 min). Using 100 fs optical pulses at 620 nm wavelength, the LT-GaAs device showed half the pulsewidth (0.87 ps FWHM) as that of the GaAs sample (1.5 ps FWHM) for sub-micrometer MSM finger widths of 300 and 100 nm, respectively (Chou et al., 1992). In addition, the bulk GaAs sample showed a 2 ps long tail potentially from photogeneration of carriers deep in the material (Chou et al., 1992).

Subsequent performance of MSM photodetectors made with LT-GaAs was investigated over a range of growth temperatures. Performance metrics varied with two parameters: carrier transit time and trapping/carrier lifetime. For best sensitivity and high-speed performance, the transit and trapping times should be equal. The results show that increasing the growth temperature causes a decrease and then an increase of two orders-of-magnitude in the dark current. At the same time, the optical response of the MSM increased by two orders-of-magnitude and then maintained that response above 350°C. The results suggest that rather than LT-GaAs at 200°C–300°C, intermediate GaAs growth temperatures of 300°C–350°C provide the best photodetector performance (Nabet et al., 1994).

These intermediate GaAs growth temperatures produced photodetectors with substantially reduced dark currents while achieving high carrier mobility (Tousley et al., 1995). Epitaxial GaAs grown at 350°C incorporated in an MSM photodetector showed a responsivity of 0.42 A/W to 800 nm excitation. With 1 mW excitation, the signal-to-dark current ratio was 7000, which was two orders-of-magnitude greater than LT-GaAs grown at 200°C. Time-response measurements show the transit time across the 5 μm gap as 31.5 and 60 ps for the electrons and holes, respectively. At higher optical power the electrical response becomes nonlinear with photocurrent due to screening and space-charge effects which reduce the carrier transit time (Tousley et al., 1995).

Even higher growth temperatures show good performance in as-grown samples. GaAs grown at 400°C and incorporated in an MSM structure shows semi-insulating behavior at low bias, but the as-grown material conducts more than the annealed material at higher bias. The authors suggest that for as-grown samples defect-assisted tunneling from metal to semiconductor is responsible for these effects. Thus, MSM photodetectors with 400°C as-grown GaAs has a unique combination of lower dark current and higher photoresponse than the annealed sample (Nabet et al., 1995).

To increase the efficiency of LT-GaAs MSM photodetectors while maintaining their high-speed performance, a thin (85 nm) LT-GaAs gating layer was integrated on top of standard-growth GaAs. In addition, the structured growth was engineered to create a thin (10 nm) GaAs channel with a vertical electric field as well as a AlGaAs heterostructure on top for increased collection efficiency, as shown in Figure 8.10. The heterostructure's Schottky contacts achieved a low dark current

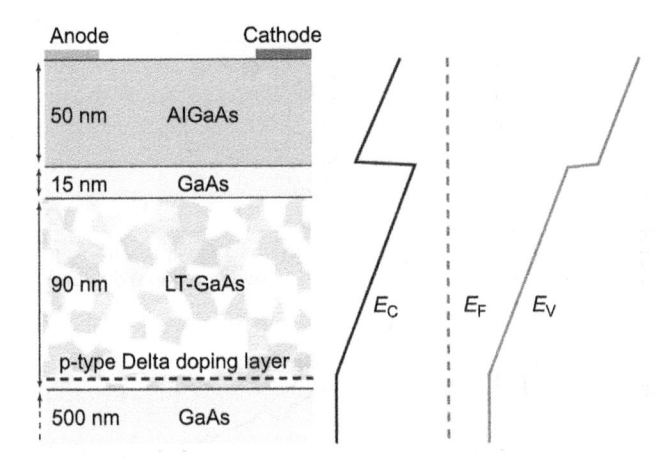

Figure 8.10 LT-GaAs detector layer structure and the associated energy band diagram, note: electrode spacing is not to scale.
Source: From Currie et al. (2011).

operation (500 pA at 10 V_{bias}), and the vertical electric field enabled efficient photoexcited carrier collection from the thin LT-GaAs and GaAs region below. This combination provided the high device responsivity (0.15 A/W, at 830 nm) with an estimated 67% internal quantum efficiency. Optoelectronic pulsed responses from 6.3 to 12 ps (FWHM) demonstrate the gating performed by the LT-GaAs for various electrode spacings, shown in Figure 8.11, as compared to 16 to 100 ps (FWHM) responses from devices without the thin LT-GaAs gating layer. These devices have very low dark currents and offer efficiencies comparable to that of standard GaAs growth, while enabling picosecond operation with >1 μm planar electrode spacing (Currie et al., 2011, 2013).

8.5.4 Waveguide photodetectors

Instead of the photoexcitation impinging normal to the photodetector as in PIN, PCS, and MSM detectors, in waveguide photodetectors the exciting photons propagate in a waveguide are coupled to one or more photodetectors while propagating in the waveguide. This occurs either by using the photodetector's absorption region as the core of the waveguide or having the guided wave periodically or continuously couple to a photodetector, for example, via evanescent coupling.

For LT-GaAs photodetectors, a 1 THz device performance was predicted by matching the microwave and optical velocities in a distributed photodetector on an optical waveguide (Böttcher and Bimberg, 1995). Implementing a traveling-wave photodetector (TWPD) geometry can overcome RC limitations in conventional (lumped-element) photodetectors, for example, PIN, PCS, and MSM. The trade-off is that TWPDs have to match the velocities of the optical and electrical waves. In

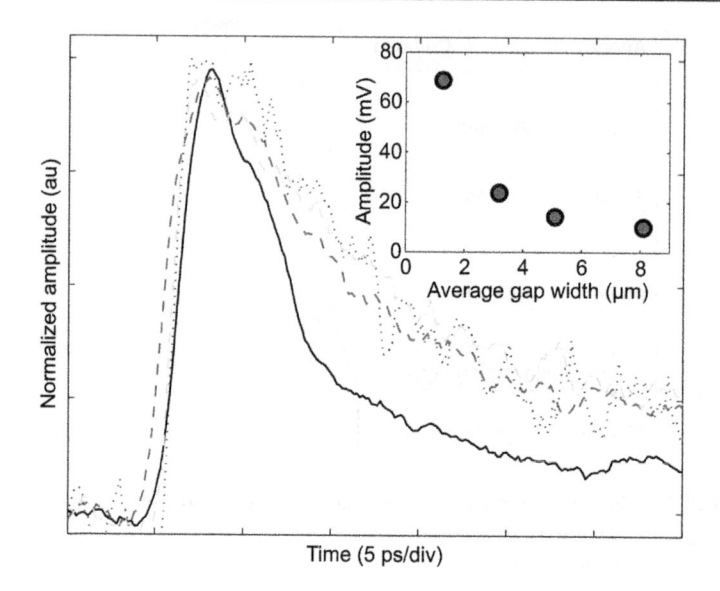

Figure 8.11 LT-GaAs MSM devices with 1.3, 3.2, 5.1, and 8.1 μm separations between cathode and anode. Their time responses were measured by electro-optic sampling, demonstrating a 6.3 ps FWHM pulsewidth for the 1.3 μm MSM while all others MSMs were approximately 12 ps. The inset shows the peak response scaled to the device's active area. *Source*: From Currie et al. (2013).

addition, for optimal performance TWPD devices must minimize both coupling and absorption losses in the optical and electrical waveguides.

Chiu et al. demonstrate a PIN TWPD in which LT-GaAs is used as the absorption layer (170 nm thick) with p- and n-layers formed by AlGaAs. For pulsed Ti: sapphire laser excitation (near 800 nm), the LT-GaAs in the TWPD provided subpicosecond performance (530 fs FWHM) while the PIN TWPD geometry enabled 8% external quantum efficiency. The high efficiency in this LT-GaAs photodetector is a tradeoff between the carrier lifetime and the transit time in the PIN structure. This TWPD combines 560 GHz bandwidth (3 dB) performance with high optoelectronic conversion efficiency (Chiu et al., 1998).

Shi et al. demonstrate slightly higher efficiency (8.1%) using MSM based TWPD (over the PIN geometry). This higher efficiency is maintained with higher optical illumination (Shi et al., 2001).

Quantum-well-based amplification is integrated into a TWPD to produce a traveling-wave amplifier photodetector, or TAP detector (Lasaosa et al., 2004). Experimental results show quantum efficiencies of 200% at 850 nm and 100% near 1550 nm (Lasaosa et al., 2004).

To reach higher-powers before saturation as well as to improve the bandwidth-efficiency product, more complex traveling-wave detector designs have been created. Distributing photodetectors in a transmission-line geometry improves the power-bandwidth product (Lin et al., 1997; Murthy et al., 2000). In addition, discrete photodetectors placed serially along a microwave transmission line allow the photocurrents to sum in phase when the velocities of the optical and electrical waves are matched (Lin et al., 1997). On a GaAs-based waveguide detector the microwave velocity is generally greater than the optical velocity, so techniques for slowing the microwave signal must be employed to match the velocities and sum the photocurrents. This allows the individual photodetectors to be kept below saturation, while producing a large current by summing the output from the photodetector array. Saturation currents of 56 mA at 50 GHz frequencies have been achieved using this method (Lin et al., 1997). These velocity-matched distributed photodetectors have a disadvantage of producing microwave power traveling in both directions on the transmission line, thus, the backwards traveling signal should be eliminated since it reduces efficiency and has the potential for signal distortion. Another design uses a tapered line distributed photodetector, where a tapered transmission line removes the backward traveling wave increasing the photodetector efficiency. This tapered transmission line geometry, therefore, can increase both the efficiency as well as improving the bandwidth, with flat spectral response from 1 to >100 GHz, and spectral content to 1 THz (Shi and Sun, 2002).

While more complex to fabricate and more difficult for light coupling, waveguide photodetection offers both high-speed and high efficiency photodetection.

8.6 Photodetector performance

The temporal pulsewidths (measured full-width at half maximum, FWHM) for LT-GaAs photodetectors cited in this chapter are plotted in Figure 8.12. The reported pulsewidths in these references vary by an order of magnitude from subpicosecond to picosecond response (350−6300 fs). No particular photodetector style (PCS, TWPD, or MSM) has a clear advantage.

These data are plotted in Figure 8.13 in terms of the reported frequency response (measured at 3 dB from the peak). Here the PCS from Zheng et al. (2003) reports a remarkable performance two times larger than any other at 1250 GHz.

For performance, including signal-to-noise ratio, dark current is an important trait in photodetectors. Figure 8.14a and b plot the reported dark currents as a function of electrode spacing as well as applied voltage. Here the MSM structures seem to have a slight edge over the PCS geometry. This is due to the Schottky contact in the MSM photodetectors. In addition, it is useful to note the work of Nabet et al. (1994) on intermediate growth temperatures, which demonstrates the influence that growth temperature has on the photodetector's dark current.

The responsivity of these devices range over several orders-of-magnitude, from nearly 0.001 to 10 A/W, as shown in Figure 8.15. While the traveling-wave amplified detectors (TAP) have some of the highest responsivities (as expected from their traveling-wave geometries), the PCS by Harmon et al. (1995) dominate below 1400 nm. This is due to a high photoconductive gain achieved by creating a large ratio of carrier lifetime to transit time in the PCS. However by 1500 nm the TAP

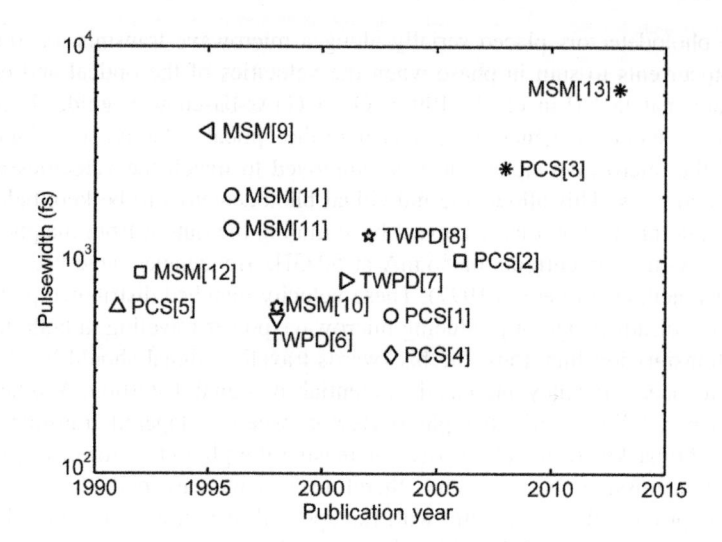

Figure 8.12 Summary of temporal pulsewidths in three types of LT-GaAs photodetectors: photoconductive (PCS), metal-semiconductor-metal (MSM), and traveling-wave (TWPD). Numbers are referenced as follows, 1: (Mikulics et al., 2003), 2: (Mikulics et al., 2006), 3: (Mikulics et al., 2008), 4: (Zheng et al., 2003), 5: (Gupta et al., 1991), 6: (Chiu et al., 1998), 7: (Shi et al., 2001), 8: (Shi et al., 2002), 9: (Boutros et al., 1995), 10: (Kordos et al., 1998), 11: (Joshi and McAdoo, 1996), 12: (Chou et al., 1992), 13: (Currie et al., 2013).

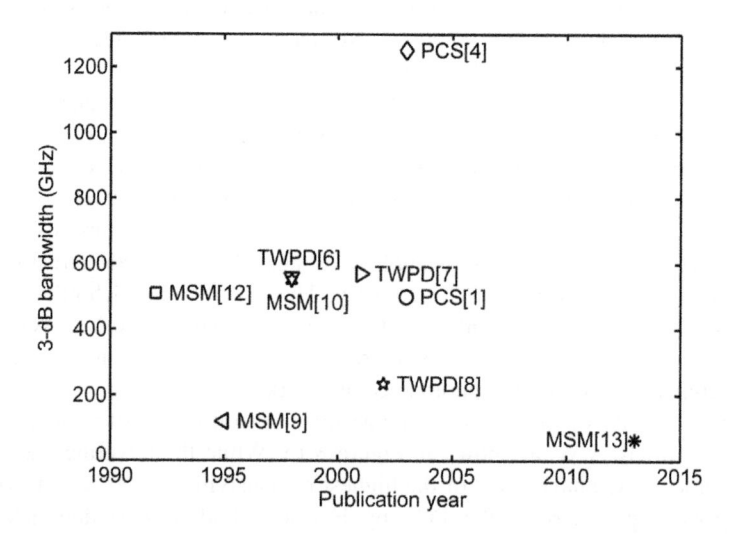

Figure 8.13 Frequency response (at 3 dB bandwidth) for some of the LT-GaAs photodetectors plotted in Figure 8.12.

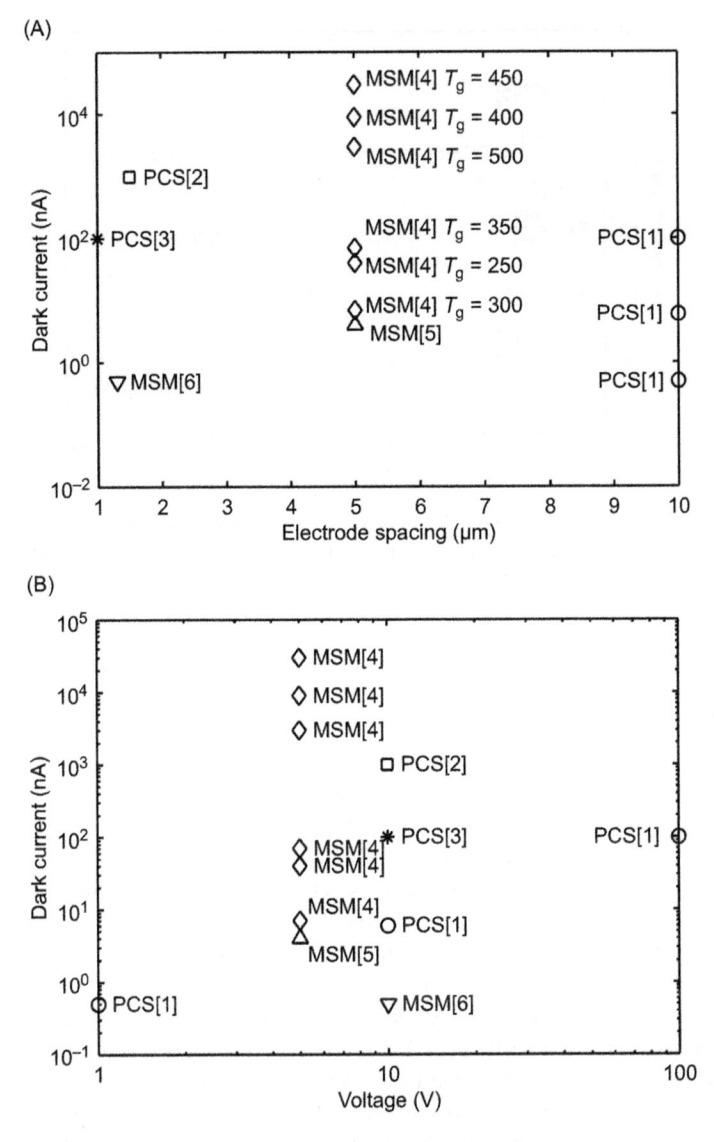

Figure 8.14 Photodetector dark current of LT-GaAs photoconductive (PCS) and metal-semiconductor-metal (MSM) devices as a function of (A) electrode spacing and (B) applied voltage. Numbers are referenced as follows, 1: (Mikulics et al., 2003), 2: (Mikulics et al., 2006), 3: (Mikulics et al., 2008), 4: (Nabet et al., 1994), 5: (Boutros et al., 1995), 6: (Currie et al., 2011).

detectors are two orders-of-magnitude better than the PCS detectors, owing to their traveling-wave structure as well as the incorporation of gain stages (Lasaosa et al., 2004). Greater efficiency was achieved by increasing the length of the traveling

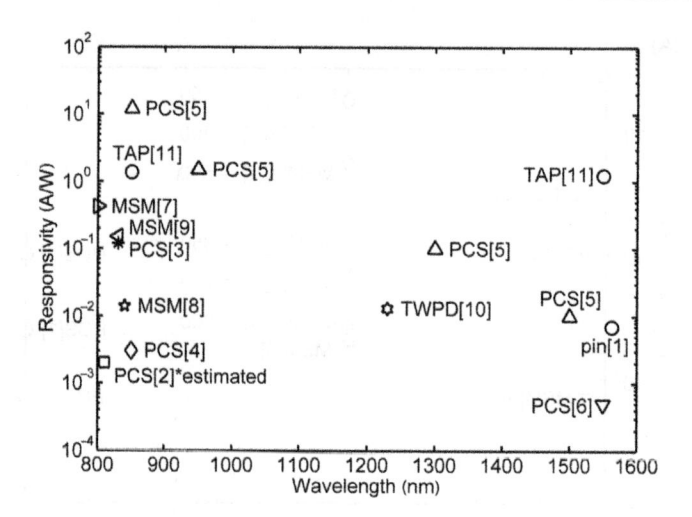

Figure 8.15 Responsivity of LT-GaAs detectors as a function of wavelength for photoconductive (PCS), metal-semiconductor-metal (MSM), traveling-wave (TWPD), and amplified traveling-wave (TAP) devices discussed in this chapter. Numbers are referenced as follows, 1: (Han et al., 2006), 2: (Mikulics et al., 2003), 3: (Mikulics et al., 2006), 4: (Mikulics et al., 2008), 5: (Harmon et al., 1995), 6: (Erlig et al., 1999), 7: (Tousley et al., 1995), 8: (Boutros et al., 1995), 9: (Currie et al., 2011), 10: (Shi et al., 2002), 11: (Lasaosa et al., 2004).

wave devices, however, increasing these devices from 40 to 70 μm increased the pulses slightly, producing lower electrical bandwidths (Shi et al., 2002).

In summarizing the performance of LT-GaAs photodetectors, a few comments should be made on response times. LT-GaAs PCS tend to show an increase in response time with increasing voltage bias. This causes the amplitude of the generated electrical transient to saturate at high bias. The result limits the THz power output from these switches. This high-bias response time increase could be due to a reduction in the carrier caption cross section due to carrier heating and Coulomb-barrier lowering, similar to an effect observed in SiO_2 (Zamdmer et al., 1999).

In contrast, epitaxially lifted-off LT-GaAs used as PCS on alternate substrates do not show an increased response time. With 10 μm spaced electrodes, no change in the 350 fs response time (1/e decay) was observed for voltages up to 120 V, corresponding to calculated electric fields greater than 100 kV/cm (Mikulics et al., 2003).

In addition, for planar nanoscale devices (∼ 100 nm contact separation), the electric field may not penetrate deep enough into the photodetector's active (absorption) region to take advantage of the material's rapid response (Joshi and McAdoo, 1996). This can be overcome by using heterostructures and/or backgating the substrate.

Finally, when using mid-gap absorption LT-GaAs for wavelengths near 1300 nm, an observed nonlinear behavior broadens the time response with increasing optical power due to potential hot-electron effects and intervalley scattering (Shi et al., 2004).

Overall, the performance of LT-GaAs photodetectors increases operational speed. Engineering modifications and clever designs have increased the collection efficiency and responsivity of these photodetectors to match that of the lower-speed GaAs rivals. This has made LT-GaAs photodetectors viable for many applications that are too fast for silicon and GaAs photodetectors.

8.7 Applications

Low-temperature grown materials have applications in fields where fast photodetection is desired. Paramount among these applications are microwave photonics and THz technology, where picosecond and subpicosecond detection is required. Before discussing this, however, there are a few other areas which benefit from application of LT-GaAs technology: sub-bandgap absorption photodetectors, nonlinear optics, and pulsed magnetic spin experiments.

8.7.1 Sub-bandgap absorption photodetectors

Extending the range of GaAs detectors into the near IR has the ability to create a single detector for this range (~500–1600 nm). LT-GaAs at 230°C shows significant absorption in the region below the bandedge of GaAs (900–1500 nm). After annealing, however, the absorption in the sub-bandgap region vanishes, due to the decreasing number of As_{Ga} antisite point defects and an increase in As nanoinclusions within stoichiometric GaAs (Pastor et al., 2013). MSM TWPDs were designed to couple 1300 nm light into an LT-GaAs waveguide where the light is absorbed and collected. These devices were fabricated with a 500 nm LT-GaAs layer guiding/absorption layer on top of a 1 μm-thick AlGaAs cladding layer for the optical waveguide. Optical pulses generated single picosecond electrical pulses with corresponding 3 dB electrical bandwidths of 234 GHz. As the length of the traveling wave device was increased from 40 to 70 μm the pulses slightly broadened producing 200 GHz electrical bandwidths. With responsivities of tens of mA/W, these devices are an order of magnitude better than conventional nonwaveguide-based MSM photodetectors (Shi et al., 2002). Unlike the work of Pastor et al. (2013), the LT-GaAs in the devices by Shi absorbs even after annealing at 600°C (Shi et al., 2002).

8.7.2 Nonlinear optics

Fast, all-optical switches can be created using the short carrier lifetimes in LT-GaAs. High-energy (1 mJ) optical pulses can create photoexcited carrier densities $>10^{19}$ cm^{-3}, which form a transient plasma and generate metal-like reflections in semiconductors. These transient reflective states exist during the subpicosecond carrier lifetime in LT-GaAs. Since GaAs and LT-GaAs transmit mid-IR wavelengths, Elezzabi et al. (1994) have used this transient reflection to generate single picosecond optical pulses from a cw, 10.6 μm laser source. Using LT-GaAs as a reflective optical switch also provides synchronization of visible and IR optical pulses. By optimizing the LT-GaAs 200 fs switching times could be generated for use at almost any wavelength in the mid- and far-IR (Elezzabi et al., 1994). This technique can be utilized for communications, as well as optical imaging and optical sensors.

8.7.3 Pulsed magnetic spin experiments

Magnetic spin dynamics are typically performed with milli-tesla fields and temporal resolution >30 ps. To reach levels beyond this linear accelerators are necessary. However, 10 mW laser pulses exciting a LT-GaAs PCS generated 20 A, 3 ps electrical pulses, thereby producing 0.6 T fields (Wang et al., 2008). With some modifications to the PCS design the pulsed field strength could increase to a few tesla. This rivals field pulses from linear accelerators ($\sim$5 T, 1 ps pulses) and provides the potential for magnetization dynamics measurements in any laboratory.

8.7.4 THz emitters and receivers

Using an epitaxial liftoff technique, LT-GaAs was transferred to sapphire and glass substrates for THz generation. Pulsed operation achieved demonstrated frequency content above 1 THz when illuminated by 150 fs optical pulses from a mode-locked Ti:sapphire laser (Heiliger et al., 1996). This provides a path for creating LT-GaAs THz sources on any material.

Materials with short carrier lifetimes are needed to maintain a high radiated THz output power. However, short carrier lifetimes reduce the material mobility and decrease the carrier collection, which limits the photogenerated current and therefore limits the THz output power (Saeedkia and Safavi-Naeini, 2008). Thus an optimal carrier lifetime needs to be determined for each design. In general, conversion efficiencies for THz generated power compared to incident optical power in photomixers is approximately 10^{-5} (Saeedkia and Safavi-Naeini, 2008). While this value is low, the following reports show several solutions to this issue.

For maximizing the THz emission from LT-GaAs devices Gregory et al. (2005) provide a detailed study. Their findings show four important areas: carrier lifetime, active area, electrode spacing, antenna design. For efficient generation in the 0.2−2 THz region, carrier lifetimes should be between 80 and 800 fs, which makes LT-GaAs a good material choice. The active area of the photodetector (photomixing region for cw-THz) is optimized by adding interdigitated electrodes. Since near-anode enhancement of the optoelectronic response occurs within 1 μm of the electrodes, the electrode spacing needs to be optimized to maximize the field in this region, producing optimal spacings between 1 and 2 μm for their devices. Finally, from 0.2 to 0.6 THz the antenna design dictated the radiated emission, while between 0.6 and 1.1 THz a resonance on the length scale of the fingers modified the emission, which then disappears above 1.3 THz, therefore requiring a more careful design of the photodetector structure as well as the antenna (Gregory et al., 2005).

In other studies, to maximize the radiated THz power from PCSs, care was taken to minimize field screening within the PCS. Two type of screening are important, screening by the: (1) space charge as well as by the (2) radiation field. Space charge screening is important but is weak during the first few picoseconds after photoexcitation, at which point screening by the generated, radiating THz field can also cause screening of the applied electric field in optoelectronic devices (Siebert et al., 2004). The radiative field screening depends on carrier mobility and plays a role in small area LT-GaAs emitters (Siebert et al., 2004). Bias-field screening times are 1–40 ns in LT-GaAs PCS for long-lived trap states (Loata et al., 2007).

In pulsed operation, using an MSM-based traveling-wave structure produced high optical to THz conversion efficiencies. At 645 GHz Lasaosa et al. (2004) demonstrated 0.11% conversion efficiency, one of the highest, with 3 mW (at ~800 nm) generating ~3 μW (at 645 GHz).

For continuous THz operation, an interdigitated MSM structure on LT-GaAs was implemented in a traveling wave geometry on a coplanar transmission line coupled to a bow-tie antenna. A THz signal was achieved by the beat frequency achieve by mixing two detuned diode lasers (in the range 780–800 nm) on the LT-GaAs photodetector. Velocity matching of the optical and electrical waveforms was achieved by adjusting the angle of incidence of the two IR beams on the LT-GaAs photodetector. Radiated microwatt level power was detected at frequencies from 100 to 1500 GHz. These power levels in the sub-millimeter and THz region are needed for local oscillators in components such as superconducting-insulator-superconducting mixers and hot-electron bolometers (Michael et al., 2004, 2005).

Separated-transport-recombination photodetectors (STR-Pds) extend the performance of the traditional PIN photodetector. Enhancements in electrical bandwidth and higher saturation current are achieved by adding a short carrier lifetime layer within the absorbing intrinsic region. LT-GaAs is a good choice for the short carrier lifetime layer since it integrates with epitaxially grown GaAs-based PIN photodetectors. Adding this layer makes the "i" region thicker, which relieves RC bandwidth limitations. At the same time this also concentrates the applied electric field in the surrounding GaAs "i" layer, which mitigates the issue of lower drift velocity (caused by the thicker "i" layer). The carrier trapping in the short lifetime LT-GaAs captures many of the carriers which limits the collection efficiency, however, this also increases the device speed. Overall, Shi et al. (2005) noticed only a 50% drop in responsivity at high-bias, due to the LT-GaAs traps. However, at low bias, the STR-PD showed improved performance compared to a traditional PIN photodetector which suffered from space-charge field screening effects causing saturation. Therefore, at low bias the STR-PD and PIN photodetectors exhibited similar responsivity. Under illumination by short optical pulses, the LT-GaAs region also provides the STR-PD with shorter pulses, thus, demonstrating enhanced performance in speed/bandwidth as well as saturation performance as compared to traditional PIN photodetectors (Shi et al., 2005).

Using LT-GaAs, a separated-transport-recombination photodetector with coplanar waveguide traveling-wave electrodes was coupled to a slot antenna to directly produce sub-THz radiation. The separated-transport-recombination photodetector design allows higher reverse bias voltages (e.g., than other LT-GaAs and uni-traveling carrier photodetectors), which provides higher THz output power before saturation. For example, the saturation fluence/current of these detectors (122 pJ/184 mA) is much larger than in a uni-traveling carrier device (1 pJ/76 mA). In addition, the separated-transport-recombination photodetector has good conversion efficiency with ~ 30 mW average optical power producing 5 µW average sub-THz power. A maximum peak power (total) was 4.5 mW, with 300 µW of power produced at 500 GHz. Integrating the slot antenna provides directional radiation without the need for an additional silicon lens making this a compact THz emitted when integrated with a laser source (Li et al., 2007).

Generating continuous-wave THz power from photonic mixers benefits from the short lifetime of LT-GaAs. However, these mixers usually have a limiting pump power density limit of ~ 1 mW/µm^2 (Michael and Mikulics, 2012). Additional saturation in these photomixers can occur due to reabsorption of the generated THz field by nanosecond-lifetime photoelectrons, whose density is proportional to the generated photocurrent (Michael and Mikulics, 2012). While this could be due to intervalley scattering or carrier field screening, Michael and Mikulics (2012) suggest that this is due to carrier effects in the underlying semi-insulating GaAs substrate. One solution is to transfer the LT-GaAs absorption region to another substrate (Michael and Mikulics, 2012).

Photomixing in a vertically integrated THz emitter provided better power capabilities up to 1.8 THz (Mouret et al., 2004). This vertical structure also removes contributions from the substrate. Spectroscopic measurements of the rotational transitions in hydrogen sulfide were measured at 3.028 THz as well as the line broadening due to air. While still requiring cryogenic detectors, the improved power capabilities of the emitter increased the system's signal-to-noise ratio for application in far-RR/THz spectroscopy (Mouret et al., 2004).

Traditional LT-GaAs photomixers generate THz emission by applying an electric field between metal contacts to accelerate carriers. The THz response is a combination of the carrier lifetime and carrier transit time between electrodes. However, designing sub-wavelength features in the contacts could enable strong near-field enhancements and can support surface plasmon polaritons (SPPs). These enhanced fields may enable higher carrier velocity, due to velocity overshoot in GaAs. In addition, the SPPs have a minimal dependence on carrier lifetime, thereby producing potential photomixers from the UV to IR (Pilla, 2007). A proposed device model of this type demonstrates orders-of-magnitude improvement in power generation efficiency at 1 and 5 THz (Pilla, 2007).

Many applications requiring operation from 100 to 5000 GHz or that require subpicosecond responses have been successfully implemented using LT-GaAs technology. The field of THz technology is thriving from the capabilities offered by LT-GaAs materials.

8.8 Conclusions and future trends

Understanding the physical mechanisms in LT-GaAs has progressed since its use for FET backgating in 1988. The physical mechanisms are now largely understood with minimal controversy over the interpretation.

Engineering device operation (largely for photodetection) has provided picosecond and subpicosecond optoelectronic devices. A variety of photodetection schemes have been created using PIN, waveguide, planar MSM, and PCS technologies to increase speed and efficiency. Further development of layered structures and heterostructures has increased device performance as well.

These gains in understanding the carrier dynamics in LT-GaAs have led to devices used for: pulsed THz generation and detection, photomixing for cw THz sources, magnetic-field pulses (0.6 T, 3 ps), and high-speed electronic circuit performance evaluation.

While the basic physics are fairly well understood, recent (Pastor et al., 2013) and future work will help to refine our knowledge, as well as to enable understanding of other low-temperature-grown materials. Meanwhile material and device engineering have focused on applications in the region between IR optics and microwaves, namely the THz spectrum. This appears to be the future for LT-GaAs photoexcitation and detection.

The future for THz emitters and detectors may be extending low-temperature materials for 1550 nm operation, as well as improved performance of LT-GaAs based detectors (e.g., efficiency and device speed—leading to THz bandwidth).

Traveling-wave LT-GaAs photomixers for THz local oscillator applications (Michael et al., 2004) along with photonic mixers for increased continuous terahertz power generation (Michael et al., 2005), and THz photonics (Saeedkia and Safavi-Naeini, 2008) are just beginning to be utilized. The future expansion of this technology will likely require and advance LT-GaAs technology.

References

Adomavičius, R., Krotkus, A., Bertulis, K., Sirutkaitis, V., Butkus, R., Piskarskas, A., 2003. Hole trapping time measurement in low-temperature-grown gallium arsenide. Appl. Phys. Lett. 83, 5304.

Auston, D.H., Lavallard, P., Sol, N., Kaplan, D., 1980. An amorphous silicon photodetector for picosecond pulses. Appl. Phys. Lett. 36, 66–68.

Bachelet, G.B., Schlüter, M., Baraff, G.A., 1983. AsGa antisite defect in GaAs. Phys. Rev. B 27, 2545–2547.

Baker, C., Gregory, I.S., Tribe, W.R., Bradley, I.V., Evans, M.J., Linfield, E.H., 2004. Highly resistive annealed low-temperature-grown InGaAs with sub-500 fs carrier lifetimes, Appl. Phys. Lett., 85. pp. 4965–4967.

Benjamin, S.D., Loka, H.S., Othonos, A., Smith, P.W.E., 1996. Ultrafast dynamics of nonlinear absorption in low-temperature-grown GaAs, Appl. Phys. Lett., 68. pp. 2544–2546.

Boutros, K.S., Roberts, J.C., Bedair, S.M., Carruthers, T.F., Frankel, M.Y., 1995. High speed metal-semiconductor-metal photodetector manufactured on GaAs by low-temperature photoassisted metalorganic chemical vapor deposition. Appl. Phys. Lett. 66, 3651–3653.

Butun, B., Biyikli, N., Kimukin, I., Aytur, O., Ozbay, E., Postigo, P.A., 2004. High-speed 1.55 μm operation of low-temperature-grown GaAs-based resonant-cavity-enhanced p–i–n photodiodes. Appl. Phys. Lett. 84, 4185–4187.

Böttcher, E.H., Bimberg, D., 1995. Millimeter wave distributed metal-semiconductor-metal photodetectors. Appl. Phys. Lett. 66, 3648–3650.

Chiu, Y.-J., Fleischer, S.B., Bowers, J.E., 1998. High-speed low-temperature-grown GaAs pin traveling-wave photodetector. Photonics Technol. Lett. IEEE 10, 1012–1014.

Chou, S.Y., Liu, Y., Khalil, W., Hsiang, T.Y., Alexandrou, S., 1992. Ultrafast nanoscale metal-semiconductor-metal photodetectors on bulk and low-temperature grown GaAs. Appl. Phys. Lett. 61, 819–821.

Currie, M., Quaranta, F., Cola, A., Gallo, E.M., Nabet, B., 2011. Low-temperature grown GaAs heterojunction metal-semiconductor-metal photodetectors improve speed and efficiency. Appl. Phys. Lett. 99, 203502.

Currie, M., Dianat, P., Persano, A., Martucci, M.C., Quaranta, F., Cola, A., 2013. Performance enhancement of a GaAs detector with a vertical field and an embedded thin low-temperature grown layer. Sensors 13, 2475–2483.

Dabrowski, J., Scheffler, M., 1989. Isolated arsenic-antisite defect in GaAs and the properties of EL2. Phys. Rev. B 40, 10391–10401.

Doany, F.E., Grischkowsky, D., Chi, C.-C., 1987. Carrier lifetime vs ion-implantation dose in silicon on sapphire. Appl. Phys. Lett. 50, 460–462.

Elezzabi, A.Y., Meyer, J., Hughes, M.K.Y., Johnson, S.R., 1994. Generation of 1-ps infrared pulses at 10.6 μm by use of low-temperature-grown GaAs as an optical semiconductor switch. Opt. Lett. 19, 898–900.

Erlig, H., Wang, S., Azfar, T., Udupa, A., Fetterman, H.R., Streit, D.C., 1999. LT-GaAs detector with 451 fs response at 1.55 μm via two-photon absorption. Electron. Lett. 35, 173–174.

Eusèbe, H., Roux, J.-F., Coutaz, J.-L., Krotkus, A., 2005. Photoconductivity sampling of low-temperature-grown Be-doped GaAs layers. J. Appl. Phys. 98, 033711.

Fukushima, S., Obata, T., Otsuka, N., 2001. Electrical properties of nearly stoichiometric GaAs grown by molecular beam epitaxy at low temperature. J. Appl. Phys. 89, 380.

Gregory, I.S., Baker, C., Tribe, W.R., Evans, M.J., Beere, H.E., Linfield, E.H., 2003. High resistivity annealed low-temperature GaAs with 100 fs lifetimes, Appl. Phys. Lett., 83. p. 4199.

Gregory, I.S., Baker, C., Tribe, W.R., Bradley, I.V., Evans, M.J., Linfield, E.H., 2005. Optimization of photomixers and antennas for continuous-wave terahertz emission, IEEE J. Quantum Electron., 41. pp. 717–728.

Grenier, P., Whitaker, J.F., 1997. Subband gap carrier dynamics in low-temperature-grown GaAs. Appl. Phys. Lett. 70, 1998–2000.

Gupta, S., Frankel, M.Y., Valdmanis, J.A., Whitaker, J.F., Mourou, G.A., Smith, F.W., 1991. Subpicosecond carrier lifetime in GaAs grown by molecular beam epitaxy at low temperatures. Appl. Phys. Lett. 59, 3276–3278.

Han, Q., Niu, Z.C., Peng, L.H., Ni, H.Q., Yang, X.H., Du, Y., 2006. High-performance 1.55 μm low-temperature-grown GaAs resonant-cavity-enhanced photodetector. Appl. Phys. Lett. 89, 131104.

Harmon, E.S., McInturff, D.T., Melloch, M.R., Woodall, J.M., 1995. Novel GaAs photodetector with gain for long wavelength detection. J. Vac. Sci. Technol. B 13, 768–770.
Heiliger, H.-M., Vossebürger, M., Roskos, H.G., Kurz, H., Hey, R., Ploog, K., 1996. Application of liftoff low-temperature-grown GaAs on transparent substrates for THz signal generation. Appl. Phys. Lett. 69, 2903–2905.
Joshi, R.P., McAdoo, J.A., 1996. Picosecond dynamic response of nanoscale low-temperature grown GaAs metal-semiconductor-metal photodetectors. Appl. Phys. Lett. 68, 1972–1974.
Kang, J.U., Frankel, M.Y., Esman, R.D., Thompson, D.A., Robinson, B.J., 1998. InGaAsP grown by He-plasma-assisted molecular beam epitaxy for 1.55 μm high speed photodetectors, Appl. Phys. Lett., 72. pp. 1278–1280.
Keil, U.D., Tautz, S., Dankowski, S.U., Kiesel, P., Döhler, G.H., 1997. Femtosecond differential transmission measurements on low temperature GaAs metal−semiconductor−metal structures. Appl. Phys. Lett. 70, 72–74.
Kordos, P., Marso, M., Forster, A., Darmo, J., Betko, J., Nimtz, G., 1997. Space-charge controlled conduction in low-temperature-grown molecular-beam epitaxial GaAs. Appl. Phys. Lett. 71, 1118–1120.
Kordos, P., Forster, A., Marso, M., Ruders, F., 1998. 550 GHz bandwidth photodetector on low-temperature grown molecular-beam epitaxial GaAs. Electron. Lett. 34, 119–120.
Kostakis, I., Saeedkia, D., Missous, M., 2012. Characterization of low temperature InGaAs-InAlAs semiconductor photo mixers at 1.55 μm wavelength illumination for terahertz generation and detection. J. Appl. Phys. 111, 103105.
Kostoulas, Y., Waxer, L.J., Walmsley, I.A., Wicks, G.W., Fauchet, P.M., 1995. Femtosecond carrier dynamics in low-temperature-grown indium phosphide. Appl. Phys. Lett. 66, 1821–1823.
Krotkus, A., Bertulis, K., Dapkus, L., Olin, U., Marcinkevičius, S., 1999. Ultrafast carrier trapping in Be-doped low-temperature-grown GaAs. Appl. Phys. Lett. 75, 3336–3338.
Krotkus, A., Bertulis, K., Kaminska, M., Korona, K., Wolos, A., Siegert, J., 2002. Be-doped low-temperature-grown GaAs. IEE Proc. Optoelectron. 149, 111–115.
Lasaosa, D., Shi, J.-W., Pasquariello, D., Gan, K.-G., Tien, M.-C., Chang, H.-H., 2004. Traveling-wave photodetectors with high power-bandwidth and gain-bandwidth product performance. IEEE J. Sel. Top Quantum Electron. 10, 728–741.
Lin, L.Y., Wu, M.C., Itoh, T., Vang, T.A., Muller, R.E., Sivco, D.L., 1997. High-power high-speed photodetectors-design, analysis, and experimental demonstration. IEEE Trans. Microwave Theory Tech. 45, 1320–1331.
Li, Y.-T., Shi, J.-W., Pan, C.-L., Chiu, C.-H., Liu, W.-S., Chen, N.-W., 2007. Sub-THz photonic-transmitters based on separated-transport-recombination photodiodes and a micromachined slot antenna. IEEE Photonics Technol. Lett. 19, 840–842.
Loata, G.C., Löffler, T., Roskos, H.G., 2007. Evidence for long-living charge carriers in electrically biased low-temperature-grown GaAs photoconductive switches. Appl. Phys. Lett. 90, 052101.
Lochtefeld, A.J., Melloch, M.R., Chang, J.C.P., Harmon, E.S., 1996. The role of point defects and arsenic precipitates in carrier trapping and recombination in low-temperature grown GaAs. Appl. Phys. Lett. 69, 1465–1467.
Look, D.C., Fang, Z.-Q., Yamamoto, H., Sizelove, J.R., Mier, M.G., Stutz, C.E., 1994. Deep traps in molecular-beam-epitaxial GaAs grown at low temperatures. J. Appl. Phys. 76, 1029–1032.

Loukakos, P.A., Kalpouzos, C., Perakis, I.E., Hatzopoulos, Z., Logaki, M., Fotakis, C., 2001. Ultrafast electron trapping times in low-temperature-grown gallium arsenide: the effect of the arsenic precipitate spacing and size. Appl. Phys. Lett. 79, 2883.

McIntosh, K.A., Nichols, K.B., Verghese, S., Brown, E.R., 1997. Investigation of ultrashort photocarrier relaxation times in low-temperature-grown GaAs. Appl. Phys. Lett. 70, 354–356.

Melloch, M.R., Otsuka, N., Woodall, J.M., Warren, A.C., Freeouf, J.L., 1990. Formation of arsenic precipitates in GaAs buffer layers grown by molecular beam epitaxy at low substrate temperatures. Appl. Phys. Lett. 57, 1531–1533.

Melloch, M.R., Woodall, J.M., Harmon, E.S., Otsuka, N., Pollak, F.H., Nolte, D.D., 1995. Low-temperature grown III-V materials. Annu. Rev. Mater. Sci. 25, 547–600.

Michael, E.A., Mikulics, M., 2012. Losses from long-living photoelectrons in terahertz-generating continuous-wave photomixers. Appl. Phys. Lett. 100, 191112.

Michael, E.A., Mikulics, M., Marso, M., Kordos, P., L'th, H., Vowinkel, B., et al., 2004. Large-area traveling-wave LT-GaAs photomixers for LO application. In: Zmuidzinas, J., Holland, W.S., Withington, S. (Eds.), Proceedings SPIE 5498, Millimeter and Submillimeter Detectors for Astronomy II. SPIE, pp. 525–536. Available from: http://doi.org/10.1117/12.551633.

Michael, E.A., Vowinkel, B., Schieder, R., Mikulics, M., Marso, M., Kordoš, P., 2005. Large-area traveling-wave photonic mixers for increased continuous terahertz power. Appl. Phys. Lett. 86, 111120.

Mikulics, M., Zheng, X., Adam, R., Sobolewski, R., Kordos, P., 2003. High-speed photoconductive switch based on low-temperature GaAs transferred on SiO_2-Si substrate. IEEE Photonics Technol. Lett. 15, 528–530.

Mikulics, M., Adam, R., Marso, M., Forster, A., Kordos, P., Luth, H., 2005. Ultrafast low-temperature-grown epitaxial GaAs photodetectors transferred on flexible plastic substrates. IEEE Photonics Technol. Lett. 17, 1725–1727.

Mikulics, M., Wu, S., Marso, M., Adam, R., Forster, A., Van Der Hart, A., 2006. Ultrafast and highly sensitive photodetectors with recessed electrodes fabricated on low-temperature-grown GaAs. IEEE Photonics Technol. Lett. 18, 820–822.

Mikulics, M., Marso, M., Wu, S., Fox, A., Lepsa, M., Grutzmacher, D., 2008. Sensitivity enhancement of metal-semiconductor-metal photodetectors on low-temperature-grown GaAs using alloyed contacts. IEEE Photonics Technol. Lett. 20, 1054–1056.

Mouret, G., Matton, S., Bocquet, R., Hindle, F., Peytavit, E., Lampin, J.F., 2004. Far-infrared cw difference-frequency generation using vertically integrated and planar low temperature grown GaAs photomixers: application to H2S rotational spectrum up to 3 THz. Appl. Phys. B 79, 725–729.

Murthy, S., Jung, T., Chau, T., Wu, M.C., Sivco, D.L., Cho, A.Y., 2000. A novel monolithic distributed traveling-wave photodetector with parallel optical feed. IEEE Photonics Technol. Lett. 12, 681–683.

Nabet, B., Paolella, A., Cooke, P., Lemuene, M.L., Moerkirk, R.P., Liou, L.-C., 1994. Intermediate temperature molecular beam-epitaxy growth for design of large-area metal-semiconductor-metal photodetectors. Appl. Phys. Lett. 64, 3151–3153.

Nabet, B., Youtz, A., Castro, F., Cooke, P., Paolella, A., 1995. Current transport in as-grown and annealed intermediate temperature molecular beam epitaxy grown GaAs. Appl. Phys. Lett. 67, 1748–1750.

Nolte, D.D., Melloch, M.R., Woodall, J.M., Ralph, S.J., 1993. Enhanced electro-optic properties of low-temperature-growth GaAs and AlGaAs. Appl. Phys. Lett. 62, 1356–1358.

Ortiz, V., Nagle, J., Lampin, J.-F., Péronne, E., Alexandrou, A., 2007. Low-temperature-grown GaAs: modeling of transient reflectivity experiments. J. Appl. Phys. 102, 043515.
Pastor, A.A., Prokhorova, U.V., Serdobintsev, P.Y., Chaldyshev, V.V., Yagovkina, M.A., 2013. Effect of annealing on the nonequilibrium carrier lifetime in GaAs grown at low temperatures. Semiconductors 47, 1137−1140.
Patkar, M.P., Chin, T.P., Woodall, J.M., Lundstrom, M.S., Melloch, M.R., 1995. Very low resistance nonalloyed ohmic contacts using low-temperature molecular beam epitaxy of GaAs. Appl. Phys. Lett. 66, 1412−1414.
Pilla, S., 2007. Enhancing the photomixing efficiency of optoelectronic devices in the terahertz regime. Appl. Phys. Lett. 90, 161119.
Prabhu, S.S., Ralph, S.E., Melloch, M.R., Harmon, E.S., 1997. Carrier dynamics of low-temperature-grown GaAs observed via THz spectroscopy. Appl. Phys. Lett. 70, 2419−2421.
Qian, L., Benjamin, S.D., Smith, P.W.E., Robinson, B.J., Thompson, D.A., 1997a. Picosecond carrier lifetime and large optical nonlinearities in InGaAsP grown by He-plasma-assisted molecular beam epitaxy, Opt. Lett., 22. pp. 108−110.
Qian, L., Benjamin, S.D., Smith, P.W.E., Robinson, B.J., Thompson, D.A., 1997b. Subpicosecond carrier lifetime in beryllium-doped InGaAsP grown by He-plasma-assisted molecular beam epitaxy, Appl. Phys. Lett., 71. pp. 1513−1515.
Qian, L., Smith, P.W.E., Matin, M.A., Pinkney, H., Robinson, B.J., Thompson, D.A., 2000. Ultrafast carrier dynamics in InGaAsP grown by He-plasma-assisted epitaxy, Opt. Commun., 185. pp. 487−492.
Saeedkia, D., Safavi-Naeini, S., 2008. Terahertz photonics: optoelectronic techniques for generation and detection of terahertz waves. J. Lightwave Technol. 26, 2409−2423.
Shi, J.-W., Sun, C.-K., 2002. Theory and design of a tapered line distributed photodetector. J. Lightwave Technol. 20, 1942−1950.
Shi, J.-W., Gan, K.-G., Chiu, Y.-J., Chen, Y.-H., Sun, C.-K., Yang, Y.-J., 2001. Metal-semiconductor-metal traveling-wave photodetectors. IEEE Photonics Technol. Lett. 13, 623−625.
Shi, J.-W., Chen, Y.-H., Gan, K.-G., Chiu, Y.-J., Sun, C.-K., Bowers, J.E., 2002. High-speed and high-power performances of LTG-GaAs based metal-semiconductor-metal traveling-wave-photodetectors in 1.3-μm wavelength regime. IEEE Photonics Technol. Lett. 14, 363−365.
Shi, J.-W., Chen, Y.-H., Gan, K.-G., Chiu, Y.-J., Bowers, J.E., Tien, M.-C., 2004. Nonlinear behaviors of low-temperature-grown GaAs-based photodetectors around 1.3-μm telecommunication wavelength. IEEE Photonics Technol. Lett. 16, 242−244.
Shi, J.-W., Hsu, H.-C., Huang, F.-H., Liu, W.-S., Chyi, J.-I., Lu, J.-Y., 2005. Separated-transport-recombination p-i-n photodiode for high-speed and high-power performance. IEEE Photonics Technol. Lett. 17, 1722−1724.
Siebert, K.J., Quast, H., Leonhardt, R., Löffler, T., Thomson, M., Bauer, T., 2002. Continuous-wave all-optoelectronic terahertz imaging, Appl. Phys. Lett., 80. pp. 3003−3005.
Siebert, K.J., Lisauskas, A., Löffler, T., Roskos, H.G., 2004. Field screening in low-temperature-grown GaAs photoconductive antennas. Jpn. J. Appl. Phys. 43, 1038.
Smith, F.W., Calawa, A.R., Chen, C.-L., Manfra, M.J., Mahoney, L.J., 1988. New MBE buffer used to eliminate backgating in GaAs MESFETs. IEEE Electron. Device Lett. 9, 77−80.

Sosnowski, T.S., Norris, T.B., Wang, H.H., Grenier, P., Whitaker, J.F., Sung, C.Y., 1997. High-carrier-density electron dynamics in low-temperature-grown GaAs. Appl. Phys. Lett. 70, 3245–3247.

Stellmacher, M., Nagle, J., Lampin, J.F., Santoro, P., Vaneecloo, J., Alexandrou, A., 2000. Dependence of the carrier lifetime on acceptor concentration in GaAs grown at low-temperature under different growth and annealing conditions. J. Appl. Phys. 88, 6026–6031.

Sun, C.-K., Chen, Y.-H., Shi, J.-W., Chiu, Y.-J., Gan, K.-G., Bowers, J.E., 2003. Electron relaxation and transport dynamics in low-temperature-grown GaAs under 1 eV optical excitation. Appl. Phys. Lett. 83, 911–913.

Takahashi, R., Kawamura, Y., Kagawa, T., Iwamura, H., 1994. Ultrafast 1.55-μm photoresponses in low-temperature-grown InGaAs/InAlAs quantum wells. Appl. Phys. Lett. 65, 1790–1792.

Takahashi, R., Kawamura, Y., Iwamura, H., 1996. Ultrafast 1.55 μm all-optical switching using low-temperature-grown multiple quantum wells. Appl. Phys. Lett. 68, 153–155.

Tousley, B.C., Davids, N., Sayles, A.N., Paolella, A., Cooke, P., Lemoune, M.L., 1995. Broad-bandwidth, high-responsivity intermediate growth temperature GaAs MSM photodetectors. IEEE Photonics Technol. Lett. 7, 1483–1485.

Wallart, X., Coinon, C., Plissard, S., Godey, S., Offranc, O., Androussi, Y., 2010. Picosecond carrier lifetime in low-temperature-grown GaAsSb. Appl. Phys. Express 3, 111202.

Wang, Z., Pietz, M., Walowski, J., Förster, A., Lepsa, M.I., Münzenberg, M., 2008. Spin dynamics triggered by subterahertz magnetic field pulses. J. Appl. Phys. 103, 123905.

Warren, A.C., Woodall, J.M., Freeouf, J.L., Grischkowsky, D., McInturff, D.T., Melloch, M.R., 1990. Arsenic precipitates and the semi-insulating properties of GaAs buffer layers grown by low-temperature molecular beam epitaxy. Appl. Phys. Lett. 57, 1331–1333.

Warren, A.C., Burroughes, J.H., Woodall, J.M., McInturff, D.T., Hodgson, R.T., Melloch, M.R., 1991. 1.3-μm PiN photodetector using GaAs with As precipitates (GaAs:As). Electron Device Lett. IEEE 12, 527–529.

Warren, A.C., Woodall, J.M., Kirchner, P.D., Yin, X., Pollak, F., Melloch, M.R., 1992. Role of excess As in low-temperature-grown GaAs. Phys. Rev. B 46, 4617.

Yablonovitch, E., Gmitter, T., Harbison, J.P., Bhat, R., 1987. Extreme selectivity in the lift-off of epitaxial GaAs films. Appl. Phys. Lett. 51, 2222–2224.

Yamamoto, H., Fang, Z.-Q., Look, D.C., 1990. Nonalloyed ohmic contacts on low-temperature molecular beam epitaxial GaAs: influence of deep donor band. Appl. Phys. Lett. 57, 1537–1539.

Zamdmer, N., Hu, Q., McIntosh, K.A., Verghese, S., 1999. Increase in response time of low-temperature-grown GaAs photoconductive switches at high voltage bias. Appl. Phys. Lett. 75, 2313–2315.

Zheng, X., Xu, Y., Sobolewski, R., Adam, R., Mikulics, M., Siegel, M., 2003. Femtosecond response of a free-standing LT-GaAs photoconductive switch. Appl. Opt. 42, 1726–1731.

Faster than electron speed: photodetectors with confined 2D charge plasma overcome transit-time limit

Bahram Nabet[1], Fabio Quaranta[2], Adriano Cola[2], Pouya Dianat[1] and Marc Currie[3]
[1]Electrical and Computer Engineering Department, Drexel University, Philadelphia, PA, United States, [2]IMM-CNR, Institute for Microelectronics and Microsystems, Unit of Lecce, National Research Council, Lecce, Italy, [3]Optical Sciences Division, Naval Research Laboratory, Washington, DC, United States

9.1 Introduction

Typical high-speed photodiodes (PDs) absorb light-producing electron and hole pairs which are then transported by an externally applied and internally aided electric field, respectively, the anode and cathode contacts. The charges collected at these contacts produce an electric signal in the external circuitry as a response to the optical stimulus. Of the several factors which determine the speed of the response of the PDs, the transport of optically generated charge carriers to contacts is an important limitation that forces a trade-off between responsivity and bandwidth, or speed of response. Figure 9.1 classifies the main trends in PD structures. The PIN PD and its avalanche photodiode version have been a staple of high-speed

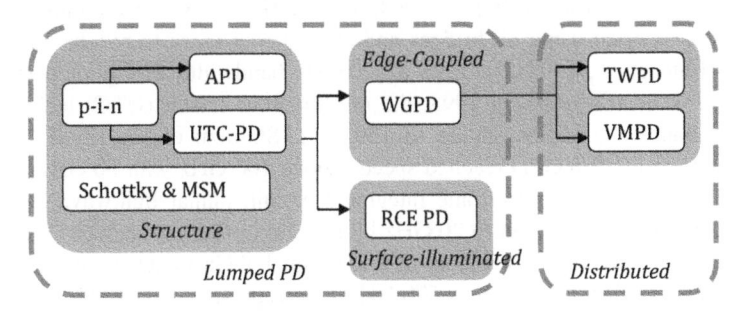

Figure 9.1 Lumped and distributed high-speed photodetectors. *APD*, Avalanche photodetector; *RCE PD*, resonant cavity enhanced photodectector; *TWPD*, traveling-wave photodetector; *UTC*, uni-traveling carrier; *VMPD*, velocity-matched photodetector; *WGPD*, waveguide photodiode.

PDs used in fiber optic communication systems since the 1980s (Sze, 1981; Auston et al., 1980; Lee, 1984; Bowers and Burrus, 1987; Kato et al., 1994; Crawford et al., 1990; Razeghi, 2000; Keiser, 2000; Donati, 2000). Recent progress in PIN diodes by some of the pioneering researchers listed above is reviewed in this book.

The study of metal-semiconductor-metal (MSM) structures emerged in the early 1970s (Lee, 1984; Sze et al., 1971) with MSM PDs being fabricated on GaAsCr (Lawton and Scavannec, 1975), amorphous silicon (Auston et al., 1980), and radiation damaged silicon-on-sapphire (Doany et al., 1987) substrates. Due to the development of monolithic fabrication techniques, the first GaAs Schottky PD above 100-GHz bandwidth was reported as early as 1983 (Wang and Bloom, 1983). A transit time-limited GaAs MSM PD was reported with a finger width of 0.75 μm and a finger spacing of 0.5-μm in 1988. This device achieved the impulse response of 4.8 ps full width at half maximum (FWHM) and a 3 dB bandwidth of 105 GHz (Patrick et al., 1988). Recombination-time-limited MSM PD on low-temperature-GaAs (LT-GaAs) with a finger width and spacing of 0.2 μm, an FWHM of 1.2 ps, and a 3 dB bandwidth of 375 GHz was reported in 1991 (Chen et al., 1991). In 1992, an MSM with a finger width and finger gap of 25 nm was reported to achieve a bandwidth of 510 GHz on LT-GaAs (Chou and Liu, 1992). Recent progress in MSM including devices employing LT-GaAs was reported (Currie, 2016).

The other branch of high-speed detectors related to the present chapter is the edge-coupled Waveguide Photodiode (WGPD) design. The first high-speed edge-coupled WGPD was demonstrated in 1986, with a bandwidth of 28 GHz and an efficiency of 25% (Bowers and Burrus, 1986). In 1995, the first traveling wave structure photodetector with a bandwidth of 176 GHz was demonstrated (Nagarajan et al., 1995) followed by a PIN LT-GaAs photodetector with a 3 dB bandwidth of 560 GHz using the LT-GaAs traveling wave technique (Chiu et al., 1998). These designs were adapted for long-haul communication purposes using InP substrates and, typically, InGaAs ternary material which allowed for the monolithic integration of photonic and high-speed electronics. In the 1990s the 16 ps response instrument limited PD was followed by the demonstration of the first long wavelength photodetector above 100 GHz bandwidth, in which the double graded layers were used to improve the device speed (Wey et al., 1991). In 1994, a 110 GHz bandwidth for mushroom-type WGPD was reported using air-bridges and an undercut mesa to reduce the RC-time constant (Kato et al., 1994). In 1995 a 120 GHz bandwidth PIN photodetector was reported (Tan et al., 1995). In 1998, the uni-traveling carrier (UTC) technique was used to enhance the bandwidth up to 150 GHz (Shimizu et al., 1998). While the WGPD and the UTC WGPD reached speeds above 100 GHz with 60 GHz packaging and 50 Gb/s receiver optoelectronic integrated circuit, planar lightwave circuit integration of WGPD was limited to 20 GHz by the end of the last century (Kato, 1999). MSM traveling wave photodetectors (TWPDs) on LT-GaAs substrate with (Fourier Transformed) bandwidth as high as 570 GHz were reported in 2001 (Shi et al., 2001). Recent progress in waveguide PDs is reported in this book by contributors with seminal contributions to the field (see Chapters 12 and 14).

An important limiting factor in the speed of the operation of these photodetector devices is, as mentioned previously, the transit time of the electron and holes to the contacts. This also sets a required minimum kinetic energy of $\frac{1}{2}m^*v_d^2$, where m^* is the carrier effective mass and v_d is the electric field-induced carrier drift velocity. This is in line with the historic perception of transistor-based microelectronics (Lilienfeld, 1933; Brattain and Bardeen, 1953) where the transport of (discrete) charge carriers whose velocity then results from their acceleration by the force of the electric field, is a requirement for the transfer of information. This (drift) velocity is derived from electrons' energy-momentum (E-K) relation and is limited to about 10^7 cm/s in high mobility semiconductors such as GaAs. Electron and hole effective masses are also derived from this relation as an important particle-like characteristic of these charge carriers.

In contrast, wave motion is built upon a dielectric function $\varepsilon(\omega)$ that characterizes the collective response of a medium to electric field excitation. The wave motion in a medium with $\varepsilon(\omega)$ can be derived from Maxwell's equations, and a dispersion relation (ω-K) can be found, from which, similar to the E-K relation, information about the group velocity of the wave traveling in the medium can be extracted. This wave (group) velocity is much higher than the electron velocity caused by the force of the electric field and may be considered the basis of electronics as envisioned by Tonks and Langmuir (1929), Debye et al. (1923), Bohm and Pines (1953), Gabor (1956), and Landau and Lifshitz (1960), among others.

The wave motion in an electron gas medium has time constants of the order of the dielectric relaxation time of the medium which is proportional to the product of the medium's permittivity ε_s and resistivity ρ, which for high charge densities is in the tens of femtoseconds range. On the other hand, time constants based on charged particle motion are much slower than these dielectric relaxation times (Haug and Koch, 2009); this is to be expected since the former can be an energy relaxation process, while the latter is due to real charge motion caused by acceleration by the force of the electric field, and deceleration due to scattering. By analogy, if the electron transport current is similar to water flow in a river, the dielectric response is the wave in a pond. In this chapter we report on a photodetector device that uses two-dimensional electron and hole gases, creating reservoirs of charge which mediate detector response optical excitation with speed and sensitivity that it is not possible to obtain with a current flow model/picture.

The reservoirs of charge produced here are those with sheets of electrons and holes whose motion is confined to two dimensions rather than the 3D motion that occurs in bulk semiconductors. The study of the interaction of electromagnetic radiation with structures that confine electrons to an interface, such as metals, was pioneered by Ritchie (1957), and verification of the collective modes of electron excitation, plasmons, appeared in 1960 (Wilberg and Ingelsfield, 1960). Besides metals, collective properties of electron gases such as at dielectric interfaces, or the surface of liquid helium have been the subject of intense interest (Ando et al., 1982), being motivated by the study of the role of surface states in metal-insulator-semiconductor devices (Brattain and Bardeen, 1953), and in the silicon metal oxide semiconductor field effect transistor (MOSFET) device (Allen et al., 1977). Collective modes of excitations, plasmons, in a two-dimensional electron gas (2DEG) were first observed for electrons in the system of electrons trapped by the image potential on liquid helium in 1976 (Grimes and Adams, 1976) and then in the inversion layer of MOSFET (Allen et al., 1977). Direct interaction of radiation with the electrons in the MOSFET inversion channel was studied as early as 1976 (Allen et al., 1976). A serious interest in the room-temperature properties of 2DEG was, however, based on the seminal work at Bell Laboratories on the "inversion channel" of AlGaAs/GaAs heterojunctions (Dingle et al., 1978) and InP (Von Klitzing et al., 1980). Though initiated for the study of the collective behavior of the electron plasma, reduced electron scattering, and higher mobility of the 2DEG compared to the bulk, it resulted in its successful incorporation as the charge transport channel of high electron mobility transistors (HEMT) (Mimura et al., 1980; Mimura, 2002; Delagebeaudeuf et al., 1980; Deal et al., 2011). Presently such transistors hold a speed of operation record at well over 650 GHz (Deal et al., 2011) that is primarily limited by the transport time of electrons in the <40 nm distance between the source and the drain electrodes. Hydrodynamic modeling of the 2DEG as a whole (Dyakonov and Shur, 1993, 1996) has shown the possibility of plasma wave propagation, and hence the much higher speed of operation, in HEMTs with lengths less than momentum relaxation distance, and resulted in a range of plasma wave electronic devices (Dyakonov and Shur, 1996; see also Chapter 6). Below we focus on a novel photodetector device with bi-layer reservoirs of confined electrons and holes, with contact separation as large as 8.5 μm—much larger than momentum relaxation distance—which in response to a 400 fs perturbation by as little as ~11,000 photons produce a short, less than 2.5–1.9 ps if corrected for testing circuitry limitations—an electric response which would take over 100 ps if it were based on charge transport. The sensitivity of the process, the mechanisms that may mediate its operation, and applications of such micro plasma-based devices will also be discussed.

9.2 Device structure

Figure 9.2A shows the layer structure of a wafer grown by metal-organic chemical vapor deposition on semi-insulating GaAs. After the growth of a buffer layer, 57 nm of $Al_{0.3}Ga_{0.7}As$ was lattice-match grown and p-type delta-doped with carbon at 2.5×10^{12} cm^{-2}. A spacer layer of AlGaAs then makes a heterojunction with an 8-nm layer of the (strained) narrow bandgap $In_{0.2}Ga_{0.8}As$. The band offsets keep the hole gas, produced from the acceptor dopants, from motion in the direction of growth, but carriers are free to move in the other two dimensions thus producing 2DHG. A 109.4-nm layer of GaAs is grown on top of InGaAs to absorb light and excite the 2D reservoirs with generated carriers, followed by a 5-nm AlGaAs spacer and a 56.4-nm AlGaAs n-type layer delta-doped with Si at 6×10^{12} cm^{-2} dopant density, to produce a 2D electron gas at this interface. Additionally, geometrically identical devices were fabricated for comparison purposes with a single 2DEG or 2DHG layer, undoped conventional structures without the two-dimensional reservoirs of charge, as well as wafers with 15 pairs of GaAs/AlGaAs, which formed a Bragg mirror for 830 nm wavelength, were also grown to produce a resonant cavity enhanced structure (Table 9.1).

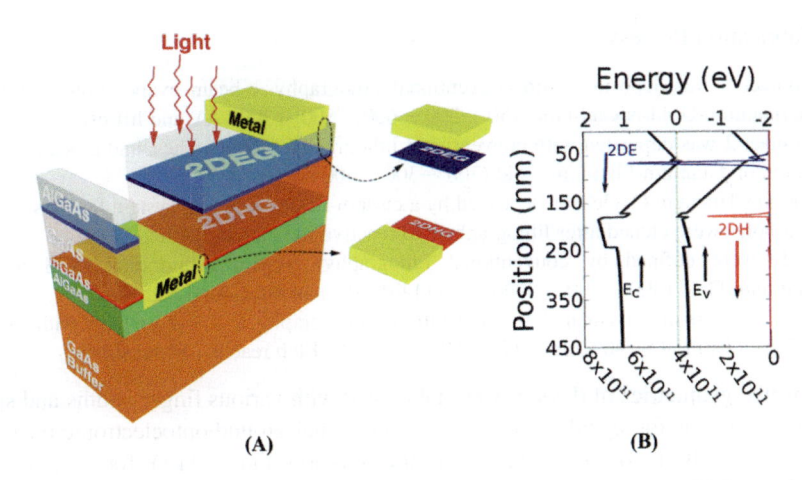

Figure 9.2 (A) Top illuminated photodetector layer structure. Two dimensional electron and hole gas reservoirs, separated by an ~110 nm GaAs region that absorbs light, are produced by modulation doping, and are separately contacted. (B) Calculated energy band diagram along the direction of growth in the middle of the device shows existence of large vertical electric field, as well as dense electron and hole gasses which can move only in two dimensions.

Table 9.1 The layer structure and thicknesses of the metal-organic chemical vapor deposition grown wafers.

Material	Thickness (Å)
GaAs	30
AlGaAs	564
Si delta doping	
AlGaAs	50
GaAs	1094
InGaAs	80
AlGaAs	50
C delta doping	
AlGaAs	573
GaAs	2000
GaAs SI substrate	

The energy band diagram (EBD) of this structure is calculated by the self-consistent solution of Poisson and Schrodinger equations and is shown in Figure 9.2B. The slope of the EBD being the electric field, this figure shows the existence of a large (>100 kV/cm) built-in *vertical* field in this structure. Electron and hole distributions are also calculated and shown, indicating the existence of relatively dense concentrations, ~6.5×10^{11} cm^{-2} electrons, and ~2.2×10^{11} cm^{-2} holes, that are verified by Hall measurement.

Fabrication Process:

- Contact 1 was deposited with conventional lithography, e-beam evaporation of Nickel-Germanium-Gold-Nickel-Gold (50 −300 −600 −300 −2000 Å), and lift-off.
- Contact 2 was deposited with conventional lithography, recess etches, and e-beam evaporation of Titanium-Platinum-Gold (300 −300 −600 Å).
- 100 nm Titanium Oxide was deposited by a custom-made sputtering system for passivation.
- Via holes were etched after lithography and reactive ion etching with CF_4 gas.
- Pads were defined by conventional lithography, e-beam evaporation of Titanium-Platinum-Gold (300 −300 −3000 Å), and lift-off in an ultrasound bath.
- Devices were mesa isolated by conventional lithography and wet etching with H_2SO_4 (95%−97%):H_2O_2 (40%):H_2O (1:8:160) $T = 20°C$. Etch reached SI substrate.

Twelve geometries of devices were fabricated with various finger widths and spacing, contact pads for signal-ground and ground-signal-ground optoelectronic measurements, as well as devices placed on transmission lines (TLs) for electro-optic sampling (EOS) measurement of the time response. Figure 9.3A shows a confocal

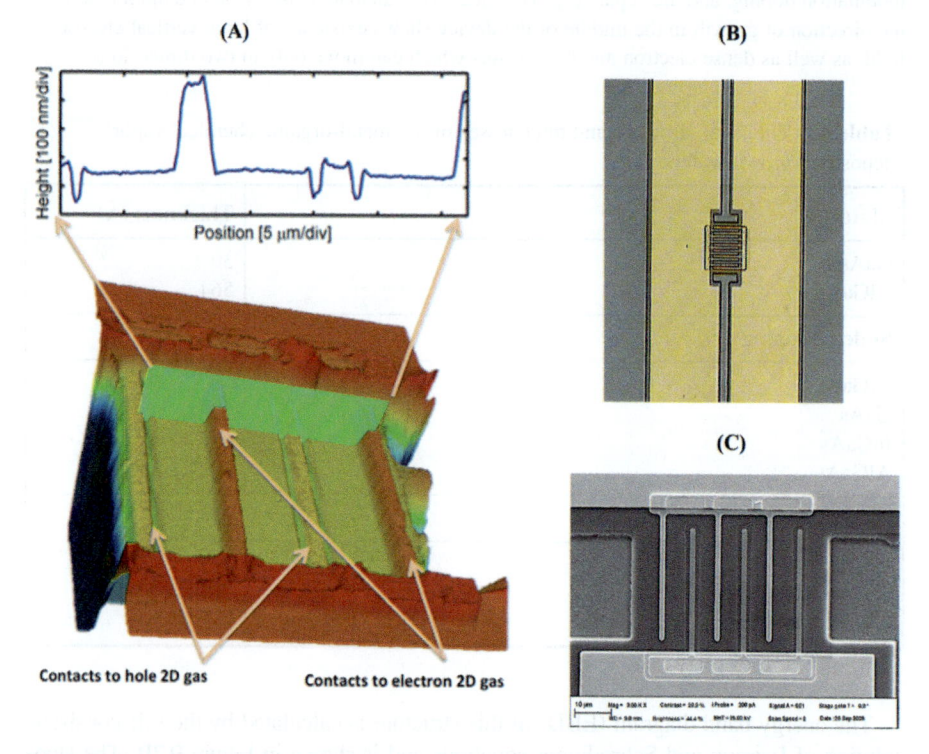

Figure 9.3 (A) A confocal microscope image of the fabricated device. Top of the figure is a cross-sectional cut showing the widths and heights of the interdigitated contacts which are separated by >8.5 μm. (B) An image of the photodetector device in a transmission line, used for electro-optic sampling measurement of the time response. (C) Close up of the fabricated device with separate contact to electron and hope charge reservoirs.

microscope image of a fabricated device. The interdigital structure reduces capacitance and ensures that the device is not RC time limited. As seen in Figure 9.3A and C, the cathode and anode separately contact 2DHG and 2DEG reservoirs, respectively. The contacts were asymmetric measuring 8.2 and 8.7 μm apart. Figure 9.3B shows the device embedded in a co-planar TL so that its time response can be measured with a high-resolution EOS technique.

This device is thus similar to the planar MSM (Sze et al., 1971) photodetector in the in-plane direction, and, due to its 2DH-insulator-2DE structure, to a (vertical) PIN (photo) diode in the direction of growth. It also has similarities to the waveguide and traveling wave, photodetectors (WGPD, TWPD) (Bowers and Burrus, 1986; Kato, 1999) with the 2D charge plasma sheets replacing the metal waveguides. It is, however, distinct from these devices in crucial characteristics. It is not the vertical PIN where the Ohmic contacts (to p^+ and n^+ doped regions) are separated by the thin intrinsic absorption region. It has the planar simplicity of top illuminated MSM but not the electric field landscape that sweeps the carriers to be collected at the Schottky contacts that are >8 μm apart; rather there exists a strong vertical field that separates the 2D charge reservoirs, with the latter mediating the electric response. And, contrary to the side-illuminated WGPD and TWPDs it does not need (long) metallic TLs although it offers the same speed of response. An interesting, somewhat similar, structure consisting of double-graphene layers has recently been proposed for resonant excitation of plasma oscillations (Ryzhii et al., 2013).

9.3 Current-voltage relationship

We have demonstrated MSM photodetector devices with an embedded single 2DEG reservoir previously (Nabet, 1997; Nabet et al., 2000, 2001, 2003; Anwar and Nabet, 2002; Chen et al., 2002, 2003, 2005; Cola et al., 2005). The dark current, being the most important source of noise in MSM and Schottky diode devices, is due to the thermionic emission current, which for bulk (3D) is given by:

$$J_{\text{thermionic}} = \int_{v_x(E_F + q\phi_B)}^{\infty} qv_x dn \qquad (9.1)$$

where v_x is the electron velocity perpendicular to the interface, dn is the product of the (3D) density of states, and the Germi function, $dn = N_{3D}(E)f(E)dE$, resulting in:

$$J_{\text{th}} = A^* T^2 \exp\left(\frac{-q\phi_{\text{bn}}}{kT}\right) \exp\left(\frac{q\Delta\phi_{\text{im}}}{kT}\right) \left[\exp\left(\frac{qV_a}{kT}\right) - 1\right] \qquad (9.2)$$

Here $A^* \equiv \frac{4\pi q m_n^* k_B^2}{h^3}$ in units of $A/(cmK^{2/3})$ is the Richardson constant, $\Delta\phi_{\text{im}}$ is the barrier lowering due to the image force, V_a is the applied voltage, and other symbols have their usual meaning.

When electrons are thermionically injected from metal to a 2D semiconductor, the current can be calculated in the same manner by considering a 2D density-of-states and integrating kinetic energies appropriately, resulting in (Anwar et al., 1999; Tait and Nabet, 2003, 2004; Ragi et al., 2005):

$$J_{2D} = J_{sat}\left[\exp\left(\frac{qV_a}{k_B T}\right) - 1\right] \tag{9.3}$$

where the reverse saturation current is:

$$J_{sat} = A_{2D}^* T^{3/2} \exp\left(-\frac{q\phi_{Bn}}{k_B T}\right) \exp\left(-\frac{E_o}{k_B T}\right) \tag{9.4}$$

and $A_{2D}^* = \frac{2q}{h^2}\sqrt{2\pi m^* k_B^3}$ is a modified Richardson constant.

Important differences thus exist when the semiconductor is reduced dimensional. First, it is observed that in the 2D system the barrier height has been increased by the value of the first confined state as indicated by the term $\exp\left(-\frac{E_o}{k_B T}\right)$. Second, the reverse saturation current J_{sat}—which is the dark current in MSM and Schottky diodes—depends on the 2D density-of-states function which is smaller than that of the bulk of the edge of the conduction band at Ec. In addition to device dimension, the prefactor in J_{sat} is based on terms due to the integration of carriers with velocity components confined to a plane, rather than a volume. Numerically, the exponential term shows a current reduction by a factor of about 12, based on a typical value of $E_o = 0.65$ meV. In the prefactor term, the constants are $A_{2D}^* = 2.6 \times 10^{-4}$ (A/cmK$^{2/3}$) and $A_{3D}^* = 8.6$ (A/cmK^2), resulting in nearly two orders of magnitude reduction in the dark current. Thirdly, the time dependence of the metal-3D Schottky diode is T^2, while it is $T^{3/2}$ for a metal-2D system. These predictions on (dark) $I-V$ relation that were originally derived for 2DEG, were later verified for physical 2D systems of graphene (Wu et al., 2008), graphene oxide (Nourbakhsh et al., 2010), and MoS$_2$ (Chen et al., 2013).

We have also reported on similar photodetectors with an embedded confined reservoir of hole gas, 2DHG (Zhao et al., 2006, 2008), as well as a novel device with both electron and hole gasses (2DEHG) (Currie et al., 2013; Nabet et al., 2014); the latter being the subject of the present chapter. Incidentally, similar electron-hole bilayer systems have been investigated in different contexts since the prediction of the formation of bound electron-hole excitons in parallel sheets of electrons and holes (Keldysh and Kozlov, 1968) which would result in important anomalous behavior, such as the formation of Bose-Einstein condensates, or possibly supersolids (Joglekar et al., 2007). However, these properties are expected to occur when the layers are much closer than the typical interparticle distance of a few nanometers (Sivan et al., 1992; Pohlt et al., 2002), which is not the case in our structures.

The current-voltage ($I-V$) relation in ambient room light (dark) and under continuous wave illumination by an 835-nm laser at different optical intensity levels are shown in Figure 9.4A and the photocurrent versus optical intensity in Figure 9.4B.

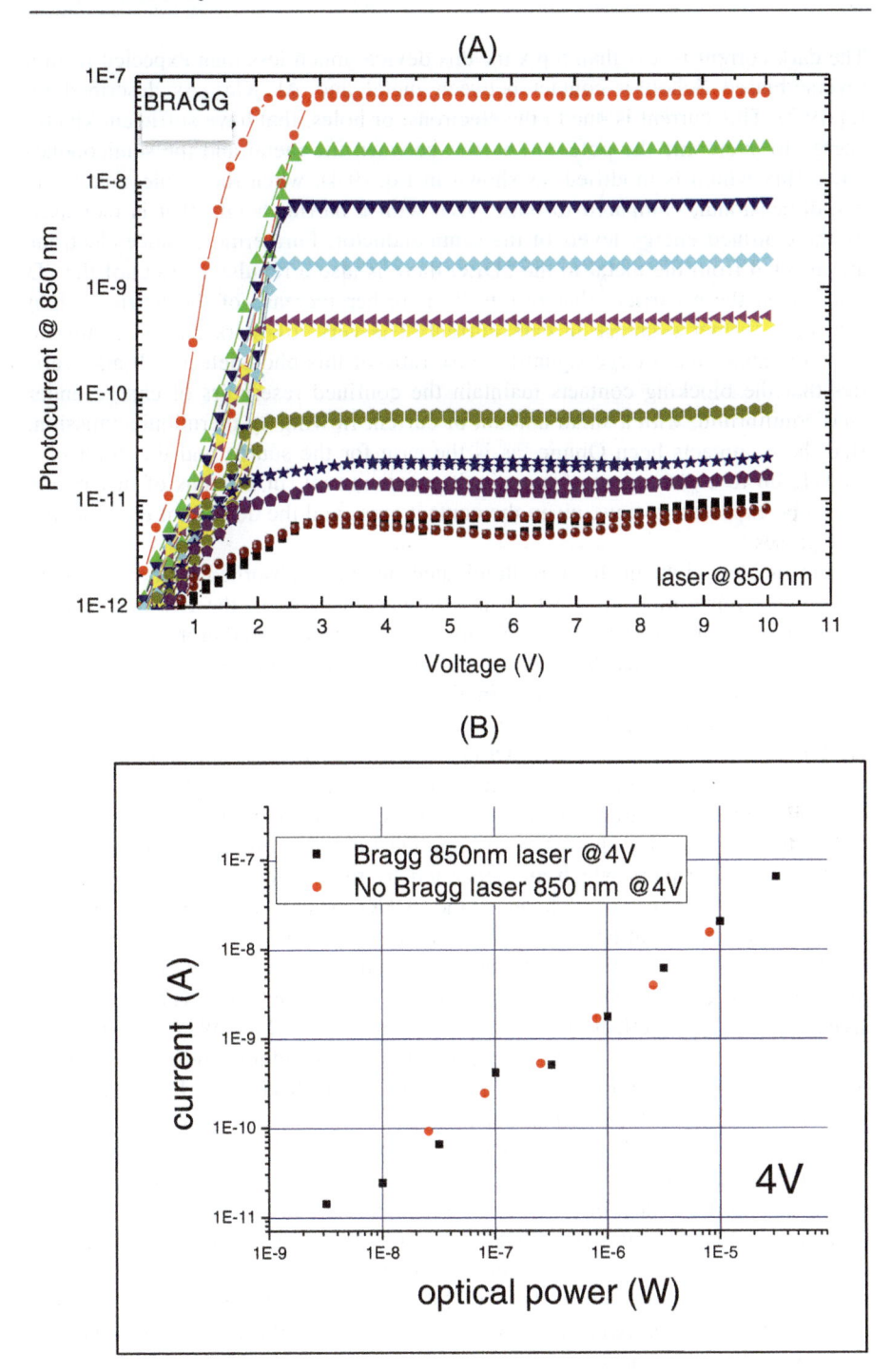

Figure 9.4 (A) Photodetector current-voltage characteristics in dark and under variable illumination intensity. (B) Photodetector responsivity is linear with illumination intensity.

The dark current is less than 5 pA for this device, much less than expected from a reverse-biased Schottky contact between metal and $Al_{0.2}Ga_{0.8}As$ described by Eq. (9.2). This current is due to the electrons, or holes, that have sufficient kinetic energy to overcome the potential barrier between the metal and the semiconductor, a flux which is modified, as shown in Eq. (9.4), when the semiconductor is two-dimensional, compared to bulk (3D), with a barrier height that is increased by the confined energy levels of the semiconductor. Furthermore, since electrons are injected from the metal to the 2DEG there is also a repulsive effect of the 2D charges on these carriers that results in a further increase of the barrier height causing a further decrease in dark current. The very low dark current results in very low noise and a large signal-to-noise ratio of this photodetector. It also verifies that the blocking contacts maintain the confined reservoirs of charge under quasiequilibrium, with a small amount of current flowing by thermionic emission. Had these contacts been Ohmic, as is the case for the source and the drain of a HEMT, up to eight orders of magnitude more current, in the tens of mA range, would be expected to flow, given the contact areas and the density of electron and hole gasses.

The incident light in this top-illuminated device is absorbed in the ~ 110 nm thick GaAs absorption layer which is sandwiched between the two-dimensional sheets of electron and hole gas reservoirs. Figure 9.5 is a simulation of the electric field distribution in this 2DEHG device under equilibrium and bias voltages of ± 0.5, ± 1.5, and ± 4 V. As seen in Figure 9.5, there is a large $> \sim 40$ kV/cm internal vertical field in the GaAs absorption region with field lines originating in 2DHG sheet and terminating in 2DEG. Under applied bias, the vertical field is seen to remain mostly unchanged. This is because the dense 2D plasma of mobile electrons and holes rearranges themselves in response to applied bias by about a Debye length, shielding the externally applied field. The incident light generates electron and hole pairs which are separated by the vertical field, with electrons moving to the 2DEG and the holes to the 2DHG, where they perturb the plasma eliciting a response that launches an external electric signal.

This device is an efficient photodetector with approximately five orders of magnitude current change caused by a 54 μW optical excitation. The DC responsivity, corrected for reflection from AlGaAs surface is 0.12 A/W. At 830 nm, the maximum responsivity of GaAs with 100% internal quantum efficiency, accounting for $\sim 30\%$ reflectivity from AlGaAs surface, is 0.47 A/W, however, for a thickness of $d = 110$ nm, and absorption coefficient $\alpha = 10^{-4}$ cm^{-1}, $[1 - \exp(\alpha d)] = 9.5\%$ of incident photons are absorbed. This DC responsivity is thus ~ 2.5 times the max responsivity expected from a 100 nm thick absorption region, mostly due to the collection of carriers generated outside this thin region, and are efficiently collected in the 2DE, 2DH reservoirs. The device is also very sensitive, with 1.2 μW of light causing a current change by a factor of over 4000 compared to the device in dark, and as low as 250 nW being detectable. The DC light responsivity and sensitivity are, however, secondary to the dynamic behavior of this device, discussed next.

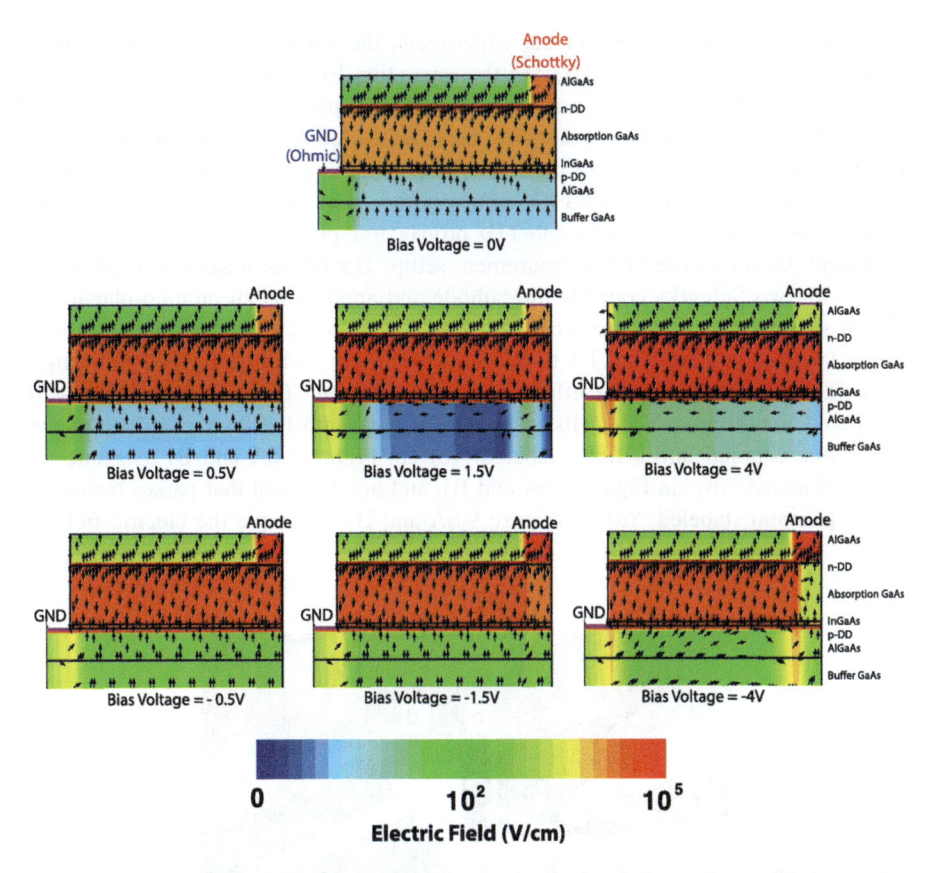

Figure 9.5 Simulation of the electric field distribution in the device under equilibrium and bias voltages of ±0.5, ±1.5, and ±4 V. The large vertical field as well as the electron and hole plasma remain unchanged under applied bias since the 2D charge plasma rearrange to shield the externally applied bias.

9.4 Time response

The dynamics of this photodetector response are probed by perturbing the device with short, 400 fs, pulses of light generated by the Ti:Sapphire laser with a center wavelength tunable from 750–1080 nm. Absorption of these pulses of light generates electron and hole pairs in the (∼110 nm thick) GaAs region. Subject to the large vertical electric field, electrons and holes separate and drift, respectively, towards the 2DEG and 2DHG reservoirs which laterally extend the long (>8 μm) distance between the contacts. High-speed testing is performed with an EOS system which can be simply thought of as an ultrafast sampling oscilloscope that uses 400 fs laser pulses to excite optoelectronic transients and then measures the electronic response by probing the refractive index change of an electro-optic crystal (e.g., $LiTaO_3$)

placed on top of the device. For our experiment, the device's electronic response is coupled to a coplanar strip (CPS) TL; the separation from the excitation fiber to the optical sampling crystal is 250 μm. Since the electro-optic crystal is sensitive to electric field (rather than voltage), sampling near the device could also probe local electric field variations, which can mask the propagating signal. Sampling the electric field 250-μm away on the CPS TL removes the local field effects, but also adds attenuation and dispersion to the sub-THz propagating pulse.

Figure 9.6 shows the EOS measurement setup. The device under test is interdigitated with two Schottky contacts for cathode and anode, that sit on a co-planar TL, shown in inset (A) of Figure 9.6A, required for EOS measurement. Devices with finger spacing ranging from 1.1 to 8.7 μm (and finger width in the 1−2 μm range) were fabricated to study the effect of transit distance. Time response data were compared for the nine wafers listed above. In our setup the laser is split into two paths: one coupled into a fiber (labeled "(c)" in Figure 9.6A and B) to excite the device (labeled "(b)" in Figure 9.6A and B), and another path that passes through a LiTaO$_3$ crystal (labeled "(d)" in Figure 9.6A and B) to sample the electric field of

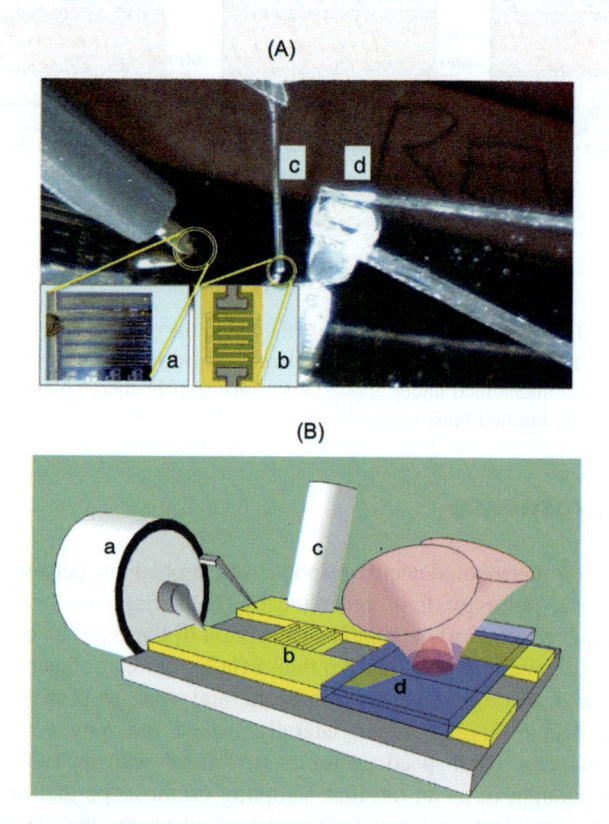

Figure 9.6 Picture (A) and sketch (B) of electro-optic sampling test set up. (a) RF probe is at one end of a transmission line in the middle of which the interdigitated device is located. (b) Picture of the fabricated device under test. (c) Pump beam is delivered by an optical fiber. (d) Probe beam sampled through the electro-optic crystal.

the propagating response. By varying the optical path of the sampling beam, the temporal response of the device is observed with a time resolution limited by the laser pulse width and the response of the electro-optic crystal. Our amplitude sensitivity is limited by the noise in our detection system. To increase the sensitivity to sub-mV levels we modulated the switching beam at 80 kHz and performed phase-sensitive detection with a lock-in amplifier on the light analyzed (with a polarizer and differential detection) from the $LiTaO_3$ crystal. In this experiment ~ 100 fs pulses from a Ti:sapphire laser with a center wavelength of 830 nm was split into the two paths. The path that is coupled into a fiber to excite the DUT (labeled "(c)" in Figure 9.6A and B) is dispersed by the fiber and exits as a 400-fs chirped optical pulse. The second laser path that passes through the $LiTaO_3$ crystal (Figure 9.6A and D) to sample the propagating electric field broadens slightly to a 150 fs optical pulse at the crystal. The photodetector's electronic response is coupled to a CPS TL. The separation from the fiber exciting the device to the optical sampling crystal is 250 μm. The device's electronic response is coupled to a CPS TL that is contacted with microwave probes (labeled "a" in Figure 9.6A and B) for DC bias and <50 GHz measurements. Bias values from -15 to $+15$ V were applied to the MSM structure with average optical powers ranging from 1/4 μW to 10 mW.

The measured time response to ~ 400 fs pulses with an average of 54 μW optical power and applied biases of 0, 1, and 2 V is shown in Figure 9.7. The same data normalized to the peak value in the inset of the figure shows pulse-width, given as the full width at half-maximum (FWHM), values of 2.9, 2.9, and 2.4 ps, respectively, with the relatively symmetrical rise and fall times. The 1.4 ps rise time of the response is longer than our EOS system response and is potentially due to TL dispersion occurring from the electrical pulse's 250 μm propagation distance. The FWHM response, however, needs further scrutiny.

9.4.1 Components of total temporal response

Several factors are involved in the temporal response measured via EOS. The overall response comprises three major elements: (1) the device response, (2) the signal propagation, and (3) the EOS system measurement. We measure the combination of these responses, resulting in a measured signal that is distorted and often results in a measured transient with a slower temporal response than the device's intrinsic speed. Our device dimensions are a fraction of the electrical transient's propagation velocity of 130 μm/ps allowing for potential temporal broadening on that time scale.

The TL adds attenuation and dispersion as the device's signal travels towards the electro-optic detection point. The TL's characteristics can be calculated, however, at frequencies exceeding 50–100 GHz; care must be taken to ensure the accuracy of calculations. Finally, the EOS system has its intrinsic response based upon parameters such as optical pulse-width, optical and electrical transit times in the crystal, optical beam size, etc. In this experiment, our EOS system has a 400 fs switching pulse and a 150 fs sampling pulse. Since the EO crystal is thin (~ 50-um) the system is primarily limited by the optical pulse-widths. As a case-in-point, the convolution of a 1500 and 400 fs Gaussian pulse (FWHM) produces a 427 fs Gaussian pulse response. This laser system response was applied to measure three

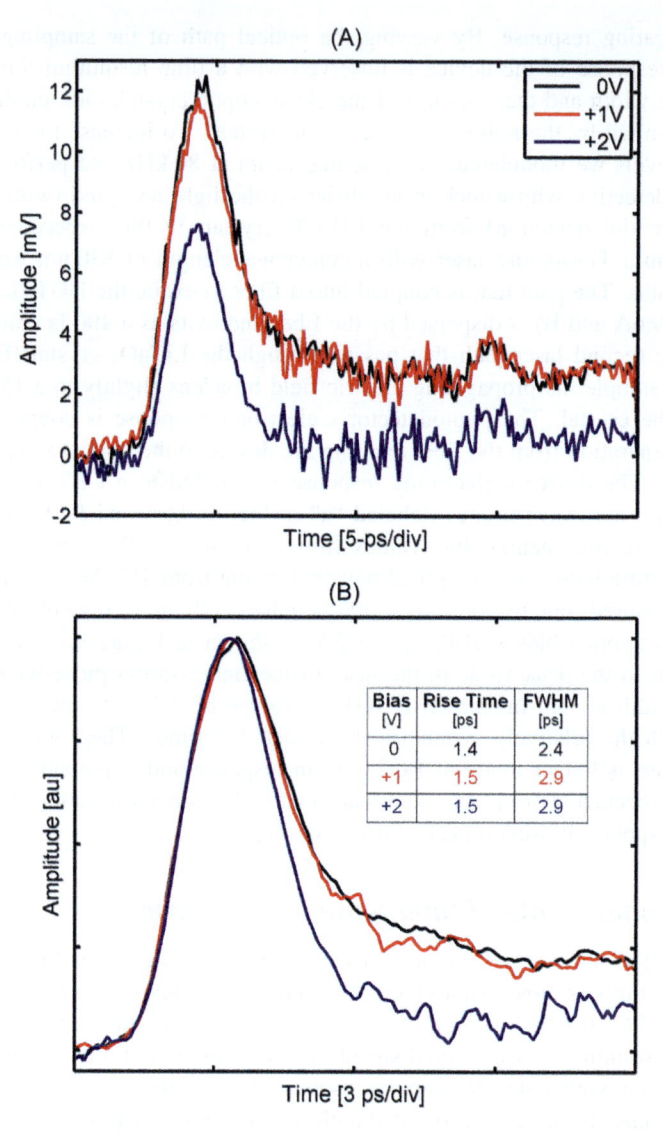

Figure 9.7 (A) Time response of the device shown in Figure 9.3C measured by electro-optic sampling technique under zero, 1 and 2 V. (B) Normalized time response of device with >8.5 μm contact separation shows full-width half-max values in the inset, which are as fast as those expected from a 100 nm device.

waveforms: (1) unit step function, (2) unit step function multiplied by an exponential decay, and (3) a 2-ps-wide Gaussian pulse. The convolution of the laser system response with these waveforms results in 0.46 ps broadened rise times and up to 0.5 ps broadened pulse-widths.

Hence the intrinsic response of the photodetector is as low as 1.9 ps for a bias voltage of 2 V, and 2.4 ps for 0 and 1 V. Such a short response was previously

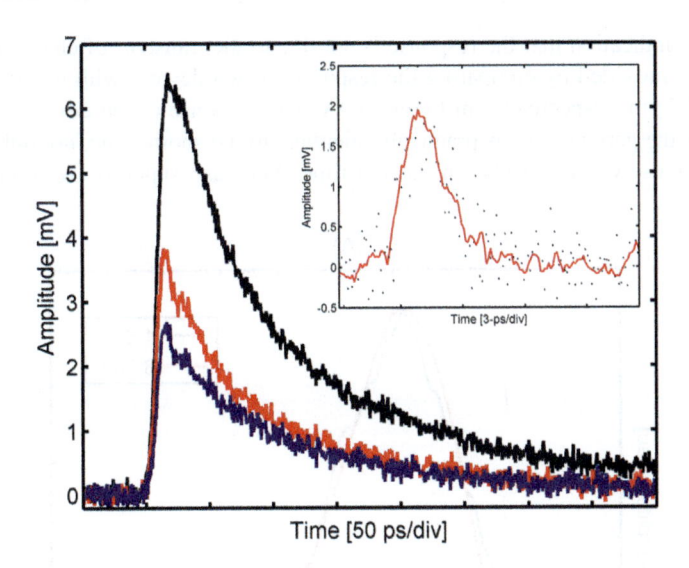

Figure 9.8 Time response of a device with 8.5 μm gap distance, but without 2D electron and 2D hole reservoirs, under 7 (blue), 9 (red), and 15 (black) V bias shows FWHM of 50, 55, and 75 ps, respectively, limited by electron transit time, with a 200–250 ps, tail that is due to the slow moving holes. Inset is time response of a 2DEHG device with similar geometry under the same optical power; it is much faster with <2.9 FWHM and ~2 ps fall time.

reported for an MSM device with tens of nanometers in finger spacing (Chen et al., 1991; Chou and Liu, 1992); it cannot be due to the transit of electrons to the anode, which in the best case of the saturation drift velocity of 10^7 cm/s would be around 80 ps, with holes taking nearly ten times longer, depending on the electric field intensity, to transit to the cathode. The time response of a similar device with 8 μm separation of contacts but without the charge reservoirs, is shown in Figure 9.8. The temporal pulse-width for 11 μW incident power is 50, 55, and 75 ps (FWHM) for respectively, 7, 9, and 15 V bias—the larger bias was chosen to assure carrier sweep out and fair comparison. The response tail, the fall time, which depends on the transport and collection of slow-moving carriers, is as high as 200–250 ps in this device. This may be contrasted with the response shown in the inset of Figure 9.8 for a device with 2DEG and 2DHG reservoirs under 7 μW of power and more than 8.2 μm cathode-anode distance, that has a pulse-width of fewer than 3 ps (FWHM)—corrected to 2.5 ps intrinsic response, and fall time of less than 2 ps. This orders-of-magnitude increase in speed is mediated by the collective response of the charge reservoirs thus circumventing the drift velocity limitations. The short tail of response shows that holes are "collected" with the same efficacy as the electrons, that is, the single carrier momentum $m^*v = \hbar k$, where m^* is the effective mass, does not limit the response, rather it works with the time constants of the collective response of the hole plasma. Indeed the speed of this large area, the high-efficiency device is still limited by the transit time of the electrons and holes, which are produced in the thin absorption region, to the 2DE and 2DH charge plasma reservoirs.

Further indication that the response is not due to the transport of charge carriers to contacts is provided by comparing the response of two devices with gap distances of 1.8 and 8.7 μm, respectively, in Figure 9.9A. The response of the device with nearly five times the gap distance is practically identical to the shorter one, not only in rising time and pulse-width but also in the fall time. This also supports the argument that

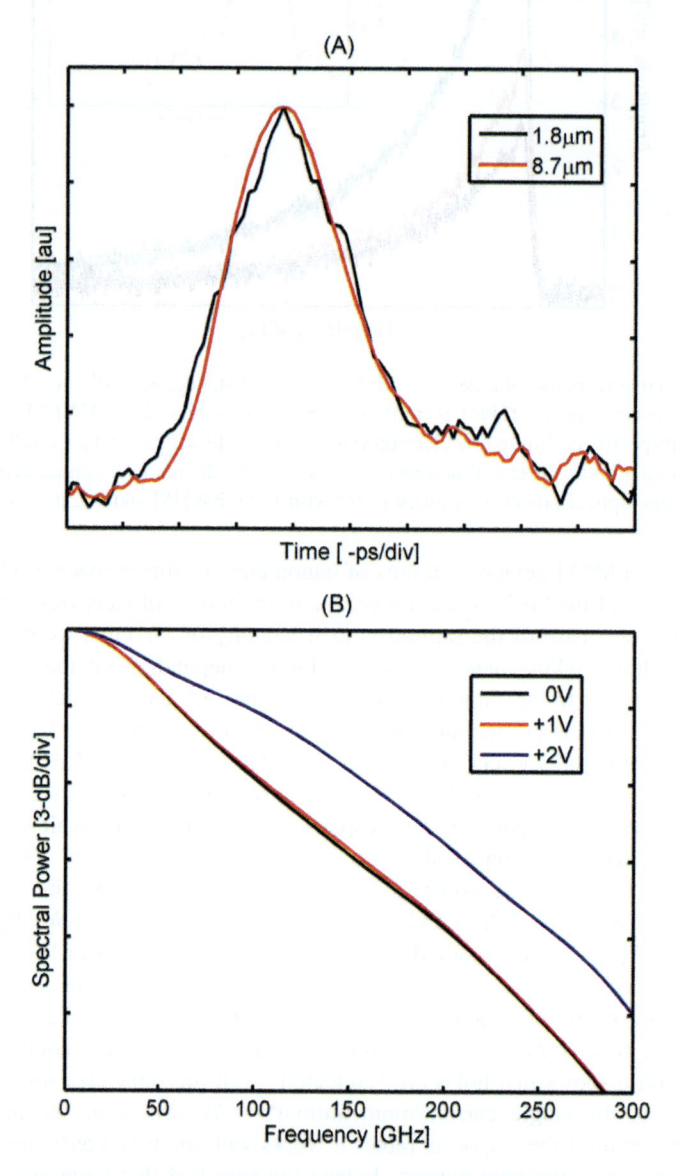

Figure 9.9 (A) Time responses for devices with 1.8 and 8.7 μm contact separations are nearly identical, and are independent of charge transport distance. (B) Power spectral response of the photodetector at three bias values derived from the as-measured time response.

the 2D hole reservoir reacts in the same manner as the 2D electron reservoir with time constants that are of the order of dielectric relaxation time, implying that the hole effective mass (used to determine the drift velocity in response to the electric field's force) is rather immaterial. This is to be expected since the effective mass is derived from the force-velocity relationship of the E-K relation, while here transfer of energy is the collective response of the medium. Additionally, the response was independent of the spatial location of the excitation beam of ~ 1 μm width; that is, the fast speed is not due to carriers being collected near one contact. The Fourier transform of the "as-measured" time response is shown in Figure 9.9B and indicates that this device is operable at several hundred GHz with no applied bias with the intrinsic bandwidth, as explained above, being much higher.

As previously noted, the key to device operation is that the dense plasma of charge is maintained in quasiequilibrium by using blocking Schottky contacts and applying relatively small biases; if a large bias is applied or perturbation by light produces a large number of carriers, device performance degrades. Figure 9.10 shows the response of a device with 1.1 μm finger spacing, under 7 μW average optical power, for a bias voltage range of ± 5 V. As seen here, the device is fastest with low or no applied bias since at higher bias values the charge reservoirs can empty, and the electric field component perpendicular to contacts weakens relative to the tangential field applied by bias. This causes a longer transit distance and higher loss in the 2D, thus degrading device response. Data were taken for various geometries and with spatial localization of the light beam. Device response was not dependent on the excitation beam's proximity to either cathode or anode, and, as Figure 9.9 shows, was independent of the distance between them. The device can operate at zero electric bias as shown. This may be explained by observing that after the photogenerated electrons and holes reach their 2DEG and 2DHG reservoirs, respectively, a potential difference

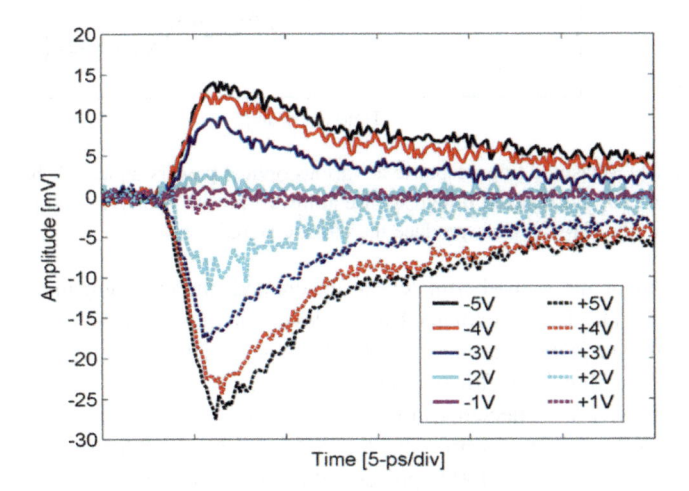

Figure 9.10 Time response data for bias voltage ranging from -5 to $+5$ V, under 7 μW optical power.

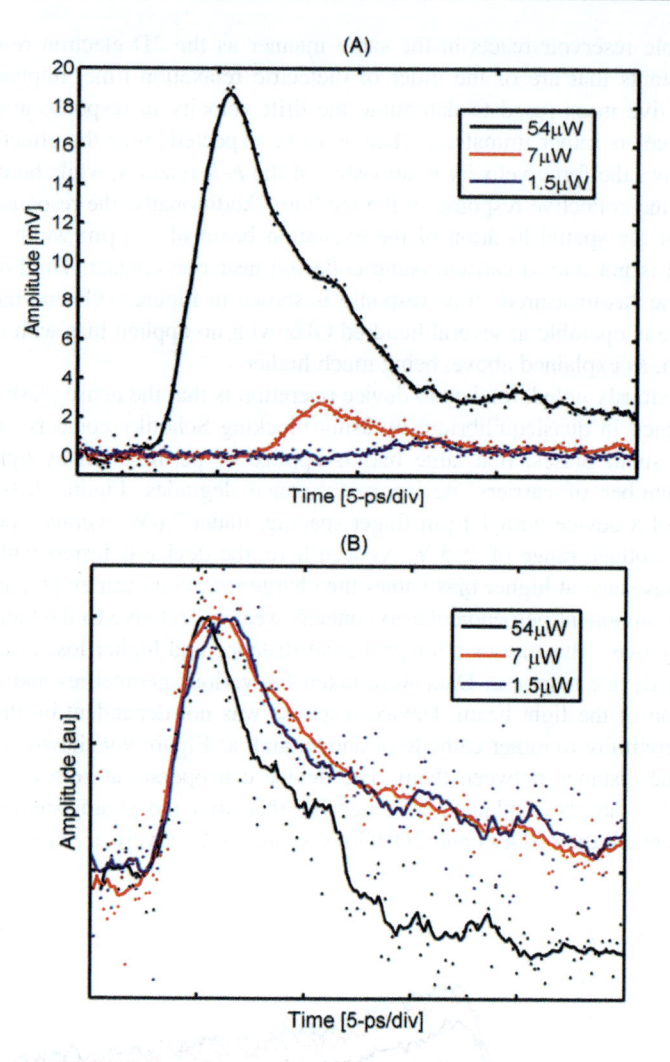

Figure 9.11 (A) Measured time response at various optical powers under 2 V bias shows high sensitivity. (B) Normalized time response at lowest power, nearly 10,500 photons that are absorbed in the GaAs region, produce the electric signal.

and separation of the quasi-Fermi levels are created, thereby launching the electric signal onto the TL. This is explained if the device is viewed in the vertical direction as 2DHG-intrinsic-2DEG, similar to a PIN detector, which has well-known photovoltaic properties, with current flowing under zero bias.

High sensitivity is expected from the picture of a reservoir being perturbed by small excitation, similar to observing the ripples caused by a drop of water on a serene lake. Response to ∼400 fs pulses with 1.5, 7, and 54 μW of average optical power under 2 V bias, shown in Figure 9.11A and B, corroborate this expectation. The 1.5 μW light pulse of 400 fs duration, repeated at 76 MHz and chopped at 50%

duty cycle, corresponds to roughly 4×10^{-14} J of energy, or equivalently, 167,000 photons at a wavelength of 830 nm. Considering 30% reflectivity from AlGaAs surface and 10% reflection by the metal electrodes, this corresponds to an incident flux of 105,000 photons. Moreover, given the absorption coefficient of GaAs, nearly 90% of these photons penetrate through the ~110 nm thick GaAs absorption layer without being absorbed. At such low flux, photon absorption is probabilistic. Furthermore, the generated electron-hole pairs (EHPs) move to charge reservoirs and will scatter and recombine during this (short) transit. The third source of loss is the (plasmon) wave loss in the >8 μm long 2D reservoirs. Lastly, the microwave signal travels 250 μm on the TL from the device to the probes, suffering its associated losses. Nevertheless, the 10,500 photons in the absorption region, produce a 6.5 ps wide and 1.5 mV tall pulse. with an identical pulse propagating in each half of the 80 Ω TL, resulting in $N = I^* dt/q = 1500$ electrons per pulse. This is remarkable considering that the device operates at room temperature, whereas such levels are typically expected from cryogenically operated detectors.

Pulses with as low as 250 nW were detected with electronic pulse-width less than 4 ps. Intensity increase results in widening of the response. This is expected since the separation of EHPs produces an electric field in the vertical direction that counters the built-in field between 2D reservoirs. Further saturation effects are expected if the EHP densities become comparable to 2DE and 2DH densities, forcing the system out of the perturbation regime. Incidentally, the delay in electric response observed in this figure is due to photon travel in the neutral density filters which are used to attenuate power, that is, EOS is measuring travel time within filters as well.

9.5 Analysis and modeling

The data shows that the collective response of the plasma overcomes charge transport limitations, however, the exact mechanism of the response of the device needs to be further discussed. We follow a two-step process to analyze device dynamics. First, the effect of the motion of optically generated carriers to charge reservoirs under the force of the electric field is calculated using the Shockley-Ramo theorem (Shockley, 1938; Ramo, 1939) which was originally proposed for one dimension, but a two-dimensional model can be derived by equating the work done by motion of charge $(-q\vec{E}dx)$ with the energy provided by the external circuit $(i(t)V\,dt)$. This allows evaluation of the current induced in the external circuit (Zhao, 2006):

$$J_{ph} = \sum_q q\vec{v}(\vec{r})\vec{E}(\vec{r})/V \tag{9.5}$$

where, $\vec{v}(\vec{r})$ is the velocity of a moving carrier; $\vec{E}$ is the component in the direction $\vec{v}$ of the electric field, which exists at the electron's instantaneous position; and V is the applied bias. The total induced current can be calculated by summing over all the charges, namely electrons and holes, in the active region. The equation involves the usual assumptions that induced currents due to magnetic effects are negligible but incorporates the displacement current since it self-consistently accounts for the electric field perturbation. The simulation starts with producing a mesh in which optically generated carriers are deposited, and the electric field is known as given in Figure 9.5. Motion of charged particles instantaneously produces a current in the external circuit, with integral of $i(t)$ being one unit of charge when EHPs reach their respective contacts, that is, are collected, which in this case are the 2DE and 2DH reservoirs. Consequently, the limiting factor in device speed of response is the transit of electrons and holes in the short ~ 100 nm thick absorption layer, since energy transfer in the long (>8.5 μm) charge reservoirs takes a significantly shorter time. The typical trade-off in thickness of absorption region and speed of response exists and can be improved by resonant cavity structures, or multilayers of 2DE/2DH structures.

Once the carriers reach the reservoirs, they cause charge imbalance, hence an internal field, resulting in a redistribution of charge and a dielectric relaxation, whose time constant is calculated from a solution of Poisson's equation, and is proportional to the dielectric constant ε_s and inversely proportional to the charge density. This dielectric relaxation time is in the range of a few femtoseconds in our case.

We model this process of the perturbation of charge reservoirs by Ensemble Monte Carlo (EMC) technique (Jacoboni and Reggiani, 1983). We start by considering the scattering rate between an electron in the 2D well with wave vector k in subband i and a second electron with wave vector k_0 in the subband j. The final states of these two electrons are k_0 and m for the first electron and $k0'$ and n for the second electron. The total scattering rate is given by (Zhao, 2006):

$$\Gamma_{im}(k) = \frac{4\pi e^4 m^*}{A\hbar^3 k^3} \sum_{k_0,j,n} f_j(k_0) \int_0^{2\pi} \frac{|F_{ijmn}(q)|^2}{(q+q_{s0})^2} d\theta \qquad (9.6)$$

where q_{s0} is the inverse screening length in two dimensions; $fi(k_0)$ is the distribution function for electrons. $Fijmn(q)$ is the form factor, given by:

$$F_{ijmn}(q) = \int_{-\infty}^{\infty} dz \int_{-\infty}^{\infty} F_i(z) F_j(z') F_m^*(z) F_n^*(z') \exp(-q|z-z'|) dz' \qquad (9.7)$$

where $Fi(z)$ is the wave function at the i-th subband. We incorporate both intra and intersubband scattering and ignore carrier motion along the direction perpendicular to the 2-D plane. Since 2DEG is degenerate, a rejection technique accounting for the Pauli exclusion principle in the EMC modeling is adopted. Once the final state is selected, during the transient phase, a random number between 0 and 1 can be used to accept or reject the transition. Screening is accounted for by a single wave-vector-independent constant. Then we introduce an ensemble of carriers, representing the photogenerated carriers injected to the 2-DEG with average energies of 150 meV above the bottom of the conduction band, calculated based on electrostatic potential simulated in Figure 9.5. Here only electron-electron scattering is

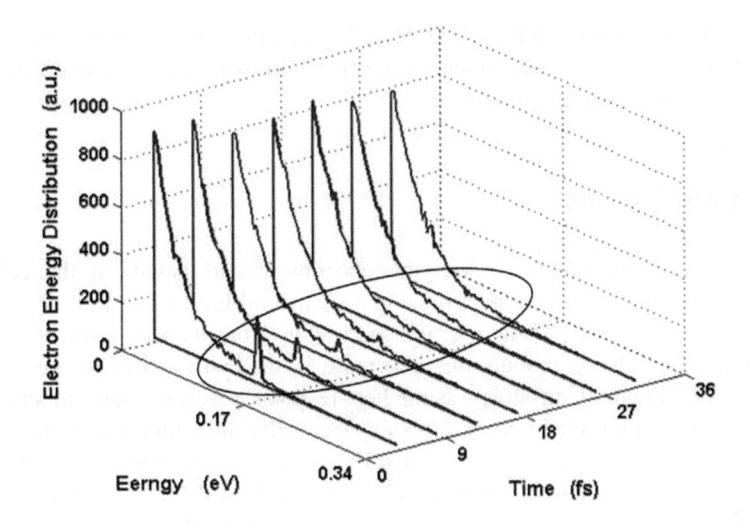

Figure 9.12 Time evolution of the energy distribution in 2DEG when perturbed by extra electrons.

considered since its rate is much larger than that of the electron-phonon scattering. The 2DEG density was assumed to be 5×10^{12} cm^{-2}.

Figure 9.12, shows the time evolution of 2DEG energy distribution when perturbed by the arrival of the optically generated electrons. Each curve represents an energy distribution function at some time, and the interval between the curves is 4.5 fs. The evolution of the nonequilibrium portion of the carriers, indicated by the solid oval in Figure 9.12, shows that these high-energy electrons reach a quasiequilibrium with the 2DEG system at around 30 fs. The time spent by nonequilibrium carriers reaching quasithermal equilibrium is the thermalization time or the energy relaxation time. This femtosecond time scale is consistent with the experimental results on carrier thermalization in a dense Fermi sea (Knox et al., 1988).

A different formalism can be applied based on a hydrodynamic description of the response of a compressible charge layer to both harmonic perturbation and a moving charge (Fetter, 1973). The advantage of this analysis is that the induced charge density and scalar potential can be calculated continuously without reverting to the two-step process of drift of carriers in the absorption region and energy relaxation in the charge layer. The shortcoming is that the response is distance dependent and in the picosecond range. Hydrodynamic modeling has also successfully been applied to short channel FETs (Dyakonov and Shur, 1993; Dyakonov and Shur, 1996), which have been shown to produce tunable detectors and emitters of THz radiation (Ryzhii et al., 2008), but apply to structures with lengths less than the momentum relaxation distance. These studies also extend the analysis of the direct interaction of radiation with the 2D gas based on polarizability (Stern, 1967), launching plasmon waves in the 2D gas, in much the same manner as performed by Allen et al. (1976) for the detection of infrared radiation. Incidentally, the possibility of excitation of edge magneto plasmons at room temperature in 2DEG for high-speed applications such as THz detection was predicted by von Klitzing et al. (2006) but requires transport under a magnetic field, which as demonstrated here is not necessary for the present device.

9.6 Conclusions

In conclusion, "electronic" devices may be constructed based on the collective response of reservoirs of charge, removing a significant limitation of the operation of devices that rely on the paradigm of transport of electrons between two contacts. Not being limited by charge motion, this also removes the constraint of the heavy effective mass of (slow-moving) holes, as the hole reservoir reacts similarly to an electron one. The device presented here is a photodetector that uses bilayers of electron and hole confined charges in a large area (~ 300 μm^2/pixel) photodetector. It dissipates nanowatts of electric power, but is operable without bias, detecting as low as tens of thousands of photons at room temperature, with speeds in hundreds of gigahertz. These properties would be expected from a device a fraction of its size if it were to be based on carrier sweep out. Beside application to optical communications, night vision, and photovoltaics in dim light, these micro plasma devices may work as detectors of other perturbation such as charged particles, or be used for direct detection of radiation such as THz. Since the reservoirs of charge can be kept near equilibrium using blocking contacts, the device operates at room temperature.

References

Allen, S.J., Tsui, D.C., Vinter, B., 1976. On the absorption of infrared radiation by electrons in semiconductor inversion layers. Solid. State Commun. 20, 425−428.

Allen, S.J., Tsui, D.C., Logan, R.A., 1977. Observation of the two-dimensional plasmon in silicon inversion layers. Phys. Rev. Lett. 38, 980−983.

Ando, T., Fowler, A.B., Stern, F., 1982. Electronic properties of two-dimensional systems. Rev. Mod. Phys. 54 (2), 437−672.

Anwar, A., Nabet, B., 2002. Barrier enhancement mechanisms in heterodimensional contacts and their effect of current transport. IEEE Trans. Microw. Theory Tech. 50 (1), 68−71.

Anwar, A., Nabet, B., Culp, J., Castro, F., 1999. Effects of electron confinement on thermionic emission current in a modulation doped heterostructure. J. Appl. Phys. 85 (5), 2663−2666.

Auston, D.H., Johnson, A.M., Smith, P.R., Bean, J.C., 1980. Picosecond optoelectronic detection, sampling, and correlation measurements in amorphous semiconductors. Appl. Phys. Lett. 37, 371−373.

Bohm, D., Pines, D., 1953. A collective description of electron interactions: III. Coulomb interactions in a degenerate electron gas. Phys. Rev. 92, 609−625.

Bowers, J.E., Burrus, C.A., 1986. High-speed zero-bias waveguide photodetectors. Electron. Lett. 22, 905−906.

Bowers, J.E., Burrus, C.A., 1987. Ultrawide-band long-wavelength p-i-n photodetectors. J. Lightwave Technol. LT-5 1339−1350.

Brattain, W.H., Bardeen, J., 1953. Surface properties of germanium. Bell Syst. Tech. J. 32, 1–41.

Chen, J.-R., Odenthal, P.M., Swartz, A.G., Floyd, G.C., Wen, H., Luo, K.Y., 2013. Control of schottky barriers in single layer MoS$_2$ transistors with ferromagnetic contacts. Nano Lett. 13, 3106–3110.

Chen, X., Nabet, B., Cola, A., Quaranta, F., Currie, M., 2002. A resonant cavity enhanced photodetector. Appl. Phys. Lett. 80 (17), 3222–3224.

Chen, X., Nabet, B., Cola, A., Quaranta, F., Currie, M., 2003. An AlGaAs/GaAs based RCE MSM photodetector with delta modulation doping. IEEE Electron. Devices Lett. 24 (5), 312–315.

Chen, X., Nabet, B., Zhao, X., Huang, H.-J., Cola, A., Quaranta, F., 2005. Optical and electrical characterization of GaAs-based high speed and high-sensitivity delta doped resonant cavity enhanced HMSM photodetector. IEEE Trans. Electron. Devices. 52 (4), 454–464.

Chen, Y., Williamson, S., Brock, T., Smith, F.W., Calawa, A.R., 1991. 375-GHz photodiode on low-temperature GaAs. Appl. Phys. Lett. 59, 1984–1986.

Chiu, Y.J., Fleischer, S.B., Bowers, J.E., 1998. High-speed lower-temperature-grown GaAs p-i-n traveling-wave photodetector. IEEE Photon. Technol. Lett. 10, 1012–1014.

Chou, S.Y., Liu, M.Y., 1992. Nanoscale tera-hertz metal-semiconductor-metal photodetectors. IEEE J. Quantum Electron. 28, 2358–2368.

Cola, Nabet, B., Chen, X., Quaranta, F., 2005. High speed heterostructure metal-semiconductor -metal photodetetors. Acta Phys. Pol. A. 107 (1), 14–25.

Crawford, D.L., Wey, Y.G., Mar, A., Bowers, J.E., Hafich, M.J., Robinson, G.Y., 1990. Highspeed InGaAs/InP p-i-n photodiodes fabricated on a semi-insulating substrate. IEEE Photon. Technol. Lett. 2, 647–649.

Currie, M., 2016. In: Nabet, B. (Ed.), In Photodetectors, first ed. Woodhead Publishing, pp. 121–155.

Currie, M., Quaranta, F., Cola, A., Dianat, P., Nabet, B., 2013. Overcoming Transit Time Limitations with Collective Excitations, FiO 2013/LS XXIX, October 6–10, 2013, Orlando, FL.

Deal, W., Mei, X.B., Leong, K.M.K.H., Radisic, V., Sarkozy, S., Lai, R., 2011. THz monolithic integrated circuits using InP high electron mobility transistors. IEEE Trans. Terahertz Sci. Technol. 1 (1), 25–33.

Debye, P., Hückel, E., Zur Theorie der Elektrolyte., I., 1923. Gefrierpunktserniedrigung und verwandte Erscheinungen [The theory of electrolytes. I. Lowering of freezing point and related phenomena]. Phys. Z. 24, 185–206.

Delagebeaudeuf, D., Delescluse, P., Etienne, P., Laviron, M., Chaplart, J., Linh, N.T., 1980. Two-dimensional electron gas MESFET structure. Electron. Lett. 16 (17), 667–668.

Dingle, R., Stormer, H.L., Gossard, A., Wiegmann, W., 1978. Electron mobilities in modulation-doped semiconductor heterojunction superlattices. Appl. Phys. Lett. 33, 665–667.

Doany, F.E., Grischkowsky, D., Chi, C.C., 1987. Carrier lifetime versus ion-implantation dose in silicon on sapphire. Appl. Phys. Lett. 50, 460–462.

Donati, S., 2000. Photodetectors: Devices, Circuits, and Applications. Prentice-Hall PTR, Upper Saddle River, NJ.

Dyakonov, M.I., Shur, M.S., 1996. Plasma wave electronics: novel terahertz devices using two dimensional electron fluid. IEEE Trans. Electron. Devices. 43 (380), 1640–1645.

Dyakonov, M.I., Shur, M.S., 1993. Shallow water analogy for a ballistic field effect transistor: new mechanism of plasma wave generation by dc current. Phys. Rev. Lett. 71, 2465.

Fetter, A.L., 1973. Electrodynamics of a layered electron gas. I. Single layer. Ann. Phys. 81, 367–393.

Gabor, D., 1956. Collective oscillations and characteristic electron energy losses. Phil. Mag. 1, 1–18.

Grimes, C.C., Adams, G., 1976. Observation of two dimensional plasmons and electron-ripplon scattering in a sheet of electrons on liquid helium. Phys. Rev. Lett. 36, 145–148.

Haug, H., Koch, S.W., 2009. Quantum Theory of the Optical and Electronic Properties of Semiconductors, fifth ed. World Scientific, Singapore.

Jacoboni, C., Reggiani, L., 1983. The Monte Carlo method for the solution of charge transport in semiconductors with applications to covalent materials. Rev. Mod. Phys. 55, 645–705.

Joglekar, Y.N., Balatsky, A.V., Das Sarma, S., 2007. Wigner supersolid of excitons in electron-hole bilayers. Phys. Rev. B. 74, 233302.

Kato, K., 1999. Ultrawide-band/high-frequency photodetectors. IEEE Trans. Microw. Theory Tech. 80 (7), 1265–1275.

Kato, Y.I.K., Kozen, A., Muramoto, Y., Nagatsuma, T., Yaita, M., 1994. 110-GHz, 50%-efficiency mushroom-mesa waveguide p-i-n photodiode for a 1.55-μm wavelength. IEEE Photonics Technol. Lett. 6, 719–721.

Keiser, G., 2000. Optical Fiber Communicationthird. McGraw-Hill, Inc., New York.

Keldysh, L.V., Kozlov, A.N., 1968. Zh. Eksp. Teor. Fiz. 54, 978.

Knox, W., Chemla, D., Livescu, G., Henry, J.E., 1988. Femtosecond carrier thermalization in dense fermi seas. Phys. Rev. Lett. 61, 1290–1293.

Landau L.D., Lifshitz, E.M., 1960. Electrodynamics of Continuous Media. Addison-Wesley, Reading, MA. 1960.

Lawton, R.A., Scavannec, A., 1975. Photoconductive detectors of fast-transition optical waveforms. Electron. Lett. 11, 74–75.

Lee, C.H., 1984. Picosecond Optoelectronic Devices. Academic, New York.

Lilienfeld, J.E., 1933. Device for Controlling Electric Current, US Patent 1,900,018.

Mimura, T., Hiyamizu, S., Fujii, T., Nanbu, K., 1980. A new field-effect transistor with selectively doped GaAs/AlxGa1-x As heterojunctions. Jpn. J. Appl. Phys. 19 (5), L225–L227.

Mimura, T., 2002. The early history of the high electron mobility transistor (HEMT). IEEE Trans. Microw. Theory Tech. 50 (3), 780–782.

Nabet, B., 1997. A heterojunction metal-semiconductor-metal photodetector. Photon. Technol. Lett. 9 (2), 223–225.

Nabet, B., Cola, A., Cataldo, A., Chen, X., Quaranta, F., 2003. High speed photodetectors based on heterostructures for opto-electronic applications. IEEE Trans. Microw. Theory Tech. 51 (10), 2063–2073.

Nabet, B., Cola, A., Quaranta, F., 2001. Heterojunction and heterodimensional optoelectronic devices. IEEE Microw. Mag. 2 (1), 40–45.

Nabet, B., Castro, F., Anwar, A., Cola, A., 2000. Heterodimensional contacts and optical detection. Int. J. High-Speed Electron. Syst. 10 (1), 375–386.

Nabet, B., Currie, M., Dianat, P., Quaranta, F., Cola, A., 2014. High-speed, high-sensitivity optoelectronic device with bilayer electron and hole charge plasma. ACS Photon.

Nagarajan, R.L., Giboney, K.S., Reynolds, T.E., Allen, S.T., Mirin, R.P., Rodwell, M.J.W., 1995. Traveling-wave photodetectors with 172-GHz bandwidth and 76-GHz bandwidth-efficiency product. IEEE Photon. Technol. Lett. 7, 412–414.

Nourbakhsh, A., Cantoro, M., Hadipour, A., Vosch, T., van der Veen, M.H., 2010. Modified, semiconducting graphene in contact with a metal: characterization of the Schottky diode. Appl. Phys. Lett. 97, 163101.
Patrick, W., van Zeghbroeck, B.J., Halbout, J.M., Vettiger, P., 1988. 105-GHz bandwidth metal-semiconductor-metal photodiode. IEEE Electron. Devices Lett. 9, 527–529.
Pohlt, M., Lynass, M., Lok, J.G.S., Dietsche, W., Klitzing, K.V., Eberl, K., 2002. Closely spaced and separately contacted two-dimensional electron and hole gases by in situ focused-ion implantation. Appl. Phys. Lett. 80 (12), 2105–2107.
Ragi, R., Romero, M., Nabet, B., 2005. Current transport in heterodimensional contacts. IEEE Trans. Electron. Devices. 52 (2), 170–175.
Ramo, S., 1939. Currents induced by electron motion. Proc. I.R.E. 27 (9), 584–585.
Razeghi, M., 2000. Optoelectronic devices based on III-V compound semiconductors which have made a major scientific and technological impact in the past 20 years. IEEE J. Sel. Top. Quantum Electron. 6, 1344–1354.
Ritchie, R.H., 1957. Plasma losses by fast electrons in thin films. Phys. Rev. 106, 874–881.
Ryzhii, V., Sato, A., Otsuji, T., Shur, M.S., 2008. Plasma mechanisms of resonant terahertz detection in a two-dimensional electron channel with split gates. J. Appl. Phys. 103, 014504.
Ryzhii, V., Ryzhii, M., Mitin, V., Shur, M.S., Satou, A., Otsuji, T., 2013. Terahertz photomixing using plasma resonances in double-graphene layer structures. J. Appl. Phys. 113, 174506.
Shi, J.-W., Gan, K.-G., Chiu, Y.-J., Chen, Y.-H., Sun, C.-K., Yang, Y.-J., 2001. Metal-semiconductor-metal traveling-wave photodetectors. IEEE Photon. Technol. Lett. 16, 623–625.
Shimizu, M.W.N., Furuta, T., Ishibashi, T., 1998. InP-InGaAs unitraveling-carrier photodiode with improved 3-dB bandwidth of over 150 GHz. IEEE Photon. Technol. Lett. 10, 412–414.
Shockley, W., 1938. Currents to conductors induced by a moving point charge. J. Appl. Phys. 9 (10), 635–636.
Sivan, U., Solomon, P.M., Shtrikman, H., 1992. Coupled electron-hole transport. Phys. Rev. Lett. 68, 1196–1198.
Stern, F., 1967. Polarizability of a two-dimensional electron gas. Phys. Rev. Lett. 18 (14), 546–548.
Sze, S.M., Coleman, D.J., Loya, A., 1971. Current transport in metal-semiconductor-metal (MSM) structures. Solid-State Electron. 14, 1209–1218.
Sze S.M., 1981. Physics of Semiconductor Devices. John Wiley & Sons, Inc., New York.
Tait, G., Nabet, B., 2003. Current transport modeling in quantum-barrier-enhanced hetero—dimensional contacts. IEEE Trans. Electron. Devices. 50 (12), 2573–2578.
Tait, G.B., Nabet, B., 2004. Physical modeling of barrier-enhanced quantum well photodetector device for optical receivers. Microw. Opt. Technol. Lett. 40 (3), 224–227.
Tan, I.H., Sun, C.K., Giboney, K.S., Bowers, J.E., Hu, E.L., Miller, B.I., 1995. 120-GHz long-wavelength low-capacitance photodetector with an air-bridged coplannar metal waveguide. IEEE Photon. Technol. Lett. 7, 1477–1479.
Tonks, L., Langmuir, I., 1929. Oscillations in ionized gases. Phys. Rev. 33, 195–210.
von Klitzing, K., Mikhailov, K.S., Smet, J.H., Kukushkin, I.V., 2006. A detector for electromagnetic radiation and a method of detecting electromagnetic radiation. US-Patent 6,987,484.
Von Klitzing, K., Englert, T., Fritzsche, D., 1980. Transport measurements on InP inversion metal-oxide semiconductor transistors. J. Appl. Phys. 51, 5893–5897.

Wang, S.Y., Bloom, D.M., 1983. 100-GHz bandwidth planar GaAs Schottky photodiode. Electron. Lett. 19.

Wey, Y.G., Crawford, D.L., Giboney, K., Bowers, J.E., Rodwell, M.J., Silvestre, P., 1991. Ultrafast graded double-heterostructure GaInAs/InP photodiode. Appl. Phys. Lett. 58, 2156–2158.

Wilberg, E., Ingelsfield, J.E., 1960. Collective modes and ground state energy of the semi-infinite electron gas. Phys. Rev. 188, 640–643.

Wu, X., Sprinkle, M., Li, X., Ming, F., Berger, C., de Heer, W.A., 2008. Epitaxial-Graphene/Graphene-Oxide junction: an essential step towards epitaxial graphene electronics. PRL 101, 026801.

Zhao, X., Cola, A., Tersigni, A., Quaranta, F., Gallo, E., Spanier, J.E., 2006. Optically modulated high sensitivity heterostructure varactor. IEEE Electron. Devices Lett. 27 (9), 710–713.

Zhao, X., Gallo, E., Cola, A., Quaranta, F., Currie, M., Spanier, J.E., 2008. Time response of two-dimensional gas-based vertical field metal−semiconductor−metal photodetectors. IEEE Trans. Electron. Devices. 55 (7), 1762–1770.

Zhao, X., 2006. Carrier transport in high speed photodetectors based on two-dimensional gas. Ph.D. dissertation. Drexel University (Chapter 5).

Plasmonic photodetectors

Arash Ahmadivand, Mustafa Karabiyik and Nezih Pala
Department of Electrical and Computer Engineering, Florida International University, Miami, FL, United States

10.1 Introduction to surface plasmon resonances

10.1.1 Physics of plasmon resonances

Coherent oscillations of free electrons at metal-dielectric interfaces or boundaries are referred to as surface plasmon resonances, which support intense electromagnetic (EM) field concentrations that can be tailored to confine the incident optical power in subwavelength dimensions at the optical frequencies (Raether, 1988; Maier, 2007; Maradudin et al., 2014). Excitation of surface plasmons with light leads to formation of surface plasmon polaritons (SPPs), a hybrid particle composed of charge oscillations coupled to EM waves. SPPs can be classified into two types, as propagating and localized. Localized SPPs involve nonpropagating excitation of conduction electrons of metal nanoparticles coupled to the photons. These modes are created by the coupling of photons to the conduction electrons in small subwavelength metal nanoparticles. Mie developed the theory of the scattering and absorption of photons from small spheres (Mie, 1908). Curved surface area of the particle exerts an effective restoring force on the driven electrons so that resonance can arise (Maradudin et al., 2014). Localized plasmon resonance can be excited by illuminating light directly onto the metal nanoparticles whereas propagating SPPs require special phase matching techniques. Shape, size, and material of nanoparticles define the optical properties of nanoparticles. By tuning the size of nanoparticle for constant shape and material, desired optical properties in transmission and reflection can be obtained.

In order to understand the properties of SPPs in various structures, one needs to understand the optical response of the metals and dielectrics. This section gives a summary of optical properties of metals, field distribution, and dispersion of SPPs. In a wide range of frequencies, optical properties of metals can be explained by a plasma model. In this plasma model, a free electron gas moves against fixed positive ion cores. Electron-electron interactions are ignored. Response of the free electron gas to the applied electric field defines the optical properties of metals. Oscillation of electrons is damped due to electron-core collisions with a characteristic collision frequency $\gamma = 1/\tau$ of the order of 100 THz, where τ is mean free time of free electron gas, which is of the order of 10^{-14} s at room temperature.

Considering the solution of equation of motion for an electron sea affected by an externally driven electric field E

$$m\ddot{x} + m\dot{x}\gamma = -eE \qquad (10.1)$$

polarization $P = nex$ caused by electron displacement, and the dielectric displacement $D = \varepsilon_0 E + P$, one can derive dielectric function of the free electron gas as $\varepsilon(\omega) = \varepsilon'(\omega) + i\varepsilon''(\omega)$ where real (ε') and imaginary (ε'') parts can be written as

$$\varepsilon'(\omega) = 1 - \frac{\omega_p^2 \tau^2}{1 + \omega^2 \tau^2} \text{ and } \varepsilon''(\omega) = \frac{\omega_p^2 \tau}{\omega(1 + \omega^2 \tau^2)} \qquad (10.2)$$

where

$$\omega_p^2 = \frac{ne^2}{\varepsilon_0} \qquad (10.3)$$

is the plasma frequency of the free electron gas. This model is known as the Drude model (Maradudin et al., 2014). The Drude model can describe the optical properties of metals for photons that have energies below the threshold of transitions between electronic bands. For some noble metals used in plasmonic applications like silver and gold, interband transitions start occurring around 2 eV (Fox, 2006) and the Drude model fails for higher energies. Similar relations can be constructed through the refractive indices of materials. Complex refractive index n is defined as $\tilde{n}(\omega) = n(\omega) + i\kappa(\omega)$. Refractive index n and absorption coefficient κ can be calculated from the dielectric functions as

$$n^2 = \frac{\varepsilon'}{2} + \frac{1}{2} + \sqrt{\varepsilon'^2 + \varepsilon''^2} \text{ and } \kappa = \frac{\varepsilon''}{2n} \qquad (10.4)$$

Refractive index approaches to zero for energies that are close to plasma energy and absorption coefficient also approaches to zero near the plasma energy and metals behave like a dielectric and becomes transparent for higher energies. Such an affect results in a drop in reflection spectrum and part of the spectrum is absorbed. This absorption determines the color of the metal.

To understand the properties of SPPs one needs to calculate the energy-momentum relation (dispersion) of SPPs. SPPs are confined at the metal-dielectric interface. Starting with Helmholtz equations at the metal-dielectric interface with a constant dielectric permittivity and proper boundary conditions, SPP dispersion relations can be derived. Let us consider the coordinate system in Figure 10.1 to define the confined wave propagation at a metal-dielectric interface. Dielectric permittivity is constant in the xy plane and changes at the interface $z = 0$ as $\varepsilon = \varepsilon_1$ for $z > 0$ and $\varepsilon = \varepsilon_2$ for $z < 0$. The wave propagates in the x direction.

Solving the Helmholtz equation

$$(\nabla^2 + k_0^2 \varepsilon)(E, H) = 0 \qquad (10.5)$$

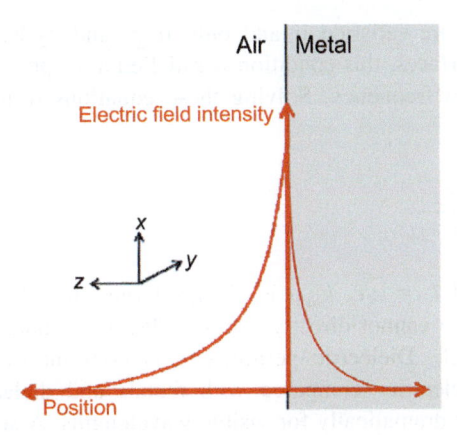

Figure 10.1 Electric field intensity profile of a SPP propagating at the metal-dielectric interface.

where $k_0 = \omega/c$ is the momentum of light in air, for TM modes of the E and H components of an EM wave propagating in x direction, one can find for $z > 0$:

$$H_y(z) = A_2 e^{i\beta x} e^{-k_2 z}$$
$$E_x(z) = \frac{iA_2}{\omega \varepsilon_0 \varepsilon_2} e^{i\beta x} e^{-k_2 z} \quad (10.6)$$
$$E_z(z) = -\frac{A_2 \beta k_2}{\omega \varepsilon_0 \varepsilon_2} e^{i\beta x} e^{-k_2 z}$$

and for $z < 0$:

$$H_y(z) = A_1 e^{i\beta x} e^{-k_1 z}$$
$$E_x(z) = \frac{iA_1}{\omega \varepsilon_0 \varepsilon_1} e^{i\beta x} e^{-k_1 z} \quad (10.7)$$
$$E_z(z) = -\frac{A_1 \beta k_1}{\omega \varepsilon_0 \varepsilon_1} e^{i\beta x} e^{-k_1 z}$$

where k_1 and k_2 are positive. Continuity at boundary conditions at $z = 0$ results

$$\begin{aligned} A_1 &= A_2 \\ \frac{k_2}{k_1} &= -\frac{\varepsilon_2}{\varepsilon_1} \\ k_1^2 &= \beta^2 - k_0^2 \varepsilon_1^2 \\ k_2^2 &= \beta^2 - k_0^2 \varepsilon_2^2 \end{aligned} \quad (10.8)$$

These equations are satisfied if and only if ε_1 and ε_2 has opposite signs. At metal-dielectric interfaces, this condition is satisfied at a specific frequency interval from zero to plasma frequency. Solving these equations results in the dispersion relation for SPPs:

$$k_{SPP} = \frac{\omega}{c} \sqrt{\frac{\varepsilon_1 \varepsilon_2}{\varepsilon_1 + \varepsilon_2}} \tag{10.9}$$

where $\beta = k_{SPP}$ and $k_0 = \omega/c$. k_{SPP} is always larger than k_0, which implies that photons in free space cannot directly excite SPPs. This condition is known as the momentum mismatch. Dielectric permittivity is constant over a wide range of wavelengths for dielectrics compared with metals and dielectric permittivity of metals changes very dramatically for visible wavelengths. A singularity is obtained when $\varepsilon_1 = -\varepsilon_2$. From the Drude model, this singularity is calculated as

$$\omega_{SP} = \frac{\omega_P}{\sqrt{1 + \varepsilon_2}} \tag{10.10}$$

This frequency is known as the surface plasmon resonance frequency. Contribution of the d orbital electrons in silver causes this frequency to shift to 3.7 eV (Maradudin et al., 2014).

Figure 10.2 represents the dispersion relation for silver-air interface. SPPs bound to the interface can be identified as the curve below the light line upto surface

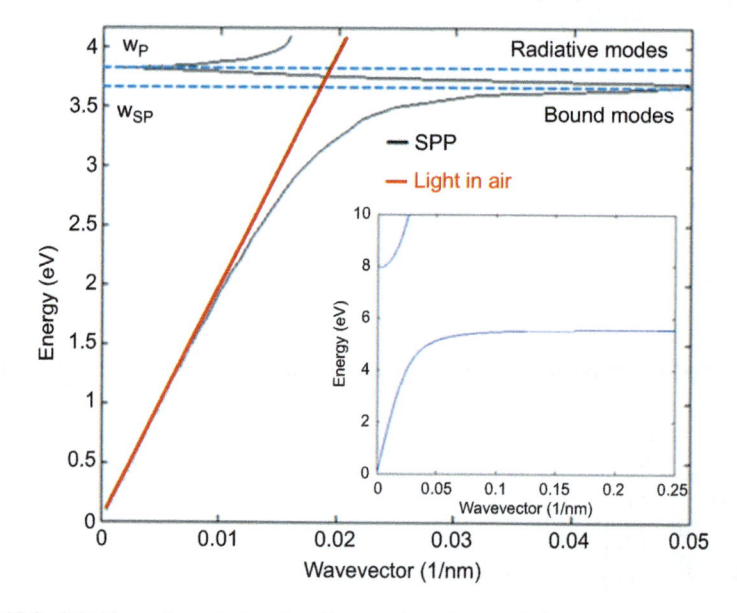

Figure 10.2 SPP dispersion relation for silver-air interface and dispersion of light in air. Dispersion of SPP is calculated with the Drude model as represented in the inset.

plasmon resonance frequency at 3.7 eV. Bound plasmons are SPPs confined at the metal-dielectric interface propagating along the interface. In the SPP bound mode regime, momentum of light in air is less than the momentum of SPP for the same energies. Because, SPPs are collective oscillations of free electrons and photons, SPPs can be visualized to have both electron and photon parts where momentum of electrons are larger than the momentum of light with the same energy. This phenomenon causes the momentum mismatch. The difference in momentum prevents the SPP excitation with incident light in air. There are several methods to compensate for the momentum mismatch which will be discussed later. Radiative modes are energetic SPPs that are above the ω_p (3.9 eV for silver) which are above the light line. In this regime, radiative bulk plasmon modes are observed and the metal behaves like a dielectric and dielectric function of the metal is positive. Light with large wave vectors can excite characteristic surface plasmon frequencies ω_{SP} around 3.7 eV. In the region between 3.7 and 3.9 eV, plasmonic modes are called quasi-bound modes where SPPs exhibit negative phase velocity. In this regime, according to the Drude model, dispersion is purely imaginary and SPPs are not allowed to propagate and this imaginary region behaves like a bandgap, but, experimentally these modes can be excited.

Electric and magnetic field distribution of SPP propagation at metal-dielectric interfaces for TM (TM represents transverse magnetic and in TM mode, there is no magnetic field in the direction of propagation) polarization is shown in Figure 10.3. Electric fields are created by positive and negative charges on the surface of the metal. The H field has only the component parallel to the interface. The E field has both parallel and perpendicular field lines with respect to the interface. E field's perpendicular components in the two media are in opposite directions and E field's parallel components in the two media are in the same direction. The boundary conditions are the continuity of $E_\parallel$ ($E_\parallel$ is the parallel component of E) and $D_\perp$ ($D_\perp$ is the perpendicular component of D and $D = \varepsilon E$). The continuity of $D_\perp$ is satisfied for positive values of ε_1 and negative values of ε_2 a condition that is only satisfied for metal-dielectric interfaces and as a result, surface plasmons exists only at the metal-dielectric interfaces. H fields should be divergence free which indicates absence of magnetic monopoles. This means that the end points of the H field lines

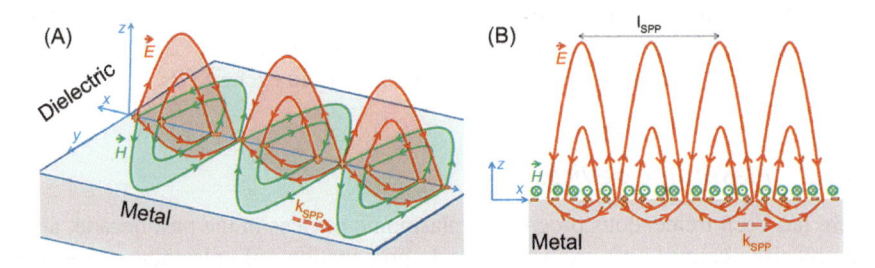

Figure 10.3 (A) Electric and magnetic field distributions of SPPs propagating at the metal-dielectric interface. (B) Cross-section view from the x-z plane, of electric and magnetic field distributions of SPP propagation at the metal-dielectric interface.

should be in the same direction. For a TE (TE represents transverse electric and in TE mode there is no electric field in the direction of propagation) polarized solutions, boundary condition for the continuity of $H_\parallel$ is violated because the H field directions are in opposite directions at the metal-dielectric interface and TE polarized SPPs cannot be excited at the metal-dielectric interfaces (Zakharian et al., 2007).

The wavelength of the SPPs is the period of the charge distribution on the surface and the associated field distributions. The real part of the dispersion relation describes the wavelength of SPP and the imaginary part of the dispersion relation describes the loss or the propagation length of SPPs. The complex wave vector of SPPs with real and imaginary parts is given as

$$k_{SPP} = k'_{SPP} + ik''_{SPP} = k_0 \sqrt{\frac{\varepsilon_d \varepsilon'_m}{\varepsilon_d + \varepsilon'_m}} + i \frac{k_0 \varepsilon''_m}{2\varepsilon'^2} \sqrt{\frac{\varepsilon_d \varepsilon'_m}{\varepsilon_d + \varepsilon'_m}} \qquad (10.11)$$

where ε'_m is the real part of dielectric function for the metal and ε_d is the dielectric function of dielectric. The complex wave vector has a wide range of usage in defining the fundamental parameters such as measuring the propagation length of SPPs along the interface which plays key role in couplers, waveguides, polarization beam splitters, etc., and defining the absorption coefficient for the interacted optical energy. The energy of distributed SPPs decays as $\exp(-2k''_{SPP}x)$ with the decay length of $L_{decay} = 1/2k''_{SPP}$. SPP wavelength can be written as

$$\lambda_{SPP} = \frac{2\pi}{k'_{SPP}} = \lambda_0 \sqrt{\frac{\varepsilon_d \varepsilon'_m}{\varepsilon_d + \varepsilon'_m}} \qquad (10.12)$$

Wavelength of the SPPs is smaller than the wavelength of the photon in air which allows us to study optics below the diffraction limit, that is, called subwavelength optics. The wavelength of light in air can be reduced to 4–5 times smaller wavelengths by exciting SPPs with energetic photons around 3.6–3.7 eV at the silver-air interface (Zakharian et al., 2007). SPPs cannot be excited with light without overcoming momentum mismatch between light in air and SPP. Methods like prism coupling or grating coupling to excite SPPs are necessary to compensate the momentum mismatch (Mie, 1908). There are also other methods like the use of trench scatterers, charged particle impact and near-field coupling (Zakharian et al., 2007). Prism coupling and grating coupling are the most common ways to excite SPPs.

10.1.2 Plasmonic devices

There has been tremendous progress in plasmonic devices in the past decade, allowing precise control of excitation and propagation of plasmon resonances. Technically, if the metallic structure is highly lossy, then an intense accumulation of surface plasmon resonant modes can be generated by an incident EM wave (Palik, 1985). This mechanism enables a wide range of applications in designing nanoscale devices (Shahbazyan and Stockman, 2013; Zayats and Maier, 2013) including antireflective surfaces (Spinelli et al., 2012; Munday and Atwater, 2011), solar energy harvesters (Linic et al., 2011; Naphade et al., 2014), all-optical biochemical and gas sensors (Chamtouri et al., 2014; Nasir et al., 2014; Golmohammadi and Ahmadivand, 2014), fast routers (Zheng et al., 2011; Chen et al., 2013a,b), plasmonic waveguides (Ahmadivand and Golmohammadi, 2014a; Oulton et al., 2008; Grandidier et al., 2009; Veronis and Fan, 2007; Hu et al., 2014), optical power splitters (Ahmadivand et al., 2012; Guo et al., 2011; Ahmadivand and Golmohammadi, 2014b), hybrid polarization beam splitters (Chen et al., 2014; Ahmadivand, 2014a), negative-index metamaterials (Atre et al., 2013; Zhang et al., 2005; Ahmadivand and Pala, 2015), visible and terahertz (THz) photodetectors (Rosenberg et al., 2010; Berry et al., 2013), photonic-plasmonic couplers (He et al., 2011; Xiao et al., 2012), and Mach-Zehnder interferometers (Hu et al., 2013; Gao et al., 2011) in a wide spectral range spanning from UV to THz frequencies. Operation of these nanosize devices strongly depends on three fundamental optical processes: (1) absorption, (2) scattering, and (3) plasmon resonance localization. Each one of these processes can be controlled and optimized based on the selected application. For instance, the absorption is important for photovoltaic devices and light harvesters (Naphade et al., 2014), while for plasmon waveguides, routers, and couplers which are operating based on field propagation, the scattering plays a key role (Oulton et al., 2008; Grandidier et al., 2009; He et al., 2011). In addition, strong localization of surface plasmon resonances have an important contribution in designing nanoscale plasmonic biochemical sensors and fast switches (Chamtouri et al., 2014; Nasir et al., 2014; Golmohammadi and Ahmadivand, 2014). As practical applications for subwavelength plasmonic devices develop, the requirements for designing and fabricating nanoscale structures with high efficiency and responsivity increases dramatically. A recent example of new technologies is molecular nanoparticles in self-assembled clusters or "oligomers" in simple and complex orientations (Prodan et al., 2003; Fan et al., 2010). Strong localization of plasmon modes in these nanoparticle aggregates can be managed and utilized in designing ultrasensitive sensors (Willet and Van Duyne, 2007; Lee and El-Sayed, 2006), metamaterials (Wu et al., 2011; Luk'yanchuk et al., 2010), and nanoswitches (Chang et al., 2012). Plasmonic photodetectors provide high responsivity and external quantum efficiency (EQE) from UV to the near infrared region (NIR) (Lee et al., 2011; Pelayo Garcia de Arquer et al., 2012). Recently, new technologies have been introduced and used to enhance the plasmon response of plasmonic structures, such as employing graphene layers or dielectric nanoparticles (Fang et al., 2012a; Wen et al., 2012).

10.2 Photodetectors

Photodetectors have emerged as an important type of electronic device to convert EM wave (photon) energy to electrical energy in a desired spectral range. The typical structure of a photodetector includes either a slab of semiconductor (photoconductors), a semiconductor/semiconductor junction (various forms of p-n diodes) or a metal/semiconductor junction (Schottky diodes) (Yang et al., 2012; Freitag et al., 2013a; Berini, 2014; Xing et al., 2012). Performance of photodetectors can be greatly improved by using complex stacks of layers or by employing sophisticated structures including plasmonic nanostructures. More specifically, plasmonic photodetectors are tailored to enhance EM wave absorption by the mediation of SPPs, typically by exciting resonant modes, and thereby enhance the photoresponse. Plasmonic photodetectors mainly include structures of noble metals to support strong plasmon resonant modes facilitating the detection mechanism which could be internal photoemission or electron-hole generation (Berini, 2014; Xing et al., 2012; Vanderlinde, 2004).

10.2.1 Semiconductor photodetectors

We will first explain the mechanism of photodetection by using a simple Schottky diode photodetector as an example which will later help to understand the contribution of plasmonic structures. Considering a Schottky diode is more instructive to describe two different photodetection mechanisms. In this example, photon detection is performed by a diode on an n-type silicon (Si) semiconductor. Figure 10.4A shows the schematic of a typical Schottky diode on a Si substrate, where two metallic contacts are placed on the top and bottom surfaces. The upper part of the Si is n-type whereas the lower part is heavily doped n^+-Si resulting a rectifying Schottky contact at the upper interface and a nonrectifying ohmic contact at the bottom interface. There are two possible photodetection mechanisms as shown in Figure 10.4B.

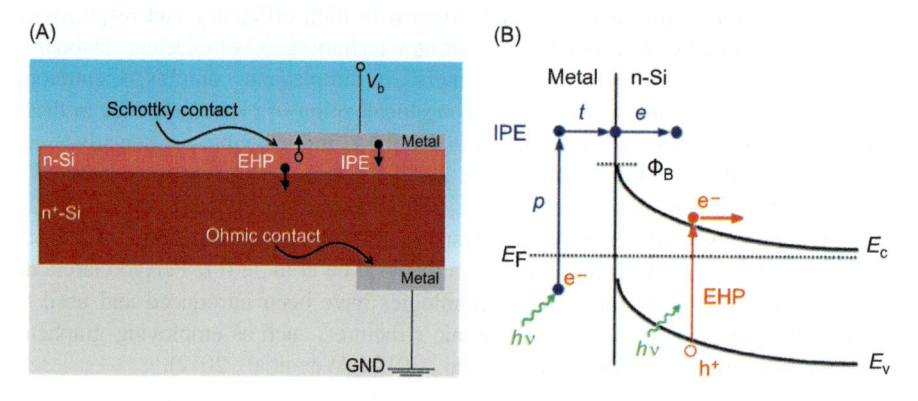

Figure 10.4 (A) A schematic diagram for the Schottky diode exposed by incoming photon energy that is operating via reverse bias. The exchange of electron-holes is illustrated inside the figure. (B) Energy band diagram of the Schottky photodiode with both electron-hole pair (EHP) generation and internal photoemission process (IPE) photodetection mechanisms. In the three-step internal photoemission mechanism. p, t, and e represent photoexcitation, transport, and emission, respectively. E_c and E_v and E_F are the conduction and valence band edges, and Fermi level, respectively, Φ_B is the Schottky barrier height and e^- and h^+ represent electrons and holes, respectively.

The first one is related to the generation of electron-hole pairs (EHP) on the semiconductor surface caused by absorption of incoming optical energy, where this energy is greater than the bandgap energy of silicon ($E_g < h\nu$). Photodetection by EHP generation process has three fundamental steps: (1) the incoming optical power is absorbed by the semiconductor and leads to creation of EHPs, (2) The applied reverse bias $V_b < 0$ to the Schottky metal contact, forms a depletion region with high electric field which causes the EHPs to be separated and carried across the absorption area, and (3) the photocurrent is formed by collecting the EHPs at the device contacts. The other possible mechanism is based on the creation of hot carriers in the metal via internal photoemission process (IPE). In this regime, the absorption of incoming optical power plays a key role. It should be underlined that in this mechanism the energy of incoming light must be greater than Schottky barrier energy (Φ_B). As shown in Figure 10.4B, this process can also described by three steps: (1) absorption of light by metallic parts leads to the photoexcitation of hot carriers (p), (2) hot carriers are transported to the Schottky contact, and (3) finally the emission and collection of hot carriers over the Schottky barrier into the semiconductor under applied a reverse bias produce photocurrent. The major difference between two processes is the ability of detection of incident energies below the bandgap energies ($E_g > h\nu$) for the last mechanism. However, creation of EHPs in the semiconductor is highly efficient and superior in comparison to the hot carrier formation and therefore typically dominates the detection process. The internal quantum efficiency (IQE) is an important parameter quantifying the performance of a photodetector. It can be defined by the number of careers that contribute in photocurrent (I_p) formation as:

$$\eta(\text{IQE}) = \frac{h\nu I_p}{Aq} \tag{10.13}$$

where q is the elemental charge and A is the absorbed optical energy using in photocurrent generation. However, for the second mechanism based on hot carrier generation, the formula above can be modified as below, which defines the IQE at the Schottky metallic contacts (Scales and Berini, 2010):

$$\eta(\text{IQE}) = 0.5 \left(1 - \sqrt{\frac{\Phi_B}{h\nu}} \right)^2 \tag{10.14}$$

To provide an example for typical Schottky photodetectors, let us consider $\Phi_B = 0.34$ and 0.72 eV for three different cases: Au/p-Si, Au/n-Si, Al/p-Si, and Al/n-Si. At the telecommunication band ($\lambda \sim 1310$ nm), the IQE can be determined in the range of 0.3%−9% (Sze, 1981). Besides the IQE, the EQE is the other fundamental parameter for a regular photodetector. To measure this parameter, an analogous method was employed, and can be defined as:

$$\eta(\text{EQE}) = \frac{h\nu I_p}{P_{inc} q} \tag{10.15}$$

in which P_{inc} is the power of the incoming power. A ratio of the EQE over the IQE indicates the optical power absorptance contributing in the performed photocurrent (A_{ph}):

$$A_{ph} = \frac{\eta(EQE)}{\eta(IQE)} \quad (10.16)$$

Finally, the responsivity of a photodetector can be defined by calculating the ratio of the generated photocurrent at the metallic contacts over the incident optical energy as:

$$R = \frac{I_p}{P_{inc}} = \frac{A_{ph}\eta(IQE)q}{h\nu} \quad (10.17)$$

Similar relations can be defined for a p-n junction photodetector where the Schottky metal is replaced with a p-type semiconductor and hence only the EHP generation mechanism occurs for photodetection.

In such photodetectors metal/semiconductor interfaces can be used for excitation of SPPs which can greatly enhance their performance. Examples of such detectors with different structures supporting SPPs will be considered in the following sections.

10.2.2 Grating-coupled plasmonic photodetectors

Grating-coupled photodetectors are widely used in plasmonic detectors for momentum matching between the incident light and the SPPs (Maier, 2007). The major advantages of this structure is easy fabrication and integration could be realized simply by depositing comb-type metallic parts on a multilayer semiconductor and/or dielectric layers (Kroo et al., 1984; Yoon et al., 2011; Rahman et al., 1991; Pala et al., 2009; Satoh and Inokawa, 2012). In grating coupler method, grating surfaces can compensate the missing momentum by supplying additional momentum to the incident light caused by the periodicity of the grating. Grating coupler allows to excite SPPs when light is incident from above the grating or below the grating. According to Figure 10.5A, SPPs can be excited using grating coupler with the condition

$$k_{SPP} = k_0 \sin\theta \pm mG \quad (10.18)$$

where $G = 2\pi/a$ is the reciprocal lattice vector of the grating with the periodicity a, θ is the angle of incidence, and m is the order of diffraction. By changing the incidence angle and wavelength of light different SPP modes can be excited.

An analogy with solid state physics and photonic crystals can be constructed to understand the dispersion relation of the SPPs in periodic medium. In solid state physics, electrons in a crystal can be considered to be in a periodic potential due to

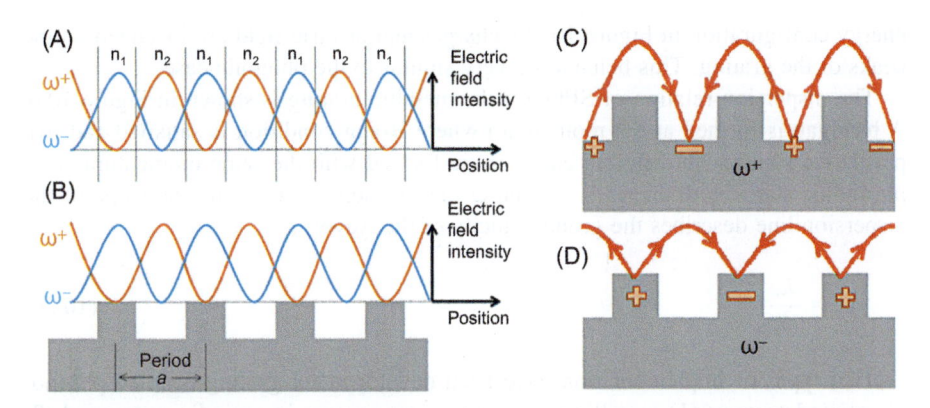

Figure 10.5 (A) Standing wave profiles of light in photonic crystal, and (B) SPP on the periodic surface. Electric field distributions for (C) ω^+ and (d) ω^- modes.

periodicity of the atoms. Electrons are not allowed to have all momentum and energy because of the periodic potential. The ranges of energies that cannot be occupied by electrons are called the forbidden band or the electronic bandgap. In the case of photonic crystals, periodic structure consisting of two materials with refractive indexes of n_1 and n_2 as seen in Figure 10.5A can also generate a photonic bandgap, provided that the Bragg condition is satisfied:

$$k_{SPP} = G/2 \tag{10.19}$$

where G is the reciprocal lattice vector. Bragg scattering of light results in both forward and backward traveling waves and interfere constructively resulting in a standing wave (Chen et al., 2013b). Light at each interface interfering constructively in the plasmonic structure forms two standing wave profiles with different energies. The maxima of the high energy (ω^+) standing wave form in the lower index material and the maxima of the low energy (ω^-) standing wave occur in the high index material. For energies between ω^- and ω^+ light interferes destructively and light propagation is not allowed. Hence, a bandgap is formed, known as the photonic bandgap. Electric field distributions at the edges of the bandgap are shown for the photonic case in Figure 10.5A. The formation of the bandgap for the plasmonic case has an analogy with the photonic case.

For SPPs, corrugations on the grating surface act as scattering centers and effective index of the plasmon changes periodically as it travels along the grating. The periodicity of the grating causes scattering of SPPs and when the Bragg condition in Eq. (10.19) is satisfied, a bandgap develops at the corresponding energy at a specific momentum. It is possible to excite SPPs with the same momentum but with different energies. Two different electric field distributions at the band edges with different energies are shown in Figure 10.5C and D. For higher energy configuration in Figure 10.5C, charges and electric field are located on the troughs. For lower energy configuration in Figure 10.5D, charges and electric field are localized on the peaks of the grating. This behavior is very similar to the photonic case.

The dispersion relation of SPPs for the metallic grating is shown in Figure 10.6. A bandgap is formed at the momentum where Bragg condition is satisfied and two plasmonic states with different energies are formed with the same momentum. Two plasmonic states with energies ω^+ and ω^- are standing waves and the slope of the dispersion line describes the group velocity of the wave:

$$v_g = \frac{d\omega}{dk} \tag{10.20}$$

Two types of implementation have been developed for grating-coupled plasmonic photodetectors: (1) metallic periodic nanowire arrays in periodic arrays and (2) corrugated gratings.

10.2.2.1 Plasmonic photodetectors with metallic nanowire grating gates

The operating mechanism plasmonic photodetectors with metallic nanowire grating gates is based on the action of nanowires as a coupler for incident waves to SPPs on detector structures (Zhang et al., 2011; Senanayake et al., 2011). Metallic wires are typically deposited in an interdigitated orientation which also play a key role in designing required contacts for the photodetector. These interdigitated contacts lead to remarkably fast-speed operation by reducing the detector capacitance and carrier transition time. Figure 10.7A shows an example of such structures, where the excitation of surface plasmons at the metal−semiconductor interface leads to a strong field enhancement near the metallic electrodes. This results in an increased absorption in the QW, allowing both fast electrical response of the photodetector and high

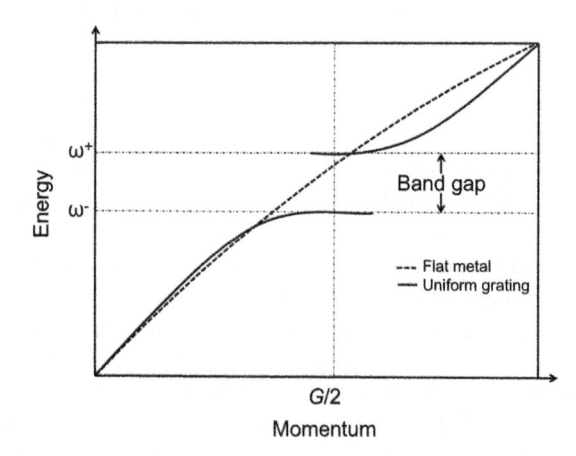

Figure 10.6 Schematic representation of the dispersion relation of SPPs for the uniform metallic grating and flat metal.

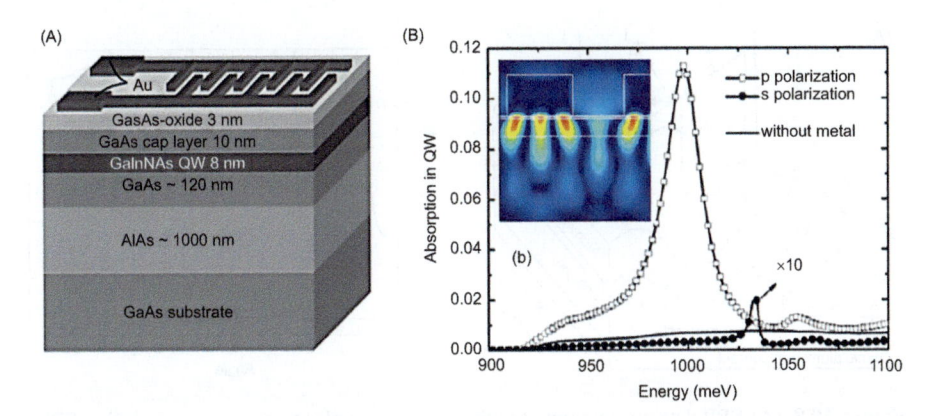

Figure 10.7 (A) Metal–semiconductor–metal (MSM) SPP photodetector based on a single GaInNAs quantum well. The interdigital metal contacts also act as a nanoscale wire grating. (B) Calculated detector absorption profile for normally-incident p- and s-polarized light and for an identical detector without the grating. The inset snapshot shows the computed magnetic field over the cross-section of about one period at 999 meV ($\lambda_0 = 1250$ nm) for p-polarized incidence (Hetterich et al., 2007).

quantum efficiencies (Hetterich et al., 2007). With a grating periodicity of 820 nm and electrode finger width of 460 nm a 16-fold increase in the absorption of p-polarized light in the QW is achieved in comparison to the case without electrodes. Shown in Figure 10.7A, the detection zone consists of an isolated GaInNAs quantum in which EHPs are generated by optical power absorption. Figure 10.7B shows the calculated absorption coefficient of the detector for the incident incoming p- and s-polarized light and for an equivalent detector without metallic contacts. It should be underlined that the peak of absorption is at 1000 meV ($\lambda_0 \sim 1250$ nm) for p-polarized incident light, which shows the strong polarization dependence of such grating gate plasmonic detectors. The inset snapshot shows a plot of the numerically calculated magnetic field over the cross-section of about one period at this peak.

10.2.2.2 Photodetectors with corrugated gratings

A corrugated photodetector comprises a semiconductor structure with a corrugated metallic grating designed to couple incident optical energy to SPPs similar to the metallic wire grating gates. An example of such devices is shown in Figure 10.8A (Berthold et al., 1986, 1987) where a structure is composed of Al electrodes on top of SiO_2/p-Si layers with a sinusoidal corrugated grating with the period of Λ and height of H, where the grating vector of the structure is given by:

$$k^G = \frac{2\pi x}{\Lambda} \tag{10.21}$$

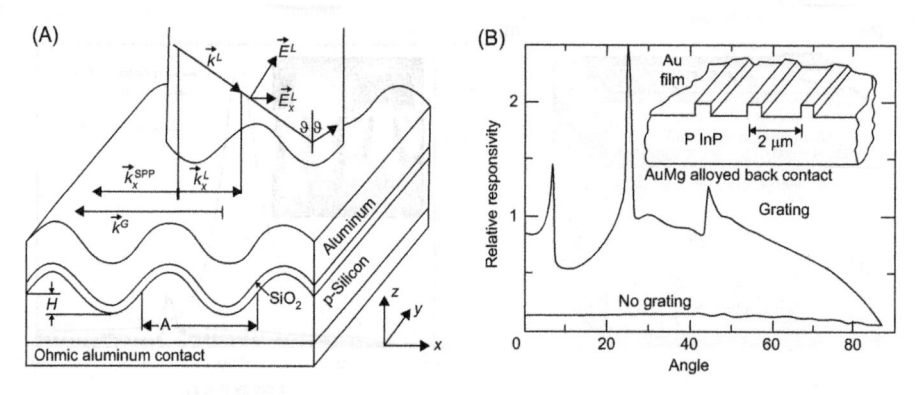

Figure 10.8 (A) SPP detector integrated with a grating coupler to excite x-propagating SPPs along the top surface of the top Al contact (Jestl et al., 1989), (B) measured responsivity for a Au/p-InP Schottky detector at $\lambda_0 = 1150$ nm with and without a grating coupler. The grating coupler excites SPPs along the air/Au interface at angles that correspond to grating orders (Brueck et al., 1985).

in which x is the x-directed unit vector. In the schematic profile, p-polarized light with a wave vector of k^L and electric field E^L is the incoming electric field to the grating with the angle of φ. The incident light strongly couples to the SPPs that are oscillating along the metal-dielectric interface, where the following equation, which is correlating with the momentum conversion, must be satisfied:

$$k_x^L + nk^G| = k_x^{SPR} \tag{10.22}$$

where k_x^L is the x-directed factor of k^L, n is an integer value, and k_x^{SPR} is the wavenumber for the SPPs. The above equation is for shallow gratings, and H is the minor perturbations to the surface of the grating. For the given example (Jestl et al., 1989), a thin Al is used for tunneling and leaking of SPPs to the p-type Si layer for detection via generation of EHPs at the operating wavelength of $\lambda_0 = 646$ nm. This type of photodetector strongly depends on the polarization direction, wavelength, and angle of incident spectrum. In addition, these grating-coupled SPPs can be used to enhance the absorptance in the metallic contact of Schottky detectors based on IPE to enhance the responsivity and quantum efficiency. Such an enhancement is clearly shown in Figure 10.8B for a Au/p-InP Schottky diode by exciting SPPs along the metal-dielectric interface at $\lambda_0 = 1150$ nm via an integrated corrugated grating (Berthold et al., 1987).

10.2.3 Plasmonic detectors with metallic nanoparticles

In recent years there has been growing interest in using metallic nanoparticles with different shapes, materials and orientations to design and fabricate highly responsive and efficient plasmonic photodetectors (Stuart and Hall, 1998; Schaadt et al., 2005). It is well-accepted that the shape and material of nanoparticles have significant impacts on defining the operation frequency of such nanoscale devices (Ahmadivand et al., 2012, 2013; Ahmadivand and Golmohammadi, 2014b,c). In this section, we will first examine the optical properties of various metallic structures by comparing the scattering and absorption features.

10.2.3.1 Metallic nanospheres

The nanosphere is the simplest type of nanoparticle with only one adjustable geometrical parameter (radius) which exhibits resonant responses under optical excitation. The spectral response of a metallic subwavelength sphere strongly depends on its material, size, and the ambience during exposure (Kelly et al., 2003). In terms of optical physics, plasmon resonances can be excited when the electron charge density oscillates coherently with the exposing optical EM field. The fundamental resonant mode of a simple spherical nanoparticle is dipolar mode with distribution of opposite charge formation at opposite spherical caps of the particle along the polarization of the illuminating electric field. This dipolar mode has a peak in the scattering cross-sectional profile that can be red- or blue-shifted along the wavelength variations. It should be underlined that the size of the nanosphere has a direct influence on excitation of dipolar and multipolar modes such as quadrupolar modes. Alterations in the refractive index of the surrounding medium also give rise to variations in the position and quality of the excited modes. In the resonance regime, the electric field in the vicinity of the sphere is strongly enhanced in comparison to the incident light. In the nonretarded limit, by assuming a nanosphere with the polarizability of α as:

$$\alpha = 4\pi r^3 \left(\frac{\varepsilon_m - \varepsilon_d}{\varepsilon_m + 2\varepsilon_d} \right) \tag{10.23}$$

where ε_m and ε_d are the permittivity of the metallic sphere and dielectric ambience, respectively. Also, r is the radius of nanosphere. Then, according to Bohren and Huffman (1983) and Jackson (2007), the scattering and absorption cross-sectional spectra are given by:

$$C_{scat} = \frac{1}{6\pi} \left(\frac{2\pi}{\lambda_0} \right)^4 |\alpha|^2 \tag{10.24}$$

$$C_{abs} = \frac{2\pi}{\lambda_0} \text{Im}\{\alpha\} \tag{10.25}$$

Figure 10.9 shows the normalized scattering cross-sectional profile ($C_{scat}/(\pi r^2)$) for a nanosphere as a function of radius in the free space ambience. As obvious it is, the peak of dipolar peak is shifted in the visible spectrum for a gold nanosphere with variant radius. This behavior of a nanosphere can be optimized by changing the shape of the nanoparticle, or the addition of an identical nanosphere in the

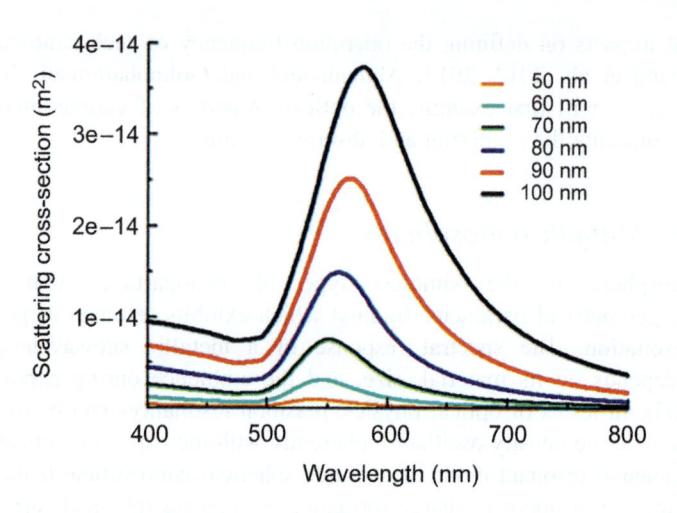

Figure 10.9 Normalized scattering cross-sectional profile for a metallic nanosphere as function of wavelength for various diameters.

proximity to enhance the plasmon resonant modes energies. It is well-understood that nanoparticles with larger numbers of revisable geometrical parameters or in other words with extra degrees of tunability can be used to shift and adjust the position of plasmon resonant peaks from the visible to the NIR spectra which is important for optical communication purposes.

In a plasmonic photodetector, the position and orientation of metallic nanoparticles are very important for the device performance. Indeed, as incoming light interacts with the nanoparticles, plasmon resonances can be excited, and thereupon, high localization and enhancement of the EM field can be achieved by the near-field coupling effect. This process also includes scattering of light depending on the substance of the particles. The field enhancement results in the generation of a larger photocurrent. Metal nanoparticles can be integrated with the photodetector by simply placing them on top of the detection medium (Stuart and Hall, 1996; Derkacs et al., 2006; Pillai et al., 2007; Sundararajan et al., 2008; Catchpole and Polman, 2008), or embedding them below or within the detection zone (Qu et al., 2011; Hyun and Lauhon, 2011; Echtermeyer et al., 2011; Ouyang et al., 2010; Pelayo Garcia de Arquer et al., 2012; Basch et al., 2012) which could be a semiconductor layer such as Si, GaAs, or organic materials. Figure 10.10 displays a schematic figure of a sample photodetector composed of Au nanospheres in linear arrays that are deposited between metallic wires embedded in a semiconductor substrate. The size of nanoparticle can be tailored to support and confine strong plasmon resonant modes from the visible to the NIR spectra. Embedding particles allows for the detection of visible and infrared spectra via IPE, since the metallic nanospheres form Schottky barriers with the surrounding semiconductor. As the plasmon resonance on a nanoparticle decays, the energy is transferred to the surrounding lattice to generate hot carriers which generate a photocurrent. During the illumination

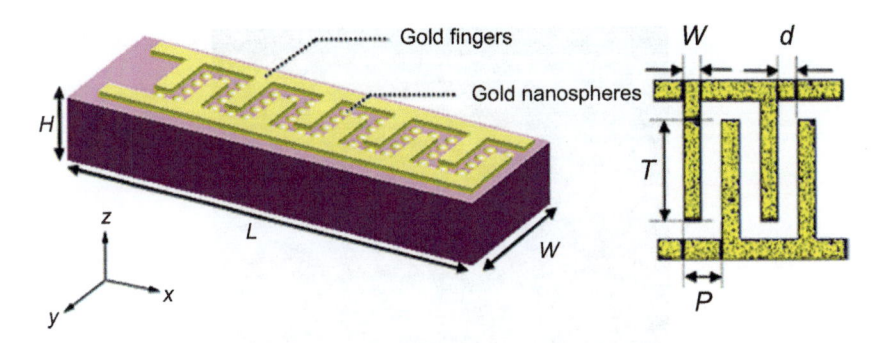

Figure 10.10 A schematic diagram of a plasmonic photodetector composed of gold fingers and nanorods on the top side with the description of geometrical sizes.

process, nanoparticles can be charged because they are floating and are not contacted to the electrode. Discharging happens via compensation charge injection from the semiconductor substrate, however, this process can slow the process of photocurrent generation (Stratton, 2007).

It has been also verified that EHPs can be generated in an Si layer without interaction of phonons via the near-field of SPPs on the embedded plasmonic nanoparticles with an energy larger than the bandgap energy of Si. It should be noted that the SPP momentum is large enough to satisfy the momentum conservation during the transition (Kirkengen et al., 2007). A big advantage of metallic nanoparticles is their easy integration with a detector via deposition onto the surface through which light penetrates (Stuart and Hall, 1996, 1998). Figure 10.11A displays a scanning electron microscope (SEM) image of Ag nanoparticles with the radii of 108 nm deposited on a LiF layer with the thickness of 30 nm, which is placed on a Si-on-insulator (SOI) detector (Stuart and Hall, 1998). These nanoparticles were formed by first depositing a thin Ag film, then annealing the film to induce islandization. Photodetection is performed by the p-n junction detectors within the thin Si slab with the overall thickness of 165 nm above the insulator but beneath the LiF and the metallic nanoparticle arrays. Figure 10.11B exhibits the experimentally measured spectral response for the proposed detector coated with Ag nanoparticles as a function of nanoparticle dimension. The obtained enhancement is defined as the ratio of the photocurrent from a detector with Ag nanoparticles to the photocurrent from a detector without particles. A strongly enhanced photocurrent is reported at $\lambda_0 = 800$ nm for the Ag nanoparticles with the size of 108 nm. This optimization in photocurrent and performance of the photodetector is correlated with the large cross-sectional spectra produced by metallic nanoparticles and particularly mediated by the dipole resonant modes excited on the nanoparticle arrays and then radiated into guided modes of the thin Si slab. The propagation direction of these modes plays a fundamental role in optimizing the performance of the structure, where the maximum absorption of optical energy is reported for the longitudinally propagated modes. On the other hand, chemically synthesized colloidal nanoparticle arrangements can also be deposited on a plasmonic detector. For instance, the plasmon

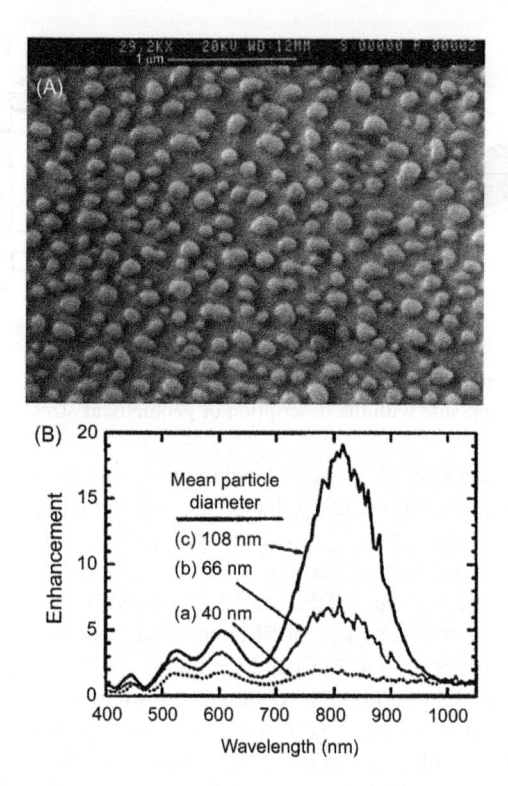

Figure 10.11 (A) Normalized scattering cross-sectional profile for a metallic nanosphere as a function of wavelength for various diameters. (B) Spectral response of p-n junction SOI photodetectors coated with Ag nanoparticles. The enhancement is defined as the ratio of the photocurrent from a detector with Ag particles to the photocurrent from an identical detector without metallic nanoparticles (Stuart and Hall, 1998).

response of photodetectors coated with Au nanosphere arrays with the radii of 25, 40, and 50 nm presented significant photocurrent enhancement at the dipolar peak positions (wavelengths) in the calculated extinction spectra (Stuart and Hall, 1998). In another example, it is shown that metallic nanoparticles can be integrated with a solar cell via deposition onto the backside of the structure (Schaadt et al., 2005). Figure 10.12A shows a thin-film solar cell including Ag nanoparticles on the back of the detection spot which was composed of a 2 μm thick layer of polysilicon. Here, Ag nanoparticles were created by deposition of Ag layer and by subsequent annealing. An SEM image of the metallic nanoparticles is shown in Figure 10.12B. The measured EQE and EQE enhancement of such a solar cell compared with various control devices are shown in Figure 10.12C. Note that in the comparative figure, the best performing cell consists of Ag nanoparticle arrays formed directly onto the Si layer with an external back surface reflector-curve labeled Ag + DP. Calculated parameters for the proposed nanostructure include a short-circuit current density of ~20 mA/cm^2 and a power conversion efficiency of 38.

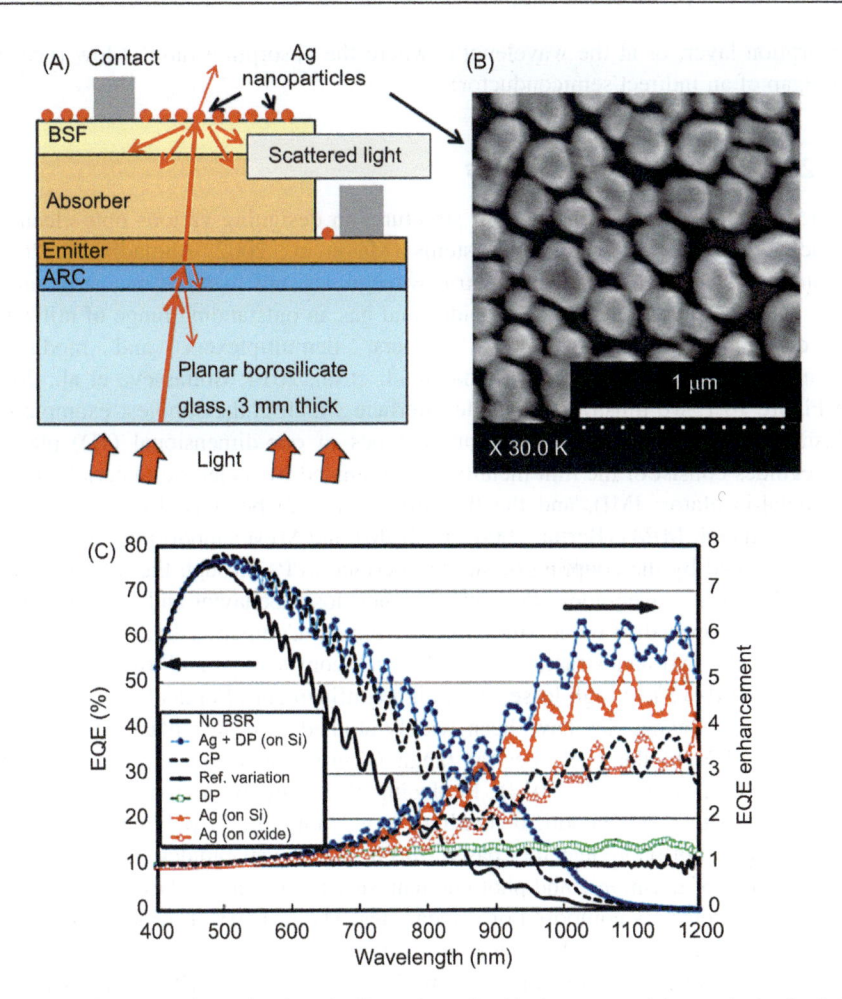

Figure 10.12 (A) A schematic of a thin-film solar cell with Ag nanoparticles on the back of the absorbing medium, where ARC is the antireflective coating, and BSF is the back surface field. The absorber consists of a 2 μm thick layer of poly-Si. A back surface reflector (BSR) painted directly onto the diodes or onto an external surface placed in proximity was also used. (B) SEM image of Ag nanoparticles on the back of a cell with the size of 150–250 nm. (C) External quantum efficiency (EQE) and enhancement of various solar cells (Ouyang et al., 2010).

So far, we have showed that plasmonic nanoparticles have made a significant contribution to the improvement of the performance of various photodetectors. The effects induced by metallic particles increases scattering of EM fields into detectors (light trapping mechanism) and the result of this process is increasing the absorption of incoming light by detector due to field enhancement. These effects are useful to increase the absorptance of detectors, specifically those based on a thin absorption layer, or at the wavelengths where the absorption ratio is low (near the bandgap of an indirect semiconductor).

10.2.4 Waveguide detectors

Dielectric waveguides are important structures in designing various opto-electronic devices, and telecommunication systems (Ma et al., 2002; Ahmadivand, 2014b; Adams, 1981). A mixture of dielectric waveguides and metallic nanostructures is well-known as a "plasmonic waveguide" and has an outstanding range of utilization in designing couplers, routers, splitters, demultiplexers, and modulators (Ahmadivand, 2012, 2014b; Golmohammadi et al., 2014; Boltasseva et al., 2005). In Figure 10.1, we illustrated a single-interface which is the simplest example of a plasmon waveguide. The other common types of one-dimensional (1D) plasmon waveguides consist of the thin metallic slab bounded by dielectric materials (insulator-metal-insulator: IMI), and the thin dielectric slab bounded by metals (metal-insulator-metal: MIM) (Berini, 2006). Both IMI and MIM support plasmon resonant modes excited by the coupling of single-interface SPPs through the thin interfering layer. However, waveguides are known for their lossy behavior and several methods have been proposed to reduce the effect of losses. Comparing both IMI and MIM structures, the IMI nanostructures provide weak confinement and low losses, and in contrast, MIMs reflect high losses with high confinement of optical energy.

Plasmonic waveguides have been widely studied to design plasmonic photodetectors for UV to the NIR spectral range (Konstantatos and Sargent, 2010; Liu et al., 2010; Butun et al., 2012). Replacing the insulator region of a plasmonic waveguide with a semiconductor medium (e.g., inorganic or polymer) is the main method to design photodetectors based on plasmonic waveguides. Integration with several types of plasmonic and photonic nanostructures is also a key feature of plasmonic photodetectors composed of waveguides. The primary absorption mechanism for the incident light is EHP generation in plasmonic waveguide photodetectors with organics or semiconductor active regions, whereas it is IPE in the ones with metal active regions (Ditlbacher et al., 2006; Neutens et al., 2009; Dufaux et al., 2010; Ren et al., 2010; Trolle and Pedersen, 2012). Thermal detection, photo-assisted transport and tunneling processes are also major photodetection mechanisms in plasmonic waveguide detectors (Colace et al., 1998; Olivieri et al., 2010; Berini et al., 2012; Akbari et al., 2013; Scales et al., 2006, 2010). Improved absorption rate, compact dimensions, and significant photocurrent via IPEs and EHPs make plasmonic waveguide photodetectors promising for several applications.

10.2.4.1 Silicon-based plasmonic waveguide photodetectors

Plasmonic waveguide detectors composed of Si-based structures have been extensively utilized to achieve efficient high-speed photodetection for the telecommunication applications with CMOS-compatible fabrication processes (Berini et al., 2012; Akbari et al., 2013; Goykhman et al., 2012). An elegant example of such devices is integrating a Schottky detector with a plasmon waveguide and employing

IPE as the detection mechanism for infrared detection below the bandgap of Si. It has been shown that Schottky photodetectors based on metallic strip waveguides (using IPE mechanism) have been utilized to locate on the n- or p-type Si layers (symmetric or antisymmetric regimes) as a cladding for photodetection (Berini et al., 2012; Akbari et al., 2013). Figure 10.13A presents a cross-sectional schematic diagram for a Germanium (Ge) on a SOI MSM photodetector with an effective device width (W) of 25 μm and length of 50 μm. The thickness of the thin Si layer between SiO_2 waveguide and Ge layer is 220 nm (Ren et al., 2010). Figure 10.13B shows the top view of the structure as a schematic diagram including interdigitated metallic electrodes. These electrodes effectively form a 1D metallic rectangular grating with finite dimensions on the top surface of the multilayer structure. Here, Al is used as a plasmonic substance due to its compatibility in Si-CMOS processes.

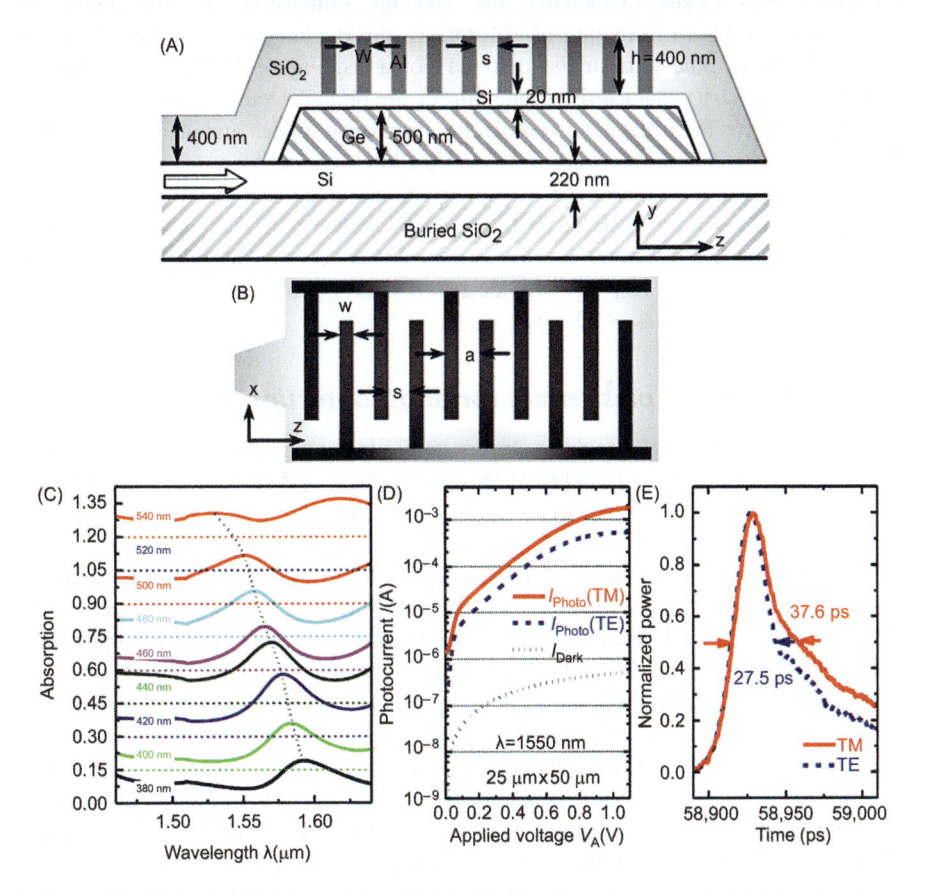

Figure 10.13 (A and B) Cross-sectional and top view schematic diagrams for the Ge on SOI MSM photodetector with interdigitated electrodes, (C) Calculated absorption spectra of the Ge region with grating periodicity 1500 nm, (D) Current-voltage characteristics, and (E) normalized temporal response of the device with TE and TM injection, respectively (Ren et al., 2010).

A crystalline Si (c-Si) thin layer with the thickness of 20 nm is located between the metal layer and the Ge section to increase the Schottky barrier and, therefore, suppress the dark current of the device. The transverse and longitudinal polarized electric/magnetic field parallel to the x-axis, and the light propagates along the z-direction. By reaching to the active condition, the Si waveguide width is adjusted with a taper to ensure efficient coupling of incidence photon from the routing Si waveguide up into the Ge region for absorption. In this photodetector, the SPPs are radiated and distributed via coupling inside the Si waveguide as a core layer and assisted by the distribution of plasmon resonances in the Al-Si interface as well as the Bragg grating. Figure 10.13C exhibits the absorption profile for the Ge layer as a function of incident beam wavelength for the examined photodetector with the periodicity of 1500 nm under transverse magnetic polarization mode excitation that is computed numerically. Changing the spacing dimensions in the range of 380–540 nm leads to formation of absorptance peaks along the $\lambda \sim 1590-1530$ nm with a dramatic blue-shift in the extreme positions. Figure 10.13D illustrates the current-voltage (I-V) characteristic diagram for the photodetector which compares the photocurrents for both magnetic and electric transverse modes and also dark current. For instance, for the applied bias of $V = 1$ V, a low dark photocurrent of 0.5 µA was defined. The normalized power as a function of simulation time delay in picoseconds is depicted in Figure 10.13E for both of the incidence polarizations at the applied bias voltage of 2 V. Herein, for transverse electric mode the full-width at half-maximum of the pulse response was determined as approximately ~ 27.5 ps, while for the wider width it was ~ 37.6 ps.

10.2.5 Graphene plasmonics for photodetection

Recently, a mixture of graphene layers and plasmonic nanoparticles has been employed as an approach to enhance the photocurrent and responsivity of photodetectors (Kim et al., 2014; Koppens et al., 2011; Schackleford et al., 2009; Furchi et al., 2012; Freitag et al., 2013b). Graphene is a promising material for new electronic devices due to its tremendously high electron mobility, atomic layer thickness, and outstanding mechanical flexibility (Kim et al., 2014; Koppens et al., 2011). The broadband absorption of graphene makes it as an attractive option for solar cells (Wang et al., 2008; Li et al., 2010), ultrafast photodetectors (Kim et al., 2014; Xia et al., 2009; Mueller et al., 2010), THz applications (Ju et al., 2011), photocatalysts (Xiang et al., 2012; Zhang et al., 2012), biological agents (Yang et al., 2013; Wu et al., 2013), photonic metamaterials (Lee et al., 2012), and DNA sensors (Liu et al., 2013). The nonlocal carrier-assisted intrinsic photoresponse in dual-gated graphene p-n junction devices has been investigated comprehensively (Lemme et al., 2011; Fang et al., 2012b). Despite the unique features of graphene, this two-dimensional (2D) monolayer shows poor absorption spectra and low quantum efficacy which limit its performance and potential to be employed in designing high-quality photonic and opto-electronic devices (Bonaccorso et al., 2010). In a recent development, a substrate of graphene layer with an array of plasmonic metallic nanostructures on top has been introduced as a graphene-plasmonic structure and widely employed to design numerous types of plasmonic devices for photocurrent generation and light harvesting applications (Grigorenko et al., 2012; Wang et al., 2012). Graphene plasmonics provides tremendous EM field confinement in subwavelength regime due to the broadband absorption property of the graphene layer during light-matter interaction (Xiang et al., 2010; Neto et al., 2009). The intensified EM field in doped graphene layers can be exploited as an alternative method instead of regular plasmon resonance excitation and localization techniques in noble metallic particles which suffer from dramatic ohmic losses and limited structural tunability (Kim et al., 2014; Koppens et al., 2011; Schackleford et al., 2009; Furchi et al., 2012; Freitag et al., 2013b). This unique ability of graphene originates from its charge carriers of zero effective mass (Dirac Fermions), which allows for long travel distances of EM waves without a significant amount of scattering (Zhao et al., 2013). Additionally, the linear dispersion of the 2D Dirac Fermions yields extraordinary wideband tunability under various implementations, while this feature can be affected by its weak interaction with incoming light due to low carrier concentrations (Jo et al., 2011). Besides, due to lack of free charges, graphene is considered as a semimetal substance. Free carriers can be induced via chemical doping or electrical gating easily due to the inherent structural nature of graphene, and obtained free carriers in graphene is almost 0.01 per atom, or doping concentration of 1×10^{13} cm^{-2}, which is dramatically lower than regular metals (Koppens et al., 2011). Thus, the 2D property and semimetallic behavior of graphene provides limited electrical tunability. However, this natural imperfection of graphene can be treated by employing metallic nanoscale structures in designing correlating nanoscale devices. For a finite graphene sheet, the absorption coefficient is given by $\alpha = e^2/\hbar c \sim 1/137$ ($\alpha\pi \approx 2.3\%$), and also, the universal optical conductivity of graphene is almost $e^2/4\hbar$ (Low and Avouris, 2014). In practical applications, graphene plasmons can be confined to volumes in the range of $\sim \alpha^{-3}$ which is much smaller than the diffraction limit allowing ultrastrong light-matter interaction (Low and Avouris, 2014; Ryzhii et al., 2008).

10.2.5.1 Graphene-plasmonic photodetectors

A blend of noble metallic nanoparticles and graphene sheets provides outstanding improvement in photocurrent efficiency and responsivity (Fang et al., 2012a; Wang et al., 2012). The type, shape, and material of employed nanoparticles in graphene-based photodetectors play a fundamental role in optimizing the spectral response. Intensifying the energy of plasmon resonant modes directly enhances the photocurrent efficiency (Berry et al., 2013; Echtermeyer et al., 2011; Schackleford et al., 2009). One of the major approaches introduced to enhance the plasmon resonance energy in metallic nanoparticles is the hybridization mechanism in molecular nanoparticles (Berry et al., 2013; Fan et al., 2010; Fang et al., 2014; Rana et al., 2009; Ferrari et al., 2006; Guo et al., 2007; Mueller et al., 2009; Capasso et al., 2000; Ahmadivand et al., 2015; Barrow et al., 2012; Hong et al., 2013). Figure 10.14A and B show the SEM images of two different types of molecular nanoparticle clusters. Figure 10.15A exhibits a schematic profile of a graphene plasmonic planar

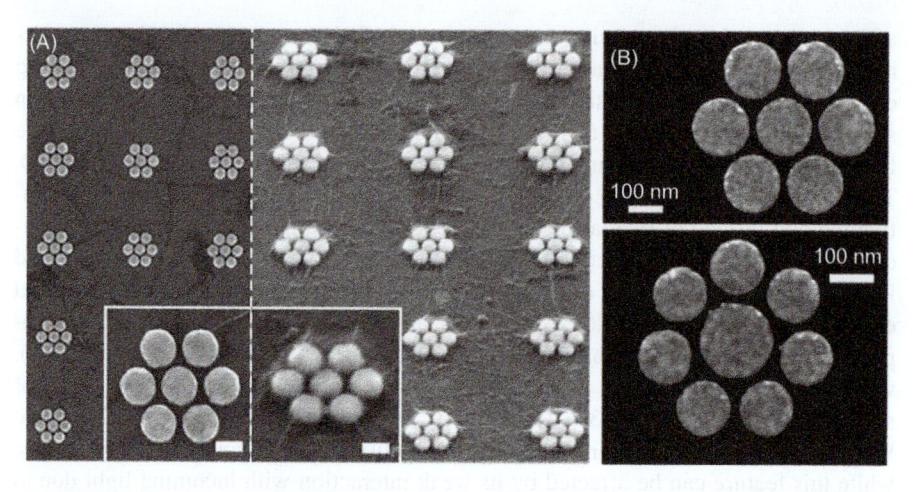

Figure 10.14 (A) SEM images for various nanoparticle clusters (Fang et al., 2012a), (B) a SEM image for a metallic symmetric heptamer and antisymmetric octamer (Fang et al., 2012a).

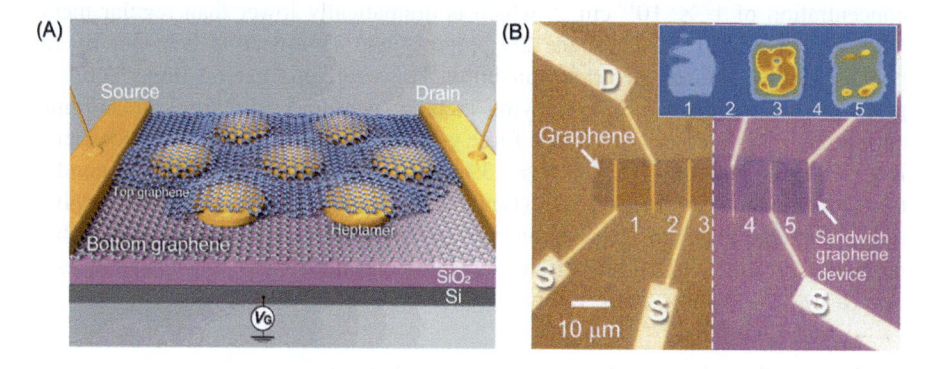

Figure 10.15 (A) A schematic diagram of a photodetector consists of an isolated gold nanodisk heptamer deposited on a multilayer substrate of silica, silicon, and graphene layers which is entirely covered with a monolayer graphene (Fang et al., 2012a). (B) SEM images of the photodetector device before (left) and after (right) deposition of cover graphene layer. The inset snapshot is the Raman mapping of the device under illuminations at $\lambda = 785$ nm (Fang et al., 2012a).

photodetector composed of multilayer substrate (SiO_2/Si/graphene) and gold heptamer arrays that are covered with a monolayer graphene and Figure 10.15B shows the SEM image of the fabricated device (Fang et al., 2012a). Due to the mechanical flexibility of the upper graphene layer, the interparticle offset gaps between nanoparticles are also covered with graphene. For this case, the radii of Au nanodisks is set to 65 nm with the offset gap distance of 15 nm to support strong plasmon resonance coupling between sub- and superradiant modes. The inset Raman image shows the spatial and spectral areas of strongest resonance coupling in the device. Considering the wavelength of the excitation light, ($\lambda = 514$ nm) any variations in the Raman intensity across the devices are due to the nonresonant behavior of the multilayer structure. On the other hand, for excitation wavelength of $\lambda = 785$ nm, significant enhancement is performed in the graphene Raman modes (see zones 3 and 5 in the SEM image). In this regime, the incoming wavelength is resonant with the induced Fano resonant mode of the heptamer (Fang et al., 2012a). Figure 10.16A illustrates the relation and dependence of resistance on the gate voltage (V_g) variations for the applied source-drain voltage as 1 mV, where the Dirac point is indicated for $V_g \sim 30$ V. The inset diagram shows the linear I-V diagram that is calculated for the gate bias in the range of 0–25 V (Fang et al., 2012a). Figure 10.16B demonstrates the measured photocurrent for the examined photodetector device with a cover of bare graphene during gate voltage variations in the range of -40 to $+40$ V. Obviously, the photocurrent becomes zero when the incoming light is moved to the center of the graphene channel. For large negative gate voltages, the peak of the photocurrent performs at the position nearby the source and drains electrodes. In contrast, for the positive gate biases, no photocurrent is observed, and increasing the gate voltage changes the direction of photocurrent and increases its magnitude (Fang et al., 2012a).

10.2.6 Plasmonic metamaterials

Metamaterials are promising artificial and engineered structures with unique properties beyond natural materials consisting of nano- and/or microscale metallic (dielectric) resonant building blocks or particles to support incoming strong EM field

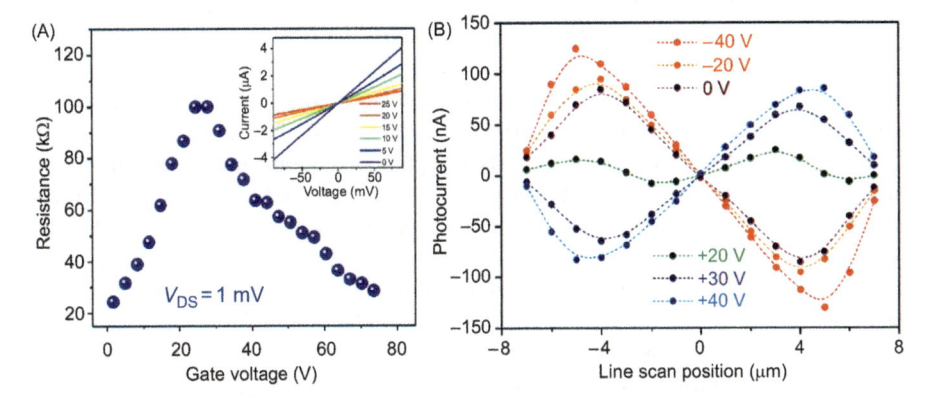

Figure 10.16 (A) Electrical transport characteristic at a drain bias of 1 mV. The inset profile indicates the I-V plot for various gate voltages V_g from 0 to 25 V. (B) Experimentally measured photocurrent for the gate bias in the range of -40 to $+40$ V. In addition, metallic nanoparticle assembly is patterned in zone 3 of the device and the incident laser power was set to 10 μW (Fang et al., 2012a).

resonances in the optical and THz frequencies (Ju et al., 2011; Boltasseva and Atwater, 2011; Hao et al., 2011; Liu et al., 2012a). Owing to several transcend features, metamaterials have been extensively employed in designing a diverse range of sensors (Liu et al., 2012b), switches (Zheludev and Kivshar, 2012), and broadband light bending (Ni et al., 2012). Recently, plasmonic metamaterials composed of noble metallic components have been introduced as promising structures for photodetection and photocurrent generation purposes (Li and Valentine, 2014; Sobhani et al., 2013). Despite having high resistive losses (ohmic losses), metallic blocks in various shapes and formations have become a traditional choice in designing metamaterials (Li and Valentine, 2014). Strong coupling of incoming optical power to the metallic structures leads to transfer of photon momentum to a free electron gas in the form of surface plasmon resonances. To mitigate the lossy behavior of plasmonic metamaterials, various techniques have been developed including combining the metamaterial with a gain medium to offset the losses regarding metallic structures (Wuestner et al., 2010). Using plasmonic substances with a negative dielectric permittivity has also been proposed (Dominguez et al., 2011). In terms of photodetection, plasmonic metamaterials typically operate on hot electron generation at the interface of metallic contact and semiconductor (metasurface) which forms a Schottky diode described in the first section of this chapter. Interaction of the incident wave of a certain frequency with the metallic blocks of a metamaterials causes excitation of SPRs. These resonance modes propagate to the metal-metasurface interface in a nonradiative fashion and generate a photocurrent at the conduction band of the metasurface. In this regime, the plasmonic photodetector's spectral response is restricted by the Schottky barrier height, instead of the energy bandgap of the metasurface (Knight et al., 2011). This mechanism allows broadband optical absorption at energies below the bandgap energy level of the metasurface. In the past few years, noticeable progresses have been reported in detection efficiency and responsivity by exploiting metallic gratings and particles (Okajima and Matsui, 2014; Dayal and Ramakrishna, 2014). On the other hand, employing multilayer metasurfaces including different surfaces with dissimilar materials with different thicknesses also provides a method to increase the absorption coefficient and photocurrent efficiency (He et al., 2015; Zhukovsky et al., 2014). Highly absorptive metamaterials for photodetection via hot electrons with high responsivity have been demonstrated with a record responsivity (Li and Valentine, 2014). Figure 10.17A illustrates a three-dimensional schematic diagram for a polarization-dependence photodetector composed of Au slabs with a thickness of 15 nm deposited on an n-type silicon as a semiconductor substrate (metasurface). Such a small thickness, which is less than the hot electron diffusion length, ensures the possibility of diffusing of hot electrons to the metal-metasurface (semiconductor) interface. To provide a strong adhesion of Au stripes to the metasurface, a thin layer of titanium (Ti) with a thickness of 1 nm is used. Illuminating the structure with a polarized light with transverse direction (0°) leads to formation of SPRs along the metallic stripes on the upper and lower resonators (see Figure 10.17B). A metamaterial structure composed of unit cell resonators (see Figure 10.17A) with

Plasmonic photodetectors

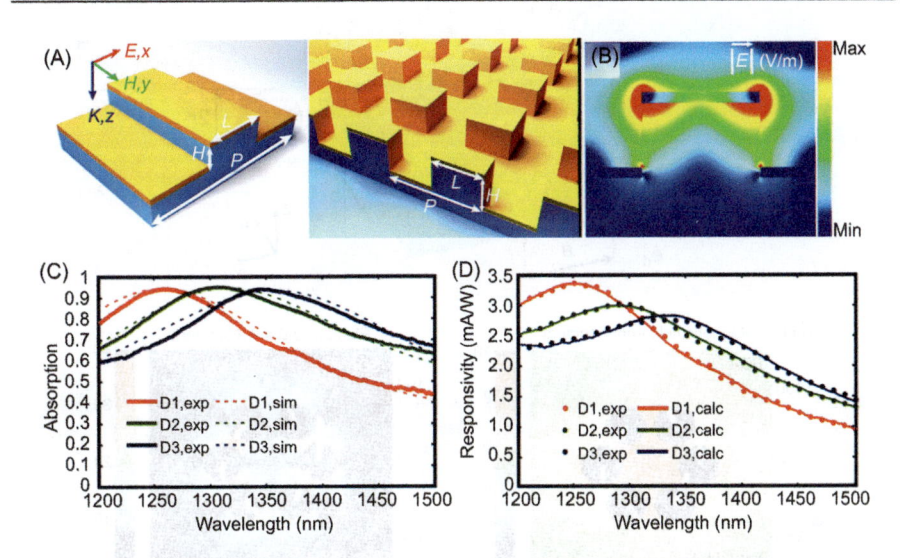

Figure 10.17 (A) Schematic diagram for a unit cell of a photodetector which shows the upper and lower Au resonators with a description of geometrical dimensions. (B) Simulation snapshot for the electric field distribution in the metamaterial and metallic slots during transverse electric illumination. (C) Numerically and experimentally measured absorption spectra for the proposed metamaterial during modifications in the geometrical dimensions of the unit cells. (D) Numerically and experimentally measured photoresponsivity for the proposed metamaterial during modifications in the geometrical dimensions of the unit cells (Li and Valentine, 2014).

the dimensions of $L = 160$, $P = 320$; $L = 170$, $P = 320$, and $L = 170$, $P = 340$ nm, and with $H = 120$ nm, exhibited the absorption spectra of D1, D2, and D3 shown in Figure 10.17C, respectively. Note that in the spectral absorption profile, the broadband absorption feature of the metamaterial in the NIR is obvious which is almost ~95% at the peak position. Figure 10.17D exhibits the photoresponsivity profile for the plasmonic photodetector that has been calculated both numerically and experimentally for the same geometrical dimensions. The figure shows the hot electron transfer efficiency in the metamaterial for the NIR spectrum. The measured responsivity at the peak positions are 3.37, 3.05, and 2.75 mA/W for D1, D2, and D3, respectively (Li and Valentine, 2014).

10.2.6.1 Plasmonic metamaterials composed of metallic assemblies

As discussed in the prior section, nanoparticle and nanocavity clusters with high plasmon resonance localization and hybridization can be used to design plasmonic photodetectors with high responsivity and giant photocurrent efficiency. Additionally, periodic arrays of nanoparticle and cavity clusters in symmetric and antisymmetric

(A)

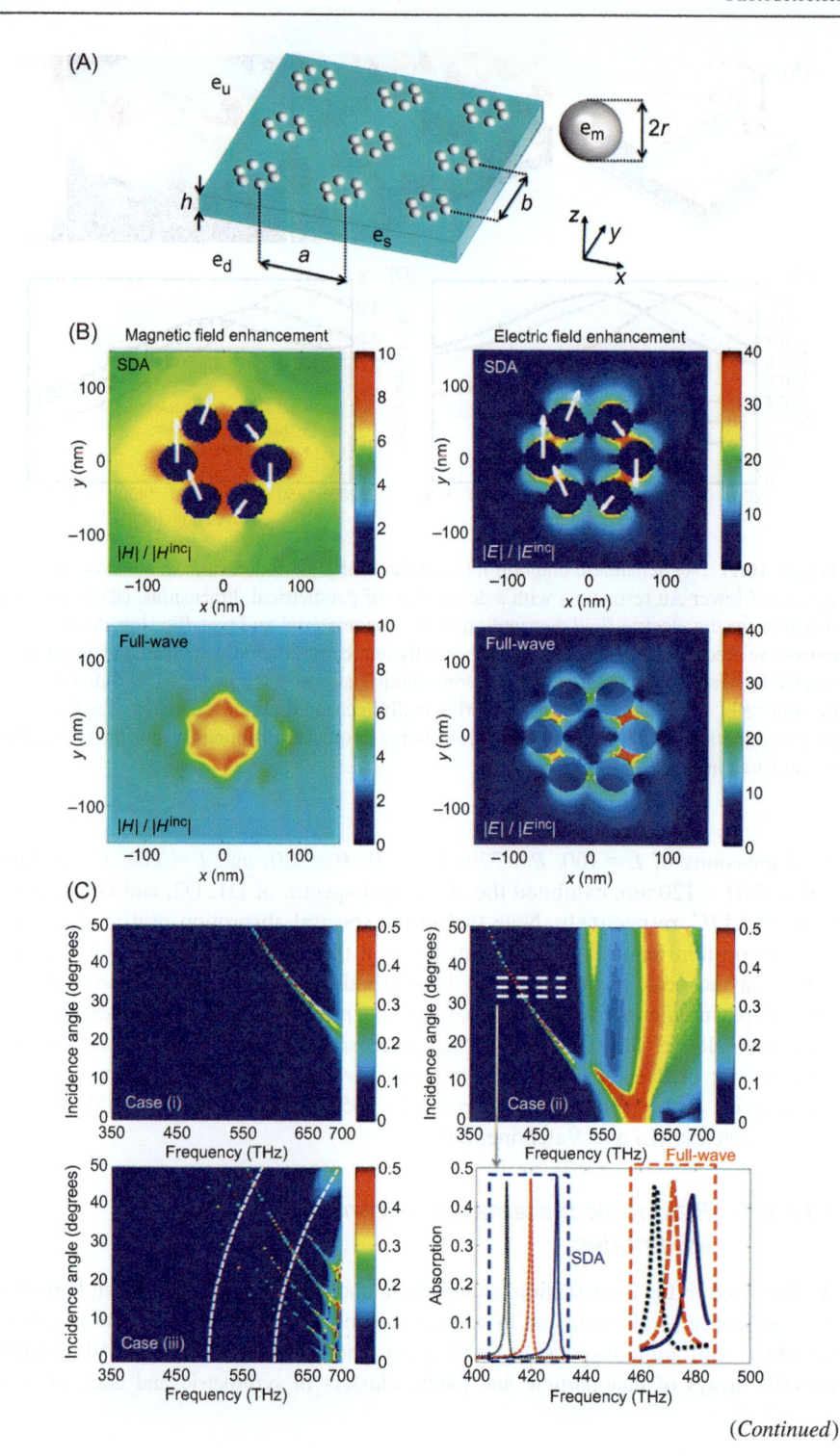

(B) Magnetic field enhancement | Electric field enhancement

SDA

Full-wave

(C)

Case (i)

Case (ii)

Case (iii)

SDA

Full-wave

orientations deposited on a metasurface can be tailored as a metamaterial structure to produce photocurrent at different spectral ranges via nonradiative hot electron transport. Figure 10.18A shows a three-dimensional schematic diagram for a metamaterial structure consisting of metallic nanosphere hexamers that are deposited on a SiO_2 surface periodically in both x and y directions (Campione et al., 2014). Having an ability to support strong plasmon resonances, this structure shows negative dielectric permittivity and magnetic permeability in the visible to the NIR spectra. Formation of dark and bright modes in nanoparticle clusters leads to possible interaction between them. Due to the destructive and weak interference between superradiant bright and subradiant dark modes in the bright energy continuum of the hexamer cluster, this structure is able to support a Fano resonant mode at the optical frequencies. Illuminating the metamaterial system by a linear EM light source, a strong hybridization of plasmon resonance modes can be realized with twofold scattering intensity due to formation of two different systems in the hexamer clusters yielding different bonding and antibonding resonant frequencies. The clusters are located 300 nm away from each other in both x and y directions to prevent strong near-field coupling between the neighboring clusters. All the dimensions are carefully selected to induce a pronounced Fano dip in a configuration (Campione et al., 2014). Figure 10.18B illustrates the numerically computed E and H field distribution and hybridization inside the clusters. The field distribution directions inside the clusters at resonance frequencies are illustrated inside the snapshots with white arrows. Figure 10.18C exhibits the snapshots for the incident angle as a function of incident frequency for different structural and chemical properties of the metasurface calculated using the single-dipole approximation. The figure also compares the behavior of the structure for three different regimes by changing the position and environmental condition for the nanoparticle arrays: Case (1) is for the array in the free space, Case (2) is for the metallic array in the SiO_2 substrate, and Case (3) is for the arrays on top of the SiO_2 layer. The absorption spectra of the nanoparticles in the hexamer-type assemblies for a metasurface under transverse polarization excitation is also presented. There are numerous other examples of metamaterials composed of nanosize clusters on metasurfaces with diverse properties and performances (Ahmadivand and Pala, 2014, 2015). Nanospirals and core-shell nanocavities are the most recent structures that have been employed in this field of optical sciences.

Figure 10.18 (A) Schematic diagram for a metamaterial consisting of metallic nanoparticle hexamers deposited on SiO_2 with description of the geometrical dimensions, (B) Magnetic and electric field enhancements in a unit cell in an $x - y$ plane. (C) Simulation diagram for the incident angle over the incident frequency: Case (1) is for the array in the free space, Case (2) is for the metallic array in the SiO_2 substrate, and Case (3) is regarding the arrays on top of the SiO_2 layer. The absorption diagram compares the SDA results (thin curves) and full-wave simulations (thick curves) for three angular cuts in the frequency range depicted by the three white dashed lines: 32.5 degrees (*solid blue*), 35 degrees (*dashed red*), and 37.5 degrees (*dotted black*) (Campione et al., 2014).

References

Adams, M.J., 1981. An Introduction to Optical Waveguides. Wiley & Sons, New York.
Ahmadivand, A., 2012. Routing properties of the T-structure based on Au/SiO$_2$ nanorings in optical nanophotonic devices. Opt. Appl. 42 (3), 659–666.
Ahmadivand, A., 2014a. Hybrid photonic-plasmonic polarization beam splitter (HPPPBS) based on metal-silica-silicon interactions. Opt. Laser Technol. 58 (4), 145–150.
Ahmadivand, A., 2014b. Ultra compact hybrid photonic-plasmonic coupler: a configuration of metal-silica-silicon. Opt. Quantum Electron. 46 (8), 1039–1048.
Ahmadivand, A., Golmohammadi, S., 2014a. Compositional arrangement of rod/shell nanoparticles: an approach to provide efficient plasmon waveguides. Opto-Electron. Rev. 22 (2), 101–108.
Ahmadivand, A., Golmohammadi, S., 2014b. Comprehensive investigation of noble metal nanoparticles shape, size, and material on the optical response of optimal plasmonic Y-splitter waveguides. Opt. Commun. 310 (1), 1–11.
Ahmadivand, A., Golmohammadi, S., 2014c. Electromagnetic plasmon propagation and coupling through the gold nanoring heptamers: a route to design optimized telecommunication photonic nanostructures. Appl. Opt. 53 (18), 3832–3840.
Ahmadivand, A., Pala, N., 2014. Plasmon response of a metal-semiconductor multilayer 4π-spiral as a negative index metamaterial. J. Nanopart. Res. 16, 2764.
Ahmadivand, A., Pala, N., 2015. Tailoring the negative-refractive-index metamaterials composed of semiconductor-metal-semiconductor gold ring/disk cavity heptamers to support strong Fano resonances in the visible spectrum. J. Opt. Soc. Am. A. 32 (2), 204–212.
Ahmadivand, A., Golmohammadi, S., Rostami, A., 2012. T- and Y-splitters based on an Au/SiO$_2$ nanoring chain an optical communication band. Appl. Opt. 51 (15), 2784–2793.
Ahmadivand, A., Golmohammadi, S., Rostami, A., 2013. Broad comparison between Au nanospheres, nanorods, and nanorings as an S-bend plasmon waveguide at optical C-band spectrum. J. Opt. Technol. 80 (2), 80–87.
Ahmadivand, A., Golmohammadi, S., Karabiyik, M., Pala, N., 2015. Fano resonances in complex plasmonic necklaces composed of gold nanodisks clusters for enhanced LSPR sensing. IEEE Sens. J. 15 (3), 1588–1594.
Akbari, A., Olivieri, A., Berini, P., 2013. Subbandgap asymmetric surface plasmon waveguide Schottky detectors on silicon. IEEE J. Sel. Top. Quantum. 19 (3), 4600209.
Atre, A.C., Garcia-Etxarri, A., Alaeian, H., Dionne, J.A., 2013. A broadband negative index metamaterial at optical frequencies. Adv. Opt. Mater. 1 (4), 350.
Barrow, S.J., Wei, X., Baldauf, S.J., Funston, A.M., Mulvaney, P., 2012. The surface plasmon modes of self-assembled gold crystals. Nat. Commun. 3 (12), 1275.
Basch, A., Beck, F.J., Söderström, T., Varlamov, S., Catchpole, K.R., 2012. Combined plasmonic and dielectric rear reflectors for enhanced photocurrent in solar cells. Appl. Phys. Lett. 100 (24), 243903.
Berini, P., 2006. Figure of merit for surface plasmon waveguides. Opt. Express. 14 (26), 13030–13042.
Berini, P., 2014. Surface plasmon photodetectors and their applications. Laser Photon. Rev. 8 (2), 197–220.
Berini, P., Olivieri, A., Chen, C., 2012. Thin Au surface plasmon waveguide Schottky detectors on p-Si. Nanotechnology. 23 (44), 444011.
Berry, C.W., Wang, N., Hashemi, M.R., Unlu, M., Jarrahi, M., 2013. Significant performance enhancement in photoconductive terahertz optoelectronics by incorporating plasmonic contact electrodes. Nat. Commun. 4 (3), 1622.

Berthold, K., Beinsting, W., Berger, R., Gornik, E., 1986. Surface plasmon enhanced quantum efficiency of metal-insulator-semiconductor junctions in the visible. Appl. Phys. Lett. 48 (8), 526.

Berthold, K., Beinsting, W., Gornik, E., 1987. Frequency- and polarization-selective Schottky detectors in the visible and near ultraviolet. Opt. Lett. 12 (2), 69–71.

Bohren, C.F., Huffman, D.R., 1983. Absorption and Scattering of Light by Small Particles. Wiley-VCH, New York, NY.

Boltasseva, A., Atwater, H.A., 2011. Low-loss plasmonic metamaterial. Science. 331, 290–291.

Boltasseva, A., Nikolajsen, T., Leosson, K., Kjaer, K., Larsen, M.S., Bozhevolnyi, S.I., 2005. Integrated optical components utilizing long-range surface plasmon polaritons. J. Lightwave Technol. 23 (1), 413–422.

Bonaccorso, F., Sun, Z., Hasan, T., Ferrari, A.C., 2010. Graphene photonics and optoelectronics. Nat. Photonics. 4 (9), 611–622.

Brueck, S.R.J., Diadiuk, V., Jones, T., Lenth, W., 1985. Enhanced quantum efficiency internal photoemission detectors by grating coupling to surface plasma waves. Appl. Phys. Lett. 46 (10), 915.

Butun, S., Cinel, N.A., Ozbay, E., 2012. LSPR enhanced MSM UV photodetectors. Nanotechnology. 23 (44), 444010.

Campione, S., Guclu, C., Ragan, R., Capolino, F., 2014. Enhanced magnetic and electric fields via Fano resonances in metasurfaces of circular clusters of plasmonic nanoparticles. ACS Photonics 1, 254–260.

Capasso, F., Gmachl, C., Paiella, R., Tredicucci, A., Hutchinson, A.L., Sivco, D.L., 2000. New frontiers in quantum cascade lasers and applications. IEEE J. Sel. Top. Quantum. 6 (6), 931–947.

Catchpole, K.R., Polman, A., 2008. Design principles for particle plasmon enhanced solar cells. Appl. Phys. Lett. 93 (19), 191113.

Chamtouri, M., Sarkar, M., Moreau, J., Besbes, M., Ghalila, H., Canva, M., 2014. Field enhancement and target localization impact on the biosensitivity of nanostructured plasmonic sensors. J. Opt. Soc. Am. B 31 (5), 1223–1231.

Chang, W.S., Lassiter, J.B., Swanglap, P., Sobhani, H., Khatua, S., Nordlander, P., 2012. A plasmonic Fano switch. Nano Lett. 12 (9), 4977–4982.

Chen, J., Li, Z., Zhang, X., Xiao, J., Gong, Q., 2013a. Submicron bidirectional all-optical plasmonic switches. Sci. Rep. 3, 1451.

Chen, S., Liu, J., Lu, H., Zhu, Y., 2013b. All-optical strong coupling switches based on coupled *meta*-atom and MIM nanocavity configuration. Plasmonics. 8 (3), 1439–1444.

Chen, J., Sun, C., Li, H., Gong, Q., 2014. Experimental demonstration of an on-chip polarization splitter in a submicron asymmetric dielectric-coated metal slit. Appl. Phys. Lett. 104 (23), 231111.

Colace, L., Masini, G., Galluzzi, F., Assanto, G., Capellini, G., Di Gaspare, L., 1998. Metal-semicondutor-metal near-infrared light detector based on epitaxial Ge/Si. Appl. Phys. Lett. 12 (24), 3175.

Dayal, G., Ramakrishna, S.A., 2014. Multipolar localized resonances for multiband metamaterial perfect absorbers. J. Opt. 16 (9), 094016.

Derkacs, D., Lim, S.H., Matheu, P., Mar, W., Yu, E.T., 2006. Improved performance of amorphous silicon solar cells via scattering from surface plasmon polaritons in nearby metallic nanoparticles. Appl. Phys. Lett. 89 (9), 093103.

Ditlbacher, H., Aussenegg, D.R., Krenn, J.R., Lamprecht, B., Jakopic, G., Leising, G., 2006. Organic diodes as monolithically integrated surface plasmon polaritons. Appl. Phys. Lett. 89 (16), 161101.

Dominguez, R.P., Tejeira, F.L., Marques, R., Gil, A.S., 2011. Metallo-dielectric core-shell nanospheres as building blocks for optical three-dimensional isotropic negative-index metamaterials. N. J. Phys. 13, 123017.

Dufaux, T., Dorfmüller, J., Vogelgesang, R., Burghard, M., Kern, K., 2010. Surface plasmon coupling to nanoscale Schottky-type electric detectors. Appl. Phys. Lett. 97 (16), 161110.

Echtermeyer, T.J., Britnell, L., Jasnos, P.K., Lombardo, A., Gorbachev, R.V., Grigorenko, A.N., 2011. Strong plasmonic enhancement of photovoltage in graphene. Nat. Commun. 2 (8), 458.

Fan, J.A., Wu, C., Bao, K., Bao, J., Bardhan, R., Halas, N.J., 2010. Self-assembled plasmonic nanoparticle clusters. Science. 328 (5982), 1135−1138.

Fang, Z., Liu, Z., Wang, Y., Ajayan, P.M., Nordlander, P., Halas, N.J., 2012a. Graphene-antenna sandwich photodetector. Nano Lett. 12 (7), 3808−3813.

Fang, Z., Wang, Y., Liu, Z., Schlather, A., Ajayan, P.M., Koppens, F.H.L., 2012b. Plasmon-induced doping of graphene. ACS Nano 6 (11), 10222−10228.

Fang, Z., Wang, Y., Schlather, A.E., Liu, Z., Ajayan, P.M., Javier, F., 2014. Garcia de Abajo, active tunable absorption enhancement with graphene nanodisk arrays. Nano Lett. 14, 299−304.

Ferrari, A.C., Meyer, J.C., Scardaci, V., Casiraghi, C., Lazzeri, M., Mauri, F., 2006. Raman spectrum of graphene and graphene layers. Phys. Rev. Lett. 97, 187401.

Fox, M., 2006. Optical Properties of Solids. Oxford University Press.

Freitag, M., Low, T., Avouris, P., 2013a. Increased responsivity of suspended graphene photodetectors. Nano Lett. 13 (4), 1644−1648.

Freitag, M., Low, T., Xia, F., Avouris, P., 2013b. Photoconductivity of biased graphene. Nat. Photonics. 7 (1), 53−59.

Furchi, M., Urich, A., Pospischil, A., Lilley, G., Unterrainer, K., Detz, H., 2012. Microcavity-integrated graphene photodetector. Nano Lett. 12 (6), 2773−2777.

Gao, Y., Gan, Q., Xin, Z., Cheng, X., Bartoli, F.J., 2011. Plasmonic Mach-Zehnder interferometer for ultrasensitive on-chip biosensing. ACS Nano 5 (12), 9836−9844.

Golmohammadi, S., Ahmadivand, A., 2014. Fano resonances in compositional clusters of aluminum nanodisks at the UV spectrum: a route to design efficient and precise biochemical sensors. Plasmonics. 9 (6), 1447−1456.

Golmohammadi, S., Khalilou, Y., Ahmadivand, A., 2014. Plasmonics: a route to design an optical demultiplexer based on gold nanorings arrays to operate at near infrared region (NIR). Opt. Commun. 321, 56−60.

Goykhman, I., Desiatov, B., Khurgin, J., Shappir, J., Levy, U., 2012. Waveguide based compact silicon Schottky photodetector with enhanced responsivity in the telecom spectral band. Opt. Express. 20 (27), 28594−28602.

Grandidier, J., Colas des Francs, G., Massenot, S., Bouhelier, A., Markey, L., Weeber, J.C., 2009. Gain-assisted propagation in a plasmonic waveguide at telecom wavelengths. Nano Lett. 9 (8), 2935−2939.

Grigorenko, A.N., Polini, M., Novoselov, K.S., 2012. Graphene plasmonics. Nat. Photonics. 6 (11), 749−758.

Guo, J., Yoon, Y., Ouyang, Y., 2007. Gate electrostatics and quantum capacitance of graphene nanoribbons. Nano Lett. 7 (7), 1935−1940.

Guo, Y., Yan, L., Pan, W., Luo, B., Wen, K., Guo, Z., 2011. A plasmonic splitter based on slot cavity. Opt. Express. 19 (15), 13831−13838.

Hao, J., Zhou, L., Qiu, M., 2011. Nearly total absorption of light and heat generation by plasmonic metamaterials. Phys. Rev. B. 83, 165107.

He, X., Yang, L., Yang, T., 2011. Optical nanofocusing by tapering coupled photonic-plasmonic waveguides. Opt. Express. 19 (14), 12865−12872.
He, X.J., Yan, S.T., Mia, Q.X., Zhang, Q.F., Jia, P., Wu, F.M., 2015. Broadband and polarization-insensitive terahertz absorber based on multilayer metamaterials. Opt. Commun. 340 (1), 44−49.
Hetterich, J., Bastian, G., Gippius, N.A., Tikhodeev, S.G., Von Plessen, G., Lemmer, U., 2007. Optimized design of plasmonic MSM photodetector. IEEE J. Quantum Electron. 43 (10), 855−859.
Hong, Y., Pourmand, M., Boriskina, S.V., Reinhard, B.M., 2013. Enhanced light focusing in self-assembled optoplasmonic clusters with subwavelength dimensions. Adv. Mater. 25 (1), 115−119.
Hu, C.C., Tsai, Y.T., Yang, W., Chau, Y.F., 2014. Effective coupling of incident light through an air region into an S-bend plasmonic Ag nanowire waveguide with relatively long propagation length. Plasmonics. 9 (3), 573−579.
Hu, H., Zeng, X., Ji, D., Zhu, L., Gan, Q., 2013. Efficient end-fire coupling of surface plasmons on flat metal surfaces for improved plasmonic Mach-Zehnder interferometers. J. Appl. Phys. 113 (5), 053101.
Hyun, J.K., Lauhon, L.J., 2011. Spatially resolved plasmonically enhanced photocurrent from Au nanoparticles on a Si nanowire. Nano Lett. 11 (7), 2731−2734.
Jackson, J.D., 2007. Classical Electrodynamics. Wiley, New York.
Jestl, M., Maran, I., Köck, A., Beinstingl, W., Gornik, E., 1989. Polarization-sensitive surface plasmon Schottky detectors. Opt. Lett. 14 (14), 719−721.
Jo, S., Ki, D.K., Jeong, D., Lee, H.J., Kettemann, S., 2011. Spin relaxation properties in graphene due to its linear dispersion. Phys. Rev. B. 84, 075453.
Ju, L., Geng, B., Horng, J., Girit, C., Martin, M., Hao, Z., 2011. Graphene plasmonic for tunable terahertz metamaterials. Nat. Nanotechnol. 6 (10), 630−634.
Kelly, K.L., Coronado, E., Zhao, L.L., Schatz, G.C., 2003. The optical properties of metal nanoparticles: the influence of size, shape, and dielectric environments. J. Phys. Chem. B 107 (3), 668−677.
Kim, J.T., Yu, Y.J., Choi, H., Choi, C.G., 2014. Graphene-based plasmonic photodetector for photonic integrated circuits. Opt. Express. 22 (1), 803−808.
Kirkengen, M., Bergli, J., Galperin, Y.M., 2007. Direct generation of charge carriers in c-Si solar cells due to embedded nanoparticles. J. Appl. Phys. 102 (9), 093713.
Knight, M.W., Sobhani, H., Nordlander, P., Halas, N.J., 2011. Photodetection with active optical antennas. Science. 332, 702−704.
Konstantatos, G., Sargent, E.H., 2010. Nanostructured materials for photon detection. Nat. Nanotechnol. 5 (6), 391−400.
Koppens, F.H.L., Chang, D.E., Javier, F., 2011. Garcia de Abajo, Graphene plasmonics: a platform for strong light-matter interactions. Nano Lett. 11 (8), 3370−3377.
Kroo, N., Szentirmay, Z.S., Felszerfalvi, J., 1984. Internal photoeffect in periodically corrugated MOM structures. Phys. Lett. A. 101 (4), 235−238.
Lee, K.S., El-Sayed, M.A., 2006. Gold and silver nanoparticles in sensing and imaging: sensitivity of plasmon response to size, shape, and metal composition. J. Phys. Chem. B 110 (39), 19220−19225.
Lee, S.C., Krishna, S., Brueck, S.R.J., 2011. Plasmonic-enhanced photodetectors for focal plane arrays. IEEE Photon. Technol. Lett. 23 (14), 935−937.
Lee, S.H., Choi, M., Kim, T.T., Lee, S., Liu, M., Yin, X., 2012. Switching terahertz waves with gate-controlled active graphene metamaterials. Nat. Mater. 11 (11), 936−941.

Lemme, M.C., Koppens, F.H.L., Falk, A.L., Rudner, M.S., Park, H., Levitov, L.S., 2011. Gate-activated photoresponse in a graphene p-n junction. Nano Lett. 11 (10), 4134–4137.

Li, W., Valentine, J., 2014. Metamaterial perfect absorber based hot electron photodetection. Nano Lett. 14 (6), 3510–3514.

Li, X., Zhu, H., Wang, K., Cao, A., Wei, J., Li, C., 2010. Graphene-on-silicon Schottky junction solar cells. Adv. Mater. 22 (25), 2743–2748.

Linic, S., Christopher, P., Ingram, D.B., 2011. Plasmonic-metal nanostructures for efficient conversion of solar to chemical energy. Nat. Mater. 10 (12), 911–921.

Liu, K., Sakurai, M., Liao, M., Aono, M., 2010. Giant improvement of the performance of ZnO nanowire photodetectors by Au nanoparticles. J. Phys. Chem. C. 114 (46), 19835–19839.

Liu, X., Gu, J., Singh, R., Ma, Y., Zhu, J., Tian, Z., 2012a. Electromagnetically induced transparency in terahertz plasmonic metamaterials via dual excitation pathways of the dark mode. Appl. Phys. Lett. 100 (13), 131101.

Liu, Z., Xia, X., Sun, Y., Yang, H., Chen, R., Liu, B., 2012b. Visible transmission response of nanoscale complementary metamaterials for sensing applications. Nanotechnology. 23 (27), 275503.

Liu, B., Sun, Z., Zhang, X., Liu, J., 2013. Mechanisms of DNA sensing on graphene oxide. Anal. Chem. 85 (16), 7987–7993.

Low, T., Avouris, P., 2014. Graphene plasmonics for terahertz to mid-infrared applications. ACS Nano 8 (2), 1086–1101.

Luk'yanchuk, B., Zheludev, N.I., Maier, S.A., Halas, N.J., Nordlander, P., Giessen, H., 2010. The Fano resonance in plasmonic nanostructures and metamaterials. Nat. Mater. 9 (9), 707–715.

Ma, H., Jen, A.K.Y., Dalton, L.R., 2002. Polymer-based optical waveguides: materials, processing, and devices. Adv. Mater. 14 (19), 1339–1365.

Maier, S.A., 2007. Plasmonics: Fundamental and Applications. Springer, New York.

Maradudin, A.A., Sambles, J.R., Barnes, W.L., 2014. Modern Plasmonics. Elsevier, Oxford.

Mie, G., 1908. Contributions to the optics of turbid media, particularly of colloidal metal solutions. Leipzig Ann. Phys. 25 (3), 377–445.

Mueller, T., Xia, F., Freitag, M., Tsang, J., Avouris, P.H., 2009. Role of contacts in graphene transistors: a scanning photocurrent study. Phys. Rev. B. 79, 245430.

Mueller, T., Xia, F., Avouris, P., 2010. Graphene photodetectors for high-speed optical communications. Nat. Photonics. 4 (5), 297–301.

Munday, J.N., Atwater, H.A., 2011. Large integrated absorption enhancement in plasmonic solar cells by combining metallic grating and antireflection coating. Nano Lett. 11 (6), 2195–2201.

Naphade, R.A., Tathavadekar, M., Jog, J.P., Agarkar, S., Ogale, S., 2014. Plasmonic light harvesting of dye sensitized solar cells by Au-nanoparticle loaded TiO_2 nanofibers. J. Mater. Chem. A. 2 (4), 975–984.

Nasir, M.E., Dickson, W., Wurtz, G.A., Wardley, W.P., Zayats, A.V., 2014. Hydrogen detected by the naked eye: optical hydrogen gas sensors based on core-shell plasmonic nanorod metamaterials. Adv. Mater. 26 (21), 3532–3537.

Neto, A.C., Guniea, F., Peres, N.M.R., Novoselov, K.S., Geim, A.K., 2009. The electronic properties of graphene. Rev. Mod. Phys. 80, 109.

Neutens, P., Van Dorpe, P., De Vlaminck, I., Lagae, L., Borghs, G., 2009. Electrical detection of confined gap plasmons in metal-insulator-metal waveguides. Nat. Photon. 3 (5), 283–286.

Ni, X., Emani, N.K., Kildishev, A.V., Boltasseva, A., Shalaev, V.M., 2012. Broadband light bending with plasmonic nanoantennas. Science. 335 (6067), 427.
Okajima, A., Matsui, T., 2014. Electron-beam induced terahertz radiation from graded metallic grating. Opt. Express. 22 (14), 17490–17496.
Olivieri, A., Akbari, A., Berini, P., 2010. Surface plasmon waveguide Schottky detectors operating near breakdown. Phys. Status Solidi-R 4 (10), 283–285.
Oulton, R.F., Sorger, V.J., Genov, D.A., Pile, D.F.P., Zhang, X., 2008. A hybrid plasmonic waveguide for subwavelength confinement and long-range propagation. Nat. Photon. 2 (8), 496–500.
Ouyang, Z., Pillai, S., Beck, F., Kunz, O., Varlamov, S., Catchpole, K.R., 2010. Effective light trapping in polycrystalline silicon thin film solar cells by means of rear localized surface plasmons. Appl. Phys. Lett. 96 (26), 261109.
Pala, R.A., White, J., Barnard, E., Liu, J., Brongersma, M.L., 2009. Design of plasmonic thin-film solar cells with broadband absorption enhancement. Adv. Mater. 21 (34), 3504–3509.
Palik, E.D., 1985. Handbook of Optical Constants of Solids. Academic Press, CA.
Pelayo Garcia de Arquer, F., Beck, F.J., Bernechea, M., Konstantatos, G., 2012. Plasmonic light trapping leads to responsivity increase in colloidal quantum dot photodetectors. Appl. Phys. Lett. 100 (4), 043101.
Pillai, S., Catchpole, K.R., Trupke, T., Green, M.A., 2007. Surface plasmon enhanced silicon solar cells. J. Appl. Phys. 101 (9), 093105.
Prodan, E., Radloff, C., Halas, N.J., Nordlander, P., 2003. A hybridization model for the plasmon response of complex nanostructures. Science. 302 (5644), 419–422.
Qu, D., Liu, F., Yu, J., Xie, W., Xu, Q., Li, X., 2011. Plasmonic core-shell gold nanoparticle enhanced optical absorption in photovoltaic devices. Appl. Phys. Lett. 98 (11), 113119.
Raether, H., 1988. Surface Plasmons on Smooth and Rough Surfaces and on Gratings. Springer, Berlin, Germany.
Rahman, M., Karakashian, A.S., Broude, S., Gladden, D., 1991. Surface-plasma-enhanced quantum efficiency of a Ag-Ti-n-GaAs grating photodiode. Appl. Opt. 30 (21), 2935–2937.
Rana, F., George, P.A., Strait, J.H., Dawlaty, J., Shivaraman, S., Chandrashekhar, M.V.S., 2009. Carrier recombination and generation rates for intravalley and intervalley phonon scattering in graphene. Phys. Rev. B. 79, 115447.
Ren, F.F., Ang, K.W., Song, J., Fang, Q., Yu, M., Lo, G.Q., 2010. Surface plasmon enhanced responsivity in a waveguided germanium metasemiconductor-metal photodetector. Appl. Phys. Lett. 97 (9), 091102.
Rosenberg, J., Shenoi, R.V., Krishna, S., Painter, O., 2010. Design of plasmonic photonic crystal cavities for polarization sensitive infrared photodetectors. Opt. Express. 18 (4), 3672–3686.
Ryzhii, V., Mitin, V., Ryzhii, M., Ryabova, N., Otsuji, T., 2008. Device model for graphene nanoribbons phototransistor. Appl. Phys. Express 1, 063002.
Satoh, H., Inokawa, H., 2012. Surface plasmon antenna with gold line and shape grating for enhanced visible light detection by a silicon-on-insulator metal-oxide-semiconductor photodiode. IEEE Trans. Nanotechnol. 11 (2), 346–351.
Scales, C., Berini, P., 2010. Thin film Schottky barrier photodetector models. IEEE J. Quantum Electron. 46 (5), 633–643.
Scales, C., Breukelaar, I., Charbonneau, R., Berini, P., 2006. Infrared performance of symmetric surface-plasmon waveguide Schottky detectors in Si. J. Light. Technol. 29 (12), 1852–1860.

Scales, C., Breukelaar, I., Berini, P., 2010. Surface-plasmon Schottky contact detector based on a symmetric metal stripe in silicon. Opt. Lett. 35 (4), 529–531.

Schaadt, D.M., Feng, B., Yu, E.T., 2005. Enhanced semiconductor optical absorption via surface plasmon excitation in metal nanoparticles. Appl. Phys. Lett. 86 (6), 063106.

Schackleford, J.A., Grote, R., Currie, M., Spanier, J.E., Nabet, B., 2009. Integrated plasmonic lens photodetector. Appl. Phys. Lett. 94 (8), 083501.

Senanayake, P., Hung, C.H., Shapiro, J., Lin, A., Liang, B., Williams, B.S., 2011. Surface plasmon-enhanced nanopillar photodetectors. Nano Lett. 11 (12), 5279–5283.

Shahbazyan, T.V., Stockman, M.I., 2013. Plasmonics: Theory and Applications. Springer, New York.

Sobhani, A., Knight, M.W., Wang, Y., Zheng, B., King, N.S., Brown, L.V., 2013. Narrowband photodetection in the near-infrared with a plasmon-induced hot electron device. Nat. Commun. 4 (3), 1643.

Spinelli, P., Verschuuren, M.A., Polman, A., 2012. Broadband omnidirectional antireflection coating based on subwavelength surface mire resonators. Nat. Commun. 3 (2), 692.

Stratton, J.A., 2007. IEEE Press Electromagnetic Theory. Wiley & Sons, New Jersey.

Stuart, H.R., Hall, D.G., 1996. Absorption enhancement in silicon-on-insulator waveguides using metal island films. Appl. Phys. Lett. 69 (16), 2327.

Stuart, H.R., Hall, D.G., 1998. Island size effects in nanoparticle-enhanced photodetectors. Appl. Phys. Lett. 73 (26), 3815.

Sundararajan, S.P., Grady, N.K., Mirin, N., Halas, N.J., 2008. Nanoparticle-induced enhancement and suppression of photocurrent in a silicon photodiode. Nano Lett. 8 (2), 624–630.

Sze, S.M., 1981. Physics of Semiconductor Devices. Wiley.

Trolle, M.L., Pedersen, T.G., 2012. Indirect optical absorption in silicon via thin-film surface plasmon. J. Appl. Phys. 112 (4), 043103.

Vanderlinde, J., 2004. Classical Electromagnetic Theory: Fundamental Theories of Physics. Springer, Kluwer Academic Publishers, USA.

Veronis, G., Fan, S., 2007. Theoretical investigation of compact couplers between dielectric slab waveguides and two-dimensional metal-dielectric-metal plasmonic waveguides. Opt. Express. 15 (3), 1211–1221.

Wang, X., Zhi, L., Müllen, K., 2008. Tranparent, conductive graphene electrodes for dye-sensitized solar cells. Nano Lett. 8 (1), 323–327.

Wang, P., Zhang, W., Liang, O., Pantoja, M., Katzer, J., Schroeder, T., 2012. Giant optical response from graphene-plasmonic system. ACS Nano 6 (7), 6244–6249.

Wen, F., Ye, J., Liu, N., Van Dorpe, P., Nordlander, P., Halas, N.J., 2012. Plasmon transmutation: inducing new modes in nanoclusters by adding dielectric nanoparticles. Nano Lett. 12 (9), 5020–5026.

Willet, K.A., Van Duyne, R.P., 2007. Localized surface plasmon resonance spectroscopy and sensing. Annu. Rev. Phys. Chem. 58, 267–297.

Wu, C., Khanikaev, A.B., Adato, R., Arju, N., Yanik, A.A., Altug, H., 2011. Fano-resonant asymmetric metamaterials for ultrasensitive spectroscopy and identification of molecular monolayers. Nat. Mater. 11 (1), 69–75.

Wu, M.C., Deokar, A.R., Liao, J.H., Shih, P.Y., Ling, Y.C., 2013. Graphene-based photothermal agent for rapid and effective killing of bacteria. ACS Nano 7 (2), 1281–1290.

Wuestner, S., Pusch, A., Tsakmakidis, K.S., Hamm, J.M., Hess, O., 2010. Overcoming losses with gain in a negative refractive index metamaterial. Phys. Rev. B. 105, 127401.

Xia, F., Mueller, T., Lin, Y.M., Garcia, A.V., Avouris, P., 2009. Ultrafast graphene photodetector. Nat. Nanotechnol. 4 (12), 839–843.

Xiang, G., Guo, H., Zhang, X., Sum, T.C., Huan, C.H.A., 2010. The physic of ultrafast saturable absorption in graphene. Opt. Express. 18 (5), 4564–4573.
Xiang, Q., Yu, J., Jaroniec, M., 2012. Graphene-based semiconductor photocatalyst. Chem. Soc. Rev. 41 (2), 782–796.
Xiao, Y.F., Liu, Y.C., Li, B.B., Chen, Y.L., Li, Y., Gong, Q., 2012. Strongly enhanced light-matter interaction in a hybrid photonic-plasmonic resonator. Phys. Rev. A. 85, 031805.
Xing, J., Zhao, C., Guo, E., Yang, F., 2012. High performance ultraviolet photodetector based on polycrystalline $SrTiO_3$ thin film. IEEE Sens. J. 12 (8), 2561–2654.
Yang, K., Feng, L., Shi, X., Liu, Z., 2013. Nano-graphene in biomedicine: theranostic applications. Chem. Soc. Rev. 42 (2), 530–547.
Yang, S., Gong, J., Deng, Y., 2012. A sandwich-structured ultraviolet photodetector driven only by opposite heterojunctions. J. Mater. Chem. 22 (28), 13899–13902.
Yoon, J.W., Park, W.J., Lee, K.J., Song, S.H., Magnusson, R., 2011. Surface plasmon mediated total absorption of light into silicon. Opt. Express. 19 (21), 20673–20680.
Zakharian, A.R., Moloney, J.V., Mansuripur, M., 2007. Surface plasmon polaritons on metallic surfaces. Opt. Express. 15 (1), 183–197.
Zayats, A.V., Maier, S.A., 2013. Active Plasmonics and Tuneable Plasmonics Metamaterials. Wiley & Sons, New Jersey.
Zhang, R., Guo, X.G., Song, C.Y., Buchanan, M., Wasilevski, Z.R., Cao, J.C., 2011. Metal-grating-coupled terahertz quantum well photodetectors. IEEE Electron. Device Lett. 32 (5), 659–661.
Zhang, N., Zhang, Y., Xu, Y.J., 2012. Recent progress on graphene photocatalysts: current status and future perspectives. Nanoscale. 4 (19), 5792–5813.
Zhang, S., Fan, W., Panoiu, N.C., Osgood, M.R.M., Brueck, S.R.J., 2005. Experimental demonstration of near-infrared negative-index metamaterial. Phys. Rev. Lett. 95, 137404.
Zhao, X., Zhang, Z., Wang, L., Xi, K., Cao, Q., Wang, D., 2013. Excellent microwave absorption property of graphene-coated Fe nanocomposites. Sci. Rep. 3, 3421.
Zheludev, N.I., Kivshar, Y.S., 2012. From metamaterials to metadevices. Nat. Mater. 11 (11), 917–924.
Zheng, Y.B., Kiraly, B., Cheunkar, S., Huang, T.J., Weiss, P.S., 2011. Incident-angle-modulated molecular plasmonic switches: a case of weak exciton-plasmon coupling. Nano Lett. 11 (5), 2061–2065.
Zhukovsky, S.V., Orlov, A.A., Babicheva, V.E., Lavrinenko, A.V., Spie, J.E., 2014. Photonic-band-gap engineering for volume plasmon polaritons in multiscale multilayer hyperbolic metamaterials. Phys. Rev. A. 90, 013801.

CMOS-integrated waveguide photodetectors for communications applications

11

Shiyang Zhu and Guo-Qiang Lo
Institute of Microelectronics, The Agency for Science, Technology and Research (A*STAR), Singapore

11.1 Introduction

Silicon photonics has widely been studied for optical telecommunications and for optical interconnects in microelectronics. The rationale is to reduce the cost of photonic systems through the integration of photonic components and electronic components/circuits on a common chip within a single complementary metal–oxide–semiconductor (CMOS) process flow (Beals et al., 2008). In the spectral range of 1.3–1.6 μm for communication applications, Si is a good waveguiding material because of its transparency and large refractive index contrast to SiO_2. Low-loss waveguides along with various passive components (e.g., power splitters, waveguide ring resonators, wavelength-division multiplexing/demultiplexing systems, grating couplers for input/output, etc.) have been demonstrated on the silicon-on-insulator (SOI) platform using the standard CMOS fabrication process (Lockwood and Pavesi, 2011). High-performance Si modulators are also demonstrated relying on the Si free-carrier plasma dispersion effect (Reed et al., 2010). However, while Si photodetectors are widely used in the visible spectral range, a defect-free bulk Si cannot perform as an efficient absorber at wavelengths longer than 1.10 μm because of its relatively large bandgap of 1.12 eV. Over the past decades, III–V compound materials such as InGaAs have been utilized for the near infrared detection due to their excellent light emission and absorption properties, high carrier mobility, and mature technology for design and fabrication of photonic devices. High-performance III–V photodetectors have been integrated in Si-waveguide circuits using the die to wafer bonding technology (the technology is also used for integration of III–V laser diodes in Si photonic circuits) (Roelkens et al., 2005). However, this hybrid integration is not a standard CMOS process and the introduction of III–V materials in the Si CMOS process flow may cause the cross-contamination issue. Therefore, monolithic integration of photodetectors in Si photonic circuits using the front-end-of-the-line (FEOL)-CMOS-compatible process is highly desired.

Germanium is an effective absorber in the 1.3–1.55 μm spectral range because of its relatively narrow direct bandgap of 0.8 eV. Ge is a CMOS-compatible material as it has been introduced in BiCMOS baseline (Lischke et al., 2015). Owing to the recently developed technology for epitaxial growth of high-quality Ge films on Si substrates, Ge-on-Si photodetectors have achieved high performances such as >50 GHz bandwidth and ~1 A/W responsivity at 1.55 μm. Thus, they have become the mainstream photodetectors for Si photonic integrated circuits (Michel et al., 2010; Liu, 2014; Wang and Lee, 2011; Ang et al., 2010b).

However, Ge is still a challenging material to integrate in a CMOS environment due to its low thermal budget constraint and its large lattice mismatch of approximately 4.2% with Si (Currie et al., 1998). To keep the advantages of low cost and easy manufacturability of Si photonic integrated circuits, all-silicon photodetectors are preferred. A number of all-silicon near-infrared photodetectors have been proposed based on different physical effects (Casalino et al., 2010b), which can be roughly cataloged to two approaches. One approach is by intentionally introducing defects or dopants into the Si lattice to stimulate defect-mediated subbandgap absorption in Si. Waveguide photodetectors with high bandwidth and large responsivity have been demonstrated using this approach. However, additional processing steps such as proton, helium, or Si^+ implantation are required and the subsequent processes are restricted to low temperature to prevent the defects being annealed out. Moreover, these photodetectors usually have a long length because the defect-mediated Si has relatively low optical absorption (Geis et al., 2009; Grote et al., 2013; Chen et al., 2015). The other approach is by employing a thin metal or metallic silicide layer to take advantage of the internal photoemission process, in which the light is absorbed by the thin metal or silicide layer and the detectable wavelength is determined by the Schottky-barrier height at the silicide/Si interface. These Schottky-barrier photodetectors (SBPDs) offer a small footprint because the metallic silicide has very large optical absorption and ease in CMOS implementation because the metallic silicide has already been used for interconnection in Si electronic circuits. However, they usually suffer from an inherent issue of low responsivity (Scales and Berini, 2010; Goykhman et al., 2012).

In this chapter, we review Si-waveguide-integrated Ge-on-Si and silicide SBPDs developed in our laboratory. For the Ge-on-Si photodetectors, the discussion first begins with the technology of selective hetero-epitaxy of Ge-on-Si, which is the basic for all Ge-on-Si devices. Then, two kinds of evanescent-coupled Ge PIN photodetectors are presented and their performance metrics are evaluated in terms of dark current, responsivity, and bandwidth and compared with those reported by the other groups. Then, a surface plasmon enhanced photodetector is introduced which can enhance the optical absorption significantly. Finally, a Ge-on-Si avalanche detector (APD) with lateral separate-absorption-charge-multiplication (SACM) configuration is introduced for large responsivity. For the silicide SBPDs, we first discuss $NiSi_2$/p-Si, $NiSi_2$/n-Si, and metal-semiconductor-metal (MSM) configurations. The tradeoff between the absorption and the internal quantum efficiency is elucidated. Then, two approaches are discussed to improve the responsivity. One approach is to use a Schottky-barrier collector phototransistor (SBCPT) configuration, which can provide a current gain. The other approach is to use silicide nanoparticles (NPs) embedded in the Si p-n junction, which can enhance both the light absorption and the internal quantum efficiency. We conclude the chapter with comparative views on the performance metrics of these photodetectors and possible solutions for the remaining technical issues.

11.2 Waveguide-integrated Ge-on-Si photodetectors

11.2.1 Selective hetero-epitaxy of Ge-on-Si and CMOS integration

High-quality Ge films grown on Si are the basic for any Ge-on-Si devices. Our laboratory has developed Ge selective epitaxial growth (SEG) technology using an ultra-high vacuum chemical vapor deposition (UHCVD) reactor to grow Ge within designed regions on Si substrates (Loh et al., 2007a,b; Ang et al., 2010a). Such a SEG technology not only eliminates the subsequent Ge etching step for patterning the Ge layer but also reduces the threading dislocation (TD) density if the selective growth region is small enough (<40 μm) (Luan et al., 1999). The SEG of Ge is started from Si (100) substrates. An SiO_2 layer is deposited using plasma-enhanced chemical vapor deposition (PECVD) and then patterned by reactive ion etching with ~10-nm SiO_2 remaining to form an oxide window for the Ge growth. The remaining SiO_2 is wet etched by dilute hydrofluoric acid (DHF, 1:200) and subsequently wet cleaned by standard SC1 ($NH_4OH:H_2O_2:H_2O = 1:2:10$ at 60°C), DHF, and isopropyl alcohol solutions. These cleaning processes will affect the Ge quality significantly. The wafers are immediately submitted to the UHCVD chamber with a base pressure of greater than 10 − 8 torr. The wafers are backed in N_2 ambient at 800°C and then ramped down and stabilized at 500°C. A thin Si-seed layer is selectively grown using SiH_4 and HCl precursors, followed by selective deposition of a thin $Si_{0.8}Ge_{0.2}$ buffer layer with a thickness of 10−15 nm using SiH_4 and GeH_4 precursors at 400°C. A graded SiGe layer with the Ge mole fraction graded from 10% to approximately 50% can also be used as the buffer layer. The wafers are then subjected to SEG of Ge seed at 400°C. The use of a low temperature growth is intended to suppress adatoms migration on Si and thus prevents the formation of three-dimensional Stranski-Krastanov growth, which allows a flat Ge surface morphology to be achieved. Upon obtaining a smooth Ge seed layer, the epitaxy process temperature is then increased to approximately 600°C to facilitate faster epitaxy growth to obtain the desired Ge thickness (Loh et al., 2007b; Sakai et al., 1997). Using this approach, high-quality Ge epilayer with a thickness of up to approximately 2 μm has been obtained. Figure 11.1 shows a 189 nm epitaxial Ge film selectively grown on a SOI substrate with 13.6 nm $Si_{0.78}Ge_{0.22}$ buffer layer, for example. The Ge layer is only grown on the Si surface (no Ge on the dielectric surface) and the Ge mesa is faceted with the (111) and (113) facets dominating when the oxide openings are aligned to a ⟨110⟩ direction. Energy dispersive X-ray spectroscopy (EDX) shows a very consistent Ge concentration of 92.1%−94% across the Ge film without any trace of Si migration at various locations. The Ge top surface has no cross-hatch pattern and is smooth. A root-mean-square (rms) roughness of approximately 0.68 nm is measured on a 5×5 μm^2 scanning area by atomic

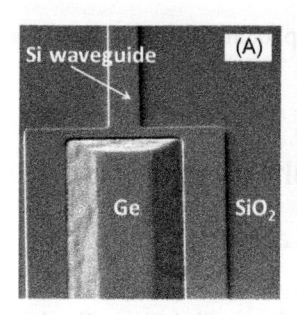

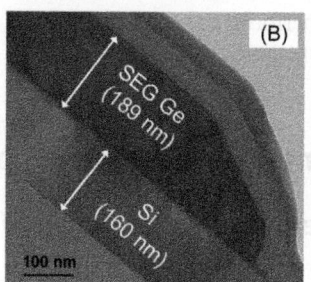

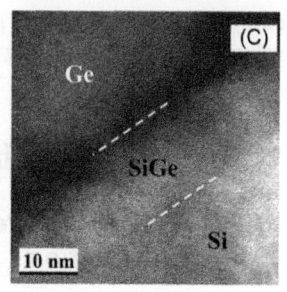

Figure 11.1 An example of SEG of Ge-on-Si (100) substrate: (A) scanning electron microscopy (SEM) image of a SEG-Ge layer in a SiO$_2$ window opened on the Si waveguide, (B) cross-sectional transmission electron microscopy (XTEM) image of a 189-nm SEG-Ge layer on a SOI substrate, and (C) high-resolution XTEM image showing a SiGe interfacial layer between the Si substrate and the SEG-Ge layer.

force microscopy. TD is evaluated by subjecting the Ge surface to a selective etchant composed of approximately 55% CrO$_3$ and approximately 49% DHF, and the estimated etch pits density is approximately 3.8×10^6 cm^{-2}. The achievement of such low defects density is predominantly due to the insertion of a thin SiGe buffer layer grown using a low temperature approach, which provides Ge dangling bonds as nucleation sites to ease the selective Ge epitaxy process. Moreover, such a SiGe buffer layer also acts as an additional interface to relieve the large lattice mismatch strain of approximately 4.2% between Ge and Si. As a result of such low defects level within the Ge film, postepitaxy Ge anneal can be skipped to reduce the thermal budget.

Since Ge is already in use at the FEOL-CMOS process, a straightforward approach is to insert the Ge epitaxy step in an established CMOS process flow after poly-gate formation and before contact module or metallization. Figure 11.2 shows the "electronic-first and photonic-last" integration approach adopted for monolithically fabricating the Ge photodetector and CMOS-integrated circuit on common SOI platform (Ang et al., 2010a). Starting with SOI wafers, channel waveguides are first formed by using the same Si mesa isolation process employed for proper electrical isolation. Using a thin sacrificial oxide, selective ion implantations are first performed to define the p-well and n-well regions of the n-MOS and p-MOS transistors, respectively, as shown in Figure 11.2A. A moderately high p-type doping is also implanted on the detector's active region, in which the dose is carefully chosen to allow low series resistance while not impacting the quality of the as-grown Ge epitaxy film. Rapid thermal annealing at 1030°C for 5 s is used to activate the dopant. After removing the sacrificial oxide, a thin gate oxide of ~ 5 nm is formed by thermal oxidation and a low-pressure chemical vapor deposition (LPCVD) polycrystalline silicon of ~ 100 nm is deposited and patterned to form the transistor gate electrodes, as shown in Figure 11.2B. A thin sidewall spacer made of conformal oxide is deposited and followed by an anisotropic reactive ion etch. Deep source/drain implants for highly-doped p^{++} and n^{++} regions are subsequently performed for the CMOS transistors and rapid thermal annealing at 1030°C for 5 s is

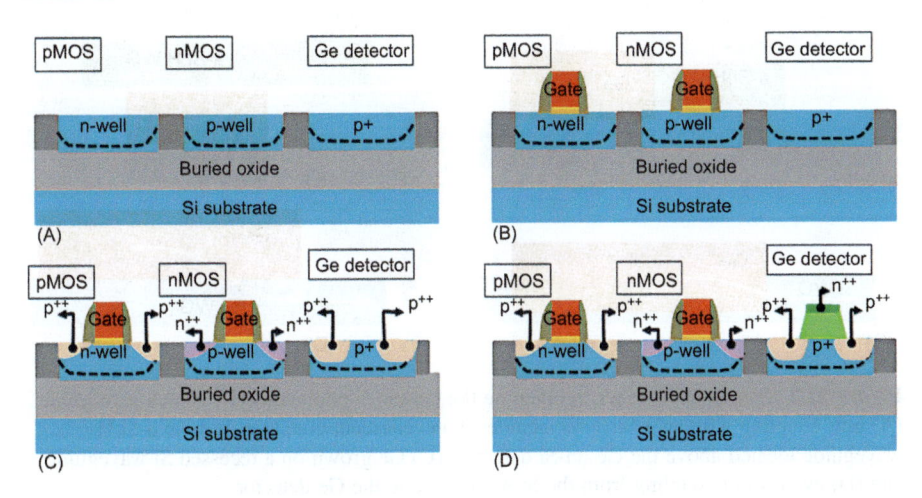

Figure 11.2 Schematic showing the "electronic-first and photonic-last" integration approach for monolithic integration of waveguide-integrated Ge-on-Si photodetectors and Si CMOS circuit on common SOI platform.

employed to activate the dopant. Simultaneously, the bottom contact of the Ge photodetector is also formed using a highly-doped p^{++} region, as shown in Figure 11.2C. A PECVD passivation oxide layer of approximately 60 nm is then deposited and patterned to open the Ge active window. SEG of Ge is performed in the UHCVD chamber. High dose selective phosphorous implant is then performed and annealed at 500°C for 5 min to form a good n-type ohmic contact for the Ge detector. Following passivation oxide deposition, contact and metallization are subsequently done to complete the device fabrication, as shown in Figure 11.2D.

11.2.2 Waveguide-integrated PIN Ge-on-Si photodetectors

There are several approaches to integrate the Ge-on-Si photodetector with a waveguide, as shown in Figure 11.3 schematically. The first approach is butt-coupling, as shown in Figure 11.3A schematically, in which light propagating along the Si channel waveguide is vertically coupled to a deposited waveguide (a hydrogenated amorphous Si (a-Si:H) waveguide is usually used due to its low optical loss and large refractive index contrast (Zhu et al., 2010), whereas other waveguides such as Si_3N_4 are also adoptable) and then butt-coupled to the Ge detector (Sun et al., 2008). Efficient butt-coupling is achieved and optical power in the waveguide is directly transferred to the detector when the mode-matching conditions for waveguide and detector modes are met. The required length is approximately 5 μm since the Ge absorption coefficient is approximately 4000 cm^{-1} at $\lambda = 1.55$ μm. The second approach is evanescent coupling, in which light propagating along the top of an a-Si:H waveguide (Ahn et al., 2007) or the bottom Si waveguide (Wang et al., 2008) is evanescently coupled to the Ge detector because Ge has a slightly larger refractive index than Si, as shown in Figure 11.3B and D, respectively. Since the

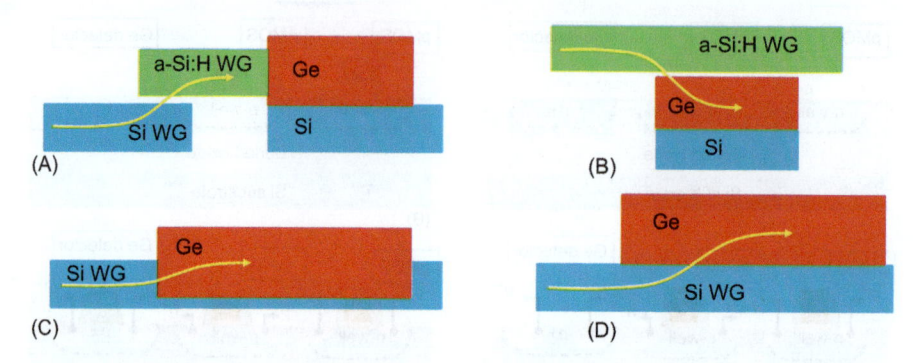

Figure 11.3 Several approaches to integrate the Ge-on-Si photodetectors with a waveguide: (A) butt-coupling with an a-Si:H waveguide; (B) evanescent coupling from an a-Si:H waveguide located above the Ge-on-Si detector; (C) Ge grown on a recessed Si waveguide; and (D) evanescent coupling from the Si waveguide to the Ge detector.

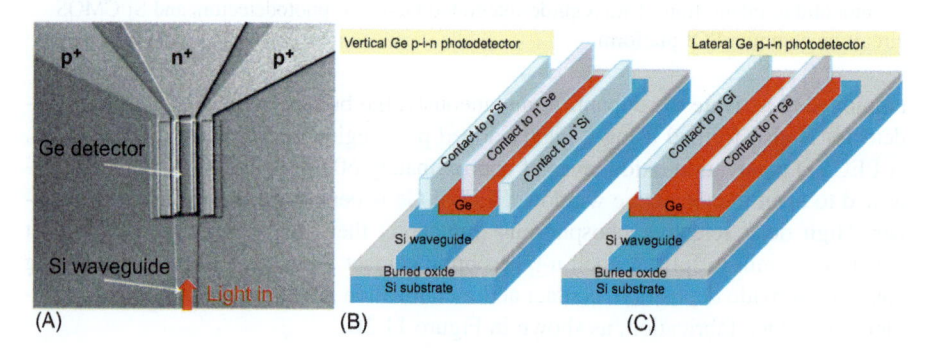

Figure 11.4 (A) SEM image of a monolithically integrated Ge PIN photodetector with a Si waveguide. (B) The design of an evanescent-coupled Ge photodetector featuring vertical PIN configuration, the width W and length L of the detector is 8 and 100 μm, respectively, and (C) the design with a lateral PIN configuration, the width W and length L of this detector is 20 and 100 μm, respectively.

optical power is slowly transferred to the high-index Ge layer in evanescent coupling, the detector requires a longer length for sufficient absorption, depending on the Ge thickness. The required length is ∼80 μm for the 0.4 μm-thick Ge layer. The third approach is that between the butt-coupling and the evanescent coupling by using a recessed Si waveguide, as shown in Figure 11.3C schematically (Vivien et al., 2009). The required length is ∼15 μm. In our laboratory, the approach of Figure 11.3D is utilized because of its ease of integration into a CMOS process.

Figure 11.4A shows the monolithically integrated Ge PIN photodetector fabricated in our laboratory. The detectors feature vertical and lateral PIN configurations, as shown schematically in Figure 11.4B and C, respectively. For a vertical PIN detector, the p^+ and n^+ junctions are formed on the Si and Ge regions, respectively, with an intrinsic Ge absorbing layer having a thickness of approximately

0.4 μm sandwiched in between these alternating junctions, as shown in Figure 11.4B. The width W and length L of the vertical PIN detector is 8 and 100 μm, respectively. Such a PIN design is often preferred over the MSM structure as it enables the achievement of low leakage current. In this design, the high absorption coefficient of the Ge layer allows the absorbing region to be kept very thin, while leverages on the detector length improve the photodetection efficiency. In another detector design featuring the lateral PIN configuration, both the p^+ and n^+ junctions are formed on the Ge region, and are separated by an intrinsic absorbing layer of approximately 0.8 μm, as shown in Figure 11.4C. The width W and length L of the lateral PIN detector is 20 and 100 μm, respectively. In this context, the insertion of a thick buried oxide serves as a cladding to confine the optical mode within the core of the channel waveguide so as to prevent leakage into the underneath Si substrate.

Figure 11.5A shows a plot of the current vs applied bias for both the vertical and lateral PIN photodetectors under dark and illumination conditions. They both exhibit excellent rectifying characteristics. At a given applied bias of -1.0 V, the dark current is measured to be ~ 0.57 μA (or ~ 0.7 nA/μm^2) in a vertical PIN detector, which is below the typical 1.0 μA generally considered to be the upper limit for high-speed receiver design. However, for a lateral PIN detector, a higher dark current of ~ 3.8 μA (or ~ 1.9 nA/μm^2) is measured. A 1.55 μm light with power of approximately 300 μW is coupled and the photocurrent is measured. Both photodetectors achieve a photocurrent of approximately 275 μA at higher biases. Figure 11.5B compares the responsivity of the detectors as a function of the applied voltages. The vertical PIN detector demonstrated a lower responsivity as compared to the lateral PIN detector for biases below -0.5 V. This could possibly be due to an enhanced carrier recombination process at the high density of defect centers near the Ge−Si hetero-junction. This is set to compromise the absolute photocurrent value of a vertical PIN detector under low field influence. However, with an increased electrostatic potential across the junction, the photo-generated carriers will be assisted across the space charge region with enhanced mobility and be collected as photocurrent before they can recombine at these recombination centers.

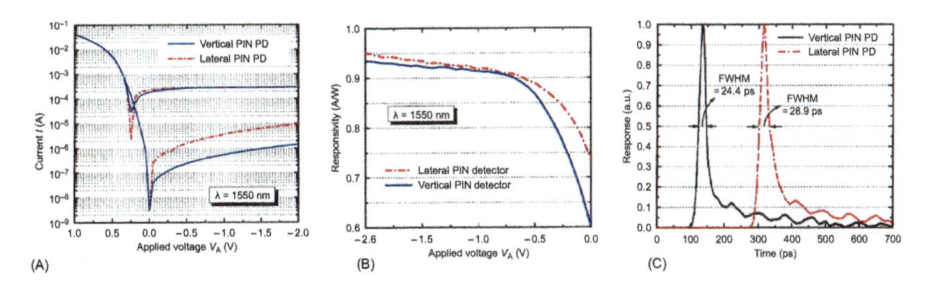

Figure 11.5 Performance metric of the fabricated Si-waveguide Ge-on-Si photodetectors with vertical and lateral PIN configurations: (A) the current-voltage characteristics of the detectors under dark and illumination conditions; (B) responsivity as a function of applied voltage measured at 1550 nm; and (C) impulse response of the photodetectors measured at an applied bias of -1.0 V and wavelength of 1550 nm.

At an applied bias larger than −1.0 V, a comparable responsivity is measured for both the vertical and lateral PIN detectors. Despite that the metallurgical junction is separated by merely 0.8 μm, a lateral PIN detector showed a high absolute responsivity of ∼0.9 A/W. Figure 11.5C shows the impulse response of the vertical and lateral PIN detectors measured at a light wavelength of 1.55 μm. The measurement is performed with a 1.55 μm pulsed laser source having an 80 fs pulse width. Devices are characterized with microwave probes and the impulse response is captured using a sampling oscilloscope. A vertical PIN detector is shown to achieve a smaller full-width at half maximum (FWHM) of ∼24.4 ps as compared to the lateral PIN detector with a slightly larger FWHM of approximately 28.9 ps. This corresponds to a −3 dB bandwidth of approximately 10.1 and 11.3 GHz, respectively. The performances of the photodetectors can be further improved by optimization of their geometry. Table 11.1 compares this works and those reported recently.

11.2.3 Surface plasmon enhanced Ge-on-Si photodetectors

Reducing the dimension of the photodetector can improve the speed and reduce the dark current. However, the length of the photodetector is limited by the light absorption efficiency. A way to enhance the light absorption is to excite surface plasmon polaritons (SPPs), which are electromagnetic waves that are strongly coupled to free electron oscillation on the metal/dielectric surfaces (Gramotnev and Bozhevolnyi, 2010). A concentric grating surface plasmon antenna of 10 μm diameter has been

Table 11.1 Summary of recently developed Si waveguide-integrated Ge-on-Si photodetectors.

Coupling scheme	Responsivity (A/W) @1.55 μm	3 dB bandwidth (GHz)	Dark current density (mA/cm^2)	Absolute dark current (μA)	References
Butt-coupling	0.8	>67	8×10^4	4 @ −1 V	Vivien et al. (2012)
Recessed-Si	1.0	42	60	4 @ −4 V	Vivien et al. (2009)
Evanescent-coupling	0.95	36	29	0.0046 @ −1 V	Liao et al. (2011)
Evanescent-coupling	0.75	60	3750	3 @ 2 V	Novack et al. (2013)
Our lab (vertical PIN)	0.9	11.3	∼0.57	0.7 @ −1 V	Ang et al. (2010a)
Our lab (lateral PIN)	0.9	10.1	∼0.7	4 @ −1 V	Ang et al. (2010a)

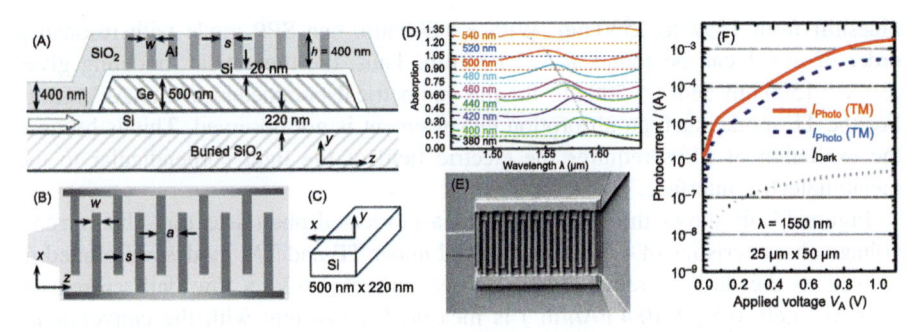

Figure 11.6 A surface plasmon enhanced waveguide Ge-on-SOI MSM photodetector: (A) schematic cross-sectional view of the device; (B) top view of the interdigitated fingers; (C) perspective view of the Si core layer with a thickness of 220 nm and a width of 500 nm where light propagates in the z-direction; (D) calculated absorption spectra of the Ge region with grating periodicity of $a = 1500$ nm and finger spacing s varying from 380 to 540 nm under TM light injection; (E) top-view SEM image of the fabricated device; and (F) current-voltage characteristics of the device with TE and TM injections.
Source: From Ren et al. (2010).

demonstrated to concentrate light into the center 300 nm-diameter Si mesa (Ishi et al., 2005). Such a structure can be used for a normal-incident Ge-on-Si photodetector (Ren et al., 2011). Here, we introduce a SPP structure for the waveguide-integrated Ge-on-Si photodetector (Ren et al., 2010). The detector has a MSM configuration with interdigitated electrodes to achieve large transverse magnetic (TM) enhancement at $\lambda = 1.55$ μm, as shown in Figure 11.6A and B schematically. The plasmon-enhancement regime is realized by properly choosing the finger width and spacing in terms of the Bloch theorem and momentum conservation. The designed Ge-on-SOI MSM photodetector has an effective device width W of 25 μm and length L of 50 μm. Figure 11.6B shows the top-view structure schematic of the interdigitated electrodes, which are defined by the figure width w and spacing s. These electrodes effectively form a one-dimensional metallic rectangular grating with finite width, length, and height on the top surface of the massive multilayer. To keep the CMOS compatibility, aluminum is used as the metal, which is not the traditional plasmonic metal. A crystalline Si interfacial layer of 20 nm is designed to be inserted between the metal layer and the Ge region so as to increase the Schottky barrier and thus suppress the dark current of device. Figure 11.6C depicts a perspective view of the waveguide Si core layer with a thickness of 220 nm and a width of 500 nm. The TE/TM-polarized light wave propagates along the z-direction. When reaching the active regime, the Si waveguide width is adjusted with a taper to ensure efficient coupling of incidence photon from the Si waveguide into the Ge region for absorption. The absorption spectra of the Ge layer for the proposed device with the fixed periodicity of 1500 nm under TM light injection are numerically calculated and the results are plotted in Figure 11.6D. The SPP modes can be detected via the maximum points located slightly after Woods−Rayleigh anomaly modes. When the spacing s is varied from 380 to 540 nm, the peaks in absorbance display a gradual blueshift from 1592 to 1530 nm. With $s = 500$ nm, one SPP mode with resonance order of $m = 1$ can be found at $\lambda = 1.55$ μm. Thus, the theoretical modeling gives $w + s = 1500$ nm and $s = 500$ nm for the geometrical parameters of the Al grating. For the TE mode, such an absorption enhancement is not observed. This is because the excitation of SPPs requires the electric field of the light perpendicular to the metal/dielectric interface.

Figure 11.6E shows the SEM image of a fabricated photodetector. The current-voltage characteristics of device illuminated under TE and TM modes are plotted in as well as the dark currents. At an applied bias of 1.0 V, a low dark current of approximately 0.5 μA (0.4 nA/μm^2) is measured, consistent with the conventional MSM Ge-on-Si photodetector (Ang et al., 2008). The optical measurement is performed by coupling a TE or TM 1.55 μm light into the Si waveguide. The incidence light power reaching the Ge detector is approximately 3 mW for TE mode and 1.5 mW for TM mode because the TE mode has an averaged insertion loss of approximately 4.5 dB and the TM mode has an averaged insertion loss of approximately 5.9 dB. At the applied bias of 1.0 V, the photocurrents of the device are measured to be 0.527 and 1.622 mA under the individual TE and TM illumination, respectively, leading to responsivities of 0.176 and 1.081 A/W. The corresponding quantum efficiencies are thus calculated to be 14.1% and 86.7%, respectively. The responsivity and quantum efficiency under TM mode overwhelm that of TE mode, which can be attributed to the excitation of SPPs.

11.2.4 Waveguide-integrated avalanche photodetectors

The responsivity of the PIN photodetectors has approached to approximately 1 A/W for the 1550 nm light detection, corresponding to approximately 90% quantum efficiency. It is well known that the APD structure may provide much higher responsivity due to its internal carrier multiplication mechanism. Si is one of the best materials for APD because of its favorable ionization coefficient ratio (Campbell, 2007; Kang et al., 2008). A normal incident Ge/Si APD with conventional SACM structure has been demonstrated recently for 1.3 μm light detection, in which an i-Si multiplication layer, a p-Si charge layer, and an i-Ge absorbing layer are epitaxially grown on an n$^+$-Si substrate sequentially (Kang et al., 2008). However, to integrate this structure in Si waveguides with evanescent-coupling scheme, the relatively thick Si epilayer for charge and multiplication will significantly degrade the coupling efficiency of light from the Si-waveguide to the Ge absorbing layer. To circumvent this issue, we introduce a novel APD structure where the p-Si charge and i-Si multiplication regions are located laterally in the Si waveguide whereas the absorbing Ge layer is directly grown on the p-Si charge region, so that the coupling efficiency keeps the same as the conventional evanescent-coupled vertical Ge/Si PIN photodiode, as shown in Figure 11.7A (Zhu et al., 2009b). The fabrication process is also similar to the conventional evanescent-coupled vertical Ge/Si PIN photodiode except the doping scheme in the Si waveguide layer. If the n-contact is applied by a positive voltage while the p-contact is grounded, the potential gradient in the Si region drops mainly along laterally whereas that in the Ge region drops

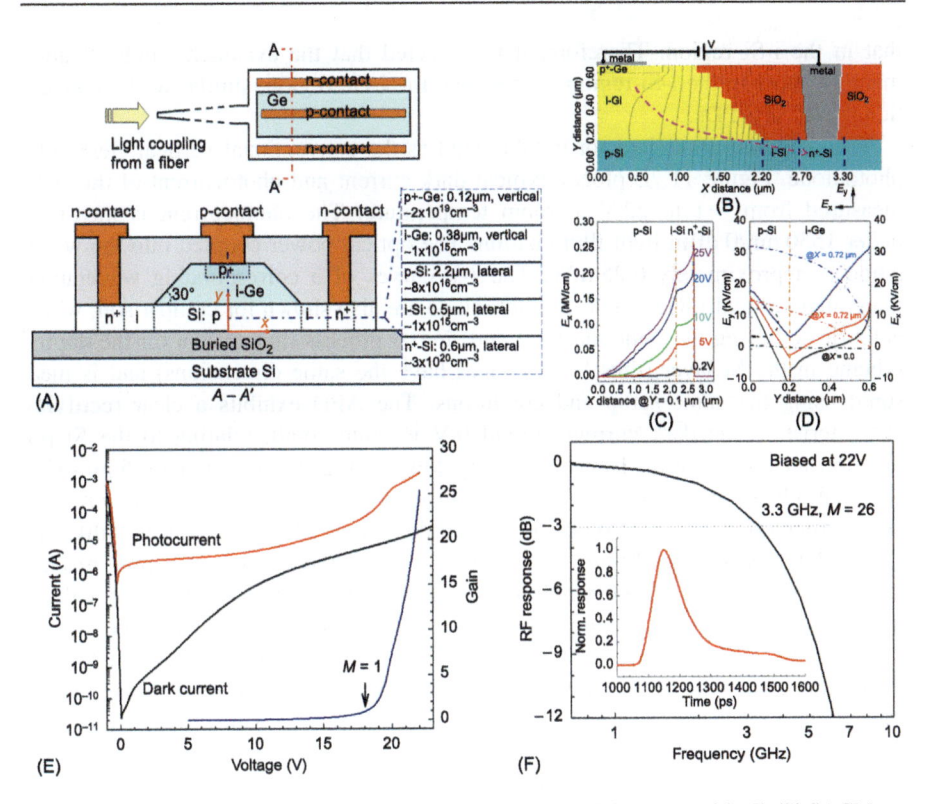

Figure 11.7 (A) Top and cross-sectional schematic views of the waveguide Ge/Si SACM APD, (B–D): the simulation results obtained from Medici: (B) the potential contours of the 25-V bias APD, $X = 0$ is at the center of device; (C) x-component electric field, E_x, along the x distance laterally at the center of Si at $Y = 0.1$ μm for the device biased at 0.2, 5, 10, 20, and 25 V; (D) x and y-component electric fields, E_x (dashed curves) and E_y (solid curves), along the Y distance vertically at $X = 0$ (the centerline), 0.44 μm, and 0.72 μm for the device at 25 V bias, (E) dark current and photocurrent of the waveguided APD under illumination of 1500-nm/0.25 mW light coupled into the waveguide, and (F) the frequency response of the waveguide APD at 22-V bias, obtained from Fourier transform of the impulse response, the inset shows the impulse response recorded from a high-speed oscilloscope.

along both laterally and vertically. It indicates that the x-component electric field (E_x) dominates and the y-component electric field (E_y) is negligible in the Si layer. Figure 11.7B–D shows the electric field distribution within the structure obtained from Medici simulation. As expected, the electric field maximizes in the i-Si multiplication region and extends into the p-Si charge region laterally with the voltage increasing. E_x and E_y in the Ge layer along the centerline are almost zero due to the symmetric structure, and increases with the distance deviating from the centerline, as shown in Figure 11.7D for the 25 V biased device. With the distance increasing from 0.44 to 0.72 μm, the maximum electric field increases from approximately 10 to 30 kV/cm. The values are about one order of magnitude smaller than that in the i-Si region. Therefore, it is expected that the avalanche multiplication mainly occurs in the i-Si region, rather than the i-Ge region, similar to the conventional SACM APDs.

The proposed photodetector is fabricated as the conventional vertical Ge/Si PIN photodiode. Figure 11.7E plots a typical dark current and photocurrent of the APD measured from −1 to 22 V at room temperature. The photocurrent is measured under 1550 nm/0 dBm light illumination. The optical power coupled into the waveguide is approximately 0.25 mW. The *I-V* curves of a corresponding waveguide-integrated vertical p-Ge/i-Ge/n-Si photodiode are also shown for comparison, which is fabricated using the same mask set and similar process flow except for the doping scheme in the Si layer (thus both devices have the same dimensions) and is measured using the same setup and conditions. The APD exhibits a clear rectifying characteristic. The dark current around 0 V is quite small, relating to the Si p-n junction leakage current. Due to the very low doping density of the 0.5 μm-wide i-Si multiplication region, this layer should be already depleted even at zero bias and the depletion region may already reach the corner of i-Ge, which is unlike the 1-D APD reported by Kang et al. (2008) where the depletion region reaches the i-Ge region through the whole depleted p-Si charge region. With the voltage increasing, the depletion region extends into the p-Si region and the above i-Ge region, both laterally to the centerline and vertically to the p^+-Ge contact, leading to the quick increase in the dark current due to the generation-recombination current at the Ge/Si interface and in the Ge layer. The increasing rate slows down beyond approximately 10−12 V probably because the major portion of i-Ge has been depleted. On the other hand, the corresponding p-Ge/i-Ge/n-Si diode at 1 V bias exhibits approximately 0.4 μA dark current, close to the approximately 10 V biased APD, reflecting the fact that the i-Ge region in p^+-Ge/i-Ge/n^+-Si diode is depleted at a relatively low voltage (around ∼1 V) and the major portion of i-Ge in APD is depleted around 10 V. The further increase in dark current with the applied voltage for both devices can be attributed to the electrical field dependent Shockley-Hall-Read effect and/or Poole-Frenkel effect.

The APD detects a photocurrent of approximately 2 μA at very low bias because the depletion has already crossed through the i-Si region to the i-Ge corner. The photocurrent increases with voltage from low bias, more rapidly from 10 V onwards, which can be attributed to both the depletion extension in the i-Ge layer and the avalanche multiplication induced by impact ionization in the i-Si region. The latter's contribution increases with voltage, especially beyond approximately 18 V. However, it is difficult to separate them out as both processes are ongoing simultaneously in the 10−20 V range due to the fact that the electrical field in i-Ge is spatial dependent, as shown in Figure 11.7D. The photocurrent reaches approximately 1.8 mA at 22 V. The internal responsivity and quantum efficiency are calculated to be approximately 7.2 A/W and 576%, respectively, indicating that the minimum avalanche gain provided by the 22 V biased APD is approximately 6. On the other hand, the Ge/Si PIN counterpart detects a photocurrent of approximately 70 μA at 1 V, and keeps almost constant up to 5 V. Therefore, our APD provides approximately 26 times larger photocurrent than the PIN counterpart, achieved only by changing the doping scheme in the Si layer. The excess noise of APD is measured under 1500 nm/0.25 mW light illumination using an HP8970B noise figure meter and a standard noise source at room temperature and at 50 and 130 MHz. The noise figure increases slowly with the voltage up to approximately 18 V and increases quickly onwards, mainly due to the occurrence of avalanche multiplication. With known photocurrent and dark current, the excess noise factor is roughly extracted to be approximately 4 at 21 V. This relatively small value indicates that the carrier impact ionization events are mainly confined in the i-Si multiplication region, rather than in the i-Ge region.

The temporal response of our APD is measured by a 1550 nm pulsed laser with 80 fs pulse width. The inset of Figure 11.7F shows an impulse response recorded from a high-speed oscilloscope for the APD probed with a microwave probe and applied at 22 V through a high-speed bias tee. Figure 11.7F shows the frequency response obtained by Fourier transform of the recorded impulse response, the 3 dB bandwidth is read to approximately 3.3 GHz. Therefore, the bandwidth-gain-product of the presented APD has reached approximately 86 GHz, which may be mainly limited by the product of resistance and capacitance and can be further improved by optimizing the structure such as the metal line for interconnect, the probe pad, the metal/Ge and metal/Si contacts, etc. and the doping concentration of the Si charge region.

11.3 Waveguide-integrated silicide Schottky-barrier photodetectors

11.3.1 NiSi₂ film and absorption

Silicide SBPDs have been used for many years to detect infrared light, which is based on the photoexcitation of metal charge carriers across the silicide/Si interface and the cut-off wavelength is mainly determined by the Schottky-barrier height (Φ_B) (Cabanski and Schulz, 1991). For operating at 1.55-μm wavelength, the conventional silicides used in the standard CMOS technology for local interconnection, that is, Ti, Co, and Ni silicides, can be adopted because they have Φ_B of 0.5–0.7 eV, thereby the corresponding cut-off wavelength of 1.8–2.4 μm. Therefore, the SBPDs are completely compatible with the standard Si-CMOS processing with no new material addition and few processing modifications. Moreover, the speed of SBPDs is inherently faster than the conventional PIN PDs. The performance of SBPDs is primarily limited by the low responsivity because only a small fraction of photoexcited carriers in the silicide layer can emit through the silicide/Si interface to be collected as a photocurrent. The internal quantum yield is low of the order of 10^{-3} as calculated by the Fowler model (Fowler, 1931). The yield can be significantly enhanced, namely a gain, if the silicide film is thin enough so that the photoexcited carriers moving away from the interface can be reflected to the emission barrier by wall or phonon scattering. However, the number of absorbed photons decreases dramatically with decreasing the silicide thickness. This conflict can be solved by using the waveguide configuration because the light is absorbed along the waveguide so that the silicide film on the waveguide can be lengthened unrestrictedly while keeping its thickness sufficiently thin. In our laboratory, ultrathin epitaxial $NiSi_2$ films are deposited on Si channel waveguides as the absorber.

Ni silicide for optical absorption is formed by a titanium-intermediated solid-state epitaxy (Nakatsuka et al., 2005; Zhu et al., 2009a). The silicide thickness is controlled by RTA after deposition of 1-nm Ti and 5-nm Ni sequentially on SOI substrates in a sputtering system: in one procedure, RTA at 450°C for 30 s, followed by selectively etching in Piranha solution at 90°C for 10 min to remove the unreacted metal; in the other procedure, first RTA at 240°C for 30 s, followed by the selectively etching, and then second RTA at 450°C for 30 s. The final Ni silicide thicknesses are 16 and 5 nm, as shown in Figure 11.8A and B, respectively. The thinner Ni silicide exhibits smoother silicide−Si interface than the thicker ones. The Ni:Si ratio of the 16 nm-thick silicide is very close to 1:2, as determined from EDX, confirming that the formed phase is $NiSi_2$, whereas the 5 nm-thick silicide is too thin for the EDX analysis. The channel SOI waveguides employed in this study have a width of 1 μm, height of 0.2 μm, and length of 5.5 mm, and have an inverted taper at both facets. The buried SiO_2 and top cladding SiO_2 are 2 and 1.5 μm thick, respectively. The excess loss (in decibel scale) induced by a 0.56 μm-wide Ni silicide layer placed on the SOI channel waveguide, as shown schematically in the inset of Figure 11.8C, is depicted as a function of its length in Figure 11.8C. The plots exhibit a good linearity, obeying an equation of Loss = $-A \times L$, where the factor $A = 4.34 \times \alpha \times \Gamma$, where α is the absorption coefficient of $NiSi_2$ and Γ is the confinement factor. A is extracted to be 5.85 and 1.58 dB/μm for the thin and the thick silicides, respectively.

11.3.2 NiSi₂/p-Si and NiSi₂/n-Si Schottky-barrier photodetectors

Figure 11.9 shows the fabricated $NiSi_2$ SBPDs. The thin $NiSi_2$ film exhibits good rectifying property on both p-Si and n-Si. The SBHs deduced from measured $I-V$

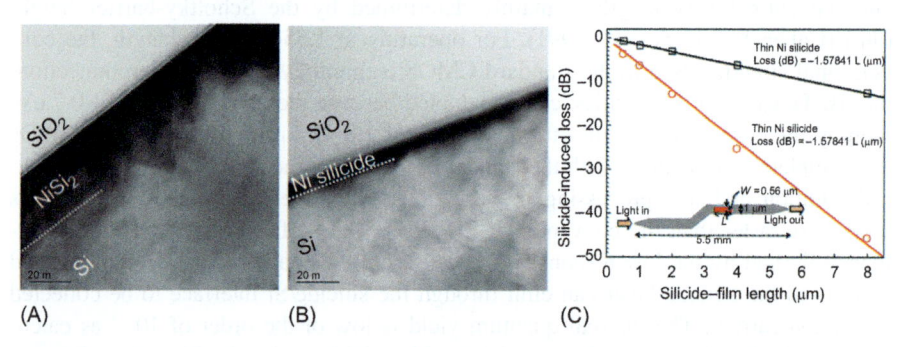

Figure 11.8 (A and B): XTEM images of Ni silicide on SOI formed by solid-state reaction of Ni(∼5 nm)/Ti(∼1 nm)/Si systems by (a) one-step RTA and (B) two-step RTA. The formed Ni silicides have thicknesses of ∼16 and ∼5 nm, respectively, (C) excess loss induced by the Ni silicide layer placed above the Si waveguide as a function of the silicide length, as shown schematically in the inset.

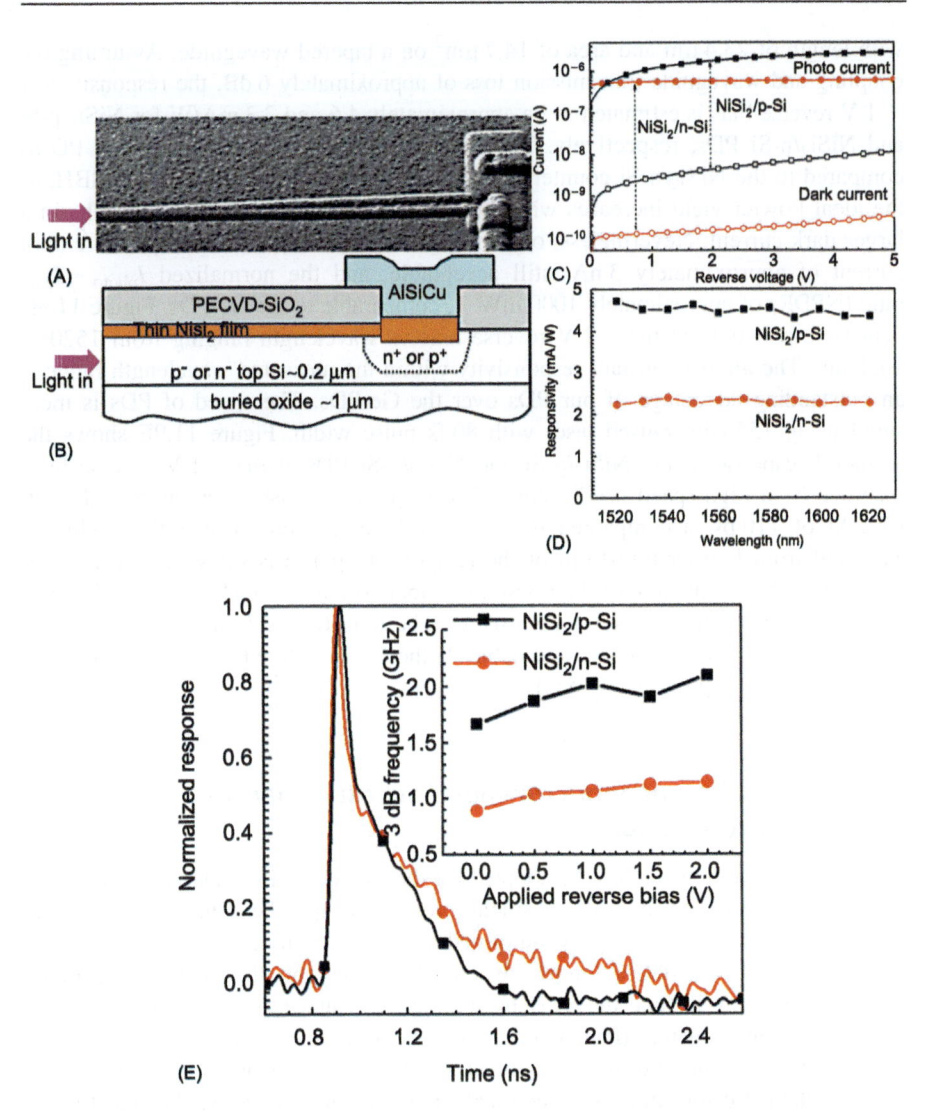

Figure 11.9 (A) Top view and (B) schematic cross-sectional view of the NiSi$_2$/Si SBPDs, (C) dark and photocurrents of NiSi$_2$/p-Si and NiSi$_2$/n-Si PDs under 1550 nm and 1 mW light coupled from an optical fiber to the Si waveguide, (D) responsivity at 1 V bias vs wavelength, and (E) impulse response at 1550 nm of NiSi$_2$/p-Si and NiSi$_2$/n-Si PDs. The inset shows 3 dB bandwidth extracted from Fourier transform of the temporal response vs the applied reverse voltage.
Source: From Zhu et al. (2008c).

curves at the small forward bias region are 0.53 and 0.62 eV for hole and electron, respectively, and the corresponding identify factor is close to unity. Figure 11.9C shows the dark and photocurrents of NiSi$_2$/n-Si and NiSi$_2$/n-Si PDs with 1550 nm, 1 mW light edge coupling from an optical fiber. The PDs have tapered NiSi$_2$ layer with length of 23.4 μm and area of 14.7 μm^2 on a tapered waveguide. Assuming the coupling and waveguide transmission loss of approximately 6 dB, the responsivity at −1 V reverse bias is estimated to be approximately 4.6 and 2.3 mA/W for NiSi$_2$/p-Si and NiSi$_2$/n-Si PDs, respectively. The larger responsivity of the NiSi$_2$/p-Si PD as compared to the NiSi$_2$/n-Si counterpart can be ascribed to the smaller hole SBH, as the ideal Fowler yield increases with the SBH decreasing, while it also results in a larger dark current. Nevertheless, our NiSi$_2$/p-Si PD has the room temperature dark current of approximately 3 nA, still acceptable, and the normalized $I_{photo} - I_{dark}$ ratio (NPDR) of approximately 1000 mW^{-1}, comparable to the Ge-PDs. Figure 11.9D depicts responsivity at the −1 V reverse bias vs wavelength ranging from 1520 to 1620 nm. The almost constant responsivity within the measured wavelength range is an outstanding advantage of our PDs over the Ge-PDs. The speed of PDs is measured using 1550 nm pulsed laser with 80 fs pulse width. Figure 11.9E shows the temporal response of the NiSi$_2$/p-Si and NiSi$_2$/n-Si PDs at the −1 V reverse bias, recorded by a high-speed oscilloscope. The response consists of a sharp peak with FWHM of 110 ps, accompanied by a longer timescale tail. The 3 dB bandwidth extracted from Fourier transform of the temporal response is shown in the inset of Figure 11.9D as a function of the revise bias: approximately 2 GHz for the NiSi$_2$/p-Si SBPD and 1 GHz for the NiSi$_2$/n-Si SBPD. The long timescale tail may arise from the response of bias tee, probe, and cable. If the tail is ignored, the 3 dB bandwidth of both SBDs is approximately 5 GHz.

11.3.3 Metal-semiconductor-metal Schottky-barrier photodetectors

A MSM configuration can also be adopted for the silicide SBPDs. Due to the elimination of heavy doping and thick silicide formation for the Ohmic contact, the fabrication process is significantly simplified and is inherently low-temperature as the highest temperature is 480°C for silicidation. More important, the MSM configuration may open the way to tune the SBH by the applied voltage as the vertical silicide−Si−silicide heterostructure (Pahun et al., 1992; Schwarz and Kanel, 1996) and to suppress the dark current by using an asymmetric MSM scheme (Chui et al., 2003).

The schematic top and cross-sectional views of the silicide MSM detector are shown in Figure 11.10A (Zhu et al., 2008b). Light propagating along the waveguide is absorbed by a 0.56 μm-wide NiSi$_2$ absorber on the SOI waveguide in contact-1. The lateral distance (finger spacing) between contact-1 and contact-2 is 2.66 μm. Figure 11.10B shows the responsivity of the fabricated silicide MSM photodetector as a function of applied voltage under 1550 nm light coupling from a fiber to the waveguide. Unlike the conventional MSM detector in which the light is mainly absorbed in the semiconductor layer, thus exhibiting nearly symmetrical response in both bias directions, the silicide Schottky-barrier MSM photodetector absorbs light along the silicide layer of contact-1. At positive bias, contact-1 is forward biased and contact-2 is reversely biased. The photoexcited holes emitted from contact-1 will recombine with the majority carriers (electrons) of n-Si and meet a barrier at

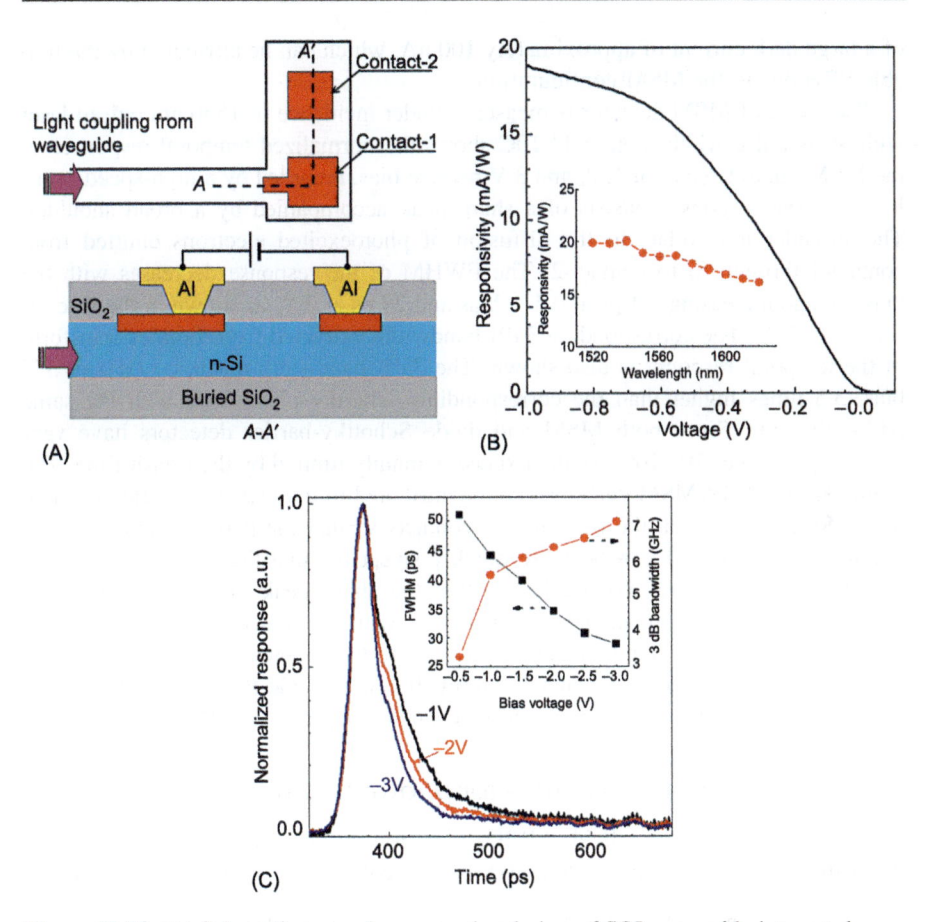

Figure 11.10 (A) Schematic top and cross-sectional view of SOI-waveguide-integrated silicide Schottky-barrier MSM PD, (B) responsivity of NiSi$_2$ MSM detector as a function of applied voltage, the inset shows the responsivity as a function of wavelength, (C) normalized temporal response of silicide MSM detector at -1, -2, and -3 V bias under 1550 nm pulse laser, the inset shows the extracted FWHM (left axis) and 3 dB bandwidth (right axis) as a function of the applied voltage.

contact-2, leading to a very small photocurrent. At negative bias, on the other hand, contact-1 is reversely biased and contact-2 is forward biased. The photoexcited electrons emitted from contact-1 meet a very small barrier at contact-2: $\Phi_B - V_2$, where V_2 is a voltage dropped at contact-2, thus it behaves somewhat like the NiSi$_2$/n-Si photodiode. We really see that the photocurrent increases with the applied voltage and tends to saturate for larger than 1 V. The photocurrent increases almost linearly with the input optical power and depends very weakly on wavelength ranging from 1520 to 1620 nm, as shown in the inset of Figure 11.10B. The responsivity at the -1 V reverse bias is approximately 19 mA/W, four times larger than the photodiode counterpart with the same NiSi$_2$ absorber, however, at the cost of a large dark current of approximately 100 nA, which can be attributed by the tunable SBH due to the MSM configuration.

The speed of MSM detector is measured under incidence of 1550 nm pulsed laser with 80 fs pulse width. Figure 11.10C shows the normalized temporal responses of the MSM photodetector at 1, 2, and 3 V reverse bias, recorded by a high-speed oscilloscope. The response consists of a sharp peak accompanied by a broad shoulder. The shoulder may relate to the diffusion of photoexcited electrons emitted from contact-1 through Si to contact-2. The FWHM of the response decreases with the bias voltage increasing: 51 ps at 0.5 V bias and 29 ps at 3 V, as shown in the inset of Figure 11.10C. The corresponding 3 dB bandwidth extracted from Fourier transform of the temporal response is also shown. The 3 dB bandwidth reaches 7 GHz at 3 V bias, 3.5 times higher than the corresponding Schottky photodiode with the same $NiSi_2$ absorber. Since both MSM and diode Schottky-barrier detectors have very small capacitance (10–100 fF), their speed is mainly limited by the transit time. The higher speed of the MSM detector may be attributed to the presence of electric field in the Si layer between the two silicide contacts so that the photoexcited electrons emitted from contact-1 can be more quickly swept through the Si layer to be collected by the contact-2. We really see that the sharp peak in the response stays the same while the shoulder becomes narrower with the bias increasing. The speed of the silicide MSM photodetector may be further increased by reducing the figure spacing between two contacts, as in the case of Ge-PDs. Table 11.2 summarizes the performances of various Si waveguide integrated SBPDs.

11.3.4 Schottky-barrier collector phototransistors

As the main issue of the silicide SBPDs is its low responsivity. A way to improve the responsivity is to utilize a Schottky-barrier collector bipolar transistor (SCBT)

Table 11.2 Summary of recently developed Si waveguide-integrated Schottky-barrier photodetectors.

Coupling scheme	Responsivity @1.55 μm	3 dB bandwidth (GHz)	Dark current	Dimension	References
Cu/p-Si	0.08 mA/W	Estimated GHz	10 nA @ −1 V	–	Casalino et al. (2010a)
Al/n-Si	0.84 mA/W	–	6 μA	1×10^{-4} cm^2	Akbari et al. (2010)
Al/p-Si	12.5 mA/W	–	0.1 μA	1×0.305 μm^2	Goykhman et al. (2012)
$NiSi_2$/p-Si (Our lab)	4.6 mA/W	2 GHz	3 nA	14.7 μm^2	Zhu et al. (2008c)
$NiSi_2$/n-Si (Our lab)	2.3 mA/W	1 GHz	0.2 nA	14.7 μm^2	Zhu et al. (2008c)
$NiSi_2$, MSM (Our lab)	19 mA/W	7 GHz @ −3 V	100 nA	2.6 μm^2	Zhu et al. (2008b)

configuration which can offer a high current gain. The SCBT has been extensively studied for electronic applications, in which the conventional p-n collector junction is replaced by a Schottky barrier with the metallic side forming the collector region (May, 1968; Jagadesh and Roy, 2005). In combination with both SBPD and SCBT, an infrared detector named SBCPT is proposed and demonstrated, in which a thin silicide layer on the SOI waveguide acts both as a metallic collector and an optical absorber, which differs from the previously reported III–V semiconductor base Schottky collector phototransistor in the light absorption (Sakai et al., 1983): the latter is in the semiconductor base region while the former is in the metallic collector region. Corresponding to p-Si and n-Si SBPDs, two kinds of SBCPTs can be designed as NPM and PNM, respectively.

Figure 11.11 shows schematically the top and cross-sectional views of a two-terminal NPM SBCPT integrated on the SOI channel waveguide (Zhu et al., 2008a).

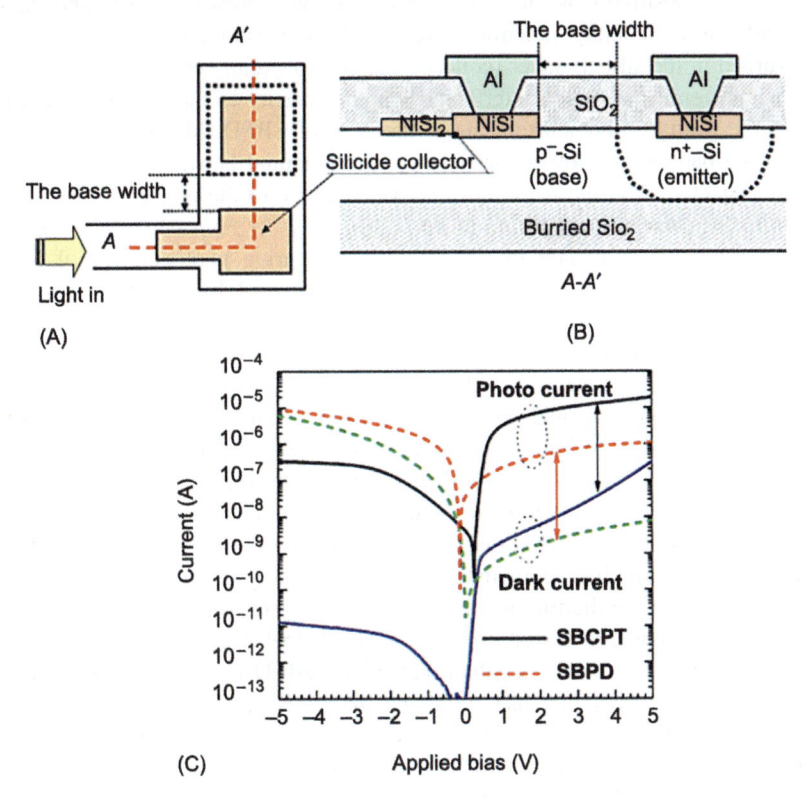

Figure 11.11 (A) Schematic top layout and (B) cross-sectional view of a SOI waveguide-based two-terminal Ni-silicide NPM SBCPT, which is exactly the same as the Ni-silicide/p-Si SBPD except the different doping types under the left electrode: n^+ for SBCPT and p^+ for SBPD, (C) current-voltage characteristics of a two-terminal NPM SBCPT and p-Si SBPD counterpart under dark environment and illuminated by ~1 mW of 1550 nm light coupled through a fiber to the waveguide.
Source: From Zhu et al. (2008a).

It contains a vertical collector-base Schottky junction and a lateral emitter-base n-p junction. Because the base is floating, the two-terminal SBCPT has apparently the same configuration as the corresponding p-Si SBPD except the doping type under the left electrode: n^+ doping to form the emitter in SBCPT whereas p^+ doping for Ohmic contact on Si in SBPD. Therefore, both NPM SBCPT and p-Si SBPD can be fabricated using the same processing flow and the same mask except the doping scheme. Figure 11.11C shows current-voltage characteristics of a fabricated two-terminal NPM SBCPT and a corresponding p-Si SBPD at dark environment and under illumination with approximately1 mW of 1550 nm light coupling from an optical fiber, from which the responsivity at 5 V bias is estimated to be approximately 8.5 mA/W for SBPD and 150 mA/W for SBCPT. Since both SBCPT and SBPD have exactly the same Ni-silicide absorber, the larger responsivity of SBCPT can be attributed to the current amplifying effect as the conventional phototransistor. For SBCPT at positive bias, the emitter p-n junction is forward biased and the collector $NiSi_2$/p-Si Schottky junction is reverse biased. Under illumination, photons are absorbed in the silicide layer (collector region) to excite hot holes and the photo-excited holes emit across the interfacial Schottky barrier to the p-Si base region to provide the primary photocurrent, as in the case of SBPD. Meanwhile, the excess holes in the base region lower the base-emitter potential barrier so that more electrons are injected from the n^+ emitter across the base to the collector. Namely, the total collector current is amplified to be $I_{SBCPT} \approx I_{ph} \cdot (\beta + 1)$, where β is the static common-emitter current gain of SBCPT. By comparing photocurrents of SBCPT and SBPD, β is roughly estimated to be approximately 12.8 at 1 V and 16.6 at 5 V. The relatively low β value of our SBCPT compared to the reported SCBTs (whose β in the range of $\sim 40-100$ (May, 1968; Jagadesh and Roy, 2005)) can be attributed to the relatively long base width of our SBCPT (~ 2.44 μm). Since β is approximately inversely proportional to the base width, β, thereby the responsivity of SBCPT, is expected to be improved simply by reducing the base width. For instance, the responsivity may achieve approximately 1 A/W by reducing the base width of our SBCPT to approximately 0.35 μm without other optimizations.

The dark current of SBPD is mainly determined by the hole SBH of the Ni-silicide/p-Si contact, which is extracted to be approximately 0.49 eV from the forward I-V curve based on the thermionic emission model, close to the reported value. The dark current of positive-biased SBCPT also originates from the reverse-biased Schottky junction but it is larger than that of the SBPD because of the amplifying effect. The dark current increases with the bias voltage due to the field induced Schottky-barrier lowering effect. The increasing rate of SBCPT is faster than that of SBPD because β also increases with the current. However, the extreme large dark current of SBCPT at large bias (~ 324 nA at 5 V, ~ 41.5 times larger than the corresponding SPD) may also arise from imperfection of the Ni-silicide/p-Si interface, especially at the thick NiSi region which suffered from dry and wet etching during fabrication. The dark current may be reduced by optimization of the processing flow. The speed of SBCPT is measured to be approximately 0.44 GHz at 5 V bias, much smaller than that of the SBPD (~ 3.7 GHz). The slow speed is a common shortcoming for the two-terminal phototransistors with the base floating because the excess holes in the base region are mainly eliminated through recombination upon illumination interrupt, which is a relatively slow process. Decreasing the base width reduces the amount of excess holes in the base region, thus may improve the speed. Moreover, the approaches developed to improve the speed of conventional heterojunction phototransistors such as by using a base terminal or substrate terminal should be also applicable to our SBCPT. In summary, the silicide SBCPT can offer a higher responsivity than the SBPD counterpart through the transistor action while maintaining the advantages of SBPD such as small footprint and ease of fabrication, but with a price of slow speed. Such a photodetector may be useful for in-line optical power monitoring in Si photonic integrated circuits (Zhu et al., 2009a).

11.3.5 Schottky-barrier detector with embedded silicide nanoparticles

Another approach to increase the responsivity is to employ silicide NPs, as shown in Figure 11.12 schematically. It is a Si p-n junction with electrically floating silicide NPs embedded in its space charge region. The concept of embedded silicide NPs for infrared detection can be traced back to the layered internal photoemission sensor reported by Fathauer et al. (Fathauer et al., 1993; Raissi, 2003). Figure 11.12C depicts the band diagram of such a p-n junction, showing that both

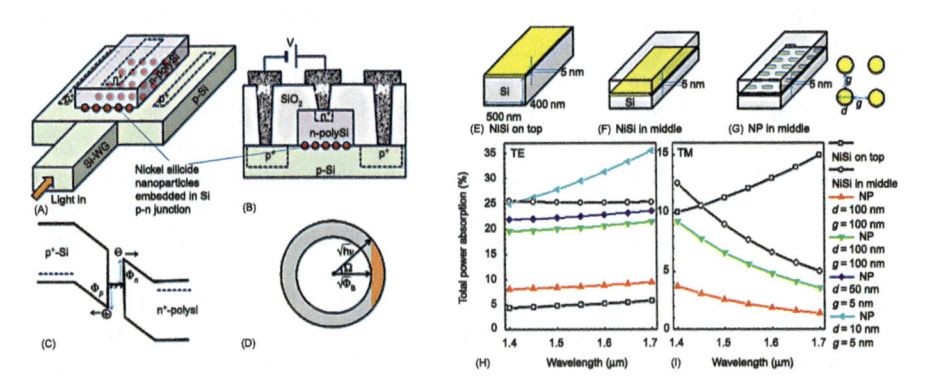

Figure 11.12 (A) Schematic of the Si waveguide-integrated photodetector based on the embedded metal silicide NPs, (B) cross-sectional view of the device, (C) energy band diagram of the Si p-n junction with embedded metal NPs under reverse bias, (D) simplified escape cone model for Schottky emission shown in momentum space, the shaded cone with angle of Ω represents a thick silicide film on Si whereas the shaded spherical ring represents a very small silicide NP surrounded by Si. (E) Schematics of Si waveguide with a 5-nm-thick NiSi placed on the top, (F) Si waveguide with a 5-nm-thick NiSi embedded in the middle, (G) Si waveguide with NiSi NPs embedded in the middle, (H) total power absorption in the above three kinds of 1 μm-long waveguides vs wavelength for TE light, and (I) total power absorption for TM light. In the case of NiSi film on the top of waveguide, the TM light can excite a SPP mode, thus being absorbed more effectively than the TE light.
Source: From Zhu et al. (2012).

electrons and holes photoexcited in the NPs are simultaneously emitted into Si over Φ_B. Because the photoexcited carriers in the silicide NPs face the silicide/Si interface at all directions, the escape cone in Figure 11.12D extends to the whole k-space sphere, therefore, $P(E)$, which is the probability of photoexcited carriers emitting from the metal silicide layer in to the Si layer, approaches to unity and η_i, which is the internal quantum efficiency, approaches to $\frac{h\nu - \Phi_B}{h\nu}$, where $h\nu$ is the photon energy.

More importantly, the light absorption can be dramatically enhanced through excitation of localized surface plasmon resonance (LSPR) on the silicide NPs. It is well known that the light absorption dominants over scattering on a metal NP with diameter well below the skin depth (~ 25 nm for common metals in the entire optical region). The absorption cross-section (C_{abs}) of a metal NP is expressed by $\frac{2\pi}{\lambda} \text{Im}\left[3V \frac{\varepsilon_m(\omega) - \varepsilon_d}{\varepsilon_m(\omega) + 2\varepsilon_d}\right]$ based on the point dipole model, where V is the particle volume, $\varepsilon_m(\omega)$ is the complex permittivity of metallic NP, and ε_d is the permittivity of the embedding medium. The C_{abs} can well excess the geometrical cross section of the metal NP when $\varepsilon_m(\omega) = -2\varepsilon_d$, known as the excitation of LSPR. To verify the above prediction in the case of metal NPs embedded in a Si waveguide, the light absorption in 1 μm-long Si waveguides with continuous silicide films or silicide NPs, as shown schematically in Figure 11.12E–G, respectively, are numerically simulated using three-dimensional full-wave finite-difference time-domain method. Here, the Si waveguide surrounded by SiO_2 is set to 0.4 μm × 0.5 μm, and the metal is 5 nm-thick NiSi silicide. The NiSi NPs are approximated as nanodisks with height of 5 nm and diameter of d, periodically arranged on a square grip with space of g. The simulation results are depicted in Figure 11.12H and I for the TE (the electric field parallel to the NiSi layer) and TM (the electric field perpendicular to the NiSi film) polarized lights, respectively. The continuous NiSi layer in the middle of Si waveguide can absorb the TE light more efficiently than that on the top of Si waveguide because the light field of TE mode maximizes in the middle of waveguide. The absorption of NPs increases with reducing d. The absorption in the condition of $d = 10$ nm and $g = 5$ nm is even larger than the continuous NiSi layer with the same thickness, which could be attributed to the excitation of LSPR, although the LSPR condition of $\varepsilon_m(\omega) = -2\varepsilon_d$ does not exactly meet ($\varepsilon_m(\omega)$ of $NiSi_2$ is $\sim -20 + i15$ at 1550 nm whereas $\varepsilon_d = 11.9$). Because $\varepsilon_m(\omega)$ of the Ni silicides depend on their phase, for example, $\sim -8.2 + i60$ for Ni_3Si, $\sim -0.88 + i29$ for Ni_2Si, and $\sim -19 + i25$ for $NiSi_2$ at 1550 nm, and several phases may coexist in the Ni silicide (relating to the fabrication condition), $\varepsilon_m(\omega)$ of the final silicide NPs may be engineered by the fabrication condition or even by adding an additional metal such as Co or Ti to form a ternary silicide. Moreover, the LSPR condition also depends on the silicide NP shape taking the higher-order LSPRs into account. Therefore, more efficient light absorption could be expected once strong LSPR is excited, which could be achieved by optimizations of fabrication parameters of silicide NPs, thus enabling to scale down the absorber length to submicron. As expected, the TM light cannot excite LSPR. Therefore, our PD is only suitable for the TE light detection.

To quickly verify the feasibility of the proposed NP-based photodetectors, a proof-of-concept device is fabricated on SOI using an existing mask set which is designed for conventional germanium PDs. Some key fabrication processes are as follows. After boron doping, the Si pattern is defined by dry etch of Si down to the buried oxide using a SiO_2 layer as hard mask. A rectangle 80 μm × 4.4 μm window is opened to expose the Si surface. Then, approximately 1 nm Ti and 3 nm Ni are deposited subsequently, followed by 200°C/30 s RTA for silicidation. After removing the unreacted metal by a 90°C Piranha solution, a 200 nm amorphous Si layer is deposited, followed by phosphorus implantation, 700°C/5 min RTA, and patterning. During this thermal process, the Ni silicide agglomerates to form NPs. Meanwhile, the a-Si layer crystallizes to form polycrystalline Si due to the well-known metal-induced crystallization effect. The p^+-Si and n^+-polycrystalline Si are contacted by Al electrodes to form a p-n diode. Figure 11.13A shows the fabricated photodetector. Figure 11.13B shows the scanning transmission electron microscopy (STEM) and Figure 11.13C shows the XTEM image. It reveals clearly that the embedded Ni-silicide film is discontinuous and Ni diffuses deeply into the single-crystalline Si layer at some spots. Meanwhile, a-Si is already crystallized to become polycrystalline Si. Figure 11.13D plots the current-voltage curve measured on the fabricated photodetectors, which exhibits a typical rectifying property of a p-n diode. The insert shows the photo- and dark currents in the semi-logarithmic scale. For optical measurement, light from a tunable laser source with power of approximately −2 dBm is coupled in the Si waveguide through a lensed polarization maintained fiber. Assuming approximately 3 dB coupling loss, the light power arrived the detector is estimated to be approximately 0.32 mW. Figure 11.13E plots the responsivity vs wavelength for quasiTE and TM modes. Unlike the conventional SB-PDs, the responsivity of this PD depends on both wavelength and polarization, and the peak responsivity for the TE light reaches approximately 30 mA/W at some specific wavelengths. The temporal responses at different biases are measured under illumination of a pulsed laser. They consist of a sharp peak with full width at high maximum of approximately 30 ps and a longer timescale tail. The 3 dB bandwidth extracted from Fourier transform is approximately 6 GHz.

The light propagating along the NP-based photodetector attenuates by absorption (by the embedded silicide NPs, the polySi layer, and the Al electrode) and scattering. The absorption by the Al electrode will not contribute to the photocurrent because the emitted hot carries are not in the space charge region. The absorption by the polySi may contribute the photocurrent, but it is weak, indicated by the fact that the propagation loss of polySi waveguide ($< \sim 0.01$ dB/μm) is quite small compared with the Si waveguide with silicide (either thin film or NPs). The responsivity of the NP-based photodetector is thus mainly limited by the ratio of the light absorbed by the embedded silicide NPs to the overall light attenuation. The silicide NPs in the SBPD are random in location and shape, and their LSPR excitation is wavelength dependent, making the responsivity also wavelength dependent. To increase the responsivity, one way is to enhance the LSPR effect, which could be achieved by optimizing the fabrication parameters and the other way is to reduce the unusable light attenuation (especially the absorption of the Al electrode), which could be achieved by optimizing the detector's configuration to move away the Al electrodes from the optical mode. Once LSPRs are excited on the silicide NPs, they

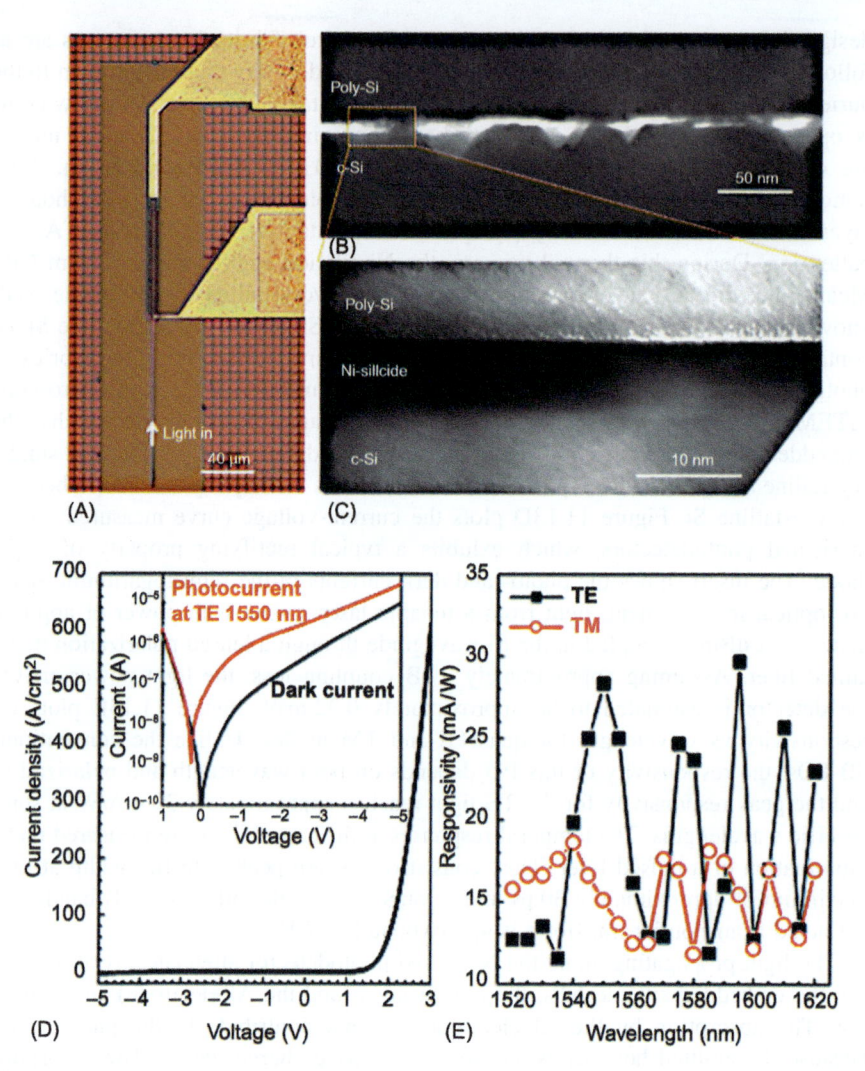

Figure 11.13 (A) Top view of the fabricated device, (B) STEM image, (C) enlarged XTEM image of the final polySi/Ni-silicide/c-Si structure, (D) *I-V* characteristic of the fabricated PD at room temperature, inset is the photo- and dark- currents of the PD, the light power arrived the detector is ∼0.32 mW, and (E) responsivity at 5 V reverse bias vs wavelength for quasi-TE and quasi-TM lights (Zhu et al., 2012).

decay into incoherent electron-hole pairs through Landau damping very quickly (i.e., in the femtosecond scale). Thus, the speed of the NP-based photodetector is still limited by the carrier transmitting time (from NPs to the electrodes) and the resistance-capacitance constant, as the conventional SB-PDs. One disadvantage of the proposed NP-based photodetector is the high dark current density (e.g., ∼2.84 A/cm^2 at 5 V bias) due to field bunching at the surfaces of NPs. Fortunately, the photodetector can be scaled down to a submicron scale, thus the dark current can be reduced to be comparable to the conventional photodetectors. For instance, the dark current will be approximately 14 nA if the active area is reduced to 0.5 μm^2.

11.4 Conclusions

In this chapter, we review the Si waveguide-integrated Ge-on-Si and metallic silicide photodetectors operating at 1.55 μm for communication applications. The Ge-on-Si photodetectors have achieved high performances. However, Ge is still a challenging material to be integrated in a CMOS environment because of its low thermal budget constraint and its large lattice mismatch with Si. The selective growth of epitaxial Ge-on-Si is an additional step in the conventional Si CMOS flow and the processes after the Ge growth may need to be changed. In contrast, the silicide SBPDs can be easily integrate in a CMOS environment without additional steps because the silicide layer already exists in a CMOS flow for interconnect. However, it usually suffers from a main drawback of low responsivity. Although several approaches can be used to improve the responsivity, they suffer from their own disadvantages and need to be further improved. At this stage, the Ge-on-Si photodetectors can be used as the mainstream photodetectors to be integrated in Si photonic integrated circuits whereas the silicide SBPDs can be used as an alternative photodetector, such as an in-line optical power monitor.

References

Ahn, D., Hong, C.Y., Liu, J., Giziewicz, W., Beals, M., Kimerling, L.C., 2007. High performance, waveguide integrated Ge photodetectors. Opt. Express. 15, 3916–3921.

Akbari, A., Tait, N., Berini, P., 2010. Surface plasmon wveguide Schottky detector. Opt. Express. 18, 8505–8514.

Ang, K.W., Yu, M.B., Zhu, S.Y., Chua, K.T., Lo, G.Q., Kwong, D.L., 2008. Novel NiGe MSM photodetector featuring asymmetrical Schottky barriers using sulfur co-implantation and segregation. IEEE Electron. Device Lett. 29, 704–707.

Ang, K.W., Liow, T.Y., Yu, M.B., Fang, Q., Song, J., Lo, G.Q., 2010a. Low thermal budget monolithic integration of evanescent-coupled Ge-on-SOI photodetector on CMOS platform. IEEE J. Sel. Top. Quantum Electron. 16, 106–113.

Ang, K.W., Lo, G.Q., Kwong, D.L., Grym, J., 2010b. Germanium photodetector technologies for optical communication applicationSemiconductor Technologies. In: Grym, J. (Ed.), Semiconductor Technologies. InTech.

Beals, M., Michel, J., Liu, J.F., Ahn, D.H., Sparacin, D., Sun, R., et al., 2008. Process flow innovations for photonic device integration in CMOS. In: Proceedings of SPIE, vol. 6898. Silicon Photonics III, paper 689804.

Cabanski, W.A., Schulz, M.J., 1991. Electronic and IR-optical properties of silicide/silicon interfaces. Infrared Phys. 32, 29–44.

Campbell, J.C., 2007. Recent advances in telecommunications avalanches photodiodes. IEEE J. Light. Technol. 25, 109–121.

Casalino, M., Sirleto, L., Iodice, M., Saffioti, N., Gioffre, M., Rendina, I., 2010a. Cu/p-Si Schottky barrier-based near infrared photodetector integrated with a silicon-on-insulator waveguide. Appl. Phys. Lett. 96.

Casalino, M., Coppola, G., Iodice, M., Rendina, I., Sirleto, L., 2010b. Near-infrared subbandgap all-silicon photodetectors: state of the art and perspectives. Sensors. 10, 10571−10600.

Chen, C.P., Driscoll, J.B., Grote, R.R., Souhan, B., Osgood, R.M., Bergman, K., 2015. Mode and polarization multiplexing in a Si photonic chip at 40 Gb/s aggregate data bandwidth. IEEE Photon. Technol. Lett. 27, 22−25.

Chui, C.O., Okyay, A.K., Saraswat, K.C., 2003. Effective dark current suppression with asymmetric MSM photodetectors in group IV semiconductors. IEEE Photon. Technol. Lett. 15, 1585−1587.

Currie, M.T., Samavedam, S.B., Langdo, T.A., Leitz, C.W., Fitzgerald, E.A., 1998. Controlling threading dislocation densities in Ge-on-Si using graded SiGe layers and chemical-mechanical polishing. Appl. Phys. Lett. 72, 1718−1720.

Fathauer, R.W., Dejewski, S.M., George, T., Jones, E.W., Krabach, T.N., Ksendzov, A., 1993. Infrared photodetectors with tailorable response due to resonant plasmon absorption in epixtaxial silicide particles embedded in silicon. Appl. Phys. Lett. 62, 1774−1776.

Fowler, R.H., 1931. The analysis of photoelectric sensitivity curves for clean metals at various temperatures. Phys. Rev. 38, 45−56.

Geis, M.W., Spector, S.J., Grein, M.E., Yoon, J.U., Lennon, D.M., Lyszczarz, T.M., 2009. Silicon waveguide infrared photodiodes with >35 GHz bandwidth and phototransistors with 50 AW-1 response. Opt. Express. 17, 5193−5204.

Goykhman, I., Desiatov, B., Khurgin, J., Shappir, J., Levy, U., 2012. Waveguide based compact silicon Schottky photodetector with enhanced responsivity in the telecom spectral band. Opt. Express. 20, 28594−28602.

Gramotnev, D.K., Bozhevolnyi, S.I., 2010. Plasmonics beyond the diffraction limit. Nat. Photonics 4, 83−91.

Grote, R.R., Padmaraju, K., Souhan, B., Driscoll, J.B., Bergman, K., Osgood, R., 2013. 10 Gb/s error-free operation of all-silicon ion-implanted-waveguide photodiodes at 1.55 μm. IEEE Photon. Technol. Lett. 25, 67−70.

Ishi, T., Fujikata, J., Makita, K., Baba, T., Ohashi, K., 2005. Si nano-photodiode with a surface plasmon antenna. Jpn. J. Appl. Phys. 44, L364−L366.

Jagadesh, M., Roy, S.D., 2005. A new high breakdown voltage lateral Schottky collector bipolar transistor on SOI: design and analysis. IEEE Trans. Electron. Devices 52, 2496−2501.

Kang, Y., Zadka, M., Litski, S., Sarid, G., Morse, M., Paniccia, M.J., 2008. Epitaxially-grown Ge/Si avalanche photodiodes for 1.3 μm light detection. Opt. Express. 16, 9365−9371.

Liao, S., Feng, N.N., Feng, D., Dong, P., Shafiiha, R., Kung, C.C., 2011. 36 GHz submicro silicon waveguide germanium photodetector. Opt. Express. 19, 10967−10972.

Lischke, S., Knoll, D., Zimmermann, L., 2015. Monolithic integration of high bandwidth waveguide coupled photodiode in a photonic BiCMOS process. In: Proceedings SPIE 9390, Next-Generation Optical Networks for Data Centers and Short-Reach Links II.

Liu, J., 2014. Monolithically integration Ge-on-Si active photonics. Photonics 1, 162−197.

Lockwood, D.J., Pavesi, L., 2011. Silicon photonics II: components and integration. In: Lockwood, D.J., Pavesi, L. (Eds.), Silicon Photonics II: Components and Integration. Springer-Verlag, Berlin Heidelberg.

Loh, T.H., Nguyen, H.S., Murthy, R., Yu, M.B., Loh, W.Y., Lo, G.Q., 2007a. Selective epitaxial germanium on silicon-on-insulator high speed photodetectors using low-temperature ultrathin Si0.8Ge0.2 buffer. Appl. Phys. Lett. 91.

Loh, W.Y., Wang, J., Ye, J.D., Yang, R., Nguyen, H.S., Chua, K.T., 2007b. Impact of local strain from selective epitaxial germanium with thin Si/SiGE buffer on high-performance p-i-n photodetectors with a low thermal budget. IEEE Electron. Device Lett. 28, 984–986.

Luan, H.C., Lim, D.R., Lee, K.K., Chen, K.M., Sandland, J.G., Wada, K., 1999. High-quality Ge epilayers on Si with low threading dislocation densities. Appl. Phys. Lett. 75, 2909–2911.

May, G.A., 1968. The Schottky-barrier-collector transistor. Solid-State Electron. 11, 613–619.

Michel, J., Liu, J., Kimerling, L.C., 2010. High-performance Ge-on-Si photodetectors. Nat. Photonics 4, 527–534.

Nakatsuka, O., Okubo, K., Tsuchiya, Y., Sakai, A., Zaima, S., Yasuda, Y., 2005. Low-temperature formation of epitaxial NiSi layers with solid-phase reaction in Ni/Ti/Si(001) systems. Jpn. J. Appl. Phys. 2945–2947.

Novack, A., Gould, M., Yang, Y., Xuan, Z., Streshinsky, M., Liu, Y., 2013. Germanium photodetector with 60 GHz bandwidth using inductive gain peaking. Opt. Express. 21, 28387–28393.

Pahun, L., Campidelli, Y., d'Avitaya, F.A., Badoz, P.A., 1992. Infrared response of Pt/Si/ErSi heterostructure: tunable internal photoemission sensor. Appl. Phys. Lett. 60, 1166–1168.

Raissi, F., 2003. A possible explanation for high quantum efficiency of PtSi/porous Si Schottky detectors. IEEE Trans. Electron. Devices 50, 1134–1137.

Reed, G.T., Mashanovich, G., Gardes, F.Y., Thomson, D.J., 2010. Silicon optical modulators. Nat. Photonics. 4, 518–526.

Ren, F.F., Ang, K.W., Song, J., Fang, Q., Yu, M., Lo, G.Q., 2010. Surface plasmon enhanced responsivity in a waveguided germanium metal-semiconductor-metal photodetector. Appl. Phys. Lett. 97.

Ren, F.F., Ang, K.W., Ye, J., Yu, M., Lo, G.Q., Kwong, D.L., 2011. Split bull's eye shaped aluminum antenna for plasmon-enhanced nanometer scale germanium photodetector. Nano Lett. 11, 1289–1293.

Roelkens, G., Brouckaert, J., Tailaert, D., Dumon, P., Bogaerts, W., Thourhout, D.V., 2005. Integration of InP/InGaAsP photodetectors onto silicon-on-insulator waveguide circuits. Opt. Express. 13, 10102–10108.

Sakai, A., Tatsumi, T., Aoyama, K., 1997. Growth of strain-relaxed Ge films on Si(001) surfaces. Appl. Phys. Lett. 71, 3510–3512.

Sakai, S., Naitoh, M., Kobayashi, M., Umeno, M., 1983. InGaAsP/InP phototransistor-based detectors. IEEE Trans. Electron. Devices 30, 404–408.

Scales, C., Berini, P., 2010. Thin-film Schottky barrier photodetector models. IEEE J. Quantum Electron. 46, 633–643.

Schwarz, C., Kanel, H.V., 1996. Tunable infrared detector with epitaxial silicide/silicon heterostructures. J. Appl. Phys. 79, 8798–8807.

Sun, R., Beals, M., Pomerene, A., Cheng, J., Hong, C.Y., Kimerling, L., 2008. Impedance matching vertical optical waveguide couplers for dense high index contrast circuits. Opt. Express. 16, 11682–11690.

Vivien, L., Osmond, J., Fedeli, J.M., Marris-Morini, D., Crozat, P., Damlencourt, J.F., 2009. 42 GHz p.i.n germanium photodetector integrated in a silicon-on-insulator waveguide. Opt. Express. 17, 6252–6257.

Vivien, L., Polzer, A., Marris-Morini, D., Osmond, J., Hartman, J.M., Crozat, P., 2012. Zero-bias 40 Gbit/s germanium waveguide photodetector on silicon. Opt. Express. 20, 1096–1101.

Wang, J., Lee, S., 2011. Ge-photodetectors for Si-based optoelectronic integration. Sensor 11, 696–718.

Wang, J., Loh, W.Y., Chua, K.T., Zang, H., Xiong, Y.Z., Loh, T.H., 2008. Evanescent-coupled Ge p-i-n photodetectors on Si-waveguide with SEG-Ge and comparative study of lateral and vertical p-i-n configurations. IEEE Electron. Device Lett. 29, 445–448.

Zhu, S.Y., Lo, G.Q., Yu, M.B., Kwong, D.L., 2008a. Low-cost and high-gain silicide Schottky-barrier collector phototransistor integrated on Si waveguide for infrared detection. Appl. Phys. Lett. 93.

Zhu, S.Y., Lo, G.Q., Kwong, D.L., 2008b. Low-cost and high-speed SOI waveguide-based silicide Schottky-barrier MSM photodetectors for broadband optical communications. IEEE Photon. Technol. Lett. 20, 1396–1398.

Zhu, S.Y., Yu, M.B., Lo, G.Q., Kwong, D.L., 2008c. Near-infrared waveguide-based nickel silicide Schottky-barrier photodetector for optical communications. Appl. Phys. Lett. 92.

Zhu, S.Y., Lo, G.Q., Yu, M.B., Kwong, D.L., 2009a. Silicide Schottky-barrier phototransistor integrated in silicon channel waveguide for in-line power monitoring. IEEE Photon. Technol. Lett. 21, 185–187.

Zhu, S.Y., Ang, K.W., Rustagi, S.C., Wang, J., Xiong, Y.Z., Lo, G.Q., 2009b. Waveguided Ge/Si avalanche photodiode with separate vertical SEG-Ge absorption, lateral Si charge and multiplication configuration. IEEE Electron. Device Lett. 30, 934–936.

Zhu, S.Y., Lo, G.Q., Kwong, D.L., 2010. Low-loss amorphous silicon wire waveguide for integrated photonics: effect of fabrication process and the thermal stability. Opt. Express. 18, 25283–25291.

Zhu, S.Y., Chu, H.S., Lo, G.Q., Kwong, D.L., 2012. Waveguide-integrated near-infrared detector with self-assembled metal silicide nanoparticles embedded in a silicon p-n junction. Appl. Phys. Lett. 100.

Photodetectors for silicon photonic integrated circuits

12

Molly Piels and John E. Bowers
Department of Electrical and Computer Engineering, University of California, Santa Barbara, CA, United States

12.1 Introduction

Silicon-based photonic components are especially attractive for realizing low-cost photonic integrated circuits (PICs) using high-volume manufacturing processes (Heck et al., 2013). Due to its transparency in the telecommunications wavelength bands near 1310 and 1550 nm, silicon is an excellent material for realizing low-loss passive optical components. For the same reason, it is not a strong candidate for sources and detectors, and photodetector fabrication requires the integration of either III/V materials or germanium if high speed and high efficiency are required. Photodetectors used in phonic integrated circuits, like photodetectors used in most other applications, typically require large bandwidth, high efficiency, and low dark current. In addition, the devices must be waveguide-integrated (rather than surface-illuminated) and the process used to fabricate the photodiode must be compatible with the processes used to fabricate other components on the chip. For many applications where PICs are a promising solution, for example microwave frequency generation, coherent receivers, and optical interconnects relying on receiverless circuit designs (Assefa et al., 2010b), the maximum output power is also an important figure of merit.

There are numerous design trade-offs between speed, efficiency, and output power. Designing for high bandwidth favors small devices for low capacitance. Small devices require abrupt absorption profiles for good efficiency, but design for high output power favors large devices with dilute absorption. Most of the work on silicon-based photodiodes to date has focused on PIN diodes. Both ultra-compact devices with abrupt absorption profiles and devices with larger active areas have been demonstrated. The results have been consistent with this trade-off: ultra-compact devices have shown the highest bandwidth-efficiency products (up to 38 GHz; Virot et al., 2013), while devices utilizing dilute absorption profiles had better power handling (up to 19 dBm output power at 1 GHz; Piels et al., 2013). Recently, photodetectors with decoupled structures, the separate absorption charge and multiplication (SACM) avalanche structure and the uni-traveling carrier (UTC) structure, have been used in both germanium (Piels and Bowers, 2014; Dai et al., 2010, 2014; Duan et al., 2013) and hybrid III/V-silicon (Beling et al., 2013b) to push performance past the limits imposed by the PIN structure.

In this chapter, we will review the status of heterogeneous integration of silicon waveguides and photodetectors. First, we will cover available fabrication technologies (both Ge and hybrid III/V-silicon). We will then discuss the design constraints that are common to all waveguide photodiodes (WGPDs) on silicon substrates. We will present an overview of demonstrated devices, and lastly conclude and show an outlook for the future.

12.2 Technology

WGPDs on silicon broadly fall into one of two categories: germanium-based and hybrid III/V-silicon. A number of groups have demonstrated Ge-based photodiodes in mature fabrication technology based on a CMOS pilot line (Assefa et al., 2010a; Marris-Morini et al., 2014; Galland, et al., 2013). In these works, the photodiodes have been cofabricated with passive optics, modulators, and in some cases transistors, but not optical sources (lasers or LEDs). On the other hand, hybrid or InGaAs-based photodiodes have generally been fabricated using technology that is further from mass-production capabilities, but that has a full library of components (competitive with InP substrate-based devices) available (Koch et al., 2013).

12.2.1 Germanium

Germanium is an appealing absorbing material for use in silicon-based PICs because it can be integrated into a CMOS pilot line relatively easily (Si/Ge alloyed contacts are already used in CMOS electronics) and because the bulk material is absorbing in the entire 1310 nm window and much of the C and L bands. There are a number of ways to integrate germanium and silicon, but selective area growth by chemical vapor deposition is the most common for WGPDs (Michel et al., 2010). The Si/Ge interface is conductive, and for vertical diodes, one contact is often composed of silicon. A typical geometry is shown in Figure 12.1A. There is a 4% lattice mismatch between germanium and silicon, which is relieved through the formation of threading defects (Hartmann et al., 2005). These threading defects form mid-gap generation-recombination centers, which increase the dark current of the detector relative to a bulk Ge diode (Mueller, 1959; Giovane et al., 2001). The threading

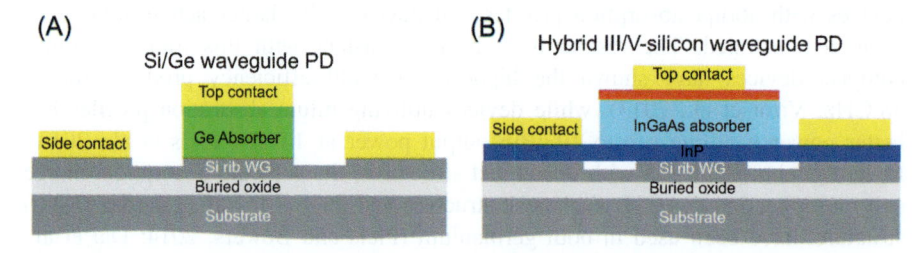

Figure 12.1 Cross-section schematics of WGPDs on (A) Si/Ge and (B) hybrid III/V-silicon. A vertical diode configuration is shown here, but the diode can also be formed laterally.

defects also pin the Fermi level in their immediate vicinity, and thus for vertical PIN diodes, the p-down configuration is preferred (Masini et al., 2001).

Germanium epitaxially grown on silicon has superior properties to bulk germanium for a C-band or L-band detector. After growth is completed, as the wafer is cooled from the growth temperature to room temperature, the thermal expansion coefficient mismatch between the two materials results in the formation of tensile strain in the germanium (Liu et al., 2004). This decreases the direct bandgap energy, pushing the direct band edge to around 0.782 eV or lower (1599 nm).

12.2.2 Hybrid III/V-silicon

The hybrid III/V-silicon platform enables the inclusion of optical gain in PICs. The III/V layer stack is usually bonded to the SOI waveguide, but can also be epitaxially grown on Si (Liu et al., 2014). For oxygen plasma enhanced bonding, lattice mismatch does not cause threading defect formation. Thus, any III/V material that can be grown on InP or other III-V substrates, including material with optical gain, can be used in a hybrid III/V-Si PIC. For low temperature oxygen plasma enhanced bonding, the Si/InP interface is not conductive, and both sides of the diode must be in the III/V layers. This results in most devices having the geometry shown in Figure 12.1B. The InP contact layer thickness affects the optical properties of the device, and is typically around 200 nm. For higher temperature bonding, conduction through the interface is possible (Hawkins et al., 1997; Tanabe et al., 2012).

It is possible to use the same epitaxial material for both a laser/amplifier and a photodiode, and this was the approach pursued by the first hybrid III/V-silicon photodiode demonstration (Park et al., 2007). However, in this case, the quantum well depth affects both amplifier performance and photodiode bandwidth, and it is difficult to optimize both simultaneously (Højfeldt and Mørk, 2002). Instead, for applications requiring high bandwidths, InGaAs lattice-matched to InP is the absorbing material of choice. Fabricating hybrid III/V-Si PICs using multiple epitaxial materials is more complex than using a single material (Chang et al., 2010), but has been successfully demonstrated (Koch et al., 2013).

The germanium absorption coefficient in the C and L-bands is a function of growth conditions, but it is typically around 3000 cm^{-1} in the C-band with a long wavelength cutoff around 1600 nm. InGaAs is direct gap and has a larger absorption coefficient in the C-band (around 9600 cm^{-1}), and absorbs well in the entire L-band independent of growth conditions. The real part of the refractive index is relatively low (about 3.6 as opposed to 4.2), which can make designing short, compact devices more difficult in the hybrid platform. However, InP offers greater flexibility in band engineering and optical matching layer design due to the availability of more mature growth technology.

12.2.3 Other technologies

There are a number of other promising technologies for fabricating photodiodes on silicon platforms. Photodiodes based on defect-enhanced absorption in silicon have been demonstrated, and are promising for monitoring purposes (Knights and Doylend, 2008). To move the Ge band edge toward longer wavelengths, Sn has been incorporated in the growth (Roucka et al., 2011), but waveguide-integrated GeSn photodiodes have yet to be demonstrated. InGaAs has been grown epitaxially on (Feng et al., 2012) and fused to (Black et al., 1997) silicon, resulting in a conductive interface. In both cases, the dark current is increased relative to low-temperature bonded and native-substrate material. For optical interconnect applications, InGaAs nanopillars grown on silicon substrates have shown good performance as both photodetectors and optical sources (Chen et al., 2011).

12.3 Optical properties of Si-based WGPDs

There are two commonly used schemes for coupling to a WGPD: butt-coupling and vertical coupling. In a butt-coupled photodiode, the absorbing region sits in a recess at the end of the input waveguide. Vertically coupled photodiodes have an absorbing region that lies on top of the input waveguide. The fabrication process for butt-coupled photodiodes is typically more complex than the fabrication of vertically coupled photodiodes and most practical when the absorbing region is grown, rather than bonded. The primary benefit of the butt-coupled photodiode is that the confinement of the optical mode in the absorbing region is very high, and thus ultra-compact devices with high efficiency and low capacitance can be fabricated. The highest bandwidth-efficiency products for WGPDs on silicon reported to-date have been for butt-coupled devices (Virot et al., 2013; DeRose et al., 2011). In addition to a less complicated fabrication process, vertically coupled photodiodes typically have a larger active device area. This makes it relatively difficult to achieve high bandwidth and low dark current, but is preferable for applications requiring high saturated output power (e.g., microwave photonics and coherent communications).

Whereas the optical design of butt-coupled detectors is straightforward (Bowers and Burrus, 1986), the optical design of vertically coupled photodetectors on silicon requires careful simulation. Cross-section schematics of both InGaAs and germanium-based WGPDs are shown in Figure 12.1. In both cases, the absorbing region has a real refractive index that is larger than the real refractive index of silicon, so the fundamental mode in the detector area has low overlap with the input mode from the input passive waveguide. Thus the coupling from the input passive waveguide is typically to higher-order modes in the photodiode area. The absorption profile depends on the confinement factor of the higher-order mode in the absorbing region and the overlap between it and the input mode, both of which are functions of the absorber thickness. The end result is that the absorption profile of the device is a strong function of absorbing region thickness. The efficiency of a device of a given length displays local maxima and minima, as shown in Figure 12.2A for germanium and Figure 12.2B for InGaAs. The simulation was done using the semivectorial beam propagation method for a TE-polarized input.

On both material platforms, the absorption profile also depends on the underlying silicon thickness, waveguide width, input polarization, and wavelength. In the

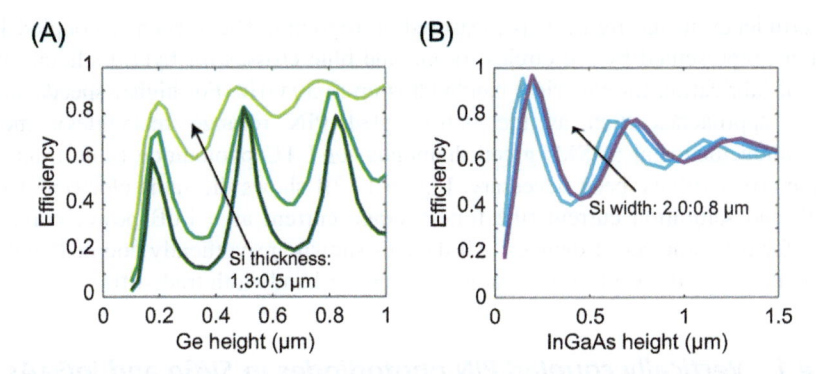

Figure 12.2 Simulated quantum efficiency of a 20-μm-long TE photodetector as a function of (A) Ge thickness and (B) InGaAs thickness for waveguide detectors on silicon. For the Si/Ge photodetector, the underlying silicon thickness is the parameter; the dependence of the absorption profile on width is minimal for easily fabricated device sizes (wider than 2 μm). For the hybrid III/V-Si detector, the parameter is the width of the silicon rib waveguide. The thickness and rib height are assumed to be 500 and 250 nm, respectively, as these dimensions are often used for this kind of device.

Si/Ge system, the locations of the maxima and minima are roughly the same for both polarizations over large optical bandwidths. Thus devices with low polarization-dependent responsivity and high efficiency over an optical bandwidth in excess of 100 nm have been demonstrated (Yin et al., 2007). On the hybrid III/V-silicon platform, performance is more sensitive to design parameters. This is for a number of reasons, primary among them is that the real part of the refractive index of InGaAs is closer to the real part of the refractive index of Si and the imaginary part is large enough to affect the confinement factor. The drawback of this sensitivity is that optical simulations of hybrid silicon devices require excellent material models in order to accurately predict performance, whereas approximate models often work well for Si/Ge detectors (Piels et al., 2013; Yin et al., 2007).

The absorption characteristics are very important to determining how design trade-offs between efficiency, bandwidth, and output power behave. For some types of waveguide detectors (e.g., butt-coupled), a minor change in absorber thickness will have either no impact or an easily mitigated impact on the absorption profile. The peak/valley behavior in Figure 12.2, on the other hand, means that for vertically coupled WGPDs on silicon, the optical and the electrical design must be done simultaneously.

12.4 Demonstrated WGPDs on silicon

This section presents the state of the art on waveguide photodetectors on silicon, organized by cross-section design. Figure 12.3A shows the electrical bandwidth and efficiency of a number of research devices on silicon. Unless otherwise indicated, the efficiency in the figure was measured at 1550 nm. The vertically coupled PIN designs, represented by red circles for Ge and blue crosses for hybrid silicon, offer ease of fabrication for electrical bandwidths up to 30 GHz. For higher speeds, alternative approaches such as the butt-coupled PIN (orange rectangles), metal-semiconductor-metal (MSM; green diamonds) or UTC photodiode (blue triangles; purple crosses) have been necessary. Figure 12.3B shows the same electrical bandwidth and saturation current (the time-average current at −1 dB power compression) for the same set of devices. All devices shown are vertically coupled, and the trend illustrates the well-known saturation current-bandwidth trade-off.

12.4.1 Vertically coupled PIN photodiodes in Si/Ge and InGaAs

The PIN diode is one of the most commonly used photodiode cross-section designs. In a PIN detector, most of the light is absorbed in the intrinsic region in the center of the device. For bandwidth, PIN design involves balancing the RC and the transit time limit. The RC limit is approximately

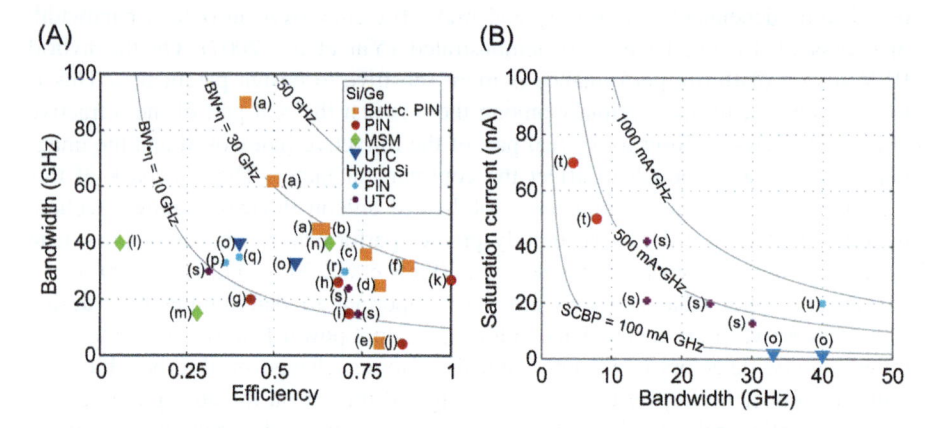

$$f_{RC} = \frac{1}{2\pi(R_s + R_L)C} \tag{12.1}$$

Figure 12.3 (A) Bandwidth-efficiency and (B) output power-bandwidth trade-offs for demonstrated waveguide devices. Devices with avalanche gain are not included in these plots. Contours for bandwidth-efficiency products of 10, 30, and 50 GHz are also shown in (A) and contours for saturation current-bandwidth products of 100, 500, and 1000 mA GHz are shown in (B). a: (Virot et al., 2013), b: (DeRose et al., 2011), c: (Liao et al., 2011), d: (Vivien et al., 2007), e: (Liu et al., 2006) (1520 nm), f: (Feng et al., 2009), g: (Wang et al., 2008), h: (Yin et al., 2007), i: (Masini et al., 2008) (divided optical bandwidth by $\sqrt{3}$), j: (Ahn et al., 2007) (divided optical bandwidth by $\sqrt{3}$), k: (Liow et al., 2013), l: (Assefa et al., 2009), m: (Assefa et al., 2013), n: (Chen and Lipson, 2009) (from pulsed measurement; $f_{3dB} = 0.312\tau_{FWHM}$ (Weingarten et al., 1988)), o- (Piels and Bowers, 2014), p: (Binetti et al., 2010), q: (Lee et al., 2013), r: (Piels et al., 2014), s: (Beling et al., 2013b), t: (Bowers et al., 2010).

where R_s is the diode series resistance, R_L is the load resistance (typically 50 Ω), and C is the diode capacitance, and the transit time limit is about (Bowers and Burrus, 1987)

$$f_\tau = \frac{0.45v}{W} \tag{12.2}$$

where v is the smaller of the saturated electron and hole velocities and W is the intrinsic region thickness. Since the diode capacitance is approximately $\varepsilon A/W$, where A is the diode area, there is an optimum intrinsic region thickness.

For vertically coupled PIN photodiodes, the highest bandwidth-efficiency products can be obtained by choosing an absorption region thickness at a peak in Figure 12.2. This is shown for a Si/Ge PIN as a surface plot in Figure 12.4A. The calculation includes the transit time and RC limits assuming a 50 Ω load, but neglects parasitic effects. The capacitance was calculated using a parallel plate model and a device width of 3 μm. The germanium region was assumed to be completely depleted for both bandwidth estimates (i.e., the thicknesses of the p- and n-contact were assumed negligible or had negligible absorption due to larger bandgap contact layers). The silicon waveguide height was 500 nm. The maximum values of bandwidth-efficiency product in Figure 12.4A are limited to below 60 GHz because the optimum thicknesses for fast absorption do not necessarily correspond to thicknesses where the RC and transit time constants have been carefully balanced. This is shown (under the same assumptions used in Figure 12.4A) in Figure 12.4B for a 30 μm long device. For maximum bandwidth, the optimum germanium region thickness is 300 nm, but for maximum efficiency, 200 and 400 nm give better performance.

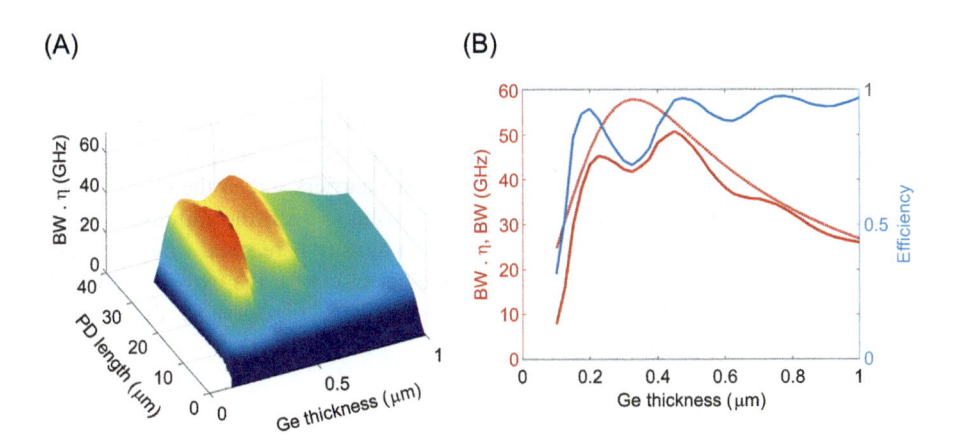

Figure 12.4 (A) Bandwidth-efficiency product for a vertically coupled Si/Ge PIN photodiode as a function of intrinsic region thickness and device length. The assumed width is 3 μm. (B) Bandwidth, efficiency, and bandwidth-efficiency product for a 30 μm × 3 μm vertically coupled Si/Ge PIN photodiode as a function of intrinsic region thickness.

Despite these difficulties, several devices with good performance have been demonstrated. The largest demonstrated bandwidth of a vertically coupled PIN photodiode on Si/Ge is 27 GHz (Liow et al., 2013). Two Intel NIP photodiodes (Yin et al., 2007) had responsivities of 0.89 and 1.16 A/W at 1550 nm and electrical bandwidths of 26 and 24.1 GHz. A Luxtera device performed similarly, with a responsivity of 0.85 A/W and bandwidth of 26 GHz (Masini et al., 2008). Finally, IME demonstrated photodetectors with a 20 GHz bandwidth and 0.54 A/W responsivity (Wang et al., 2008). The same group has recently demonstrated detectors with improved responsivity and the same bandwidth using (low-field) avalanche multiplication (Liow et al., 2013), and extended the 3 dB bandwidth by using inductive gain peaking (Novack et al., 2013). It has proven difficult to increase the bandwidth of a vertically coupled germanium PIN beyond 30 GHz in a 50 Ω environment while maintaining high efficiency.

The design space and results for hybrid III/V-silicon PIN photodetectors are similar to those for Si/Ge photodetectors. Optically, although the absorption profile can be altered by changing the width of the underlying silicon, achieving a large change in confinement factor requires a very narrow waveguide, which in turn requires lithography with higher resolution than what has historically been used to fabricate this kind of device. Electrically, the transit time-limited bandwidth of an InGaAs-based PIN detector is slightly lower and the RC-limited bandwidth is slightly higher than the same quantities for an equivalent Si/Ge PIN. This is because the hole velocity in InGaAs is slower than the carrier velocities in Ge, increasing the transit time, and the dielectric constant is lower, which decreases the capacitance. The performance of demonstrated devices is also similar to the performance of Si/Ge PINs; a number of different devices have been demonstrated, and the bandwidths were all around 30 GHz (Binetti et al., 2010; Lee et al., 2013; Piels et al., 2014).

Both Si/Ge and hybrid III/V-silicon PIN detectors have been investigated for use in high-power applications. The main limitations on output power are the active area of the device and the maximum current density the cross-section can sustain before the internal field collapses (Williams and Esman, 1999). In the intrinsic region of a photodetector, under low-injection conditions, there is a roughly constant electric field due to the applied bias that separates the charge carriers. As the current density in the intrinsic region increases, the carriers screen the electric field. Under high injection, the field distribution redistributes with the minimum occurring in the intrinsic region. The maximum current density is reached when the minimum of the electric field drops below the value necessary for the carriers to be able to maintain their saturation velocities. For a PIN photodetector, this can be expressed as (Piels, 2013)

$$J_{max} = \frac{6\varepsilon v_n v_p}{W^2(v_n + v_p)}(V_{bi} + V_{PD} - E_{crit}W) \qquad (12.3)$$

where ε is the dielectric constant of the intrinsic region, v_n and v_p are the saturated electron and hole velocities, W is the intrinsic region width, V_{bi} is the diode built-in voltage, V_{PD} is the voltage drop across the device due to the load resistance and applied bias, and E_{crit} is the electric field at which the carrier velocities saturate. The factor $6\varepsilon v_n v_p/W_i^2(v_n+v_p)$ determines how the saturation current scales with bias voltage, and should be as large as possible for high-power conversion efficiency (Tulchinsky et al., 2008). In germanium, it is 2.6e − 5 A/V/cm², while for InGaAs it is 2.3e − 5 A/V/cm², and so we expect slightly better power handling from a Si/Ge PIN than from an InGaAs PIN with the same dimensions at the same bias voltage. However, the breakdown field of InGaAs is about twice that of Ge, so in the absence of a system limit on bias voltage (and neglecting thermal effects), the InGaAs device would have a larger saturated output power.

12.4.2 Butt-coupled PIN photodiodes in Ge

One approach to increasing the bandwidth beyond 30 GHz is to use a butt-coupled or nearly butt-coupled design (Virot et al., 2013; DeRose et al., 2011; Liao et al., 2011; Vivien et al., 2007; Feng et al., 2009). The confinement factor in the germanium for these photodiodes is nearly 100% regardless of the total germanium thickness used. As a result, most of the light can be absorbed by very short (less than 10 μm long) devices. The highest bandwidths for WGPDs on silicon to date have been reported for ultra-compact butt-coupled PIN detectors.

The optical characteristics of the device are relatively insensitive to the germanium thickness and device width, so a large degree of electrical optimization is possible. Due to their small size, the capacitance of ultra-compact devices is usually very small, regardless of the intrinsic region thickness. This enables the use of very thin intrinsic regions for decreased transit time and operating voltage. There is a significant fabrication challenge in getting the contact resistance low enough for high bandwidth in a 50 Ω environment, but in large part these are being advocated for applications where a larger series resistance may be acceptable. The primary disadvantage of such ultra-compact designs comes in power handling. Due to the small active device area, the maximum output power is expected to be low.

12.4.3 Metal-semiconductor-metal photodetectors

Another way to increase the bandwidth is to use a cross-section design with lower capacitance per unit area. MSM detectors have this property, and have been demonstrated on Si/Ge. IBM successfully integrated a germanium-based photodiode with a 38 GHz bandwidth into a CMOS process flow, though the responsivity at 1550 nm was only 0.07 A/W (Assefa et al., 2009). The same group integrated similar devices with higher responsivity and lower bandwidth with TIAs (Assefa et al., 2013). They also showed that it is possible to use such a structure in avalanche mode at low (1.5 V) bias, which yields a significant sensitivity improvement, while maintaining a 30 GHz bandwidth (Assefa et al., 2010c). Chen and Lipson demonstrated a device with 40 GHz bandwidth and higher (0.35 A/W) responsivity fabricated using wafer bonding (Chen and Lipson, 2009). For both devices, because the germanium is not grown on silicon using CVD, the responsivity begins to roll off at relatively short wavelengths. In general, MSM devices can have lower capacitance per unit area than PIN detectors because the depletion region only occupies a fraction of the device area. One consequence of this is that the saturation current density is also decreased. In addition, the dark current of MSM detectors is often higher than the dark current of PIN diode-based devices.

12.4.4 Separate absorption charge and multiplication avalanche photodiodes

For the lowest noise receivers, photodetectors with gain are attractive, using either avalanche or photoconductive gain. Silicon is an excellent material for avalanche gain due to the low electron and hole ionization coefficient ratio ($k < 0.1$), which allows for high gain-bandwidth products and low excess noise factors. SACM avalanche photodetectors (APDs) using silicon avalanche regions have been demonstrated using both III-V (Hawkins et al., 1997) and Ge (Dai et al., 2010) absorbing regions. In the surface-normal configuration, gain-bandwidth products up to 840 GHz (Sfar Zaoui et al., 2009) have been demonstrated in for Si/Ge devices and up to 315 GHz (Hawkins et al., 1997) have been demonstrated for III-V/Si ones. As waveguide detectors, the highest gain-bandwidth product was at least 380 GHz (Duan et al., 2013) (20 GHz, gain > 19).

12.4.5 Si/Ge uni-traveling carrier photodiodes

The UTC cross-section is an alternative way to push the bandwidth of a waveguide Si/Ge photodiode beyond 30 GHz in a vertically coupled configuration without decreasing the active device area. A UTC is a decoupled structure where the absorption occurs in a doped layer and carriers are collected through a depleted layer in the silicon. As a result, the germanium thickness can be chosen for optimal coupling from the silicon waveguide without affecting the capacitance, and vertically coupled devices with relatively large footprints and fast transit times can be fabricated without sacrificing RC performance. Recently, Si/Ge UTCs with a bandwidth of 40 GHz and responsivity of 0.5 A/W have been demonstrated (Piels and Bowers, 2014).

The cross-section and band diagram of the devices in Piels and Bowers (2014) are shown in Figure 12.5. The transit time is dominated by minority electron transport through the absorber and collector (holes in the absorber move to screen the minority charges), which is the origin of the term *uni-traveling* (Ishibashi et al., 1997). In the absorber, photogenerated carriers move toward the collector by a combination of diffusion and drift; the absorber is doped on a grade to produce a small electric field and decrease the transit time. In the collector, injected minority electrons form a drift current (from the electric field due to the bias voltage) and are collected at the n-contact. Unlike PIN diodes in Si/Ge, UTC detectors perform better when they are p-side up. The threading defects that form at the Si/Ge heterointerface pin the Fermi level in that region, but if the germanium is doped sufficiently

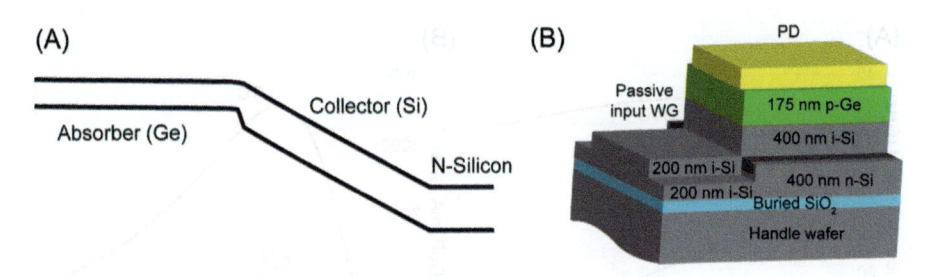

Figure 12.5 (A) Band diagram and (B) cross-section schematic of a waveguide Si/Ge uni-traveling carrier photodetector.
Source: Reprinted with permission from Piels and Bowers (2014).

highly, this does not result in the formation of a large barrier (Piels and Bowers, 2014).

UTC photodiodes were first demonstrated in III/V materials, where they have been shown to have superior bandwidth and power handling due to the large electron velocity in InP relative to the hole velocity. In group IV materials, electron and hole velocities are similar, and the benefit of the UTC is instead that it allows us to choose a capacitance per unit area independently from the absorbing region thickness. For the absorption peak at 200 nm Ge thickness, this results in a higher bandwidth-efficiency product. In a UTC photodiode, the transit time is dominated by the electron transport properties and the capacitance is dominated by the collector thickness. The material constants in Eqs (1.1) and (1.2) change, and Eq. (1.3) becomes (Mishra and Singh, 2008)

$$J_{max} = \frac{2\varepsilon v_n}{W^2}(V_{bi} + V_{PD} - E_{crit}W) \quad (12.4)$$

where ε now refers to the dielectric constant of the silicon collector and W is its thickness.

Figure 12.6A shows idealized design curves for bandwidth and efficiency for PIN and UTC photodiodes assuming a 200 nm thick absorber and a 3 μm wide mesa. Parasitic effects (e.g., pad capacitance and diode series resistance) are neglected in the simulation. The collector thickness of the UTC was chosen separately for each detector length to maximize the bandwidth-efficiency product. As the figure shows, for even moderate efficiency, the UTC out-performs the PIN. Figure 12.6B shows the calculated saturation current-bandwidth product for 3 μm × 30 μm PIN and UTC photodiodes as a function of intrinsic region thickness (Ge for the PIN, Si for the UTC). The assumed bias voltage is half the breakdown voltage of the diode, and the illumination profile was assumed uniform (this will result in an optimistic estimate of the saturation current). Parasitic effects were again ignored. The UTC has both higher saturation current and higher bandwidth, which leads to higher estimated saturation-current bandwidth products.

In Piels and Bowers (2014), high-speed (>33 GHz) waveguide-type Ge/Si UTC photodiodes with dilute absorption profiles were demonstrated with high

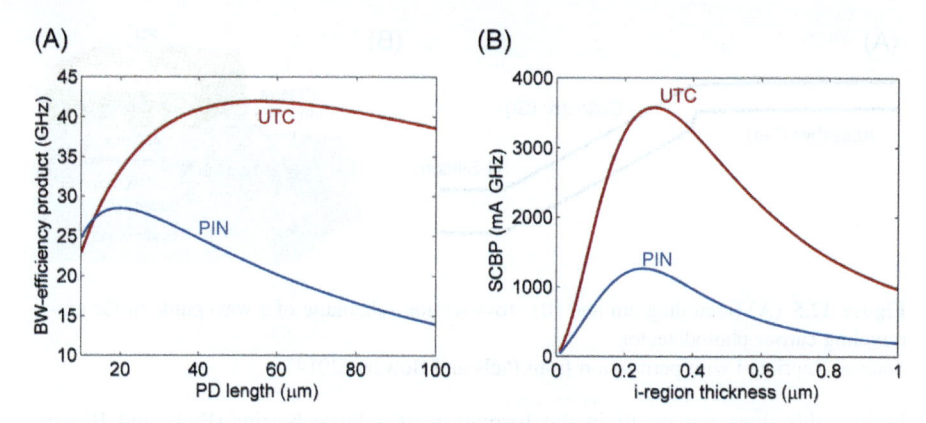

Figure 12.6 (A) BW-efficiency product and (B) saturation current-bandwidth product (SCBP) for ideal PIN and UTC photodiodes assuming 200 nm Ge thickness and a 3-μm wide mesa.

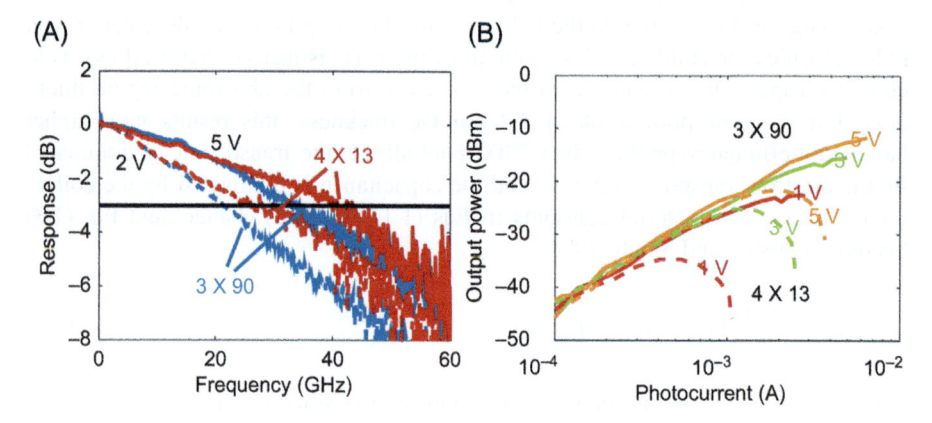

Figure 12.7 (A) Bandwidth and (B) power handling of Si/Ge UTC photodetectors at 30 GHz. *Source*: Reprinted with permission from Piels and Bowers (2014).

responsivities (>0.5 A/W) and high 1 dB-compression (>1.5 mA). Figure 12.7A shows the frequency responses of a 3 μm × 90 μm and a 4 μm × 13 μm device at −2 and −5 V bias. The frequency response was measured with an Agilent lightwave component analyzer at 1550 nm and includes the effect of the probe pad impedance. At −5 V bias, the 3 dB electrical bandwidth of the 3 μm × 90 μm detector was 33 GHz while for the 4 μm × 13 μm detector, it was 40 GHz. The corresponding optical bandwidths (i.e., 10 $\log_{10}(S_{21})$) of the two devices are 54 and 56 GHz, respectively.

The large-signal saturation characteristics of the 40 GHz 4 μm × 13 μm and the 33 GHz 3 μm × 90 μm devices discussed above are shown in Figure 12.7B. An 80% modulation depth tone fixed at 30 GHz was generated using the standard heterodyne technique with two free-running lasers at 1537 nm. The RF power was measured on an electrical spectrum analyzer. The loss of the cables was measured with a network analyzer and subtracted from the data. In both cases, the −1 dB compression current is around 2 mA, which corresponds to an output power around −20 dBm. However, the longer device has larger maximum output power since the output power continues to increase after the −1 dB compression current is reached. This is because the back of the device continues to operate in the linear regime even when the front of the device is compressed, whereas for the shorter device, the current is more uniformly distributed. This leads to a sharp decrease in output power beyond the −1 dB compression current for the shorter device, in contrast to a slow increase in output power for the longer one. The maximum output power of the 4 μm × 13 μm detector at 30 GHz is −21.4 dBm, while it is −11.7 dBm for the longer one. Si/Ge UTCs are thus promising for use in high-power high-speed applications. At present, the primary limitation to both bandwidth and power handling is carrier transport through the heterojunction; improved performance can be expected as the growth technology matures.

12.4.6 Hybrid III/V-silicon uni-traveling carrier photodiodes

UTC designs can also be used to increase the bandwidth and output power of hybrid III/V-silicon photodiodes. Figure 12.8 shows a cross-section of a fabricated device (Beling et al., 2013b); it is based on a surface-normal structure that had high bandwidth and high saturated output power (Li et al., 2011). The modified uni-traveling carrier (MUTC) differs from the UTC in that part of the absorbing region closest to the collector is left undoped. The bias voltage thus induces an electric field in the area of the heterojunction, which improves the carrier transport

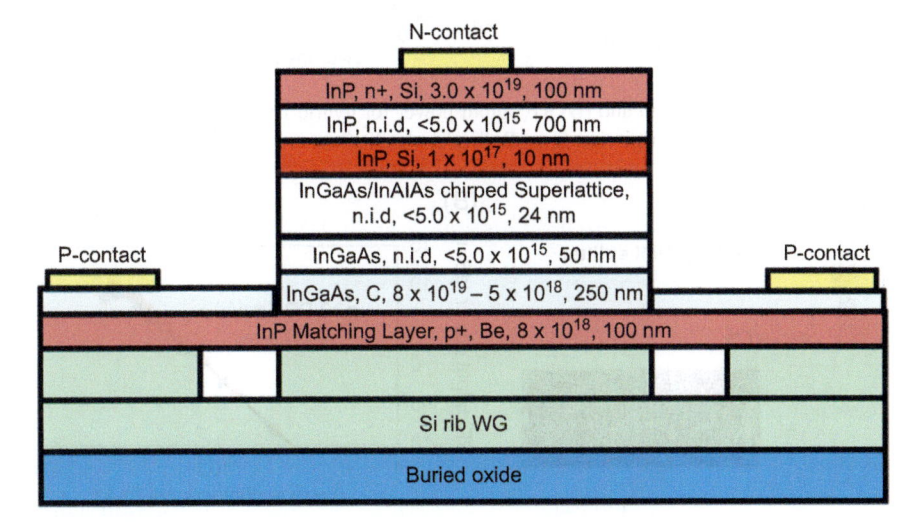

Figure 12.8 Layer stack and cross-section schematic of the hybrid III/V-silicon MUTC photodiode. Doping concentrations are in cm^{-3}.
Source: Reprinted with permission from Beling et al. (2013a).

properties, increasing linearity and bandwidth. The ability to do band engineering is a strong difference between the hybrid III/V-Si platform and Si/Ge platform, and we see much better performance for HSP UTCs than for Si/Ge UTCs.

Beling et al. (2013a,b) report on high-speed high-power waveguide MUTC PDs on the hybrid silicon platform with internal responsivities up to 0.85 A/W. Figure 12.9A shows the measured bandwidths of three devices with different lengths (25, 50, and 100 μm) and a 14 μm wide mesa. The bandwidth was measured using an optical heterodyne setup with a modulation depth close to 100%. For the shortest (25 μm) device, the 3 dB bandwidth was 30 GHz. Figure 12.9B shows the RF output power and compression of this device as a function of time-average photocurrent for the shortest device at 30 GHz. The maximum at 30 GHz was 3.9 dBm, while at 20 GHz it was 5.6 dBm.

To increase the RF output power, photodiode arrays were designed and fabricated. Optically, the input was distributed to multiple diodes via a multimode interferometer as shown in Figure 12.10A. Electrically, the signals were summed in

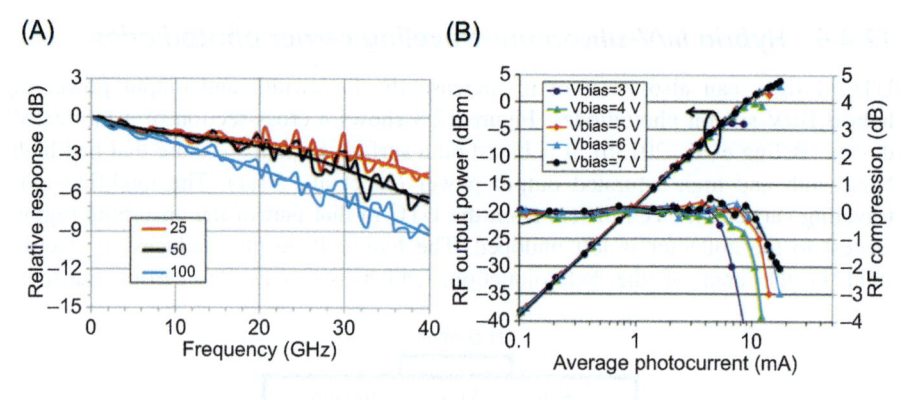

Figure 12.9 (A) Bandwidth and (B) RF output power for hybrid III/V-silicon modified UTCs.
Source: Reprinted with permission from Beling et al. (2013a).

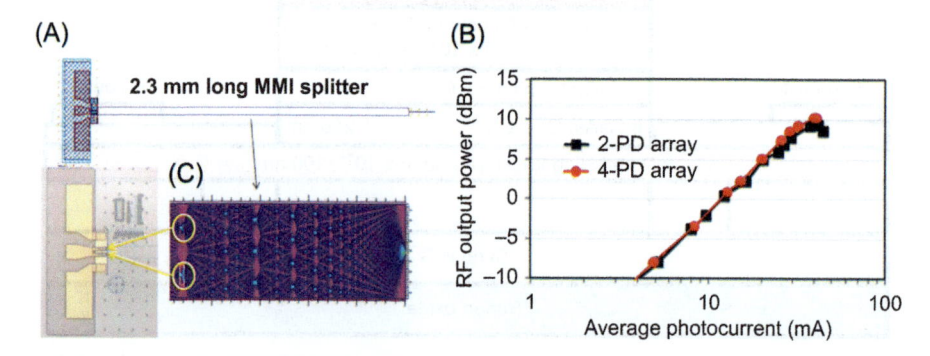

Figure 12.10 MMI-fed photodiode array (A) operating principle and (B) RF output power at 20 GHz as a function of average photocurrent for arrays of 10 μm × 37 μm photodetectors.
Source: Reprinted with permission from Beling et al. (2013a).

parallel at the probe pads. At 20 GHz, a two-photodiode array achieved an RF output power of 9.3 dBm, while a four-detector array achieved an RF output power of 10.2 dBm. The performance was limited by the available input power; in principle, higher output power is possible. For both arrays, the RF output power exceeded that of the single photodiodes discussed above, which indicates the promise of this approach for increasing the output power available from a PIC.

12.5 Conclusions and future outlook

We have presented an overview of photodetectors heterogeneously integrated on silicon-on-insulator. For very high-speed applications where output power is not of great concern, ultra-compact butt-coupled Si/Ge PINs have the best bandwidth and efficiency. For microwave photonics and coherent communication applications requiring large photocurrents, vertically coupled UTC detectors, either on the hybrid III/V-silicon platform or in Si/Ge, are preferable. Both of these technologies are in their initial stages of development, and can be further improved. The band engineering currently available in III/V (that has been critical to enhancing III/V MUTC performance) can be applied to the Si/Ge UTC as germanium-on-silicon growth technology matures. Advanced optical designs (Beling et al., 2013a,b) can be applied to both platforms, further increasing output power.

References

Ahn, D., Hong, C.-Y., Liu, J., Giziewicz, W., Beals, M., Kimerling, L.C., 2007. High performance, waveguide integrated Ge photodetectors. Opt. Express. 15 (7), 3916–3921.

Assefa, S., Xia, F., Bedell, S.W., Zhang, Y., Topuria, T., Rice, P.M., et al., 2009. CMOS-integrated 40 GHz germanium waveguide photodetector for on-chip optical interconnects. In: Optical Fiber Communication Conference and National Fiber Optic Engineers Conference, San Diego, CA.

Assefa, S., Xia, F., Bedell, S.W., Zhang, Y., Topuria, T., Rice, P.M., 2010a. CMOS-integrated high-speed MSM germanium waveguide photodetector. Opt. Express. 18 (5), 4986–4999.

Assefa, S., Xia, F., Green, W., Schow, C., Rylyakov, A., Vlasov, Y., 2010b. CMOS-integrated optical receivers for on-chip interconnects. IEEE J. Sel. Top. Quantum Electron. 16 (5), 1376–1385.

Assefa, S., Xia, F., Vlasov, Y.A., 2010c. Reinventing germanium avalanche photodetector for nanophotonic on-chip optical interconnects. Nature 464, 80–84.

Assefa, S., Pan, H., Shank, S., Green, W., Rylyakov, A., Schow, C., et al., 2013. Monolithically integrated silicon nanophotonics receiver in 90 nm CMOS technology node. In: Optical Fiber Communication Conference/National Fiber Optic Engineers Conference, Anaheim, CA.

Beling, A., Cross, A., Piels, M., Peters, J., Zhou, Q., Bowers, J., 2013a. InP-based waveguide photodiodes heterogeneously integrated on silicon-on-insulator for photonic microwave generation. Opt. Express. 21 (22), 25901–25906.

Beling, A., Cross, A.S., Piels, M., Peters, J., Fu, Y., Zhou, Q., et al., 2013b. High-power high-speed waveguide photodiodes and photodiode arrays heterogeneously integrated on silicon-on-insulator. In: Optical Fiber Communication Conference/National Fiber Optic Engineers Conference, Anaheim, CA.
Binetti, P., Leitjens, X.J.M., de Vries, T., Oei, Y.S., Di Cioccio, L., Fédéli, J.M., 2010. InP/InGaAs photodetector on SOI photonic circuitry. IEEE Photonics J. 2 (3), 299–305.
Black, A., Hawkins, A.R., Margalit, N.M., Babic, D.I., Holmes, A.L., Chang, Y.-L., 1997. Wafer fusion: materials issues and device results. IEEE J. Sel. Top. Quantum Electron. 3 (3), 943–951.
Bowers, J.E., Burrus, C.A., 1986. High speed zero bias waveguide photodetectors. Electron. Lett. 22 (17), 905.
Bowers, J.E., Burrus, C., 1987. Ultrawide-band long-wavelength p-i-n photodetectors. J. Light. Technol LT-5 (10), 1339–1350.
Bowers, J.E., Piels, M., Ramaswamy, A., Yin, T., 2010. High power waveguide Ge/Si photodiodes. In: Annual Meeting of the Electrochemical Society, Las Vegas, NV.
Chang, H., Kuo, Y., Jones, R., Barkai, A., Bowers, J.E., 2010. Integrated hybrid silicon triplexer. Opt. Express. 18 (23), 23891–23899.
Chen, L., Lipson, M., 2009. Ultra-low capacitance and high speed germanium photodetectors on silicon. Opt. Express. 17 (10), 7901–7906.
Chen, R., Tran, T.-T.D., Ng, K.W., Ko, S.W., Chuang, L.C., Sedgwick, F.G., 2011. Nanolasers grown on silicon. Nat. Photonics. 5, 170–175.
Dai, D., Chen, H.-W., Bowers, J.E., Kang, Y., Morse, M., Paniccia, M., 2010. Equivalent circuit model of a waveguide-type Ge/Si avalanche photodetector. Phys. Status Solidi C. 7 (10), 2532–2535.
Dai, D., Piels, M., Bowers, J.E., 2014. Monolithic Germanium/Silicon photodetectors with decoupled structures: resonant APDs and UTC photodiodes. J. Sel. Top. Quantum Electron. 20 (6).
DeRose, C.T., Trotter, D.C., Zortman, W.A., Starbuck, A.L., Fisher, M., Watts, M.R., 2011. Ultra compact 45 GHz CMOS compatible germanium waveguide photodiode with low dark current. Opt. Express. 19 (25), 24897–24904.
Duan, N., Liow, T.-Y., Lim, A., Ding, L., Lo, G., 2013. High speed waveguide-integrated Ge/Si avalanche photodetector. In: Optical Fiber Communication Conference, Anaheim, CA.
Feng, D., Liao, S., Dong, P., Feng, N.-N., Liang, H., Zheng, D., 2009. High-speed Ge photodetector monolithically integrated with large cross-section silicon-on-insulator waveguide. Appl. Phys. Lett. 95 (26), 261105–261105-3.
Feng, S., Geng, Y., Lau, K.M., Poon, A.W., 2012. Epitaxial III-V-on-silicon waveguide butt-coupled photodetectors. Opt. Lett. 37 (19), 4035–4037.
Galland,C., Novack,A., Liu,Y., Ding,R., Gould,M., Baehr-Jones,T., 2013. A CMOS-Compatible Silicon Photonic Platform for High-Speed Integrated Opto-Electronics. Proceedings Volume 8767, Integrated Photonics: Materials, Devices, and Applications II; 87670G (2013) https://doi.org/10.1117/12.2017053.
Giovane, L.M., Luan, H.-C., Agarwal, A., Kimerling, L.C., 2001. Correlation between leakage current density and threading dislocation density in SiGe p-i-n diodes grown on relaxed graded buffer layers. Appl. Phys. Lett. 78, 541–543.
Hartmann, J., Damlencourt, J.-F., Bogumilowicz, Y., Holliger, P., Rolland, G., Billon, T., 2005. Reduced pressure-chemical vapor deposition of intrinsic and doped Ge layers on Si(001) for microelectronics and optoelectronics purposes. J. Cryst. Growth 274 (1–2), 90–99.

Hawkins, A.R., Wu, W., Abraham, P., Streubel, K., Bowers, J.E., 1997. High gain-bandwidth-product silicon heterointerface photodetector. Appl. Phys. Lett. 70, 303–305.

Heck, M.J.R., Bauters, J.F., Davenport, M.L., Doylend, J.K., Jain, S., Kurczveil, G., 2013. Hybrid silicon photonic integrated circuit technology. IEEE J. Sel. Top. Quantum Electron. 19 (4).

Højfeldt, S., Mørk, J., 2002. Modeling of carrier dynamics in quantum-well electroabsorption modulators. IEEE J. Sel. Top. Quantum Electron. 8 (6), 1265–1276.

Ishibashi, T., Shimizu, N., Kodama, S., Ito, H., Nagatsuma, T., Furuta, T., 1997. Uni-traveling-carrier photodiodes. In: Ultrafast Electronics and Optoelectronics, Incline Village, NV.

Knights, A.P., Doylend, J.K., 2008. Silicon photonics – recent advances in device developmentAdvances in Information Optics and Photonics. SPIE Press.

Koch, B., Norberg, E., Kim, B., Hutchinson, J., Shin, J., Fish, G., et al., 2013. Integrated silicon photonic laser sources for telecom and datacom. In: Optical Fiber Communication Conference/National Fiber Optic Engineers Conference, Anaheim, CA.

Lee, B., Rylyakov, A., Proesel, J., Baks, C., Rimolo-Donadio, R., Schow, C., et al., 2013. 60-Gb/s receiver employing heterogeneously integrated silicon waveguide coupled photodetector. CLEO, 2013 Postdeadline, San Jose, CA.

Li, Z., Fu, Y., Piels, M., Pan, H., Beling, A., Bowers, J.E., 2011. High-power high-linearity flip-chip bonded modified uni-traveling carrier photodiode. Opt. Express. 19 (26), B385–B390.

Liao, S., Feng, N.-N., Feng, D., Dong, P., Shafiiha, R., Kung, C.-C., 2011. 36 GHz submicron silicon waveguide germanium photodetector. Opt. Express. 19 (11), 10967–10972.

Liow, T.-Y., Lim, A.E., Duan, N., Yu, M., Lo, G., 2013. Waveguide germanium photodetector with high bandwidth and high L-band responsivity. In: Optical Fiber Communication Conference/National Fiber Optic Engineers Conference, Anaheim, CA.

Liu, J., Cannon, D.D., Wada, K., Ishikawa, Y., Danielson, D.T., Jongthammanurak, S., 2004. Deformation potential constants of biaxially tensile stressed Ge epitaxial films on Si (100). Phys. Rev. B. 70 (15), 155309.

Liu, J., Ahn, D., Hong, C.Y., Pan, D., Jongthammanurak, S., Beals, M., et al., 2006. Waveguide integrated Ge p-i-n photodetectors on a silicon-oninsulator platform. In: Optics Valley of China International Symposium on Optoelectronics. 10.1109/OVCISO.2006.302697.

Liu, A.Y., Zhang, C., Norman, J., Snyder, A., Lubyshev, D., Fastenau, J.M., 2014. High performance continuous wave 1.3 μm quantum dot lasers on silicon. Appl. Phys. Lett. 104, 041104.

Marris-Morini, D., Virot, L., Baudot, C., Fédéli, J., Rasigade, G., Perez-Galacho, D., 2014. A 40 Gbit/s optical link on a 300-mm silicon platform. Opt. Express. 22 (6), 6674–6679.

Masini, C., Colace, L., Assanto, G., Luan, H.-C., Kimerling, L.C., 2001. High performance p-i-n Ge on Si photodetectors for the near infrared: from model to demonstration. IEEE Trans. Electron. Devices 48 (6), 1092–1096.

Masini, G., Sahni, S., Capellini, G., Witzens, J., Gunn, C., 2008. High-speed near infrared optical receivers based on Ge waveguide photodetectors integrated in a CMOS process. Adv. Opt. Technol. 2008, 196572.

Michel, J., Liu, J., Kimerling, L.C., 2010. High-performance Ge-on-Si photodetectors. Nat. Photonics. 4 (8), 527–534.

Mishra, U.K., Singh, J., 2008. Semiconductor Device Physics and Design. Springer-Verlag, AA Dordrecht, the Netherlands.

Mueller, R.K., 1959. Dislocation acceptor levels in germanium. J. Appl. Phys. 30 (12), 2015–2016.

Novack, A., Gould, M., Yang, Y., Xuan, Z., Streshinsky, M., Liu, Y., 2013. Germanium photodetector with 60 GHz bandwidth using inductive gain peaking. Opt. Express. 21 (23), 28387−28393.
Park, H., Fang, A., Jones, R., Cohen, O., Raday, O., Sysak, M., 2007. A hybrid AlGaInAs-silicon evanescent waveguide photodetector. Opt. Express. 15 (6), 6044−6052.
Piels, M., 2013. Si/Ge photodiodes for coherent and analog communication. University of California, Santa Barbara, CA.
Piels, M., Bowers, J., 2014. 40 GHz Si/Ge uni-traveling carrier waveguide photodiode. J. Light. Technol. 32 (20), 3502−3508.
Piels, M., Ramaswamy, A., Bowers, J., 2013. Nonlinear modeling of waveguide photodetectors. Opt. Express. 21, 15634−15644.
Piels, M., Bauters, J.F., Davenport, M.L., Heck, M.J.R., Bowers, J.E., 2014. Low-loss silicon nitride AWG demultiplexer heterogeneously integrated with hybrid III-V/Silicon photodetectors. J. Light. Technol. 32 (4), 817−823.
Roucka, R., Mathews, J., Weng, C., Beeler, R., Tolle, J., Menéndez, J., 2011. Development of high performance near IR photodiodes: a novel chemistry based approach to Ge-Sn devices integrated on silicon. IEEE J. Quantum Electron. 47 (2), 213−222.
Sfar Zaoui, W., Chen, H.-W., Bowers, J.E., Bowers, J.E., Kang, Y., Morse, M., 2009. Frequency response and bandwidth enhancement in Ge/Si avalanche photodiodes with over 840 GHz gain-bandwidth-product. Opt. Express. 17 (15).
Tanabe, K., Watanabe, K., Arakawa, Y., 2012. III-V/Si hybrid photonic devices by direct fusion bonding. Sci. Rep. 2, 349.
Tulchinsky, D.A., Boos, J.B., Park, D., Goetz, P.G., Rabinovich, W.S., Williams, K.J., 2008. High-current photodetectors as efficient, linear, and high-power RF output stages. J. Light. Technol. 26 (4), 408−416.
Virot, L., Vivien, L., Fédéli, J., Bogumilowicz, Y., Hartmann, J., Boeuf, F., 2013. High-performance waveguide-integrated germanium PIN photodiodes for optical communication applications. Photonics Res. 1, 140−147.
Vivien, L., Rouviere, M., Fédéli, J.-M., Marris-Morini, D., Damlencourt, J.F., Mangeney, J., 2007. High speed and high responsivity germanium photodetector integrated in a silicon-on-insulator microwaveguide. Opt. Express. 15 (15), 9843−9848.
Wang, J., Loh, W.-Y., Chua, K., Zhang, H., Xiong, Y., Loh, T.H., 2008. Evanescent-coupled Ge p-i-n photodetectors on Si-waveguide with SEG-Ge and comparative study of lateral and vertical p-i-n configurations. IEEE Electron. Device Lett. 29 (5), 445−448.
Weingarten, K., Rodwell, M., Bloom, D., 1988. Picosecond optical sampling of GaAs integrated circuits. IEEE J. Quantum Electron. 24 (2), 198−220.
Williams, K.J., Esman, R., 1999. Design considerations for high-current photodetectors. J. Light. Technol. 17 (8), 1443−1454.
Yin, T., Cohen, R., Morse, M.M., Sarid, G., Chetrit, Y., Rubin, D., 2007. 31 GHz Ge n-i-p waveguide photodetectors on silicon-on-insulator substrate. Opt. Express. 15 (21), 13965−13971.

Efficient surface nano-textured CMOS-compatible photodiodes for Optical Interconnects

Soroush Ghandiparsi, Ahmed S. Mayet, Cesar Bartolo-Perez and M. Saif Islam
Department of Electrical and Computer Engineering University of California, Davis, CA, United States

13.1 Introduction

13.1.1 Global IP traffic trend and forecast

Connected sensors (IoT) (Hassan, 2019; Laghari et al., 2022), monitoring, streaming media, next generation of wireless networks, cloud computing (Varghese and Buyya, 2018; El Srouji, 2022), remote learning, virtual reality, online gaming, and other emerging online applications has led to an explosion of online data. Annual data traffic is expected to increase by over 20x by 2030. The global data center construction market size is projected to surpass around US$ 369.6 billion by 2030 and register growth at a compound annual growth rate (CAGR) of 6.7% from 2022 to 2030 (Mozo et al., 2018; Bilal et al., 2014).

Most of the global internet traffic (more than 85%) has originated or terminated in a data center since 2008. Datacenter traffic will continue to dominate internet traffic in the near future. Still, the nature of data center traffic is undergoing a fundamental transformation brought about by cloud applications, services, and infrastructure. Based on the CAGR forecast provided by CISCO, more than 95% of data center traffic reached cloud traffic in 2021 (Bilal, et al., 2014; Yap et al., 2017; Meeker and Wu, 2018) (Figure 13.1).

13.1.2 Optical communication in datacenters

A data center is a massively parallel supercomputing infrastructure, including clusters with several thousands of servers interconnected together in each cluster. The early small-scale data center networks used copper-based interconnects. Still, as bandwidth requirements scaled, fiber-optics-based optical interconnects were introduced and quickly established as the most cost-effective, deployable solution to scale out the data center. Presently, high bandwidth and low power

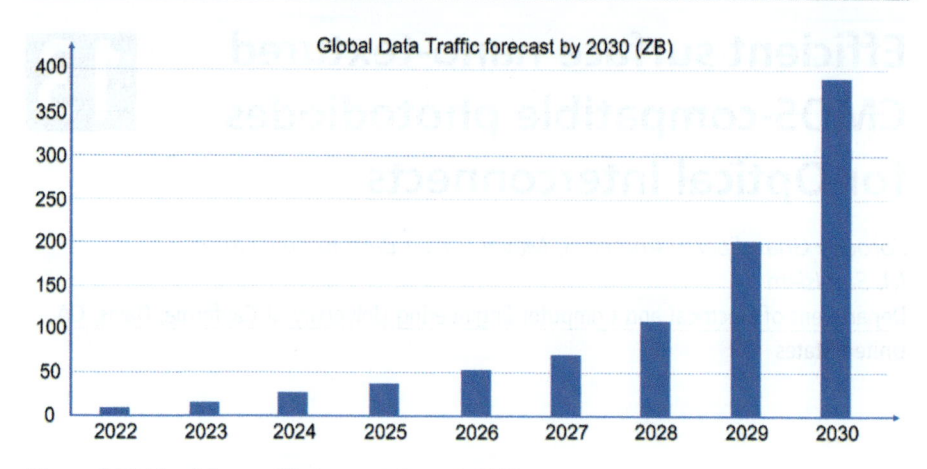

Figure 13.1 Total data traffic forecast through 2030.
Source: "Impact of AI on electronics and semiconductor industries", IBS, April 2020.

optical interconnects have been ubiquitously used in data centers for any link distance beyond a few meters.

To handle higher data volume, higher speed infrastructure is needed to sustain data movement. Currently, 56 and 112 Gbps SerDes IP that supports PAM-4 modulation enables 400 Gbps Ethernet connectivity in hyperscale data centers, and it is expected to speed up to 800 Gbps in the future (Figure 13.2). Leading Ethernet switch vendors are already developing 800 Gbps switches based on 112G SerDes IP, with plans to introduce 1.6 Tbps Ethernet (using a faster, next-generation SerDes) within the next few years to meet the demands of increasing data volumes. Data communication between servers within a rack is managed by the Top-of-Rack (ToR) switch and network interface cards within each server. The most common interface speed in cloud data centers at this level has been 25 Gbps for the past few years. However, as infrastructure speeds increase to 400 Gbps, Ethernet speed within the rack is increasing to 100 Gbps.

The most important consideration for interconnecting is the bandwidth cost. To lower the total interconnection cost, various technologies are developed for different segments of the network. Fiber-based optical interconnects are used for interconnection between the TOR switch and the edge switch, as well as between the edge aggregation switch and the spine switches. Commercial systems are using multimode fibers (MMFs) for short-reach applications as the preferred medium of transmission due to their relative ease of use compared to single-mode fibers (SMFs). In short-reach applications such as the datacenters, the constraints due to the cost and complexity of deployment drive the choice of associated technology. vertical-cavity surface-emitting laser (VCSEL)-based MMF optical links offer a cost-effective solution to the growing demand for bandwidth today with the large core MMF relaxing the requirement of high-end optics at the transmitter and receiver. For a link distance of less than 500 m, VCSEL and MMF-based technologies have proven to

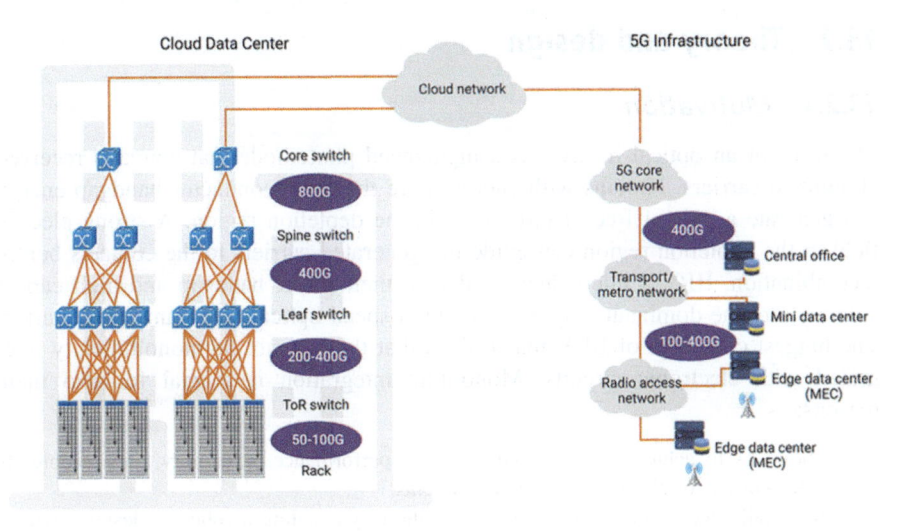

Figure 13.2 Hyperscale data center infrastructures are transitioning to 400 + GbE.

give the best overall link cost (transceiver cost plus fiber cost). Beyond 100 m, however, more expensive SMF transmission technologies usually have to be chosen due to the following reasons:

1. The bandwidth of commercially available VCSELs has been limited (about 20 GHz).
2. The bandwidth of an MMF reduces as the distance increases [e.g., the BW is limited to about 20 GHz for 100 m of optical multimode (OM3) fiber].
3. The higher cost of an MMF, although the price of a VCSEL transceiver, is significantly lower than an edge-emitting laser-based SMF transceiver (Zhou et al., 2018; Garlinska et al., 2020; Plant et al., 2017; Mellette et al., 2017; Eriksson et al., 2017; Sato 2018; Wang et al., 2022; Yoo et al., 2012; Ben Yoo et al., 2018; Gerstel et al., 2012).

13.1.3 Chapter outline

In this chapter, we present new methods to improve the performance of high-speed silicon-based photodiodes for data centers' interconnects. Efficient structures for coupling the incident light to the thin absorbing layer are investigated to increase the responsivity while reaching higher bandwidths. The advantages of developing high-speed silicon photodiodes and the challenges that hindered their application in high-speed applications are discussed. A vertical all-silicon pin photodiode based on the light-trapping theory optimized for short-reach ($\lambda = 850$ nm) communication is demonstrated. Complete optical, DC, RF, and bit error rate (BER) characterizations are presented, and the design challenges are explored. Based on the measurement results, an optical-electrical photodiode model is developed. Employing the PD model, we provide feasible solutions to reach the design targets.

13.2 Theory and design

13.2.1 Motivation

The heart of an optical receiver is a high-speed photodiode that converts received photons to carriers. Photons with energy more than semiconductor bandgap energy can generate a pair of free electron-hole in the depletion region. A strong electric field in the depletion region can guide the generated carriers to the contacts before recombination. III/V semiconductors, due to their direct bandgap and high carrier velocity, are the dominant players in the high-speed optical communication market. The biggest challenge of III-V materials is that they cannot be monolithically integrated with electronic circuits. Monolithic integration is crucial for two main reasons:

1. It allows us to achieve the required levels of performance, scalability, and complexity simultaneously for electronic–photonic systems.
2. Substantially accelerates system-level innovation by enabling a cohesive design environment to realize the systems on a chip goal.

Silicon is the most desirable material to bridge the gap between photonics to electronics in terms of CMOS integration. Replacing III/V material with Silicon-based materials is not as easy as we think. Silicon is an indirect material, and its carrier mobility is much lower than III-V material. A single silicon photodiode could not pass a 10 GHz limit, while a low responsivity is a challenging performance parameter to be overcome (Jalali and Fathpour, 2006; Thomson et al., 2016; Soref, 2006; Siew et al., 2021; Margalit et al., 2021; Chen et al., 2018; Cyriac et al., 2022; Siviero et al., 2022).

What is the main challenge with silicon photodiodes? Figure 13.3 is a comparison between a 30 μm GaAs photodiode and a similar size silicon photodiode at an 850 nm window. We can see the optical absorption and 3 dB bandwidth in one graph.

Optical absorption exponentially depends on absorber thickness, and the silicon absorption coefficient is very low close to its bandgap. Therefore, a very thick silicon layer is needed to reach an acceptable level of optical absorption. On the other hand, the bandwidth of the photodiode decreases with a larger RC time constant. Thus, there is a critical trade-off between bandwidth and absorption (responsivity). As shown in Figure 13.3, a photodiode with less than a 3 μm absorbing layer could reach 25G bandwidth, but the absorption will be limited to less than 20% (Cansizoglu et al., 2018a; Gao et al., 2017a; Cansizoglu et al., 2018b; Sentieri et al., 2020).

13.2.2 Light trapping theory and background

Light trapping is the capturing of as many photons as possible from an impinging electromagnetic (E-M) wave to generate heat or charge carriers, excitons, and other absorption mechanisms. Trapping of photons of different wavelengths is vital to many applications, including sensing, photovoltaics, photo-electrochemistry, solar

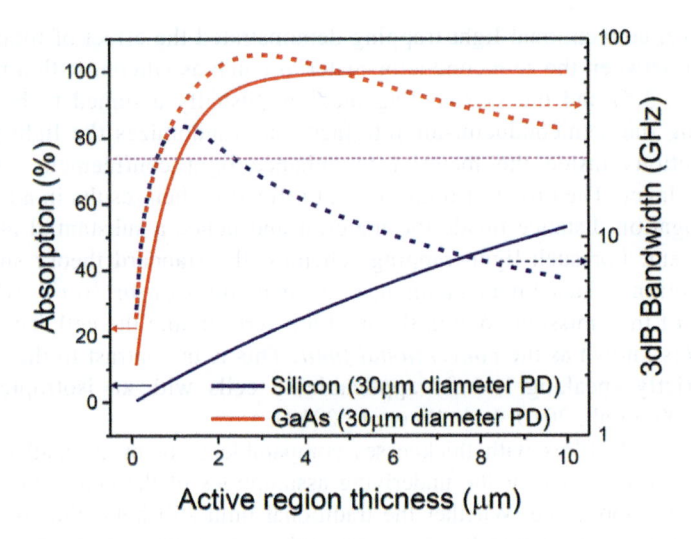

Figure 13.3 Photodiode absorption and bandwidth versus active region thickness. There is a trade-off between 3 dB bandwidth (dash line) and optical absorption (solid line). To overcome the weak absorption in Si while the bandwidth is kept in the Gb region, new solutions need to be offered.

fuel production, and thermal photovoltaics. Some discussions break light trapping into light capture and light trapping. With the thin devices and, in particular, the thin photon absorbers used in high-speed detectors and advanced solar cells, it is somewhat artificial to separate light capture and light trapping (Pala et al., 2013; Yu et al., 2010; Wang et al., 2012; Isabella et al., 2018).

The optical phenomena that provide the essential tools for light trapping are listed as:

1. **Interference:** The phenomenon where two or more E-M waves exist at a point constructively or destructively add together to some degree at that point.
2. **Scattering:** The result of impinging E-M waves bouncing off objects by being absorbed and emitted (Material property dependent). In general, it can be elastic or inelastic.
3. **Reflection:** The result of some portion of an impinging E-M wave being scattered backward (Material property dependent). They are generally taken as elastic.
4. **Diffraction:** The result of impinging E-M waves bouncing off objects, waves change direction, and constructively interfering in certain specific directions. It is taken as elastic.
5. **Plasmonics:** The result of an impinging E-M wave being absorbed by the extremely numerous electrons of metal, thereby exciting an oscillating plasma. This plasma dissipates energy through electron collisions and reradiates an E-M scattered wave (Material property dependent). Overall, it is taken as inelastic.
6. **Refraction:** The result of an impinging E-M wave changing direction and wavelength due to a change in the transmission medium through which it is passing (Material property dependent). It is taken as elastic.

The theory of light trapping was initially developed for conventional solar cells where the light-absorbing film is typically many wavelengths thick. From the ray-optics perspective, conventional light trapping demonstrated the effect of total internal reflection between the semiconductor material (such as silicon, with a refractive index $n \sim 3.5$) and the surrounding medium (usually assumed to be air). By roughening the semiconductor-air interface, one randomizes the light propagation directions inside the material to enhance light confinement within the absorbing layer. The effect of total internal reflection then results in a much longer propagation distance inside the material and hence a substantial absorption enhancement. For such light-trapping schemes, the standard theory shows that the absorption enhancement factor has an upper limit of $4n^2/sin^2\theta$, where θ is the angle of the emission cone in the medium surrounding the cell. This limit of $4n^2/sin^2\theta$ is known as the *conventional limit*. This is in contrast to the $4n^2$ limit, which strictly speaking is only applicable to cells with an isotropic angular response (Yu et al., 2010; Isabella et al., 2018).

For nanoscale films with thicknesses comparable to or even smaller than the wavelength scale, some of the underlying assumptions of the conventional theory are no longer applicable. Whether the traditional limit still holds thus becomes an essential open question that has been pursued numerically and experimentally. It has shown the conventional limit ($4n^2/sin^2\theta$) is only correct in bulk structures. In the nanophotonic regime, the absorption enhancement factor can go far beyond this limit with proper design.

Consider a high-index thin-film active layer with a high reflectivity mirror at the bottom and air on top. Such a film supports guided optical modes. In the limit where the absorption of the active layer is weak, these guided modes typically have a propagation distance along with the film that is much longer than the thickness of the film. Light trapping is accomplished by coupling the incident plane waves into these guided modes, with either a grating with periodicity L (Figure 13.4A) or random Lambertian roughness. It is well known that a system with random roughness can be understood by taking the $L \to \infty$ limit of the periodic system. As long as L is chosen to be sufficiently large, for example, at least comparable to the free space

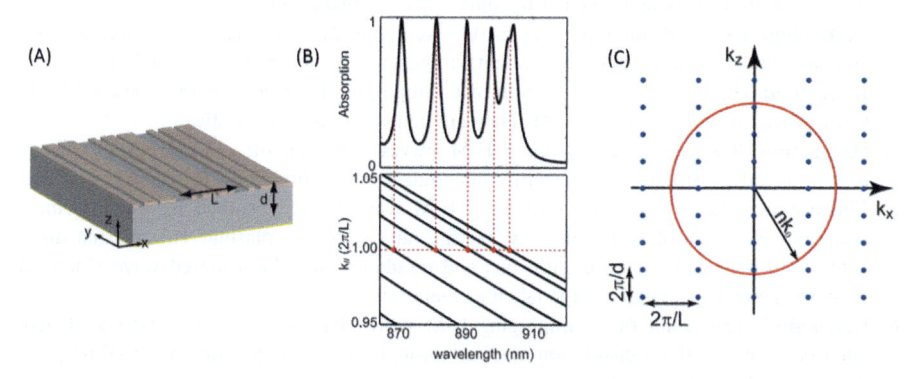

Figure 13.4 (A) Simple light trapping (grating) structure with dimensions comparable to the wavelength, (B) absorption enhancement in thin-film by applying surface photon trapping structure, (C) guided lateral modes based on the structure parameters.

wavelength of the incident light, each incident plane wave can couple into at least one guided mode. By the same argument, such a guided mode can couple to external plane waves, creating a guided resonance.

A typical absorption spectrum for such a film is reproduced in Figure 13.4B. The absorption spectrum consists of multiple peaks, each corresponding to a guided resonance. The absorption is strongly enhanced in the vicinity of each resonance. However, compared to the broad solar spectrum, each individual resonance has a very narrow spectral width. Consequently, to enhance absorption over a substantial portion of the solar spectrum, one must rely upon a collection of these peaks. Motivated by this observation, we develop a statistical temporal coupled-mode theory that describes the aggregate contributions from all resonances.

The behavior of an individual guided resonance, when excited by an incident plane wave, is described by the temporal coupled-mode theory equation (Yu et al., 2010; Yu and Fan 2011):

$$\frac{d}{dt}a = \left(j\omega_0 - \frac{N\gamma_e + \gamma_i}{2}\right)a + j\sqrt{\gamma_e}S \qquad (13.1)$$

Here a is the resonance amplitude, normalized such that $|a|^2$ is the energy/unit area in the film, ω_0 is the resonance frequency, and γ_i is the intrinsic loss rate of the resonance due to material absorption. S is the amplitude of the incident plane wave, with $|S|^2$ corresponding to its intensity. We refer to a plane wave that couples to the resonance as a channel. γ_e is the leakage rate of the resonance to the channel that carries the incident wave. In general, the grating may phase-match the resonance to other plane-wave channels as well.

For light incidents from the normal direction, the periodicity results in the excitation of other plane waves with $k_x = 0$, $\pm \frac{2\pi}{L}$, $\pm \frac{4\pi}{L}$, ... in the free space above the waveguide. Since the plane waves are propagating modes in vacuum space, it is required to satisfy $k_x \leq k_0$, where k_0 is the wavevector of the incident light. These two requirements completely specify the number of channels available in the k-space (Figure 13.4C). In addition, consider the case $L \geq \lambda$. The spacing between the channels is $\frac{2\pi}{L} \leq \frac{2\pi}{\lambda} = k_0$. The total number of channels (Figure 2.1B) at a wavelength (λ) thus becomes (Yu et al., 2010; Wang et al., 2012):

$$N = \frac{2k_0}{2\pi/L} = \frac{2L}{\lambda} \qquad (13.2)$$

In the frequency range $[\omega, \Delta\omega]$, the total number of guided resonances supported by the film is (Figure 13.4C):

$$M = \frac{2n^2\omega\pi}{c^2}\left(\frac{L}{2\pi}\right)\left(\frac{d}{2\pi}\right)\Delta\omega. \qquad (13.3)$$

As it is shown in the absorption graph, resonance occurs when a specific lateral mode is excited considering the designed grating period. For more complex structures such as 2D nanostructures, a numerical calculation is needed. To overcome the weak absorption in thin-film silicon while the 3 dB bandwidth has remained in the Gb regime, the PIN Si PD is integrated with an optimized light-trapping structure (Cansizoglu et al., 2018b; Saive, 2021; Mokkapati and Catchpole, 2012; Peter Amalathas and Alkaisi, 2019; Martins et al., 2012; Bhattacharya and John, 2020; Qiao et al., 2020; Rim et al., 2007; Zhou and Biswas, 2008; Saini and Nair, 2019; Huo and Konstantatos, 2018; Razmjooei and Magnusson, 2020; Cesar et al., 2021).

13.3 Vertical PIN silicon-based photodiode for short-reach communication

13.3.1 Device design

The devices include a nip and pin mesa structure on top of a silicon-on-insulator (SOI) wafer, over a 3 μm SiO$_2$ layer. Micro- and nanoscale holes with different diameters/periods were etched through the mesa to a depth sufficient to reach the bottom n^{++} layer (Figure 13.5). Holes were etched into the active region of the photodiode as square or hexagonal lattices and were fabricated as either uniform cylindrical shapes, inverted pyramids, and funnel shapes with tapered sidewalls.

The photodiode structure was grown epitaxially on an SOI substrate with a 0.25 μm device layer (p-Si). The thin absorption region was designed to comprise a 2 μm-thick i-Si layer to minimize the transit time for electrons and holes. Lattice-matched 0.2 μm p^{++} was used as the bottom p-contact layer. A 0.3 μm P-doped n^{++} thin layer served as the top *n*-ohmic contact. High doping decreases the minority carrier lifetime and minimizes the diffusion of photocarriers generated in the n and p layers into the high field i-region, as well as reducing the series resistance (Cansizoglu et al., 2018a; Gao et al., 2017a; Lv et al., 2018; Casalino et al., 2016; Liu et al., 2021; Ackert et al., 2015; Kyomasu, 1995; Beling and Campbell, 2014).

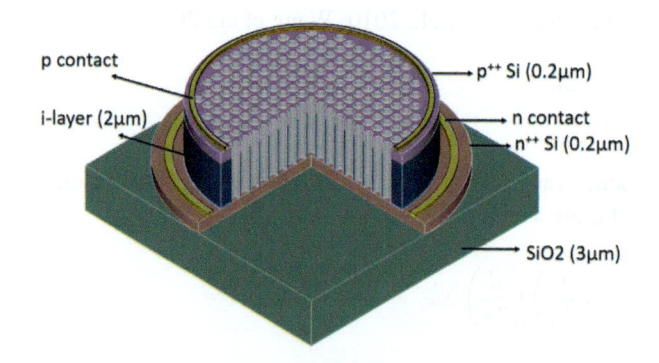

Figure 13.5 Schematic of PIN-designed photodiode integrated with light trapping micro/nanohole array on SOI wafer.

13.3.2 Simulation and optimization

To find the optimized structure that reaches the design targets of maximum quantum efficiency (QE) (close to unity) at 850 nm wavelength and reaches more than 25 GHz 3 dB bandwidth, the structures are simulated using different parameters. The simulation is conducted via the Lumerical optical [Finite-Difference Time-Domain (FDTD)] and Device [Finite Element (FE)] methods to reach the best structure in terms of layers thickness, doping profile of each layer, and applied bias. The results of optical simulation (optical generation) are imported to the Device module to simulate the bandwidth and output current.

By employing the FDTD method, the light-matter interaction via solving Maxwell equations is studied numerically in the structures consisting of different micro/nanohole designs. A plane-wave source with a Gaussian profile is used as a normal incident light source. Periodic boundary conditions are used for the x-y direction, and a perfectly matched layer (PML) is used for Z-direction.

The simulation result shows that the integration of a micro/nanohole array could enhance the photon absorption in thin-film silicon effectively. The light confinement and lateral propagation in cylindrical, inverted pyramid and funnel-shaped structures are shown in Figure 13.6A–C. The highest absorption enhancement occurred in the funnel-shaped structure. Figure 13.6D shows the absorption in the funnel-shaped structures with different sidewall angles (Gao et al., 2017a; Razmjooei and Magnusson, 2020; Fard et al., 2017; Gao et al., 2017b; Devine et al., 2021; Ghandiparsi et al., 2019; Razmjooei et al., 2021; Razmjooei and Magnusson, 2022; Ahamad et al., 2020).

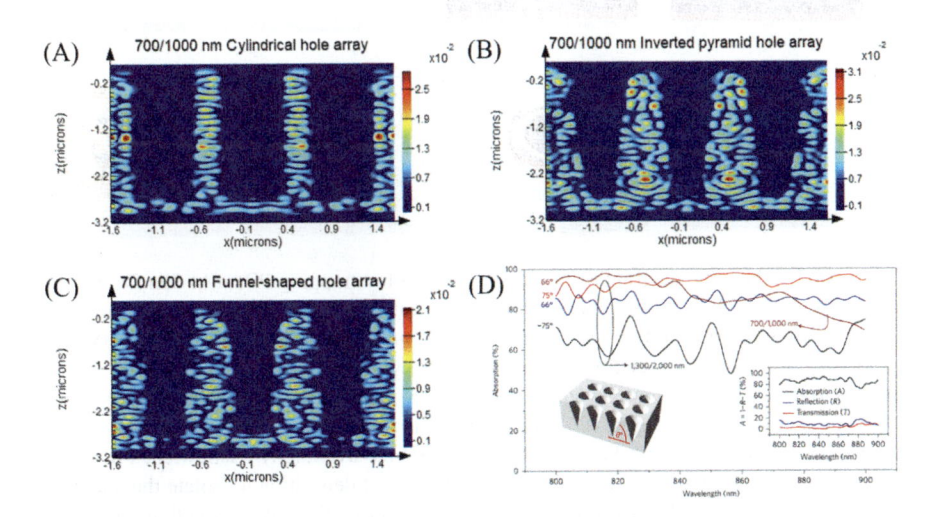

Figure 13.6 Optical confinement in (A) cylindrical, (B) inverted pyramid, and (C) funnel-shaped light-trapping structures (FDTD simulation), (D) optical absorption enhancement in funnel-shaped structure with different side-wall angles.

13.3.3 Fabrication

After evaluating the simulation results, 2D hexagonal and square periodic nanohole patterns with different diameters/periods (d/p) of 1300/2000 nm and 700/1000 nm were chosen to fabricate Si PDs. The bottom contact layer, an active region, and top contact were epitaxially grown using Chemical Vapor Deposition (CVD) method. To boost the speed of the device, p- and n-layer doped to the level of 5×10^{20} cm^{-3} (boron-doped) and 10^{19} cm^{-3} (As-doped) were grown to minimize the recombination time of diffused minority carriers in contact layers. The active layer (i-layer) was slightly unintentionally doped to n-type characteristics ($\leq 5 \times 10^{16}$ cm^{-3}) during the growth process.

The schematics of the fabrication processes of the photon-trapping photodiodes (PDs) are shown in Figure 13.7. There are two types of nanoholes (cylindrical and tapered) in our PDs design, and their fabrication processes are slightly different (Gao et al., 2017a; Liu et al., 2021; Ackert et al., 2015; Gao et al., 2017b; Yang et al., 2019; Hossain et al., 2018).

Deep reaction ion etching (DRIE) is a highly anisotropic etching process used to create high aspect ratio vertical holes, and it utilizes the Bosch process, which alternates between two modes to achieve holes with almost 90 degrees anisotropic etching (Sato, 2018). One mode is an anisotropic etching of Si by sulfur hexafluoride

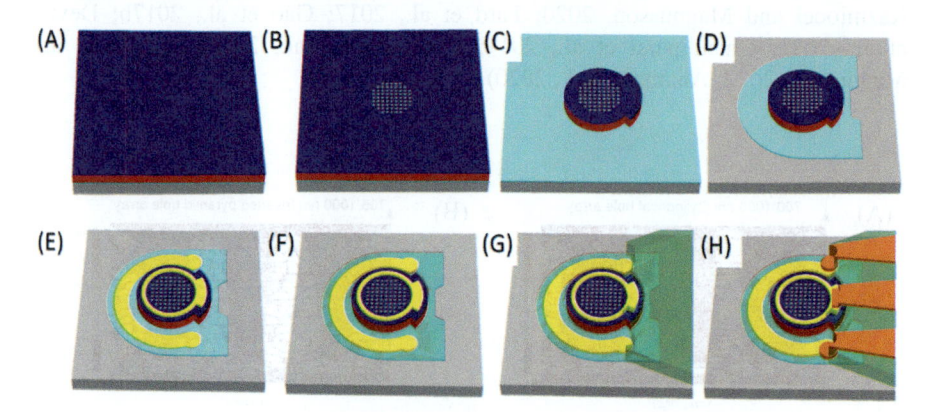

Figure 13.7 Schematic diagram of fabricating the photon trapping PDs. (A) Starting wafer (*gray*: SOI wafer substrate; turquoise: p-type layer, composed of 0.2 μm of SiGeB and 0.25 μm p-Si SOI device layer; *red*: 2 μm i-Si layer; *blue*: 0.2 μm n-Si layer). (B) DUV photolithography and holes etch to create tapered or cylindrical holes with diameters ranging from 600 to 1500 nm in a square (shown here) or hexagonal (not shown here) lattice. (C) N-mesa etches to p-Si layer. (D) P-mesa etches to the substrate layer. (E) Ohmic metal deposition (100 nm Al, 20 nm Pt) followed by HF dip passivation. (F) Sandwiched insulation layer (150 nm Si$_3$N$_4$/300 nm SiO$_2$/150 nm Si$_3$N$_4$) PECVD deposition to isolate the *n* and p mesas. (Semitransparent brownish layer represents this insulation layer on both *n*-mesa sidewall and the top surface of p-mesa with contacts opening). (G) Polyimide planarization (semitransparent green color). (H) Coplanar waveguides (CPWs) metal deposition (*brown color*).

(SF$_6$) gas, and the other mode is passivation/deposition by octafluorocyclobutane (C$_4$F$_8$) gas to protect the sidewalls (Figure 13.8A). Since the spacing between the holes in our PDs is very small (ranging from 270 to 700 nm), it is not sufficient to create tapered etched holes with microns deep by the DRIE method. The RIE process was employed to etch gradual funnel-shaped holes. In our RIE etching system, HBr and Cl$_2$ gases were used to etch Si, and it usually produces almost vertical holes with an angle of 83 − 87 degrees. To create a tapered sidewall profile, one can start with a near-vertical hole etch and then convert it into a tapered one after an additional mask-less etching step that eroded the silicon structure laterally while etching along the vertical direction (Figure 13.8B) (Cansizoglu et al., 2018a; Gao et al., 2017a, 2017b Fard et al., 2017; Laermer and Urban, 2003; Marty et al., 2005; Csutak et al., 2002; Bartolo-Perez et al., 2021; Cansizoglu et al., 2019).

After DRIE etching of holes and mesas, an extremely high leakage current was observed in our devices as shown in Figure 13.9A (*black curve*). This high leakage can be attributed to the electrically active surface sites arising from dangling bonds and defects created after the DRIE process. H (Hydrogen) termination of the Si surface is a well-known method to decrease the surface recombination velocity of Si. After dipping our devices in 1:10, HF: H$_2$O solution for 10 s, the leakage current was suppressed to the nA levels, whereas it was in the mA range. Slightly higher current in surface textured devices compared to the flat-top device after HF passivation can be attributed to the high surface area of the hole-based structures (Gao et al., 2017b; Mayet et al., 2018; Ahmed et al., 2016).

In addition to hydrogen passivation, several other methods have been applied to inhibit device degradation induced by surface damage from the dry etch, including thermal oxidation, subsequent oxide removal, and low ion energy etch. All the passivation methods lowered the leakage current drastically, and some of these methods even reduced the leakage current by more than four orders of magnitude, as shown in Figure 13.9A. Oxidation and hydrogen passivation with HF dip reduced

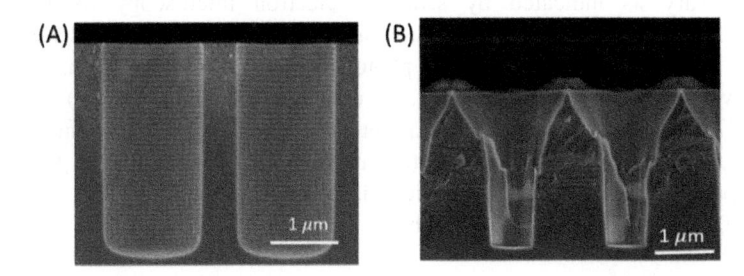

Figure 13.8 (A) SEM image of cross-sectional cylindrical holes etched by DRIE, showing holes of 1300 nm in diameter and 2000 nm in a period in a square lattice, (B) SEM image of the cross-section of the tapered holes by two steps RIE etch. The resist (*brown color*) on the spacing of the holes formed pyramid-shaped islands of less than 250 nm thick, and thus facilitated the lateral Si etch. The design of the holes is 1500 nm in diameter and 2000 nm in the period, and the holes are in a square lattice. The angle of the opening of the holes is around 60 degrees; the angle of the sidewall at the bottom side is around 84 degrees.

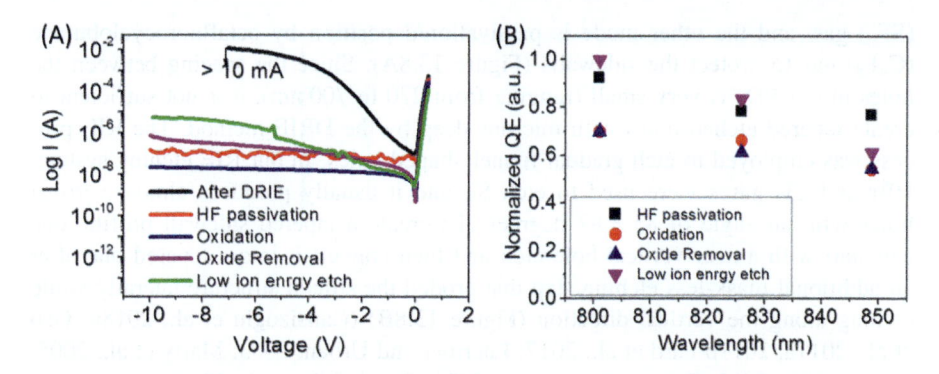

Figure 13.9 Comparison of different passivation methods applied for inhibiting leakage current. (A) Leakage current before and after hydrogen passivation, oxidation, oxide removal, and low ion energy etch. (B) Normalized QE was measured at three different wavelengths (800, 826, and 848 nm) after each passivation under 10 V reverse bias for PDs with cylindrical holes (d/p:1500/2000 nm).

the leakage current to the nA range, whereas it was around $\sim$ µA range after low ion energy etch. QE measurements were conducted using a supercontinuum laser and a tunable filter that transmits a band of wavelengths with a 1 nm width blocking the adjacent wavelength. For accurate calibration, we also used 5 discrete fibers coupled to lasers at 780, 800, 826, 848, and 940 nm. The highest efficiency of $\sim 65\%$ (-10 V) was achieved from a PD with hexagonally packed holes of 700 nm diameter and 1000 nm period at 850 nm (Mayet et al., 2018).

The doping profile shows that there are transition regions at both n^+ and p^+ boundaries with the i-layer. This contributed to a reduced effective absorbing medium of less than 2 µm. DRIE and RIE techniques were used to form cylindrical or funnel-shaped micro/nanoholes, respectively. The diameters and sidewall angles of holes vary as indicated by scanning electron microscopy (SEM) images (Figure 13.10A–D). After mesa isolation, aluminum (Al) ring contacts were deposited on n^+ and p^+ regions to facilitate photogenerated carrier collection. To conduct wafer-level high-speed measurements, Al coplanar waveguide (CPW) is designed with 50 Ω characteristic impedance. An optical image of the 30 µm diameter device that was used for DC and RF characterizations is depicted in Figure 13.10E. The doping profile of the fabricated devices is shown in Figure 13.10F (Gao et al., 2017a, 2017b; Cansizoglu et al., 2018b).

13.3.4 Experimental results

QE measurements were conducted using a supercontinuum laser and a tunable filter that transmits a band of wavelengths with a 1 nm width blocking the adjacent wavelength. For accurate calibration, we also used 5 discrete fibers coupled to lasers at 780, 800, 826, 848, and 940 nm, and delivered light to the devices by a SMF probe

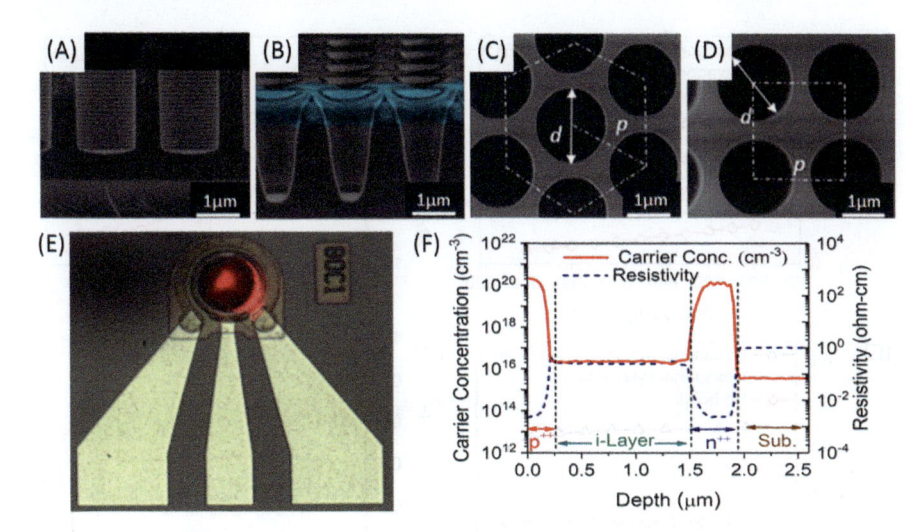

Figure 13.10 Micrograph picture of fabricated devices, (A) hexagonal pattern (*top*), (B) square pattern (*top*), (C) cylindrical (*cross-section*), (D) funnel-shaped (*cross-section*), (E) complete photodiode under laser illumination, (F) doping profile of fabricated devices.

on a probe station. The external quantum efficiency (EQE) characterization is extended to wavelengths that range from 800 to 1100 nm.

Measurement shows that absorption is enhanced for arrays with $d/p = 700/1000$ nm holes (larger number of holes per area) combined with the silicon dioxide layer that helps to confine light into the absorption layer. To compare the performance of the devices with and without micro/nanoholes, we fabricated control devices without holes in the same processing steps. The devices with micro/nanoholes have exhibited strong coupling of the vertically oriented incident plane waves into a multitude of lateral modes in the active region. The EQE of both flat and patterned ($d/p = 700/1000$ nm) devices is presented in Figure 13.11A. The devices exhibited more than 55% EQE at 850 nm wavelength for the PDs with funnel-shaped holes ($\theta = 61$ degrees) that are arranged in hexagonal arrays and more than 37% EQE for devices with cylindrical holes. By contrast, the devices without integrated micro/nanoholes exhibited less than 9% EQE. Therefore, a significant absorption enhancement of four times for cylindrical holes and six times for funnel-shaped holes is achieved compared to the devices with flat surfaces (Figure 13.11A, *blue* and *green* colors) (Cansizoglu et al., 2018b; Ghandiparsi et al., 2019, 2018).

To evaluate the absorption efficiency of the light-trapping structures the enhanced absorption coefficient (α_{eff}) is calculated via:

$$\alpha_{eff.} = -\ln(1-A)/L \tag{13.4}$$

Figure 13.11B compares the effective absorption coefficient (α_{eff}) of our devices with bulk silicon (αSi) and GaAs (αGaAs), which is the dominant material used in high-speed high-efficient PDs due to its direct bandgap. It implies that designed Si

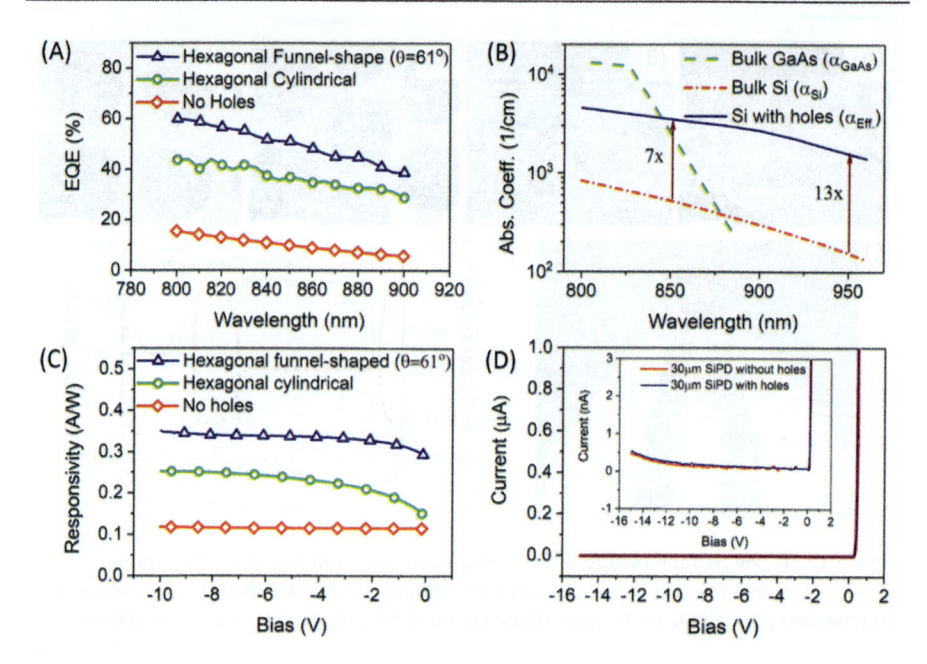

Figure 13.11 DC characterization of 30 μm devices, (A) external quantum efficiency (EQE), (B) absorption coefficient of bulk Si, GaAs, and fabricated device active region, (C) responsivity, (D) I−V characteristics (*dark current*).

PIN PDs integrated with surface nanohole arrays have a broader absorption spectrum compared to GaAs-PD, which makes the device suitable for transceivers design for short wavelength division multiplexing. Moreover, to reach the same optical absorption (55%) in PDs with flat surfaces, 13 times thicker absorbing region is required. Responsivity—the capability of Si PIN PD to convert incident illumination power into electrical current—as a function of reverse bias is shown in Figure 13.11C for $\lambda = 850$ nm. Due to higher optical absorption in the devices with integrated micro/nanohole array, the responsivity is enhanced to 0.33 A/W (at 3 V) compared to 0.11 A/W exhibited by flat devices. We also observed that Figure 13.11C the responsivity is almost voltage-independent for higher reverse bias after 3 V.

The results of the DC measurement of the fabricated devices with 30 μm diameters on the same wafer with identical physical parameters are shown in Figure 13.11D. The Current-Voltage (I-V) measurement was done using a semiconductor parameter analyzer (Agilent 4156b). The devices with and without micro/nanoholes were measured with less than 1 nA dark current. However, the higher dark current of surface patterned devices is due to the surface damage that occurred during the DRIE process. The series resistance of 30 μm devices with surface nanoholes was measured to be ∼83 Ω using I−V characteristics in the forwarding bias regime.

Wafer-level RF measurement was conducted using a microwave probe (GSG, Cascade) for devices with a 30 μm diameter containing hexagonal holes arrays with d/p = 700/1000 nm. A mode-locked pulsed laser (Calmar Laser Inc.) with 850 nm center frequency, 20 MHz duty-cycle, and the sub-picosecond (200 fs) pulse width was used on a probe station as a surface illuminating light source. The FWHM of the laser pulse delivered to the PD surface was increased to ~5 ps due to propagation through a few meters of fiber. The high-speed measurement setup is presented in Figure 13.12. The light beam was aligned close to normal to the surface to reach maximum photocurrent. The device output signal was recorded by a sampling oscilloscope with a 3 dB bandwidth of 20 GHz (DSA8300, Tektronix) (Ghandiparsi et al., 2019, 2018).

Using the RF measurement setup, the impulse (narrow Gaussian pulse with FWHM < 5 ps) response of a 30 μm device with the highest responsivity (0.35 A/W) is measured at different biases. Figure 13.13A shows the normalized impulse response at −3 V (*green*), −10 V (*blue*), and −15 V (*red*), respectively. Although there is a small difference in FWHM at different biases, the fall time (tail) changes considerably with bias. The fall time strongly depends on the slow carriers dynamic, which differs with applied bias. By increasing the reverse bias from −3 V to −15 V, the depletion region covers a larger part of the absorbing region and is enabled to extract the photogenerated carriers more efficiently. To understand the dynamics of the fabricated PD more in detail it is needed to find the de-convolve impulse response from laser and scope effect.

Applying a computer algorithm, we generated a PD model that matches the impulse response at each bias. Later, using an extrapolation tail mechanism, the PD fall time is fitted -to reach a mathematical explanation and remove the effect of laser- and added to the PD model to generate the deconvolved impulse response. Figure 13.13B−D demonstrates the impulse response and deconvolved impulse response by eliminating the laser and scope effect at −3, −10, and −15 V, respectively.

Deconvolved impulse responses convolved with 2E10 random bits at a different frequency (symbol rate) and passed through a 3rd order Butterworth filter with a 0.75 × Bit rate.

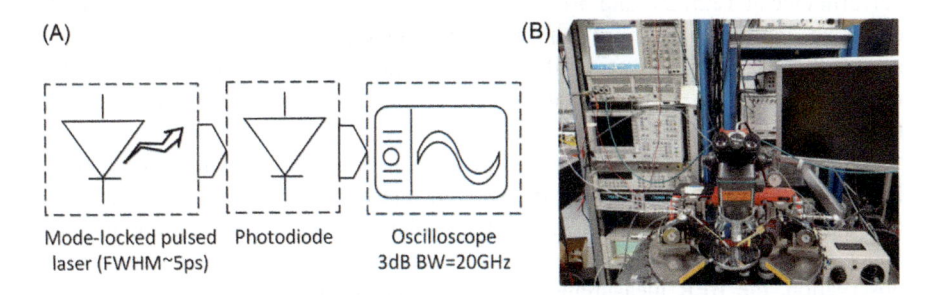

Figure 13.12 (A) Block diagram of high-speed measurement setup including mode-locked pulsed laser, photodiode, and scope, (B) lab. experimental setup for RF and BER measurement.

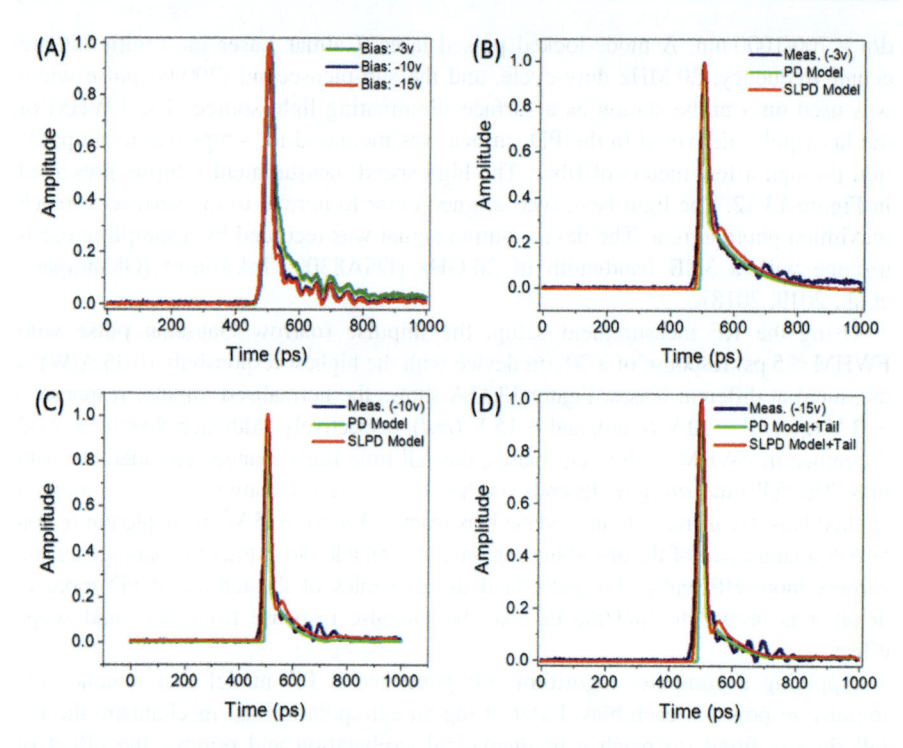

Figure 13.13 (A) Impulse response measurement of 30 μm device at −3, −10, and −15 V, deconvolved impulse response and effect of laser-scope with extrapolated tail, (B) −3 V, (C) −10 V, (D) −15 V.

To evaluate the high-speed performance of fabricated PD the simulated eye diagrams using measured impulse responses are shown in Figure 13.14. For commercial applications, the standard bias is around 3 V. As is shown in Figure 13.14A−C, the fabricated PD is capable of handling up to 8 Gb/s and with some modifications could reach 10 Gb/s. Figure 13.14D−F demonstrate the PD performance at 12.5, 25, and 50 Gb/s under −10 V bias. It can realize that the PD could be applicable in more than 12.5 Gb/s bit rates. Figure 13.14G−I demonstrates the PD operation at similar bit rates under −15 V bias. The PD performance improved at −15 V bias due to the extended depletion region compared to the −10 V bias.

To evaluate the device's high-speed performance, a measurement setup including Bit Error Rate Test, commercial transceiver, and digital sampling scope is developed. Measured eye diagrams for a 30 μm diameter Si PD at different bit rates under various reverse biases are shown in Figure 13.15. Based on our previous observation, the BER measurements confirm that the PD performance is strongly dependent on the bias and depletion region thickness, consequently. The Si PD measurements at 1 Gb/s under −3, −10, −15, and −15 V biases, are shown in Figure 13.15A−D.

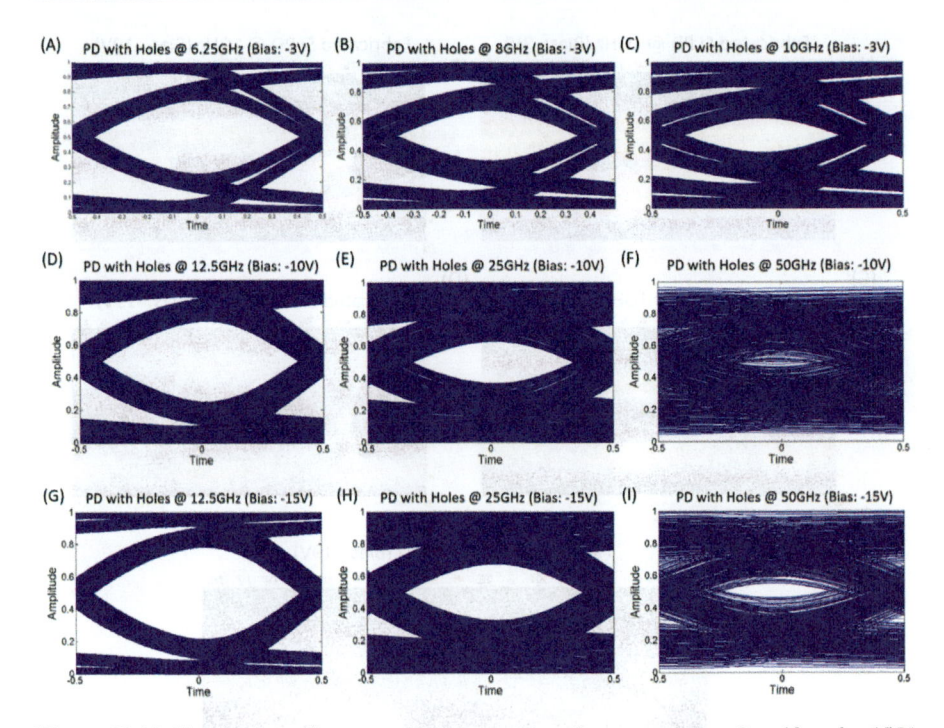

Figure 13.14 Simulated eye diagrams using measured impulse responses at -3, -10, and -15 V.

It can be seen that the device's high-speed performance is dependent on voltage (as discussed before). At -3 V, the PD could not show more than 1 GHz bandwidth. To realize the best performance the reversed bias reached -19 V, where we made sure that the absorbing region is completely depleted. A measured eye diagram at 5 Gb/s is shown in Figure 13.15E. Although the eye diagram confirms the device is not suitable for 5 Gb/s and higher, it shows a promising result that with several materials and fabrication modifications reaching higher bandwidth is quite feasible.

13.3.5 Equivalent photodiode model (system verification)

The PD model (Figure 13.16) includes a photocurrent generation mechanism block and the equivalent circuit.

To obtain accurate Si PIN PD model parameters the effect of the on-wafer probe has been included in the equivalent circuit model. Therefore, PD equivalent circuit consists of junction capacitance (C_j), series resistance (R_s), and large shunt resistance (R_j); it is connected to bonding capacitance (C_b) and 50 Ω load resistance (R_L). The series resistance and combination of junction-bonding capacitors ($C_j + C_b$) are obtained from the device I-V (forward bias) and C-V (reverse bias) measurements, respectively.

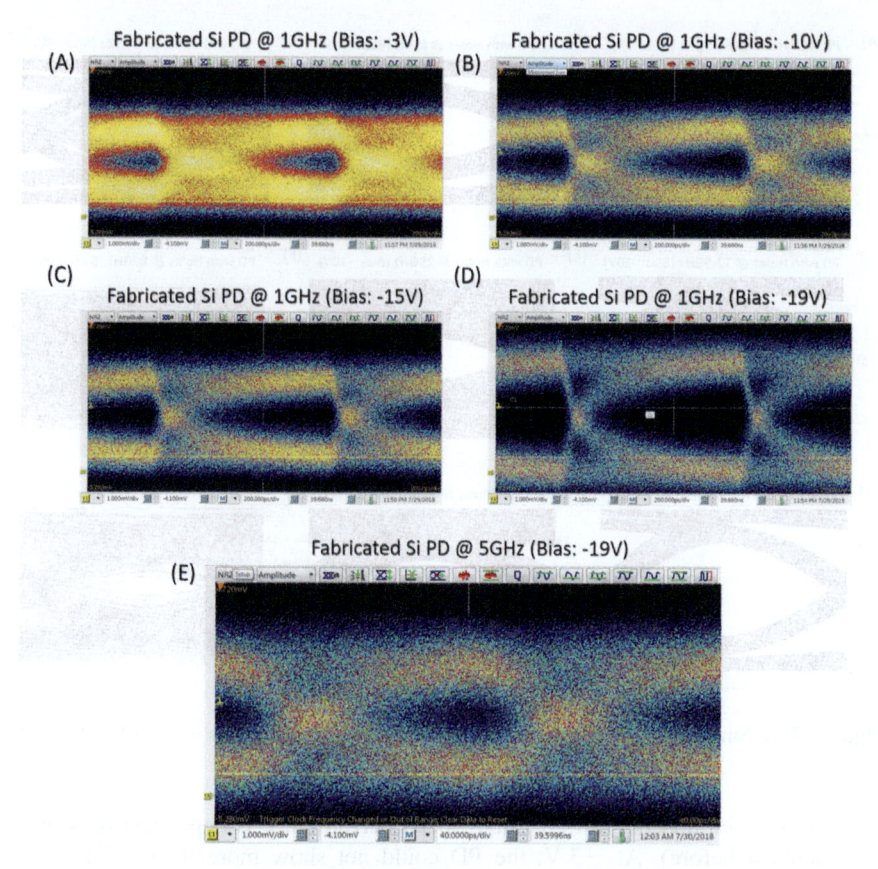

Figure 13.15 Eye-diagram measurement of fabricated 30 μm Silicon PD at (A) 1 GHz under −3 V bias, (B) 1 GHz under −10 V bias, (C) 1 GHz under −15 V bias, (D) 1 GHz under −19 V bias, (E) 5 GHz under −19 V bias.

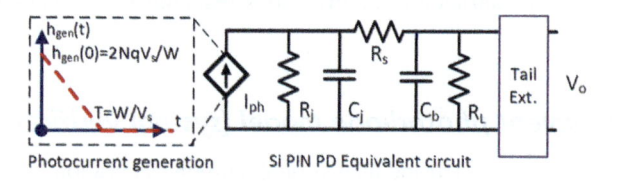

Figure 13.16 Photodetector circuit model, including the photocurrent generation process and the PD equivalent circuit. The circuit parameters are extracted from the I-V and C-V measurements.

Photocurrent generation in the depletion region has been modeled based on the assumption of uniform electron-hole generation along the depletion region with a triangular impulse response $h_{gen}(t)$ versus time (Thomson et al., 2016) ($h_{gen}(0) = 2NqV_s/W$)

where N is electron-hole count, q is the electron charge, and V_s is the generated carriers saturation velocity in depletion region (W).

Capacitance measurement (at 1 MHz) was conducted for devices with various diameters (20–80 μm) using a high precision LCR meter (Agilent 4284A). The applied electric field over device contacts controls the thickness of the depletion region formed in the i-layer. Therefore, lower PD capacitance in higher reverse voltages is the consequence of increased depletion region.

In order to find the individual equivalent circuit capacitances (C_j and C_b) and the depletion region width (W), an extrapolation method was employed to calculate C_j and C_b using capacitance measurements of devices with various areas (A) at a given voltage. Assuming that the surface nanostructure has an insignificant effect on the device capacitance, the following equation reflects the linear interpolation of the measured capacitance area:

$$C = C_j + C_b = \left(\varepsilon_{Si}\frac{A}{W}\right) + C_b \tag{13.5}$$

Where ε_{Si} is the silicon dielectric constant. The linear interpolation slope (ε_{Si}/W) and intercept (C_b) are presented in Figure 13.17 for 3, 10, and 15 V biases.

The device equivalent circuit parameters and the depletion region thickness extracted from I-V and C-V measurements are presented in Table 13.1 for 30 μm devices.

Unintended dopant diffusion from n^+ and p^+ regions to the active layer during epitaxial growth caused a smaller i-layer thickness relative to the original design value of 2 μm. An optimized growth process to ensure abrupt junctions (p-i and i-n) can inhibit the dopant diffusion issue and effectively restore the active region to 2 μm.

The measured impulse response of the 30 μm device is recorded in the 800 ps (0.8 ns) window with a 1 ps (1 GHz) resolution shown in Figure 13.18A. Due to the

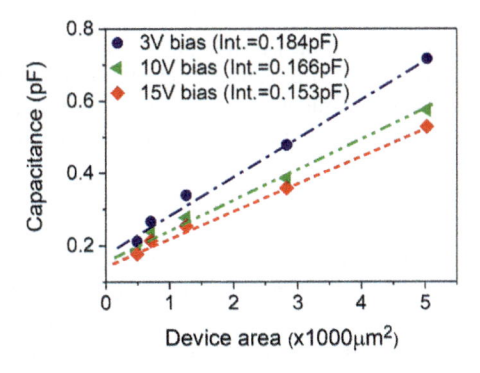

Figure 13.17 Extraction of junction (C_j) capacitance, bonding (C_b) capacitance, depletion region width (W) using a linear interpolation of devices with different capacitance at certain voltages, 3, 10, and 15 V.

Table 13.1 Measured 30 μm device parameters versus bias.

Bias	R_s (Ω)	C_j (pF)	C_b (pF)	W (μm)
3 V	83	0.084	0.18	0.9
10 V	83	0.075	0.17	1.3
15 V	83	0.06	0.15	1.4

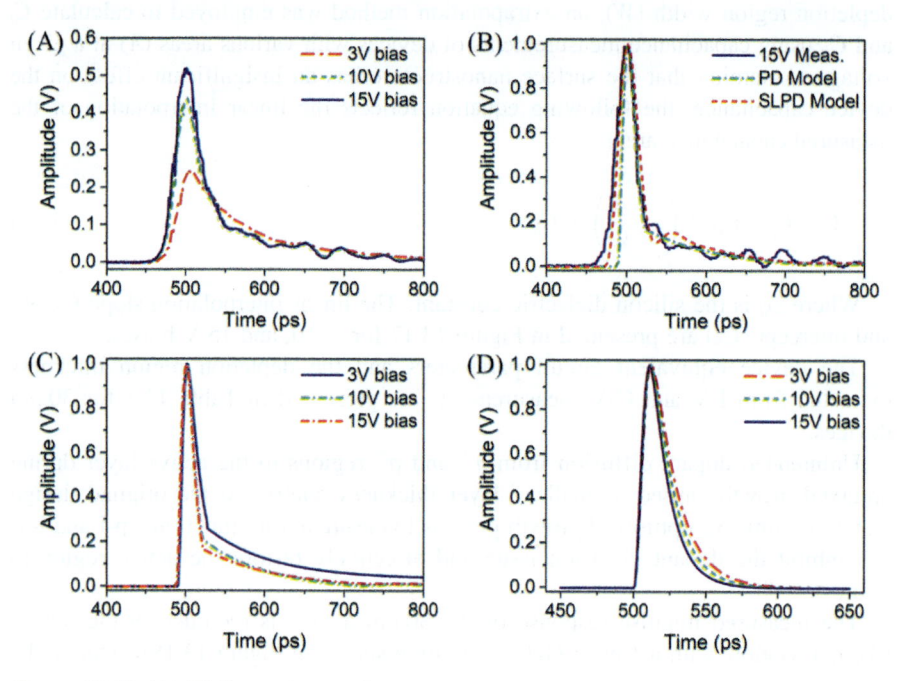

Figure 13.18 (A) High-speed on-wafer measurement using a mode-locked pulsed laser at 850 nm with FWHM = 5 ps fabricated PD with CPW contacts, sampling oscilloscope with a 3 dB bandwidth equal to 20 GHz. (B) Measured impulse response at different voltage biases. FWHM is measured at 29 ps (*navy*), 31 ps (*green*), and 34 ps (*red*) at 15, 10, and 3 V, respectively. (C) Comparison of measured (*navy*), PD model (*green*), and Scope-PD-laser (SLPD) (*red*) impulse response at 15 V, (D) PD model at 3 V (*red*), 10 V (*green*), and 15 V (*navy*).

variation of depletion region width at different biases, as discussed previously, the impulse response is also voltage-dependent. The FWHM of Si PIN PD at 3, 10, and 15 V biases were measured to be 34, 31, and 29 ps. Based on the design considerations, the tail followed by fall time was expected to be much smaller than the observed results. However, the additional tail after the pulse fall time is due to the slow diffusion of photogenerated minority carriers at the p-i and i-n interfaces caused by unintended dopant diffusion, as was described earlier.

The effect of the measured unexpected extra tail is imported to the PD model using a nonlinear fitting algorithm (Figure 13.18B, *green color*). Also, the effect of the measurement equipment, including sampling oscilloscope (3 dB bandwidth: 20 GHz modeled as a third-order Butterworth filter) and pulsed optical laser (FWHM: 5 ps) on the measurement was combined with the PD model to form the Scope-Laser-PD model (SLPD model). There is a strong agreement between the SLPD model (Figure 13.18B, *red color*) and measured impulse response (Figure 13.18B, *navy color*).

The deconvolved impulse response for the fabricated 30 μm device at 15, 10, and 3 V are shown in Figure 13.18C with FWHM of ∼19, ∼21, and ∼23 ps, respectively. The corresponding 3 dB bandwidth is calculated to be 3.5 GHz (15 V), 3 GHz (10 V), and 2 GHz (3 V). The impulse response for future devices designed with abrupt junctions is shown in Figure 13.18D with similar FWHM to the current devices. A considerably higher 3 dB bandwidth of 13 GHz (15 V), 11 GHz (10 V), and 10 GHz (3 V) due to a smaller tail.

The performance of an end-to-end optical link (Figure 13.19), including the PD models derived in Figure 13.18D, has been evaluated using computer simulations. The optical link is comprised of a transmitter (T_x), MMF, and a Receiver (R_x) unit. The T_x block includes a random bit source (RBS), pulse-shaping filter, 2-tap pre-emphasis feed-forward equalizer (FFE), and an MM-VCSEL. A train of 4000 random NRZ (Non-Return to Zero) or PAM-4 (Four-Level Pulse Amplitude Modulation) is generated (by RBS) and convolved with pulse shaping low-pass filter $T_r = 12$ ps (is 10%–90% rise time) and with MM-VCSEL with 3 dB bandwidth of 25 GHz. Both pulse shaping filters and VCSEL filters are modeled as Gaussian filters.

On the receiver (Rx) side, the VCSEL output is convolved with the impulse response of the Si PIN PD models at different biases. The output of the PD model is applied to a trans-impedance amplifier (TIA), which is modeled as a third-order Butterworth filter with a 3 dB bandwidth of 0.75 × Bit Rate (Siew et al., 2021). Further, a de-emphasis of FFE was implemented in the optical link to overcome the channel bandwidth limits (Ghandiparsi et al., 2019; Afzal et al., 2021, 2022; Li et al., 2017, 2019; Ozolins et al., 2017; Marufuzzaman et al., 2018; Szilagyi et al., 2019; Ahmed et al., 2018).

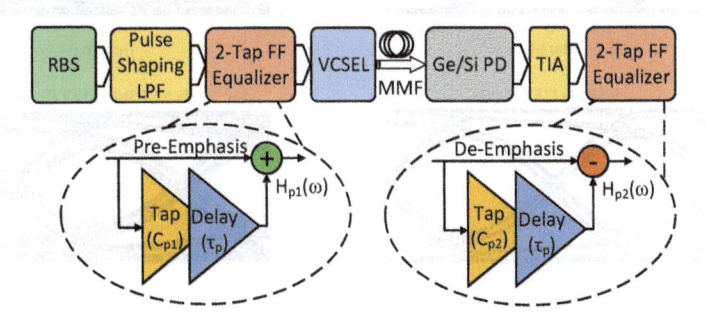

Figure 13.19 The optical link includes the transmitter unit, an optical fiber transmission line, and the receiver unit.

The performance of the fabricated device (at 3, 10, and 15 V biases) in a full optical link (discussed above) was evaluated at the receiver output without equalizers. Considering the PD model at 3 V bias the eye diagrams for 4 and 8 Gb/s with 0.78 and 0.47 eye-opening are shown in Figure 13.20A and B. Applying higher voltages effectively improves the vertical eye-opening to 0.81 and 0.61 at 6 and 10 Gb/s (Figure 13.20C and D), 0.65 and 0.54 at 8 and 12.5 Gb/s (Figure 13.20E and F) for 10 and 15 V biases, respectively.

The random bit source rise time, VCSEL laser limited bandwidth (25 GHz), and the PD bandwidth limitations are the most effective pulse distortion sources for the optical link considered in this simulation.

A set of 2-tap pre/de-emphasis FFEs are employed in both the transmitter and receiver ends to improve the signal quality. The pair of 2-tap preemphasis FFE is chosen in the way that the input pulse peak to the VCSEL driver is approximately equal to the original pulse peak without equalization (Li et al., 2017, 2021; Afzal et al., 2022; Badal et al., 2019). The transfer functions of the 2-tap pre/de-emphasis FFE are given by:

$$H_{p-pre}(\omega) = \frac{1 + C_{p1} \, e^{-j\omega\tau_p}}{1 + C_{p1}} \tag{13.6}$$

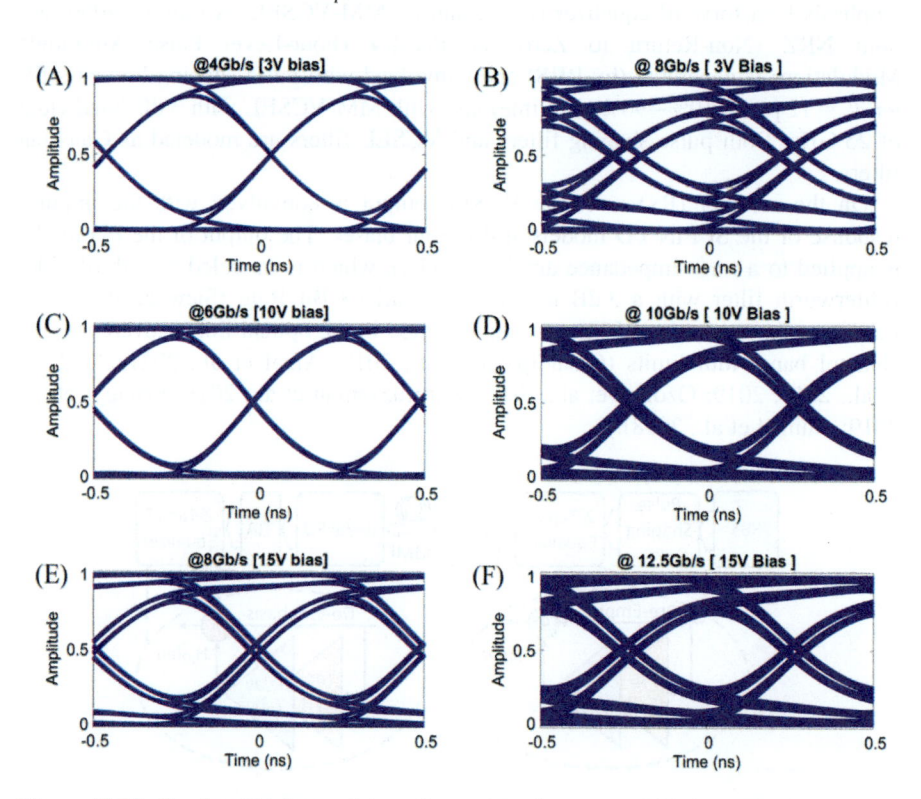

Figure 13.20 Simulated high-speed (eye-diagram) performance of fabricated devices using the PD equivalent circuit model.

$$H_{p-pre}(\omega) = \frac{1 - C_{p1}\, e^{-j\omega \tau_p}}{1 - C_{p1}} \tag{13.7}$$

where C_{p1} and C_{p2} are the tap coefficients and t_p is the tap delay.

For the 50 Gb/s bit rate, the tap delay (t_p) was chosen to be 10 ps (1/2 × bit time) to obtain equalized eye diagrams at the filter output with impulse response at 3 V bias. To obtain the maximum eye-opening, a series of simulations are conducted to reach the optimum tap coefficients. Using equal tap coefficients $C_{p1} = C_{p2} = 0.7$, the vertical opening of ∼0.67 (Figure 13.21A) was reached. By taking larger tap coefficients ($C_{p1} = C_{p2} = 0.73$), a reasonable vertical eye-opening of ∼0.74 has been achieved for the impulse response at 10 V bias (Figure 13.21B). Applying the impulse response at 15 V bias (FWHM = 19 ps) and optimized equalizer parameters ($t_p = 10$ ps, $C_{p1} = C_{p2} = 0.72$) the simulated eye diagram reached ∼0.81 (Figure 13.21C) vertical opening. Furthermore, for 100 Gb/s PAM-4-bit rate the vertical eye-opening of ∼0.56 (Figure 13.21D) was obtained using the optimized pre/de-emphasis equalizer ($t_p = 5$ ps, $C_{p1} = C_{p2} = 0.76$) (Ghandiparsi et al., 2019; Kim and Buckwalter 2012; Morero et al., 2016; Lambrecht et al., 2019).

13.3.6 Discussion and future roadmap

There is a discrepancy between the designed *pin* structure and measured fabricated device parameters. In this section, a complete theoretical and numerical study is conducted in order to address the longer fall time (in terms of bandwidth) and lower bandwidth of fabricated devices. Theoretically, a long fall time of the impulse response (Figure 13.22A) is the result of slow carriers' diffusion from a part of an

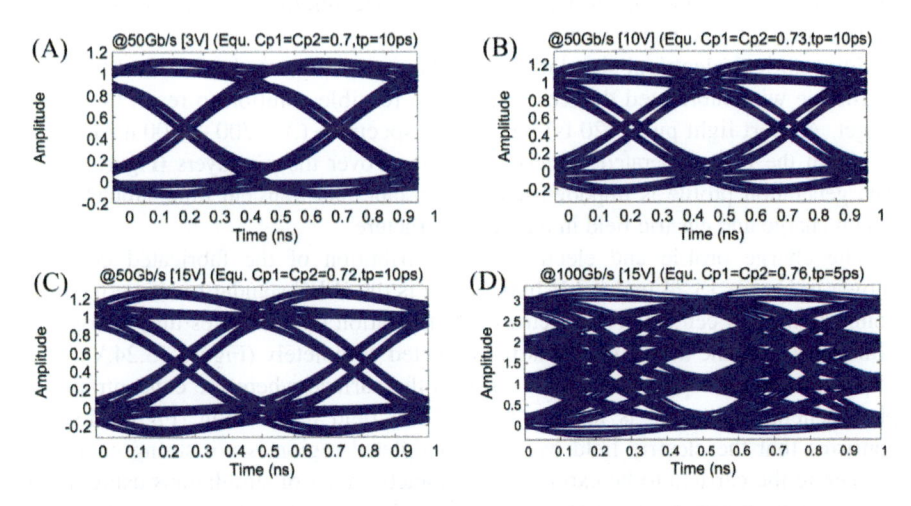

Figure 13.21 Link equalized eye diagrams with 2-tap pre/de-emphasis FFE for future devices with an abrupt junction.

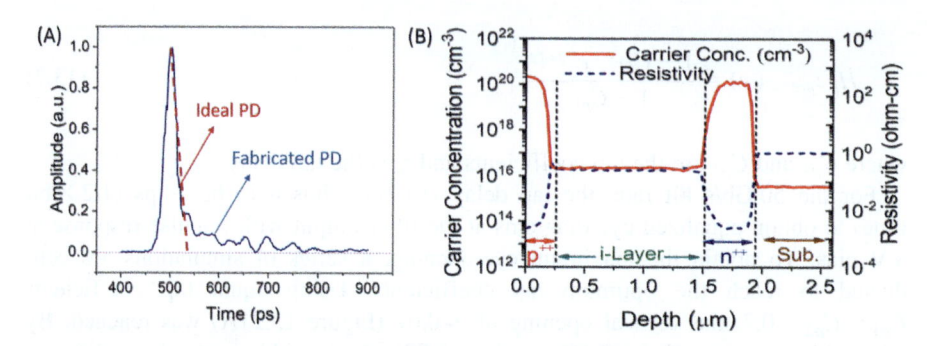

Figure 13.22 Compare fabricated PD impulse with design targets; (A) fabricated PD demonstrated long fall time due to slow carriers' diffusion from the active layer non-depleted region, (B) fabricated device doping profile.

active region that is not depleted completely under a certain reverse bias (3 V). Therefore, as mentioned before, we had to increase the reverse bias to higher levels (say 15 V) to reach higher bandwidth by pushing the borders of the depletion region.

To reach a desirable structure that can satisfy our goals, a set of effective parameters such as layers' doping profile (Figure 13.22B), material characteristics, and physical properties are used to solve diffusion equation 3.5 in a 30 μm device.

$$\frac{\partial n_p}{\partial t} = n_p \mu_n \frac{\partial E}{\partial x} + \mu_n E \frac{\partial n_p}{\partial x} + D_n \frac{\partial^2 n_p}{\partial x^2} + G_n - \frac{n_p - n_{p0}}{\tau_n} \qquad (13.8)$$

Where n_p, μ_n, D_n, and G_n are minority carrier concentration, carriers' mobility, diffusion constant, and generation rate, respectively.

Joint optical-electrical simulations of 30 μm are conducted to (a) find the exact challenge with fabricated devices and offer a feasible solution to reach the design target. A short light pulse (20 fs) with broad-spectrum ($\lambda = 700-1100$ nm) is used to obtain the photogenerated carriers' distribution over the *pin* layers (Figure 13.23). The generation profile is imported into electrostatic and transient time simulations to obtain charge and electric field in the device structure.

The charge profile and electric field distribution of the fabricated device is shown in Figure 13.24 under 3 V bias, which is the IEEE standard voltage for interconnects in data centers. The electrostatic simulation demonstrates that a large part (almost 2/3) of the active region is not depleted completely (Figure 13.24A). More important, the undepleted region is close to the surface where the concentration of photogenerated carriers is higher. The electric field distribution (Figure 13.24B) confirms that the electric field over the undepleted region is not strong enough to accelerate the carriers to be extracted by contacts. A set of simulations using different parameters is conducted to reach the maximum depleted region in the active layer and high electric field, consequently. We found that by reducing the i-layer

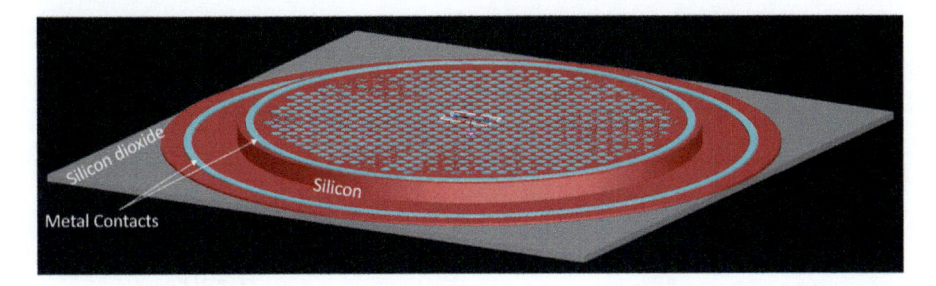

Efficient surface nano-textured CMOS-compatible photodiodes for Optical Interconnects   461

Figure 13.23 A complete 30 μm optical and device simulation setup. Using the optical simulation, photocarrier generation in pin layers using a pulsed plane wave is calculated. The photocarriers' distribution was imported into a device simulation setup to calculate electrostatic and transient time simulations.

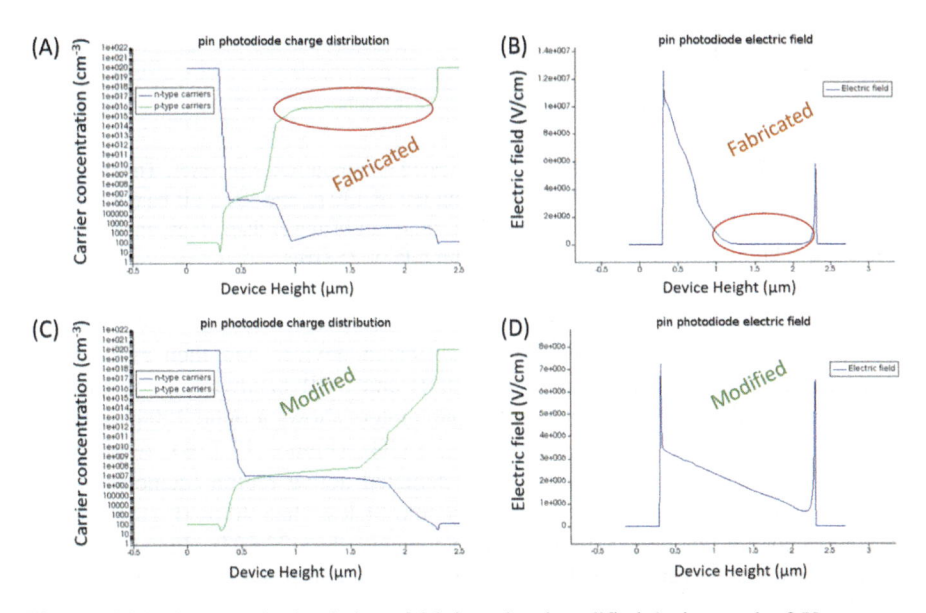

Figure 13.24 Electrostatic simulation of fabricated and modified devices under 3 V reverse bias. By lowering i-layer doping, the depletion region covered the whole absorbing region.

doping to 2 orders lower than the fabricated device, the depletion regions at both sides (i-n and i-p junctions) would extend and cover the whole absorbing region.

In the modified structure, the charge simulation (Figure 13.24C and D) shows less than 15% of the absorbing region is not completely depleted and carrier concentration in the depleted region lowered to less than 10^7 (cm^{-3}). The optimum design could reach more than 10^6 V/cm all over the absorbing region which is very close to the required electric field and capable of accelerating generated carriers up to saturation velocity.

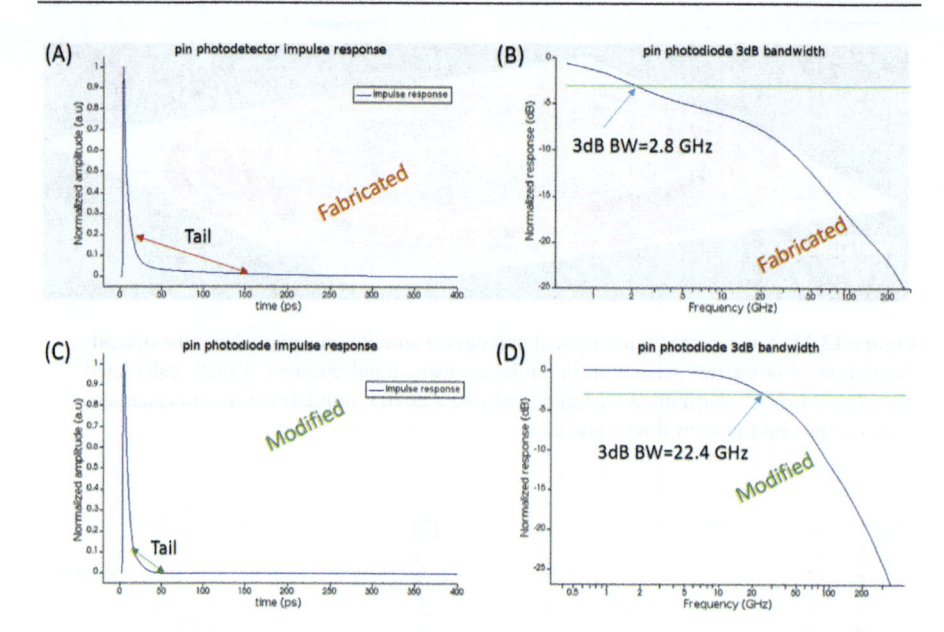

Figure 13.25 Transient time simulation of fabricated and modified structures. The impulse response of the fabricated device shows: (A) long tail of 130 ps and (B) 3 dB bandwidth of 2.8 GHz. The modified device could reach (C) a short tail of fewer than 30 ps, and (D) a 3 dB bandwidth of 22.4 GHz that guarantees a bit rate of 25 GB/s.

Next step, transient time simulations were conducted to confirm the calculated results with the measured results from fabricated devices and then evaluate the modified structure in terms of fall time (tail) and 3 dB bandwidth. The structure that is simulated based on fabricated PD parameters shows around 120 ps fall time (Figure 13.25A) and 3 dB bandwidth of 2.8 GHz (Figure 13.25B), respectively. The modified device due to its complete depleted active region (demonstrated in the last section) shows a very short falling time of less than 30 ps (Figure 13.25C) that elevated the device's 3 dB bandwidth to more than 22 GHz (Figure 13.25D). This is a great achievement that confirms the device can reach our target bit rate of 25 GB/s.

References

Ackert, J.J., et al., 2015. High-speed detection at two micrometres with monolithic silicon photodiodes. Nat. Photonics 9 (6), 393–396.

Afzal, H., et al., 2021. An mm-wave scalable PLL-coupled array for phased-array applications in 65-nm CMOS. IEEE Trans. Microw. Theory Tech. 69 (2), 1439–1452.

Afzal, H., Li, C., Momeni, O., 2022. A 17 Gb/s 10.7 pJ/b 4FSK transceiver system for point to Point communication in 65 nm CMOS. 2022 IEEE Radio Frequency Integrated Circuits Symposium (RFIC). IEEE.

Ahamad, A., et al., 2020. Smart nanophotonics silicon spectrometer array for hyperspectral imaging. Conference on Lasers and Electro-Optics. Optica Publishing Group, Washington, DC.

Ahmed, S.M., et al., 2016. Inhibiting device degradation induced by surface damages during top-down fabrication of semiconductor devices with micro/nano-scale pillars and holes. Proc. SPIE.

Ahmed, M.G., et al., 2018. A 12-Gb/s −16.8-dBm OMA sensitivity 23-mW optical receiver in 65-nm CMOS. IEEE J. Solid-State Circuits 53 (2), 445−457.

Badal, M.T.I., et al., 2019. Advancement of CMOS transimpedance amplifier for optical receiver. Trans. Electr. Electron. Mater. 20 (2), 73−84.

Bartolo-Perez, C., et al., 2021. Maximizing absorption in photon-trapping ultrafast silicon photodetectors. Adv. Photonics Res. 2 (6), 2000190.

Beling, A., Campbell, J.C., 2014. High-Speed photodiodes. IEEE J. Sel. Top. Quantum Electron. 20 (6), 57−63.

Ben Yoo, S., Proietti, R., Grani, P., 2018. Photonics in data centers. Optical Switching in Next Generation Data Centers. Springer, pp. 3−21.

Bhattacharya, S., John, S., 2020. Photonic crystal light trapping: beyond 30% conversion efficiency for silicon photovoltaics. APL Photonics 5 (2), 020902.

Bilal, K., et al., 2014. Trends and challenges in cloud datacenters. IEEE Cloud Comput. 1 (1), 10−20.

Cansizoglu, H., et al., 2018a. A new paradigm in high-speed and high-efficiency silicon photodiodes for communication—Part II: device and vlsi integration challenges for low-dimensional structures. IEEE Trans. Electron. Devices 65 (2), 382−391.

Cansizoglu, H., et al., 2018b. A new paradigm in high-speed and high-efficiency silicon photodiodes for communication—Part I: enhancing photon−material interactions via low-dimensional structures. IEEE Trans. Electron. Devices 65 (2), 372−381.

Cansizoglu, H., et al., 2019. Dramatically enhanced efficiency in ultra-fast silicon MSM photodiodes via light trapping structures. IEEE Photonics Technol. Lett. 31 (20), 1619−1622.

Casalino, M., et al., 2016. State-of-the-art all-silicon sub-bandgap photodetectors at telecom and datacom wavelengths. Laser Photonics Rev. 10 (6), 895−921.

Cesar, B.-P., et al., 2021. Controlling the photon absorption characteristics in avalanche photodetectors for high resolution biomedical imaging. Proc. SPIE.

Chen, X., et al., 2018. The emergence of silicon photonics as a flexible technology platform. Proc. IEEE 106 (12), 2101−2116.

Csutak, S.M., et al., 2002. CMOS-compatible high-speed planar silicon photodiodes fabricated on SOI substrates. IEEE J. Quantum Electron. 38 (2), 193−196.

Cyriac, S.L., et al., 2022. Emerging trends in nano structured silicon detectors for neutron spectroscopy. Silicon 14 (4), 1331−1337.

Devine, E.P., et al., 2021. Single microhole per Pixel in CMOS image sensors with enhanced Optical sensitivity in near-infrared. IEEE Sens. J. 21 (9), 10556−10562.

El Srouji, L., et al., 2022. Photonic and optoelectronic neuromorphic computing. APL Photonics 7 (5), 051101.

Eriksson, T.A., et al., 2017. 56 Gbaud probabilistically shaped PAM8 for data center interconnects. 2017 European Conference on Optical Communication (ECOC). IEEE.

Fard, M.M.P., et al., 2017. High-speed grating-assisted all-silicon photodetectors for 850 nm applications. Opt. Express 25 (5), 5107−5118.

Gao, Y., et al., 2017a. Photon-trapping microstructures enable high-speed high-efficiency silicon photodiodes. Nat. Photonics 11 (5), 301−308.

Gao, Y., et al., 2017b. High speed surface illuminated si photodiode using microstructured holes for absorption enhancements at 900–1000 nm wavelength. ACS Photonics 4 (8), 2053–2060.

Garlinska, M., et al., 2020. From mirrors to free-space optical communication—historical aspects in data transmission. Future Internet 12 (11), 179.

Gerstel, O., et al., 2012. Elastic optical networking: a new dawn for the optical layer? IEEE Commun. Mag. 50 (2), s12–s20.

Ghandiparsi, S., et al., 2019. High-speed high-efficiency photon-trapping broadband silicon PIN photodiodes for short-reach optical interconnects in data centers. J. Lightwave Technol. 37 (23), 5748–5755.

Ghandiparsi, S., et al., 2018. High-speed high-efficiency broadband silicon photodiodes for short-reach optical interconnects in data centers. 2018 Optical Fiber Communications Conference and Exposition (OFC). Optica Publishing Group.

Hassan, W.H., 2019. Current research on internet of things (Iot) security: a survey. Computer Netw. 148, 283–294.

Hossain, M., et al., 2018. Transparent, flexible silicon nanostructured wire networks with seamless junctions for high-performance photodetector applications. ACS Nano 12 (5), 4727–4735.

Huo, N., Konstantatos, G., 2018. Recent progress and future prospects of 2D-based photodetectors. Adv. Mater. 30 (51), 1801164.

Isabella, O., et al., 2018. Advanced light trapping scheme in decoupled front and rear textured thin-film silicon solar cells. Sol. Energy 162, 344–356.

Jalali, B., Fathpour, S., 2006. Silicon photonics. J. Lightwave Technol. 24 (12), 4600–4615.

Kim, J., Buckwalter, J.F., 2012. A 40-Gb/s optical transceiver front-end in 45 nm SOI CMOS. IEEE J. Solid-State Circuits 47 (3), 615–626.

Kyomasu, M., 1995. Development of an integrated high speed silicon PIN photodiode sensor. IEEE Trans. Electron. Devices 42 (6), 1093–1099.

Laermer, F., Urban, A., 2003. Challenges, developments and applications of silicon deep reactive ion etching. Microelectron. Eng. 67, 349–355.

Laghari, A.A., et al., 2022. A review and state of art of internet of things (IoT). Arch. Comput. Methods Eng. 29 (3), 1395–1413.

Lambrecht, J., et al., 2019. 90-Gb/s NRZ optical receiver in silicon using a fully differential transimpedance amplifier. J. Lightwave Technol. 37 (9), 1964–1973.

Liu, J., Cristoloveanu, S., Wan, J., 2021. A review on the recent progress of silicon-on-insulator-based photodetectors. Phys. Status Solidi (A) 218 (14), 2000751.

Li, Z., et al., 2017. Investigation on the equalization techniques for 10G-class optics enabled 25G-EPON. Opt. Express 25 (14), 16228–16234.

Li, D., et al., 2019. Low-noise broadband CMOS TIA based on multi-stage stagger-tuned Amplifier for high-speed high-sensitivity optical communication. IEEE Trans. Circuits Syst. I: Regul. Pap. 66 (10), 3676–3689.

Li, H., et al., 2021. 11.6 A 100Gb/s-8.3dBm-sensitivity PAM-4 optical receiver with integrated TIA, FFE and direct-feedback DFE in 28nm CMOS. 2021 IEEE International Solid- State Circuits Conference (ISSCC).

Lv, J., et al., 2018. Review application of nanostructured black silicon. Nanoscale Res. Lett. 13 (1), 1–10.

Margalit, N., et al., 2021. Perspective on the future of silicon photonics and electronics. Appl. Phys. Lett. 118 (22), 220501.

Martins, E.R., et al., 2012. Engineering gratings for light trapping in photovoltaics: the supercell concept. Phys. Rev. B 86 (4), 041404.

Marty, F., et al., 2005. Advanced etching of silicon based on deep reactive ion etching for silicon high aspect ratio microstructures and three-dimensional micro-and nanostructures. Microelectron. J. 36 (7), 673–677.
Marufuzzaman, M., et al., 2018. Design of low-cost transimpedance amplifier for optical receiver. Trans. Electr. Electron. Mater. 19 (1), 7–13.
Mayet, A.S., et al., 2018. Surface passivation of silicon photonic devices with high surface-to-volume-ratio nanostructures. J. Opt. Soc. Am. B 35 (5), 1059–1065.
Meeker, M., Wu, L., 2018. Internet Trends 2018. Kleiner Perkins.
Mellette, W.M., et al., 2017. Rotornet: a scalable, low-complexity, optical datacenter network. Proceedings of the Conference of the ACM Special Interest Group on Data Communication. ACM.
Mokkapati, S., Catchpole, K., 2012. Nanophotonic light trapping in solar cells. J. Appl. Phys. 112 (10), 101101.
Morero, D.A., et al., 2016. Design tradeoffs and challenges in practical coherent optical transceiver implementations. J. Lightwave Technol. 34 (1), 121–136.
Mozo, A., Ordozgoiti, B., Gómez-Canaval, S., 2018. Forecasting short-term data center network traffic load with convolutional neural networks. PLoS One 13 (2), e0191939.
Ozolins, O., et al., 2017. 100 Gbaud 4PAM Link for high speed optical interconnects. 2017 European Conference on Optical Communication (ECOC). ECOC.
Pala, R.A., et al., 2013. Optimization of non-periodic plasmonic light-trapping layers for thin-film solar cells. Nat. Commun. 4 (1), 1–7.
Peter Amalathas, A., Alkaisi, M.M., 2019. Nanostructures for light trapping in thin film solar cells. Micromachines 10 (9), 619.
Plant, D.V., Morsy-Osman, M., Chagnon, M., 2017. Optical communication systems for data-center networks. Optical Fiber Communication Conference. Optical Society of America.
Qiao, F., et al., 2020. Light trapping structures and plasmons synergistically enhance the photovoltaic performance of full-spectrum solar cells. Nanoscale 12 (3), 1269–1280.
Razmjooei, N., Magnusson, R., 2020. Properties of resonant photonic lattices at the lattice-particle Mie scattering wavelength. Frontiers in Optics/Laser Science. Optica Publishing Group, Washington, DC.
Razmjooei, N., et al., 2021. Resonant reflection by microsphere arrays with AR-quenched Mie scattering. Opt. Express 29 (12), 19183–19192.
Razmjooei, N., Magnusson, R., 2022. Experimental band flip and band closure in guided-mode resonant optical lattices. Opt. Lett. 47 (13), 3363–3366.
Rim, S.-B., et al., 2007. An effective light trapping configuration for thin-film solar cells. Appl. Phys. Lett. 91 (24), 243501.
Saini, S.K., Nair, R.V., 2019. Quantitative analysis of gradient effective refractive index in silicon nanowires for broadband light trapping and anti-reflective properties. J. Appl. Phys. 125 (10), 103102.
Saive, R., 2021. Light trapping in thin silicon solar cells: a review on fundamentals and technologies. Prog. Photovolt: Res. Appl. 29 (10), 1125–1137.
Sato, K.-i, 2018. Realization and application of large-scale fast optical circuit switch for data center networking. J. Lightwave Technol. 36 (7), 1411–1419.
Sentieri, E., et al., 2020. 12.2 A 4-Channel 200Gb/s PAM-4 BiCMOS transceiver with silicon photonics front-ends for gigabit ethernet applications. 2020 IEEE International Solid-State Circuits Conference (ISSCC).
Siew, S.Y., et al., 2021. Review of silicon photonics technology and platform development. J. Lightwave Technol. 39 (13), 4374–4389.

Siviero, F., et al., 2022. Optimization of the gain layer design of ultra-fast silicon detectors. Nucl. Instrum. Methods Phys. Res. Sect. A: Accelerators, Spectrometers, Detect. Associated Equip. 1033, 166739.

Soref, R., 2006. The past, present, and future of silicon photonics. IEEE J Sel Top Quantum Electron 12 (6), 1678–1687.

Szilagyi, L., et al., 2019. A 53-Gbit/s optical receiver frontend with 0.65 pJ/bit in 28-nm bulk-CMOS. IEEE J. Solid-State Circuits 54 (3), 845–855.

Thomson, D., et al., 2016. Roadmap on silicon photonics. J. Opt. 18 (7), 073003.

Varghese, B., Buyya, R., 2018. Next generation cloud computing: new trends and research directions. Future Gener. Comput Syst. 79, 849–861.

Wang, F., et al., 2022. Hybrid optical-electrical data center networking: challenges and solutions for bandwidth resource optimization. IEEE Commun. Mag.

Wang, K.X., et al., 2012. Absorption enhancement in ultrathin crystalline silicon solar cells with antireflection and light-trapping nanocone gratings. Nano Lett. 12 (3), 1616–1619.

Yang, W., et al., 2019. Silicon-compatible photodetectors: trends to monolithically integrate photosensors with chip technology. Adv. Funct. Mater. 29 (18), 1808182.

Yap, K.-K., et al., 2017. Taking the edge off with espresso: scale, reliability and programmability for global internet peering. Proceedings of the Conference of the ACM Special Interest Group on Data Communication. ACM.

Yoo, S.B., Yin, Y., Wen, K., 2012. Intra and inter datacenter networking: the role of optical packet switching and flexible bandwidth optical networking. 2012 16th International Conference on Optical Network Design and Modelling (ONDM). IEEE.

Yu, Z., Fan, S., 2011. Angular constraint on light-trapping absorption enhancement in solar cells. Appl. Phys. Lett. 98 (1), 011106.

Yu, Z., Raman, A., Fan, S., 2010. Fundamental limit of nanophotonic light trapping in solar cells. Proc Natl Acad Sci 107 (41), 17491–17496.

Zhou, Z., et al., 2018. Development trends in silicon photonics for data centers. Optical Fiber Technol. 44, 13–23.

Zhou, D., Biswas, R., 2008. Photonic crystal enhanced light-trapping in thin film solar cells. J. Appl. Phys. 103 (9), 093102.

Photodetectors for microwave photonics

14

Tadao Nagatsuma
Graduate School of Engineering Science, Osaka University, Osaka, Japan

14.1 Signal generation

Research on exploring terahertz (THz) waves, which cover the frequency range from 100 GHz to 10 THz, have lately increased since the nature of these electromagnetic waves is suited to spectroscopic sensing as well as to ultra-broadband wireless communications (Siegel, 2002; Tonouchi, 2007; Nagatsuma, 2011). One of the obstacles to developing applications of THz waves is a lack of solid-state signal sources (Mann, 2007).

For the generation of THz waves, photonic techniques are considered to be superior to conventional techniques based on electronic devices with respect to wide frequency bandwidth, tenability, and stability. Moreover, the use of optical fiber cables enables us to distribute high-frequency (RF) signals over long distances instead of metallic transmission media such as coaxial cables and hollow waveguides (Nagatsuma, 2009).

14.1.1 Schemes

Figure 14.1 shows a schematic of CW THz-signal generation based on optical-to-terahertz conversion. First, intensity-modulated optical signals, whose envelope is sinusoidal at a designated THz frequency, are generated with use of light waves at different wavelengths, λ_1 and λ_2. Then, these two-wavelength of lights are injected to conversion media such as nonlinear electro-optical (EO) materials, PCs, and photodiodes (PD), which leads to the generation of THz waves at a frequency given by $f_{RF} = c\Delta\lambda/\lambda^2$,

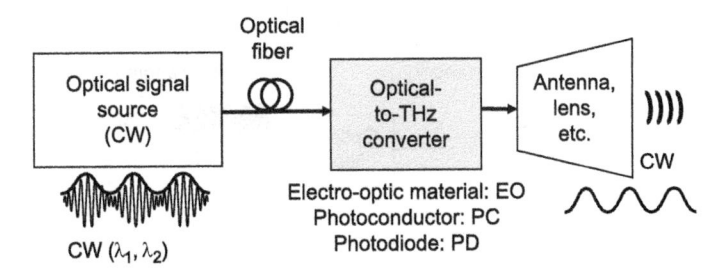

Figure 14.1 Schematic diagram of photonic CW MMW and THz-signal generation.

where $\Delta\lambda$ is a difference in wavelength of lights, and c is a velocity of light. The converted signals are finally radiated into free space by an antenna, a lens, etc.

14.1.2 Optical signal generation

Typical optical signal sources are depicted in Figure 14.2A and B; an optical heterodyning technique using two frequency-tunable laser diodes (LDs), and using the combination of an optical frequency comb generator and filters, respectively. In the latter case, two wavelength lights are filtered from the optical filters, and this offers both wideband frequency tunability and excellent stability. When so narrow linewidth or frequency resolution in the spectroscopy and/or imaging is not required, we can use two wavelength light filtered from the optical noise source such as ASE noise in optical amplifiers as shown in Figure 14.2C.

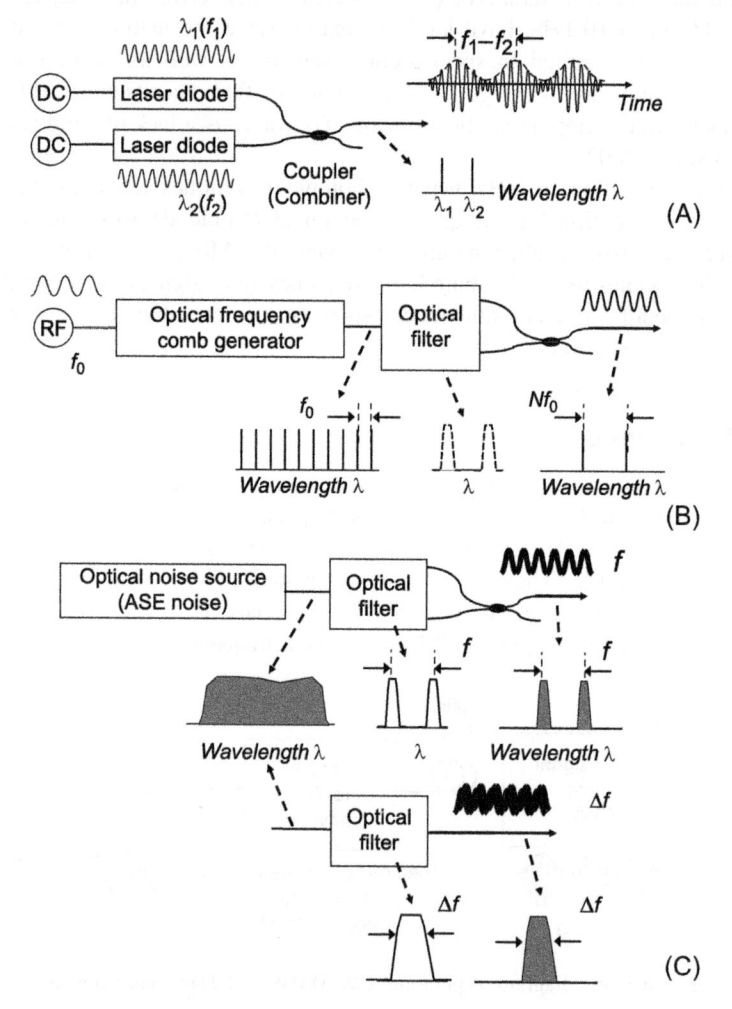

Figure 14.2 Schematic diagram of optical signal sources.

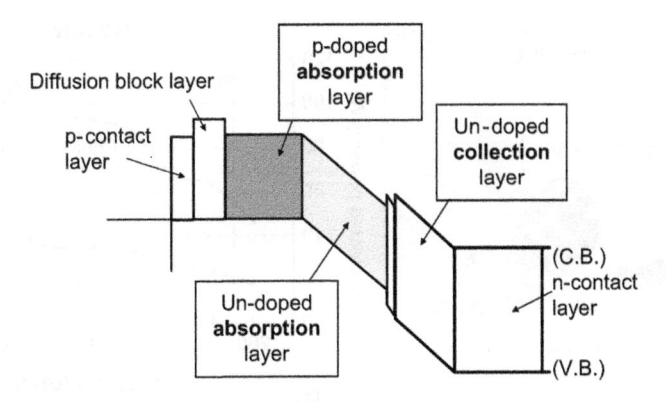

Figure 14.3 Block diagram of the modified UTC-PD. *UTC-PD*, Uni-traveling-carrier photodiode.

Among the above-mentioned three types of optical-to-electrical conversion media, the PD is highly advantageous with respect to the conversion efficiency. In addition to the operation at long optical wavelengths (1.3–1.55 μm), large bandwidth and high output current are required for practical applications. Among various types of long-wavelength PD technologies, a uni-traveling-carrier photodiode (UTC-PD) and its derivatives have exhibited the highest output powers at frequencies from 100 GHz to 1 THz, with improvement in layer and device (Ito et al., 2005; Beck et al., 2008; Nagatsuma et al., 2009; Shi et al., 2012; Rouvalis et al., 2012). Figure 14.3 shows the band diagram of the PD optimized for higher power and efficiency operation, which is a modification of the original UTC-PD.

14.1.3 Fundamental characteristics

The PD chips are usually packaged into the module with a rectangular waveguide output port as shown in Figure 14.4A. Figure 14.4B shows the frequency dependence of the output power generated from the module. The 3 dB bandwidth is 140 GHz (from 270 to 410 GHz). The peak output power was 110 μW at 380 GHz for a photocurrent of 10 mA with a bias voltage of 1.1 V. The output power could be further increased to over 500 μW by increasing the photocurrent up to 20 mA with responsivity of 0.22 A/W (Wakatsuki et al., 2008).

To increase the output power to more than 1 mW, one of the practical approaches is a power-combining technique using an array of PDs as shown in Figure 14.5A. With two PDs, the output power of greater than 1 mW has been obtained at the photocurrent of 18 mA per PD at 300 GHz (Figure 14.5B) (Song et al., 2012).

At frequencies of over 300 GHz extending to 1 THz or higher, an antenna-integrated PD is more efficient, and semispherical silicon lens is often used to collimate a beam radiated from a planar antenna such as bow-tie, log-periodic and dipole antennas (Ishibashi et al., 2013). Photos of such a quasioptical module and its characteristics are shown in Figure 14.6A and B, respectively.

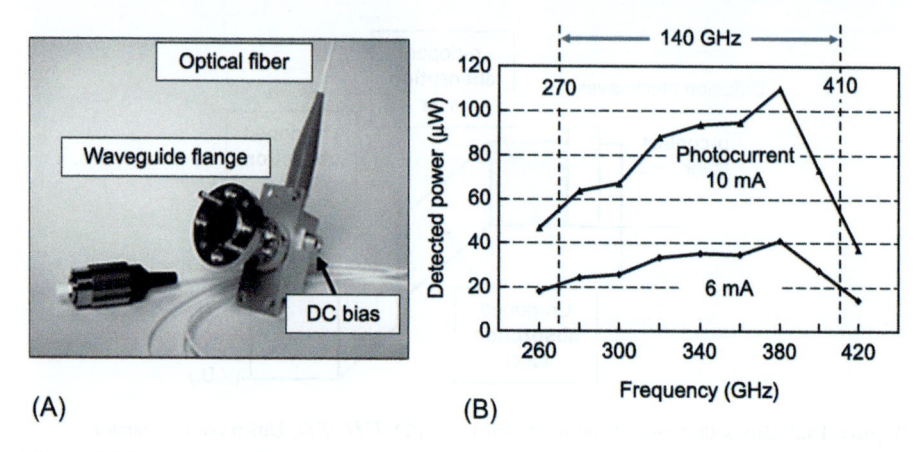

Figure 14.4 (A) Photo of waveguide-mounted PD module. (B) Output power characteristics.

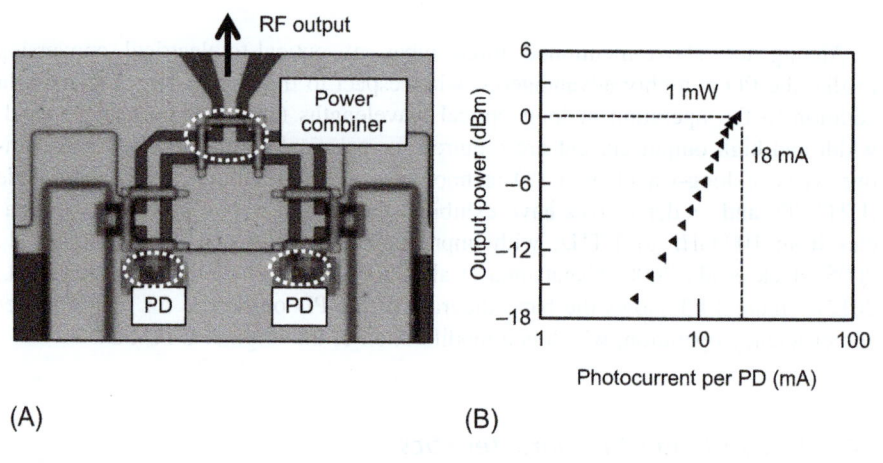

Figure 14.5 (A) Photo of array PDs. (B) Output power characteristics.

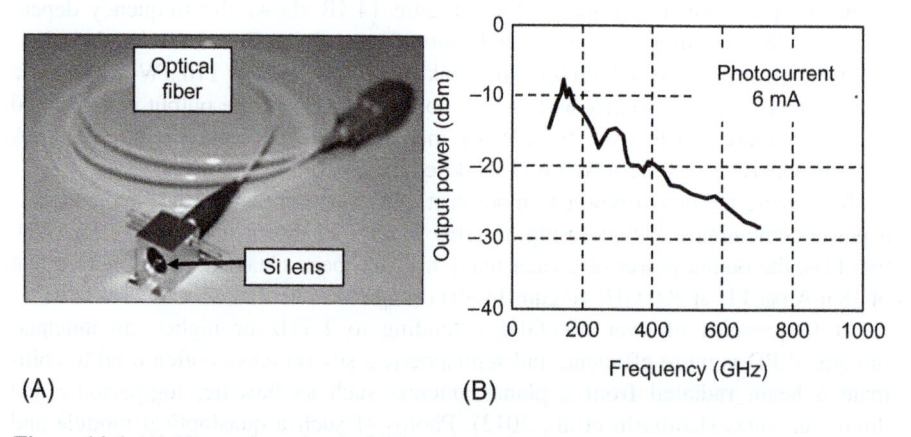

Figure 14.6 (A) Photo of quasioptical PD module. (B) Output power characteristics.

14.2 Signal detection

14.2.1 Schemes

As for the signal detection, there are several choices in the THz regions, such as "direct detection" using Schottky barrier diodes (SBDs) or bolometers and "heterodyne detection" by combining mixers and local oscillators (LOs) (Figure 14.7). There are electronic and photonic mixers as well as electronic and photonic LOs. We can choose the best one depending on required performance in each application.

Direct detection (Figure 14.7A) using SBDs or bolometers is the most widely used technique for the detection of amplitude or power of THz electromagnetic waves. The cutoff frequency of the SBD can be made more than 10 THz with GaAs materials (Peatman and Crowe, 1990; Hesler and Crowe, 2007) and more than 1 THz even with silicon CMOS (Sankaran and O, 2005). However, detection based on the plasma-wave phenomenon enables a sensitive reception of signals whose frequencies are much higher than cut-off frequencies of transistors (Blin et al., 2013; Otsuji and Shur, 2014). Heterodyne detection with the SBD mixer and a LO signal source (Figure 14.6B) provides higher sensitivity and phase information of THz waves, which is important for network analyzers, radars, etc. The sensitivity can be further enhanced when the superconductor-insulator-superconductor mixer or superconducting hot electron bolometer mixer is used. Another advantage of the superconducting mixer is that it requires an extremely small LO power (10 nW–1 μW). Choice between the SBD and bolometer should be made considering the sensitivity, the response speed, and the operation temperature.

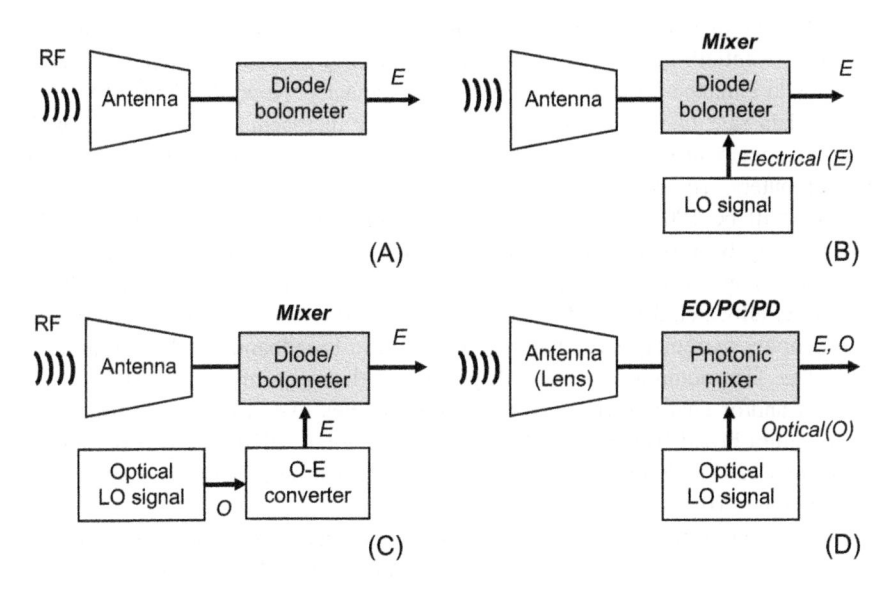

Figure 14.7 Configurations of THz detection system. O, Optical signal; E, electrical signal.

14.2.2 Photonic local oscillator for mixers

Figure 14.7C and D show heterodyne detection systems which use photonically generated THz LO signal sources. An advantage of the scheme of Figure 14.7C is that we can not only deliver THz LO signals with optical fiber cables at some distances, but also increase a receiver bandwidth because of the inherently wider frequency tunability in photonic signal generation (Takano et al., 2003; Kohjiro et al., 2008; Shimizu et al., 2012).

14.2.3 Photonic mixers

Together with photonic mixers, the heterodyne system of Figure 14.7D provides the largest bandwidth, which is useful in particular for spectroscopy applications. Typical photonic mixers are PCs, EO materials, and PDs, where nonlinear interaction between RF signals and optical signals is a key to enhance a sensitivity.

14.3 Applications

This section will describe how effectively the photonics technologies are introduced in Microwave Photonics applications in the millimeter- and terahertz-wave regions, presenting some of our recent applications, in particular based on "continuous wave (CW)" signals, such as wireless communications, spectroscopy, imaging and measurement.

14.3.1 Wireless communications

In the history of wireless communications technologies initiated by G. Marconi in the early twentieth century, we have been exploring higher and higher carrier frequencies to enhance a speed and/or a channel capacity. Now, the demand for much greater data rate of wireless technologies is ever increasing in accordance with a rapid advancement of mobile networks and rich contents handled by the networks and computers. The prospective data rate for wireless communications in the marketplace will be 100 Gbit/s within 10 years. Against this background, researchers have recently been seeking a use of radio waves whose frequency is over 275 GHz for ultrahigh-speed wireless links, since the frequency bands from 275 to 3000 GHz are not yet allocated for specific active services in the world, and there is a possibility to employ extremely large bandwidths for ultra-broadband wireless communications. The THz communications will promise a data rate of over 100 Gbit/s using low-cost and/or energy-efficient modulation schemes like ASK and QPSK because of their ultra-broad bandwidths (Federici and Moeller, 2010; Kleine-Ostmann and Nagatsuma, 2011; Song and Nagatsuma, 2011; Kürner, 2012; Li et al., 2013; Koenig et al., 2013; Ducournau et al., 2015).

Introduction of photonics technologies for signal generation, modulation and detection is very effective not only to enhance the bandwidth and/or the data rate, but also to combine fiber-optic (wired) and wireless networks. In addition, the photonics-based approach is expected to bring such ultra-high data rate wireless technologies to potential users and to meet and explore real-world applications at the earliest opportunity as a technology driver (Nagatsuma et al., 2013; Takahashi et al., 2014).

Figure 14.8 illustrates an example of the application schemes with photonics-based approach, showing how the photonic RF signal generation can be employed together with fiber-optic links. In addition to the wired link using the light wave at a wavelength of λ_1, we can transmit the same data with the wireless link by introducing another wavelength (λ_2) of light wave and mixing the two wavelengths of light in the RF PD. The PD generates RF signals, whose frequency can be determined the difference in the wavelength of the two light waves, as described in Section 14.1.1. As for the receiver, we can use, for example, a simple diode such as a SBD for the demodulation based on the square-law detection in the case of the ASK (OOK) data format. Thus, the receiver becomes more cost-effective and energy-efficient, if we can make use of a wide bandwidth lying over the THz frequency region.

Figure 14.9 shows a detailed experimental setup, where a photonics-based ASK modulation in the transmitter and a direct detection in the receiver are employed at 300-GHz band system. The link distance is 1–2 m with 40 dBi antennas (horn antenna and dielectric lens) for the transmitter and receiver. Figure 14.10A and B show a photo of the experimental setup, and eye diagrams demodulated at the receiver, respectively. Clear eye diagrams are obtained up to 50 Gbit/s, and an error-free (BER $< 10^{-11}$) transmission has been confirmed up to 40 Gbit/s (Nagatsuma, 2014a,b).

To extend a transmission distance to over 100 m, for instance, the most effective ways are an increase in the output power of the transmitter, and in the sensitivity of the receiver (Nagatsuma and Kato, 2014). The deployment of PD arrays is promising as for the former countermeasure (Nagatsuma, 2014a,b), while coherent detection techniques together with frequency- and phase-stabilized carrier signal sources are employed as the latter approach (Yoshimizu et al., 2013a,b). Figure 14.11 shows

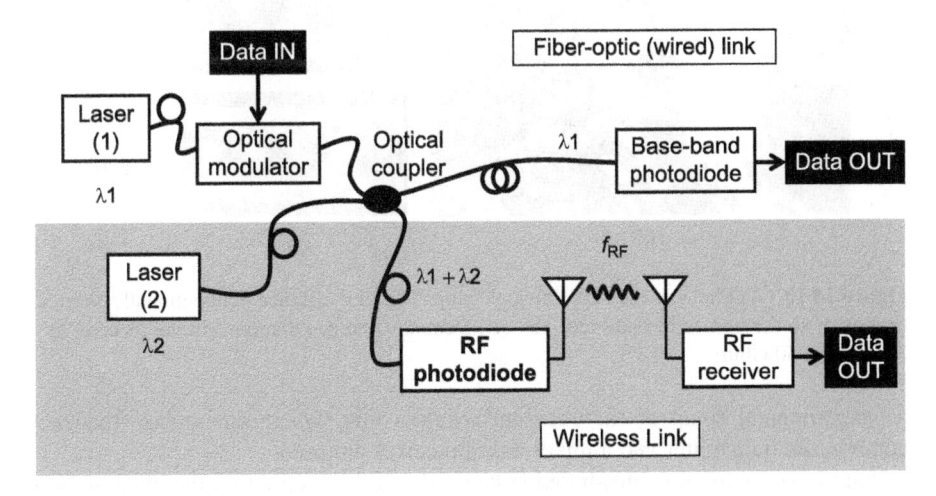

Figure 14.8 System concept of wired and wireless convergence.

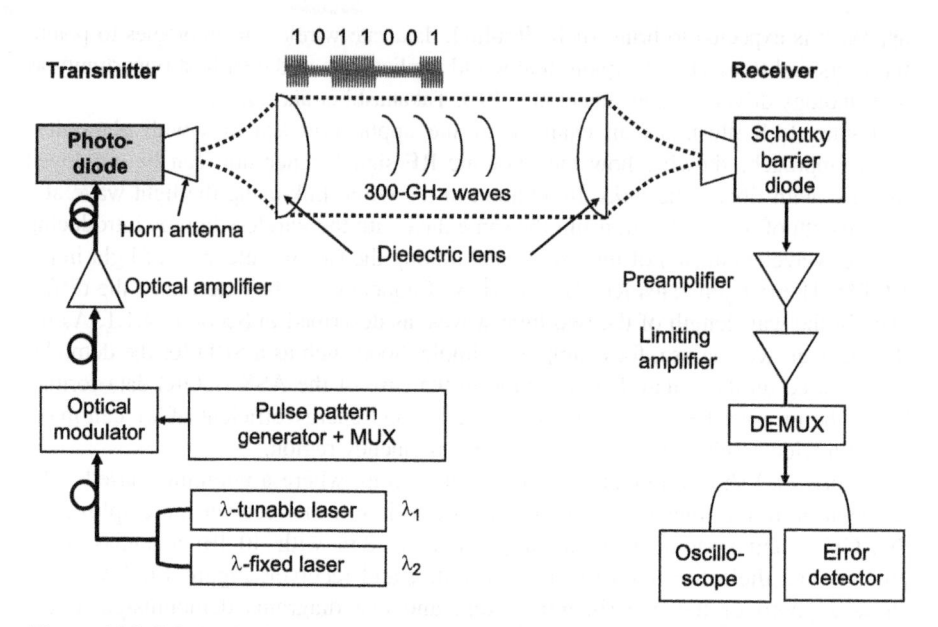

Figure 14.9 Schematic diagram of experimental setup to evaluate transmission characteristics of 300-GHz wireless link based on ASK modulation and direct detection.

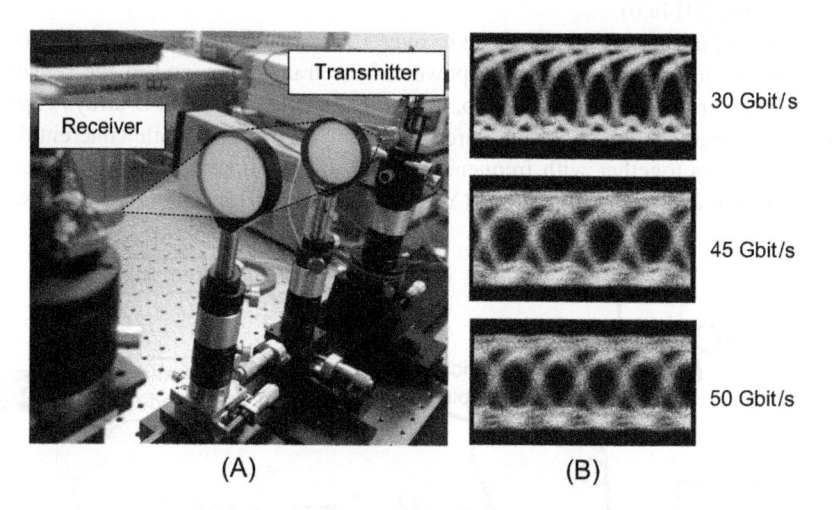

Figure 14.10 (A) Photo of the experimental setup. The link distance is intentionally shorten to show both the transmitter and receiver. (B) Demodulated eye diagrams in the receiver at 30, 45, and 50 Gbit/s.

an experimental setup of the coherent wireless link, where an optical frequency comb-based transmitter and a mixer-based receiver are used.

Future works in photonics-based systems should be focused on the integration to compete with electronics with respect to cost and size (Carpintero et al., 2014).

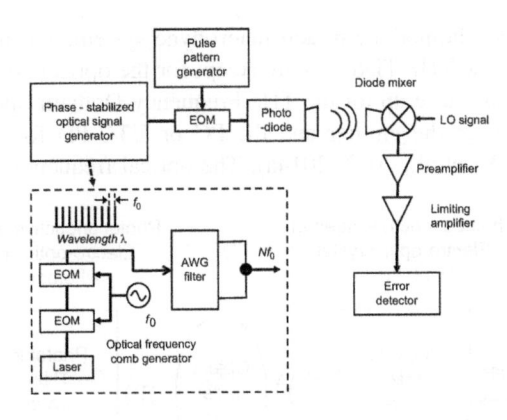

Figure 14.11 Experimental setup of coherent systems.

14.3.2 Spectroscopy

One of the interesting aspects of THz-waves is in their interaction with matters via the motion of groups of relatively large molecules such as biological molecules like proteins and DNA, and chemicals. Pharmaceutical inspection of tablets with THz-waves is expected to have a large market opportunity (Brock et al., 2013). Other potential profitable applications for THz spectroscopy would be targeted to material inspection and evaluation during or after manufacturing process of semiconductors, solar cells, polymeric films, dielectric composite, plastics, paint, etc. (Robin and Bouyé, 2014).

Figure 14.12 shows a typical configuration of THz spectroscopy system, by combining THz sources and detectors (Figure 14.7D). Laser-pulse-assisted THz-wave technology has proven to be powerful and useful in spectroscopy applications because of its unprecedented capability to control ultrashort timing. In particular, the THz time-domain spectroscopy (THz-TDS) system has been established as a laboratory standard for THz spectroscopy and is commercially available from a number of companies (Withayachumnankul and Naftaly, 2014). In the THz-TDS system, frequency characteristics are obtained by Fourier transforming the time-domain data, and the frequency resolution is limited by the time window given by the optical delay shown in Figure 14.12. The typical frequency resolution of the THz-TDS system is around 100 MHz–1 GHz and is determined by the repetition frequency of the pulse laser and the scan length of the optical delay line.

Recently, spectroscopy systems based on CW technology, which use monochromatic sources with an accurate frequency control capability, have attracted great interest (Demers et al., 2007; Stanze et al., 2011; Göbel et al., 2013). The CW source-based systems provide a higher signal-to-noise ratio and spectral resolution. When the frequency band of interest is targeted for the specific absorption line of the objects being tested, CW systems with the selected frequency-scan length and resolution are more practical in terms of data acquisition time as well as system cost. A CW spectroscopy system with a photonic THz-wave emitter and detector is often referred to as a homodyne or self-heterodyne system, and its configuration is the same as that of the THz-TDS system except for the optical signal source.

Figure 14.13 shows a setup for the THz-Frequency-Domain Spectroscopy system using the UTC-PD for the emitter and the PC or UTC-PD for the detector (Kim et al., 2013; Hisatake et al., 2013, 2014a). The optical frequency/phase shifter (FS)

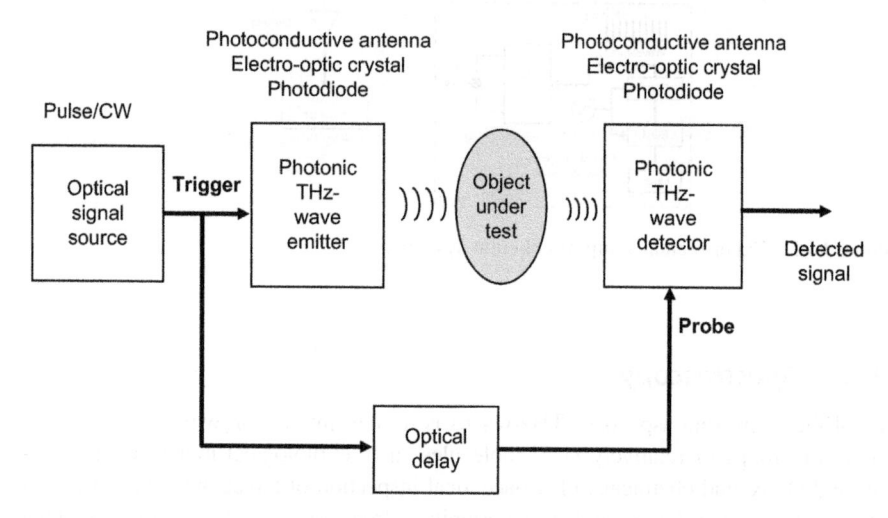

Figure 14.12 Configuration of THz spectroscopy/imaging system based on photonic techniques.

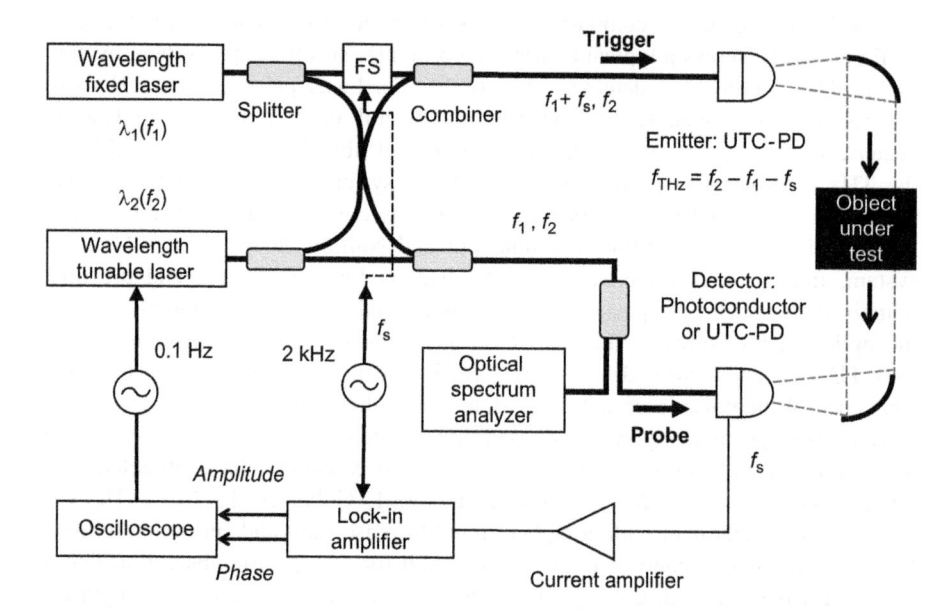

Figure 14.13 Experimental setup for the frequency-domain spectroscopy. *FS*, Electro-optic frequency shifter.

enables us to accurately measure both the amplitude and phase. Since two LDs are free-running, we monitor each wavelength by the optical spectrum analyzer or the wave meter. Currently, the experimental standard deviation of the mean for the frequency and the phase are is about 50 MHz, and 0.37 degrees, respectively (Hisatake and Nagatsuma, 2014).

14.3.3 Electric-field measurement

The same configuration of Figure 14.13 is applicable to the characterization of antennas by using EO sensors in place of photoconductive detectors or PDs. Figure 14.14A shows a close-up view of the measurement, where the fiber-fed EO sensor of Figure 14.14B is placed very close to the aperture of a W-band horn antenna to detect electric fields radiated from the antenna. The EO sensor consists of a 1-mm cubic EO crystal, which has a Pockels effect, a collimating lens, and a polarization maintaining fiber cable, as shown in Figure 14.14C. Typical EO crystals are GaAs, CdTe, and ZnTe (Yang et al., 2001; Togo et al., 2007).

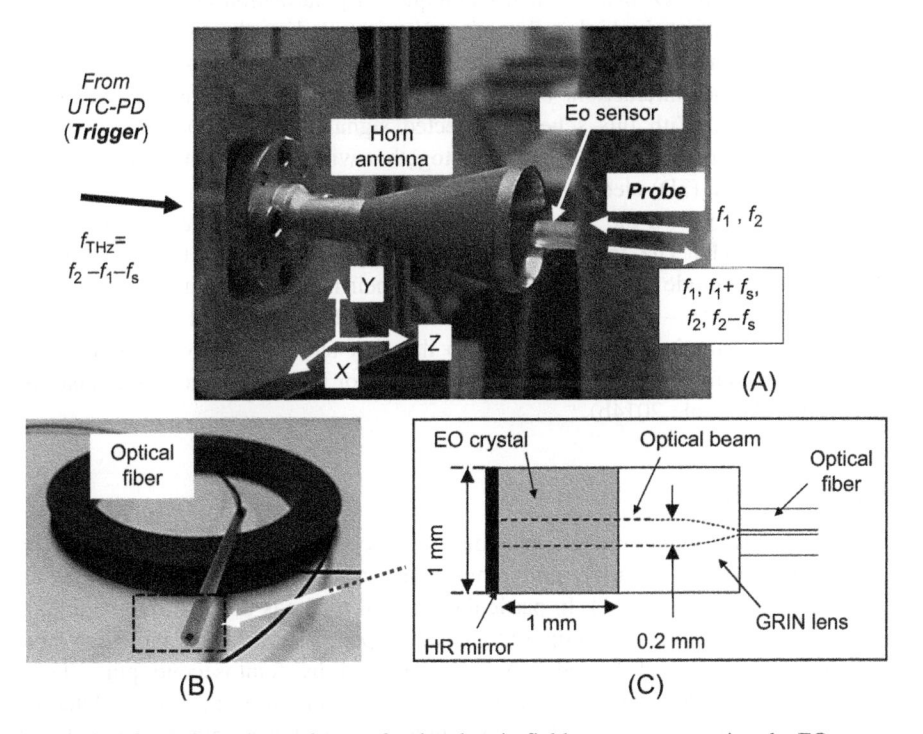

Figure 14.14 (A) Experimental setup for the electric-field measurement using the EO sensor. The EO sensor is placed close to the antenna aperture. Overall optics and signal detection schemes are the same with those of Figure 14.13. (B) Photo of fiber-fed EO sensor. (C) Detailed structure of the EO sensor. *HR*, High reflectivity.

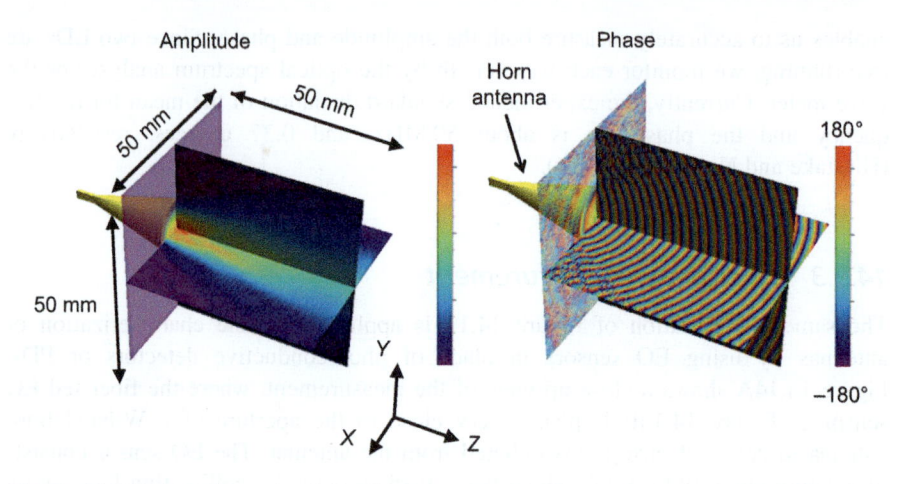

Figure 14.15 Three-dimensional distributions of the freely propagating wave measured at 125 GHz.

In conventional EO measurement techniques, a polarization change of the laser beam reflected off at the high-reflectivity mirror coated on the EO crystal is measured. On the contrary, we detect side-band optical signals, $f_1 + f_s, f_2 - f_s$, generated through the interaction between the THz wave and the optical beam (Figure 14.14A). With this scheme, detected signal intensity does not fluctuate, even when the optical fiber cable moves together with the EO sensor to perform a three-dimensional (3D) electric-field mapping.

Figure 14.15 shows the 3D distributions of the freely propagating wave measured at 125 GHz. The measured data were acquired with the time constant of 30 Ms. The amplitude data were normalized to their maximum values in each plane. The scan area ($20\lambda = 50$ mm) has been limited by the mechanical stage. The measured field distributions agreed well with those of simulations even for the near-field measurement, which verifies that our EO sensor is noninvasive in the measurement (Hisatake et al., 2014b).

14.3.4 Imaging

THz waves have been applied to imaging of objects in various fields such as medical, pharmaceutical, security, and other nondestructive inspection (Chan et al., 2007; Ajito and Ueno, 2011). In particular, several methods of THz 3D imaging, or THz tomography have been proposed and developed. Most of them including commercial ones employ THz pulse waves generated by femtosecond pulse lasers (Fitzgerald et al., 2005; Takayanagi et al., 2009; Imamura et al., 2010). Although THz tomography systems based on THz pulses have high output power and high spatial resolution, their drawbacks are size, cost, and complexity of the system.

Tomographic imaging based on CWs is expected to overcome the above issues and to become more widespread in industrial applications (Cooper et al., 2008; Quast

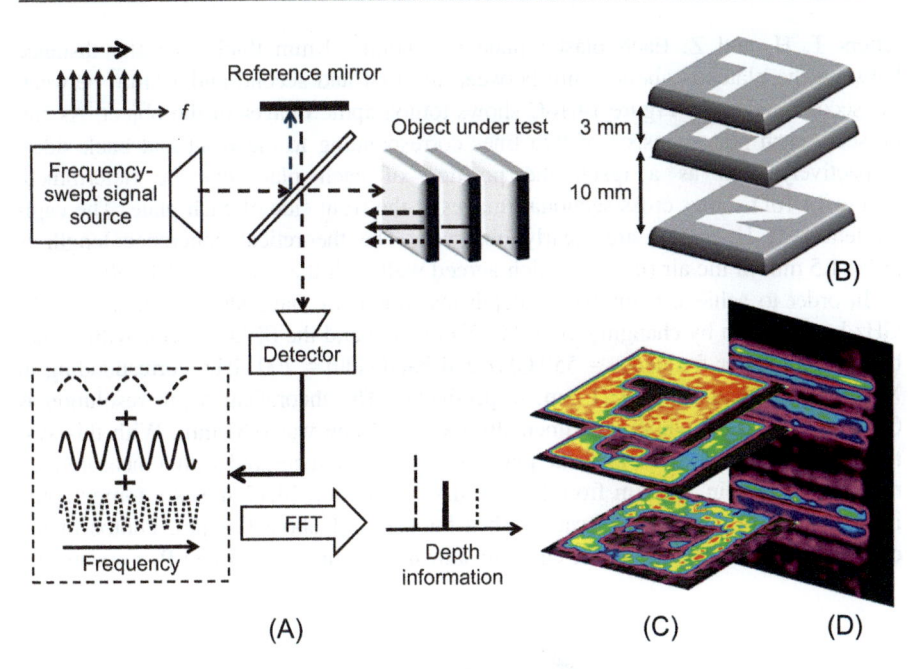

Figure 14.16 (A) Block diagram of the swept-source THz OCT system. (B) Object under test: three plastic plates which have hollow holes in the shape of the capital letters T, H, and Z. (C) Tomographic image of the object measured with 300-GHz band system. (D) Cross-sectional images at each depth position.

and Loeffler, 2009; Recur et al., 2011). We have developed THz tomographic imaging systems using optical coherence tomography (OCT) techniques in the frequency band from 300 to 800 GHz (Isogawa et al., 2012; Ikeo et al., 2012, 2013; Nagatsuma et al., 2014). In our system, photonic generation of such wideband signals is one of the most important enabling technologies. We have examined two approaches; one is the time-domain OCT with use of incoherent noise source and the other is the frequency-domain OCT with frequency-swept monochromatic source.

Figure 14.16A shows a block diagram of the THz frequency-domain (swept-source: SS) OCT system. It consists of a Mickelson interferometer and a frequency-swept THz source. In order to generate frequency-swept THz signals, we employ a photomixing technique using two laser sources and the UTC-PD module as is used in communications and spectroscopy systems: one is a fixed light source and the other is a tunable source. In the SS-OCT system, the spatial (depth) information is obtained by Fourier transforming the frequency domain interference signals. The theoretical value of depth resolution, Δz, is given as $2 \ln 2 \lambda_c^2/(\pi \Delta \lambda)$, where λ_c is the center wavelength of the THz source and $\Delta \lambda$ is the spectral width.

By using 300-GHz band system (center frequency = 325 GHz and bandwidth = 87.4 GHz, corresponding to $\lambda_c = 0.92$ mm and $\Delta \lambda = 0.25$ mm, respectively), we conducted a tomographic imaging of object as shown in Figure 14.16B−D. The object consists of three plastic plates which have hollow holes in the shape of the capital letters T, H, and Z. Each plastic plate is 50 mm^2, 1 mm thick, and the distance between the plates is about 3 mm between the first and second and 10 mm between the second and third. Figure 14.16C shows tomographic images of the object. As can be seen, each plate consists of two lines corresponding to the front and back sides, respectively, and as a result the position of each plate can be determined. Figure 14.16D shows cross-sectional images at the front side of each plate. The capital letters T, H, and Z are clearly observed. The theoretical coherence length is $\Delta z = 1.5$ mm in the air ($n = 1$), which agreed well with the experimental value.

In order to achieve a submicron depth resolution, we have also developed a 600-GHz band system by changing the UTC-PD module and the SBD detector with higher bandwidth (center frequency = 550 GHz and bandwidth = 200 GHz, corresponding to $\lambda_c = 0.55$ mm and $\Delta \lambda = 0.21$ mm, respectively). The theoretical depth resolution is 0.64 mm in the air, while experimentally measured one was 0.61 mm. With this system, 0.37 mm thickness of a plastic plate ($n = 1.65$) was measured with a measurement repeatability within ± 3 μm from the mean value. Figure 14.17 shows a tomographic imaging of a driver's license card with a thickness of about 0.8 mm. Photos of the card and internal structure of the card are shown in Figure 14.17A and B, respectively.

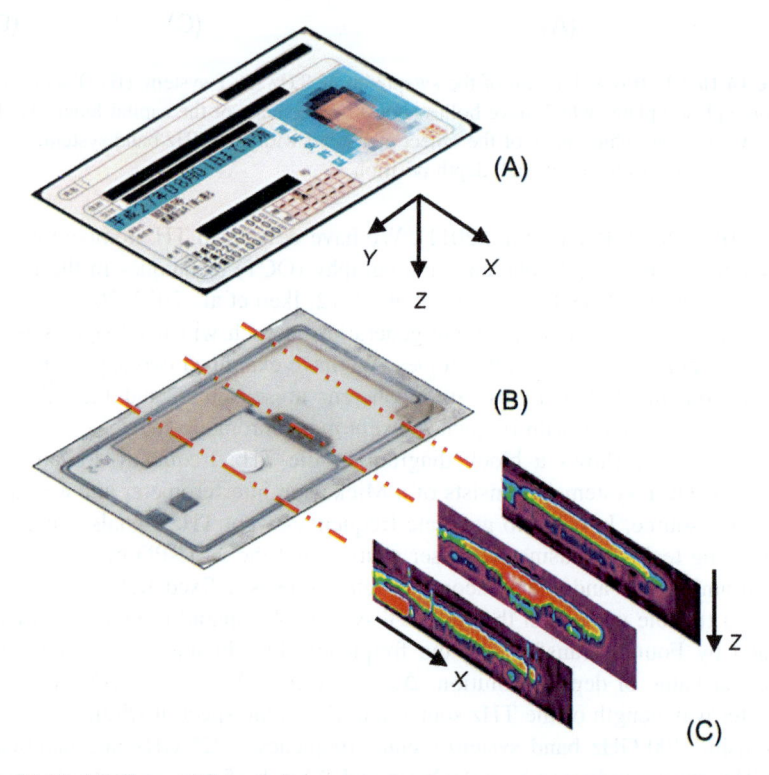

Figure 14.17 (A) Photo of driver's license card to be measured. (B) Internal structure of the card consisting of electronic components made of metals. (C) Cross-sectional images taken for each line indicated in (B).

Cross-sectional images were taken as shown in Figure 14.17C for each line indicated in Figure 14.17B (Nagatsuma et al., 2014).

References

Ajito, K., Ueno, Y., 2011. THz chemical imaging for biological applications. IEEE Terahertz Sci. Technol. 1 (1), 293–300.

Beck, A., Ducournau, G., Zaknoune, M., Peytavit, E., Akalin, T., Lampin, J.-F., 2008. High-efficiency uni-travelling-carrier photomixer at 1.55 μm and spectroscopy applications up to 1.4 THz. Electron. Lett. 44 (22), 1320–1322.

Blin, S., Tohme, L., Coquillat, D., Horiguchi, S., Minamikata, Y., Hisatake, S., 2013. Wireless communication at 310 GHz using GaAs high-electron-mobility transistors for detection. J. Commun. Netw. 15 (6), 559–568.

Brock, D., Zeitler, J.A., Funke, A., Knop, K., Kleinebudde, P., 2013. Evaluation of critical process parameters for intra-tablet coating uniformity using terahertz pulsed imaging. Eur. J. Pharm. Biopharm. 85 (3), 1122–1129.

Carpintero, G., Balakier, K., Yang, Z., Guzmán, R.C., Corradi, A., Jimenez, A., 2014. Microwave photonic integrated circuits for millimeter-wave wireless communications. IEEE J. Light. Technol. 32 (20), 3495–3501.

Chan, W.L., Deibel, J., Mittleman, D.M., 2007. Imaging with terahertz radiation. Rep. Progr. Phys. 70, 1325–1379.

Cooper, K.B., Dengler, R.J., Llombart, N., Bryllent, T., Chattopadhyay, G., Schlecht, E., 2008. Penetrating 3-D imaging at 4- and 25-m range using a submillimeter-wave radar. IEEE Trans. Microw. Theory Tech. 56, 2771–2778.

Demers, J.R., Logan Jr., R.T., Brown, E.R., 2007. An optically integrated coherent frequency-domain THz spectrometer with signal-to-noise ratio up to 80 dB. In: Tech. Dig. IEEE International Topical Meeting on Microwave Photonics (MWP2007), Victoria, pp. 92–95.

Ducournau, G., Szriftgiser, P., Pavanello, F., Peytavit, E., Zaknoune, M., Bacquet, D., 2015. THz communications using photonics and electronic devices: the race to data-rate. J. Infrared Millim. Terahertz Waves 36 (2), 198–220.

Federici, J., Moeller, L., 2010. Review of terahertz and subterahertz wireless communications. J. Appl. Phys. 107 (11), 111101.

Fitzgerald, A.J., Cole, B.E., Taday, P.F., 2005. Nondestructive analysis of tablet coating thickness using terahertz pulsed imaging. J. Pharm. Sci. 94, 177–183.

Göbel, T., Stanze, D., Globisch, B., Dietz, R.J.B., Roehle, H., Schell, M., 2013. Telecom technology based continuous wave terahertz photomixing system with 105 dB signal-to-noise ratio and 3.5 terahertz bandwidth. Opt. Lett. 38 (20), 4197–4199.

Hesler, J.L., Crowe, T.W., 2007. NEP and responsivity of THz zero-bias Schottky diode detectors (IRMMW-THz). In: Joint 32nd International Conference on Infrared and Millimeter Waves, 2007 and the 2007 15th International Conference on Terahertz Electronics, Sendai, pp. 844–845.

Hisatake, S., Nagatsuma, T., 2014. Precise terahertz-wave phase measurement based on photonics technology. In: 39th International Conference on Infrared, Millimeter, and Terahertz waves (IRMMW-THz2014), T2/C-13.7.

Hisatake, S., Kitahara, G., Ajito, K., Fukada, Y., Yoshimoto, N., Nagatsuma, T., 2013. Phase-sensitive terahertz self-heterodyne system based on photodiode and low-temperature-grown GaAs photoconductor at 1.55 μm. IEEE. Sens. J. 13 (1), 31–36.

Hisatake, S., Kim, J.-Y., Ajito, K., Nagatsuma, T., 2014a. Self-heterodyne spectrometer using uni-traveling-carrier photodiodes for terahertz-wave generators and optoelectronic mixers. J. Light. Technol. 32 (20), 3683–3689.

Hisatake, S., Nguyen Pham, H.H., Nagatsuma, T., 2014b. Visualization of the spatial–temporal evolution of continuous electromagnetic waves in the terahertz range based on photonics technology. Optica. 1 (6), 365–371.

Ikeo, T., Isogawa, T., Ajito, K., Kukutsu, N., Nagatsuma, T., 2012. Terahertz imaging using swept source optical-coherence-tomography techniques. In: Tech. Dig. IEEE International Topical Meeting on Microwave Photonics (MWP2012), Session 8, Noordwijk.

Ikeo, T., Isogawa, T., Nagatsuma, T., 2013. Three dimensional millimeter- and terahertz-wave imaging based on optical coherence tomography. IEICE Trans. Electron. E96-C (10), 1210–1217.

Imamura, M., Nishina, S., Irisawa, A., Yamashita, T., Kato, E., 2010. 3D imaging and analysis system using terahertz waves. In: Proceedings of International Conference on Infrared, Millimeter, and Terahertz Waves (IRMMW-THz 2010), We-B3.1, Rome.

Ishibashi, T., Muramoto, Y., Yoshimatsu, T., Ito, H., 2013. Continuous THz wave generation by photodiodes up to 2.5 THz. In: Proceedings of International Conference on Infrared, Millimeter, and Terahertz Waves (IRMMW-THz 2013), We2-5.

Isogawa, T., Kumashiro, T., Song, H.-J., Ajito, K., Kukutsu, N., Iwatsuki, K., 2012. Tomographic imaging using photonically generated low-coherence terahertz noise sources. IEEE Trans. Terahertz Sci. Technol. 2 (5), 485–492.

Ito, H., Furuta, T., Nakajima, F., Yoshino, K., Ishibashi, T., 2005. Photonic generation of continuous THz wave using uni-traveling-carrier photodiode. IEEE J. Lightwave Technol. 23 (12), 4016–4021.

Kim, J.-Y., Song, H.-J., Ajito, K., Yaita, M., Kukutsu, N., 2013. Continuous-wave THz homodyne spectroscopy and imaging system with electro-optical phase modulation for high dynamic range. IEEE Trans. Terahertz Sci. Technol. 3 (2), 158–164.

Kleine-Ostmann, T., Nagatsuma, T., 2011. A review on terahertz communications research. J. Infrared Millim. Terahertz Waves 32 (2), 143–171.

Koenig, S., Lopez-Diaz, D., Antes, J., Henneberger, R., Schmogrow, R., Hillerkuss, D., et al., 2013. 100 Gbit/s wireless link with mm-wave photonics. In: Tech. Dig. Optical Fiber Communication Conference and Exposition and the National Fiber Optics Engineers Conference (OFC/NFOEC 2013), post deadline paper.

Kohjiro, S., Kikuchi, K., Maezawa, M., Furuta, T., Wakatsuki, A., Ito, H., 2008. A 0.2–0.5 THz single-band heterodyne receiver based on a photonic local oscillator and a superconductor-insulator-superconductor mixer. Appl. Phys. Lett. 93, 093508.

Kürner, T., 2012. Towards future THz communications systems. Terahertz Sci. Technol. 5 (1), 11–17.

Li, X., Yu, J., Zhang, J., Dong, Z., Li, F., Chi, N., 2013. A 400G optical wireless integration delivery system. Opt. Express. 21 (16), 187894.

Mann, C., 2007. Practical challenges for the commercialization of terahertz electronics. In: IEEE MTT-S International Microwave Symposium (IMS 2007), Honolulu, pp. 1705–1708.

Nagatsuma, T., 2009. Generating millimeter and terahertz waves. IEEE Microw. Mag. 10 (4), 64–74.

Nagatsuma, T., 2011. Terahertz technologies: present and future. IEICE Electron. Express. 8 (14), 1127–1142.

Nagatsuma, T., 2014a. Breakthroughs in photonics 2013: THz communications based on photonics. IEEE Photonics J. 6 (2), 14263570.
Nagatsuma, T., 2014b. 300-GHz-band wireless communications with high-power photonic sources. In: 31th URSI General Assembly and Scientific Symposium, DFC01.3, Beijing.
Nagatsuma, T., Kato, K., 2014. Photonically-assisted 300-GHz wireless link for real-time 100-Gbit/s transmission. In: IEEE MTT-S International Microwave Symposium (IMS 2014), WE1H-5, Tampa.
Nagatsuma, T., Ito, H., Ishibashi, T., 2009. High-power RF photodiodes and their applications. Laser Photon. Rev. 3 (1−2), 123−137.
Nagatsuma, T., Horiguchi, S., Minamikata, Y., Yoshimizu, Y., Hisatake, S., Kuwano, S., 2013. Terahertz communications based on photonics technologies. Opt. Express. 21 (20), 23736.
Nagatsuma, T., Nishii, H., Ikeo, T., 2014. Terahertz imaging based on optical coherence tomography. OSA J. Photonics Res. 2 (4), B64−B69.
Otsuji, T., Shur, M.S., 2014. Terahertz plasmonics: good results and great expectations. IEEE Microw. Mag. 15 (7), 43−50.
Peatman, W.C.B., Crowe, T.W., 1990. Design and fabrication of 0.5 micron GaAs Schottky barrier diodes for low-noise terahertz receiver applications. Int. J. Infrared Millim. Waves. 11 (3), 355−365.
Quast, H., Loeffler, T., 2009. 3D-terahertz-tomography for material inspection and security. In: Proceedings of International Conference on Infrared, Millimeter, and Terahertz Waves (IRMMW-THz 2009), T3D02.0311, Busan.
Recur, B., Younus, A., Salort, S., Mounaix, P., Chassagne, B., Desbarats, P., 2011. Investigation on reconstruction methods applied to 3D terahertz computed tomography. Opt. Express. 19, 5106−5117.
Robin, T., Bouyé, C., 2014. THz technologies offer varied options for industry. Ind. Photonics. 1 (2), 11−13.
Rouvalis, E., Renaud, C.C., Moodie, D., Robertson, M.J., Seeds, A.J., 2012. Continuous wave terahertz generation from ultra-fast InP-based photodiodes. IEEE Trans. Microw. Theory Tech. 60, 509−517.
Sankaran, S., O, K.K., 2005. Schottky barrier diodes for millimeter wave detection in a foundry CMOS process. IEEE Electron. Device Lett. 26 (7), 492−494.
Shi, J.-W., Kuo, F.-M., Bowers, J.E., 2012. Design and analysis of ultra-high speed near-ballistic uni-traveling-carrier photodiodes under a 50 Ω load for high-power performance. IEEE Photon. Technol. Lett. 24 (7), 533−535.
Shimizu, N., Kukutsu, N., Wakatsuki, A., Muramoto, Y., 2012. Remote detection of hazardous gases in full-scale simulated fire by using terahertz electromagnetic waves. NTT Tech. Rev. 10 (2), 1−6.
Siegel, P.H., 2002. Terahertz technology. IEEE Trans. Microw. Theory Tech. 50 (3), 910−928.
Song, H.-J., Nagatsuma, T., 2011. Present and future of terahertz communications. IEEE Trans. Terahertz Sci. Technol. 1 (1), 256−264.
Song, H.-J., Ajito, K., Muramoto, Y., Wakatsuki, A., Nagatsuma, T., Kukutsu, N., 2012. Uni-travelling-carrier photodiode module generating 300 GHz power greater than 1 mW. IEEE Microw. Wirel. Compon. Lett. 22 (7), 363−365.
Stanze, D., Deninger, A., Roggenbuck, A., Schindler, S., Schlak, M., Sartorius, B., 2011. Compact cw terahertz spectrometer pumped at 1.5 μm wavelength. J. Infrared Millim. Terahertz Waves 32 (2), 225−232.

Takahashi, H., Hirata, A., Ajito, K., Hisatake, S., Nagatsuma, T., 2014. 10-Gbit/s close-proximity wireless system meeting the regulation for extremely low-power radio stations. IEICE Electron. Express. 11 (3), 20130989.

Takano, S., Ueda, A., Yamamoto, T., Asayama, S., Sekimoto, Y., Noguchi, T., 2003. The first radioastronomical observation with photonic local oscillator. Publ. Astron. Soc. Japan. 55 (4), L53−L56.

Takayanagi, J., Jinno, H., Ichino, S., Suizu, K., Yamashita, M., Ouchi, T., 2009. High-resolution time-of-flight terahertz tomography using a femtosecond fiber laser. Opt. Express. 17, 7549−7555.

Togo, H., Shimizu, N., Nagatsuma, T., 2007. Near-field mapping system using fiber-based electro-optic probe for specific absorption rate measurement. IEICE Trans. Electron. E90-C (2), 436−442.

Tonouchi, M., 2007. Cutting-edge terahertz technology. Nat. Photonics. 1 (2), 97−105.

Wakatsuki, A., Furuta, T., Muramoto, Y., Yoshimatsu, T., Ito, H., 2008. High-power and broadband sub-terahertz wave generation using a J-band photomixer module with rectangular-waveguide output port. In: Proceedings of International Conference on Infrared, Millimeter, and Terahertz Waves, pp. 1999-1−1999-2.

Withayachumnankul, W., Naftaly, M., 2014. Fundamentals of measurement in terahertz time-domain spectroscopy. J. Infrared Millim. Terahertz Waves 35 (8), 610−637.

Yang, K., Katehi, L.P., Whitaker, J.F., 2001. Electric field mapping system using an optical-fiber based electrooptic probe. IEEE Microw. Wirel. Compon. Lett. 11 (4), 164−166.

Yoshimizu, Y., Hisatake, S., Kuwano, S., Terada, J., Yoshimoto, N., Nagatsuma, T., 2013a. Wireless transmission using coherent terahertz wave with phase stabilization. IEICE Electron. Express. 10 (18), 20130578.

Yoshimizu, Y., Hisatake, S., Kuwano, S., Terada, J., Yoshimoto, N., Nagatsuma, T., 2013b. Generation of coherent sub-terahertz carrier with phase stabilization for wireless communications. J. Commun. Netw. 15 (6), 569−575.

Index

Note: Page numbers followed by "*f*" and "*t*" refer to figures and tables, respectively.

A
Active illumination based ToF 3D imaging, 62–66
Active organic pixels, 121
Active quenching and reset (AQR), 61
Aerosol jet printing, 81–82
Analog SiPM (aSiPM), 61
Analog-to-digital conversion (ADC), 49
Analogous method, 361–362
Annealing process, 297–298
Antiphase boundaries (APBs), 204–205
Antireflection (AR), 174
Antisite defects and carrier traps, 295–297
Array, homogeneity of SPAD parameters in, 60
Arsenic defects, 295
ASE noise, 468
Avalanche photodetectors (APDs), 44, 236–238, 249–256, 392–393, 426
Avalanche photodiodes (APD), 205, 327–328. *See also* Photodiodes (PDs)
to SPAD, 44–49

B
Back-end-of-line (BEOL), 242–243
Back-side illuminated SPAD (BackSPAD), 52–53
Bandgap energy, 440
Bandwidth (BW), 84, 241–242
bandwidth-efficiency, 215
cost, 438–439
Biological molecules, 475
Bis (2-methyl-8-quinolinato) – 4- phenylphenolate aluminum (BAlq), 104
Bit-error-rate (BER), 235, 439
Blocking layers (BLs), 78

Bolometers, 471
Boltzmann constant, 29, 31–32
Boron nitride (BN), 283
Bosch process, 446–447
Bose-Einstein statistics, 29
Bow-tie antennas, 469
Bragg condition, 362–363
Bridging process, 143–144
Buffer layer, 293
Bulk heterojunctions (BHJs), 76
Bulk thermal generation, 89–90
Butt-coupled PIN photodiodes in Ge, 427
Butt-waveguide-coupling, direct, 243–244

C
Calcium fluoride (CaF_2), 276
Carbon nanotube (CNT), 93–94
Carrier lifetime, 300–303
Catalytic nanoparticles, 141
Center-of-Gravity determination, 39–40
Central Limit Theorem, 11
Charge carrier narrowing (CCN), 102
Charge plasma, 333
Charge transfer (CT), 76
Charge transport channel, 329–330
Chemical analysis, 267–268
Chemical vapor deposition method (CVD method), 155–156, 446
Chip-integrated silicon-germanium photodetectors
  contemporary advances and state-of-the-art silicon-germanium photodetectors, 245–256
  photodetection material systems, 236–240
  processing methods and integration opportunities, 240–245
Coating technique, 79
Coaxial cables, 467

Coefficients of thermal expansion (CTE), 204–205
Coherent oscillations, 353
Complementary metal–oxide–semiconductor (CMOS), 200, 233–234, 391
　compatibility, 398–400
　fabrication
　　issues, 49–57
　　process, 391
　selective hetero-epitaxy of Ge-on-Si and, 393–395
　SPAD arrays in CMOS technology, 61–62
Complementary metal–oxide–semiconductor Image Sensor (CIS), 52
Compound annual growth rate (CAGR), 437
Computer algorithm, 451
Conductivity, 16
Continuous-wave (cw), 293
　modulated active illumination configurations, 63
　signals, 472
Conventional EO measurement techniques, 478
Conventional limit, 441–442
Coplanar strip (CPS), 337–338
Coplanar waveguide (CPW), 448
Coulomb-barrier lowering, 303
Crystalline semiconductors, 236
Current-voltage (I-V), 372, 450
　relationship, 333–336

D

Dark count rate (DCR), 49, 58, 62
Dark currents, reduction of, 86–91
Dark shot noise, 43
Data center, 437
Deep reaction ion etching (DRIE), 446–447
Deep-level transient spectroscopy (DLTS), 210
Deep-UV range (DUV range), 103
Dense wavelength division multiplexing (DWDM systems), 254
Deposition techniques, 77–82
　aerosol-jet printing, 81–82
　ink-jet printing, 79–81
　spin-coating, 78–79
　spray-coating, 82
　vacuum thermal evaporation, 77–78
Detector-waveguide integration, 241–242
Device design, 444
Device structure, 330–333
Dielectric function, 329, 353–354
Dielectric waveguides, 372
Diffraction, 441
Digital Photon-Counting Devices (DPC), 48
Digital SiPM (dSiPM), 61
Dilute hydrofluoric acid (DHF), 393–394
4,7-diphenyl-1,10-phenanthroline-(bathophenanthroline) (Bphen), 104
Dipolar mode, 367
Dipole antennas, 469
Direct detection, 471
Direct nanowire integration, 141–144
Direct-growth photodetectors, demonstrations of, 147–148
Direct-ToF (dToF), 63
Dispersion relation, 358
Drop-on-demand approach (DOD approach), 79–81
Drude model, 353–354

E

Effective noise bandwidth, 25–27
Electric fields, 357–358
　measurement, 477–478
Electrical bias, 32–33
Electrical circuits, 24
Electrical interconnects, 233–234
Electrical properties of quantum dots, 206–213
　in-plane, 212–213
　stark effect in, 213
　in vertical device structures, 207–211
Electro-optic sampling (EOS), 308, 332–333
Electromagnetic (EM), 353
　spectrum, 267
　wave, 440–441
Electron affinity (EA), 75
Electron dynamics, 302–303
Electron speed
　analysis and modeling, 345–348
　current-voltage relationship, 333–336
　device structure, 330–333
　time response, 337–345

Index

Electron-electron interactions, 353–354
Electron-hole pairs (EHPs), 42–43, 344–345, 359–360
Electronic bandgap, 362–363
Electronic noise, 30, 86–89
Electronics measurement, 24–25
Enabling Technologies and Innovation (ETI), 34
Energy band diagram (EBD), 331
Energy dispersive X-ray spectroscopy (EDX), 393–394
Energy relaxation process, 329
Ensemble Monte Carlo technique (EMC technique), 346
Equivalent circuit model, 31, 33
Equivalent photodiode model, 453–459
Evanescent light coupling, 242–243
Exciton binding energy, 14
Extended sources, 7–9
External Quantum Efficiency (EQE), 84, 215–219, 358–359, 448–449

F

Fabrication process, 332, 400–402, 446–448
Feed-forward equalizer (FFE), 457
Femtosecond transmission experiments, 304
FEOL-CMOS process, 393–394
Fiber optic communication systems, 327–328
Field-of-view (FOV), 7–9
5th generation (5G), 199
  cellular networks, 234–235
Fill factor (FF), 60, 65
Finite element (FE), 445
Finite-difference time-domain (FDTD), 445
Flicker noise, 29–30
Fluorescence lifetime imaging spectroscopy (FLIM), 61
Fluorine-doped-tin-oxide (FTO), 156–157
Forbidden band, 362–363
4-level pulse amplitude modulation (4-PAM), 254
Fourier transform, 22
Free electron gas, 353–354
Frequency-dependent noise sources, 26
Frequency-dependent responsivity, 13–14
Frequency-swept THz source, 479
Frequency-tunable laser diodes (Frequency-tunable LDs), 468
Front-end-of-the-line (FEOL), 391
Full-width-at-half-maximum (FWHM), 9–10, 99–100, 313, 328, 339, 393–394
Fundamental sampling theorem, 10–11, 28

G

GaAs-on-V-grooves-Si (GoVS), 205–206
Gallium arsenide (GaAs), 238, 293
Gallium oxide (Ga_2O_3), 158, 283
  growth process of, 159–161
  nanowires photodetectors, 158–159
Gamma rays, 39–40
Gaussian distribution, 11
Geiger-mode APDs (GAPDs), 44–45
Geiger-mode limitation, 47–48
Geiger-mode operation, 44–49
Generation-recombination noise, 29
Germanium (Ge), 234–235, 391–392, 420–421
  photodetectors, 238–239
  surface morphology, 393–394
Germanium-Tin (GeSn), 234–235
  photodetectors, 239
  photodiodes, 421–422
Glancing angle deposition technique (GLAD technique), 166
Global communications, 233–234
Global IP traffic trend and forecast, 437
  total data traffic forecast through 2030, 438f
Graphene plasmonics
  for photodetection, 374–377
  photodetectors, 375–377
Grating-coupled plasmonic photodetectors, 362–366
  photodetectors with corrugated gratings, 365–366
  plasmonic photodetectors with metallic nanowire grating gates, 364–365
Green's function, 296
Group-IV photonics, 234–235
Growth methods, 241
Growth temperature, 294–295

H

Helmholtz equations, 354

Hetero-epitaxial growth focuses, 201–202
Heterodyne detection systems, 471–472
Heteroepitaxial, 140–141
Heterogeneous integration, 220–221
Heterojunction, 431–432
Hexagonal boron nitride (h-BN), 286
High electron mobility transistors (HEMT), 329–330
High-speed InAs quantum dot
  photodetectors for data/telecom
  electrical and photoelectrical properties of quantum dots, 206–213
  epitaxial growth advances, 201–206
  InAs quantum dots photodetectors, 221–222
  photodetectors for optical communication, 213–215
  trends and performance, 215–221
High-speed photodiodes, 327–328
High-speed testing, 337–338
High-speed transceivers, 235
Highest occupied molecular orbital (HOMO), 74
Hollow waveguides, 467
Homo-junction photodetectors, 247
Homodyne system, 475–476
Hot-electron effect, 304
Hybrid contacts for print transferred nanowires, 172
Hybrid III/V-silicon, 421
  uni-traveling carrier photodiodes, 431–433
Hybrid imagers, 117–118
Hydrodynamic modeling, 329–330
Hydrogenated microcrystalline silicon, 140

I

III/V compound photodetectors, 238
III–nitride-based ultraviolet photodetectors, 272–279
Image intensifiers (I2), 40
Imaging, 472, 478–481
  applications, 115–121
    active organic imagers integrating passive pixels, 117–121
    active organic pixels, 121
    passive organic imager, 116–117
Impediments on contact formation, 170–171
Indirect-ToF (iToF), 63

Indium arsenide (InAs)
  quantum dots, 201–206
  quantum dot-based photodetectors, 215–222
Indium gallium arsenide (InGaAs), 238, 391, 426–427
  InGaAs-based photodiodes, 420
  vertically coupled PIN photodiodes in Si/Ge and, 424–427
Indium gallium arsenide phosphide (InGaAsP), 238
Indium phosphide (InP), 238
Indium tin oxide (ITO), 83–84, 157
Indium zinc oxide (IZO), 283
Infrared devices (IR devices), 200
Ink-jet printing technique, 79–81
Inkjet-printed organic photodetectors, 108–111
Inorganic crystalline semiconductors, 14–16
Inorganic semiconductor photodetectors, 1–2
Insulator-metal-insulator (IMI), 372
Integrated transceivers, 113–115
Integration opportunities, 240–245
  detector-waveguide integration, 241–242
  growth methods, 241
  on-chip schemes for waveguide coupling, 242–244
  Si-complementary metal-oxide-semiconductor and Si-foundry integration, 244–245
Intense pulse light (IPL), 155–156
Interference, 441
Internal optoelectronic filter (IOEF), 101–102
Internal photoemission process (IPE), 359–360, 392
Internal quantum efficiency (IQE), 359–360
International system of units (SI), 4
Internet of Things, 199
Ionization potential (IP), 75
Irradiance, 6
Isolation process, 394–395

L

Lambertian source, 7–8
Langmuir-Blodgett technique (LB technique), 143–144

Laser-pulse-assisted THz-wave technology, 475
Lateral topology, 84
Lattice mismatched systems, 201–202
Lead (Pb), 40
Light Detection And Ranging (LiDAR), 60
Light sources, spectral characteristics of, 9–10
Light trapping theory, 440–444
Light-emitting diodes (LEDs), 3–4, 150–151
Local oscillators (LOs), 471
Local Oxidation of Silicon (LOCOS), 56–57
Localized surface plasmon resonance (LSPR), 412
Lock-in measurement techniques, 26–27
Log-periodic antennas, 469
Low viscosity materials, 81–82
Low-pressure chemical vapor deposition (LPCVD), 394–395
Low-temperature grown gallium arsenide (LT-GaAs), 293, 328
  applications, 317–320
    nonlinear optics, 317–318
    pulsed magnetic spin experiments, 318
    sub-bandgap absorption photodetectors, 317
    THz emitters and receivers, 318–320
  attributes of low-temperature-grown photodetectors, 294–304
  material systems, 305
  photodetector
    performance, 313–317
    technologies, 306–313
  principle of operation for, 306
Low-temperature-grown photodetectors
  annealing, 297–298
  antisite defects and carrier traps, 295–297
  attributes, 294–304
  carrier lifetime, 300–303
  growth temperature, 294–295
  mobility and resistivity, 298–300
  optical absorption at mid-gap states, 303–304
Lowest unoccupied molecular orbital (LUMO), 74

M

Mach-Zehnder interferometers, 358–359
Magnetic field, 357–358
Material systems, 236, 305
Maxwell's equations, 329, 445
Metal organic chemical vapor deposition (MOCVD), 305
Metal oxide nanowires-based photodetectors, 169–170
Metal oxide semiconductor field effect transistor (MOSFET), 329–330
Metal-insulator-metal (MIM), 372
Metal-insulator-semiconductor (MIS), 276
Metal-semiconductor-metal (MSM), 214, 270, 306, 328, 392–393
  diodes, 245–246
  photodetectors, 309–311, 427–428
  Schottky-barrier photodetectors, 406–408
Metallic silicide, 392
Metallic transmission media, 467
Metals, 353–354
  electrodes, 299–300
  metal-organic semiconductor contacts, 75
  nanoparticles, 368–369
Metasurfaces, 379–381
Metrology of thin-film photodetectors, 3
  basic radiometry, 3–10
  electrical signals, 14–20
    photogenerated current, 14–19
    photogenerated voltage, 19–20
  ideal measurement of physical quantities, 10–12
  measurement and noise, 21–30
    effective noise bandwidth, 25–27
    generation-recombination noise, 29
    measuring electronics, 24–25
    1/f or flicker noise, 29–30
    shot noise, 28–29
    sources of noise, 27
    thermal noise, 27–28
  NEP, 20–21
  photodiodes, 30–34
  responsivity, 12–14
    frequency-dependent responsivity, 13–14
  specific detectivity, 30
Mickelson interferometer, 479
Micropillars (MPs), 152
Microwalls (MWs), 152

Microwave photonics
  applications, 472–481
    electric-field measurement, 477–478
    imaging, 478–481
    spectroscopy, 475–477
    wireless communications, 472–474
  signal detection, 471–472
    photonic local oscillator for mixers, 472
    photonic mixers, 472
    schemes, 471
  signal generation, 467–470
    fundamental characteristics, 469–470
    optical signal generation, 468–469
    schemes, 467–468
Mobility and resistivity, 298–300
Modified uni-traveling carrier (MUTC), 426
Molecular beam epitaxy (MBE), 295
Molecular solids, 74
Monochromator, 9–10
Monolithic integration, 440
Motivation, 440
Multi project wafer (MPW), 242–243
Multi-channel plates (MCP), 40
Multimode fibers (MMFs), 438–439
Multiplication avalanche photodiodes, 428

N

N,N'-bis (naphthalene-1-yl)-N,N'-bis (phenyl) benzidine (NPB), 104
N,N'-diphenyl-N,N'-bis (3-methyl-phenyl) $(1,1'$-biphenyl$)$ $-$ $4,4'$-diamine (TPD), 103–104
Nanocrystals (NCs), 3, 93–94
Nanoparticles (NPs), 392–393
Nanowires (NWs), 3, 139
  control of nanowires doping, 172–174
  development of integrated multispectral nanowires photodetectors, 177–178
  device demonstrations, 147–170
    demonstrations of direct-growth photodetectors, 147–148
    photoactive oxide devices, 150–170
    plasmonic photodetectors, 150
    waveguide coupled photodetectors, 148–150
  device design challenges, 170–177
    additional challenges, 177
    control of nanowires doping, 172–174
    hybrid contacts for print transferred nanowires, 172
    impediments on contact formation, 170–171
    nanowire photo-trapping enhancement, 174–177
  photodetectors, 139
    direct nanowire integration, 141–144
    fabrication themes, 141–147
    transfer printing of horizontally oriented semiconductor nanowires, 145–147
    transfer-printing/pick-and-place techniques, 144–145
    transfer printing of horizontally oriented semiconductor, 145–147
Narrowband-absorption-type (NBA), 100–101
National Nuclear Security Administration (NNSA), 34
Near infra-red (NIR), 37, 57, 73, 150–151
  photodetectors, 91–94
  region, 358–359
Near-UV range (NUV range), 103
$NiSi_2$ film and absorption, 403–404
$NiSi_2$/n-Si Schottky-barrier photodetectors, 404–406
Noise, sources of, 27
Noise equivalent power (NEP), 2, 20–21, 86–89, 270–271
Nonlinear electro-optical (Nonlinear EO), 467–468
Nonlinear optics, 317–318
Normalized photocurrent-to-dark current ratio (NPDR), 270
Nyquist-Johnson noise, 27–28
Nyquist-Shannon theorem, 10–11

O

Oligomers, 358–359
On-chip schemes for waveguide coupling, 242–244
  direct butt-waveguide-coupling, 243–244
  evanescent light coupling, 242–243
On-off keying signaling (OOK signaling), 254
One-dimensional nanowires (1D-NWs), 139
1/f noise, 29–30
Optical absorption, 440

Index   491

at mid-gap states, 303–304
Optical coherence tomography (OCT), 478–479
Optical communication, 235
Optical interconnects
  global IP traffic trend and forecast, 437
  optical communication in datacenters, 437–439
  theory and design, 440–444
    light trapping theory and background, 440–444
    motivation, 440
    vertical PIN silicon-based photodiode for short-reach communication, 444–462
Optical multimode (OM3), 439
Optical noise source, 468
Optical photodetectors, 235
Optical power function, 9
Optical signal generation, 468–469
Optical technologies, 233–234
Optimization, 445
Optoelectronics, 234–235
  devices, 221
  time, 303–304
Organic electronics, 122–123
Organic field-effect transistors, 73
Organic light-emitting diodes (OLEDs), 73
Organic memories, 73
Organic photodetectors
  all-printed organic photodetectors fabricated of scalable solution-based processes, 108–113
    fabricated by spray-coating, 111–113
    inkjet-printed, 108–111
  applications of, 113–123
    imaging applications, 115–121
    integrated transceivers, 113–115
    other applications, 121–123
  device structure and operation mechanisms, 83–91
    photoconductors, 86
    photodiodes operation mechanism, 84–85
    reduction of dark currents, 86–91
    organic semiconductors, 73–82
    photoactive materials and detectors for different spectral regions, 91–108
Organic photodiodes (OPDs), 18–19

Organic photovoltaics, 73
Organic semiconductors, 73–82
  charge photogeneration, 75–77
  deposition techniques, 77–82
Organic sensors, 73
Oscillation, 353–354

P
P-type/intrinsic/n-type (PIN), 207
  diodes, 246–249
  photodiodes, 306–307, 421–422
Parseval's theorem, 23, 26
Passive pixels, 117
  active organic imagers integrating, 117–121
  all-organic pulse oximeter, 122f
Pauli exclusion principle, 346–347
Perfectly matched layer (PML), 445
Performance metrics, 311–313
Perovskites, 3
Persistent photoconductivity (PPC), 273–274
Photoactive materials, 79
  and detectors for different spectral regions, 91–108
    UV detectors, 103–108
    visible and NIR photodetectors, 91–94
    wavelength selective OPDs, 94–103
Photoactive oxide devices, 150–170
  Ga_2O_3 nanowires photodetectors, 158–159
  growth process of Ga_2O_3 nanowires, 159–161
  metal oxide nanowires-based photodetectors, 169–170
  silver catalyst, 162–164
  SnO_2 nanowires-based photodetectors, 164–167
  TiO_2 nanowires-based photodetectors, 167–169
  ZnO nanowires photodetectors, 151–158
Photoconductive gain, 86
Photoconductive switches (PCSs), 293, 308–309
Photoconductors, 83, 86, 275–276
  organic and hybrid photoconductors, 87t
Photoconversion layer (PCL), 101–102
Photocurrent (PC), 206–207
  generation, 454–455
Photocurrent spectroscopy (PS), 207–208

Photocurrent-to-dark current ratio (PDCR), 270
Photodetection efficiency (PDE), 49−52
Photodetection material systems, 236−240
 germanium photodetectors, 238−239
 germanium-tin photodetectors, 239
 III/V compound photodetectors, 238
 new materials beyond semiconductors, 239−240
 silicon photodetectors, 236−238
Photodetectors (PDs), 2−6, 13−14, 37, 83, 139, 199, 235, 248−249, 275−276, 359−381
 chips, 469
 with corrugated gratings, 365−366
 fabrication, 419
 graphene plasmonics for photodetection, 374−377
 grating-coupled plasmonic, 362−366
 metrics, 2−3
 parameters, 270−272
 plasmonic detectors with metallic nanoparticles, 366−372
 plasmonic metamaterials, 377−381
 semiconductor, 360−362
 technologies, 306−313
  MSM photodetectors, 309−311
  performance metrics, 311−313
  photoconductive switches, 308−309
  PIN photodiodes, 306−307
  waveguide detectors, 372−374
Photodiodes (PDs), 30−34, 42−43, 83, 327−328, 446, 467−468
 operation mechanism, 84−85
Photoelectrical properties of quantum dots, 206−213
 electrical properties of quantum dots in vertical device structures, 207−211
 in-plane electrical properties of quantum dots, 212−213
 stark effect in quantum dots, 213
Photogenerated current, 14−19
Photogenerated voltage, 19−20
Photoluminescence (PL), 207−208
Photomultiplier tubes (PMT), 37
Photon detection efficiency (PDE), 46−47, 58, 62
Photon shot noise, 43
Photonic integrated circuits (PICs), 200, 419

Photonics
 bandgap, 362−363
 local oscillator for mixers, 472
 mixers, 472
 systems, 391
 technologies, 472−473
Photons, 440
Photosensing arrays, 115−116
Phototubes, 38−39
Physics of plasmon resonances, 353−358
 electric field intensity profile of SPP propagating, 355f
Pick-and-place techniques, 144−145
Piezoelectric transduction, 79−81
Pixel cells, 48
Planar antenna, 469
Planck's constant, 18−19
Plank's law, 267
Plasma-enhanced chemical vapor deposition (PECVD), 393−394
Plasmonic detectors with metallic nanoparticles, 366−372
 metallic nanospheres, 367−372
Plasmonic devices, 358−359
Plasmonic metamaterials, 377−381
 metallic assemblies, 379−381
Plasmonic modes, 356−357
Plasmonic photodetectors, 150. *See also* Waveguide photodetectors
 with metallic nanowire grating gates, 364−365
 photodetectors, 359−381
 surface plasmon resonances, 353−359
Plasmonics, 360−361, 441
Point sources, 5−7
Poisson distribution, 28, 58−59
Poisson equations, 331
Poisson statistics, 29
Poly dithienobenzodithiophene-co-diketopyrrolopyrrolebithiophene (PDPDBD), 86
Poly(3,4-ethylene dioxythiophene) polystyrene sulfonate, 75
Poly(9,9-dihexylfluorene-2,7-diyl) (PFH), 104
Poly(9,9'-dioctylfluorene-co-benzothiadiazole) (F8BT), 108−109

Poly(9,9′-dioctylfluorinene-co-bis-N,N' (4-butylphenyl)-bis-N, N'-phenyl-1,4 phenylenediamine) (PFB), 108–109
Poly(N-vinylcarbazole) (PVK), 104
Polyethylene naphthalate (PEN), 117–118
Polyethyleneimine (PEIE), 83–84
Polymeric optical fibers (POF), 114–115
Poole-Frenkel effect, 402
Positron-Emission Tomography, 48
Power density spectrum, 22
Print transfer process, 172
Printed electronics, 79–81
Pulsed magnetic spin experiments, 318
Pulsed modulated active illumination configuration (PM active illumination configuration), 63
Pulses, 345
Pump-probe techniques, 300

Q

Quantum confined Stark effect (QCSE), 213
Quantum dots (QDs), 199
  electrical and photoelectrical properties of, 206–213
  in-plane electrical properties of, 212–213
  stark effect in, 213
Quantum efficiency (QE), 43, 270, 445
Quantum wells (QWs), 199
  quantum-well-based amplification, 312
Quasibound modes, 356–357
Quenching resistor limits, 45–46

R

Radiant energy, 4
Radiometers, 8–9
Radiometry, 3–10
  extended sources, 7–9
  point sources, 5–7
  spectral characteristics of light sources, 9–10
Read noise, 43
Readout electronics (ROIC), 62
Recombination process, 393–394
Reflection, 441
Refraction, 441
Refractive index, 354
Research and development (R&D), 235–236
Resistance-capacitance (RC), 246–247
  delay, 246–247

RC-limited bandwidth, 426
Resonant-cavity-enhanced PDs (RCE PDs,), 215
Responsivity (R), 270
Reverse-biased p-n junction, 44
Roll-to-roll process, 77–78
Room temperature, 274–275
Root-mean-square (rms), 393–394

S

Scalar potential, 347–348
Scanning electron microscopy (SEM), 369–370, 448
Scanning transmission electron microscopy (STEM), 412–413
Scattering, 441
Schottky barrier diodes (SBDs), 471, 480–481
Schottky barriers, 274–275, 293–294
  energy, 359–360
Schottky diodes, 236–238, 359–360
  devices, 333
Schottky-barrier collector bipolar transistor (SCBT), 408–409
Schottky-barrier collector phototransistor (SBCPT), 392–393, 408–411
Schottky-barrier detector with embedded silicide nanoparticles, 411–415
Schottky-barrier photodetectors (SBPDs), 392. See also Organic photodetectors
  metal-semiconductor-metal SBPDs, 406–408
  $NiSi_2$/p-Si and $NiSi_2$/n-S, 404–406
  waveguide-integrated silicide, 403–415
Schrodinger equations, 331
Secondary emission, 38–39
Selective epitaxial growth (SEG), 393–394
Selective hetero-epitaxy of Ge-on-Si, 393–395
Self-heterodyne system, 475–476
Semiconductors, 3, 309–310, 440
  chips, 233–234
  photodetectors, 1, 12, 14, 17–18, 360–362
Sense-node (SN), 37, 43
Separate absorption, charge, and multiplications (SACM), 219, 252, 392–393, 419
Separate absorption charge, 428

Separated-transport-recombination photodetectors (STR-Pds), 319
Shockley diode model (S-model), 31–32
Shockley-Hall-Read effect, 402
Shockley-Ramo theorem, 345–346
Shockley-Read-Hall model, 302
Shot noise, 28–29
Signal-to-noise ratio (SNR), 11, 86–89, 270–271
Silicon (Si), 234–235, 440
   optical properties of Si-based WGPDs, 422–423
   Si-complementary metal-oxide-semiconductor, 244–245
   Si-foundry integration, 244–245
   Si/Ge, 426–427
     and InGaAs, vertically coupled PIN photodiodes in, 424–427
     Si/Ge uni-traveling carrier photodiodes, 428–431
   silicon-based photonic components, 419
   single-photon counting in, 42–57
   waveguide photodiodes on, 423–433
     Butt-coupled PIN photodiodes in Ge, 427
     hybrid III/V-silicon uni-traveling carrier photodiodes, 431–433
     metal-semiconductor-metal photodetectors, 427–428
     separate absorption charge and multiplication avalanche photodiodes, 428
     Si/Ge uni-traveling carrier photodiodes, 428–431
     vertically coupled PIN photodiodes in Si/Ge and InGaAs, 424–427
Silicon carbide (SiC), 268–270
   SiC-based photodetectors, 272
   SIC-based ultraviolet photodetectors, 279–282
Silicon nitride (SiN), 234–235
Silicon oxide, 40
Silicon photodetectors, 236–238
Silicon photodiodes (SiPDs), 18–19, 440
Silicon photomultipliers (SiPM), 46–48
Silicon photonics, 391
   integrated circuits
     optical properties of Si-based WGPDs, 422–423
     technology, 420–422
     waveguide photodiodes on silicon, 423–433
Silicon-on-insulator (SOI), 235, 369–370, 391, 444
Silver catalyst, 162–164
Simulation, 445
Single-mode fibers (SMFs), 438–439
Single-photon avalanche diode (SPAD), 45–46
   active illumination based ToF 3D imaging and ranging with, 62–66
   arrays in CMOS technology, 61–62
   figures of merit and general aspects of SPAD characterization, 57–60
     afterpulsing probability, 58–59
     crosstalk, 59
     dark count rate, 58
     fill factor, 60
     general aspects of SPAD characterization, 60
     homogeneity of SPAD parameters in array, 60
     photon detection efficiency, 58
     SPAD dead time, 59
     timing jitter, 59
   single-photon counting in silicon, 42–57
     avalanche photodiodes to SPAD, 44–49
     fabrication issues in CMOS-based SPAD arrays, 49–57
Single-photon counting (SPC), 45–46
   in silicon, 42–57
6G cellular networks, 234–235
Solar energy harvesters, 358–359
Solution-based process, 77–78
Sony/Philips Digital Interface Format standard (S/PDIF), 115
Space charge limited (SCL), 84–85
Space-charge region (SCR), 42–43, 49–52
Spectroscopy, 472, 475–477
Spin-coating technique, 78–79
Spray-coating, 111–112
   fabricated by, 111–113
   organic photoactive blends, 119
   technique, 82
Sputtering system, 404
SS-OCT system, 479

State-of-the-art silicon-germanium photodetectors, 245–256
  avalanche photodiodes, 249–256
  metal-semiconductor-metal diodes, 245–246
  PIN diodes, 246–249
Stranski-Krastanov InAs QDs, 200
Sub-bandgap absorption photodetectors, 317
Submonolayer growth (SML growth), 204
Superconducting-insulator-superconducting mixers, 319
Surface plasmon enhanced Ge-on-Si photodetectors, 398–400
Surface plasmon polaritons (SPPs), 320, 353, 398–400
Surface plasmon resonances, 353–359
  physics of plasmon resonances, 353–358
  plasmonic devices, 358–359
Surface recombination, 300
Synthetic organic compounds, 3

T

Technology, 420–422
  germanium, 420–421
  hybrid III/V-silicon, 421
Temporal coherence, 4
Temporal coupled-mode theory, 443
Terahertz (THz), 467
  detectors, 475
  emitters and receivers, 318–320
  frequency-domain, 479
  photodetectors, 358–359
  sources, 475
  waves, 467
Terahertz time-domain spectroscopy system (THz-TDS system), 475
Terahertz-Frequency-Domain Spectroscopy system, 476–477
2-(4-tertbutylphenyl) − 5-(4-biphenylyl) − 1,3,4-oxadiazole (PBD), 104
Thermal noise, 27–28
Thermal oxidation, 159–161
Thermally stimulated current (TSC), 212
Thin film transistors (TFTs), 117–118
Thin-film photodetectors, 3
Third-order optical aberrations, 7
Threading dislocations (TDs), 204–205, 393–394
Three-dimension (3D), 478

electric-field mapping, 478
imaging, 49
  active illumination based ToF 3D imaging and ranging with SPAD, 62–66
  systems, 62
Stranski-Krastanov growth, 393–394
300-GHz band system, 479–480
Through silicon vias (TSVs), 62
Time response, 337–345
  components of total temporal response, 339–345
Time-averaged autocorrelation function, 22
Time-of-flight (TOF), 49, 62
  active illumination based ToF 3D imaging and ranging with SPAD, 62–66
Time-to-digital-converters (TDC), 61, 65
Timing jitter, 59
Tin dioxide (SnO_2), 164
  nanowires-based photodetectors, 164–167
Tomographic imaging, 478–479
Top-of-Rack (ToR), 438
Trans-impedance-amplifier (TIA), 61, 457
Transfer-printing techniques, 144–145
Transient spectroscopy, 296–297
Transmission lines (TLs), 332–333
Transverse Electric (TE), 208–209
Transverse magnetic (TM), 398–400
Traveling wave photodetectors (TWPDs), 328
Traveling-wave amplified detectors (TAP), 313–316
Traveling-wave photodetector (TWPD), 311–312
4,4′,4″-tri-(2-methylphenylphenylamino) triphenylaine (m-MTDATA), 104
Tris (8-hydroxyquinoline) gallium (Gaq_3), 104
1,3,5-tris (N-phenyl-benzimidazol-2-yl) benzene (TPBi), 104
Two-dimensional electron and hole gasses (2DEHG), 334
Two-dimensional electron gas (2DEG), 272–273, 329–330
Two-dimensional semiconductors (2-D semiconductors), 3

U

Ultra-high vacuum chemical vapor deposition (UHCVD), 393–394
Ultrasonic systems, 82
Ultraviolet (UV), 37, 150–151
 detectors, 103–108
 detectors for harsh environments
  III–nitride-based ultraviolet photodetectors, 272–279
  photodetector parameters, 270–272
  SIC-based ultraviolet photodetectors, 279–282
  types of ultraviolet photodetectors, 283–286
 to NIR spectral range, 372
Uni-traveling carrier (UTC), 328, 419, 428
Uni-traveling-carrier photodiode (UTC-PD), 428–431, 469

V

Vacuum thermal evaporation technique, 77–78
Vapor-liquid-solid (VLS), 140
Vertical PIN silicon-based photodiode, 444–462
 device design, 444
 equivalent photodiode model, 453–459
 experimental results, 448–453
 fabrication, 446–448
 simulation and optimization, 445
Vertical-cavity surface-emitting laser (VCSEL), 438–439
Visible (VIS) range, 37

W

Wave motion, 329
Waveguide (WG), 205–206
 coupled photodetectors, 148–150
 detectors, 372–374
  silicon-based plasmonic waveguide photodetectors, 372–374

Waveguide photodiode (WGPD), 328, 420
waveguide-integrated Ge-on-Si photodetectors, 393–403
waveguide-integrated silicide Schottky-barrier photodetectors, 403–415
Waveguide-integrated Ge-on-Si photodetectors, 393–403
 selective hetero-epitaxy of Ge-on-Si and CMOS integration, 393–395
 surface plasmon enhanced Ge-on-Si photodetectors, 398–400
 waveguide-integrated avalanche photodetectors, 400–403
 waveguide-integrated PIN Ge-on-Si photodetectors, 395–398
Waveguide-integrated PIN Ge-on-Si photodetectors, 395–398
 Si waveguide-integrated Ge-on-Si photodetectors, 398t
Waveguide-integrated silicide Schottky-barrier photodetectors, 403–415
 metal-semiconductor-metal Schottky-barrier photodetectors, 406–408
 $NiSi_2$ film and absorption, 403–404
 $NiSi_2$/p-Si and $NiSi_2$/n-Si Schottky-barrier photodetectors, 404–406
 Schottky-barrier collector phototransistors, 408–411
 Schottky-barrier detector with embedded silicide nanoparticles, 411–415
Wavelength, 358
 selective OPDs, 94–103
Wetting layer (WL), 202
Wiener–Khintchine theorem, 22
Wireless communications, 472–474

Z

0-dimensional semiconductors (0D), 200
Zinc oxide (ZnO), 150–151, 268–270, 283
 nanowires photodetectors, 151–158

CPI Antony Rowe
Eastbourne, UK
February 22, 2023